SYMMETRIC MARKOV PROCESSES, TIME CHANGE, AND BOUNDARY THEORY

London Mathematical Society Monographs

The London Mathematical Society Monographs Series was established in 1968. Since that time it has published outstanding volumes that have been critically acclaimed by the mathematics community. The aim of this series is to publish authoritative accounts of current research in mathematics and high-quality expository works bringing the reader to the frontiers of research. Of particular interest are topics that have developed rapidly in the last ten years but that have reached a certain level of maturity. Clarity of exposition is important and each book should be accessible to those commencing work in its field. The original series was founded in 1968 by the Society and Academic Press; the second series was launched by the Society and Oxford University Press in 1983. In January 2003, the Society and Princeton University Press united to expand the number of books published annually and to make the series more international in scope.

Vol. 35, *Symmetric Markov Processes, Time Change, and Boundary Theory*, by Zhen-Qing Chen and Masatoshi Fukushima
Vol. 34, *Log-Gases and Random Matrices*, by Peter J. Forrester
Vol. 33, *Prime-Detecting Sieves*, by Glyn Harman
Vol. 32, *The Geometry and Topology of Coxeter Groups*, by Michael W. Davis
Vol. 31, *Analysis of Heat Equations on Domains*, by El Maati Ouhabaz

SYMMETRIC MARKOV PROCESSES, TIME CHANGE, AND BOUNDARY THEORY

Zhen-Qing Chen
Masatoshi Fukushima

PRINCETON UNIVERSITY PRESS
PRINCETON AND OXFORD

Published by Princeton University Press, 41 William Street, Princeton, New Jersey 08540
In the United Kingdom: Princeton University Press, 6 Oxford Street, Woodstock, Oxfordshire OX20 1TW

ISBN: 978-0-691-13605-9

Library of Congress Control Number: 2011934606

British Library Cataloging-in-Publication Data is available

This book has been composed in Times Roman
Printed on acid-free paper ∞
press.princeton.edu

Typeset by S R Nova Pvt Ltd, Bangalore, India
Printed in the United States of America

10 9 8 7 6 5 4 3 2 1

To Chen, Minda, and Angela

Z.-Q. C.

To Masako

M. F.

Contents

Notation

1. $\mathbb{R}$: set of all real numbers, $\mathbb{R}_+$: set of all non-negative real numbers, $\mathbb{Q}$: set of all rational numbers, $\mathbb{Q}_+$: set of all non-negative rational numbers, $\mathbb{Z}$: set of all integers, $\mathbb{N}$: set of all natural numbers
2. A $[-\infty, \infty]$-valued function is called a *numerical function*. An $\mathbb{R}$-valued function is called a *real function*. For a family $\mathcal{H}$ of numerical functions, $b\mathcal{H}$ (resp. $\mathcal{H}_+$) denotes the family of all bounded (resp. $[0, \infty]$-valued) functions in $\mathcal{H}$. Sometimes we also denote $b\mathcal{H}$ by $\mathcal{H}_b$.
3. For a, $b \in [-\infty, \infty]$, $a \vee b$ and $a \wedge b$ denote the larger and smaller one, respectively. We let $a^+ = a \vee 0$ and $a^- := (-a) \vee 0$. For numerical functions f, g, we define $(f \vee g)(x) = f(x) \vee g(x)$, $f^+(x) := (f(x))^+$, and so on.
4. If f is a measurable map from a measurable space $(E, \mathcal{B})$ to another measurable space $(F, \mathcal{G})$, we write $f \in \mathcal{B}/\mathcal{G}$. If in particular f is a measurable numerical function on a measurable space $(E, \mathcal{B}(E))$, we write $f \in \mathcal{B}(E)$ by regarding $\mathcal{B}(E)$ as the space of all $\mathcal{B}(E)$-measurable functions on E. Further, $f \in b\mathcal{B}(E)$ ($f \in \mathcal{B}_+(E)$) indicates that f is a bounded (non-negative) measurable function on E.
5. The symmetric difference $(A \setminus B) \cup (B \setminus A)$ of sets A and B is denoted by $A \Delta B$.
6. The characteristic function (indicator function) $\mathbf{1}_B$ of a set B is defined by

$$\mathbf{1}_B(x) = \begin{cases} 1 & x \in B \\ 0 & x \notin B. \end{cases}$$

7. $\mathbb{R}^n$ denotes the n-dimensional Euclidean space for $n \in \mathbb{N}$. $\mathbb{R}^1$ is denoted by $\mathbb{R}$. For $x \in \mathbb{R}^n$, A, $B \subset \mathbb{R}^n$, we define $A - x = \{y - x : y \in A\}$ and $A - B = \{y - x : y \in A, \ x \in B\}$.
8. For $\ell \in \mathbb{N}$, the contractive real functions φ^ℓ and φ_ℓ are defined by

$$\varphi^\ell(t) = ((-\ell) \vee t) \wedge \ell, \quad t \in \mathbb{R},$$

$$\varphi_\ell(t) = t - ((-1/\ell) \vee t) \wedge (1/\ell), \quad t \in \mathbb{R}.$$

9. $f = g \ [m]$ means that $f = g$ m-a.e.

10. $A \subset B$ m-a.e. (respectively, $A = B$ m-a.e.) means that $m(A \setminus B) = 0$ (respectively, $m(A \Delta B) = 0$).
11. For $p \geq 1$, the L^p-norm in the space $L^p(E; m)$ is denoted as $\|f\|_p = (\int_E |f(x)|^p m(dx))^{1/p}$.
12. δ_a denotes the probability measure concentrated at a point a.
13. For a measure μ and a non-negative function $f \in \mathcal{B}_+$, the measure $A \mapsto \int_A f(x)\mu(dx)$ is denoted by $f \cdot \mu$ or by $f\mu$.
14. For a measure μ and a function f on E, the integral $\int_E f d\mu$ is denoted by $\langle \mu, f \rangle$ or by $\langle f, \mu \rangle$ or by $\mu(f)$.
15. $\sigma\{\,\cdot\,\}$: the smallest σ-field making the family of sets or functions $\{\,\cdot\,\}$ measurable. Sometimes we also use $\sigma(\,\cdot\,)$ for $\sigma\{\,\cdot\,\}$.
16. For a Markov process X, $\mathbf{P}_x$-a.s. means that $\mathbf{P}_x$-a.e. on Ω. Further, *almost surely* or *a.s.* means that $\mathbf{P}_x$-a.s. for every $x \in E$.
17. $C(E)$ denotes the space of all continuous real functions on a topological space E. When E is locally compact, $C_c(E)$ ($C_\infty(E)$) denotes the family of all functions in $C(E)$ with compact support (vanishing at infinity). For a metric space (E, d), we denote by $C_u(E)$ the collection of all d-uniformly continuous real functions on E.
18. We use $:=$ as a means of definition. For a Markov process X and a subset A on its state space, we use $\sigma_A := \inf\{t > 0 : X_t \in A\}$ and $\tau_A := \inf\{t > 0 : X_t \notin A\}$ to denote the first hitting time of A and the first exit time from A, respectively, by X. The entrance time of A by X is defined to be $\dot{\sigma}_A := \inf\{t \geq 0 : X_t \in A\}$. We use the convention that $\inf \emptyset = \infty$.

Preface

The seminal paper of A. Beurling and J. Deny on the theory of Dirichlet spaces appeared in 1959, just half a century ago. Since then the theory of Dirichlet spaces has been growing in close relationship with the theory of Markov processes, especially symmetric Markov processes. The scope of the original Dirichlet space theory has since been greatly expanded both in theory and applications. Books bearing the term "Dirichlet forms" or "symmetric Markov processes" in the title continue to be published. For instance, the following volumes, listed in the chronological order, are devoted to this theory: [138, 63, 140, 64, 15, 119, 47, 73, 100, 141].

In 1960's and 1970's, the study of the Beurling-Deny theory was motivated by a desire to comprehend and develop the boundary theory for Markov processes as is evident in the papers by M. Fukushima, H. Kunita, M. L. Silverstein, and Y. LeJan and in the two books of Silverstein. Indeed, the concept of the reflected Dirichlet space and the space of functions with finite Douglas integrals involved there were the outgrowth of the idea in the preceding works by W. Feller in 1957 and J. L. Doob in 1962 reinterpreted in terms of Dirichlet forms. But the study in this direction was left halfway until Z.-Q. Chen gave an appropriate reformulation of the reflected Dirichlet space in 1992.

Over the last 30 years, time changes of symmetric Markov processes have been extensively studied. It is well understood now that time change of a symmetric Markov process by means of a positive continuous additive functional (PCAF in abbreviation) with full support corresponds precisely to the replacement of the symmetrizing measure while keeping the extended Dirichlet space invariant. The relevant stochastic calculus is also well developed accompanied by basic decomposition theorems of (not necessarily positive) additive functionals. However, the intrinsic Beurling-Deny decomposition of the trace Dirichlet form or characterization of the time-changed process by a non–fully supported PCAF in terms of a (generalized) Douglas integral has been obtained only quite recently; this is the topic of Chapter 5 of this book.

The notion of the quasi-regular Dirichlet form due to S. Albeverio, Z.-M. Ma, and M. Röckner and a related result of P. J. Fitzsimmons have enabled us to reduce the study of a general symmetric (not necessarily Borel) right process to the study of a symmetric Borel special standard process. It is established by Z.-Q. Chen, Z.-M. Ma, and M. Röckner that a Dirichlet form is quasi-regular if and only if it is quasi-homeomorphic to a regular Dirichlet form on a locally compact separable metric space. This quasi-homeomorphism allows one to transfer problems concerning those quasi-regular Dirichlet forms and symmetric right processes on (possibly infinite dimensional) general Hausdorff topological spaces to problems for regular Dirichlet forms and symmetric Hunt processes on locally compact separable metric spaces.

The development of the Dirichlet form theory benefits from its interaction with other areas of probability theory such as the theory of general Markov processes, martingale theory, and stochastic analysis, and with analytic potential theory, harmonic analysis, Riemannian geometry, theory of function spaces, partial differential equations and pseudo-differential operators, and mathematical physics. On the other hand, Dirichlet form theory has wide range of applications to these fields. For example, it is an effective tool in studying various probabilistic models as well as analytic problems with non-smooth data or in non-smooth media, such as reflecting Brownian motion on non-smooth domains, Brownian motion with random obstacles, diffusions and analysis on fractals, Markov processes and analysis on metric measure spaces, and diffusion processes and differential analysis on path spaces or loop spaces over a compact Riemannian manifold. It is a powerful machinery in studying various stochastic differential equations in infinite dimensional spaces, stochastic partial differential equations, various models in statistical physics such as quantum field theory and interacting particle systems. It also provides a probabilistic means to study various problems in partial differential equations with singular coefficients, analytic potential theory, and theory of function spaces.

The aim of this book is twofold. First, it gives a systematic introduction to the essential ingredients of both the probabilistic part and the analytic part of the theory of quasi-regular Dirichlet form. This is done in the first four and a half chapters of the book, where the theory of quasi-regular Dirichlet form and that of regular Dirichlet form are developed in a unified way. Second, it presents some recent developments of the theory in the last two and a half chapters. Its aim is, along with the characterization of the trace Dirichlet form by the Douglas integral, to give a comprehensive account of the reflected Dirichlet space and then to show the important role they play in the recent development of the boundary theory of symmetric Markov processes.

We strived to make the contents of this book self-contained so that it, especially its first four and a half chapters, can be used as a textbook for advanced graduate students. Chapters 2, 3, 5, 6, and 7 contain many examples illustrating the theory presented. Exercises given throughout the book are an integral part of the book. The solutions to these exercises are given in Appendix B.

The rest of the book is organized as follows. In Chapter 1, we introduce the concepts of Dirichlet form on $L^2(E, \mathcal{B}; m)$ and its extended Dirichlet space, where $(E, \mathcal{B})$ is a measurable space without any topological assumption imposed on E. In the remaining sections of Chapter 1, we give a quick introduction to the basic theory of quasi-regular Dirichlet forms, where E is assumed to be a topological Hausdorff space with the Borel σ-field $\mathcal{B}(E)$ being generated by the continuous functions on E. After the concept of a quasi-regular Dirichlet form is introduced, it is shown that every quasi-regular Dirichlet form is quasi-homeomorphic to a regular Dirichlet space on a locally compact separable metric space.

In Chapter 2, we investigate the transience and recurrence of the semigroups associated with general Dirichlet forms. Analytic potential theory for regular Dirichlet forms, such as capacity, smooth measures, and their potentials, are studied in Section 2.3. Various equivalent characterizations of the local property for a quasi-regular Dirichlet form are given in Section 2.4. Some basic examples of Dirichlet forms corresponding to symmetric Markov processes are presented in Section 2.2, including symmetric pure jump step processes, symmetric Lévy processes, one-dimensional diffusions, multidimensional Brownian motions, and Brownian motions on manifolds. An example of a quasi-regular but not regular Dirichlet form on $\mathbb{R}^n$ is given in Example 5.1.11.

Probabilistic potential theory of symmetric Markov processes and its relationship to analytic potential theory of the associated Dirichlet forms are presented in Chapter 3.

In Chapter 4, additive functionals of symmetric Markov processes are studied. In particular, the one-to-one correspondence between positive continuous additive functionals and the smooth measures is established. Fukushima decomposition, which serves as a counterpart of Itô's formula for symmetric Markov processes, is presented in Section 4.2. It plays an important role in analyzing the sample path properties of the processes as well as in other areas of the Dirichlet form theory. Beurling-Deny decomposition of a regular Dirichlet form is derived in Section 4.3 by utilizing martingale additive functionals.

The first half of Chapter 5 is devoted to a study of time changes of symmetric Markov processes, their Dirichlet form characterization, and applications. In the second half, Feller measures are introduced. They are used to characterize trace Dirichlet forms and to identify jump and killing measures of the time-changed process.

The reflected Dirichlet space of a Dirichlet form is introduced and investigated in Chapter 6. It is first introduced under the regular Dirichlet form setting. The transient and recurrent cases are treated separately. It is then extended to the quasi-regular Dirichlet form setting by using quasi-homeomorphism. Concrete examples of reflected Dirichlet spaces are exhibited for a number of Dirichlet forms including most of those appearing in Section 2.2. The important role that reflected Dirichlet spaces would play in the boundary theory for symmetric Markov processes is indicated by the fact that the active reflected Dirichlet form is the maximal Silverstein extension of the Dirichlet form.

In Chapter 7, we present some recent developments of boundary theory of symmetric Markov processes, emphasizing the role of reflected Dirichlet spaces and the function spaces of finite Douglas integrals. In the second half of Chapter 7, we develop the theory when the boundary is countable and give many concrete illustrative examples.

In Appendix A, we present basic materials on (not necessarily symmetric) right processes that are utilized in the text.

For readers' convenience, an index of some useful results is provided in the *Catalogue of Some Useful Theorems* at the end of the book.

The material starting from Section 5.5 to the end of Chapter 7 appears here for the first time in a book. Except for a few sections, most results in other parts of the book are not new. However, their proofs and presentations can be found to be novel in many places. As compared to the book by M. Fukushima, Y. Oshima, and M. Takeda published in 1994 [73], the approach presented in this book is more probabilistic and the framework of a quasi-regular Dirichlet form is employed for the first time in parallel with a regular one. On the other hand, [73] contains more detailed expositions of analytic properties of a regular Dirichlet form as well as some other relevant topics including a construction of an associated Hunt process and a Girsanov-type transformation. We refer readers to the book by Z. M. Ma and M. Röckner [119] published in 1992 for further readings on quasi-regular (non-symmetric) Dirichlet forms, in particular for basic concrete examples of quasi-regular Dirichlet forms in infinite dimensions that are not touched upon in the present volume.

The materials in Sections 3.5 and 5.4 and Appendix A owe a lot to the work by P. J. Fitzsimmons, R. K. Getoor, P. A. Meyer, and M. J. Sharpe on right processes, additive functionals, and energy functionals.

The notes at the end of this book provide information on other closely related books, sources of materials, and related literature.

In the bibliography, we list only literature that is directly linked to the topics of this book. But we admit with apology that it is still far from being complete partly due to a great diversity and vast literature of related areas.

In August 2003, the second-named author gave an invited lecture series of London Mathematical Society at the University of Wales Swansea, kindly arranged by N. Jacob. The lecture notes contained a time-change theory as well as a preliminary account of the Douglas integrals for diffusions. They eventually grew into the present book, in collaboration with the first author. We are indebted to N. Jacob for creating an opportunity to write the preliminary version of the book.

Thanks are due to M. Takeda for allowing us to use several ingredients from the recently published Japanese book by Fukushima and Takeda [74].

We are grateful to K. Burdzy, W. T. Fan, M. Hino, N. Kajino, P. Kim, K. Kuwae, S. Lou, and Y. Oshima for reading parts of the manuscript and providing us with helpful comments and lists of typos. We thank the reviewers of this book for their helpful comments. We also thank Vickie Kearn and Stefani Wexler at Princeton University Press for their truly kind assistance during the preparation of this manuscript.

The research of the authors was supported in part by NSF grants in the United States and a Grant-in-Aid for Scientific Research in Japan.

Zhen-Qing Chen
Seattle

Masatoshi Fukushima
Osaka

Chapter One

SYMMETRIC MARKOVIAN SEMIGROUPS AND DIRICHLET FORMS

1.1. DIRICHLET FORMS AND EXTENDED DIRICHLET SPACES

The concepts of Dirichlet form and Dirichlet space were introduced in 1959 by A. Beurling and J. Deny [8] and the concept of the extended Dirichlet space was given in 1974 by M. L. Silverstein [138]. They all assumed that the underlying state space E is a locally compact separable metric space. Concrete examples of Dirichlet forms (bilinear form, weak solution formulations) have appeared frequently in the theory of partial differential equations and Riemannian geometry. However, the theory of Dirichlet forms goes far beyond these.

In this section, we work with a σ-finite measure space $(E, \mathcal{B}(E), m)$ without any topological assumption on E and establish the correspondence of the above-mentioned notions to the semigroups of symmetric Markovian linear operators. The present arguments are a little longer than the usual ones under the topological assumption found in [39] and [73, §1.4] but they are quite elementary in nature.

Only at the end of this section, we shall assume that E is a Hausdorff topological space and consider the semigroups and Dirichlet forms generated by symmetric Markovian transition kernels on E.

Let $(E, \mathcal{B}(E))$ be a measurable space and m a σ-finite measure on it. Let $\mathcal{B}^m(E)$ be the completion of $\mathcal{B}(E)$ with respect to m. Numerical functions f, g on E are said to be m-equivalent ($f = g$ $[m]$ in notation) if $m(\{x \in E : f(x) \neq g(x)\}) = 0$. For $p \geq 1$ and a numerical function $f \in \mathcal{B}^m(E)$, we put

$$\|f\|_p = \left(\int_E |f(x)|^p m(dx) \right)^{1/p}.$$

The family of all m-equivalence classes of $f \in \mathcal{B}^m(E)$ with $\|f\|_p < \infty$ is denoted by $L^p(E; m)$, which is a Banach space with norm $\|\cdot\|_p$, namely, a complete normed linear space. We denote by $L^\infty(E; m)$ the family of all m-equivalence classes of $f \in \mathcal{B}^m(E)$ which are bounded m-a.e. on E. $L^\infty(E; m)$ is a Banach space with norm

$$\|f\|_\infty := \inf_{N:\, m(N)=0} \sup_{x \in E \setminus N} |f(x)|.$$

Note that $L^2(E; m)$ is a real Hilbert space with inner product

$$(f, g) = \int_E f(x)g(x)m(dx), \qquad f, g \in L^2(E; m).$$

For a moment, let us consider an abstract real Hilbert space H with inner product $(\cdot, \cdot)$. $\sqrt{(f,f)}$ for $f \in H$ is denoted by $\|f\|_H$. As is summarized in Section A.4, there are mutual one-to-one correspondences among four objects on the Hilbert space H: the family of all closed symmetric forms $(\mathcal{E}, \mathcal{D}(\mathcal{E}))$, the family of all strongly continuous contraction semigroups $\{T_t; t \geq 0\}$, the family of all strongly continuous contraction resolvents $\{R_\alpha; \alpha > 0\}$, and the family of all non-positive definite self-adjoint operators A. Here we mention the correspondences among the first three objects only.

$\mathcal{E}$ or $(\mathcal{E}, \mathcal{D}(\mathcal{E}))$ is said to be a *symmetric form* on H if $\mathcal{D}(\mathcal{E})$ is a dense linear subspace of H and $\mathcal{E}$ is a non-negative definite symmetric bilinear form defined on $\mathcal{D}(\mathcal{E}) \times \mathcal{D}(\mathcal{E})$ in the sense that for every $f, g, h \in \mathcal{D}(\mathcal{E})$ and $a, b \in \mathbb{R}$

$$\mathcal{E}(f, g) = \mathcal{E}(g, f), \quad \mathcal{E}(f, f) \geq 0, \quad \text{and}$$

$$\mathcal{E}(af + bg, h) = a\mathcal{E}(f, h) + b\mathcal{E}(g, h).$$

For $\alpha > 0$, we define

$$\mathcal{E}_\alpha(f, g) = \mathcal{E}(f, g) + \alpha(f, g), \quad f, g \in \mathcal{D}(\mathcal{E}).$$

We call a symmetric form $(\mathcal{E}, \mathcal{D}(\mathcal{E}))$ on H *closed* if $\mathcal{D}(\mathcal{E})$ is complete with norm $\sqrt{\mathcal{E}_1(f,f)}$. $\mathcal{D}(\mathcal{E})$ is then a real Hilbert space with inner product $\mathcal{E}_\alpha$ for each $\alpha > 0$.

A family of symmetric linear operators $\{T_t; t > 0\}$ on H is called a *strongly continuous contraction semigroup* if, for any $f \in H$,

$$T_s T_t f = T_{s+t} f, \quad \|T_t f\|_H \leq \|f\|_H, \quad \lim_{t \downarrow 0} \|T_t f - f\|_H = 0.$$

We call a family of symmetric linear operators $\{G_\alpha; \alpha > 0\}$ on H a *strongly continuous contraction resolvent* if for every $\alpha, \beta > 0$ and $f \in H$,

$$G_\alpha f - G_\beta f + (\alpha - \beta) G_\alpha G_\beta f = 0, \quad \alpha \|G_\alpha f\|_H \leq \|f\|_H,$$

$$\lim_{\alpha \to \infty} \|\alpha G_\alpha f - f\|_H = 0.$$

The semigroup $\{T_t; t \geq 0\}$ and the resolvent $\{G_\alpha; \alpha > 0\}$ as above correspond to each other by the next two equations:

$$G_\alpha f = \int_0^\infty e^{-\alpha t} T_t f dt, \quad f \in H, \tag{1.1.1}$$

the integral on the right hand side being defined in Bochner's sense, and

$$T_t f = \lim_{\beta \to \infty} e^{-t\beta} \sum_{n=0}^{\infty} \frac{(t\beta)^n}{n!} (\beta G_\beta)^n f, \quad f \in H. \tag{1.1.2}$$

$\{G_\alpha; \alpha > 0\}$ determined by (1.1.1) from $\{T_t; t > 0\}$ is called the *resolvent of* $\{T_t; t \geq 0\}$.

Given a strongly continuous contraction symmetric semigroup $\{T_t; t > 0\}$ on H, for each $t > 0$,

$$\mathcal{E}^{(t)}(f, g) := \frac{1}{t}(f - T_t f, g), \quad f, g \in H \tag{1.1.3}$$

defines a symmetric form $\mathcal{E}^{(t)}$ on H with domain H. For each $f \in H$, $\mathcal{E}^{(t)}(f, f)$ is non-negative and increasing as $t > 0$ decreases (this can be shown, for example, by using spectral representation of $\{T_t; t > 0\}$). We may then set

$$\mathcal{D}(\mathcal{E}) = \{f \in H\colon \lim_{t \downarrow 0} \mathcal{E}^{(t)}(f, f) < \infty\}, \tag{1.1.4}$$

$$\mathcal{E}(f, g) = \lim_{t \downarrow 0} \mathcal{E}^{(t)}(f, g), \qquad f, g \in \mathcal{D}(\mathcal{E}), \tag{1.1.5}$$

which becomes a closed symmetric form on H called the *closed symmetric form of the semigroup* $\{T_t; t > 0\}$. We call $\mathcal{E}^{(t)}$ of (1.1.3) the *approximating form* of $\mathcal{E}$.

Conversely, suppose that we are given a closed symmetric form $(\mathcal{E}, \mathcal{D}(\mathcal{E}))$ on H. For each $\alpha > 0, f \in H$ and $v \in \mathcal{D}(\mathcal{E})$, we have

$$|(f, v)| \leq \|f\|_2 \|v\|_2 \leq (1/\alpha)^{1/2} \|f\|_2 \sqrt{\mathcal{E}_\alpha(v, v)},$$

which means that $\Phi(v) = (f, v)$ is a bounded linear functional on the Hilbert space $(\mathcal{D}(\mathcal{E}), \mathcal{E}_\alpha)$. By the Riesz representation theorem, there exists a unique element of $\mathcal{D}(\mathcal{E})$ denoted by $G_\alpha f$ such that for every $f \in H$ and $v \in \mathcal{D}(\mathcal{E})$,

$$G_\alpha f \in \mathcal{D}(\mathcal{E}) \quad \text{and} \quad \mathcal{E}_\alpha(G_\alpha f, v) = (f, v). \tag{1.1.6}$$

$\{G_\alpha; \alpha > 0\}$ so defined is a strongly continuous contraction resolvent on H, which in turn determines a strongly continuous contraction semigroup

$\{T_t; t > 0\}$ on H by (1.1.2). They are called the *resolvent* and *semigroup generated by the closed symmetric form* $(\mathcal{E}, \mathcal{D}(\mathcal{E}))$, respectively.

The above-mentioned correspondences from $\{T_t; t > 0\}$ to $(\mathcal{E}, \mathcal{D}(\mathcal{E}))$ and from $(\mathcal{E}, \mathcal{D}(\mathcal{E}))$ to $\{T_t; t > 0\}$ are mutually reciprocal.

From now on, we shall take as H the space $L^2(E; m)$ on a σ-finite measure space $(E, \mathcal{B}(E), m)$. In this book, we need to consider extensions of the domain $\mathcal{D}(\mathcal{E})$ of a closed symmetric form $\mathcal{E}$ on $L^2(E; m)$. For this purpose, we shall designate $\mathcal{D}(\mathcal{E})$ by $\mathcal{F}$ so that a closed symmetric form on $L^2(E; m)$ will be denoted by $(\mathcal{E}, \mathcal{F})$. We now proceed to introduce the notions of Dirichlet form and extended Dirichlet space.

DEFINITION 1.1.1. For $1 \le p \le \infty$, a linear operator L on $L^p(E; m)$ with domain of definition $\mathcal{D}(L)$ is called *Markovian* if

$$f \in \mathcal{D}(L) \textit{ with } 0 \le f \le 1\ [m] \quad \Longrightarrow \quad 0 \le Lf \le 1\ [m].$$

A real function φ, namely, a mapping from $\mathbb{R}$ to $\mathbb{R}$, is said to be a *normal contraction* if

$$\varphi(0) = 0 \quad \text{and} \quad |\varphi(s) - \varphi(t)| \le |s - t| \text{ for every } s, t \in \mathbb{R}.$$

A function defined by $\varphi(t) = (0 \vee t) \wedge 1$, $t \in \mathbb{R}$, is a normal contraction which is called the *unit contraction*. For any $\varepsilon > 0$, a real function φ_ε satisfying the next condition is a normal contraction:

$$\begin{aligned} &\varphi_\varepsilon(t) = t \text{ for } t \in [0,1]; \quad -\varepsilon \le \varphi_\varepsilon(t) \le 1 + \varepsilon \text{ for } t \in \mathbb{R}, \\ &0 \le \varphi_\varepsilon(t) - \varphi_\varepsilon(s) \le t - s \quad \text{for } s < t. \end{aligned} \tag{1.1.7}$$

DEFINITION 1.1.2. A symmetric form $(\mathcal{E}, \mathcal{D}(\mathcal{E}))$ on $L^2(E; m)$ is called *Markovian* if, for any $\varepsilon > 0$, there exists a real function φ_ε satisfying (1.1.7) and

$$f \in \mathcal{D}(\mathcal{E}) \quad \Longrightarrow \quad g := \varphi_\varepsilon \circ f \in \mathcal{D}(\mathcal{E}) \text{ with } \mathcal{E}(g, g) \le \mathcal{E}(f, f). \tag{1.1.8}$$

A closed symmetric form $(\mathcal{E}, \mathcal{F})$ on $L^2(E; m)$ is called a *Dirichlet form* if it is Markovian. In this case, the domain $\mathcal{F}$ is said to be a *Dirichlet space*.

THEOREM 1.1.3. *Let* $(\mathcal{E}, \mathcal{F})$ *be a closed symmetric form on* $L^2(E; m)$ *and* $\{T_t\}_{t>0}$, $\{G_\alpha\}_{\alpha>0}$ *be the strongly continuous contraction semigroup and resolvent on* $L^2(E; m)$ *generated by* $(\mathcal{E}, \mathcal{F})$, *respectively. Then the following conditions are mutually equivalent:*

(a) T_t *is Markovian for each* $t > 0$.

(b) αG_α *is Markovian for each* $\alpha > 0$.

(c) $(\mathcal{E}, \mathcal{F})$ *is a Dirichlet form on* $L^2(E; m)$.

(d) *The unit contraction operates on* $(\mathcal{E},\mathcal{F})$*:*

$$f \in \mathcal{F} \implies g := (0 \vee f) \wedge 1 \in \mathcal{F} \text{ and } \mathcal{E}(g,g) \le \mathcal{E}(f,f).$$

(e) *Every normal contraction operates on* $(\mathcal{E},\mathcal{F})$*: for any normal contraction* φ

$$f \in \mathcal{F} \implies g = \varphi \circ f \in \mathcal{F} \text{ and } \mathcal{E}(g,g) \le \mathcal{E}(f,f).$$

Proof. The implications **(a)** $\Rightarrow$ **(b)** and **(b)** $\Rightarrow$ **(a)** follow from (1.1.1) and (1.1.2), respectively. The implication **(e)** $\Rightarrow$ **(d)** $\Rightarrow$ **(c)** is obvious.

(c) $\Rightarrow$ **(b)**: We fix an $\alpha > 0$ and a function $f \in L^2(E;m)$ with $0 \le f \le 1$ $[m]$, and introduce a quadratic form on $\mathcal{F}$ by

$$\Phi(v) = \mathcal{E}(v,v) + \alpha\left(v - \frac{f}{\alpha}, v - \frac{f}{\alpha}\right), \quad v \in \mathcal{F}.$$

It follows from (1.1.6) that

$$\Phi(G_\alpha f) + \mathcal{E}_\alpha(G_\alpha f - v, G_\alpha f - v) = \Phi(v), \quad v \in \mathcal{F},$$

namely, $G_\alpha f$ is a unique element of $\mathcal{F}$ minimizing $\Phi(v)$. Suppose $(\mathcal{E},\mathcal{F})$ is a Dirichlet form on $L^2(E;m)$. There exists then for any $\varepsilon > 0$ a real function φ_ε satisfying (1.1.7) and (1.1.8). We let $\widetilde{\varphi}_\varepsilon(t) = (1/\alpha)\varphi_{\alpha\varepsilon}(\alpha t)$, $u = \widetilde{\varphi}_\varepsilon \circ G_\alpha f$ to obtain

$$u \in \mathcal{F} \text{ and } \mathcal{E}(u,u) \le \mathcal{E}(G_\alpha f, G_\alpha f).$$

Since $|\widetilde{\varphi}_\varepsilon(t) - s| \le |t-s|$ for every $s \in [0,1/\alpha]$ and $t \in \mathbb{R}$, we have $|u(x) - f(x)/\alpha| \le |G_\alpha f(x) - f(x)/\alpha|$ $[m]$ and $(u - f/\alpha, u - f/\alpha) \le (G_\alpha f - f/\alpha, G_\alpha f - f/\alpha)$. Therefore, $\Phi(u) \le \Phi(G_\alpha f)$ and consequently $u = G_\alpha f$ $[m]$, which means that $-\varepsilon \le G_\alpha f \le 1/\alpha + \varepsilon$ $[m]$. Letting $\varepsilon \to 0$, we get **(b)**.

It remains to prove the implication **(a)** $\Rightarrow$ **(e)**, which will follow from a more general theorem formulated below. □

In what follows, we occasionally use for a symmetric form $(\mathcal{E},\mathcal{F})$ on $L^2(E;m)$ the notations

$$\|f\|_{\mathcal{E}} := \sqrt{\mathcal{E}(f,f)}, \quad \|f\|_{\mathcal{E}_\alpha} := \sqrt{\mathcal{E}_\alpha(f,f)}, \quad f \in \mathcal{F},\ \alpha > 0$$

DEFINITION 1.1.4. Let $(\mathcal{E},\mathcal{F})$ be a closed symmetric form on $L^2(E;m)$. Denote by $\mathcal{F}_e$ the totality of m-equivalence classes of all m-measurable functions f on E such that $|f| < \infty$ $[m]$ and there exists an $\mathcal{E}$-Cauchy sequence $\{f_n, n \ge 1\} \subset \mathcal{F}$ such that $\lim_{n\to\infty} f_n = f$ m-a.e on E. $\{f_n\} \subset \mathcal{F}$ in the above is called an *approximating sequence of* $f \in \mathcal{F}_e$. We call the space $\mathcal{F}_e$ the *extended space* attached to $(\mathcal{E},\mathcal{F})$. When the latter is a Dirichlet form on $L^2(E;m)$, the space $\mathcal{F}_e$ will be called its *extended Dirichlet space*.

THEOREM 1.1.5. *Let $(\mathcal{E},\mathcal{F})$ be a closed symmetric form on $L^2(E;m)$ and $\mathcal{F}_e$ be the extended space attached to it. If the semigroup $\{T_t; t>0\}$ generated by $(\mathcal{E},\mathcal{F})$ is Markovian, then the following are true:*
(i) *For any $f\in\mathcal{F}_e$ and for any approximating sequence $\{f_n\}\subset\mathcal{F}$ of f, the limit $\mathcal{E}(f,f)=\lim_{n\to\infty}\mathcal{E}(f_n,f_n)$ exists independently of the choice of an approximating sequence $\{f_n\}$ of f.*
(ii) *Every normal contraction operates on $(\mathcal{F}_e,\mathcal{E})$: for any normal contraction φ*

$$f\in\mathcal{F}_e \implies g:=\varphi\circ f\in\mathcal{F}_e,\quad \mathcal{E}(g,g)\le\mathcal{E}(f,f).$$

(iii) *$\mathcal{F}=\mathcal{F}_e\cap L^2(E;m)$. In particular, $(\mathcal{E},\mathcal{F})$ is a Dirichlet form on $L^2(E;m)$.*

Assertion (ii) of this theorem implies the implication **(a)** $\Rightarrow$ **(e)** in Theorem 1.1.3, completing the proof of Theorem 1.1.3.

For $f,g\in\mathcal{F}_e$, clearly both $f+g$ and $f-g$ are in $\mathcal{F}_e$. Define $\mathcal{E}(f,g)=\frac{1}{4}(\mathcal{E}(f+g,f+g)-\mathcal{E}(f-g,f-g))$, which is a symmetric bilinear form over $\mathcal{F}_e$. $(\mathcal{E},\mathcal{F}_e)$ is called the *extended Dirichlet form* of $(\mathcal{E},\mathcal{F})$.

If a given closed symmetric form $(\mathcal{E},\mathcal{F})$ on $L^2(E;m)$ is a Dirichlet form, then the corresponding semigroup $\{T_t; t>0\}$ is Markovian by virtue of the already proven implication **(c)** $\Rightarrow$ **(a)** of Theorem 1.1.3. So the extended Dirichlet space $\mathcal{F}_e$ satisfies all properties mentioned in Theorem 1.1.5.

Before giving the proof of Theorem 1.1.5, we shall fix a Markovian contractive symmetric linear operator T on $L^2(E;m)$ and make some preliminary observations on T.

By the linearity and the Markovian property of T on $L^2(E;m)\cap L^\infty(E;m)$,

$$f_1,\,f_2\in L^2\cap L^\infty,\ 0\le f_1\le f_2\ [m] \Longrightarrow 0\le Tf_1\le Tf_2\le \|f_2\|_\infty\ [m].$$

Due to the σ-finiteness of m, we can construct a Borel function $\eta\in L^1(E;m)$ which is strictly positive on E. If we put $\eta_n(x)=(n\eta(x))\wedge 1$, then $0<\eta_n\le 1$, $\eta_n\uparrow 1$, $n\to\infty$. Hence we can define an extension of T from $L^2(E;m)\cap L^\infty(E;m)$ to $L^\infty(E;m)$ as follows:

$$\begin{cases} Tf(x):=\lim_{n\to\infty}T(f\cdot\eta_n)(x)\ [m], & f\in L^\infty_+(E;m),\\ Tf:=Tf^+-Tf^-, & f\in L^\infty(E;m).\end{cases}\tag{1.1.9}$$

By the symmetry of T, $(g,T(f\cdot\eta_n))=(Tg,f\cdot\eta_n)$ for $g\in bL^1(E;m)$. Letting $n\to\infty$, we see that the function $Tf, f\in L^\infty(E;m)$, defined by (1.1.9), satisfies the identity

$$\langle g,Tf\rangle=\langle Tg,f\rangle\quad\text{for every } g\in bL^1(E;m),\tag{1.1.10}$$

where $\langle g,\ f\rangle$ denotes the integral $\int_E gfdm$ for $g\in L^1(E;m)$, $f\in L^\infty(E;m)$. Consequently, Tf is uniquely determined up to the m-equivalence for

$f \in L^\infty(E; m)$. T becomes a Markovian linear operator on $L^\infty(E; m)$ and satisfies

$$f_n,\ f \in L_+^\infty(E; m),\ f_n \uparrow f\ [m] \Longrightarrow \lim_{n\to\infty} Tf_n = Tf\ [m]. \tag{1.1.11}$$

Further, if a sequence $\{f_n\} \subset L^\infty(E; m)$ is uniformly bounded and converges to f m-a.e. as $n \to \infty$,

$$\lim_{n\to\infty} \langle g, Tf_n\rangle = \langle g, Tf\rangle \quad \text{for every } g \in bL^1(E; m). \tag{1.1.12}$$

Lemma 1.1.6. (i) *For any $g \in L^\infty(E; m)$,*

$$T(g^2) - 2gTg + g^2T1 \geq 0 \quad [m].$$

(ii) *For any $g \in L^\infty(E; m)$, define*

$$\mathcal{A}_T(g) = \frac{1}{2}\int_E \left(T(g^2) - 2gTg + g^2T1\right) dm + \int_E g^2(1 - T1)dm. \tag{1.1.13}$$

It holds for $g \in L^2(E; m) \cap L^\infty(E; m)$ that

$$\mathcal{A}_T(g) = (g - Tg, g). \tag{1.1.14}$$

(iii) *For any $g \in L^\infty(E; m)$ and for any normal contraction φ,*

$$\mathcal{A}_T(\varphi \circ g) \leq \mathcal{A}_T(g). \tag{1.1.15}$$

(iv) *For any $f, g \in L^\infty(E; m)$,*

$$\mathcal{A}_T(f + g)^{1/2} \leq \mathcal{A}_T(f)^{1/2} + \mathcal{A}_T(g)^{1/2}. \tag{1.1.16}$$

Proof. (i) For a simple function on E expressed by

$$s = \sum_{i=1}^n a_i \mathbf{1}_{B_i}, \tag{1.1.17}$$

where $a_i \in \mathbb{R}$, $B_i \in \mathcal{B}(E)$ with $B_i \cap B_j = \emptyset$ for $i \neq j$ and $\cup_{i=1}^n B_i = E$, we have

$$T(g^2) - 2sTg + s^2T1 = \sum_{i=1}^n \mathbf{1}_{B_i} T\left((g - a_i)^2\right) \geq 0 \quad [m]. \tag{1.1.18}$$

Hence it suffices to choose an increasing sequence of simple functions $\{s_\ell, \ell \geq 1\}$ of this type such that

$$\lim_{\ell\to\infty} s_\ell = g \quad [m].$$

(ii) Recall that $\{\eta_n, n \geq 1\}$ is an increasing sequence of positive functions that is defined preceding (1.1.9). For $g \in L^2 \cap L^\infty$, we have $(Tg^2, \eta_n) = (g^2, T\eta_n)$

by the symmetry of T. By letting $n \to \infty$, we get $\int_E Tg^2 dm = \int_E g^2 T1 dm < \infty$ and $(g - Tg, g) = \frac{1}{2}\int_E \left(2g^2T1 - 2gTg\right) dm + \int_E g^2(1 - T1)dm = \mathcal{A}_T(g)$.
(iii) For $g \in L^\infty(E; m)$ and $k = 1, 2, \ldots$, we put

$$\mathcal{A}_T^k(g) = \frac{1}{2}\langle T(g^2) - 2gTg + g^2T1, \eta_k\rangle + \langle g^2(1 - T1), \eta_k\rangle.$$

When g is a simple function of the type (1.1.17),

$$\mathcal{A}_T^k(s) = \frac{1}{2}\sum_{1\le i,j\le n}(a_i - a_j)^2 J_{ij}^k + \sum_{1\le i\le n} a_i^2\kappa_i^k.$$

Here $J_{ij}^k = \int_E (T\mathbf{1}_{B_i})\mathbf{1}_{B_j}\eta_k dm$, $\kappa_i^k = \int_E \mathbf{1}_{B_i}(1 - T1)\eta_k dm$.
For any normal contraction φ, it holds that

$$(\varphi(a_i) - \varphi(a_j))^2 \le (a_i - a_j)^2 \quad \text{and} \quad \varphi(a_i)^2 \le a_i^2.$$

Thus for a simple function s, $\mathcal{A}_T^k(\varphi \circ s) \le \mathcal{A}_T^k(s)$. For any $g \in L^\infty(E; m)$, we can take uniformly bounded simple functions s_ℓ with $\lim_{\ell\to\infty} s_\ell = g$ $[m]$ to obtain $\mathcal{A}_T^k(\varphi \circ s_\ell) \le \mathcal{A}_T^k(s_\ell)$. Letting $\ell \to \infty$ and then $k \to \infty$, we have by (1.1.12) that (1.1.15) holds.
(iv) It suffices to show the triangular inequality (1.1.16) for $\mathcal{A}_T^k$ for each fixed k instead of $\mathcal{A}_T$. Since $0 \le \mathcal{A}_T^k(g) < \infty$, $g \in L^\infty(E; m)$, the bilinear form defined by

$$\mathcal{A}_T^k(f, g) = \frac{1}{4}\left(\mathcal{A}_T^k(f + g) - \mathcal{A}_T^k(f - g)\right), \quad f, g \in L^\infty(E; m),$$

satisfies the Schwarz inequality

$$|\mathcal{A}_T^k(f, g)| \le \mathcal{A}_T^k(f)^{1/2} \cdot \mathcal{A}_T^k(g)^{1/2},$$

from which follows the desired triangular inequality. □

Let $\{\varphi^\ell, \ \ell > 0\}$ be a specific family of normal contractions defined by

$$\varphi^\ell(t) = ((-\ell) \vee t) \wedge \ell, \quad t \in \mathbb{R}. \tag{1.1.19}$$

For any m-measurable function g on E with $|g| < \infty$ $[m]$, $\mathcal{A}_T(\varphi^\ell \circ g)$ is increasing as ℓ increases. This is clear from $\varphi^\ell \circ (\varphi^{\ell+1} \circ g) = \varphi^\ell \circ g$ and Lemma 1.1.6(iii). We can then extend $\mathcal{A}_T(g)$ to g by letting

$$\mathcal{A}_T(g) = \lim_{\ell\to\infty} \mathcal{A}_T(\varphi^\ell \circ g) \ (\le \infty). \tag{1.1.20}$$

Lemma 1.1.7. (i) *For* $g \in L^2(E; m)$, $\mathcal{A}_T(g) = (g - Tg, g)$.
(ii) **(Fatou's property)** *For any m-measurable functions g_n, g on E with* $|g_n| < \infty$, $|g| < \infty$ $[m]$, $\lim_{n\to\infty} g_n = g$ $[m]$,

$$\mathcal{A}_T(g) \le \liminf_{n\to\infty} \mathcal{A}_T(g_n). \tag{1.1.21}$$

(iii) *For any m-measurable function g on E with* $|g| < \infty$ $[m]$ *and for any normal contraction* φ, $\mathcal{A}_T(\varphi \circ g) \leq \mathcal{A}_T(g)$.
(iv) *The triangular inequality (1.1.16) holds for every m-measurable functions f and g that are finite m-a.e. on E.*

Proof. (i) follows from Lemma 1.1.6(ii) and the contraction property of T on $L^2(E; m)$.
(ii) We first give a proof when $|g_n| \leq M$, $|g| \leq M$ for some M and $\lim_{n\to\infty} g_n = g$ $[m]$. From the linearity, the Markovian property of T on $L^\infty(E; m)$, and (1.1.11), we have for $b \in \mathbb{R}$

$$T((g-b)^2) = \lim_{j\to\infty} T\left(\inf_{n\geq j}(g_n - b)^2\right) \leq \liminf_n T((g_n - b)^2).$$

Since the identity (1.1.18) holds when s is a simple function like (1.1.17), we get from the above inequality

$$0 \leq T(g^2) - 2sTg + s^2T1 \leq \liminf_n (T(g_n^2) - 2sTg_n + s^2T1).$$

On the other hand,

$$\begin{aligned} &|\left(T(g_n^2) - 2g_nTg_n + g_n^2T1\right) - \left(T(g_n^2) - 2sTg_n + s^2T1\right)| \\ &\leq 2|Tg_n|\,|g_n - s| + |g_n^2 - s^2|\,T1 \leq 2M|g_n - s| + |g_n^2 - s^2| \end{aligned}$$

hence

$$\begin{aligned} 0 &\leq T(g^2) - 2sTg + s^2T1 \\ &\leq \liminf_n \left(T(g_n^2) - 2g_nTg_n + g_n^2T1\right) + 2M|g-s| + |g^2 - s^2|. \end{aligned}$$

Taking a sequence of simple functions s such that $s \to g$ $[m]$,

$$0 \leq T(g^2) - 2gTg + g^2T1 \leq \liminf_n \left(T(g_n^2) - 2g_nTg_n + g_n^2T1\right).$$

Integrating both sides with respect to m and taking the defining formula (1.1.13) into account, we arrive at the desired (1.1.21) using the Fatou's lemma in the Lebesgue integration theory.

When g_n and g are not necessarily uniformly bounded, we can use the results obtained above to get

$$\mathcal{A}_T(\varphi^\ell \circ g) \leq \liminf_n \mathcal{A}_T(\varphi^\ell \circ g_n) \leq \liminf_n \mathcal{A}_T(g_n).$$

By letting $\ell \to \infty$, we still have the inequality (1.1.21).

(iii) It holds for $f = \varphi \circ g$, $f_\ell = \varphi \circ \varphi^\ell \circ g$ that $\lim_{\ell\to\infty} f_\ell = f$. Equations (1.1.15) and (1.1.21) then lead us to

$$\mathcal{A}_T(f) \le \liminf_{\ell\to\infty} \mathcal{A}_T(f_\ell) \le \liminf_{\ell\to\infty} \mathcal{A}_T(\varphi^\ell \circ g) = \mathcal{A}_T(g).$$

(iv) If we let $f_n := \varphi^n \circ f$ and $g_n := \varphi^n \circ g$, then $\lim_{n\to\infty}(f_n + g_n) = f + g[m]$ so that (1.1.16) and (1.1.21) yield

$$\begin{aligned}\mathcal{A}_T(f+g)^{1/2} &\le \liminf_{n\to\infty} \mathcal{A}_T(f_n+g_n)^{1/2} \\ &\le \lim_{n\to\infty} \left(\mathcal{A}_T(f_n)^{1/2} + \mathcal{A}_T(g_n)^{1/2}\right) = \mathcal{A}_T(f)^{1/2} + \mathcal{A}_T(g)^{1/2}.\end{aligned}$$

□

Proof of Theorem 1.1.5. (i) For any $f \in \mathcal{F}_e$, take its approximating sequence $\{f_n\} \subset \mathcal{F}$. f_n being $\mathcal{E}$-Cauchy, the triangular inequality guarantees the existence of the limit $\mathcal{E}(f,f) = \lim_{n\to\infty} \mathcal{E}(f_n,f_n)$. Let us prove that

$$\frac{1}{t}\mathcal{A}_{T_t}(f) \uparrow \mathcal{E}(f,f) \quad \text{as } t \downarrow 0, \tag{1.1.22}$$

which in particular implies that $\mathcal{E}(f,f)$ does not depend the choice of the approximating sequence.

Since $f - f_\ell \in \mathcal{F}_e$ for each ℓ and $\{f_n - f_\ell\} \subset \mathcal{F}$ is its approximating sequence, we have from Lemma 1.1.7 and (1.1.4)

$$\frac{1}{t}\mathcal{A}_{T_t}(f - f_\ell) \le \liminf_{n\to\infty} \frac{1}{t}\mathcal{A}_{T_t}(f_n - f_\ell) \le \lim_{n\to\infty} \|f_n - f_\ell\|_{\mathcal{E}}^2.$$

Therefore, $\lim_{\ell\to\infty} \mathcal{A}_{T_t}(f - f_\ell) = 0$, and by the triangular inequality $\mathcal{A}_{T_t}(f) = \lim_{\ell\to\infty} \mathcal{A}_{T_t}(f_\ell)$, which particularly implies that $\frac{1}{t}\mathcal{A}_{T_t}(f)$ increases as t decreases to 0. Since $\lim_{t\downarrow 0} \frac{1}{t}\mathcal{A}_{T_t}(f_\ell) = \|f_\ell\|_{\mathcal{E}}^2$, we can get from the triangular inequality and the inequality obtained above that

$$\left|\lim_{t\downarrow 0} \sqrt{\frac{1}{t}\mathcal{A}_{T_t}(f)} - \|f_\ell\|_{\mathcal{E}}\right| \le \lim_{t\downarrow 0} \sqrt{\frac{1}{t}\mathcal{A}_{T_t}(f - f_\ell)} \le \lim_{n\to\infty} \|f_n - f_\ell\|_{\mathcal{E}}.$$

The last term tends to 0 as $\ell \to \infty$. The proof of (1.1.22) is complete.

(ii) For any $f \in \mathcal{F}_e$ and any normal contraction φ, we are led from Lemma 1.1.7(iii) and (1.1.22) to

$$\frac{1}{t}\mathcal{A}_{T_t}(\varphi \circ f) \le \frac{1}{t}\mathcal{A}_{T_t}(f) \le \mathcal{E}(f,f) \quad \text{for every } t > 0.$$

Hence it suffices to show $\varphi \circ f \in \mathcal{F}_e$. For an approximating sequence $\{f_n\} \subset \mathcal{F}$ of f, we obtain by Lemma 1.1.7 and (1.1.4)

$$\frac{1}{t}\mathcal{A}_{T_t}(\varphi \circ f_n) \le \frac{1}{t}\mathcal{A}_{T_t}(f_n) \le \mathcal{E}(f_n,f_n).$$

Thus $\varphi \circ f_n \in \mathcal{F}$ with $\mathcal{E}(\varphi \circ f_n, \varphi \circ f_n) \le \mathcal{E}(f_n, f_n)$. This means that $\varphi \circ f_n$ are elements of $\mathcal{F}$ with uniformly bounded $\mathcal{E}$-norm. Therefore, the Cesàro mean $g_k = (1/k)\sum_{j=1}^k \varphi \circ f_{n_j}$ of its suitable subsequence $\{f_{n_j}\}$ is an $\mathcal{E}$-Cauchy sequence by Theorem A.4.1. Since $\lim_{k\to\infty} g_k = \varphi \circ f$ $[m]$, we arrive at $\varphi \circ f \in \mathcal{F}_e$.
(iii) The first identity follows from (1.1.4), Lemma 1.1.7, and (1.1.22). Since every normal contraction operates on $(\mathcal{E}, \mathcal{F})$ by (ii), $(\mathcal{E}, \mathcal{F})$ is Markovian, namely, a Dirichlet form. □

Remark 1.1.8. Property (1.1.22) in particular implies that if $\{f_k, k \ge 1\} \subset \mathcal{F}$ is an $\mathcal{E}$-Cauchy sequence and $f_k \to 0$ $[m]$, then $\mathcal{E}(f_k, f_k) \to 0$. □

COROLLARY 1.1.9. (Fatou's lemma) *Suppose $\{f_k, k \ge 1\} \subset \mathcal{F}_e$ and $f \in \mathcal{F}_e$. If $f_k \to f$ $[m]$, then*

$$\mathcal{E}(f, f) \le \liminf_{k\to\infty} \mathcal{E}(f_k, f_k).$$

Proof. It follows from (1.1.21) and (1.1.22) that

$$\mathcal{E}(f, f) \le \liminf_{t\to 0} \liminf_{k\to\infty} \frac{1}{t} \mathcal{A}_{T_t}(f_k, f_k) \le \liminf_{k\to\infty} \mathcal{E}(f_k, f_k).$$

□

In the remainder of this section, $(\mathcal{E}, \mathcal{F})$ is a Dirichlet form on $L^2(E; m)$.

Exercise 1.1.10. Show that for $f, g \in \mathcal{F}_e \cap L^\infty(E; m)$, $f \cdot g \in \mathcal{F}_e$ and $\|f \cdot g\|_{\mathcal{E}} \le \|g\|_\infty \cdot \|f\|_{\mathcal{E}} + \|f\|_\infty \cdot \|g\|_{\mathcal{E}}$.

We state two lemmas for later use.

LEMMA 1.1.11. (i) *Let $\{\psi_\ell\}_{\ell \ge 1}$ be a sequence of normal contractions satisfying $\lim_{\ell\to\infty} \psi_\ell(t) = t$ for every $t \in \mathbb{R}$. Then $\lim_{\ell\to\infty} \|\psi_\ell(f) - f\|_{\mathcal{E}_1} = 0$ for any $f \in \mathcal{F}$.*
(ii) *Suppose $\{f_n\} \subset \mathcal{F}$ is $\mathcal{E}_1$-convergent to $f \in \mathcal{F}$. Then, for any normal contraction φ, $\{\varphi(f_n)\}$ is $\mathcal{E}_1$-weakly convergent to $\varphi(f)$. If further $\varphi(f) = f$, then the convergence is $\mathcal{E}_1$-strong.*

Proof. (i) If we let $\psi_\ell(f) = f_\ell$, then $f_\ell \in \mathcal{F}$ and $\|f_\ell\|_{\mathcal{E}_1}$ is uniformly dominated by $\|f\|_{\mathcal{E}_1}$. Since $G_1(L^2)$ is $\mathcal{E}_1$-dense in $\mathcal{F}$ by (1.1.6) and $\mathcal{E}_1(f_\ell, G_1 g) = (f_\ell, g) \to (f, g) = \mathcal{E}_1(f, G_1 g)$ for every $g \in L^2$, we can conclude that f_ℓ converges as $\ell \to \infty$ to f weakly in $(\mathcal{F}, \mathcal{E}_1)$. But $\|f_\ell - f\|_{\mathcal{E}_1}^2 \le 2\|f\|_{\mathcal{E}_1}^2 - 2\mathcal{E}_1(f_\ell, f)$ means that the convergence is strong as well.

(ii) $\mathcal{E}_1$-norm of $\varphi(f_n)$ is uniformly bounded and, for any $g \in L^2(E;m)$, $\mathcal{E}_1(G_1 g, \varphi(f_n) - \varphi(f)) = (g, \varphi(f_n) - \varphi(f)) \to 0$, $n \to \infty$. Hence the first assertion follows. If $\varphi(f) = f$, then as $n \to \infty$,

$$\mathcal{E}_1(\varphi(f_n) - f, \varphi(f_n) - f) \leq \mathcal{E}_1(f_n, f_n) + \mathcal{E}_1(f, f) - 2\mathcal{E}_1(f, \varphi(f_n)) \to 0.$$

□

LEMMA 1.1.12. *Let f be an m-measurable function on E with $|f| < \infty$ $[m]$. If, for the contractions φ^ℓ of (1.1.19), $f_\ell := \varphi^\ell \circ f \in \mathcal{F}_e$ for every $\ell \geq 1$, and $\sup_\ell \|f_\ell\|_{\mathcal{E}} < \infty$, then $f \in \mathcal{F}_e$.*

Proof. Without loss of generality, we assume that f is non-negative. For each ℓ, choose an approximating sequence $f_{\ell,k} \in \mathcal{F}$ for f_ℓ such that $\sup_k \|f_{\ell,k}\|_{\mathcal{E}}^2 \leq \|f_\ell\|_{\mathcal{E}}^2 + 1$. We put $v_{\ell,k} = f_{\ell,k}^+ \wedge f_\ell$. Then $v_{\ell,k} \in \mathcal{F}_e \cap L^2(E;m) = \mathcal{F}$ and it converges to f_ℓ m-a.e. as $k \to \infty$ for each ℓ. Furthermore,

$$\|v_{\ell,k}\|_{\mathcal{E}}^2 \leq \|f_{\ell,k}\|_{\mathcal{E}}^2 + \|f_\ell\|_{\mathcal{E}}^2 \leq 2 \sup_\ell \|f_\ell\|_{\mathcal{E}}^2 + 1 < \infty.$$

Take a strictly positive m-measurable function g with $\int_E g\,dm \leq 1$ and put $\widetilde{g}(x) = g(x)/(f(x) \vee 1)$. Since $0 \leq v_{\ell,k}(x) \leq f_\ell(x)$ for $x \in E$, $v_{\ell,k}$ is convergent to f_ℓ in $L^1(E;\widetilde{g}dx)$ as $k \to \infty$ and the latter converges to f in $L^1(E;\widetilde{g}dx)$ as $\ell \to \infty$. Hence $w_\ell = v_{\ell,k_\ell}$ converges to f in $L^1(E;\widetilde{g}dx)$ as well as m-a.e. on E if we choose a suitable subsequence $\{k_\ell\}$ of $\{k\}$. According to the boundedness of $\|v_{\ell,k}\|_{\mathcal{E}}$ obtained above, we can conclude that $f \in \mathcal{F}_e$ admits the Cesàro mean of a subsequence of $w_\ell \in \mathcal{F}$ as its approximating sequence by Theorem A.4.1. □

A numerical function $K(x,B)$ of two variables $x \in E$, $B \in \mathcal{B}(E)$, is said to be a *kernel* on the measurable space $(E, \mathcal{B}(E))$ if, for each fixed $x \in E$, it is a (positive) measure in B and, for each fixed $B \in \mathcal{B}(E)$, it is a $\mathcal{B}(E)$-measurable function in x. We then put

$$Kf(x) := \int_E f(y)K(x,dy), \quad x \in E. \tag{1.1.23}$$

$Kf \in \mathcal{B}_+(E)$ for $f \in \mathcal{B}_+(E)$ because the latter is an increasing limit of simple functions. A kernel K is called *Markovian* if $K(x,E) \leq 1$ for every $x \in E$. A Markovian kernel K defines a linear operator on the space of bounded $\mathcal{B}(E)$-measurable functions by (1.1.23). A Markovian kernel K on E is said to be *conservative* or a *probability kernel* if $K(x,\cdot)$ is a probability measure for each $x \in E$.

We call a kernel $K(x,\cdot)$ (or an operator K) on $(E,\mathcal{B}(E))$ *m-symmetric* if

$$\int_E (Kf)(x)g(x)m(dx) = \int_E f(x)(Kg)(x)m(dx) \quad \text{for } f,g \in \mathcal{B}_+(E). \tag{1.1.24}$$

Let K be an m-symmetric Markovian kernel on $(E,\mathcal{B}(E))$ and $f \in b\mathcal{B}(E) \cap L^2(E;m)$. We then have from (1.1.23) $(Kf)^2(x) \le (Kf^2)(x)$, which yields by integrating with respect to m and using (1.1.24) the contraction property

$$\|Kf\|_2^2 \le \int_E K1(x)f(x)^2m(dx) \le \|f\|_2^2.$$

This means that K can be regarded as a bounded linear operator on the space of m-essentially bounded m-measurable functions in $L^2(E;m)$, which is dense in $L^2(E;m)$. Hence K is uniquely extended to a linear contraction symmetric operator on $L^2(E;m)$.

So far we have assumed that $(E,\mathcal{B}(E))$ is only a measurable space. In the rest of this section, we assume that E is a Hausdorff topological space. In this case, we shall use the notation $\mathcal{B}(E)$ exclusively for the Borel field, namely, the σ-field of subsets of E generated by open sets. The space of $\mathcal{B}(E)$-measurable real-valued functions will be denoted by $\mathcal{B}(E)$. We sometimes need to consider a larger σ-field $\mathcal{B}^*(E)$; the family of universally measurable subsets of E: $\mathcal{B}^*(E) = \bigcap_{\mu\in\mathcal{P}(E)} \mathcal{B}^\mu(E)$, where $\mathcal{P}(E)$ denotes the family of all probability measures on E and $\mathcal{B}^\mu(E)$ is the completion of $\mathcal{B}(E)$ with respect to $\mu \in \mathcal{P}(E)$.

DEFINITION 1.1.13. (i) A family $\{P_t; t \ge 0\}$ is called a *transition function* on $(E,\mathcal{B}(E))$ (resp. $(E,\mathcal{B}^*(E))$) if P_t is a Markovian kernel on $(E,\mathcal{B}(E))$(resp. $(E,\mathcal{B}^*(E))$) for each $t \ge 0$ and the following four conditions are satisfied:

- **(t.1)** $P_sP_tf = P_{s+t}f$ for $s,t \ge 0$ and $f \in b\mathcal{B}(E)$ (resp. $f \in b\mathcal{B}^*(E)$). Here $P_tf(x) := \int_E f(y)P_t(x,dy)$.
- **(t.2)** For each $B \in \mathcal{B}(E)$, $P_t(x,B)$ is $\mathcal{B}([0,\infty)) \times \mathcal{B}(E)$-measurable (resp. $\mathcal{B}([0,\infty)) \times \mathcal{B}^*(E)$-measurable) in two variables $(t,x) \in [0,\infty) \times E$.
- **(t.3)** For each $x \in E$, $P_0(x,\cdot) = \delta_x(\cdot)$, where δ_x denotes the unit mass concentrated at the one-point set $\{x\}$.
- **(t.4)** $\lim_{t\downarrow 0} P_tf(x) = f(x)$ for any $f \in bC(E)$ and $x \in E$.

A transition function $\{P_t; t \ge 0\}$ is called a *transition probability* if P_t is conservative for every $t > 0$.

(ii) A family $\{R_\alpha; \alpha > 0\}$ is called a *resolvent kernel* on $(E,\mathcal{B}(E))$ (resp. $(E,\mathcal{B}^*(E))$) if, for each $\alpha > 0$, αR_α is a Markovian kernel on $(E,\mathcal{B}(E))$

(resp. $(E, \mathcal{B}^*(E))$) and

$$R_\alpha f - R_\beta f + (\alpha - \beta) R_\alpha R_\beta f = 0, \quad \alpha, \beta > 0, \; f \in b\mathcal{B}(E). \tag{1.1.25}$$

$$\lim_{\alpha \to \infty} \alpha R_\alpha f(x) = f(x), \quad x \in E, \quad f \in bC(E). \tag{1.1.26}$$

Property **(t.1)** is called the *semigroup property* or *Chapman-Kolmogorov equation*. Identity (1.1.25) is called the *resolvent equation*. For a transition function $\{P_t; t \geq 0\}$ on $(E, \mathcal{B}(E))$ (resp. $(E, \mathcal{B}^*(E))$), it is easy to verify that

$$R_\alpha f(x) = \int_0^\infty e^{-\alpha t} P_t f(x) dt, \quad \alpha > 0, \quad f \in \mathcal{B}(E), \tag{1.1.27}$$

determines uniquely a resolvent on $(E, \mathcal{B}(E))$, (resp. $(E, \mathcal{B}^*(E))$), which is called the *resolvent kernel of the transition function* $\{P_t; t \geq 0\}$.

A topological space E is called a *Lusin space* (resp. *Radon space*) if it is homeomorphic to a Borel (resp. universally measurable) subset of a compact metric space F. For a topological space E, a measure m on $(E, \mathcal{B}(E))$ is said to be *regular* if, for any $B \in \mathcal{B}(E)$, $m(B) = \inf\{m(U) \colon B \subset U, U \text{ open}\} = \sup\{m(K) \colon K \subset B, K \text{ compact}\}$. Any Radon measure on a locally compact separable metric space is regular. Any finite measure on a Lusin space or on a Radon space is regular.

LEMMA 1.1.14. *Let $\{P_t; t \geq 0\}$ be a family of Markovian kernels on a Lusin space E equipped with the Borel field $\mathcal{B}(E)$ or on a Radon space equipped with the σ-field $\mathcal{B}^*(E)$ of its universally measurable subsets.*
(i) *Suppose $\{P_t; t \geq 0\}$ satisfies* **(t.1), (t.3)** *and*

> **(t.4)′** *For every $f \in bC(E)$, $P_t f(x)$ is right continuous in $t \in [0, \infty)$ for each $x \in E$.*

Then $\{P_t; t \geq 0\}$ is a transition function.

(ii) *Suppose $\{P_t; t \geq 0\}$ satisfies* **(t.1), (t.4)** *and, for a σ-finite measure m on E, $\{P_t; t \geq 0\}$ is m-symmetric in the sense that P_t is m-symmetric for each $t > 0$. Let T_t be the symmetric linear operator on $L^2(E; m)$ uniquely determined by P_t. Then $\{T_t; t \geq 0\}$ is a strongly continuous contraction semigroup on $L^2(E; m)$.*

Proof. We give a proof for a family $\{P_t; t \geq 0\}$ of Markovian kernels on a Lusin space $(E, \mathcal{B}(E))$. The proof for a Radon space $(E, \mathcal{B}^*(E))$ is the same.
(i) It suffices to establish **(t.2)**. Let H be the collection of functions in $b\mathcal{B}(E)$ such that $P_t f(x)$ is measurable in two variables (t, x). H is then a linear space closed under the operation of taking uniformly bounded increasing limits. By **(t.4)′**, it holds that $bC(E) \subset H$. Hence **(t.2)** follows from Proposition A.1.3.

(ii) We may assume that E is a Borel subset of a compact metric space (F, d) and identify $L^2(E; m)$ with $L^2(F; m)$ by setting $m(F \setminus E) = 0$. We first show that $bC(F) \cap L^2(F; m)$ is dense in $L^2(F; m)$. Since m is σ-finite, it suffices to assume that m is a finite measure and that the indicator function of a set $B \in \mathcal{B}(F)$ can be L^2-approximated. For any ε, there exist a compact set K and an open set U such that $K \subset B \subset U$, $m(U \setminus K) < \varepsilon$. If we let $g(x) = d(x, U^c)/(d(x, U^c) + d(x, K))$, $x \in F$, then $g \in bC(F) \cap L^2(F; m)$ and $\|g - \mathbf{1}_B\|_2 < \sqrt{\varepsilon}$.

For any $f \in L^2(E; m)$ and $\varepsilon > 0$, take a function $g \in bC(F) \cap L^2(E; m)$ such that $\|f - g\|_2 < \varepsilon$. Because of the contraction property of $\{T_t; t > 0\}$, we then have $\|T_t f - f\|_2 \le \|P_t g - g\|_2 + 2\varepsilon$. Further,

$$\|P_t g - g\|_2^2 \le 2\|g\|_2^2 - 2(g, P_t g),$$

which tends to 0 as $t \downarrow 0$ by **(t.4)** and the Lebesgue-dominated convergence theorem. □

By virtue of Lemma 1.1.14, any m-symmetric transition function $\{P_t; t \ge 0\}$ on a Lusin space $(E, \mathcal{B}(E))$ or a Radon space $(E, \mathcal{B}^*(E))$ determines a unique strongly continuous contraction semigroup $\{T_t; t \ge 0\}$ on $L^2(E; m)$, which in turn decides a Dirichlet form $(\mathcal{E}, \mathcal{F})$ on $L^2(E; m)$ according to Theorem 1.1.3. $(\mathcal{E}, \mathcal{F})$ is called the *Dirichlet form of the transition function* $\{P_t; t \ge 0\}$. In this case, the resolvent $\{G_\alpha; \alpha > 0\}$ of $\{T_t; t \ge 0\}$ is the unique extension of the resolvent kernel $\{R_\alpha; \alpha > 0\}$ of $\{P_t; t \ge 0\}$ from $b\mathcal{B}(E) \cap L^2(E; m)$ to $L^2(E; m)$. Moreover, we have from (1.1.6) that for $f \in b\mathcal{B}(E) \cap L^2(E; m)$,

$$R_\alpha f \in \mathcal{F} \quad \text{with} \quad \mathcal{E}_\alpha(R_\alpha f, v) = (f, v) \ \text{ for every } v \in \mathcal{F}. \tag{1.1.28}$$

Conversely, if the resolvent kernel $\{R_\alpha; \alpha > 0\}$ of a transition function $\{P_t; t \ge 0\}$ satisfies (1.1.28) for a Dirichlet form $(\mathcal{E}, \mathcal{F})$ on $L^2(E; m)$, then $\{P_t; t \ge 0\}$ is m-symmetric and its Dirichlet form coincides with $(\mathcal{E}, \mathcal{F})$.

In the rest of this chapter, we give a quick introduction to the basic theory of quasi-regular Dirichlet forms. The importance of a quasi-regular Dirichlet form is that they are in one-to-one correspondence with symmetric Markov processes having some nice properties. We will show that any quasi-regular Dirichlet form is quasi-homeomorphic to a regular Dirichlet form on a locally compact separable metric space. Thus the study of quasi-regular Dirichlet forms can be reduced to that of regular Dirichlet forms.

1.2. EXCESSIVE FUNCTIONS AND CAPACITIES

In this section, let E be a Hausdorff topological space with the Borel σ-field $\mathcal{B}(E)$ being assumed to be generated by the continuous functions on

E and m be a σ-finite measure with $\text{supp}[m] = E$. Here for a measure ν on E, its *support* $\text{supp}[\nu]$ is by definition the smallest closed set outside which ν vanishes. Let $(\mathcal{E}, \mathcal{F})$ be a symmetric Dirichlet form on $L^2(E; m)$, and $\{T_t; t \geq 0\}$ and $\{G_\alpha; \alpha > 0\}$ be its associated semigroup and resolvents on $L^2(E; m)$.

DEFINITION 1.2.1. For $\alpha > 0$, $u \in L^2(E; m)$ is called *α-excessive* if $e^{-\alpha t} T_t u \leq u$ m-a.e. for every $t > 0$.

Remark 1.2.2. (i) If u is α-excessive, then $u \geq 0$. This is because

$$\|e^{-\alpha t} T_t u\|_2 = e^{-\alpha t} \|T_t u\|_2 \leq e^{-\alpha t} \|u\|_2$$

and so $u \geq \lim_{t \to \infty} e^{-\alpha t} T_t u = 0$.
(ii) The constant function 1 is α-excessive if $m(E) < \infty$. For $f \in L^2_+(E; m)$, $G_\alpha f$ is α-excessive.
(iii) If $u_1 \geq 0$, $u_2 \geq 0$ are α-excessive functions, then so are $u_1 \wedge u_2$ and $u_1 \wedge 1$. □

LEMMA 1.2.3. *Let $u \in L^2_+(E; m)$ be α-excessive for $\alpha > 0$. Assume there is $v \in \mathcal{F}$ such that $u \leq v$. Then $u \in \mathcal{F}$ and $\mathcal{E}_\alpha(u, u) \leq \mathcal{E}_\alpha(v, v)$.*

Proof. By the symmetry and contraction property of T_t in $L^2(E; m)$, for each $t > 0$, $(f, g - e^{-\alpha t} T_t g)$ is a non-negative symmetric bilinear form on $L^2(E; m)$. So it satisfies the following Cauchy-Schwarz inequality:

$$|(f, g - e^{-\alpha t} T_t g)| \leq (f, f - e^{-\alpha t} T_t f)^{1/2} \cdot (g, g - e^{-\alpha t} T_t g)^{1/2}.$$

Thus we have by the α-excessiveness of u,

$$(u - e^{-\alpha t} T_t u, u) \leq (u - e^{-\alpha t} T_t u, v) \leq (u, u - e^{-\alpha t} T_t u)^{1/2} \cdot (v, v - e^{-\alpha t} T_t v)^{1/2},$$

and so

$$(u - e^{-\alpha t} T_t u, u) \leq (v, v - e^{-\alpha t} T_t v).$$

It follows then that

$$\begin{aligned} \lim_{t \to 0} \frac{1}{t}(u - T_t u, u) &= \lim_{t \to 0} \frac{1}{t}(u - e^{-\alpha t} T_t u, u) + \lim_{t \to 0} \frac{1}{t}(e^{-\alpha t} - 1)(T_t u, u) \\ &\leq \lim_{t \to 0} \frac{1}{t}(v - e^{-\alpha t} T_t v, v) - \alpha(u, u) \\ &= \mathcal{E}(v, v) + \alpha(v, v) - \alpha(u, u) < \infty. \end{aligned}$$

We conclude from (1.1.4)–(1.1.5) that $u \in \mathcal{F}$ with $\mathcal{E}_\alpha(u, u) \leq \mathcal{E}_\alpha(v, v)$. □

LEMMA 1.2.4. *The following statements are equivalent for $u \in \mathcal{F}$ and $\alpha > 0$:*
(i) *u is α-excessive.*
(ii) *$\mathcal{E}_\alpha(u, v) \geq 0$ for every non-negative $v \in \mathcal{F}$.*

Proof. (i) $\Rightarrow$ (ii): It follow from (1.1.5) that

$$0 \leq \frac{1}{t}(u - e^{-\alpha t}T_t u, v) = \frac{1}{t}(u - T_t u, v) + \frac{1-e^{-\alpha t}}{t}(T_t u, v) \to \mathcal{E}_\alpha(u, v) \quad (1.2.1)$$

as $t \downarrow 0$.
(ii) $\Rightarrow$ (i): For $v \in L^2_+(E; m)$ and $t > 0$, since

$$G_\alpha v - e^{-\alpha t}T_t G_\alpha v = \int_0^t e^{-\alpha s}T_s v ds \geq 0,$$

we have

$$(u - e^{-\alpha t}T_t u,\ v) = (u,\ v - e^{-\alpha t}T_t v) = \mathcal{E}_\alpha(u,\ G_\alpha(v - e^{-\alpha t}T_t v)) \geq 0.$$

This implies that $u - e^{-\alpha t}T_t u \geq 0$ and so (i) holds. □

For a closed subset F of E, define

$$\mathcal{F}_F := \{f \in \mathcal{F} : f = 0\ m\text{–a.e. } on\ E \setminus F\}. \quad (1.2.2)$$

THEOREM 1.2.5. *Let $\alpha > 0$ and f be a non-negative function defined on E. For an open set D, denote $\mathcal{L}_{D,f} = \{u \in \mathcal{F} : u \geq f$ m-a.e. on $D\}$. Suppose $\mathcal{L}_{D,f} \neq \emptyset$. Then*
(i) *there is a unique $f_D \in \mathcal{L}_{D,f}$ such that*

$$\mathcal{E}_\alpha(u, u) \geq \mathcal{E}_\alpha(f_D, f_D) \quad \text{for every } u \in \mathcal{L}_{D,f}.$$

(ii) *f_D is the unique function in $\mathcal{L}_{D,f}$ such that*

$$\mathcal{E}_\alpha(u, f_D) \geq \mathcal{E}_\alpha(f_D, f_D) \quad \text{for every } u \in \mathcal{L}_{D,f}.$$

(iii) *$\mathcal{E}_\alpha(f_D, v) \geq 0$ for every $v \in \mathcal{F}$ with $v \geq 0$ m-a.e. on D. In particular, f_D is α-excessive and $\mathcal{E}_\alpha(f_D, v) = 0$ for every $v \in \mathcal{F}_{D^c}$.*
(iv) *$f_D \leq f$ if and only if $f_D \wedge f$ is an α-excessive function. In this case, $f_D = f$ m-a.e. on D. f_D is the minimum element among α-excessive functions in $\mathcal{L}_{D,f}$ in the sense that, if $u \in \mathcal{L}_{D,f}$ is α-excessive, then $f_D \leq u$.*
(v) *If open sets $D_1 \subset D_2$ and $\mathcal{L}_{D_2,f} \neq \emptyset$, then $f_{D_1} \leq f_{D_2}$ and*

$$\mathcal{E}_\alpha(f_{D_1}, f_{D_1}) \leq \mathcal{E}_\alpha(f_{D_2}, f_{D_2}).$$

(vi) *For open sets $D_1 \subset D_2$, if $f \wedge f_{D_2}$ is an α-excessive function, then $(f_{D_2})_{D_1} = f_{D_1}$. If further $f \wedge f_{D_1}$ is α-excessive, then*

$$\mathcal{E}_\alpha(f_{D_1}, f_{D_2}) = \mathcal{E}_\alpha(f_{D_1}, f_{D_1}).$$

(vii) *For open sets $D_1 \subset D_2$, $(f_{D_1})_{D_2} = f_{D_1}$.*

Proof. (i) Because $\mathcal{L}_{D,f}$ is a closed convex set in the Hilbert space $(\mathcal{F}, \mathcal{E}_\alpha)$, it has a unique minimizer f_D.
(ii) For every $u \in \mathcal{L}_{D,f}$ and $0 < \varepsilon < 1$, $f_D + \varepsilon(u - f_D) = (1-\varepsilon)f_D + \varepsilon u \in \mathcal{L}_{D,f}$ and so $\mathcal{E}_\alpha(f_D + \varepsilon(u - f_D), f_D + \varepsilon(u - f_D)) \geq \mathcal{E}_\alpha(f_D, f_D)$. This implies that $\mathcal{E}_\alpha(f_D, u - f_D) \geq 0$. Now suppose $v \in \mathcal{L}_{D,f}$ is another function such that for every $u \in \mathcal{L}_{D,f}$, $\mathcal{E}_\alpha(v, u - v) \geq 0$. As $f_D \in \mathcal{L}_{D,f}$, $\mathcal{E}_\alpha(v, f_D - v) \geq 0$. But with $\mathcal{E}_\alpha(f_D, v - f_D) \geq 0$, we have $\mathcal{E}_\alpha(f_D - v, f_D - v) \leq 0$. Therefore, $v = f_D$.
(iii) For any $v \in \mathcal{F}$ with $v \geq 0$ m-a.e. on D, $f_D + \varepsilon v \in \mathcal{L}_{D,f}$ for every $\varepsilon > 0$. One immediately deduces from $\mathcal{E}_\alpha(f_D + \varepsilon v, f_D + \varepsilon v) \geq \mathcal{E}_\alpha(f_D, f_D)$ that $\mathcal{E}_\alpha(f_D, v) \geq 0$.
(iv) This follows immediately from (iii) and Lemma 1.2.3.
(v) The first part follows from (iv). The second part follows from (i).
(vi) By (iv), $f = f_{D_2}$ on D_2 and hence by definition, $(f_{D_2})_{D_1} = f_{D_1}$. The second assertion follows from (iii) and (iv).
(vii) For every $u \in \mathcal{L}_{D_2, f_{D_1}}$, $\mathcal{E}_\alpha(f_{D_1}, u - f_{D_1}) \geq 0$ by (iii). We therefore have by (ii) that $f_{D_1} = (f_{D_1})_{D_2}$. □

The function f_D is called the *α-reduced function* of f on D.

Remark 1.2.6. (i) If f is α-excessive in $\mathcal{F}$, then f_D is the $\mathcal{E}_\alpha$-orthogonal projection of f into the $\mathcal{E}_\alpha$-orthogonal complement of $\mathcal{F}_{D^c}$. This is because $f = (f - f_D) + f_D$, where $f - f_D \in \mathcal{F}_{D^c}$ by Theorem 1.2.5(iv) and f_D is $\mathcal{E}_\alpha$-orthogonal to $\mathcal{F}_{D^c}$ by Theorem 1.2.5(iii).
(ii) By (iii) and (iv) of Theorem 1.2.5, if $g \in \mathcal{F}$ is α-excessive, then $\mathcal{E}_\alpha(f_D, g) = \mathcal{E}_\alpha(f_D, g_D)$. □

DEFINITION 1.2.7. ((h, α)-capacity) Fix $\alpha > 0$. Let $h \geq 0$ be a function on E satisfying one of the following two conditions:
(i) $h \in \mathcal{F}$ and h is α-excessive;
(ii) $h \wedge h_D$ is a α-excessive function for every open set $D \subset E$ with $\mathcal{L}_{D,h} \neq \emptyset$. (This is equivalent to, by Theorem 1.2.5(iv), that $h \geq h_D$ for every open set $D \subset E$ with $\mathcal{L}_{D,h} \neq \emptyset$). Define for open subset $D \subset E$,

$$\mathrm{Cap}_{h,\alpha}(D) := \begin{cases} \mathcal{E}_\alpha(h_D, h_D) & \text{if } \mathcal{L}_{D,h} \neq \emptyset, \\ +\infty & \text{otherwise,} \end{cases} \tag{1.2.3}$$

and for an arbitrary subset $A \subset E$,

$$\mathrm{Cap}_{h,\alpha}(A) := \inf\left\{\mathrm{Cap}_{h,\alpha}(D) : \text{ open set } D \supset A\right\}. \tag{1.2.4}$$

Remark 1.2.8. (i) Important cases are $h = 1$ and $h = G_\alpha\varphi$ for some strictly positive $\varphi \in L^2(E;m)$.
(ii) Under either of conditions (i) and (ii), $h = h_D\ [m]$ on D whenever $\mathcal{L}_{D,h} \neq \emptyset$.
(iii) When $h > 0\ [m]$ on E, then $\mathrm{Cap}_{h,\alpha}(A) = 0$ implies that $m(A) = 0$.
(iv) If $0 \le h^{(1)} \le h^{(2)}$ are two functions satisfying either condition (i) or (ii) in Definition 1.2.7, we have by Theorem 1.2.5(iv) that $h_D^{(1)} \le h_D^{(2)}$ for any open set D with $\mathcal{L}_{D,h^{(2)}} \neq \emptyset$. Therefore $\mathrm{Cap}_{h^{(1)},\alpha}(D) \le \mathrm{Cap}_{h^{(2)},\alpha}(D)$ by Lemma 1.2.3.
(v) We shall use the following comparison in $\alpha > 0$ for the capacity: if h_1 is 1-excessive, h_2 is 2-excessive, and $h_2 \le h_1$, then

$$\mathrm{Cap}_{h_2,2}(A) \le 2\mathrm{Cap}_{h_1,1}(A), \quad A \subset E.$$

In fact, we have for an open set D,

$$\begin{aligned}\mathrm{Cap}_{h_2,2}(D) &= \inf_{u\in\mathcal{F},\ u\ge h_2 \text{ on } D} \mathcal{E}_2(u,u) \le \inf_{u\in\mathcal{F},\ u\ge h_1 \text{ on } D} 2\mathcal{E}_1(u,u)\\ &\le 2\mathrm{Cap}_{h_1,1}(D).\end{aligned}$$

□

In the remainder of this section $h \ge 0$ is a non-trivial function on E satisfying one of the conditions (i) or (ii) in Definition 1.2.7.

THEOREM 1.2.9. (i) *For open sets* $D_1 \subset D_2$, $\mathrm{Cap}_{h,\alpha}(D_1) \le \mathrm{Cap}_{h,\alpha}(D_2)$.
(ii) *For open sets* D_1 *and* D_2,

$$\mathrm{Cap}_{h,\alpha}(D_1 \cup D_2) + \mathrm{Cap}_{h,\alpha}(D_1 \cap D_2) \le \mathrm{Cap}_{h,\alpha}(D_1) + \mathrm{Cap}_{h,\alpha}(D_2).$$

(iii) *For any increasing sequence of open sets* $\{D_k, k \ge 1\}$,

$$\mathrm{Cap}_{h,\alpha}(\cup_{k\ge1} D_k) = \sup_{k\ge1} \mathrm{Cap}_{h,\alpha}(D_k).$$

(iv) *For any decreasing sequence of open sets* $\{D_k, k \ge 1\}$ *with* $\mathcal{L}_{D_1,h} \neq \emptyset$, $\{h_{D_k}; k \ge 1\}$ *is decreasing to as well as* $\mathcal{E}_\alpha$*-convergent to a function* $h_\infty \in \mathcal{F}$, *and* $\inf_{k\ge1} \mathrm{Cap}_{h,\alpha}(D_k) = \mathcal{E}_\alpha(h_\infty, h_\infty)$.

Proof. (i) follows from Theorem 1.2.5(v).

(ii) Without loss of generality, we may assume $\mathrm{Cap}_{h,\alpha}(D_i) < \infty$ for $i = 1, 2$. By the property of h_D,

$$\begin{aligned}
&\mathrm{Cap}_{h,\alpha}(D_1 \cup D_2) + \mathrm{Cap}_{h,\alpha}(D_1 \cap D_2) \\
&\quad \le \mathcal{E}_\alpha(h_{D_1} \vee h_{D_2},\, h_{D_1} \vee h_{D_2}) + \mathcal{E}_\alpha(h_{D_1} \wedge h_{D_2},\, h_{D_1} \wedge h_{D_2}) \\
&\quad = \tfrac{1}{2}\mathcal{E}_\alpha(h_{D_1} + h_{D_2},\, h_{D_1} + h_{D_2}) + \tfrac{1}{2}\mathcal{E}_\alpha(|h_{D_1} - h_{D_2}|,\, |h_{D_1} - h_{D_2}|) \\
&\quad \le \tfrac{1}{2}\mathcal{E}_\alpha(h_{D_1} + h_{D_2},\, h_{D_1} + h_{D_2}) + \tfrac{1}{2}\mathcal{E}_\alpha(h_{D_1} - h_{D_2},\, h_{D_1} - h_{D_2}) \\
&\quad = \mathcal{E}_\alpha(h_{D_1},\, h_{D_1}) + \mathcal{E}_\alpha(h_{D_2},\, h_{D_2}) \\
&\quad = \mathrm{Cap}_{h,\alpha}(D_1) + \mathrm{Cap}_{h,\alpha}(D_2).
\end{aligned}$$

(iii) Without loss of generality, assume that $\sup_{k\ge 1} \mathrm{Cap}_{h,\alpha}(D_k) < \infty$. For $j > k$, we have from Theorem 1.2.5(vi)

$$\begin{aligned}
&\mathcal{E}_\alpha(h_{D_j} - h_{D_k},\, h_{D_j} - h_{D_k}) \\
&\quad = \mathcal{E}_\alpha(h_{D_j}, h_{D_j}) - 2\mathcal{E}_\alpha(h_{D_j}, h_{D_k}) + \mathcal{E}_\alpha(h_{D_k}, h_{D_k}) \\
&\quad = \mathcal{E}_\alpha(h_{D_j}, h_{D_j}) - \mathcal{E}_\alpha(h_{D_k}, h_{D_k}) \\
&\quad = \mathrm{Cap}_{h,\alpha}(D_j) - \mathrm{Cap}_{h,\alpha}(D_k) \to 0 \qquad \text{as } j, k \to \infty.
\end{aligned}$$

So h_{D_k} is $\mathcal{E}_\alpha$-convergent to some $h_\infty \in \mathcal{F}$. As $h_\infty = h_k = h\ [m]$ on D_k, we have $h_\infty = h\ [m]$ on $\cup_{k\ge 1} D_k$. For $v \in \mathcal{L}_{\cup_{k\ge 1} D_k, h}$, by Theorem 1.2.5(ii),

$$\mathcal{E}_\alpha(h_\infty, v) = \lim_{k\to\infty} \mathcal{E}_\alpha(h_{D_k}, v) \ge \lim_{k\to\infty} \mathcal{E}_\alpha(h_{D_k}, h_{D_k}) = \mathcal{E}_\alpha(h_\infty, h_\infty).$$

By Theorem 1.2.5(ii) again, $h_\infty = h_{\cup_{k\ge 1} D_k}$ and therefore

$$\begin{aligned}
\sup_{k\ge 1} \mathrm{Cap}_{h,\alpha}(D_k) &= \lim_{k\to\infty} \mathrm{Cap}_{h,\alpha}(D_k) = \lim_{k\to\infty} \mathcal{E}_\alpha(h_{D_k}, h_{D_k}) \\
&= \mathcal{E}_\alpha(h_\infty, h_\infty) = \mathrm{Cap}_{h,\alpha}(\cup_{k\ge 1} D_k).
\end{aligned}$$

(iv) $\{h_{D_k}\}$ is decreasing by Theorem 1.2.5(v). For $j > k$, we have from Theorem 1.2.5(vi)

$$\begin{aligned}
&\mathcal{E}_\alpha(h_{D_j} - h_{D_k},\, h_{D_j} - h_{D_k}) \\
&\quad = \mathcal{E}_\alpha(h_{D_j}, h_{D_j}) - 2\mathcal{E}_\alpha(h_{D_j}, h_{D_k}) + \mathcal{E}_\alpha(h_{D_k}, h_{D_k}) \\
&\quad = \mathcal{E}_\alpha(h_{D_k}, h_{D_k}) - \mathcal{E}_\alpha(h_{D_j}, h_{D_j}),
\end{aligned}$$

which leads us to (iv). □

Observe that the proof of Theorem 1.2.9(iii) shows that h_{D_k} converges to $h_{\cup_{k\ge 1} D_k}$ both monotonously and in $(\mathcal{F}, \mathcal{E}_\alpha)$.

THEOREM 1.2.10. $\mathrm{Cap}_{h,\alpha}$ *is a Choquet $\mathcal{K}$-capacity, where $\mathcal{K}$ denotes all the compact subsets of E; that is,*
(i) *For any subsets $A \subset B$, $\mathrm{Cap}_{h,\alpha}(A) \leq \mathrm{Cap}_{h,\alpha}(B)$;*
(ii) *For any increasing sequence of subsets $\{A_j, j \geq 1\}$,*

$$\mathrm{Cap}_{h,\alpha}(\cup_{j\geq 1} A_j) = \sup_{j\geq 1} \mathrm{Cap}_{h,\alpha}(A_j);$$

(iii) *For any decreasing sequence of compact subsets $\{K_j, j \geq 1\}$,*

$$\mathrm{Cap}_{h,\alpha}(\cap_{j\geq 1} K_j) = \inf_{j\geq 1} \mathrm{Cap}_{h,\alpha}(K_j).$$

Proof. (i) follows immediately from Theorem 1.2.9(i) and the definition of $\mathrm{Cap}_{h,\alpha}$.
(ii) Without loss of generality, we may assume that $\mathrm{Cap}_{h,\alpha}(A_j) < \infty$ for every $j \geq 1$. In view of (i), it suffices to show

$$\mathrm{Cap}_{h,\alpha}(\cup_{j\geq 1} A_j) \leq \sup_{j\geq 1} \mathrm{Cap}_{h,\alpha}(A_j).$$

For any $\varepsilon > 0$, let an open set $O_j \supset A_j$ be such that $\mathrm{Cap}_{h,\alpha}(O_j) < \mathrm{Cap}_{h,\alpha}(A_j) + 2^{-j}\varepsilon$. Define $D_j := \cup_{k=1}^{j} O_k$. Then $\{D_j, j \geq 1\}$ is an increasing sequence of open sets. We claim that

$$\mathrm{Cap}_{h,\alpha}(D_j) \leq \mathrm{Cap}_{h,\alpha}(A_j) + (1 - 2^{-j})\varepsilon \qquad \text{for every } j \geq 1. \tag{1.2.5}$$

We prove this by induction. Clearly this is true for $j = 1$. Suppose it is true for $j \geq 1$. Since $D_{j+1} = D_j \cup O_{j+1}$, we have by Theorem 1.2.9(ii),

$$\mathrm{Cap}_{h,\alpha}(D_{j+1}) + \mathrm{Cap}_{h,\alpha}(D_j \cap O_{j+1}) \leq \mathrm{Cap}_{h,\alpha}(D_j) + \mathrm{Cap}_{h,\alpha}(O_{j+1}).$$

But as $A_j \subset D_j \cap O_{j+1}$, we have

$$\begin{aligned}
\mathrm{Cap}_{h,\alpha}(D_{j+1}) &\leq \mathrm{Cap}_{h,\alpha}(D_j) + \mathrm{Cap}_{h,\alpha}(O_{j+1}) - \mathrm{Cap}_{h,\alpha}(A_j) \\
&\leq \mathrm{Cap}_{h,\alpha}(O_{j+1}) + (1 - 2^{-j})\varepsilon \\
&\leq \mathrm{Cap}_{h,\alpha}(A_{j+1}) + 2^{-j-1}\varepsilon + (1 - 2^{-j})\varepsilon \\
&= \mathrm{Cap}_{h,\alpha}(A_{j+1}) + (1 - 2^{-j-1})\varepsilon.
\end{aligned}$$

This proves the claim (1.2.5). Therefore, we have

$$\begin{aligned}
\mathrm{Cap}_{h,\alpha}(\cup_{j\geq 1} A_j) &\leq \mathrm{Cap}_{h,\alpha}(\cup_{j\geq 1} D_j) = \sup_{j\geq 1} \mathrm{Cap}_{h,\alpha}(D_j) \\
&\leq \sup_{j\geq 1} \mathrm{Cap}_{h,\alpha}(A_j) + \varepsilon.
\end{aligned}$$

Passing $\varepsilon \downarrow 0$, we get $\mathrm{Cap}_{h,\alpha}(\cup_{j\geq 1} A_j) \leq \sup_{j\geq 1} \mathrm{Cap}_{h,\alpha}(A_j)$, and so (ii) is established.

(iii) It suffices to show that $\mathrm{Cap}_{h,\alpha}(\cap_{j\geq 1}K_j) \geq \inf_{j\geq 1}\mathrm{Cap}_{h,\alpha}(K_j)$. We may assume that $\mathrm{Cap}_{h,\alpha}(\cap_{j\geq 1}K_j) < \infty$. For any $\varepsilon > 0$, let D be an open set such that $D \supset \cap_{j\geq 1}K_j$ and $\mathrm{Cap}_{h,\alpha}(D) < \mathrm{Cap}_{h,\alpha}(\cap_{j\geq 1}K_j) + \varepsilon$. Since K_j is compact for every $j \geq 1$, there is $n \geq 1$ such that $D \supset K_n$. Therefore, $\mathrm{Cap}_{h,\alpha}(D) \geq \mathrm{Cap}_{h,\alpha}(K_n) \geq \inf_{j\geq 1}\mathrm{Cap}_{h,\alpha}(K_j)$. This yields that, after letting $\varepsilon \downarrow 0$, $\mathrm{Cap}_{h,\alpha}(\cap_{j\geq 1}K_j) \geq \inf_{j\geq 1}\mathrm{Cap}_{h,\alpha}(K_j)$. □

DEFINITION 1.2.11. A subset $A \subset E$ is said to be *C-capacitable* for a set function C on E if

$$C(A) = \sup_{\substack{K\subset A\\ K:\ \text{compact}}} C(K).$$

The celebrated Choquet's theorem says that every $\mathcal{K}$-analytic set is C-capacitable for any Choquet $\mathcal{K}$-capacity C (cf. [37, III: 28]). It is known that any Borel subset of a compact metric space is $\mathcal{K}$-analytic (cf. [37, III: 7,13]). In particular, for a Lusin space E, it holds from Theorem 1.2.10 that

$$\mathrm{Cap}_{h,\alpha}(B) = \sup_{K\subset B,\ K\ \text{compact}} \mathrm{Cap}_{h,\alpha}(K) \quad \text{for } B \in \mathcal{B}(E). \tag{1.2.6}$$

DEFINITION 1.2.12. (i) An increasing sequence $\{F_k, k \geq 1\}$ of closed sets of E is an *$\mathcal{E}$-nest* if $\cup_{k\geq 1}\mathcal{F}_{F_k}$ is $\mathcal{E}_1$-dense in $\mathcal{F}$, where $\mathcal{E}_1 = \mathcal{E} + (\ ,\ )_{L^2(E;m)}$.
(ii) A subset N of E is *$\mathcal{E}$-polar* if there is an $\mathcal{E}$-nest $\{F_k, k \geq 1\}$ such that $N \subset \cap_{k\geq 1}(E \setminus F_k)$.
(iii) A statement depending on $x \in A$ is said to hold *$\mathcal{E}$-quasi-everywhere* ($\mathcal{E}$-q.e. in abbreviation) on A if there is an $\mathcal{E}$-polar set $N \subset A$ such that the statement is true for every $x \in A \setminus N$.
(iv) A function f on E is said to be *$\mathcal{E}$-quasi-continuous* if there is an $\mathcal{E}$-nest $\{F_k, k \geq 1\}$ such that $f|_{F_k}$ is finite and continuous on F_k for each $k \geq 1$, which will be denoted in abbreviation as $f \in C(\{F_k\})$.
(v) An increasing sequence $\{F_k\}$ of closed sets of E is *$\mathrm{Cap}_{h,\alpha}$-nest* if

$$\lim_{k\to\infty} \mathrm{Cap}_{h,\alpha}(E \setminus F_k) = 0.$$

(vi) A subset N of E is *$\mathrm{Cap}_{h,\alpha}$-polar* if $\mathrm{Cap}_{h,\alpha}(N) = 0$.

Obviously, if $\{F_n, n \geq 1\}$ of E is an $\mathcal{E}$-nest, then so is $\{K_n, n \geq 1\}$ where $K_n = \mathrm{supp}[\mathbf{1}_{F_n} \cdot m]$. Since $\mathcal{F}$ is a dense linear subspace in $L^2(E;m)$, every $\mathcal{E}$-polar set is m-null.

THEOREM 1.2.13. *Fix an arbitrary $\alpha > 0$ and let $h = G_\alpha\varphi$ for some strictly positive $\varphi \in L^2(E;m)$. Let $\{F_k, k \geq 1\}$ be an increasing sequence of closed subsets. Then*

(i) $\{F_k, k \geq 1\}$ *is an* $\mathcal{E}$*-nest if and only if it is a* $\mathrm{Cap}_{h,\alpha}$*-nest.*
(ii) *A set* $N \subset E$ *is* $\mathcal{E}$*-polar if and only if it is* $\mathrm{Cap}_{h,\alpha}$*-polar.*
(iii) *If* $\{F_k^1, k \geq 1\}$ *and* $\{F_k^2, k \geq 1\}$ *are two* $\mathcal{E}$*-nests, then* $\{F_k^1 \cap F_k^2, k \geq 1\}$ *is also an* $\mathcal{E}$*-nest.*

Proof. Let $h_k := h_{F_k^c}$. By Theorem 1.2.9(iv), h_k is decreasing to as well as $\mathcal{E}_\alpha$-convergent to some non-negative $h_\infty \in \mathcal{F}$ and

$$\lim_{k\to\infty} \mathrm{Cap}_{h,\alpha}(F_k^c) = \mathcal{E}_\alpha(h_\infty, h_\infty). \tag{1.2.7}$$

In particular, for every $v \in \cup_{k\geq 1}\mathcal{F}_{F_k}$, by Theorem 1.2.5(iii),

$$\mathcal{E}_\alpha(h_\infty, v) = \lim_{k\to\infty} \mathcal{E}_\alpha(h_k, v) = 0. \tag{1.2.8}$$

Now suppose $\{F_k, k \geq 1\}$ is an $\mathcal{E}$-nest. Then by (1.2.8), $h_\infty = 0$ and so $\lim_{k\to\infty} \mathrm{Cap}_{h,\alpha}(F_k^c) = 0$ by (1.2.7).

Conversely, suppose that $\lim_{k\to\infty} \mathrm{Cap}_{h,\alpha}(F_k^c) = 0$. Then $h_\infty = 0$ by (1.2.7). For any α-excessive function $v \in \mathcal{F}$, denote $v_{F_k^c}$ by v_k so $v - v_k \in \mathcal{F}_{F_k}$. By the same reasoning as above for h, we see that v_k is decreasing to as well as $\mathcal{E}_\alpha$-convergent to some $v_\infty \in \mathcal{F}$. By Remark 1.2.6(i),

$$\begin{aligned}\int_E \varphi(x)v_\infty(x)m(dx) = \mathcal{E}_\alpha(h, v_\infty) &= \lim_{k\to\infty} \mathcal{E}_\alpha(h, v_k)\\ &= \lim_{k\to\infty} \mathcal{E}_\alpha(h_k, v) = \mathcal{E}_\alpha(h_\infty, v) = 0.\end{aligned}$$

This implies that $v_\infty = 0$ $[m]$ on E as $\varphi > 0$ $[m]$ on E. Therefore, $v = \lim_{k\to\infty}(v - v_k)$ is in the $\mathcal{E}_\alpha$-completion of $\cup_{k\geq 1}\mathcal{F}_{F_k}$. Since $G_\alpha L^2(E; m)$ is $\mathcal{E}_\alpha$ dense in $\mathcal{F}$, we have that $\cup_{k\geq 1}\mathcal{F}_{F_k}$ is $\mathcal{E}_\alpha$-dense (and hence $\mathcal{E}_1$-dense) in $\mathcal{F}$; that is, $\{F_k, k \geq 1\}$ is an $\mathcal{E}$-nest.

The second assertion of the theorem is immediate from the first. The third follows from the first and Theorem 1.2.9. □

THEOREM 1.2.14. *Suppose that* $h > 0$ *is a function that satisfies one of the conditions in Definition 1.2.7 for* $\alpha = 1$*. Suppose there is an increasing sequence of open sets* $\{D_k, k \geq 1\}$ *of finite* $(h, 1)$*-capacity such that* $\overline{D}_k \subset D_{k+1}$*,* $k \geq 1$*, and* $\{\overline{D}_k, k \geq 1\}$ *constitutes an* $\mathcal{E}$*-nest.*
(i) *An increasing sequence* $\{F_k, k \geq 1\}$ *of closed subsets of* E *is an* $\mathcal{E}$*-nest if and only if* $\lim_{k\to\infty} \mathrm{Cap}_{h,1}(D_n \setminus F_k) = 0$ *for every* $n \geq 1$*.*
(ii) *A set* $N \subset E$ *is* $\mathcal{E}$*-polar if and only if it is* $\mathrm{Cap}_{h,1}$*-polar.*

Proof. (i) *Proof of the "only if" part:* Suppose $\{F_k, k \geq 1\}$ is an $\mathcal{E}$-nest. For a fixed n, let $g_k = h_{D_n\setminus F_k}$. By Theorem 1.2.9(iv), $\{g_k, k \geq 1\}$ is then decreasing to and $\mathcal{E}_1$-convergent to a function $g_\infty \in \mathcal{F}$. By Theorem 1.2.5(iii), g_k is $\mathcal{E}_1$-orthogonal to the space $\mathcal{F}_{\{D_n\setminus F_\ell\}^c} \supset \mathcal{F}_{F_\ell}$ for any $k \geq \ell$. Hence g_∞ is

$\mathcal{E}_1$-orthogonal to $\cup_\ell \mathcal{F}_{F_\ell}$, which is $\mathcal{E}_1$-dense in $\mathcal{F}$. Thus $g_\infty = 0$ and so for each fixed $n \geq 1$, $\mathrm{Cap}_{h,1}(D_n \setminus F_k) = \mathcal{E}_1(g_k, g_k) \to 0$ as $k \to \infty$.
Proof of the "if" part: By the assumption and Theorem 1.2.5(iv),

$$h_1 := \sum_{n=1}^{\infty} 2^{-n} \|h_{D_n}\|_2^{-1} h_{D_n}$$

is a 1-excessive function in $L^2(E; m)$ with $0 < h_1 \leq \|h_{D_1}\|_2^{-1} h$ $[m]$ on E. Let $h_2 := G_2 h_1$. Clearly $h_2 \leq h_1 \leq \|h_{D_1}\|_2^{-1} h$. As h_1 is 1-excessive, h_2 is 2-excessive, and $h_2 \leq h_1$, we see from Remark 1.2.8(iv)–(v) that, for any open set D,

$$\mathrm{Cap}_{h_2,2}(D) \leq 2\mathrm{Cap}_{h_1,1}(D) \leq 2\|h_{D_1}\|_2^{-2}\mathrm{Cap}_{h,1}(D). \tag{1.2.9}$$

Now suppose that $\lim_{k\to\infty} \mathrm{Cap}_{h,1}(D_n \setminus F_k) = 0$ for every $n \geq 1$. Since $F_k^c \subset \overline{D}_n^c \cup (D_{n+1} \setminus F_k)$, we see by Theorem 1.2.9(ii) that

$$\mathrm{Cap}_{h_2,2}(F_k^c) \leq \mathrm{Cap}_{h_2,2}(\overline{D}_n^c) + \mathrm{Cap}(D_{n+1} \setminus F_k).$$

By noting (1.2.9) and Theorem 1.2.13, we let $k \to \infty$ and then $n \to \infty$ to get $\lim_{k\to\infty} \mathrm{Cap}_{h_2,2}(F_k^c) = 0$, which means that $\{F_k, k \geq 1\}$ is an $\mathcal{E}$-nest by Theorem 1.2.13.
(ii) If N is $\mathcal{E}$-polar, then it is a subset of $\cap_k (E \setminus F_k)$ for some $\mathcal{E}$-nest $\{F_k\}$. By (i), $\mathrm{Cap}_{h,1}(D_n \cap N) = 0$ for each n, and by letting $n \to \infty$, we get $\mathrm{Cap}_{h,1}(N) = 0$ on account of Theorem 1.2.10(ii). Conversely, suppose $\mathrm{Cap}_{h,1}(N) = 0$. For the 2-excessive function $h_2 := G_2 h_1$ defined in the "if" part of the proof of (i), the inequality (1.2.9) holds for any set D and so $\mathrm{Cap}_{h_2,2}(N) = 0$, which implies that N is $\mathcal{E}$-polar in view of Theorem 1.2.13. □

Remark 1.2.15. Let $h > 0$ be a function that satisfies one of the conditions in Definition 1.2.7.
(i) Any $\mathrm{Cap}_{h,1}$-nest is an $\mathcal{E}$-nest.
(ii) If $h \in \mathcal{F}$, then one can take $D_n = E$, $n \geq 1$, in Theorem 1.2.14 and hence a $\mathrm{Cap}_{h,1}$-nest becomes a synonym of an $\mathcal{E}$-nest.
(iii) $h = 1$ is an important case for Theorem 1.2.14. □

1.3. QUASI-REGULAR DIRICHLET FORMS

We maintain the same assumptions on $(E, \mathcal{B}(E), m)$ as in the preceding section. Let $(\mathcal{E}, \mathcal{F})$ be a Dirichlet form on $L^2(E; m)$.

LEMMA 1.3.1. *Let S be a countable family of $\mathcal{E}$-quasi-continuous functions on E. Then there is an $\mathcal{E}$-nest $\{F_k, k \geq 1\}$ such that $S \subset C(\{F_k\})$.*

Proof. Fix some $\varphi \in L^2(E;m)$ with $0 < \varphi \leq 1$ and set $h = G_1\varphi$. Spell out $S = \{f_k, k \geq 1\}$. For each $n \geq 1$, there is an $\mathcal{E}$-nest $\{F_{n,k}, k \geq 1\}$ such that $f_n \in C(\{F_{n,k}\})$ and $\mathrm{Cap}_{h,1}(F_{n,k}^c) \leq 2^{-nk}$. Define $F_k := \cap_{n\geq 1} F_{n,k}$, which is closed. By Theorems 1.2.9 and 1.2.10,

$$\mathrm{Cap}_{h,1}(F_k^c) = \mathrm{Cap}_{h,1}(\cup_{n\geq 1} F_{n,k}^c) \leq \sum_{n\geq 1} \mathrm{Cap}_{h,1}(F_{n,k}^c) \leq 2^{-k} \quad \text{for } k \geq 1.$$

So $\{F_k, k \geq 1\}$ is an $\mathcal{E}$-nest by Theorem 1.2.13 and clearly $S \subset C(\{F_k\})$. □

THEOREM 1.3.2. *Let $h = G_1\varphi$ for some $\varphi \in L^2(E;m)$ with $0 < \varphi \leq 1$. Suppose that $u \in \mathcal{F}$ has an $\mathcal{E}$-quasi-continuous m-version $\widetilde{u}$. Then*

$$\mathrm{Cap}_{h,1}(|\widetilde{u}| > \lambda) \leq \mathcal{E}_1(u,u)/\lambda^2 \quad \textit{for every } \lambda > 0.$$

Proof. Let $\{F_k, k \geq 1\}$ be an $\mathcal{E}$-nest such that $\widetilde{u} \in C(\{F_k\})$. For $\lambda > 0$, let $D_k := \{x \in F_k : |\widetilde{u}(x)| > \lambda\} \cup F_k^c$, which is an open subset of E. Let $u_k := \lambda^{-1}|\widetilde{u}| + h_{F_k^c} \in \mathcal{F}$. Since $0 < h \leq 1$ $[m]$ on E, $u_k \geq h$ on D_k. Thus

$$\begin{aligned} \mathrm{Cap}_{h,1}(|\widetilde{u}| > \lambda) &\leq \mathrm{Cap}_{h,1}(D_k) \leq \mathcal{E}_1(u_k, u_k) \\ &\leq \lambda^{-2}\mathcal{E}_1(|u|,|u|) + 2\lambda^{-1}\mathcal{E}_1(|u|, h_{F_k^c}) + \mathrm{Cap}_{h,1}(F_k^c). \end{aligned}$$

It follows then $\mathrm{Cap}_{h,1}(|\widetilde{u}| > \lambda) \leq \limsup_{k\to\infty} \mathrm{Cap}_{h,1}(D_k) \leq \mathcal{E}_1(u,u)/\lambda^2$. □

THEOREM 1.3.3. *Suppose each $u_k \in \mathcal{F}$ has an $\mathcal{E}$-quasi-continuous m-version $\widetilde{u}_k$ and that u_k converges to u in $(\mathcal{F}, \mathcal{E}_1)$ as $k \to \infty$. Then there exists a subsequence $\{u_{n_k}, k \geq 1\}$ such that $\widetilde{u}_{n_k}$ converges to an $\mathcal{E}$-quasi-continuous m-version $\widetilde{u}$ of u quasi uniformly; that is, there is an $\mathcal{E}$-nest $\{F_k, k \geq 1\}$ such that $\{\widetilde{u}, \widetilde{u}_{n_j}, j \geq 1\} \subset C(\{F_k\})$ and $\widetilde{u}_{n_k}$ converges to $\widetilde{u}$ uniformly on each F_k.*

Proof. Taking a subsequence if necessary, we may assume that $\mathcal{E}_1(u_{k+1} - u_k, u_{k+1} - u_k) < 2^{-3k}$ for every $k \geq 1$. Define

$$A_k := \left\{x \in E : |\widetilde{u}_{k+1}(x) - \widetilde{u}_k(x)| > 2^{-k}\right\}.$$

By Theorem 1.3.2, $\mathrm{Cap}_{h,1}(A_k) \leq 2^{2k}\mathcal{E}_1(u_{k+1} - u_k, u_{k+1} - u_k) < 2^{-k}$. Let $\{E_\ell, \ell \geq 1\}$ be an $\mathcal{E}$-nest such that $\{\widetilde{u}_k, k \geq 1\} \subset C(\{E_k\})$. Define $F_k := E_k \cap \left(\cap_{l\geq k} A_l^c\right)$, which is closed. Since

$$\mathrm{Cap}_{h,1}(F_k^c) \leq \mathrm{Cap}_{h,1}(E_k^c) + \sum_{l\geq k} \mathrm{Cap}_{h,1}(A_l) < \mathrm{Cap}_{h,1}(E_k^c) + 2^{-k+1},$$

which tends to 0 as $k \to \infty$, $\{F_k, k \geq 1\}$ is an $\mathcal{E}$-nest and clearly $\widetilde{u}_n$ converges to some $\widetilde{u}$ uniformly on each F_k. Thus $\widetilde{u} \in C(\{F_k\})$ and therefore it is an $\mathcal{E}$-quasi-continuous m-version of u. □

DEFINITION 1.3.4. An $\mathcal{E}$-nest $\{F_k, k \geq 1\}$ is called *m-regular* if for each $k \geq 1$, $\text{supp}[\mathbf{1}_{F_k} m] = F_k$; that is, for every $x \in F_k$ and every neighborhood $U(x)$ of x, $m(U(x) \cap F_k) > 0$.

DEFINITION 1.3.5. A Hausdorff topological space is called a *Lindelöf space* if every open covering of the space has a countable subcover.

Lindelöf theorem asserts that a topological space is Lindelöf if its topology has a countable base (see, e.g., [103, p. 49]). On a Lindelöf space, the topological support of any σ-finite measure μ is well defined to the smallest closed set F that μ does not charge on its complement F^c.

LEMMA 1.3.6. *Let $\{F_k, k \geq 1\}$ be an $\mathcal{E}$-nest. Suppose that the relative topology on each F_k is Lindelöf. Let $\widehat{F}_k = \text{supp}[\mathbf{1}_{F_k} m]$. Then $\{\widehat{F}_k, k \geq 1\}$ is an m-regular $\mathcal{E}$-nest.*

Proof. Let $h := G_1\varphi$ for some $\varphi \in L^2(E; m)$ such that $0 < \varphi \leq 1$. Note that

$$F_k \setminus \widehat{F}_k = \{x \in F_k : \text{ there is an open neighborhood } U(x) \text{ of } x \\ \text{such that } m(U(x) \cap F_k) = 0\}$$

By the Lindelöf property, $m(F_k \setminus \widehat{F}_k) = 0$. Thus $\mathcal{L}_{F_k^c, h} = \mathcal{L}_{\widehat{F}_k^c, h}$ and therefore $\text{Cap}_{h,1}(\widehat{F}_k^c) = \text{Cap}_{h,1}(F_k^c)$. This proves that $\{\widehat{F}_k, k \geq 1\}$ is an m-regular $\mathcal{E}$-nest. □

THEOREM 1.3.7. *Suppose $\{F_k, k \geq 1\}$ is an m-regular $\mathcal{E}$-nest and $f \in C(\{F_k\})$. If $f \geq 0$ $[m]$ on an open set D, then $f(x) \geq 0$ for every $x \in D \cap \left(\cup_{k \geq 1} F_k\right)$; i.e., $f \geq 0$ $\mathcal{E}$-q.e. on D.*

Proof. Since $f \geq 0$ $[m]$ on D and f is continuous on each F_k, $f \geq 0$ on $F_k \cap D$ due to the assumption $F_k = \text{supp}[\mathbf{1}_{F_k} m]$. Thus $f \geq 0$ on $\cup_{k \geq 1} (F_k \cap D)$. □

DEFINITION 1.3.8. A Dirichlet form $(\mathcal{F}, \mathcal{E})$ on $L^2(E; m)$ is called *quasi-regular* if:
(i) there exists an $\mathcal{E}$-nest $\{F_k, k \geq 1\}$ consisting of compact sets;
(ii) there exists an $\mathcal{E}_1$-dense subset of $\mathcal{F}$ whose elements have $\mathcal{E}$-quasi-continuous m-versions;
(iii) there exists $\{f_k, k \geq 1\} \subset \mathcal{F}$ having $\mathcal{E}$-quasi-continuous m-versions $\{\widetilde{f}_k, k \geq 1\}$ and an $\mathcal{E}$-polar set $N \subset E$ such that $\{\widetilde{f}_k : k \geq 1\}$ separates the points of $E \setminus N$.

Remark 1.3.9. (i) Part (i) of Definition 1.3.8 implies that, for any α-excessive function h in $\mathcal{F}$ with $\alpha > 0$, $\text{Cap}_{h,1}$ is tight; that is, there is an increasing sequence of compact sets $\{K_j, j \geq 1\}$ such that $\lim_{j\to\infty} \text{Cap}_{h,1}(E \setminus K_j) = 0$.
(ii) Part (ii) of Definition 1.3.8 implies by Theorem 1.3.3 that every function in $\mathcal{F}$ has an $\mathcal{E}$-quasi-continuous m-version, which will be denoted by $\widetilde{f}$.
(iii) We may assume, by parts (i) and (iii) of Definition 1.3.8 together with Theorem 1.2.13 and Lemma 1.3.1, that there is an $\mathcal{E}$-nest $\{F_k, k \geq 1\}$ consisting of compact sets so that $\{\widetilde{f}_k, k \geq 1\} \subset C(\{F_k\})$ and $\{\widetilde{f}_k, k \geq 1\}$ separates points of $\cup_k F_k$. Define

$$\rho(x, y) := \sum_{j=1}^{\infty} 2^{-j} \left(|\widetilde{f}_j(x) - \widetilde{f}_j(y)| \wedge 1 \right) \qquad \text{for } x, y \in \cup_{k\geq 1} F_k.$$

Then $\rho(x, y)$ is a (separating) metric on each F_k, which by the compactness of F_k is compatible with the original topology on F_k inherited from E. Since the topology induced by ρ on each F_k has countable base, F_k is a separable metric space. Hence $L^2(E; m) = L^2(\cup_{j\geq 1} F_j; m)$ is separable and therefore so is $(\mathcal{F}, \mathcal{E}_1)$. Thus Definition 1.3.8(ii) can be replaced by

(ii). There exists an $\mathcal{E}_1$-dense countable subset $\{u_k, k \geq 1\}$ of $\mathcal{F}$ whose elements have $\mathcal{E}$-quasi-continuous m-version.

(iv) By the Lindelöf theorem, each compact set F_k in (iii) is Lindelöf. Hence for a quasi-regular Dirichlet form $(\mathcal{E}, \mathcal{F})$, if $f \geq 0$ $[m]$ on an open set D and if f is $\mathcal{E}$-quasi-continuous, then $f \geq 0$ $\mathcal{E}$-q.e. on D by Lemma 1.3.6 and Theorem 1.3.7.
(v) By Corollary 2 on p.12 of [136] $Y := \cup_{k\geq 1} F_k$ is a Lusin space (i.e., it is homeomorphic to a Borel subset of a compact metric space). Since $L^2(E; m)$ can be identified with $L^2(Y; m)$, when dealing with quasi-regular Dirichlet forms, we can assume that E is a topological Lusin space. □

For an m-measurable function f defined and finite m-a.e. on E, the *support* of f is defined to be the support of the measure $f \cdot m$. When f is continuous, the support of f is just the closure of the set $\{f \neq 0\}$. When E is a locally compact separable metric space, we shall denote by $C_c(E)$ the family of all continuous functions on E with compact support, and by $C_\infty(E)$ the family of all continuous functions f on E which vanishes at ∞, namely, there exists for any $\varepsilon > 0$ a compact set with $|f(x)| < \varepsilon$ for every $x \in E \setminus K$. $C_\infty(E)$ is a Banach space with respect to the uniform norm $\|f\|_\infty = \sup_{x\in E} |f(x)|$.

DEFINITION 1.3.10. A Dirichlet form $(\mathcal{E}, \mathcal{F})$ on $L^2(E; m)$ is said to be *regular* if
(i) E is a locally compact separable metric space and m is a Radon measure on E with full support;

(ii) $\mathcal{F} \cap C_c(E)$ is $\mathcal{E}_1$-dense in $\mathcal{F}$;
(iii) $\mathcal{F} \cap C_c(E)$ is uniformly dense in $C_c(E)$.

Remark 1.3.11. (i) By Stone-Weierstrass theorem (cf. [58, Theorem 4.45]), (iii) in Definition 1.3.10 is equivalent to
(iii′) $\mathcal{F} \cap C_c(E)$ separates the points of E.
(ii) Clearly, a regular Dirichlet form is quasi-regular. □

LEMMA 1.3.12. *Let E be a locally compact separable metric space and m a Radon measure on E with full support. Suppose that $(\mathcal{E}, \mathcal{F})$ is a Dirichlet form on $L^2(E; m)$. If the space $\mathcal{F} \cap C_\infty(E)$ is dense both in $(\mathcal{F}, \|\cdot\|_{\mathcal{E}_1})$ and in $(C_\infty(E), \|\cdot\|_\infty)$, then $(\mathcal{E}, \mathcal{F})$ is regular.*

Proof. For a fixed $f \in \mathcal{F} \cap C_\infty(E)$, we consider its composition $f_\ell = \varphi_\ell \circ f$ with specific normal contractions defined by

$$\varphi_\ell(t) := t - ((-1/\ell) \vee t) \wedge (1/\ell), \qquad t \in \mathbb{R},\ \ell \geq 1, \tag{1.3.1}$$

then

$$f_\ell \in \mathcal{F} \cap C_c(E) \text{ and } \|f_\ell - f\|_\infty \leq \frac{1}{\ell} \text{ for } \ell \geq 1,$$

and we see that $\mathcal{F} \cap C_c(E)$ is dense in the space $(C_c(E), \|\cdot\|_\infty)$. On the other hand, Lemma 1.1.11 implies that f_ℓ is $\mathcal{E}_1$-convergent to f and hence $\mathcal{F} \cap C_c(E)$ is dense in $(\mathcal{F}, \mathcal{E}_1)$. □

Exercise 1.3.13. Let $(\mathcal{E}, \mathcal{F})$ be a regular Dirichlet form on $L^2(E; m)$. Show that, for any $f \in C_c(E)$, there exist $f_n \in \mathcal{F} \cap C_c(E)$ such that $\text{supp}[f_n] \subset \text{supp}[f]$ for every $n \geq 1$, and f_n converges to f uniformly on E as $n \to \infty$.

For a regular Dirichlet form $(\mathcal{E}, \mathcal{F})$ on $L^2(E; m)$, it is customary to use 1-capacity denoted by Cap_1, that is, (h, α)-capacity with $h = 1$ and $\alpha = 1$. This is because in this case $\text{Cap}_1(D) < \infty$ for every relatively compact open subset $D \subset E$. Note that since E is a locally compact separable metric space, there is a sequence of relatively compact open subsets $\{D_k, k \geq 1\}$ with $\overline{D}_k \subset D_{k+1}$, $k \geq 1$, and $\cup_{k \geq 1} D_k = E$. Thus Theorem 1.2.14 is applicable with $h = 1$. In particular, we have the following, which gives the equivalence of $\mathcal{E}$-polar set, $\mathcal{E}$-nest, and $\mathcal{E}$-quasi-continuity with the notions of set of capacity zero, generalized nest, and quasi continuity, respectively, defined in the book [73].

THEOREM 1.3.14. *Suppose that $(\mathcal{E}, \mathcal{F})$ is a regular Dirichlet form on $L^2(E; m)$. Then*
(i) *A subset set of E is $\mathcal{E}$-polar if and only if it is* Cap_1*-polar.*

(ii) *An increasing sequence of closed subsets* $\{F_j, j \geq 1\}$ *is an* $\mathcal{E}$*-nest if and only if* $\lim_{j\to\infty} \text{Cap}_1(K \setminus F_j) = 0$ *for every compact set* $K \subset E$.
(iii) *A function* f *is* $\mathcal{E}$*-quasi-continuous if and only if for every* $\varepsilon > 0$, *there is an open set* $D \subset E$ *with* $\text{Cap}_1(D) < \varepsilon$ *such that* $f\big|_{E\setminus D}$ *is finite and continuous or, equivalently, there exists a* Cap_1*-nest* $\{F_k\}$ *such that* $f \in C(\{F_k\})$.

Proof. (i) and (ii) follow immediately from Theorem 1.2.14.

For (iii), if f is $\mathcal{E}$-quasi-continuous, then, in view of Theorem 1.2.13, there is an $\mathcal{E}$-nest $\{F_k, k \geq 1\}$ consisting of closed sets so that $f \in C(\{F_k\})$. Let $\{D_k, k \geq 1\}$ be an increasing sequence of relatively compact open subsets with $\cup_{k\geq 1} D_k = E$ and $\text{Cap}(D_k) < \infty$, $k \geq 1$. By Theorem 1.2.14, for every $\varepsilon > 0$ and $n \geq 1$, there is an integer $k_n \geq 1$ so that $\text{Cap}_1(D_n \setminus F_{k_n}) < 2^{-n-1}\varepsilon$. Define $D := \cup_{n\geq 1}(D_n \setminus F_{k_n})$, which is an open set with $\text{Cap}_1(D) \leq \sum_{n\geq 1} \text{Cap}_1(D_n \setminus F_{k_n}) < \varepsilon$.

It is easy to check that $f\big|_{E\setminus D}$ is continuous. Indeed, as for every $x_0 \in E \setminus D$ there is some $r_0 > 0$ and $n_0 \geq 1$ so that $B(x_0, r_0) \subset D_{n_0}$ and that $E \setminus D = \cap_{n\geq 1}(D_n^c \cup F_{k_n})$, we have $B(x_0, r_0) \cap (E \setminus D) \subset B(x_0, r_0) \cap F_{k_{n_0}}$. It follows that $f|_{E\setminus D}$ is continuous at x_0.

The sufficiency in (iii) is obvious because any Cap_1-nest is an $\mathcal{E}$-nest by Remark 1.2.15. □

The last assertion of the above theorem can be strengthened as follows.

Let $E_\partial = E \cup \{\partial\}$ be the one-point compactification of the locally compact metric space E. For a closed set $F \subset E$, we regard $F \cup \{\partial\}$ as a topological subspace of E_∂. For an increasing sequence $\{F_k\}$ of closed sets, we denote by $C_\infty(\{F_k\})$ the collection of functions f on E such that, if f is extended to E_∂ by setting $f(\partial) = 0$, then $f\big|_{F_k\cup\{\partial\}}$ is finite and continuous for each k. Obviously the space $C_\infty(E)$ is contained in $C_\infty(\{F_k\})$.

Suppose, for a function f on E, there exists an $\mathcal{E}$-nest $\{F_k\}$ such that $f \in C_\infty(\{F_k\})$. Then f is said to be *quasi continuous in the restricted sense* relative to the $\mathcal{E}$-nest $\{F_k\}$.

Lemma 1.3.15. *If* $(\mathcal{E}, \mathcal{F})$ *is a regular Dirichlet form on* $L^2(E; m)$, *then each element* $f \in \mathcal{F}$ *admits an m-version* $\widetilde{f}$ *which is quasi continuous in the restricted sense relative to a* Cap_1*-nest.*

Proof. For $f \in \mathcal{F} \cap C_c(E)$ and $\lambda > 0$, the set $D_\lambda = \{x \in E : |f(x)| > \lambda\}$ is an open set with $f/\lambda \in \mathcal{L}_{D_\lambda,1}$ so that

$$\text{Cap}_1(D_\lambda) \leq \mathcal{E}_1(f, f)/\lambda^2, \tag{1.3.2}$$

which yields the above assertion as in Theorem 1.3.3 because $\mathcal{F} \cap C_c(E)$ is $\mathcal{E}_1$-dense in $\mathcal{F}$. □

Exercise 1.3.16. Let $(\mathcal{E}, \mathcal{F})$ be a regular Dirichlet form. Show that the statement in Theorem 1.3.3 holds with an $\mathcal{E}$-nest $\{F_k\}$ and the space $C(\{F_k\})$ being replaced by a Cap_1-nest $\{F_k\}$ and $C_\infty(\{F_k\})$, respectively.

We give the following definition for future use.

DEFINITION 1.3.17. Let E be a locally compact separable metric space, m be a Radon measure on E with $\mathrm{supp}[m] = E$, and $(\mathcal{E}, \mathcal{F})$ be a Dirichlet form on $L^2(E; m)$.

(i) $\mathcal{C} \subset \mathcal{F} \cap C_c(E)$ is said to be a *core* of $(\mathcal{E}, \mathcal{F})$ if $\mathcal{C}$ is dense both in $(\mathcal{F}, \|\cdot\|_{\mathcal{E}_1})$ and in $(C_c(E), \|\cdot\|_\infty)$. Clearly the Dirichlet form $(\mathcal{E}, \mathcal{F})$ is regular if it has a core.

(ii) A core $\mathcal{C}$ is said to be *standard* if it is a dense linear subspace of $C_c(E)$, and for any $\varepsilon > 0$, there exists a normal contraction φ_ε of (1.1.7) such that $\varphi_\varepsilon(\mathcal{C}) \subset \mathcal{C}$.

(iii) A standard core $\mathcal{C}$ is said to be *special* if $\mathcal{C}$ is a dense subalgebra of $C_c(E)$, and for any compact set K and relatively compact open set G with $K \subset G$, there exists $f \in \mathcal{C}_+$ such that $f = 1$ on K and $f = 0$ on $E \setminus G$.

(iv) $(\mathcal{E}, \mathcal{F})$ is called *local* if $\mathcal{E}(f, g) = 0$ whenever $f,\ g \in \mathcal{F}$ have disjoint compact supports.

(v) $(\mathcal{E}, \mathcal{F})$ is called *strongly local* if $\mathcal{E}(f, g) = 0$ whenever $f \in \mathcal{F}$ has a compact support and $g \in \mathcal{F}$ is constant on a neighborhood of the support of f.

1.4. QUASI-HOMEOMORPHISM OF DIRICHLET SPACES

Suppose $(\mathcal{E}, \mathcal{F})$ is a Dirichlet form on $L^2(E; m)$. Let $(\widehat{E}, \mathcal{B}(\widehat{E}))$ be a second measurable space and $j : (E, \mathcal{B}(E)) \to (\widehat{E}, \mathcal{B}(\widehat{E}))$ be a measurable map. Define $\widehat{m} := m \circ j^{-1}$, the push forward measure of m under map j; that is, for $A \in \mathcal{B}(\widehat{E})$, $\widehat{m}(A) = m(j^{-1}(A))$. Then $j^* : L^2(\widehat{E}; \widehat{m}) \to L^2(E; m)$ is an isometry, where $j^*\widehat{f} := \widehat{f} \circ j$ for $\widehat{f} \in L^2(\widehat{E}; \widehat{m})$. $j^*L^2(\widehat{E}; \widehat{m})$ is, in general, a closed subspace of $L^2(E; m)$. Define $\widehat{\mathcal{F}} := \{\widehat{f} \in L^2(\widehat{E}; \widehat{m}) : j^*\widehat{f} \in \mathcal{F}\}$ and

$$\widehat{\mathcal{E}}(\widehat{f}, \widehat{g}) := \mathcal{E}(j^*\widehat{f}, j^*\widehat{g}) \qquad \text{for } \widehat{f}, \widehat{g} \in \widehat{\mathcal{F}}.$$

Clearly $(\widehat{\mathcal{E}}, \widehat{\mathcal{F}})$ is a closed form on $L^2(\widehat{E}; \widehat{m})$. If j^* maps $L^2(\widehat{E}, \widehat{m})$ onto $L^2(E; m)$, then $(\widehat{\mathcal{E}}, \widehat{\mathcal{F}})$ is a Dirichlet form on $L^2(\widehat{E}; \widehat{m})$, which is called the *image Dirichlet form* of $(\mathcal{E}, \mathcal{F})$ under j. We denote in the sequel $(\widehat{\mathcal{E}}, \widehat{\mathcal{F}})$ as $j(\mathcal{E}, \mathcal{F})$.

DEFINITION 1.4.1. Given two Dirichlet forms $(\mathcal{E}, \mathcal{F})$ and $(\widehat{\mathcal{F}}, \widehat{\mathcal{E}})$ on $L^2(E; m)$ and $L^2(\widehat{E}; \widehat{m})$, respectively, where E and $\widehat{E}$ are two Hausdorff topological spaces and m and $\widehat{m}$ are σ-finite measures on E and $\widehat{E}$ respectively with $\mathrm{supp}[m] = E$ and $\mathrm{supp}[\widehat{m}] = \widehat{E}$. The Dirichlet form $(\mathcal{E}, \mathcal{F})$ is said to be *quasi-homeomorphic*

to $(\widehat{\mathcal{E}}, \widehat{\mathcal{F}})$ if there is an $\mathcal{E}$-nest $\{F_n\}_{n\geq 1}$ and an $\widehat{\mathcal{E}}$-nest $\{\widehat{F}_n\}_{n\geq 1}$ and a map $j\colon \cup_{k\geq 1} F_k \to \cup_{k\geq 1}\widehat{F}_k$ such that
(a) j is a topological homeomorphism from F_k onto $\widehat{F}_k$ for each $k \geq 1$.
(b) $\widehat{m} = m \circ j^{-1}$.
(c) $(\widehat{\mathcal{E}}, \widehat{\mathcal{F}}) = j(\mathcal{E}, \mathcal{F})$; that is, $(\widehat{\mathcal{E}}, \widehat{\mathcal{F}})$ is the image Dirichlet form of $(\mathcal{E}, \mathcal{F})$ under map j.

For every function $\widehat{f}$ on $\widehat{E}$, $j^*\widehat{f}$ is uniquely defined on E modulo an m-null set and j^* is an isometry from $L^2(\widehat{E}; \widehat{m})$ onto $L^2(E; m)$.

Exercise 1.4.2. Suppose two Dirichlet form $(\mathcal{E}, \mathcal{F})$ on $L^2(E; m)$ and $(\widehat{\mathcal{E}}, \widehat{\mathcal{F}})$ on $L^2(\widehat{E}; \widehat{m})$ are quasi-homeomorphic by a map j as in the above definition. Prove that j is quasi notion preserving in the following sense:
(i) An increasing sequence $\{E_k\}$ of closed subsets of E is an $\mathcal{E}$-nest if and only if $\{j(E_k \cap F_k)\}$ is an $\widehat{\mathcal{E}}$-nest.
(ii) $N \subset E$ is $\mathcal{E}$-polar if and only if $j\left((\cup_{k\geq 1} F_k) \cap N\right)$ is $\widehat{\mathcal{E}}$-polar.
(iii) A function f defined $\mathcal{E}$-q.e. on E is $\mathcal{E}$-quasi-continuous if and only if $f \circ j^{-1}$ is $\widehat{\mathcal{E}}$-quasi-continuous.

The following theorem gives an important connection between quasi-regular Dirichlet forms and regular Dirichlet forms, which enables us to transfer known results for regular Dirichlet forms to quasi-regular Dirichlet forms.

THEOREM 1.4.3. *A Dirichlet space $(\mathcal{E}, \mathcal{F})$ on $L^2(E; m)$ is quasi-regular if and only if $(\mathcal{E}, \mathcal{F})$ is quasi-homeomorphic to a regular Dirichlet space on a locally compact separable metric space.*

Proof. The "if" part is trivial. We only need to show the "only if" part. Take a strictly positive bounded $\varphi \in L^1(E; m)$ and let $h = G_1\varphi$, which is strictly positive m-a.e. on E and is in $\mathcal{F}$. Since $(\mathcal{E}, \mathcal{F})$ is quasi-regular on $L^2(E; m)$, by Theorem 1.3.3, h has an $\mathcal{E}$-quasi-continuous m-version $\widetilde{h}$. We claim that there is an $\mathcal{E}$-nest $\{K_j, j \geq 1\}$ consisting of compact sets so that $\widetilde{h} \in C(\{K_j\})$ and $\widetilde{h} \geq 1/j$ on each K_j. By Theorems 1.2.13 and 1.3.3, there is an $\mathcal{E}$-nest $\{\widetilde{K}_j, j \geq 1\}$ consisting of compact sets so that $\widetilde{h} \in C(\{\widetilde{K}_j\})$. For each $j \geq 1$, define $K_j = \widetilde{K}_j \cap \{\widetilde{h} \geq 1/j\}$, which is compact. We show that $\{K_j, j \geq 1\}$ is an $\mathcal{E}$-nest. Note that $K_j^c = \widetilde{K}_j^c \cup \{x \in \widetilde{K}_j : \widetilde{h}(x) < 1/j\}$ and $v_j := h_{\widetilde{K}_j^c} + (1/j) \wedge h \in \mathcal{L}_{K_j^c, h}$. Thus

$$
\begin{aligned}
\lim_{j\to\infty} \mathrm{Cap}_{h,1}(K_j^c) &\leq \lim_{j\to\infty} \mathcal{E}_1\left(h_{\widetilde{K}_j^c} + (1/j) \wedge h,\ h_{\widetilde{K}_j^c} + (1/j) \wedge h\right) \\
&\leq 2 \lim_{j\to\infty} \mathcal{E}_1\left(h_{\widetilde{K}_j^c}, h_{\widetilde{K}_j^c}\right) + 2 \lim_{j\to\infty} \mathcal{E}_1\left((1/j) \wedge h, (1/j) \wedge h\right) \\
&= 0,
\end{aligned}
$$

where in the last equality we used Lemma 1.1.11(i) applied to normal contractions $\psi_j(t) := t - ((-1/j) \vee t) \wedge (1/j)$. This establishes that $\{K_j, j \geq\}$ is an $\mathcal{E}$-nest. Observe that for $f \in L^2(E; m)$, $\psi_j(f) \in L^1(E; m)$. So by Lemma 1.1.11 and Remark 1.3.9, there exists a countable $\mathcal{E}_1$-dense set $B_0 = \{f_n, n \geq 1\}$ of bounded $\mathcal{E}$-quasi-continuous functions in $\mathcal{F} \cap L^1(E; m)$ such that
(i) $\widetilde{h} \in B_0$ and B_0 is an algebra over the rational numbers,
(ii) There is an $\mathcal{E}$-nest $\{F_k, k \geq 1\}$ consisting of compact sets such that $B_0 \subset C(\{F_k\})$ and B_0 separates points of $\cup_{k\geq 1} F_k$ and $\widetilde{h} \geq 1/k$ on F_k.

We make functions in B_0 to take value zero on $E \setminus \cup_{k\geq 1} F_k$. Define $B := \overline{B_0}^{\|\cdot\|_\infty}$, which is a commutative Banach algebra. We now construct a regular Dirichlet form $(\widehat{\mathcal{E}}, \widehat{\mathcal{F}})$ on a locally compact separable metric space $\widehat{E}$ via the Gelfand transform which will be quasi-homeomorphic to $(\mathcal{E}, \mathcal{F})$.

Step 1. Construct a locally compact separable metric space $\widehat{E}$.

Let $\widehat{E}$ be a collection of non-trivial real-valued functionals γ on B which satisfy for $f, g \in B$ and for rational numbers a and b,

(a) $|\gamma(f)| \leq \|f\|_\infty$,
(b) $\gamma(fg) = \gamma(f)\gamma(g)$,
(c) $\gamma(af + bg) = a\gamma(f) + b\gamma(g)$.

We equip $\widehat{E}$ with the weakest topology so that the function $\Phi_f : \gamma \mapsto \gamma(f)$ is continuous for every $f \in B$. It is well-known that $\widehat{E}$ is a separable locally compact Hausdorff space which is compact if and only if $1 \in B$, and $\{\Phi_f, f \in B\} \subset C_\infty(\widehat{E})$. The topological space $\widehat{E}$ is metrizable with metric δ defined by

$$\delta(\gamma, \eta) := \sum_{n\geq 1} 2^{-n} \left(|\gamma(f_n) - \eta(f_n)| \wedge 1\right), \qquad \gamma, \eta \in \widehat{E}.$$

Let j be the unique map from $\cup_{k\geq 1} F_k$ into $\widehat{E}$ such that

$$(jx)(f) := f(x) \qquad \text{for } f \in B \text{ and } x \in \cup_{k\geq 1} F_k.$$

By (ii) above, j is a continuous one-to one map on each F_k. Hence $\widehat{F}_k := j(F_k)$ is compact in $\widehat{E}$ and j is a topological homeomorphism from F_k onto $\widehat{F}_k$ for every $k \geq 1$. Note that $j : \cup_{k\geq 1} F_k \to \widehat{E}$ is Borel measurable and $m(E \setminus \cup_{k\geq 1} F_k) = 0$. Define $\widehat{m} := m \circ j^{-1}$. Clearly $\widehat{m}(\widehat{E} \setminus \cup_{k\geq 1} \widehat{F}_k) = 0$. It follows from the m-integrability of functions in B_0 that $\widehat{m}$ is a Radon measure, and it is easy to check that $\text{supp}[\widehat{m}] = \widehat{E}$ (see [138, p. 23]). Since B_0 is dense in $L^2(E; m)$, j^* is a unitary map from $L^2(\widehat{E}; \widehat{m})$ onto $L^2(E; m)$.

Step 2. Φ maps B onto $C_\infty(\widehat{E})$.

For $f \in B$, $\Phi_f \in C_\infty(\widehat{E})$, where $\Phi_f(\gamma) = \gamma(f)$. Clearly, $\|\Phi_f\|_\infty = \|f\|_\infty$. So $\Phi(B)$ is closed under uniform norm. Since $h \in B$ and $\Phi(B)$ is an algebra of real-valued functions that vanish at infinity and separates points in $\widehat{E}$ with $\Phi(\widehat{h}) > 0$

on $\widehat{E}$, by Stone-Weierstrass theorem (cf. [58, Theorem 4.52]), $\Phi(B) = C_\infty(\widehat{E})$.

Step 3. The image Dirichlet form $j(\mathcal{E}, \mathcal{F})$ is regular on $L^2(\widehat{E}; \widehat{m})$.

Let $(\widehat{\mathcal{E}}, \widehat{\mathcal{F}}) := j(\mathcal{E}, \mathcal{F})$. Then $(\widehat{\mathcal{E}}, \widehat{\mathcal{F}})$ is a Dirichlet form on $L^2(\widehat{E}; \widehat{m})$ as j^* is an isometry from $L^2(\widehat{E}; \widehat{m})$ onto $L^2(E; m)$. Since $\widehat{\mathcal{F}} \cap C_\infty(\widehat{E}) \supset \Phi(B_0)$ and the latter is uniformly dense in $C_\infty(\widehat{E})$ and $\widehat{\mathcal{E}}_1$-dense in $\widehat{\mathcal{F}}$, $(\widehat{\mathcal{E}}, \widehat{\mathcal{F}})$ is a regular Dirichlet form on $L^2(\widehat{E}; \widehat{m})$. Since $j^* \widehat{\mathcal{F}}_{\widehat{F}_k} = \mathcal{F}_{F_k}$ for every $k \geq 1$, $\{\widehat{F}_k, k \geq 1\}$ is an $\widehat{\mathcal{E}}$-nest and therefore j is a quasi-homeomorphism from $(\mathcal{E}, \mathcal{F})$ to $(\widehat{\mathcal{E}}, \widehat{\mathcal{F}})$. This completes the proof of the theorem. □

1.5. SYMMETRIC RIGHT PROCESSES AND QUASI-REGULAR DIRICHLET FORMS

THEOREM 1.5.1. *Let $(\mathcal{E}, \mathcal{F})$ be a regular Dirichlet form on $L^2(E; m)$, where E is a locally compact separable metric space and m is a Radon measure on E with full support. There exists then a Hunt process X on E with an m-symmetric transition function so that $(\mathcal{E}, \mathcal{F})$ is the Dirichlet form of the transition function of X.*

This theorem was proved by the second-named author [62] in 1971. A rather different proof from that of [62] is presented in [73]. Theorem A.1.37 of Appendix A on the Feller semigroup and this theorem constitute basic existence theorems of Hunt processes on locally compact spaces.

We can now combine Theorem 1.4.3 with Theorem 1.5.1 to show that there is a nice Markov process called an m-tight special Borel standard process associated with every quasi-regular Dirichlet form. See Section A.1.3 for the definition of a right process and a special Borel standard process.

Let $(E, \mathcal{B}^*(E))$ be a Radon space, m be a σ-finite measure on it, and X a right process on it. If the transition function $\{P_t; t \geq 0\}$ is m-symmetric, we say that X is *m-symmetric*. In this case, the Dirichlet form $(\mathcal{E}, \mathcal{F})$ on $L^2(E; m)$ of $\{P_t; t \geq 0\}$ is called the *Dirichlet form* of the m-symmetric right process X. We say further that X is *properly associated with* $(\mathcal{E}, \mathcal{F})$ if $P_t f$ is an $\mathcal{E}$-quasi-continuous m-version of $T_t f$ for any $f \in \mathcal{B}(E) \cap L^2(E; m)$ and $t > 0$, where $\{T_t; t > 0\}$ is the $L^2(E; m)$-semigroup generated by $(\mathcal{E}, \mathcal{F})$.

A right process X is called *m-tight* if there is an increasing sequence of compact sets $\{K_j, j \geq 1\}$ so that $\mathbf{P}_m(\lim_{j\to\infty} \tau_{K_j} < \zeta) = 0$. Here $\tau_{K_j} := \inf\{t \geq 0 : X_t \notin K_j\}$ is the first exit time from K_j by X and ζ is the lifetime of X.

THEOREM 1.5.2. *Suppose that $(\mathcal{E}, \mathcal{F})$ is a quasi-regular Dirichlet form on $L^2(E; m)$, where E is a Hausdorff topological space such that the Borel σ-field $\mathcal{B}(E)$ is generated by the continuous functions on E. Then there is an $\mathcal{E}$-polar Borel set $N \subset E$ and an m-symmetric, m-tight, special Borel standard process X on $E \setminus N$ that is properly associated with $(\mathcal{E}, \mathcal{F})$.*

Proof. By Theorem 1.4.3, $(\mathcal{E}, \mathcal{F})$ is quasi-homeomorphic to an $\widehat{m}$-symmetric regular Dirichlet form $(\widehat{\mathcal{E}}, \widehat{\mathcal{F}})$ on a locally compact separable metric space $\widehat{E}$ through quasi-homeomorphism j. More precisely, $\widehat{m} = m \circ j^{-1}$, $(\widehat{\mathcal{E}}, \widehat{\mathcal{F}}) = j(\mathcal{E}, \mathcal{F})$, and there are $\mathcal{E}$-nest $\{F_k, k \geq 1\}$ and $\widehat{\mathcal{E}}$-nest $\{\widehat{F}_k, k \geq 1\}$ so that j is a topological homeomorphism from F_k onto $\widehat{F}_k$ for every $k \geq 1$. j is a one-to-one map from $E_1 = \cup_{k=1}^{\infty} F_k$ onto $\widehat{E}_1 = \cup_{k=1}^{\infty} \widehat{F}_k$ and it can be extended to a one-to-one map from $E_1 \cup \{\partial\}$ onto $\widehat{E}_1 \cup \{\widehat{\partial}\}$, where ∂ is an extra point adjoined to E and $\widehat{\partial}$ is the point at infinity of $\widehat{E}$. On account of Theorem 1.2.13 and Theorem 1.3.14, we may and do assume that each $\widehat{F}_k$ is compact (consequently, each F_k is compact) by taking an intersection with another $\widehat{\mathcal{E}}$-nest if necessary.

By virtue of Theorem 1.5.1, there is an $\widehat{m}$-symmetric Hunt process

$$\widehat{X} = (\widehat{\Omega}, \{\widehat{\mathfrak{F}}_t\}, \widehat{\zeta}, \widehat{X}_t, \widehat{\mathbf{P}}_{\widehat{x}})$$

on $\widehat{E}$ such that $(\widehat{\mathcal{E}}, \widehat{\mathcal{F}})$ is the Dirichlet form of $\widehat{X}$ on $L^2(\widehat{E}, \widehat{m})$. We shall make use of some theorems in Section 3.1 concerning the relations between $\widehat{X}$ and $\widehat{\mathcal{E}}$. (This is the only proof in the book that uses forward references.) In view of Proposition 3.1.9, $\widehat{X}$ is automatically properly associated with $(\widehat{\mathcal{E}}, \widehat{\mathcal{F}})$.

Denote by $\widehat{\tau}_{\widehat{F}_k}$ the first exit time from $\widehat{F}_k$ of $\widehat{X}$. By Theorem 3.1.4 and Theorem 3.1.5, there exists a Borel set $\widehat{N}$ containing $\widehat{E} \setminus \widehat{E}_1$ such that $\widehat{m}(\widehat{N}) = 0$ and $\widehat{\mathbf{P}}_{\widehat{x}}(\widehat{\Lambda}) = 1$ for every $\widehat{x} \in \widehat{E} \setminus \widehat{N}$, where

$$\widehat{\Lambda} = \left\{\widehat{\omega} \in \widehat{\Omega} : \lim_{k \to \infty} \widehat{\tau}_{\widehat{F}_k} = \widehat{\zeta},\ \widehat{X}_t,\ \widehat{X}_{t-} \in \widehat{E}_{\widehat{\partial}} \setminus \widehat{N} \text{ for every } t \geq 0\right\}.$$

The above set $\widehat{N}$ is called a *Borel properly exceptional set* for the Hunt process $\widehat{X}$.

We define an $\mathcal{E}$-polar Borel set $N \subset E$ by $E \setminus N = j^{-1}(\widehat{E}_1 \setminus \widehat{N})$. We let $\Omega = \widehat{\Lambda}$, $\mathfrak{F}_t = \widehat{\mathfrak{F}}_t \cap \widehat{\Lambda}$, $t \in [0, \infty]$, and denote an element of Ω (resp. $\mathfrak{F}_\infty$) by ω (resp. Γ). Finally we define $X = (\Omega, \{\mathfrak{F}_t\}, X_t, \zeta, \mathbf{P}_x)$ by

$$X_t(\omega) := j^{-1}(\widehat{X}_t(\omega)) \quad \text{and} \quad \zeta(\omega) := \widehat{\zeta}(\omega) \qquad \text{for } \omega \in \Omega \text{ and } t \geq 0,$$

and

$$\mathbf{P}_x(\Gamma) := \widehat{\mathbf{P}}_{j(x)}(\Gamma) \quad \text{for } x \in E \setminus N \text{ and } \Gamma \in \mathfrak{F}_\infty.$$

Observe that $\tau_{F_k} = \widehat{\tau}_{\widehat{F}_k}$ for every $k \geq 1$, where τ_{F_k} is the first exit time from F_k by X. It is straightforward to check that X is an m-symmetric, m-tight, special Borel standard process on $E \setminus N$ properly associated with $(\mathcal{E}, \mathcal{F})$. □

As will be shown in Theorems 3.1.12 and 3.1.13, the Hunt process (respectively, m-symmetric right process) associated with a regular Dirichlet form (respectively, quasi-regular Dirichlet form) is unique in distribution. Moreover, it will be shown in Theorem 3.1.13 that for a quasi-regular Dirichlet form $(\mathcal{E}, \mathcal{F})$, $\{F_k, k \geq 1\}$ is an $\mathcal{E}$-nest if and only if

$$\lim_{k\to\infty} \tau_{F_k} = \zeta \qquad \mathbf{P}_x\text{-a.s. for } \mathcal{E}\text{-q.e. } x \in E,$$

where ζ is the lifetime of for the right process X associated with $(\mathcal{E}, \mathcal{F})$. Thus quasi-homeomorphism is not only an isometry at the Dirichlet form level but also a topological isometry at the process level up to its lifetime, as $j : F_k \mapsto \widehat{F}_k$ is a topological homeomorphism. In view of Theorem 1.5.2, we can assume without loss of generality, in most of the rest of the book, that the Dirichlet form $(\mathcal{E}, \mathcal{F})$ is regular, as corresponding results for quasi-regular Dirichlet forms can be easily deduced via quasi-homeomorphism.

In fact, the quasi-regularity of a Dirichlet form is not only sufficient but also necessary for the association of an m-tight special Borel standard process. More generally the following theorem holds:

THEOREM 1.5.3. *Let E be a Radon space and m be a σ-finite measure on E with full support. Suppose that X is an m-symmetric and m-tight right process on E. Then the Dirichlet form $(\mathcal{E}, \mathcal{F})$ on $L^2(E; m)$ of X is quasi-regular and X is properly associated with $(\mathcal{E}, \mathcal{F})$.*

If in particular E is a Lusin space, then the Dirichlet form of an m-symmetric right process is quasi-regular and X is properly associated with $(\mathcal{E}, \mathcal{F})$.

The second statement follows from the first because an m-symmetric right process on a Lusin space E is necessarily m-tight (see [119, Theorem IV.1.15]).

When X is an m-tight, m-special Borel standard process on E, this result was proved by S. Albeverio and Z. M. Ma [2] in 1991 and its proof can be found in the book by Z. M. Ma and M. Röckner [119, Theorem IV.5.1] under a more general assumption on the state space E. The result under the current condition follows from the aforementioned result in [119] together with a result of P. J. Fitzsimmons [53, Theorem 3.22], who showed that the restriction of X on the complement of some m-inessential set is an m-special standard process and the Borel measurability assumption on the transition function can be weakened to the universal measurability. As a matter of fact, the stated results in [119] and [53] are formulated for a more general (not necessarily symmetric) sectorial Dirichlet form $(\mathcal{E}, \mathcal{F})$. Moreover, Theorem 1.4.3 also holds for more general sectorial Dirichlet forms; see [31].

It is important to consider a general right process in applications as it is invariant under variety of transformations (for example, time change, killing, h-transformations) while Borel measurability of the transition function is not. However, we shall prove in Theorem 3.1.13 of Section 3.1 that any m-symmetric right process properly associated with a quasi-regular Dirichlet form is, when restricted to the complement of an m-inessential set, a Borel special standard process properly associated with the form. This combined with Theorem 1.5.3 means that any m-tight m-symmetric right process on a Radon space or any m-symmetric right process on a Lusin space can always be modified to be a Borel special standard process (see Corollary 3.1.14 below).

We end this chapter by noting that any quasi-regular Dirichlet form $(\mathcal{E}, \mathcal{F})$ on $L^2(E; m)$ admits the following expression in terms of an associated m-symmetric right process $X = (X_t, \mathbf{P}_x, \zeta)$ on E: for any $f \in \mathcal{F}_e$,

$$\mathcal{E}(f,f) = \lim_{t \to 0} \frac{1}{2t} \left(\mathbf{E}_m \left[(f(X_t) - f(X_0))^2; t < \zeta \right] + 2 \int_E f(x)^2 \mathbf{P}_x(\zeta \geq t) m(dx) \right), \tag{1.5.1}$$

where $(\mathcal{F}_e, \mathcal{E})$ is the extended Dirichlet space of $(\mathcal{E}, \mathcal{F})$. To see this, let $\{T_t; t > 0\}$ be the semigroup on $L^\infty(E; m)$ determined by the transition function of X. Then by (1.1.13), for $f \in L^\infty(E; m)$,

$$\mathcal{A}_{T_t}(f,f) = \frac{1}{2} \mathbf{E}_m \left[(f(X_t) - f(X_0))^2; t < \zeta \right] + \int_E f(x)^2 \mathbf{P}_x(\zeta \geq t) m(dx).$$

In view of (1.1.20), (1.1.22),

$$\mathcal{E}(f,f) = \lim_{t \to 0} \frac{1}{t} \mathcal{A}_{T_t}(f,f) = \lim_{t \to 0} \lim_{\ell \to \infty} \frac{1}{t} \mathcal{A}_{T_t}(\varphi^\ell \circ f, \varphi^\ell \circ f), \quad f \in \mathcal{F}_e,$$

for the normal contraction φ^ℓ defined by (1.1.19), and consequently, we get (1.5.1) by the monotone convergence theorem.

Chapter Two

BASIC PROPERTIES AND EXAMPLES OF DIRICHLET FORMS

2.1. TRANSIENCE, RECURRENCE, AND IRREDUCIBILITY

In this section, we introduce the concepts of the transience, recurrence, and irreducibility of the semigroup for general Markovian symmetric operators and present their characterizations by means of the associated Dirichlet form as well as the associated extended Dirichlet space. These notions are invariant under the time changes of the associated Markov process. We shall also examine them in the concrete examples of the next section.

As in Section 1.1, we let $(E, \mathcal{B}(E), m)$ be a σ-finite measure space and consider a strongly continuous contraction semigroup $\{T_t; t > 0\}$ of symmetric Markovian operators and a Dirichlet form $(\mathcal{E}, \mathcal{F})$ on $L^2(E; m)$, which are mutually related according to Theorem 1.1.3. Denote by $\{G_\alpha; \alpha > 0\}$ the resolvent of $\{T_t; t > 0\}$. For $f \in L^2(E; m)$, we define

$$S_t f = \int_0^t T_s f ds, \qquad t > 0, \tag{2.1.1}$$

where the integral is in the sense of Bochner. S_t is a linear operator on $L^2(E; m)$ and satisfies $\|S_t f\|_2 \leq t\|f\|_2$ for $f \in L^2(E; m)$.

Take $f \in L^2(E; m) \cap L^1(E; m)$. Choose $B_n \in \mathcal{B}(E)$ with $m(B_n) < \infty$ and $B_n \uparrow E$. By the symmetry and the Markov property of T_t,

$$\int_{B_n} |T_t f(x)| m(dx) \leq (T_t|f|, \mathbf{1}_{B_n}) = (|f|, T_t \mathbf{1}_{B_n}) \leq \int_E |f(x)| m(dx).$$

Letting $n \to \infty$, we get $\|T_t f\|_1 \leq \|f\|_1$. Similarly, we get $\|S_t f\|_1 \leq t\|f\|_1$. Hence both $\{T_t, t \geq 0\}$ and $\{S_t, t \geq 0\}$ can be extended to linear operators on $L^1(E; m)$ satisfying for $f \in L^1(E; m)$,

$$T_s T_t f = T_{s+t} f, \quad \|T_t f\|_1 \leq \|f\|_1 \quad \text{and} \quad \|S_t f\|_1 \leq t\|f\|_1.$$

Moreover, T_t and $\frac{1}{t} S_t$ are Markovian. The same can be said about $\{G_\alpha\}$. A concrete way to extend T_t is to define $T_t f = \lim_{\ell \to \infty} T_t(\varphi^\ell \circ f)$ for $f \in L^1(E; m)$, by the function $\varphi^\ell(t)$ of (1.1.19).

Since S_t and G_α so extended enjoy the positivity and the monotonicity for $f \in L^1_+(E;m)$: for $t > s > 0$ and $\beta > \alpha > 0$,

$$0 \leq S_s f \leq S_t f \ [m] \quad \text{and} \quad 0 \leq G_\beta f \leq G_\alpha f \ [m],$$

we can define for $f \in L^1_+(E;m)$ a function $Gf(x)$ by

$$Gf = \lim_{N\to\infty} S_N f = \lim_{N\to\infty} G_{1/N} f \quad [m], \tag{2.1.2}$$

which is uniquely up to the m-equivalence.

The operation that maps $f \in L^1_+(E;m)$ to Gf is called the *potential operator* of the semigroup $\{T_t; t > 0\}$. When E is a Hausdorff topological space and $\{T_t; t > 0\}$ is determined by an m-symmetric transition function $\{P_t; t \geq 0\}$ as in Lemma 1.1.14, then for $f \in L^1_+(E;m) \cap \mathcal{B}(E)$, Gf has an m-equivalent version Rf defined by

$$Rf(x) = \int_0^\infty P_t f(x) dt, \quad x \in E. \tag{2.1.3}$$

This operator R is said to be the *potential operator* or *0-order resolvent kernel* of the transition function $\{P_t; t \geq 0\}$.

DEFINITION 2.1.1. (i) $\{T_t; t > 0\}$ is called *transient* if $Gg < \infty \ [m]$ for some $g \in L^1_+(E;m)$ with $g > 0 \ [m]$.
(ii) $\{T_t; t > 0\}$ is called *recurrent* if, for any $f \in L^1_+(E;m)$, Gf is either ∞ or $0 \ [m]$, namely, $m\{x \in E : 0 < Gf(x) < \infty\} = 0$.
(iii) $A \in \mathcal{B}^m(E)$ is called $\{T_t\}$*-invariant* if $T_t(\mathbf{1}_{A^c} f) = 0$ m-a.e. on A for every $t > 0$ and $f \in L^2(E;m)$.
(iv) $\{T_t; t > 0\}$ is called *irreducible* if any $\{T_t\}$-invariant set A is trivial in the sense that either $m(A) = 0$ or $m(A^c) = 0$.

Inequality (2.1.4) in the first part of next lemma is called *Hopf's maximal inequality*.

LEMMA 2.1.2. (i) *For $f \in L^1(E;m)$ and $t > 0$, consider the set $E_t = \{x \in E : \sup_n S_{nt} f(x) > 0\}$. Then*

$$\int_{E_t} S_t f(x) m(dx) \geq 0. \tag{2.1.4}$$

(ii) *For $f \in L^1_+(E;m)$ and $B \in \mathcal{B}(E)$,*

$$\liminf_{t\downarrow 0} \frac{1}{t} \int_B S_t f dm \geq \int_B f dm. \tag{2.1.5}$$

Proof. (i) If we let

$$E_t^n := \left\{ x \in E : \max_{1 \le k \le n} S_{kt} f(x) > 0 \right\}$$

$$= \left\{ x \in E : \max_{1 \le k \le n} (S_{kt} f)^+(x) > 0 \right\},$$

then, for $x \in E_t^n$,

$$S_t f(x) + \max_{1 \le k \le n} (S_{(k+1)t} f - S_t f)^+(x) \ge \max_{1 \le k \le n} (S_{kt} f)^+(x).$$

On the other hand, since $S_{(k+1)t} f - S_t f = T_t(S_{kt} f)$ and T_t is positivity preserving (namely, sending a non-negative function to a non-negative function) linear operator, we have

$$\max_{1 \le k \le n} (S_{(k+1)t} f - S_t f)^+(x) \le T_t \left(\max_{1 \le k \le n} (S_{kt} f)^+ \right)(x).$$

Combining the above two inequalities, we obtain

$$\int_{E_t^n} S_t f(x) m(dx)$$

$$\ge \int_{E_t^n} \left(\max_{1 \le k \le n} (S_{kt} f)^+(x) - T_t\Big(\max_{1 \le k \le n} (S_{kt} f)^+ \Big)(x) \right) m(dx)$$

$$\ge \left\| \max_{1 \le k \le n} (S_{kt} f)^+ \right\|_1 - \left\| T_t\Big(\max_{1 \le k \le n} (S_{kt} f)^+ \Big) \right\|_1 \ge 0.$$

(ii) Since $\lim_{t \to 0+} \| \frac{1}{t} S_t g - g \|_2 = 0$ for $g \in L^2(E; m)$, we can get, by choosing $B_N \in \mathcal{B}(E)$ with $m(B_N) < \infty$ and $B_N \uparrow E$,

$$\frac{1}{t} \int_B S_t f dm \ge \left(\mathbf{1}_{B \cap B_N}, \frac{1}{t} S_t (f \wedge N) \right) \to (\mathbf{1}_{B \cap B_N}, f \wedge N), \ t \downarrow 0.$$

It suffices then to let $N \to \infty$. □

PROPOSITION 2.1.3. (i) $\{T_t; t > 0\}$ *is transient if and only if* $Gf < \infty$ $[m]$ *for every* $f \in L_+^1(E; m)$.
(ii) *The recurrence of* $\{T_t; t > 0\}$ *is equivalent to either of the following two conditions:*

$$Gf = \infty \, [m] \quad \textit{for every } f \in L_+^1(E; m) \textit{ with } f > 0 \, [m]. \tag{2.1.6}$$

$$\textit{There is some } g \in L_+^1(E; m) \textit{ so that } Gg = \infty \, [m]. \tag{2.1.7}$$

(iii) *If* $\{T_t; t > 0\}$ *is irreducible, then it is either transient or recurrent.*

(iv) $\{T_t; t > 0\}$ *is irreducible and recurrent if and only if, for any* $f \in L^1_+(E; m)$, *either* $Gf = \infty\ [m]$ *or* $Gf = 0\ [m]$.

Proof. (i) Assuming the transience, we take a function g satisfying condition (i) in Definition 2.1.1. For any $f \in L^1_+(E; m)$, put $B = \{x \in E : Gf(x) = \infty\}$. It suffices to show that $m(B) = 0$. For any $a > 0,\ h > 0$, we have for the set

$$A = \left\{x \in E : \sup_n S_{nh}(f - ag)(x) > 0\right\},$$

the inequality $\int_A S_h(f - ag)dm \geq 0$ by virtue of (2.1.4). Since $B \subset A$, m-a.e.,

$$h \int_E fdm \geq \int_A S_h fdm \geq a \int_A S_h gdm \geq a \int_B S_h gdm,$$

and hence

$$\frac{1}{a} \int_E fdm \geq \frac{1}{h} \int_B S_h gdm, \tag{2.1.8}$$

which, combined with (2.1.5), implies $\int_B gdm \leq \frac{1}{a}\|f\|_1$. By letting $a \to \infty$, we get $\int_B gdm = 0$ and accordingly $m(B) = 0$.
(ii) If we put $B = \{x \in E : Gf(x) = 0\}$ in (2.1.5), then we get $f = 0$ m-a.e. on B and hence the recurrence implies (2.1.6), which in turn implies (2.1.7) trivially. For the function g satisfying (2.1.7) and $f \in L^1_+(E; m)$, we put $A := \{x \in E : \sup_n S_n(g - af)(x) > 0\}$ for $a > 0,\ h > 0$, and $B := \{x \in E : Gf(x) < \infty\}$. We then obtain, by (2.1.4) and the same argument as in the proof of (i), the inequality (2.1.8) with f and g being interchanged. By letting $a \to \infty$, we get $S_h f = 0$ and hence $Gf = 0$ on the set B. This shows $\{T_t; t > 0\}$ is recurrent.
(iii) For $g \in L^1_+(E; m)$, the set $B = \{x \in E : Gg(x) = \infty\}$ is $\{T_t; t > 0\}$-invariant. To see this, put $C_n = \{x \in E_n : Gg(x) \leq n\}$, $g_\ell = g \wedge \ell$ for $E_n \in \mathcal{B}(E)$ such that $m(E_n) < \infty$, $E_n \uparrow E$. Then $C_n \uparrow B^c$ and, by the symmetry of T_t, we have for any $f \in L^2_+(E; m)$

$$\begin{aligned}(T_t(\mathbf{1}_{C_n} f), G_{1/\ell} g_\ell) &= (\mathbf{1}_{C_n} f, T_t G_{1/\ell} g_\ell) \leq (\mathbf{1}_{C_n} f, Gg) \\ &\leq n(f, \mathbf{1}_{C_n}) < \infty.\end{aligned}$$

By letting $\ell \to \infty$, we get $(T_t(\mathbf{1}_{C_n} f), Gg) < \infty$ and consequently $\mathbf{1}_B \cdot T_t(\mathbf{1}_{C_n} f) = 0\ [m]$. We finally let $n \to \infty$ to conclude $\mathbf{1}_B \cdot T_t(\mathbf{1}_{B^c} f) = 0\ [m]$.

Therefore, under the irreducible assumption, the above set B for a strictly positive $g \in L^1(E; m)$ satisfies either $m(B) = 0$ or $m(B^c) = 0$. $\{T_t; t > 0\}$ is transient in the former case and recurrent in the latter case by (ii).
(iv) Under the irreducibility, the above set B in the proof of (iii) for $g \in L^1_+(E; m)$ satisfies either $m(B) = 0$ or $m(B^c) = 0$. If we assume the recurrence additionally, then $Gg = 0\ [m]$ in the former case.

Conversely, assume the condition in (iv). Then $\{T_t; t > 0\}$ is obviously recurrent. If $A \in \mathcal{B}(E)$ is $\{T_t\}$-invariant, then $G(\mathbf{1}_{A^c}f) = 0$ m-a.e. on A for a strictly positive $f \in L^1(E; m) \cap L^2(E; m)$. If both $m(A)$ and $m(A^c)$ are positive, then $m(g > 0) > 0$ and $m(Gg = 0) > 0$ for $g = \mathbf{1}_{A^c}f$. By the assumption, $Gg = 0$ $[m]$, which forces $m(g > 0) = 0$ by (2.1.5), a contradiction. □

We now give a transience characterization in terms of the Dirichlet form.

LEMMA 2.1.4. (i) *For any $g \in L^2(E; m)$ and $t > 0$, $S_t g \in \mathcal{F}$ and*

$$\mathcal{E}(S_t g, u) = (g - T_t g, u) \quad \text{for every } u \in \mathcal{F}. \tag{2.1.9}$$

(ii) *For any non-negative $g \in L^1(E; m) \cap L^2(E; m)$,*

$$\sup_{u \in \mathcal{F}} \frac{(|u|, g)}{\sqrt{\mathcal{E}(u, u)}} = \sqrt{\int_E g \cdot Gg dm} \ (\leq +\infty). \tag{2.1.10}$$

Proof. (i) Since $S_t g - T_s S_t g = -\int_t^{t+s} T_v g dv + \int_0^s T_v g dv$ and $\frac{1}{s}(S_t g - T_s S_t g, S_t g)$ converges as $s \downarrow 0$ to a finite limit $(g, S_t g) - (T_t g, S_t g)$, we get the first conclusion. The same computation gives (2.1.9).
(ii) Denote by c the left hand side of (2.1.10). Suppose $c < \infty$, then $(S_t g, g) \leq c\sqrt{\mathcal{E}(S_t g, S_t g)} \leq c\sqrt{(S_t g, g)}$ by (i) and so $\sqrt{(S_t g, g)} \leq c$. Letting $t \uparrow \infty$, we obtain $\sqrt{\int_E gGg dm} \leq c$. Conversely, suppose that the right hand side of (2.1.10) is finite. Since $\int_E gGg dm = \int_0^\infty (T_s g, g) ds$ and $(T_s g, g) = (T_{s/2} g, T_{s/2} g)$ is non-increasing as s increases, we must have $\lim_{s \uparrow \infty} (T_s g, g) = 0$. By (i), we then have for any $u \in \mathcal{F}$,

$$\begin{aligned}(|u|, g) &= \mathcal{E}(|u|, S_t g) + (|u|, T_t g) \leq \|S_t g\|_{\mathcal{E}} \ \|u\|_{\mathcal{E}} + \|T_t g\|_2 \ \|u\|_2 \\ &\leq \sqrt{(S_t g, g)} \ \|u\|_{\mathcal{E}} + \sqrt{(T_{2t} g, g)} \ \|u\|_2 \to \sqrt{\int_E gGg dm} \ \|u\|_{\mathcal{E}}\end{aligned}$$

as $t \uparrow \infty$. □

THEOREM 2.1.5. (i) *$\{T_t; t > 0\}$ is transient if and only if there exists a bounded m-integrable function g strictly positive on E such that*

$$\int_E |u(x)| g(x) m(dx) \leq \|u\|_{\mathcal{E}} \quad \text{for every } u \in \mathcal{F}. \tag{2.1.11}$$

(ii) *Suppose $\{T_t; t > 0\}$ is transient. Then the inequality* (2.1.11) *holds for every $u \in \mathcal{F}_e$. Furthermore, the extended Dirichlet space $\mathcal{F}_e$ is a real Hilbert space with inner product $\mathcal{E}$.*

Proof. (i) If (2.1.11) holds, then the right hand side of (2.1.10) is no larger than 1. Consequently, $Gg < \infty$ $[m]$; namely, $\{T_t; t > 0\}$ is transient. Conversely, if $\{T_t; t > 0\}$ is transient, then the condition of Proposition 2.1.3(i) is fulfilled. By taking a strictly positive bounded function h on E with $\int_E hdm = 1$ and putting $g = h/(Gh \vee 1)$, we have $0 < g \leq h$ and

$$\int_E g \cdot Ggdm \leq \int_E h \cdot Ggdm \leq \int_E Gh \cdot (h/Gh)dm = \int_E hdm = 1,$$

which, combined with (2.1.10), leads us to (2.1.11).
(ii) Suppose $\{T_t; t > 0\}$ is transient. The first assertion is obvious due to the definition of the extended Dirichlet space $\mathcal{F}_e$ and indeed $\mathcal{F}_e$ is continuously embedded into the space $L^1(E; g \cdot m)$.

As for the second assertion, it suffices to prove the completeness of the space $(\mathcal{F}_e, \mathcal{E})$. Observe that, for $u \in \mathcal{F}_e$ and its approximating functions $\{u_n\} \subset \mathcal{F}$, $\|u - u_n\|_{\mathcal{E}} = \lim_{\ell\to\infty} \|u_\ell - u_n\|_{\mathcal{E}}$ and hence $\{u_n\}$ is $\mathcal{E}$-convergent to u. Now take any $\mathcal{E}$-Cauchy sequence $\{u_n\}$ in $\mathcal{F}_e$. Choose $\{v_n\} \subset \mathcal{F}$ with $\lim_{n\to\infty} \|u_n - v_n\|_{\mathcal{E}} \to 0$. Then $\{v_n\}$ is $\mathcal{E}$-Cauchy and hence $L^1(E; g \cdot m)$-Cauchy by virtue of (2.1.11). There is subsequence n_k so that v_{n_k} converges m-a.e. to a function $u \in L^1(E; g \cdot m)$ as $k \to \infty$. Then $u \in \mathcal{F}_e$ and

$$\|u_n - u\|_{\mathcal{E}} \leq \|u_n - u_{n_k}\|_{\mathcal{E}} + \|u_{n_k} - v_{n_k}\|_{\mathcal{E}} + \|v_{n_k} - u\|_{\mathcal{E}}.$$

Letting $k \to \infty$ and then $n \to \infty$, we see that u_n is $\mathcal{E}$-convergent to u as $n \to \infty$. □

A Dirichlet form possessing the property of Theorem 2.1.5(i) is called *transient* and a function g appearing there will be called a *reference function* for the transient Dirichlet form. A strictly positive bounded m-integrable function g on E will be called a *reference function* of a transient semigroup $\{T_t; t > 0\}$ if $\int_E g \cdot Ggdm \leq 1$. A function g is a reference function of a transient semigroup if it is so for the associated Dirichlet form.

We next formulate the restriction of the Dirichlet form to a $\{T_t\}$-invariant set.

Proposition 2.1.6. *$A \in \mathcal{B}^m(E)$ is $\{T_t\}$-invariant if and only if so is A^c. Furthermore, the following conditions for $A \in \mathcal{B}^m(E)$ are mutually equivalent.*

(i) *A is $\{T_t\}$-invariant.*
(ii) *$T_t(\mathbf{1}_A f) = \mathbf{1}_A T_t f$ for every $t > 0$ and $f \in L^2(E; m)$.*
(iii) *$G_\alpha(\mathbf{1}_A f) = \mathbf{1}_A G_\alpha f$ for every $\alpha > 0$ and $f \in L^2(E; m)$.*
(iv) *For any $f \in \mathcal{F}$, $\mathbf{1}_A \cdot f \in \mathcal{F}$ and*

$$\mathcal{E}(f, g) = \mathcal{E}(\mathbf{1}_A f, \mathbf{1}_A g) + \mathcal{E}(\mathbf{1}_{A^c} f, \mathbf{1}_{A^c} g), \quad f, g \in \mathcal{F}. \tag{2.1.12}$$

(v) *For any $f \in \mathcal{F}_e$, $\mathbf{1}_A \cdot f \in \mathcal{F}_e$ and*

$$\mathcal{E}(f, g) = \mathcal{E}(\mathbf{1}_A f, \mathbf{1}_A g) + \mathcal{E}(\mathbf{1}_{A^c} f, \mathbf{1}_{A^c} g), \quad f, g \in \mathcal{F}_e. \tag{2.1.13}$$

Proof. Due to the symmetry of T_t, $(\mathbf{1}_A f, T_t(\mathbf{1}_{A^c} g)) = (T_t(\mathbf{1}_A f), \mathbf{1}_{A^c} g)$, for $f, g \in L^2(E; m)$ and consequently the $\{T_t\}$-invariance of A is equivalent to that of A^c. Therefore, if (i) holds, then

$$T_t(\mathbf{1}_A f) = \mathbf{1}_A T_t(\mathbf{1}_A f) = \mathbf{1}_A(T_t f - T_t(\mathbf{1}_{A^c} f)) = \mathbf{1}_A T_t f,$$

namely, (ii) is valid. The implications (ii) $\Rightarrow$ (i) and (ii) $\Leftrightarrow$ (iii) are obvious.

If we assume (i), then

$$(f, (I - T_t)f) = (\mathbf{1}_A f, (I - T_t)(\mathbf{1}_A f)) + (\mathbf{1}_{A^c} f, (I - T_t)(\mathbf{1}_{A^c} f))$$

and hence (iv) follows from (1.1.4), (1.1.5). We next assume (iv). Comparing the equations obtained by substituting $\mathbf{1}_A f$, $\mathbf{1}_A g$ for f, g in (2.1.12), respectively, we get $\mathcal{E}(\mathbf{1}_A f, g) = \mathcal{E}(\mathbf{1}_A f, \mathbf{1}_A g) = \mathcal{E}(f, \mathbf{1}_A g)$ for $f, g \in \mathcal{F}$, and accordingly we have for any $f \in L^2(E; m)$, $g \in \mathcal{F}$,

$$\begin{aligned}\mathcal{E}_\alpha(G_\alpha(\mathbf{1}_A f), g) &= (\mathbf{1}_A f, g) = (f, \mathbf{1}_A g) \\ &= \mathcal{E}_\alpha(G_\alpha f, \mathbf{1}_A g) = \mathcal{E}_\alpha(\mathbf{1}_A G_\alpha f, g),\end{aligned}$$

yielding (iii).

The implication (iv) $\Rightarrow$ (v) follows from Theorem 1.1.5(i). The converse is trivial. □

We call a Dirichlet form or an extended Dirichlet space *irreducible* if, for any $A \in \mathcal{B}^m(E)$ satisfying condition (iv) or (v) of Proposition 2.1.6, either $m(A) = 0$ or $m(A^c) = 0$.

Let $A \subset E$ be an m-measurable $\{T_t\}$-invariant set. The restrictions of a function f and a measure m on E to A will be denoted by $f|_A$ and $m|_A$, respectively. If we let

$$\mathcal{F}^A = \{f|_A : f \in \mathcal{F}\}, \quad \mathcal{E}^A(f|_A, g|_A) = \mathcal{E}(\mathbf{1}_A f, \mathbf{1}_A g), \quad f, g \in \mathcal{F}, \tag{2.1.14}$$

then $(\mathcal{E}^A, \mathcal{F}^A)$ is a closed symmetric form on $L^2(A; m_A)$. The semigroup $\{T_t^A; t \geq 0\}$ and the resolvent $\{G_\alpha^A; \alpha > 0\}$ generated by it can be verified to satisfy

$$T_t^A(f|_A) = T_t(\mathbf{1}_A f)\big|_A, \quad G_\alpha^A(f|_A) = G_\alpha(\mathbf{1}_A f)\big|_A, \quad f \in L^2(E; m), \tag{2.1.15}$$

and accordingly, $(\mathcal{E}^A, \mathcal{F}^A)$ becomes a Dirichlet form on $L^2(A; m_A)$. We call this the *restriction of the Dirichlet form* $(\mathcal{E}, \mathcal{F})$ to the $\{T_t\}$-invariant set A.

Let $(\mathcal{F}_e^A, \mathcal{E}^A)$ be the space defined by (2.1.14) with $\mathcal{F}_e$ in place of $\mathcal{F}$. $(\mathcal{F}_e^A, \mathcal{E}^A)$ is called the *restriction of the extended Dirichlet space* $(\mathcal{F}_e, \mathcal{E})$ to A. $(\mathcal{F}_e^A, \mathcal{E}^A)$ is easily seen to be the extended Dirichlet space of $(\mathcal{E}^A, \mathcal{F}^A)$.

We now turn to a recurrence characterization. We employ a simple perturbation method for Dirichlet forms. We choose a function η on E satisfying

$$\eta \in L^1(E; m) \cap L^\infty(E; m), \quad \eta > 0 \ [m], \tag{2.1.16}$$

and put

$$\mathcal{E}^\eta(f,g) = \mathcal{E}(f,g) + (f,g)_{\eta\cdot m}, \quad f,g \in \mathcal{F}, \tag{2.1.17}$$

where $(f,g)_{\eta\cdot m} = \int_E f(x)g(x)\eta(x)m(dx)$. Since

$$\mathcal{E}_1(f,f) \le \mathcal{E}^\eta(f,f) + (f,f) \le \mathcal{E}(f,f) + (1 + \|\eta\|_\infty)(f,f), \quad f \in \mathcal{F},$$

$(\mathcal{E}^\eta, \mathcal{F})$ is clearly a Dirichlet form on $L^2(E;m)$. The quantities related to this Dirichlet form will be designated by a superscript η.

LEMMA 2.1.7. *Suppose* $\{T_t; t > 0\}$ *is recurrent. If we let* $f_n = G^\eta_{1/n}\eta$, *then* $f_n \in \mathcal{F}$, $0 \le f_n \uparrow 1$ $[m]$ *as* $n \to \infty$, *and* $\lim_{n\to\infty} \mathcal{E}(f_n,f_n) = 0$.

Proof. For any $f \in L^2(E;m)$, $g \in \mathcal{F}$ and $\alpha > 0$, it holds that

$$\mathcal{E}_\alpha(G^\eta_\alpha f, g) = \mathcal{E}^\eta_\alpha(G^\eta_\alpha f, g) - (G^\eta_\alpha f, g)_{\eta\cdot m} = (f - \eta G^\eta_\alpha f, g). \tag{2.1.18}$$

Hence

$$G^\eta_\alpha f = G_\alpha(f - \eta G^\eta_\alpha f). \tag{2.1.19}$$

On the other hand, for any $\varepsilon > 0$, $(\mathcal{E}, \mathcal{F})$ can be viewed as a Dirichlet form on $L^2(E; (\eta + \varepsilon)m)(= L^2(E;m))$ and we have the identity

$$\begin{aligned}
&\mathcal{E}(G^\eta_\varepsilon(\varepsilon f + f\eta), g) + (G^\eta_\varepsilon(\varepsilon f + f\eta), g)_{(\eta+\varepsilon)m} \\
&\quad = \mathcal{E}^\eta_\varepsilon(G^\eta_\varepsilon(\varepsilon f + f\eta), g) = (\varepsilon f + f\eta, g) = (f,g)_{(\eta+\varepsilon)m},
\end{aligned}$$

which means that $G^\eta_\varepsilon(\varepsilon f + f\eta)$ is nothing but the 1-order resolvent of f generated by the Dirichlet form $(\mathcal{E}, \mathcal{F})$ on $L^2(E; (\eta + \varepsilon)m)$. Due to the Markov property of the resolvent (Theorem 1.1.3,) we have $0 \le G^\eta_\varepsilon(\varepsilon f + f\eta) \le 1$ for any $f \in \mathcal{F}$ with $0 \le f \le 1$. Since G^η_ε is positivity preserving, it holds then that $0 \le G^\eta_\varepsilon(f\eta) \le 1$. By letting $\varepsilon \downarrow 0$ and then $f \uparrow 1$, we have

$$0 \le G^\eta\eta \le 1 \quad [m]. \tag{2.1.20}$$

Taking $f = \eta$ in (2.1.20), we have by (2.1.20) that

$$0 \le G(\eta(1 - G^\eta\eta)) \le \lim_{\alpha\downarrow 0} G(\eta - \eta G^\eta_\alpha \eta) = G^\eta\eta \le 1 \quad [m]. \tag{2.1.21}$$

Since $\{T_t; t > 0\}$ is recurrent and $\eta > 0$ on E, (2.1.21) implies $G^\eta\eta = 1$ $[m]$ on account of (2.1.5). We now let $f_n = G^\eta_{1/n}\eta$. Then $0 \le f_n \uparrow 1$, $n \to \infty$, and we have from (2.1.18)

$$\begin{aligned}
\mathcal{E}(f_n,f_n) &\le \mathcal{E}_{1/n}(f_n,f_n) = (\eta(1 - f_n), f_n) \\
&\le \int_E \eta(1 - f_n)dm \to 0 \quad \text{as } n \to \infty.
\end{aligned}$$

□

THEOREM 2.1.8. *The following are mutually equivalent:*

(i) $\{T_t; t > 0\}$ *is recurrent.*
(ii) *There exists a sequence* $\{f_n\} \subset \mathcal{F}$ *such that* $\lim_{n\to\infty} f_n = 1$ $[m]$ *and* $\lim_{n\to\infty} \mathcal{E}(f_n, f_n) = 0$.
(iii) $1 \in \mathcal{F}_e$ *and* $\mathcal{E}(1,1) = 0$.

Proof. By the definition of the extended Dirichlet space $(\mathcal{F}_e, \mathcal{E})$, (ii) and (iii) are equivalent. The implication (i) $\Rightarrow$ (ii) has been proved by the preceding lemma.

Assume that (ii) holds. If (i) is false, then (2.1.6) does not hold and there exists a function $g \in L^1_+(E; m)$ with $g > 0$ $[m]$ such that the set $A = \{x \in E : Gg(x) < \infty\}$ satisfies $m(A) > 0$. In view of the first statement in Proposition 2.1.6 and the proof of Proposition 2.1.3(iii), A is $\{T_t\}$-invariant so that we may consider the restriction $(\mathcal{E}^A, \mathcal{F}_A)$ of the Dirichlet form $(\mathcal{E}, \mathcal{F})$ to the set A defined by (2.1.14).

(2.1.15) then implies $G^A g|_A < \infty$ $[m_A]$ and Relation (2.1.15) then implies $G^A g|_A < \infty$ $[m_A]$ and, accordingly, $(\mathcal{E}^A, \mathcal{F}^A)$ is transient as a Dirichlet form on $L^2(A; m_A)$. By virtue of Theorem 2.1.5, there exists an m_A-integrable bounded function h on A with $h > 0$ $[m_A]$ such that

$$\int_A |f| h dm \le \|\mathbf{1}_A f\|_{\mathcal{E}} \le \|f\|_{\mathcal{E}} \quad \text{for every } f \in \mathcal{F}.$$

Applying the above to f_n, we have from the assumption (ii) and Fatou's lemma that $\int_A hdm = 0$. This contradiction establishes the implication that (ii) $\Rightarrow$ (i). □

Based on this theorem, a Dirichlet form or an extended Dirichlet space satisfying condition (ii) or (iii) of Theorem 2.1.8, respectively, can be called *recurrent*.

We now give characterizations of the transience in terms of the extended Dirichlet space.

THEOREM 2.1.9. *The following conditions are mutually equivalent:*

(i) $\{T_t; t > 0\}$ *is transient.*
(ii) *The extended Dirichlet space* $(\mathcal{F}_e, \mathcal{E})$ *is a real Hilbert space.*
(iii) $f = 0$ $[m]$ *for every* $f \in \mathcal{F}_e$ *with* $\mathcal{E}(f, f) = 0$.

Proof. The implication (i) $\Rightarrow$ (ii) has already been established in Theorem 2.1.5. The implication (ii) $\Rightarrow$ (iii) is trivial.

Assume (iii) holds. If (i) were not true, (2.1.6) of Proposition 2.1.3(ii) fails and thus there exists a function $g \in L^1_+(E; m)$ so that the set $A = \{x \in E : Gg(x) = \infty\}$ has $m(A) > 0$. By the proof of Proposition 2.1.3(iii), A is $\{T_t\}$-invariant and we may consider the restricted Dirichlet form

$(\mathcal{E}^A, \mathcal{F}_A)$ to A defined by (2.1.14). By (2.1.15) and Proposition 2.1.6, $G^A(g|_A) = G(\mathbf{1}_A g)|_A = (Gg)|_A = \infty\ [m_A]$. So by (2.1.7), $\{T_t^A; t > 0\}$ is recurrent. Hence Theorem 2.1.8 yields that there is a sequence $\{f_n\} \subset \mathcal{F}$ such that

$$\lim_{n\to\infty} \mathbf{1}_A f_n = \mathbf{1}_A \quad \text{and} \quad \lim_{n\to\infty} \mathcal{E}(\mathbf{1}_A f_n, \mathbf{1}_A f_n) = 0.$$

Since $\mathbf{1}_A f_n \in \mathcal{F}$ for each $n \geq 1$, the above means $\mathbf{1}_A \in \mathcal{F}_e$ with $\mathcal{E}(\mathbf{1}_A, \mathbf{1}_A) = 0$, a contradiction to the condition (iii). □

As was mentioned in Section 1.1, we can make use of an increasing sequence of functions $\{\eta_n; n \geq 1\} \subset L^1_+(E; m)$ that increases to 1 to deduce from the operator T_t on $L^2(E; m)$ a unique Markovian linear operator T_t on $L^\infty(E; m)$ by (1.1.9). If $T_t 1 = 1\ [m]$ for some (and hence for all) $t > 0$, then we call the semigroup $\{T_t; t > 0\}$ or the Dirichlet form $(\mathcal{E}, \mathcal{F})$ *conservative.*

PROPOSITION 2.1.10. *If $\{T_t; t > 0\}$ is recurrent, then it is conservative. If $\cup_\ell \{T_\ell 1 < 1\} = E\ [m]$, then $\{T_t; t > 0\}$ is transient.*

Proof. If $f \in L^1(E; m) \cap L^\infty(E; m)$, $f > 0\ [m]$, then for every $t > 0$,

$$\begin{aligned}(S_N f, \eta_n - T_t\eta_n) &= \left(f, \int_0^N T_s(\eta_n - T_t\eta_n)ds\right) \\ &= \left(f, \int_0^t T_s\eta_n ds - \int_N^{N+t} T_s\eta_n ds\right) \leq \left(f, \int_0^t T_s\eta_n ds\right) \leq t\int_E f dm.\end{aligned}$$

By letting $n \to \infty$ and then $N \to \infty$, we obtain

$$\int_E Gf(x)(1 - T_t 1(x))m(dx) \leq t\int_E f dm < \infty,$$

which implies the first assertion (with the aid of Proposition 2.1.3(ii)) as well as the second one. □

The next theorem gives a criterion for a recurrent Dirichlet form to be irreducible. According to Theorem 2.1.8, the recurrence of the Dirichlet form $(\mathcal{E}, \mathcal{F})$ is equivalent to the following property when $m(E) < \infty$:

$$1 \in \mathcal{F} \quad \text{and} \quad \mathcal{E}(1, 1) = 0. \tag{2.1.22}$$

THEOREM 2.1.11. *Suppose that $m(E) < \infty$ and that a Dirichlet form $(\mathcal{E}, \mathcal{F})$ is recurrent. Then the following conditions are mutually equivalent.*

(i) *$(\mathcal{E}, \mathcal{F})$ is irreducible.*
(ii) *f is constant m-a.e. for every $f \in \mathcal{F}$ with $\mathcal{E}(f, f) = 0$.*
(iii) *f is constant m-a.e. for every $f \in \mathcal{F}_e$ with $\mathcal{E}(f, f) = 0$.*
(iv) *f is constant m-a.e. for every $f \in L^2(E; m)$ with $T_t f = f$ for every $t > 0$.*

Proof. (i) $\Rightarrow$ (ii). We first note that f is $\{T_t\}$-invariant whenever $f \in \mathcal{F}$ with $\mathcal{E}(f,f) = 0$. Indeed, by the Cauchy-Schwarz inequality holding for the non-negative definite symmetric form $\mathcal{E}$, we have $\mathcal{E}(f,g) = 0$ and hence $\mathcal{E}_\alpha(f,g) = \alpha(f,g)$ for every $g \in \mathcal{F}$ and $\alpha > 0$. Therefore by (1.1.6), $\alpha G_\alpha f = f$ for every $\alpha > 0$, and so by (1.1.2) $T_t f = f$ for every $t > 0$.

Assume that $f \in \mathcal{F}$ with $\mathcal{E}(f,f) = 0$. By (2.1.22), it holds that for any $\lambda \in \mathbb{R}$ that $f - \lambda \in \mathcal{F}$ and $\mathcal{E}(f-\lambda, f-\lambda) = 0$. Since $\varphi^+(t) = t \vee 0$, $t \in \mathbb{R}$, is a normal contraction, $f_\lambda = \varphi^+ \circ (f - \lambda) \in \mathcal{F}$ with $\mathcal{E}(f_\lambda, f_\lambda) = 0$. Consequently, f_λ is $\{T_t\}$-invariant.

Put $B_\lambda = \{x \in E : f_\lambda(x) = 0\}$. Since

$$(\mathbf{1}_{B_\lambda}, T_t(\mathbf{1}_{B_\lambda^c} f_\lambda)) = (\mathbf{1}_{B_\lambda}, T_t(f_\lambda)) = (\mathbf{1}_{B_\lambda}, f_\lambda) = 0,$$

we have by the Markovian property of T_t,

$$(\mathbf{1}_{B_\lambda}, T_t(\mathbf{1}_{B_\lambda^c} \mathbf{1}_{\{f_\lambda \geq 1/n\}})) = 0 \quad \text{for every } n \geq 1.$$

By letting $n \to \infty$, we get $\mathbf{1}_{B_\lambda} T_t(\mathbf{1}_{B_\lambda^c}) = 0$ $[m]$, which gives the $\{T_t\}$-invariance of the set B_λ.

By the irreducibility assumption (i), either $m(B_\lambda) = 0$ or $m(B_\lambda^c) = 0$. If we let $\lambda_0 = \sup\{\lambda : m(B_\lambda) = 0\}$, then, for any $\lambda > \lambda_0$, $m(B_\lambda) \neq 0$ and hence $m(B_\lambda^c) = 0$, namely, $m(\{f > \lambda_0\}) = 0$. On the other hand, we have, for any $\lambda < \lambda_0$, $m(B_\lambda) = 0$ and $m(\{f < \lambda_0\}) = 0$. We have derived $f = \lambda_0$ $[m]$.
(ii) $\Rightarrow$ (iv). If $f \in L^2(E;m)$ is $\{T_t\}$-invariant, then $f \in \mathcal{F}$ with $\mathcal{E}(f,f) = 0$ in view of (1.1.4) and (1.1.5).
(iv) $\Rightarrow$ (i). If $A \in \mathcal{B}^m(E)$ is T_t-invariant, then, by Proposition 2.1.6 and Proposition 2.1.10, we have $T_t \mathbf{1}_A = \mathbf{1}_A T_t 1 = \mathbf{1}_A$. Therefore either $m(A) = 0$ or $m(A^c) = 0$ by (iv). This shows that $(\mathcal{E}, \mathcal{F})$ is irreducible.
(ii) $\Rightarrow$ (iii). If $f \in \mathcal{F}_e$ with $\mathcal{E}(f,f) = 0$, then by Theorem 1.1.5(ii), $f_\ell = \varphi^\ell \circ f \in \mathcal{F} = \mathcal{F}_e \cap L^2(E;m)$ and $\mathcal{E}(f_\ell, f_\ell) = 0$. Here φ^ℓ is the normal contraction given by (1.1.19). By the assumption (ii), f_ℓ is constant m-a.e. Since $\ell \in \mathbb{N}$ is arbitrary, f is constant m-a.e. The converse implication is trivial. □

The finiteness assumption of $m(E)$ in the above theorem will be removed in Theorem 5.2.16 under the (quasi) regularity assumption on the Dirichlet form $(\mathcal{E}, \mathcal{F})$.

In the remainder of this section, we consider a relation of the extended Dirichlet space of a transient semigroup to the potential operator G and give a characterization of the former by means of the latter.

The function Gf has been defined for $f \in L^1_+(E;m)$ by (2.1.2). We extend the potential operator G for any non-negative m-measurable function f on E by

$$Gf = \lim_{n \to \infty} G(f \wedge (n\eta)) \quad [m], \tag{2.1.23}$$

where η is a fixed strictly positive bounded m-integrable function on E. We note that for $f \in L^1_+(E;m)$, the right hand side of (2.1.23) coincides with $Gf(x)$ of (2.1.2) m-a.e. due to the exchangeability of monotone limits. We also note that when E is a Hausdorff topological space and $\{T_t; t > 0\}$ is determined by an m-symmetric transition function, Gf defined by (2.1.23) has $R\widetilde{f}$ of (2.1.3) as its version for any Borel version $\widetilde{f}$ of f.

THEOREM 2.1.12. *Assume that $\{T_t\}$ is transient.*
(i) *If a non-negative m-measurable function f has $\int_E f \cdot Gf dm < \infty$, then*

$$Gf \in \mathcal{F}_e \quad \textit{and} \quad f \cdot v \in L^1(E; m) \qquad \textit{for every } v \in \mathcal{F}_e, \tag{2.1.24}$$

$$\mathcal{E}(Gf, v) = \int_E f \cdot v dm \qquad \textit{for every } v \in \mathcal{F}_e. \tag{2.1.25}$$

Furthermore, the function Gf does not depend on the choice of η in its definition (2.1.23) *up to m-equivalence.*
(ii) *For a reference function g of the transient Dirichlet form $(\mathcal{E}, \mathcal{F})$, let $\mathcal{L} = \{f = h \cdot g : h \in b\mathcal{B}(E)\}$. Then $\int_E f \cdot Gf dm < \infty$ for any $f \in \mathcal{L}_+$ and $G(\mathcal{L})$ is a dense linear subspace of the extended Dirichlet space $(\mathcal{F}_e, \mathcal{E})$.*

Proof. (i) First consider a non-negative bounded m-integrable function f with $\int_E f \cdot Gf dm < \infty$. By (2.1.9), we have for $t > t'$

$$\|S_t f - S_{t'} f\|_{\mathcal{E}}^2 \leq (S_t f, f) - (S_{t'} f, f) + (S_{t'} f, T_t f - T_{t'} f),$$

which converges to zero as $t, t' \to \infty$ because $(S_t f, f) \to \int_E f \cdot Gf dm$ and $(S_{t'} f, T_t f) = \int_t^{t+t'} (T_u f, f) du \to 0$. Hence $\{S_n f\}$ is an $\mathcal{E}$-Cauchy sequence which converges to Gf m-a.e., and so $Gf \in \mathcal{F}_e$.

We saw in the proof of Lemma 2.1.4(ii) that $\|T_t f\|_2^2 = (T_{2t} f, f) \to 0$ as $t \to \infty$ for a function f as above. Hence by letting $t \to \infty$ in (2.1.9) with $g = f$ and $u = v \in \mathcal{F}$, we arrive at the identity (2.1.25) for $v \in \mathcal{F}$.

Next take $v \in \mathcal{F}_e$ with its approximating sequence $\{v_n\} \subset \mathcal{F}$. Since $\int |v_n - v_m| f dm = \mathcal{E}(Gf, |v_n - v_m|) \leq \|Gf\|_{\mathcal{E}} \|v_n - v_m\|_{\mathcal{E}}$, we see that $\{v_n\}$ is $L^1(E; f \cdot m)$-Cauchy and converges to v m-a.e. Therefore we have the second property of (2.1.24). By letting $n \to \infty$ in (2.1.25) for $v = v_n$, we get the same equation for $v \in \mathcal{F}_e$.

Now consider a non-negative m-measurable function f with $\int_E f \cdot Gf dm < \infty$. Put $f_n = f \wedge (n\eta) \in bL^1_+(E; m)$. Then, by what has just been proved, we have

$$\|Gf_n - Gf_k\|_{\mathcal{E}}^2 \leq \int_E f_n \cdot Gf_n dm - \int_E f_k \cdot Gf_k dm, \qquad n > k,$$

which converges to zero as $n, k \to \infty$. Therefore we are led to (2.1.24) and (2.1.25) for f from those for f_n.

Equation (2.1.25) in particular implies that the function $Gf \in \mathcal{F}_e$ depends only on f and does not depend on a particular choice of η in (2.1.23).

(ii) The first assertion is immediate from (2.1.10). The second is also clear because any $f \in \mathcal{L}$ satisfies (2.1.25). □

Exercise 2.1.13. Assume $\{T_t\}$ is transient. Prove the following.

(i) If $u_n \in \mathcal{F}_e$ is $\mathcal{E}$-convergent to $u \in \mathcal{F}_e$ as $n \to \infty$ and a real-valued function φ is a normal contraction, then $\varphi(u_n)$ converges to $\varphi(u)$ $\mathcal{E}$-weakly as $n \to \infty$. If, in addition, $\varphi(u) = u$, then the convergence is $\mathcal{E}$-strong.
(ii) Let $\{\varphi_\ell\}_{\ell \geq 1}$ be a sequence of normal contractions satisfying $\lim_{\ell\to\infty} \varphi_\ell(t) = t$. Then for any $u \in \mathcal{F}_e$, $\lim_{\ell\to\infty} \|\varphi_\ell(u) - u\|_{\mathcal{E}} = 0$.

Now we present a useful characterization for a given function space to be the extended Dirichlet space of a transient Markovian semigroup.

THEOREM 2.1.14. *Suppose that $\{T_t; t > 0\}$ is transient and $(\mathcal{G}, a)$ is a real Hilbert space satisfying the following conditions.*

(i) *$\mathcal{G}$ is a collection of m-equivalence class of $\mathcal{B}^m(E)$-measurable functions on E.*
(ii) *There exists a linear subspace $\mathcal{L}$ of $L^1(E; m)$ such that*

(a) $f \in \mathcal{L} \Longrightarrow |f| \in \mathcal{L}$.
(b) *If an m-measurable function v satisfies $f \cdot v \in L^1(E; m)$ and $\int_E f v dm = 0$ for any $f \in \mathcal{L}_+$, then $v = 0$ $[m]$.*

(iii) *For any $f \in \mathcal{L}_+$ and any $v \in \mathcal{G}$,*

$$Gf \in \mathcal{G} \quad \text{and} \quad a(Gf, v) = \int_E f v dm. \tag{2.1.26}$$

Then $(\mathcal{G}, a) = (\mathcal{F}_e, \mathcal{E})$ the extended Dirichlet space of $\{T_t; t > 0\}$.

Proof. Put $G(\mathcal{L}) = \{Gf_1 - Gf_2 : f_1, f_2 \in \mathcal{L}_+\}$. Then by (ii)(b) and (iii), $G(\mathcal{L})$ is a dense linear subspace of $(\mathcal{G}, a)$ and

$$\int_E f \cdot Gf dm = a(Gf, Gf) < \infty \quad \text{for every } f \in \mathcal{L}_+.$$

Hence, by virtue of Theorem 2.1.12, for every $f \in \mathcal{L}_+$ and $v \in \mathcal{F}_e$,

$$Gf \in \mathcal{F}_e, \quad f \cdot v \in L^1(E; m) \quad \text{and} \quad \mathcal{E}(Gf, v) = \int_E f v dm.$$

By the assumption (ii), we then see that $G(\mathcal{L})$ is dense in $(\mathcal{F}_e, \mathcal{E})$. Since $a = \mathcal{E}$ on $\mathcal{G}(\mathcal{L})$, we get the desired conclusion. □

A typical example of the space $\mathcal{L}$ satisfying conditions of the above theorem is $\mathcal{L} = \{f \cdot g : f \in b\mathcal{B}(E)\}$, where g is a reference function for the transient semigroup $\{T_t; t > 0\}$.

Finally we present a useful lemma which will reduce some arguments for a general (not necessarily transient) Dirichlet form $\mathcal{E}$ to those for transient ones. Let us consider a collection of functions

$$\mathcal{K}_0 = \left\{ g : g \in b\mathcal{B}(E) \text{ strictly positive } m\text{-a.e. with } \int_E g dm \leq 1 \right\}. \tag{2.1.27}$$

For $g \in \mathcal{K}_0$, define

$$\mathcal{E}^g(u, v) = \mathcal{E}(u, v) + (u, v)_{g \cdot m}, \quad u, v \in \mathcal{F}, \tag{2.1.28}$$

where $(u, v)_{g \cdot m} = \int_E u(x)v(x)g(x)m(dx)$.

Since

$$\mathcal{E}_1(u, u) \leq \mathcal{E}^g(u, u) + (u, u) \leq \mathcal{E}(u, u) + (\|g\|_{L^\infty} + 1)(u, u),$$

$(\mathcal{E}^g, \mathcal{F})$ is a Dirichlet form on $L^2(E; m)$ and the norm $\sqrt{\mathcal{E}_1^g}$ is equivalent to $\sqrt{\mathcal{E}_1}$ on $\mathcal{F}$. In particular, $(\mathcal{E}^g, \mathcal{F})$ shares the same quasi notions with $(\mathcal{E}, \mathcal{F})$. Furthermore, $(\mathcal{E}^g, \mathcal{F})$ is transient and possesses g as a reference function because

$$\int_E |u| g dm \leq \sqrt{(u, u)_{g \cdot m}} \leq \sqrt{\mathcal{E}^g(u, u)}, \quad u \in \mathcal{F}.$$

Denote by $(\mathcal{F}_e^g, \mathcal{E}^g)$ the extended Dirichlet space of $(\mathcal{F}, \mathcal{E}^g)$.

Lemma 2.1.15. *For any $u \in \mathcal{F}_e$ and for any $\mathcal{E}$-approximating sequence $\{u_n, n \geq 1\} \subset \mathcal{F}$ for u, there exists $g \in \mathcal{K}_0$ such that $u \in \mathcal{F}_e^g$ and $\{u_n\}$ is an $\mathcal{E}^g$-approximating sequence for u.*

Proof. Take a bounded strictly positive Borel function f with $\int_E f dm \leq 1$ and put $g = f / (\sup_n u_n(x)^2 \vee 1)$. Then $g \in \mathcal{K}_0$ and u_n is $L^2(E; g \cdot m)$-convergent to u. □

2.2. BASIC EXAMPLES

In this section, we present some basic examples of Dirichlet forms. Special attention will be paid to their transience, recurrence, and irreducibility as well as explicit expressions of the corresponding extended Dirichlet spaces.

2.2.1. Pure Jump Step Processes

Let E be a locally compact separable metric space and $Q(x, dy)$ be a probability kernel on $(E, \mathcal{B}(E))$. We assume that $Q(x, \{x\}) = 0$ for every $x \in E$. Let $\lambda(x)$ be a Borel measurable function on E such that $0 < \lambda(x) < \infty$. From these two objects one can construct a Markov process X on E in the following way. If X starts from $x_0 \in E$, it remains there for an exponentially distributed holding time T_1 with parameter $\lambda(x_0)$, then it jumps to some x_1 according to the probability distribution $Q(x_0, dy)$; it remains at x_1 for an exponentially distributed holding time T_2 with parameter $\lambda(x_1)$, which is independent of T_1, before jumping to x_2 according to $Q(x_1, dy)$, and so on. From the probabilistic role they play, sometimes we call $Q(x, dy)$ the *road map* and $\lambda(x)$ the *speed function* of the process X.

Suppose now $\lambda(x) \equiv 1$ on E. Denote by $P_t^{(k)}(x, dy)$ the distribution of X_t that starts from x and there are exactly k jumps that occurred during the time period $[0, t]$. Then

$$P_t^{(k)}(x, dy) = \frac{t^k}{k!} e^{-t}\, Q^{(k)}(x, dy), \qquad t > 0,\ k \geq 1,$$

where $Q^{(k)}$ is the kth iterated kernel of the probability kernel Q. So the transition function of X is given by

$$P_t(x, dy) = \sum_{k=0}^{\infty} P_t^{(k)}(x, dy) = \sum_{k=0}^{\infty} \frac{t^k}{k!} e^{-t}\, Q^{(k)}(x, dy)$$

with the convention that $Q^0(x, dy) = \delta_{\{x\}}(dy)$, the Dirac measure at x.

When $\lambda(x)$ is not a constant function, it is difficult to write out the transition function explicitly. The process X, which was rigorously constructed in Section 12 of Chapter 1 in [13], is called the *regular step process* there and it is shown to be a standard process. It was proved there that the resolvent kernel $\{R_\alpha; \alpha > 0\}$ of X_t is given by

$$\begin{aligned} R_\alpha f(x) &= \sum_{k=0}^{\infty} \int_E \frac{f(y)}{\alpha + \lambda(y)} Q_\alpha^{(k)}(x, dy) \\ &= \frac{f(x)}{\alpha + \lambda(x)} + \frac{\lambda(x)}{\alpha + \lambda(x)} \int_E R_\alpha f(y) Q(x, dy) \end{aligned} \tag{2.2.1}$$

for $f \in \mathcal{B}_+(E)$, where $Q_\alpha(x, dy) := \frac{\lambda(x)}{\alpha + \lambda(x)} Q(x, dy)$.

We now assume that there exists a σ-finite measure m_0 on E with $\text{supp}[m_0] = E$ such that

$$Q(x, dy) m_0(dx) = Q(y, dx) m_0(dy). \tag{2.2.2}$$

We call m_0 a *symmetrizing measure* of the road map Q. Define

$$m(dx) = \frac{1}{\lambda(x)} m_0(dx). \tag{2.2.3}$$

The density function $\frac{1}{\lambda(x)}$ of m is the mean holding time of the process X at x so we may call m the *speed measure* for X. We are going to show that under the condition (2.2.2) the process X is m-symmetric. For $f \in \mathcal{B}_+(E)$, let $Q_\alpha^{(k)} f(x) := \int_E f(y) Q_\alpha^{(k)}(x, dy)$.

LEMMA 2.2.1. *For $f, g \in \mathcal{B}_+(E)$, $\alpha > 0$ and $k \geq 0$,*

$$\int_E f(x) Q_\alpha^{(k)} \left(\frac{g}{\alpha + \lambda} \right)(x) m(dx) = \int_E g(x) Q_\alpha^{(k)} \left(\frac{f}{\alpha + \lambda} \right)(x) m(dx). \tag{2.2.4}$$

Proof. We prove it by mathematical induction on k. When $k = 0$, (2.2.4) is obviously true. For $k = 1$, by (2.2.2),

$$\begin{aligned}
&\int_E f(x) Q_\alpha \left(\frac{g}{\alpha + \lambda} \right)(x) m(dx) \\
&= \int_{E \times E} \frac{f(x)}{\alpha + \lambda(x)} \frac{g(y)}{\alpha + \lambda(y)} \lambda(x) Q(x, dy) m(dx) \\
&= \int_{E \times E} \frac{f(x)}{\alpha + \lambda(x)} \frac{g(y)}{\alpha + \lambda(y)} \lambda(y) Q(y, dx) m(dy) \\
&= \int_E g(x) Q_\alpha \left(\frac{f}{\alpha + \lambda} \right) m(dx).
\end{aligned}$$

So (2.2.4) holds for $k = 1$. Now suppose that (2.2.4) holds for $k = j$, then

$$\begin{aligned}
&\int_E f(x) Q_\alpha^{(j+1)} \left(\frac{g}{\alpha + \lambda} \right)(x) m(dx) \\
&= \int_E f(x) Q_\alpha \left(\frac{\alpha + \lambda}{\alpha + \lambda} Q_\alpha^{(j)} \left(\frac{g}{\alpha + \lambda} \right) \right)(x) m(dx) \\
&= \int_E Q_\alpha \left(\frac{f}{\alpha + \lambda} \right)(x) \cdot (\alpha + \lambda(x)) Q_\alpha^{(j)} \left(\frac{g}{\alpha + \lambda} \right)(x) m(dx) \\
&= \int_E g(x) Q_\alpha^{(j)} \left(Q_\alpha \left(\frac{f}{\alpha + \lambda} \right) \right)(x) m(dx) \\
&= \int_E g(x) Q_\alpha^{(j+1)} \left(\frac{f}{\alpha + \lambda} \right)(x) m(dx).
\end{aligned}$$

So (2.2.4) holds for $k = j + 1$ and the lemma is proved. □

It follows from (2.2.1) and (2.2.4) that for $f, g \in \mathcal{B}_+(E)$ and $\alpha > 0$,

$$\int_E f(x)R_\alpha g(x)m(dx) = \int_E g(x)R_\alpha f(x)m(dx).$$

Consequently, the strong Markov process X is m-symmetric. Next we are going to figure out its associated Dirichlet form $(\mathcal{E}, \mathcal{F})$ on $L^2(E;m)$. We know from Section 1.1 that each αR_α extends uniquely to a Markovian contraction operator αG_α in $L^2(E;m)$ and $\{G_\alpha; \alpha > 0\}$ forms a strongly continuous resolvent in $L^2(E;m)$.

THEOREM 2.2.2. *Assume that the speed function λ is bounded and that* (2.2.2) *holds. Then the Dirichlet form $(\mathcal{E}, \mathcal{F})$ of the regular step process X on $L^2(E;m)$ is given by*

$$\mathcal{F} = L^2(E;m),$$
$$\mathcal{E}(u,u) = \frac{1}{2}\int_{E\times E} (u(x) - u(y))^2 Q(x,dy)m_0(dx) \quad \textit{for } u \in L^2(E;m).$$

Proof. By (2.2.2) and the Cauchy-Schwarz inequality, for $f \in L^2(E;m)$,

$$\begin{aligned}\int_{E\times E} |f(x)f(y)|\lambda(x)Q(x,dy)m(dx) &\le \int_{E\times E} f(x)^2\lambda(x)Q(x,dy)m(dx)\\ &= \int_E f(x)^2\lambda(x)m(dx) < \infty\end{aligned}$$

and, since $\alpha G_\alpha f$ converges to f in $L^2(E;m)$ as $\alpha \to \infty$,

$$\begin{aligned}&\lim_{\alpha\to\infty}\left|\int_{E\times E} f(x)\left(\frac{\alpha}{\alpha+\lambda(x)}(\alpha G_\alpha f(y) - f(y))\right)\lambda(x)Q(x,dy)m(dx)\right|\\
&\le \limsup_{\alpha\to\infty}\left(\int_{E\times E} f(x)^2\lambda(x)Q(x,dy)m(dx)\right)^{1/2}\\
&\quad\cdot\left(\int_{E\times E} (\alpha G_\alpha f(y) - f(y))^2\,\lambda(x)Q(x,dy)m(dx)\right)^{1/2}\\
&\le c\limsup_{\alpha\to\infty}\left(\int_{E\times E} (\alpha G_\alpha f(x) - f(x))^2\,\lambda(x)Q(x,dy)m(dx)\right)^{1/2}\\
&= c\limsup_{\alpha\to\infty}\left(\int_E (\alpha G_\alpha f(x) - f(x))^2\,\lambda(x)m(dx)\right)^{1/2} = 0.\end{aligned}$$

These together with (2.2.1) and the dominated convergence theorem yield

$$\lim_{\alpha\to\infty} \alpha(f - \alpha R_\alpha f, f)_{L^2(E;m)}$$

$$= \lim_{\alpha\to\infty} \alpha \left(\frac{\lambda}{\alpha+\lambda} \left\{ f - \alpha \int_E G_\alpha f(y) Q(\cdot, dy) \right\}, f \right)$$

$$= \int_E f(x)^2 \lambda(x) m(dx) - \lim_{\alpha\to\infty} \int_{E\times E} \frac{\alpha}{\alpha+\lambda(x)} f(x) f(y) \lambda(x) Q(x, dy) m(dx)$$

$$= \int_E f(x)^2 m_0(dx) - \int_{E\times E} f(x) f(y) Q(x, dy) m_0(dx)$$

$$= \frac{1}{2} \int_{E\times E} (f(x) - f(y))^2 Q(x, dy) m_0(dx).$$

This proves the theorem. □

When the road map $Q(x, dy)$ is a Markovian kernel, that is, when $Q(x, dy)$ is a measure on E with $Q(x, E) \leq 1$ for every $x \in E$, we can extend it to be a probability kernel on $E_\partial := E \cup \{\partial\}$, the one-point compactification of E, by setting $Q(x, \{\partial\}) := 1 - Q(x, E)$ and $Q(\partial, \{\partial\}) := 1$. Extend the definition of λ to E_∂ by setting $\lambda(\partial) = 0$. Then by the same argument as in Section 12 of Chapter 1 in [13], there exists a regular step process X on E with ∂ as the trap. Under the hypothesis of (2.2.2), X is still an m-symmetric standard process on E. In this case, we have for $f \in L^2(E; m)$

$$\lim_{\alpha\to\infty} \alpha(f - \alpha R_\alpha f, f)_{L^2(E;m)} = \frac{1}{2} \int_{E\times E} (f(x) - f(y))^2 Q(x, dy) m_0(dx)$$
$$+ \int_E f(x)^2 (1 - Q(x, E)) m_0(dx),$$

provided that λ is bounded. Thus we have the following.

THEOREM 2.2.3. *Assume that $Q(x, dy)$ is a Markovian kernel, λ is bounded, and* (2.2.2) *holds. Then the Dirichlet form $(\mathcal{E}, \mathcal{F})$ of the regular step process X on $L^2(E; m)$ is given by*

$$\mathcal{F} = L^2(E; m),$$

$$\mathcal{E}(u, u) = \frac{1}{2} \int_{E\times E} (u(x) - u(y))^2 Q(x, dy) m_0(dx)$$
$$+ \int_E u(x)^2 (1 - Q(x, E)) m_0(dx) \quad \text{for } u \in \mathcal{F}. \tag{2.2.5}$$

Remark 2.2.4. Notice that only the road map Q and its symmetrizing measure m_0 are involved in the description (2.2.5) of the Dirichlet form $\mathcal{E}$. The role of λ is reflected in the reference measure m.

Note further that in Theorems 2.2.2 and 2.2.3 the speed function λ is assumed to be bounded. For an unbounded speed function, one can still obtain an explicit characterization of the Dirichlet form of X by time change. Assume that the symmetrizing measure m_0 of the road map $Q(x, dy)$ is a positive Radon measure and that λ is locally bounded away from zero in the sense that $\inf_{x\in K} \lambda(x) > 0$ for any compact set $K \subset E$. The speed measure m defined by (2.2.3) is then a positive Radon measure again. Let X be the m-symmetric regular step process corresponding to (Q, λ) and $(\mathcal{E}, \mathcal{F})$ be the Dirichlet form of X on $L^2(E; m)$.

Let Z be the m_0-symmetric regular step process on E corresponding to $(Q, 1)$. Denote by $(\mathcal{E}^Z, \mathcal{F}^Z)$ and $\mathcal{F}_e^Z$ the Dirichlet form of Z on $L^2(E; m_0)$ and its extended Dirichlet space, respectively. Since Theorem 2.2.3 is applicable to $(\mathcal{E}^Z, \mathcal{F}^Z)$, we see that $\mathcal{F}^Z = L^2(E; m_0)$ and $\mathcal{E}^Z$ is given by (2.2.5). In particular, $(\mathcal{E}^Z, \mathcal{F}^Z)$ is a regular Dirichlet form on $L^2(E; m_0)$. Its extended Dirichlet space $\mathcal{F}_e^Z$ is a linear subspace of $\mathcal{G}$ containing $L^2(E; m_0)$, where

$$\mathcal{G} = \left\{ u : |u| < \infty \ m_0\text{-a.e. with } \int_{E\times E} (u(x) - u(y))^2 Q(x, dy) m_0(dx) \right.$$
$$\left. + \int_E u(x)^2 (1 - Q(x, E)) m_0(dx) < \infty \right\}. \tag{2.2.6}$$

It is clear that X is a time change of Z by using the new time clock $\tau_t := \inf\{s : A_s > t\}$ with $A_s = \int_0^s \frac{1}{\lambda(Z_r)} dr$; that is, $X_t = Z_{\tau_t}$. By invoking (5.2.17) of Section 5.2, we can conclude that

$$\begin{cases} \mathcal{F} = \mathcal{F}_e^Z \cap L^2(E; m), \\ \mathcal{E}(u, u) = \dfrac{1}{2} \displaystyle\int_{E\times E} (u(x) - u(y))^2 Q(x, dy) m_0(dx) \\ \qquad\qquad + \displaystyle\int_E u(x)^2 (1 - Q(x, E)) m_0(dx) \qquad \text{for } u \in \mathcal{F}. \end{cases} \tag{2.2.7}$$

As we shall see in Section 6.5, the function space $\mathcal{G}$ is the reflected Dirichlet space of $(\mathcal{E}^Z, \mathcal{F}^Z)$ that will be introduced in Chapter 6. We shall also see there that, when Q is a probability kernel, the identity $\mathcal{F}_e^Z = \mathcal{G}$ holds if and only if Z is recurrent. Exercise 2.2.5 concerns a simple example of Z, a (continuous time) symmetric simple random walk on $\mathbb{Z}^n$, which is recurrent when $n \leq 2$ and transient when $n \geq 3$. □

2.2.2. Translation-Invariant Dirichlet Forms

Let $\mathbb{R}^n$ be the n-dimensional Euclidean space and $L^p(\mathbb{R}^n)$ the L^p space with respect to the Lebesgue measure dx on $\mathbb{R}^n$. The inner product in $L^2(\mathbb{R}^n)$ is denoted by $\langle \cdot, \cdot \rangle$. For $x = (x_1, \ldots, x_n)$ and $y = (y_1, \ldots, y_n)$ in $\mathbb{R}^n$, we adopt the notations

$$\langle x, y\rangle := \sum_{n=1}^{n} x_i y_i, \qquad |x| := \sqrt{\langle x, x\rangle}.$$

Define the *convolution* of two finite measures μ and ν on $\mathbb{R}^n$ by

$$\mu * \nu(B) = \int_{\mathbb{R}^n\times\mathbb{R}^n} \mathbf{1}_B(x+y)\mu(dx)\nu(dy), \quad B \in \mathcal{B}(\mathbb{R}^n).$$

We call a system of probability measures $\{\nu_t, t \geq 0\}$ on $\mathbb{R}^n$ a *continuous symmetric convolution semigroup* on $\mathbb{R}^n$ if $\nu_0 = \delta_0$, $\nu_t * \nu_s = \nu_{t+s}$ for $t, s > 0$, $\nu_t(A) = \nu_t(-A)$ for $t > 0$ and $A \in \mathcal{B}(\mathbb{R}^n)$, and $\lim_{t\downarrow 0}\nu_t = \delta_0$ weakly. Here δ_0 is the Dirac measure concentrated at the origin $\{0\}$.

The celebrated Lévy-Khinchin formula[1] for such $\{\nu_t; t \geq 0\}$ reads as follows:

$$\widehat{\nu_t}(x)\left(:= \int_{\mathbb{R}^n} e^{i\langle x,y\rangle}\nu_t(dy)\right) = e^{-t\psi(x)} \tag{2.2.8}$$

$$\psi(x) = \frac{1}{2}\langle Sx, x\rangle + \int_{\mathbb{R}^n\setminus\{0\}} (1 - \cos\langle x, y\rangle)J(dy), \tag{2.2.9}$$

where S is a (constant) non-negative definite $n \times n$ symmetric matrix and J is a symmetric measure on $\mathbb{R}^n \setminus \{0\}$ with $\int_{\mathbb{R}^n\setminus\{0\}} \frac{|x|^2}{1+|x|^2} J(dx) < \infty$. $\psi(x)$ and J are called the *Lévy exponent* and the *Lévy measure*, respectively, for $\{\nu_t, t \geq 0\}$. We extend J to $\mathbb{R}^n$ by setting $J(\{0\}) = 0$.

Let $\{\nu_t; t \geq 0\}$ be a continuous symmetric convolution semigroup on $\mathbb{R}^n$ with the Lévy exponent $\psi(x)$ given by (2.2.9). We put

$$P_t(x, B) = \nu_t(B - x), \quad t \geq 0,\ x \in \mathbb{R}^n,\ B \in \mathcal{B}(\mathbb{R}^n). \tag{2.2.10}$$

Then $P_t f(x) = \int_{\mathbb{R}^n} f(x+y)\nu_t(dy)$ for $t \geq 0$ and $x \in \mathbb{R}^n$, and we readily see that $\{P_t; t \geq 0\}$ is a transition probability on $\mathbb{R}^n$ in the sense of Definition 1.1.13 which is also symmetric with respect to dx.

Let $(\mathcal{E}, \mathcal{F})$ be the Dirichlet form on $L^2(\mathbb{R}^n)$ of $\{P_t; t \geq 0\}$. It is defined by (1.1.4) and (1.1.5) by the strongly continuous contraction semigroup $\{T_t; t > 0\}$ of Markovian symmetric operators on $L^2(\mathbb{R}^n)$ determined by $\{P_t; t \geq 0\}$ according to Lemma 1.1.14. Since the resolvent $\{R_\alpha; \alpha > 0\}$ of $\{P_t; t \geq 0\}$ admits the expression

$$R_\alpha f(x) = \int_{\mathbb{R}^n} f(x+y)w_\alpha(dy), \quad w_\alpha(B) = \int_0^\infty e^{-\alpha t}\nu_t(B)dt$$

for $B \in \mathcal{B}(\mathbb{R}^n)$, we see that $R_\alpha(C_c(\mathbb{R}^n))$ is not only a $\|\cdot\|_\infty$-dense subset of $C_\infty(\mathbb{R}^n)$ but also an $\mathcal{E}_1$-dense subset of $\mathcal{F}$ on account of equation (1.1.28).

[1] Cf. [132].

Hence the Dirichlet form $(\mathcal{E}, \mathcal{F})$ is regular in the sense of Definition 1.3.10 by virtue of Lemma 1.3.12. The associated Hunt process X on $\mathbb{R}^n$ is called a *symmetric Lévy process*. The process X has independent stationary increments and hence can start from every point in $\mathbb{R}^n$.

We shall show that

$$\begin{cases} \mathcal{F} = \{u \in L^2(\mathbb{R}^n) : \int_{\mathbb{R}^n} |\widehat{u}(x)|^2 \psi(x)dx < \infty\}, \\ \mathcal{E}(u, v) = \int_{\mathbb{R}^n} \widehat{u}(x)\overline{\widehat{v}}(x)\psi(x)dx \quad \text{for } u, v \in \mathcal{F}. \end{cases} \tag{2.2.11}$$

Here the *Fourier transform* $\widehat{f}$ of a function f is defined first by

$$\widehat{f}(x) = (2\pi)^{-\frac{n}{2}} \int_{\mathbb{R}^n} e^{i(x,y)} f(y)dy, \quad f \in \mathcal{S},$$

where $\mathcal{S}$ is the space of rapidly decreasing functions, namely, infinitely differentiable functions on $\mathbb{R}^n$ whose derivatives of any order are bounded when multiplied by polynomials. The definition of $\widehat{f}$ is then extended to any $f \in L^2(\mathbb{R}^n)$ so that the Parseval formula

$$(f, g) = \int_{\mathbb{R}^n} \widehat{f}(y)\overline{\widehat{g}}(y)dy, \quad f, g \in L^2(\mathbb{R}^n),$$

remains valid.

Since $\widehat{P_t u}(x) = \widehat{\nu}_t(x)\widehat{u}(x) = e^{-t\psi(x)}\widehat{u}(x), \ x \in \mathbb{R}^n, \ u \in \mathcal{S}$, we have by the Parseval formula

$$\frac{1}{t}(u - P_t u, u) = \int_{\mathbb{R}^n} |\widehat{u}(x)|^2 \frac{1 - e^{-t\psi(x)}}{t} dx, \quad t > 0, \ u \in \mathcal{S}, \tag{2.2.12}$$

which readily extends to any $u \in L^2(\mathbb{R}^n)$ by choosing $u_k \in \mathcal{S}$ converging to u in $L^2(\mathbb{R}^n)$ as $k \to \infty$. The right hand side of (2.2.12) for $u \in L^2(\mathbb{R}^n)$ increases to $\int_{\mathbb{R}^n} |\widehat{u}(x)|^2 \psi(x)dx (\leq \infty)$ as $t \downarrow 0$ and so we get (2.2.11). Obviously the Dirichlet form (2.2.11) is *translation-invariant* in the sense that if $u \in \mathcal{F}$, then for any $y \in \mathbb{R}^n$, the translated function $x \mapsto u_y(x) := u(x - y)$ is in $\mathcal{F}$ and $\|u_y\|_{\mathcal{E}} = \|u\|_{\mathcal{E}}$.

When S is the identity matrix and J vanishes in (2.2.9), namely, when $\psi(x) = \frac{1}{2}|x|^2$, the associated convolution semigroup is

$$\nu_t(dx) = g_t(x)dx, \quad g_t(x) = \frac{1}{(2\pi t)^{n/2}} e^{-\frac{|x|^2}{2t}}, \tag{2.2.13}$$

which corresponds to the transition probability of the n-dimensional standard Brownian motion by (2.2.10). The associated Dirichlet form (2.2.11) on $L^2(\mathbb{R}^n)$ can be written as

$$(\mathcal{E}, \mathcal{F}) = \left(\tfrac{1}{2}\mathbf{D}, H^1(\mathbb{R}^n)\right), \tag{2.2.14}$$

where

$$
\begin{cases}
\mathbf{D}(f,g) = \displaystyle\sum_{i=1}^{n} \int_{\mathbb{R}^n} \frac{\partial f(x)}{\partial x_i} \frac{\partial g(x)}{\partial x_i} dx, \\
H^1(\mathbb{R}^n) = \left\{ f \in L^2(\mathbb{R}^n) : \dfrac{\partial f}{\partial x_i} \in L^2(\mathbb{R}^n) \quad \text{for } 1 \le i \le d \right\},
\end{cases}
\tag{2.2.15}
$$

where $\frac{\partial}{\partial x_i}$ is the derivative in the sense of the Schwartz distribution. $\mathbf{D}(f,g)$ is the *Dirichlet integral* of functions f, g, and $H^1(\mathbb{R}^d)$ is called the *Sobolev space* on $\mathbb{R}^d$ of order 1. A. Beurling had this example in mind when he gave the name "Dirichlet space" to a function space on which every normal contraction operates.

When S in the formula (2.2.9) vanishes, we can also derive from (2.2.11) the following expression of the Dirichlet form $(\mathcal{E}, \mathcal{F})$ in terms of the Lévy measure J:

$$
\begin{cases}
\mathcal{E}(f,g) = \dfrac{1}{2} \displaystyle\int_{\mathbb{R}^n \times \mathbb{R}^n} (f(x+y) - f(x))(g(x+y) - g(x)) J(dy)dx, \\
\mathcal{F} = \left\{ f \in L^2(\mathbb{R}^n) : \displaystyle\int_{\mathbb{R}^n \times \mathbb{R}^n} (f(x+y) - f(x))^2 J(dy)dx < \infty \right\}.
\end{cases}
\tag{2.2.16}
$$

In fact, we have from (2.2.9) and (2.2.11)

$$
\mathcal{E}(f,f) = \int_{\mathbb{R}^n \times \mathbb{R}^n} |\widehat{f}(x)|^2 (1 - \cos\langle x, y\rangle) J(dy)dx.
$$

On the other hand, for each $y \in \mathbb{R}^n$, the Fourier transform of the function $g_y(x) = f(x+y) - f(x)$ is $\widehat{g}_y(x) = \widehat{f}(x)(e^{-i\langle x,y\rangle} - 1)$ and the Parseval formula gives

$$
\int_{\mathbb{R}^n \times \mathbb{R}^n} g_y(x)^2 J(dy)dx = 2 \int_{\mathbb{R}^n \times \mathbb{R}^n} |\widehat{f}(x)|^2 (1 - \cos\langle x, y\rangle) dx J(dy).
$$

The transience of a symmetric Markovian semigroup $\{T_t; t > 0\}$ on an L^2-space was defined in Definition 2.1.1. In the present case where $T_t = P_t$ determined by (2.2.10), the following five conditions can be verified to be mutually equivalent (cf. [73, Example 1.5.2]):

(1) $\{T_t; t > 0\}$ is transient.

(2) $w(K) < \infty$ for any compact $K \subset \mathbb{R}^n$, where

$$
w(B) = \int_0^\infty \nu_t(B)dt, \qquad B \in \mathcal{B}(\mathbb{R}^n).
$$

(3) $\int_0^\infty (P_t f, f)dt < \infty$ for any non-negative $f \in C_c(\mathbb{R}^n)$.

(4) For any compact $K \subset \mathbb{R}^n$, there exists a constant C_K such that

$$\int_K |u(x)|dx \le C_K \|u\|_{\mathcal{E}} \quad \text{for every } u \in \mathcal{F}. \tag{2.2.17}$$

(5) $\psi(x)^{-1}$ is locally integrable on $\mathbb{R}^n$.

Exercise 2.2.5. Consider the special case where $S = 0$ and J is a symmetric probability measure on $\mathbb{R}^n \setminus \{0\}$ in (2.2.9). The associated transition semigroup $\{T_t; t > 0\}$ defined by (2.2.10) corresponds to a compound Poisson process which is a special case of the regular step process X of Section 2.2.1 with $E = \mathbb{R}^n$, $\lambda \equiv 1$, $Q(x, B) = J(B - x)$ and m_0 being the Lebesgue measure (cf. [132]). Let J be a discrete distribution on $\mathbb{Z}^n$ with $J(\{\mathbf{e}_k\}) = J(\{-\mathbf{e}_k\}) = \frac{1}{2n}$ where $\mathbf{e}_k$ denotes a vector with 1 in the kth coordinate and 0 elsewhere. Show that $\{T_t; t > 0\}$ is transient if and only if $n \ge 3$.

The restriction of X to its invariant set $\mathbb{Z}^n$ is the (continuous-time) *symmetric simple random walk* on $\mathbb{Z}^n$.

Assume that $\{T_t; t > 0\}$ is transient. Then all the conditions **(2)**–**(5)** are fulfilled. Let us show that the extended Dirichlet space $(\mathcal{F}_e, \mathcal{E})$ of the Dirichlet form (2.2.11) admits an explicit expression[2]

$$\begin{cases} \mathcal{F}_e = \left\{ u \in L^1_{\text{loc}}(\mathbb{R}^n) \cap \mathcal{S}' : \widehat{u} \in L^2(\mathbb{R}^n; \psi \cdot dx) \right\}, \\ \mathcal{E}(u, v) = \displaystyle\int_{\mathbb{R}^n} \widehat{u}(x)\overline{\widehat{v}}(x)\psi(x)dx, \quad u, v \in \mathcal{F}_e, \end{cases} \tag{2.2.18}$$

where $\mathcal{S}'$ denotes the space of tempered distributions, namely, continuous linear functionals on the space $\mathcal{S}$, and, for $u \in \mathcal{S}'$, its Fourier transform $\widehat{u} \in \mathcal{S}'$ is defined by the formula $\langle \widehat{u}, \phi \rangle = \langle u, \widehat{\phi} \rangle$ for every $\phi \in \mathcal{S}$. Notice that a positive Radon measure μ on $\mathbb{R}^n$ can be considered as an element of $\mathcal{S}'$ if and only if $\int_{\mathbb{R}^n} (1 + |x|^2)^{-m} \mu(dx) < \infty$ for some $m \in \mathbb{N}$. For $\phi \in \mathcal{S}$, define $\check{\phi}$ by $\check{\phi}(x) := \phi(-x)$ for $x \in \mathbb{R}^n$, and denote by $\widetilde{\phi}$ the Fourier transform of $\check{\phi}$. It is easy to see that ϕ is the Fourier transform of $\widetilde{\phi}$.

By integrating the identity (2.2.8) in t and using the Fubini theorem, we get

$$\int_{\mathbb{R}^n} \psi(x)^{-1} \phi * \check{\phi}(x) dx = \int_{\mathbb{R}^n} |\widehat{\phi}|^2(y) w(dy) \ge 0 \quad \text{for } \phi \in C_c^\infty(\mathbb{R}^n),$$

where w is the measure defined in **(2)**. This means that the Radon measure $\mu_0(dx) = \psi(x)^{-1}dx$ is of positive type and hence is the Fourier transform of a tempered distribution.[3] Accordingly, $\mu_0 \in \mathcal{S}'$, which in turn implies that, for

[2] Due to J. Deny [39].

[3] Cf. L. Schwartz [135].

any $v \in L^2(\mathbb{R}^n; \psi \cdot dx)$, $v \cdot dx$ is tempered because

$$\int_{\mathbb{R}^n} \frac{|v(x)|}{(1+|x|^2)^{m/2}} dx \leq \|v\|_{L^2(\mathbb{R}^n;\psi\cdot dx)} \left(\int_{\mathbb{R}^n} (1+|x|^2)^{-m} \mu_0(dx) \right)^{1/2}.$$

Thus $L^2(\mathbb{R}^n; \psi \cdot dx) \subset \mathcal{S}'$. Further we see from the above estimate that if v_ℓ converges to v in $L^2(\mathbb{R}^n; \psi \cdot dx)$ as $\ell \to \infty$, then $\lim_{\ell\to\infty}\langle v_\ell, \phi\rangle = \langle v, \phi\rangle$ for every $\phi \in \mathcal{S}$.

Take any $u \in \mathcal{F}_e$. There exists then a sequence $\{u_\ell\}_{\ell\geq 1} \subset \mathcal{F}$ which is $\mathcal{E}$-convergent as well as a.e. on $\mathbb{R}^n$ to u. Notice that, on account of **(4)**, u_ℓ then converges to u in the space $L^1_{\text{loc}}(\mathbb{R}^n)$ as $\ell \to \infty$. By (2.2.11), $\{\widehat{u_\ell}\}_{\ell\geq 1}$ is convergent in $L^2(\mathbb{R}^n; \psi \cdot dx)$ to a function $v \in L^2(\mathbb{R}^n; \psi \cdot dx)$ and $\|u\|_{\mathcal{E}} = \|v\|_{L^2(\mathbb{R}^n;\psi\cdot dx)}$. Since $v \in \mathcal{S}'$ by the above observation, v is the Fourier transform of $\widetilde{v} \in \mathcal{S}'$ and moreover we have

$$\lim_{\ell\to\infty} \langle u_\ell, \widehat{\phi}\rangle = \lim_{\ell\to\infty} \langle \widehat{u}_\ell, \phi\rangle = \langle v, \phi\rangle = \langle \widetilde{v}, \widehat{\phi}\rangle \quad \text{for } \phi \in \mathcal{S}.$$

Hence $\widetilde{v} = u$ and we get $v = \widehat{u}$ and $\|u\|_{\mathcal{E}} = \|\widehat{u}\|_{L^2(\mathbb{R}^n;\psi\cdot dx)}$. Thus the inclusion $\subset$ holds in (2.2.18). The converse inclusion is readily verifiable.

An explicit expression of the extended Dirichlet space $(\mathcal{F}_e, \mathcal{E})$ in the recurrence case will be given in Section 6.5.

The continuous symmetric convolution semigroup corresponding to $\psi(x) = |x|^\alpha$ with $0 < \alpha \leq 2$ is called the *rotation-invariant stable semigroup* of index α, which corresponds in the formula (2.2.9) to[4]

$$S = 0 \quad \text{and} \quad J(dy) = \mathcal{A}(n, -\alpha)|y|^{-n-\alpha} dy, \tag{2.2.19}$$

where $\mathcal{A}(n, -\alpha) := \frac{\alpha 2^{\alpha-1}\Gamma((\alpha+n)/2)}{\pi^{n/2}\Gamma((2-\alpha)/2)}$. The corresponding Lévy process X on $\mathbb{R}^n$ is called a *rotationally symmetric α-stable process*. According to **(5)**, the associated L^2-semigroup is transient if and only if $\alpha < n$. In this case, its extended Dirichlet space is described as (2.2.18) with $\psi(x) = |x|^\alpha$. We shall now present two other characterizations of its extended Dirichlet space in terms of the Riesz convolution kernels.

In what follows, we shall fix α with $0 < \alpha \leq 2$ and $\alpha < n$, the rotation-invariant stable semigroup $\{\nu_t, t \geq 0\}$ of index α and the corresponding extended Dirichlet space $(\mathcal{F}_e, \mathcal{E})$. The first characterization is as follows:

$$\begin{cases} \mathcal{F}_e = \left\{u = I_{\alpha/2} * f : f \in L^2(\mathbb{R}^n)\right\}, \\ \mathcal{E}(u, v) = \displaystyle\int_{\mathbb{R}^n} f(x)g(x)dx \quad \text{for } u = I_{\alpha/2} * f,\ v = I_{\alpha/2} * g, \end{cases} \tag{2.2.20}$$

where I_α denotes the *Riesz convolution kernel* of index α given by

$$I_\alpha(x) = \gamma_{n,\alpha}\, |x|^{-(n-\alpha)} \quad \text{with} \quad \gamma_{n,\alpha} = \frac{\Gamma((n-\alpha)/2)}{2^\alpha \pi^{n/2}\Gamma(\alpha/2)}. \tag{2.2.21}$$

So $\mathcal{F}_e$ is the space of all Riesz potentials of index $\alpha/2$ of L^2-functions.

---

[4] Cf. [132, E18.9].

Characterization (2.2.20) is readily obtained from

$$\begin{cases} I_\beta * f \in L^1_{\text{loc}}(\mathbb{R}^n) \cap \mathcal{S}', \\ \widehat{I_\beta * f} = |x|^{-\beta}\widehat{f}, \end{cases} \tag{2.2.22}$$

for $\beta < n/2$ and $f \in L^2(\mathbb{R}^n)$. Indeed, using (2.2.22), it follows from (2.2.18) with $\psi(x) = |x|^\alpha$ that $I_{\alpha/2} * f \in \mathcal{F}_e$ for $f \in L^2(\mathbb{R}^n)$ and

$$\mathcal{E}(u, v) = \int_{\mathbb{R}^n} |x|^{-\alpha}\widehat{f}(x)\overline{\widehat{g}}(x)|x|^\alpha dx = (f, g), \ u = I_{\alpha/2} * f, \ v = I_{\alpha/2} * g,$$

which mean that the space defined by the right hand sides of (2.2.20) is a closed subspace of $(\mathcal{F}_e, \mathcal{E})$. Suppose $u \in \mathcal{F}_e$ is $\mathcal{E}$-orthogonal to this subspace, then

$$\int_{\mathbb{R}^n} \widehat{u}(x)\overline{\widehat{g}}(x)|x|^{\alpha/2}dx = 0 \quad \text{for } g \in L^2(\mathbb{R}^n).$$

Choosing a complex function $g \in L^2(\mathbb{R}^n)$ such that $\widehat{g}(x) = \widehat{u}(x)|x|^{\alpha/2} \in L^2(\mathbb{R}^n)$, we have $\widehat{u}(x)|x|^{\alpha/2} = 0$ a.e. and consequently $u = 0$.

To prove (2.2.22), let $B = \{x \in \mathbb{R}^n : |x| < 1\}$ be the unit open ball in $\mathbb{R}^n$ and define $K_1 := I_\beta \cdot \mathbf{1}_B$ and $K_2 := I_\beta - K_1$. For $\beta < n/2$, $K_2 \in L^2(\mathbb{R}^n)$. Thus for $f \in L^2(\mathbb{R}^n)$, $K_1 * f \in L^2(\mathbb{R}^2)$ and $K_2 * f \in C_\infty(\mathbb{R}^n)$. This proves the first half of (2.2.22).

On the other hand, for $\delta > 0$, since the Fourier transform of the function $\exp(-\frac{\delta}{2}|x|^2)$ is $\delta^{-n/2}\exp(-\frac{|x|^2}{2\delta})$, we have

$$\int_{\mathbb{R}^n} e^{-\frac{\delta}{2}|x|^2}\widehat{\phi}(x)dx = \delta^{-n/2}\int_{\mathbb{R}^n} e^{-\frac{|x|^2}{2\delta}}\phi(x)dx, \quad \phi \in \mathcal{S}.$$

Multiply the both hand sides by $\delta^{\frac{n-\beta}{2}-1}$ and then integrate in δ over $[0, \infty)$ to get

$$\int_{\mathbb{R}^n} I_\beta(x)\widehat{\phi}(x)dx = (2\pi)^{-\frac{n}{2}}\int_{\mathbb{R}^n} |x|^{-\beta}\phi(x)dx, \quad \phi \in \mathcal{S}, \ \beta < n. \tag{2.2.23}$$

Take $\phi(x) = e^{-i\langle y,x\rangle}\psi(-x)$, then $\widehat{\phi}(x) = \widehat{\psi}(y - x)$. Multiply the resulting equation by $f(y)$ ($f \in L^2(\mathbb{R}^n)$) and then integrate in y on $\mathbb{R}^n$ to get the second half of (2.2.22), namely,

$$\int_{\mathbb{R}^n} (I_\beta * f)(x)\widehat{\psi}(x)dx = \int_{\mathbb{R}^n} |x|^{-\beta}\widehat{f}(x)\psi(x)dx, \quad \psi \in \mathcal{S}, \ \beta < n/2.$$

As a direct application of (2.2.20), we can rewrite the *Sobolev inequality*[5]

$$\|I_{\alpha/2} * f\|_{L^{p_0}(\mathbb{R}^n)} \leq C\|f\|_{L^2(\mathbb{R}^n)},$$

[5] Cf. E. M. Stein [142, Chap. V].

where $1/p_0 = 1/2 - \alpha/2d$, to obtain

$$\|u\|_{L^{p_0}(\mathbb{R}^n)} \leq \|u\|_{\mathcal{E}}, \qquad u \in \mathcal{F}_e, \tag{2.2.24}$$

where C is a positive constant. Inequality (2.2.24) is stronger than (2.2.17). In fact, it implies that the space $(\mathcal{F}_e, \mathcal{E})$ is continuously embedded into the space $L^2_{\rm loc}(\mathbb{R}^n)$.

We next observe that the Radon measure w defined in **(2)** admits the Riesz kernel of index α as a density function, namely, $w(dx) = I_\alpha(x)dx$, because we have for any $\phi \in \mathcal{S}$

$$\begin{aligned}\int_{\mathbb{R}^n} \widehat{\phi}(x)w(dx) &= (2\pi)^{-n/2}\int_{\mathbb{R}^n}\phi(x)\int_0^\infty e^{-t|x|^\alpha}dtdx\\ &= (2\pi)^{-n/2}\int_{\mathbb{R}^n}|x|^{-\alpha}\phi(x)dx,\end{aligned}$$

which coincides with the left hand side of (2.2.23).

Therefore, for non-negative $f \in C_c(\mathbb{R}^n)$, its Riesz potential $I_\alpha * f$ is a version of Rf defined by (2.1.3) for the transition function (2.2.10) for $\{\nu_t; t \geq 0\}$. By virtue of Theorem 2.1.12, it holds that for every $f \in C_c(\mathbb{R}^n)$ and $v \in \mathcal{F}_e$,

$$I_\alpha * f \in \mathcal{F}_e \quad \text{and} \quad \mathcal{E}(I_\alpha * f, v) = \int_{\mathbb{R}^n} f(x)v(x)dx. \tag{2.2.25}$$

Furthermore, by taking $C_c(\mathbb{R}^n)$ as the space $\mathcal{L}$ in Theorem 2.1.14, we can conclude that the extended Dirichlet space $(\mathcal{F}_e, \mathcal{E})$ is characterized as a Hilbert space consisting of equivalence classes (in the sense of a.e.) of measurable functions on $\mathbb{R}^n$ for which (2.2.25) holds.

We also note here that the irreducibility of the present semigroup $\{T_t; t > 0\}$ can be deduced as follows. Suppose that a measurable set $A \subset \mathbb{R}^n$ is $\{T_t\}$-invariant. By Proposition 2.1.6, $\mathbf{1}_{A^c}R_\beta(\mathbf{1}_A f) = 0$ a.e. for any non-negative $f \in C_c(\mathbb{R}^n)$ and $\beta > 0$. Letting $\beta \to 0$, we get $\mathbf{1}_{A^c}R(\mathbf{1}_A f) = 0$ a.e., which forces A to be either $\mathbb{R}^n$ or $\emptyset$ a.e. because R has a strictly positive convolution kernel I_α.

When $\psi(x) = \frac{1}{2}|x|^2$ and $n \geq 3$, we have $w(dx) = 2I_2(x)dx$ so that the associated potential operator R is given by the *Newtonian convolution kernel*

$$2I_2(x) = \frac{\Gamma(\frac{n}{2} - 1)}{2\pi^{n/2}}|x|^{2-n}. \tag{2.2.26}$$

The extended Dirichlet space of the Dirichlet form (2.2.14) can be characterized by (2.2.25) using the Newtonian potentials of C_c functions.

We now present a brief discussion of symmetric convolution semigroups on the unit circle and their associated Dirichlet forms. The definition of a continuous symmetric convolution semigroup $\{\mu_t, t > 0\}$ on the unit circle $\mathbb{S}$ is analogous to that on $\mathbb{R}^n$ above. We can identify $\mathbb{S}$ with $[0, 2\pi)$. By the Lévy-Khinchin formula [78], $\{\mu_t, t > 0\}$ is characterized by

$$\widehat{\mu}_t(k) := \int_0^{2\pi} e^{ikx} \mu_t(dx) = e^{-ta_k}, \qquad k \in \mathbb{Z},$$

with

$$a_k = \beta k^2 + \int_0^{2\pi} (1 - \cos(kx))J(dx), \tag{2.2.27}$$

where $\beta \geq 0$ is a constant and J is a non-negative Radon measure on the circle $\mathbb{S} = [0, 2\pi)$ satisfying $\int_0^{2\pi} \frac{1}{\sin^2(x/2)} J(dx) < \infty$. Let $\{T_t; t > 0\}$ be the L^2-strongly continuous semigroup generated by $\{\mu_t, t > 0\}$, that is, $T_t f := \int_0^{2\pi} f(x+y)\mu_t(dy)$. For $f \in L^2([0, 2\pi); dx)$, let $\widehat{f}(k) := \frac{1}{2\pi} \int_0^{2\pi} e^{-ikx} f(x)dx$, $k \in \mathbb{Z}$, denote its Fourier coefficients. Then by Parseval's formula,

$$\frac{1}{t}(f - T_t f, f) = \frac{2\pi}{t} \sum_{k \in \mathbb{Z}} |\widehat{f}(k)|^2 (1 - \widehat{\mu}_t(k)) = \frac{2\pi}{t} \sum_{k \in \mathbb{Z}} |\widehat{f}(k)|^2 (1 - e^{-ta_k}),$$

which increases to $2\pi \sum_{k \in \mathbb{Z}} a_k |\widehat{f}(k)|^2$. Hence the corresponding Dirichlet space is

$$\begin{cases} \mathcal{F} = \left\{ f \in L^2([0, 2\pi); dx) : \sum_{k \in \mathbb{Z}} a_k |\widehat{f}(k)|^2 < \infty \right\} \\ \mathcal{E}(f, g) = 2\pi \sum_{k \in \mathbb{Z}} a_k \widehat{f}(k) \overline{\widehat{g}(k)} \quad \text{for } f, g \in \mathcal{F}. \end{cases}$$

It is easy to see that if we take $\beta = 0$ and

$$J(dx) = c \sum_{k \in \mathbb{Z}} |x + 2k\pi|^{-\alpha - 1} dx$$

with $\alpha \in (0, 2)$ in (2.2.27), then $a_k = c_0 |k|^\alpha$ for $k \in \mathbb{Z}$. The measures $\{\mu_t, t > 0\}$ are called the symmetric stable semigroup of index α on the unit circle. As can be easily checked, the corresponding Dirichlet form is regular. It uniquely determines a Hunt process which is called the symmetric α-stable process on the unit circle. When $\alpha = 1$, it is called the symmetric Cauchy process on the unit circle.

2.2.3. One-Dimensional Strongly Local Dirichlet Forms

The second-order ordinary differential operator $\mathcal{A}u(x) = a(x)u''(x) + b(x)u'(x)$ with real-valued functions $a > 0$ and b can be converted into Feller's canonical form $\frac{d}{dm}\frac{du}{ds}$ with

$$ds = e^{-B(x)}dx, \quad dm = \frac{e^{B(x)}}{a(x)}dx, \quad B(x) = \int_{x_0}^{x} \frac{b(y)}{a(y)}dy.$$

Further, formal computation gives

$$-\int \mathcal{A}u \cdot v\, dm = -\int v \cdot d\frac{du}{ds} = \int \frac{du}{ds}\frac{dv}{ds}ds.$$

This example leads us to the following general consideration.

Let $I \subset \mathbb{R}$ be a one-dimensional open interval and $s = s(x)$, $x \in I$, be a strictly increasing continuous function on I. Following [95], we call s a *canonical scale*. Consider the space

$$\mathcal{F}^{(s)} = \{u : u \text{ is absolutely continuous on } I \text{ with respect to } ds$$
$$\text{and } \mathcal{E}^{(s)}(u,u) < \infty\}, \tag{2.2.28}$$

where

$$\mathcal{E}^{(s)}(u,v) = \int_I \frac{du}{ds}(x)\frac{dv}{ds}(x)ds(x). \tag{2.2.29}$$

Let m be a positive Radon measure on I with full support. Such a measure on an interval will be called a *canonical measure*. The aim of this subsection is to study the function space $(\mathcal{E}, \mathcal{F})$ defined by

$$\mathcal{E} = \mathcal{E}^{(s)}, \qquad \mathcal{F} = \mathcal{F}^{(s)} \cap L^2(I;m). \tag{2.2.30}$$

First of all, we make a simple observation. From an elementary equality

$$u(z) - u(y) = \int_z^y \frac{du}{ds}(x)ds(x), \quad y, z \in I,$$

we get for $u \in \mathcal{F}^{(s)}$,

$$(u(z) - u(y))^2 \le |s(z) - s(y)|\mathcal{E}^{(s)}(u,u), \quad y, z \in I, \tag{2.2.31}$$

which also implies that for any compact set $K \subset I$, there exists a positive constant C_K with

$$\sup_{x \in K} u(x)^2 \le C_K \mathcal{E}_1^{(s)}(u,u). \tag{2.2.32}$$

In fact, if $[\alpha, \beta] \subset I$, we get from (2.2.31)

$$\sup_{\alpha \le y \le \beta} u(y)^2 \le 2(s(\beta) - s(\alpha))\mathcal{E}^{(s)}(u,u) + 2u(x)^2, \quad \alpha \le x \le \beta.$$

Integrating both sides by $dm(x)$ on $[\alpha, \beta]$, we arrive at (2.2.32).

For any $\ell \in \mathbb{N}$, we consider the normal contraction φ_ℓ defined by (1.3.1).

LEMMA 2.2.6. (i) *Assume that* $\{u_n, n \geq 1\} \subset \mathcal{F}^{(s)}$ *is* $\mathcal{E}^{(s)}$*-Cauchy and that* u_n *converges to a function* u *m-a.e. Then* $u \in \mathcal{F}^{(s)}$ *and* u_n *is* $\mathcal{E}^{(s)}$*-convergent to* u.
(ii) *For any* $u \in \mathcal{F}^{(s)}$, *there exists a subsequence* $\{\ell_n; n \geq 1\}$ *such that, if we denote by* $\{u_n; n \geq 1\}$ *the Cesàro mean sequence of* $\{\varphi_{\ell_k}(u); k \geq 1\}$, *then* $u_n \in \mathcal{F}^{(s)}$ *and* u_n *is* $\mathcal{E}^{(s)}$*-convergent to* u.

Proof. (i) If $\{u_n\}$ is $\mathcal{E}^{(s)}$-Cauchy, $\{\frac{du_n}{ds}\}$ is $L^2(I; ds)$-convergent to some $v \in L^2(I; ds)$. Since u_n is convergent to u m-a.e., we see from (2.2.31) that $\{u_n\}$ is convergent uniformly on each compact subinterval of I to a continuous version of u and hence

$$\int_I v(x)\psi(s(x))ds(x) = -\lim_{n\to\infty}\int_I u_n(x)\psi'(s(x))ds(x)$$
$$= -\int_I u(x)\psi'(s(x))ds(x),$$

for any $\psi \in C_c^1(J)$, where $J = s(I)$ and $C_c^k(J)$ denotes the space of k times continuously differentiable functions on J with compact support. This means that $u \in \mathcal{F}^{(s)}$ with $\frac{du}{ds} = v$ and u_n is $\mathcal{E}^{(s)}$-convergent to u.

(ii) For $u \in \mathcal{F}^{(s)}$ and $\ell \in \mathbb{N}$,

$$\varphi_\ell(u) \in \mathcal{F}^{(s)}, \quad \mathcal{E}^{(s)}(\varphi_\ell(u), \varphi_\ell(u)) \leq \mathcal{E}^{(s)}(u, u).$$

Hence by the Banach-Saks Theorem A.4.1, we can find a subsequence $\{\ell_n\}$ such that, for the Cesàro mean u_n of $\{\varphi_{\ell_n}(u)\}$, $\frac{du_n}{ds}$ is $L^2(I; ds)$-convergent to some $v \in L^2(I; ds)$. Since u_n converges to u pointwisely, the same argument as in the proof of (i) works to get the desired conclusion. □

LEMMA 2.2.7. *The bilinear form* $(\mathcal{E}, \mathcal{F})$ *defined by* (2.2.30) *is a Dirichlet form on* $L^2(I; m)$.

Proof. The closedness of the bilinear form $(\mathcal{E}, \mathcal{F})$ follows from Lemma 2.2.6(i) immediately. $\mathcal{F}^{(s)}$ is dense in $L^2(I; m)$ as it contains the space $\{u(s) : u \in C_c^1(s(I))\}$. For any $\varepsilon > 0$, choose a smooth function φ_ε satisfying (1.1.7) to obtain

$$\mathcal{E}^{(s)}(\varphi_\varepsilon(u), \varphi_\varepsilon(u)) = \int_{\mathbb{R}} \varphi_\varepsilon'(u(x))^2 \left(\frac{du}{ds}(x)\right)^2 ds(x) \leq \mathcal{E}^{(s)}(u, u),$$

the Markovian property of $(\mathcal{E}, \mathcal{F})$. □

Let $I = (r_1, r_2)$. We say that the boundary r_1 (respectively, r_2) is *approachable* if

$$-\infty < s(r_1) \qquad \text{(respectively, } s(r_2) < \infty).$$

By (2.2.31), we see that if r_1 is approachable, then for any $u \in \mathcal{F}^{(s)}$, the limit $u(r_1) = \lim_{x\to r_1, x\in I} u(x)$ exists, $u \in C([r_1, r_2))$, and (2.2.31) holds for any

$a, b \in [r_1, r_2)$. The same statement holds for r_2. In particular, if both r_1 and r_2 are approachable, then

$$\mathcal{F}^{(s)} \subset C([r_1, r_2]), \qquad 1,\ s \in \mathcal{F}^{(s)}, \tag{2.2.33}$$

and consequently $\mathcal{F}^{(s)}$ is a uniformly dense subalgebra of $C([r_1, r_2])$.

The point r_1 is called *regular* if

$$r_1 \text{ approachable} \quad \text{and} \quad m((r_1, c)) < \infty \quad \text{for every } c \in (r_1, r_2).$$

The regularity of r_2 is defined similarly.

We now study the properties of the Dirichlet form $(\mathcal{E}, \mathcal{F})$ defined by (2.2.30) in three cases separately.

Proposition 2.2.8. *Suppose both r_1 and r_2 are regular. Then*

$$\mathcal{F} = \mathcal{F}_e = \mathcal{F}^{(s)}, \tag{2.2.34}$$

and $(\mathcal{E}, \mathcal{F})$ is a regular, strongly local, recurrent, and irreducible Dirichlet form on $L^2([r_1, r_2]; \mathbf{1}_I \cdot m)$.

Proof. Property (2.2.33) and the finiteness of $m(I)$ imply that $\mathcal{F}$ coincides with $\mathcal{F}^{(s)}$, which also equals $\mathcal{F}_e$ by virtue of Lemma 2.2.6. In view of Lemma 2.2.7 and (2.2.33), $(\mathcal{E}, \mathcal{F})$ is a regular Dirichlet form on $L^2([r_1, r_2];\ \mathbf{1}_I \cdot m)$. Since $1 \in \mathcal{F}$ and $\mathcal{E}(1, 1) = 0$, it is recurrent by Theorem 2.1.8. It is obvious that $(\mathcal{E}, \mathcal{F})$ is strongly local. Suppose there is a Borel set $A \subset [r_1, r_2]$ such that $\mathbf{1}_A$ equals some function $u \in \mathcal{F}$ m-a.e. Since u is continuous and takes values 0 or 1 only, $u^{-1}(\{1\})$ is a closed and open subset of $[r_1, r_2]$ and hence either A or A^c is m-negligible, yielding the irreducibility by virtue of Proposition 2.1.6. □

Proposition 2.2.9. *Assume that both r_1 and r_2 are approachable but non-regular. If we let*

$$\mathcal{F}_0^{(s)} = \left\{u \in \mathcal{F}^{(s)} : u(r_1) = 0 = u(r_2)\right\}, \tag{2.2.35}$$

then

$$\mathcal{F} \subset \mathcal{F}_e = \mathcal{F}_0^{(s)}, \tag{2.2.36}$$

and $(\mathcal{E}, \mathcal{F})$ is a regular, strongly local, transient, and irreducible Dirichlet form on $L^2(I; m)$.

Proof. Since m diverges in neighborhoods of r_1 and r_2, we see from (2.2.33) that $\mathcal{F} \subset \mathcal{F}_0^{(s)}$.

Take any $u \in \mathcal{F}_0^{(s)}$. For any $\ell \in \mathbb{N}$, $\varphi_\ell(u)$ has compact support in I and so $\varphi_\ell(u) \in \mathcal{F}$. Hence the functions u_n in Lemma 2.2.6(ii) are in $\mathcal{F}$ and converge to u both pointwise and in $\mathcal{E}^{(s)}$-metric. This implies that $u \in \mathcal{F}_e$. So we have established $\mathcal{F}_0^{(s)} \subset \mathcal{F}_e$. Conversely, assume $u_n \in \mathcal{F}(\subset \mathcal{F}_0^{(s)})$ is $\mathcal{E}^{(s)}$-Cauchy and $u_n \to u$ m-a.e. Then $u \in \mathcal{F}^{(s)}$ and u_n is $\mathcal{E}^{(s)}$-convergent to u by Lemma 2.2.6. Since inequality (2.2.31) holds on $[r_1, r_2]$,

$$u(r_i) = \lim_{n\to\infty} u_n(r_i) = 0, \quad i = 1, 2.$$

Consequently, $u \in \mathcal{F}_0^{(s)}$. We have thus established that $\mathcal{F}_0^{(s)} = \mathcal{F}_e$.

For $u \in \mathcal{F}$, the functions u_n of Lemma 2.2.6(ii) are in $\mathcal{F} \cap C_c(I)$ and $\mathcal{E}_1$-convergent to u. Since $\mathcal{F}^{(s)}$ is dense in $C([r_1, r_2])$, $\mathcal{F} \subset \mathcal{F}_0^{(s)}$ is dense in $C_\infty(I)$ in view of Lemma 1.3.12, proving the regularity of the space $(\mathcal{E}, \mathcal{F})$. Its strongly local property is obvious. Since, for $u \in \mathcal{F}_e$, $\mathcal{E}(u, u) = 0$ implies $u = 0$, the transience follows from Theorem 2.1.9. The irreducibility can be proved in the same way as the proof of the preceding theorem because, for any $a < b$ with $[a, b] \subset I$, $\mathcal{F}$ admits a function taking value 1 on $[a, b]$. □

PROPOSITION 2.2.10. *Suppose both r_1 and r_2 are non-approachable. Then*

$$\mathcal{F}_e = \mathcal{F}^{(s)} \tag{2.2.37}$$

and $(\mathcal{E}, \mathcal{F})$ is a regular, strongly local, recurrent, and irreducible Dirichlet form on $L^2(I; m)$.

Proof. Lemma 2.2.6(i) implies the inclusion $\mathcal{F}_e \subset \mathcal{F}^{(s)}$. To prove the converse, take any $u \in \mathcal{F}^{(s)}$. Since the truncation $\varphi^\ell \circ u$ is $\mathcal{E}^{(s)}$-convergent to u as $\ell \to \infty$, in view of Lemma 1.1.12, we may assume without loss of generality that u is bounded, that is, $|u| \leq M$ for some constant M.

We consider a sequence of functions $\psi_n \in C_c^1(\mathbb{R}_+)$ such that

$$\begin{cases} \psi_n(x) = 1 & \text{for } 0 \leq x < n; \qquad \psi_n(x) = 0 \ \text{ for } x > 2n+1; \\ |\psi_n'(x)| \leq \frac{1}{n}, & n \leq x \leq 2n+1; \quad 0 \leq \psi_n(x) \leq 1,\ x \in \mathbb{R}_+. \end{cases} \tag{2.2.38}$$

Put $u_n(x) = u(x) \cdot \psi_n(|s(x)|)$ for $x \in I$. Then, owing to the non-approachability of the both boundaries, $u_n \in \mathcal{F}^{(s)} \cap C_c(I) (\subset \mathcal{F})$ and

$$\begin{aligned} &\mathcal{E}^{(s)}(u - u_n, u - u_n) \\ &\leq 2 \int_I \left(\frac{du}{ds}\right)^2 (1 - \psi_n(|s(x)|)^2 ds(x) + 2 \int_I u^2(x) (\psi_n'(|s(x)|)^2 ds(x) \\ &\leq 2 \int_{|s(x)| \geq n} \left(\frac{du}{ds}\right)^2 ds(x) + 2M^2 \int_{n \leq |s(x)| < 2n+1} (\psi_n'(|s(x)|)^2 ds(x) \\ &\leq 2 \int_{|s(x)| \geq n} \left(\frac{du}{ds}\right)^2 ds(x) + 4M^2 \frac{n+1}{n^2} \to 0 \quad \text{as } n \to \infty. \end{aligned}$$

This shows that $\{u_n\} \subset \mathcal{F}$ is $\mathcal{E}^{(s)}$-Cauchy. Since u_n converges to u pointwise, we get $u \in \mathcal{F}_e$.

For any bounded $u \in \mathcal{F}$, the same functions $\{u_n, n \geq 1\}$ as above are in $\mathcal{F} \cap C_c(I)$ and $\mathcal{E}_1$-convergent to u as $n \to \infty$. The family of such functions is easily seen to be uniformly dense in $C_c(I)$. Hence $(\mathcal{E}, \mathcal{F})$ is regular. Since $1 \in \mathcal{F}^{(s)} = \mathcal{F}_e$, $\mathcal{E}(1,1) = 0$, $(\mathcal{E}, \mathcal{F})$ is recurrent. Its strong locality is obvious. The irreducibility can be shown as in the preceding theorem. □

THEOREM 2.2.11. *Let $(\mathcal{E}, \mathcal{F})$ be the Dirichlet form on $L^2(I; m)$ defined by* (2.2.30) *and $\mathcal{F}_e$ be its extended Dirichlet space.*
(i) *Denote by I^* the interval obtained from $I = (r_1, r_2)$ by adding the boundary point r_i to it when r_i is regular $(i = 1, 2)$. Then $(\mathcal{E}, \mathcal{F})$ is a regular, strongly local, and irreducible Dirichlet form on $L^2(I^*; m)(= L^2(I; m))$, m being extended to I^* by $m(\cdot) = m(\cdot \cap I)$.*
(ii) *It holds that*

$$\mathcal{F}_e = \{u \in \mathcal{F}^{(s)} : u(r_i) = 0 \text{ if } r_i \text{ is approachable but non-regular}, \quad i = 1, 2\}. \tag{2.2.39}$$

In particular, $(\mathcal{E}, \mathcal{F})$ is transient if and only if either r_1 or r_2 is approachable and non-regular. Otherwise it is recurrent.

Assertion (i) and (2.2.39) of this theorem have been proved in several cases in the preceding three propositions, and they can be shown similarly in the remaining cases. The proof of the second assertion of (ii) is similar.

The inequalities (2.2.31) and (2.2.32) are extended to $a, b \in I^*$ and to any compact set $K \subset I^*$, respectively. Combining (2.2.32) with (2.3.1) and (2.3.2) we see that the capacity of each single point is positive and uniformly bounded away from zero on each compact set; that is, for every $a \in I^*$ and every compact subset $K \subset I^*$,

$$\text{Cap}_1(\{a\}) > 0 \quad \text{and} \quad \inf_{a \in K} \text{Cap}_1(\{a\}) > 0. \tag{2.2.40}$$

Consequently, a function is quasi continuous if and only if it is continuous.

In the transient case, the inequality (2.1.11) for $(\mathcal{F}_e, \mathcal{E})$ can be also derived directly from (2.2.31). Suppose for instance that $s(r_2) < \infty$ and $m(c, r_2) = \infty$ for some $c \in I$. Then

$$|u(x)| \leq \sqrt{|s(x) - s(r_2)|}\sqrt{\mathcal{E}^{(s)}(u, u)}, \qquad x \in I, \ u \in \mathcal{F}_e.$$

Consequently, inequality (2.1.11) holds for every $u \in \mathcal{F}_e$ and for any m-integrable strictly positive bounded function g with

$$\int_I \sqrt{|s(x) - s(r_2)|}g(x)dm(x) \leq 1.$$

When $s(x) = x$ and m is the Lebesgue measure on I, then the space (2.2.28) (resp. (2.2.30)) is denoted by $\mathrm{BL}(I)$ (resp. $H^1(I)$) and called the space of BL-functions (resp. the Sobolev space of order 1). The space $\mathcal{F}_e$ is then denoted by $H_e^1(I)$.

2.2.4. Space of BL Functions

Let D be a domain in the n-dimensional Euclidean space $\mathbb{R}^n$. For $p \geq 1$, the L^p-space of real-valued functions on D with respect to the Lebesgue measure dx will be denoted as $L^p(D)$. We focus our attention on the space

$$\mathrm{BL}(D) = \left\{ T : \frac{\partial T}{\partial x_i} \in L^2(D),\ 1 \leq i \leq n \right\} \tag{2.2.41}$$

of Schwartz distributions T. Any distribution $T \in \mathrm{BL}(D)$ can be identified with a function in $L^2_{\mathrm{loc}}(D)$[6] so that

$$\mathrm{BL}(D) = \left\{ u \in L^2_{\mathrm{loc}}(D) : \frac{\partial u}{\partial x_i} \in L^2(D),\ 1 \leq i \leq n \right\}, \tag{2.2.42}$$

where the derivatives are taken in the Schwartz distribution sense. Members in $\mathrm{BL}(D)$ are called BL(Beppo-Levi)-*functions* on D. For $u, v \in \mathrm{BL}(D)$, we put

$$\mathbf{D}(u, v) = \sum_{i=1}^{n} \int_D \frac{\partial u}{\partial x_i} \frac{\partial v}{\partial x_i} dx, \tag{2.2.43}$$

The space $\mathrm{BL}(D)$ is known to enjoy the following properties (cf. [40], [121]):

(BL.1) The quotient space $\dot{\mathrm{BL}}(D)$ of $\mathrm{BL}(D)$ by the subspace of constant functions is a Hilbert space with inner product $\mathbf{D}$. Any $\mathbf{D}$-Cauchy sequence $u_n \in \mathrm{BL}(D)$ admits $u \in \mathrm{BL}(D)$ and constants c_n such that u_n is $\mathbf{D}$-convergent to u and $u_n + c_n$ is L^2_{loc}-convergent to u.

(BL.2) A function u on D is in $\mathrm{BL}(D)$ if and only if, for each i ($1 \leq i \leq n$), there is a version $u^{(i)}$ of u such that it is absolutely continuous on almost all straight lines parallel to x_i-axis and the derivative $\partial u^{(i)}/\partial x_i$ in the ordinary sense (which exists a.e. on D) is in $L^2(D)$. In this case, the ordinary derivatives coincide with the distribution derivatives of u.

The *Sobolev space* of order 1 on the domain $D \subset \mathbb{R}^n$ is defined by

$$H^1(D) = \mathrm{BL}(D) \cap L^2(D). \tag{2.2.44}$$

---

[6] Cf. L. Schwartz [135], J. Deny and J. L. Lions [40].

Then

$$(\mathcal{E}, \mathcal{F}) = \left(\tfrac{1}{2}\mathbf{D}, H^1(D)\right) \tag{2.2.45}$$

is a closed symmetric form on $L^2(D)$. To see this, suppose $\{u_n\} \subset \mathcal{F}$ is $\mathcal{E}_1$-Cauchy. Then $\frac{\partial u_n}{\partial x_i}$ is $L^2(D)$-convergent to some $v_i \in L^2(D)$ for each $1 \le i \le n$ and u_n is $L^2(D)$-convergent to some $u \in L^2(D)$. Then for any $f \in C_c^\infty(D)$,

$$(v_i, f) = \lim_{n\to\infty}\left(\frac{\partial u_n}{\partial x_i}, f\right) = -\lim_{n\to\infty}\left(u_n, \frac{\partial f}{\partial x_i}\right) = -\left(u, \frac{\partial f}{\partial x_i}\right).$$

Hence $v_i = \frac{\partial u}{\partial x_i}$ for $1 \le i \le n$. It is a Dirichlet form on $L^2(D)$ because its Markov property can be verified in the same way as the proof of Lemma 2.2.7 by using the property (BL.2), which also implies the strongly local property of (2.2.45).

Denote by $H_e^1(D)$ the extended Dirichlet space of (2.2.45). We shall be concerned with its relationship to the space $\mathrm{BL}(D)$. Let us first consider the case that $D = \mathbb{R}^n$. The Dirichlet form

$$(\mathcal{E}, \mathcal{F}) = \left(\tfrac{1}{2}\mathbf{D}, H^1(\mathbb{R}^n)\right) \tag{2.2.46}$$

on $L^2(\mathbb{R}^n)$ has already appeared in (2.2.14) of Subsection 2.2.2, where we saw that it is a regular Dirichlet form, namely, the space $H^1(\mathbb{R}^n) \cap C_c(\mathbb{R}^n)$ is a core of it. We can further see that it has $C_c^\infty(\mathbb{R}^n)$ as a special standard core by making convolutions with mollifiers $\varepsilon^{-n}\varphi(x/\varepsilon)$, $\varepsilon > 0$, where φ is a non-negative infinitely differentiable function supported by $B = \{|x| < 1\}$ with $\int_B \varphi(x)dx = 1$.

The Dirichlet form (2.2.46) on $L^2(\mathbb{R}^n)$ is associated with the transition density $g_t(x)$ of (2.2.13). Since, for $x \neq 0$, $\int_0^\infty g_t(x)dt$ is divergent when $n = 1$, 2, but convergent and equal to the Newtonian kernel (2.2.26) when $n \ge 3$, the corresponding L^2-semigroup is recurrent in the former case and transient in the latter case.

When $n \ge 3$, the extended Dirichlet space $(H_e^1(\mathbb{R}^n), \mathcal{E})$ of (2.2.46) is a real Hilbert space on account of Theorem 2.1.5 and, in particular,

$$u \in H_e^1(\mathbb{R}^n) \text{ with } \mathcal{E}(u,u) = 0 \implies u = 0. \tag{2.2.47}$$

THEOREM 2.2.12. *Assume that $n \ge 3$. Then $H_e^1(\mathbb{R}^n) \subset \mathrm{BL}(\mathbb{R}^n)$, $\mathcal{E}(u,u) = \mathbf{D}(u,u)$ for $u \in H_e^1(\mathbb{R}^n)$, and $\mathrm{BL}(\mathbb{R}^n)$ is the linear space spanned by $H_e^1(\mathbb{R}^n)$ and constant functions. The space $(H_e^1(\mathbb{R}^n), \mathcal{E})$ is isometric with the space $(\dot{\mathrm{BL}}(\mathbb{R}^n), \frac{1}{2}\mathbf{D})$ by the canonical map $\mathrm{BL}(\mathbb{R}^n) \to \dot{\mathrm{BL}}(\mathbb{R}^n)$. Furthermore,*

$$H_e^1(\mathbb{R}^n) = \{u \in \mathrm{BL}(\mathbb{R}^n) \cap \mathcal{S}' : \widehat{u} \in L^1_{\mathrm{loc}}(\mathbb{R}^n)\}. \tag{2.2.48}$$

Proof. For $u \in H_e^1(\mathbb{R}^n)$, there is a sequence $\{u_n\} \subset H^1(\mathbb{R}^n)$ which is $\mathbf{D}$-Cauchy and convergent to u a.e. $\mathbf{D}(u_n, u_n)$ then converges to $\mathcal{E}(u,u)$. By (BL.1), there exist $v \in \mathrm{BL}(\mathbb{R}^n)$ and constants c_n such that $\{u_n\}$ is $\mathbf{D}$-convergent to

v and the sequence $\{u_n + c_n\}$ is convergent to v in $L^2_{\rm loc}(\mathbb{R}^n)$. By choosing a subsequence if necessary, we may assume that the latter sequence converges to v a.e. Then $\lim_{n\to\infty} c_n = c$ exists, $u = v - c$ and, consequently, $u \in {\rm BL}(\mathbb{R}^n)$ and $\mathcal{E}(u,u) = \mathbf{D}(u,u)$. Further, we see from (2.2.47) that the Hilbert space $(H^1_e(\mathbb{R}^n), \mathcal{E})$ is isometrically embedded into a closed subspace of $(\dot{\rm BL}(\mathbb{R}^n), \frac{1}{2}\mathbf{D})$ by the canonical map ${\rm BL}(\mathbb{R}^n) \mapsto \dot{\rm BL}(\mathbb{R}^n)$. We denote by $\dot{u}$ the equivalence class represented by $u \in {\rm BL}(\mathbb{R}^n)$.

If $\dot{u} \in \dot{\rm BL}(\mathbb{R}^n)$ is $\mathbf{D}$-orthogonal to this closed subspace, then, since $C_c^\infty(\mathbb{R}^n) \subset H^1(\mathbb{R}^n)$, we have

$$(\Delta u, f) = -\mathbf{D}(u,f) = 0 \quad \text{for every } f \in C_c^\infty(\mathbb{R}^n),$$

which implies that $\Delta u = 0$, namely, (a version of) u is harmonic on $\mathbb{R}^n$. Since the ordinary derivatives $\frac{\partial u}{\partial x_i}$, $1 \le i \le n$, are also harmonic on $\mathbb{R}^n$, we get from the mean-value theorem the estimate

$$\left|\frac{\partial u}{\partial x_i}(x)\right| \le \frac{1}{|B_r(x)|}\int_{B_r(x)} \left|\frac{\partial u}{\partial x_i}\right| dx \le \left(\frac{1}{|B_r(x)|}\mathbf{D}(u,u)\right)^{1/2}, \qquad x \in \mathbb{R}^n,$$

where $B_r(x)$ is the ball of radius r centered at x and $|B_r(x)|$ denotes its volume. By letting $r \to \infty$, we see that all derivatives of u vanish and hence u is constant, and consequently $\dot{u}$ is the 0 element of $\dot{\rm BL}(\mathbb{R}^n)$.

Identity (2.2.48) can be derived from the identity (2.2.18) with $\psi(x) = \frac{1}{2}|x|^2$, namely,

$$\begin{cases} H^1_e(\mathbb{R}^n) = \left\{u \in L^1_{\rm loc}(\mathbb{R}^n) \cap \mathcal{S}' : \widehat{u} \in L^2(\mathbb{R}^n; |x|^2 dx)\right\} \\ \mathcal{E}(u,v) = \dfrac{1}{2}\displaystyle\int_{\mathbb{R}^n} \widehat{u}(x)\overline{\widehat{v}}(x)|x|^2 dx, \quad u, v \in H^1_e(\mathbb{R}^n). \end{cases} \tag{2.2.49}$$

Indeed, from the Plancherel theorem[7], we see for $u \in \mathcal{S}'$ that $u \in {\rm BL}(\mathbb{R}^n)$ if and only if the distributions $x_i\widehat{u}$, $1 \le i \le n$, are functions in $L^2(\mathbb{R}^n)$. On the other hand, the last condition combined with the condition that $\widehat{u}$ is a function in $L^1_{\rm loc}(\mathbb{R}^n)$ can be seen to be equivalent to that $\widehat{u}$ is a function in $L^2(\mathbb{R}^n; |x|^2 dx)$. □

THEOREM 2.2.13. *Assume that $n \le 2$. Then, for any domain $D \subset \mathbb{R}^n$, the Dirichlet form $(\frac{1}{2}\mathbf{D}, H^1(D))$ on $L^2(D)$ is recurrent. Furthermore, denoting its extended Dirichlet space by $(\mathcal{E}, H^1_e(D))$, it holds that*

$$(\mathcal{E}, H^1_e(D)) = \left(\tfrac{1}{2}\mathbf{D}, {\rm BL}(D)\right). \tag{2.2.50}$$

Proof. When $n = 1, 2$, the first statement is true for $D = \mathbb{R}^n$ and, by restricting to D those functions in $H^1(\mathbb{R}^n)$ appearing in the recurrence criterion in

[7] K. Yosida [154].

Theorem 2.1.8(ii), we see that the same criterion holds for the Dirichlet form (2.2.45) on $L^2(D)$.

When $n = 1$, the identity (2.2.50) is established in Theorem 2.2.11 under a more general context. A similar method to the proof of Proposition 2.2.10 works for $n = 2$.

Consider any domain $D \subset \mathbb{R}^2$ and take any function $u \in \mathrm{BL}(D)$ such that $|u| \leq \ell$ for some constant. Let $\psi_n \in C_c^1(\mathbb{R}_+)$ be functions satisfying properties (2.2.38) and put $u_n(x) = u(x)\psi_n(|x|)$ for $x \in D$. Then $u_n \in \mathrm{BL}(D) \cap L^2(D) = H^1(D)$ and

$$\begin{aligned}\mathbf{D}(u_n, u_n) &\leq 2\int_D |\nabla u|^2(x)\psi_n(|x|)^2 dx + 2\int_D u^2(x)\psi_n'(|x|)^2 dx \\ &\leq 2\mathbf{D}(u,u) + 2\ell^2 \int_{\{x\in\mathbb{R}^2 : |x|\leq 2n+1\}} \psi_n'(r)^2 r dr d\theta \\ &\leq 2\mathbf{D}(u,u) + \frac{2\ell^2\pi(2n+1)^2}{n^2} \leq 2\mathbf{D}(u,u) + 18\ell^2\pi.\end{aligned}$$

Hence by Theorem A.4.1, a Cesàro mean of a subsequence of $\{u_n\}$ is $\mathbf{D}$-convergent. Since u_n converges to u pointwise, we conclude that $u \in H_e^1(D)$ and $\mathcal{E}(u,u) = \frac{1}{2}\mathbf{D}(u,u)$.

Next take any $u \in \mathrm{BL}(D)$ and put $u_\ell = \varphi^\ell \circ u$, $\ell \in \mathbb{N}$, for the normal contraction φ^ℓ of (1.1.19). By (BL.2), $u_\ell \in \mathrm{BL}(D)$ and

$$\|u - u_\ell\|_D^2 = \int_{\{|u|>\ell\}} |\nabla u|^2 dx \to 0, \quad \ell \to \infty.$$

So in particular $\|u_\ell\|_{\mathbf{D}}$ is bounded. We have just shown that $u_\ell \in H_e^1(D)$ with $\mathcal{E}(u_\ell, u_\ell) = \frac{1}{2}\mathbf{D}(u_\ell, u_\ell)$. Thus we conclude by Lemma 1.1.12 that $u \in H_e^1(D)$. □

We next consider, on a general domain $D \subset \mathbb{R}^n$, a finite measure $m(dx) = m(x)dx$ with a density function $m(x)$ satisfying

$$m(x) > 0 \text{ for every } x \in D, \quad m \in bC(D) \cap L^1(D) \tag{2.2.51}$$

and an associated form

$$(\mathcal{E}, \mathcal{F}) = \left(\frac{1}{2}\mathbf{D}, \mathrm{BL}(D) \cap L^2(D;m)\right), \tag{2.2.52}$$

which is obtained just by replacing $L^2(D)$ with $L^2(D;m)$ in (2.2.44) and (2.2.45). Since the convergence in $L^2(D;m)$ implies the convergence in $L^2_{\mathrm{loc}}(D)$, (2.2.52) can be readily seen to be a Dirichlet form on $L^2(D;m)$.

THEOREM 2.2.14. *The Dirichlet form (2.2.52) on $L^2(D;m)$ is recurrent and irreducible. Its extended Dirichlet space $(\mathcal{E},\mathcal{F}_e)$ coincides with the space $(\frac{1}{2}\mathbf{D}, \mathrm{BL}(D))$.*

Proof. Since m is assumed to be a finite measure on D, the present Dirichlet form enjoys the recurrence condition (2.1.22) as well as the condition (ii) of Theorem 2.1.11. According to the same theorem, it is irreducible and we further have

$$u \in \mathcal{F}_e \text{ with } \mathcal{E}(u,u) = 0 \implies u \text{ is constant } a.e. \tag{2.2.53}$$

Denote by $\dot{\mathcal{F}}_e$ the quotient space of $\mathcal{F}_e$ by the subspace of constant functions. Just as in the proof of the preceding theorem but using (2.2.53) in place of (2.2.47), we conclude that the space $(\dot{\mathcal{F}}_e, \mathcal{E})$ is isometrically embedded into the space $(\dot{\mathrm{BL}}(D), \frac{1}{2}\mathbf{D})$.

For any $u \in \mathrm{BL}(D)$, the functions $u_\ell \in \mathrm{BL}(D)$, $\ell \geq 1$, defined as in the last part of the proof of the preceding theorem is $\mathbf{D}$-convergent to u. Since $u_\ell \in \mathcal{F} = \mathrm{BL}(D) \cap L^2(D;m)$ and u_ℓ converges to u pointwise, u must be an element of $\mathcal{F}_e$. Hence the above isometric embedding is an onto map and $\mathcal{F}_e = \mathrm{BL}(D)$. □

When the Lebesgue measure of the domain D is finite, then we can take $m(dx)$ to be the Lebesgue measure in (2.2.52) in reducing $\mathcal{F}$ to $H^1(D)$. Hence

COROLLARY 2.2.15. *If the domain D is of finite Lebesgue measure, then $(\frac{1}{2}\mathbf{D}, H^1(D))$ is irreducible recurrent and*

$$H^1_e(D) = \mathrm{BL}(D). \tag{2.2.54}$$

Recall that the Dirichlet form (2.2.46) on $L^2(\mathbb{R}^n)$ is regular and possesses $C_c^\infty(\mathbb{R}^n)$ as its core. Therefore each function $u \in H^1(\mathbb{R}^n)$ admits a quasi continuous version $\widetilde{u}$ based on the capacity Cap_1 on $\mathbb{R}^n$ evaluated by (1.2.3)–(1.2.4) with $h = 1$ and $\alpha = 1$ there using the form (2.2.46). In what follows, "q.e." means "except for an $\mathcal{E}$-polar set $N \subset \mathbb{R}^n$" (or, equivalently, in view of Theorem 1.3.14, except for a set $N \subset \mathbb{R}^n$ with $\mathrm{Cap}_1(N) = 0$).

The Dirichlet form $(\frac{1}{2}\mathbf{D}, H^1(D))$ on $L^2(D)$ is not necessarily regular. But if the boundary ∂D is so regular that $C_c^\infty(\overline{D})$, the space of functions in $C_c^\infty(\mathbb{R}^n)$ restricted to $\overline{D}$, is $\mathcal{E}_1$-dense in $H^1(D)$, then (2.2.45) can be regarded as a regular Dirichlet form on $L^2(\overline{D}; \mathbf{1}_D \cdot dx)(= L^2(D))$. Such a property holds, for example, when the domain D *has continuous boundary* in the following sense:[8] any $x \in \partial D$ has a neighborhood U such that

$$D \cap U = \{(x_1, \ldots, x_n) : x_n > F(x_1, \ldots, x_{n-1})\} \cap U \tag{2.2.55}$$

[8] Cf. Theorem 2 on page 14 of V. G. Maz'ja [121].

in some coordinate $(x_1, \dots, x_n)$ and with a continuous function F. In this case, $(\frac{1}{2}\mathbf{D}, H^1(D))$ can be regarded as a regular, strongly local Dirichlet form on $L^2(\overline{D})(= L^2(\overline{D}; \mathbf{1}_D \cdot dx))$ rather than on $L^2(D)$.

A counterexample concerning the above property is provided by a planar domain $D = \{(r, \theta) : 1 < r < 2, 0 < \theta < 2\pi\}$, which is obtained from the annulus $D_1 = \{(r, \theta) : 1 < r < 2, 0 \le \theta < 2\pi\}$ by removing a segment on the positive x-axis. The function $u(r, \theta) = \theta$ is in BL(D) but not in BL(D_1) because it violates the property (BL.2) on D_1. Hence it cannot be $\mathbf{D}$-approximated by functions in $C_c^\infty(\overline{D}) = C_c^\infty(\overline{D}_1)$.

Finally we consider the very special case where D is the unit disk $\mathbb{D} = \{x \in \mathbb{R}^2 : |x| < 1\}$ in $\mathbb{R}^2$. Then the *Poincaré inequality*

$$\|u - u_{\mathbb{D}}\|_2 \le C\|u\|_{\mathbf{D}} \quad \text{for } u \in C^1(\overline{\mathbb{D}}) \tag{2.2.56}$$

is known to be true for some constant $C > 0$, where $u_{\mathbb{D}} := \frac{1}{\pi}\int_{\mathbb{D}} u(x)dx$. In fact, (2.2.56) holds for every $u \in H_e^1(D)$. This is because for $u \in H_e^1(D)$, we can find an approximating sequence $\{u_n\}$ for u from $C^1(\overline{D})$. Then $u_n - (u_n)_{\mathbb{D}}$ is L^2-convergent and u_n converges to u a.e. In particular,

$$H^1(\mathbb{D}) = H_e^1(\mathbb{D}) = \mathrm{BL}(\mathbb{D}). \tag{2.2.57}$$

The second identity follows from Corollary 2.2.15.

The boundary of $\mathbb{D}$, the unit circle, is denoted by T and parameterized by $T = \{\theta : 0 \le \theta < 2\pi\}$. Let us introduce a Dirichlet form $(\mathbf{C}, \mathcal{G})$ on $L^2(T)$ by

$$\begin{cases} \mathbf{C}(\varphi, \psi) = \dfrac{1}{2}\displaystyle\int_{T\times T} (\varphi(\theta) - \varphi(\theta'))(\psi(\theta) - \psi(\theta'))U(\theta - \theta')d\theta d\theta' \\ \mathcal{G} = \{\varphi \in L^2(T) : \mathbf{C}(\varphi, \varphi) < \infty\}, \end{cases} \tag{2.2.58}$$

where

$$U(\theta) = (4\pi(1 - \cos\theta))^{-1}. \tag{2.2.59}$$

The above integral is called the *Douglas integral*.[9] We shall show that

$$\begin{cases} \frac{1}{2}\mathbf{D}(\mathbf{H}\varphi, \mathbf{H}\psi) = \mathbf{C}(\varphi, \psi) \\ \big\{\varphi \in L^2(T) : \mathbf{H}\varphi \in \mathrm{BL}(\mathbb{D})\big\} = \mathcal{G}, \end{cases} \tag{2.2.60}$$

where $\mathbf{H}\varphi(x) = \int_T K(x, \theta')\varphi(\theta')d\theta'$, $x \in \mathbb{D}$, is the integral with the Poisson kernel

$$K(x, \theta') = \frac{1}{2\pi}\frac{1 - \rho^2}{1 - 2\rho\cos(\theta - \theta') + \rho^2}, \quad x = \rho e^{i\theta}. \tag{2.2.61}$$

---

[9] Cf. J. Douglas [42] and J. L. Doob [41].

To see (2.2.60), we express $\varphi \in L^2(T)$ by the Fourier series $\varphi(\xi) = \sum_{\nu=-\infty}^{\infty} c_\nu e^{i\nu\xi}$, $c_\nu = \frac{1}{2\pi}\int_0^{2\pi} e^{-i\nu\theta'}\varphi(\theta')d\theta'$. A change of variables $\eta = \theta - \theta'$, $\xi = \theta'$and an evaluation of residue yield

$$\begin{aligned}\mathbf{C}(\varphi,\varphi) &= \frac{1}{8\pi}\int_0^{2\pi}\int_0^{2\pi}(\varphi(\xi+\eta)-\varphi(\xi))^2(1-\cos\eta)^{-1}d\xi d\eta\\ &= \frac{1}{2}\sum_\nu |c_\nu|^2\int_0^{2\pi}(1-\cos\nu\eta)(1-\cos\eta)^{-1}d\eta\\ &= \pi\sum_\nu |c_\nu|^2|\nu|(\le\infty).\end{aligned}$$

Now take $\varphi \in L^2(T)$. Using Green's formula on each disk of radius ρ,

$$\frac{1}{2}\mathbf{D}(\mathbf{H}\varphi,\mathbf{H}\varphi) = \lim_{\rho\uparrow 1}\frac{1}{2}\int_0^{2\pi}\mathbf{H}\varphi(\rho e^{i\theta})\frac{\partial(\mathbf{H}\varphi)}{\partial\rho}(\rho e^{i\theta})\rho d\theta.$$

On the other hand, the expression $K(\rho e^{i\theta},\theta') = \frac{1}{2\pi}\sum_\nu \rho^{|\nu|}e^{i\nu(\theta-\theta')}$ leads us to $\mathbf{H}\varphi(\rho e^{i\theta}) = \sum_\nu c_\nu\rho^{|\nu|}e^{i\nu\theta}$, which converges uniformly in θ and in $\rho \le \rho_0$ for each fixed $\rho_0 < 1$. Hence

$$\frac{1}{2}\mathbf{D}(\mathbf{H}\varphi,\mathbf{H}\varphi) = \lim_{\rho\uparrow 1}\pi\sum_{\nu=-\infty}^{\infty}|\nu|\rho^{2|\nu|}|c_\nu|^2 = \mathbf{C}(\varphi,\varphi)(\le\infty).$$

We note the relation

$$-\frac{1}{2}\frac{d}{d\rho}K(\rho e^{i\theta},\theta')\Big|_{\rho=1-} = U(\theta-\theta'). \tag{2.2.62}$$

As we shall see in Example (**3**°) of Section 5.3, $(\mathbf{C},\mathcal{G})$ is the Dirichlet form of the reflecting Brownian motion X on $\overline{\mathbb{D}}$ time-changed by its local time on $\partial\mathbb{D}$ corresponding to the Lebesgue surface measure of $\partial\mathbb{D}$. In Section 7.2 we shall encounter a general situation where the trace Dirichlet form of a symmetric Markov process has the Douglas integral representation analogous to (2.2.58).

2.2.5. Brownian Motions on Manifolds

Let (M,g) be an n-dimensional Riemannian manifold. A Brownian motion $B = \{B_t, t \ge 0\}$ is the minimal M-valued strong Markov process with continuous sample paths with infinitesimal generator $\frac{1}{2}\Delta_g$, where Δ_g is the Laplace-Beltrami operator on M. There are several ways to construct a Brownian motion on M. One can solve stochastic differential equations in local coordinates and then piece together the resulting diffusions in local coordinates. This was done by R. Gangolli in [79]. It can also be constructed by solving an M-valued stochastic differential equation using Itô's map (see [91]). In this subsection, we illustrate that Brownian motion on M can be constructed through a regular Dirichlet form via Theorem 1.5.1.

For $p \in M$, g_p is an inner product on the tangent space $T_p(M)$ of M at p, which is of dimension n. Hence g_p establishes a one-to-one correspondence between $T_p(M)$ and its dual $T_p^*(M)$, the co-tangent space at p. For every $f \in C(M)$, df, the C^∞ co-vector field, is in $T^*(M)$. It uniquely determines an $X_f \in T(M)$ by $g(X_f, Y) = \langle df, Y\rangle = Yf$ for every $Y \in T(M)$. In local coordinates,

$$df = \sum_{i=1}^n \frac{\partial f}{\partial x_i} dx_i \quad \text{and} \quad X_f = \sum_{i=1}^n \left(\sum_{j=1}^n g^{ij} \frac{\partial f}{\partial x_j} \right) \frac{\partial}{\partial x_i},$$

where (g_{ij}) is the matrix of the inner product g in these local coordinates and (g^{ij}) is its inverse matrix. Let $V(dp)$ be the volume element of (M, g) and define, for $u, v \in C_c^\infty(M)$,

$$\mathcal{E}(u, v) = \frac{1}{2} \int_M g_p(X_u, X_v) V(dp).$$

In local coordinates, $V(dx) = \sqrt{g(x)}dx$, where $g(x)$ is the determinant of $(g_{ij}(x))$ and, with $\nabla u(x) := (\frac{\partial u(x)}{\partial x_1}, \dots, \frac{\partial u(x)}{\partial x_n})$ denoting the (Euclidean) gradient of u,

$$\begin{aligned} g_p(X_u, X_v)V(dp) &= ((g^{ij})\nabla u)(x) \cdot (g_{ij}(x))((g^{ij})\nabla v)(x) \sqrt{g(x)}dx \\ &= \nabla u(x) \cdot (g^{ij}(x)) \nabla v(x) \sqrt{g(x)}dx. \end{aligned}$$

In the same way as the proof of Lemma 2.2.7, the Markovian property of the symmetric form $(\mathcal{E}, C_c^\infty(M))$ in the sense of Definition 1.1.2 is readily verifiable. Note that for $u, v \in C_c^\infty(M)$, $\mathcal{E}(u, v) = (-\frac{1}{2}\Delta_g u, v)_{L^2(M;\, dV)}$. This can be easily seen since in local coordinates, the Laplace-Beltrami differential operator Δ_g on (M, g) has the expression

$$\Delta_g = \frac{1}{\sqrt{g(x)}} \sum_{i,j=1}^n \frac{\partial}{\partial x_i} \left(\sqrt{g(x)} g^{ij}(x) \frac{\partial}{\partial x_j} \right).$$

As, for $\{u, v_k, k \geq 1\} \subset C_c^\infty(M)$ with $v_k \to 0$ in $L^2(M; V(dp))$,

$$\lim_{k\to\infty} \mathcal{E}(u, v_k) = \lim_{k\to\infty} \left(-\frac{1}{2} \Delta_g u, v \right)_{L^2(M;\, dV)} = 0,$$

the quadratic form $(\mathcal{E}, C_c^\infty(M))$ is closable.

Define $H_0^1(M)$ to be the closure of $C_c^\infty(M)$ under the inner product $\mathcal{E}_1 := \mathcal{E} + (\cdot, \cdot)_{L^2(M;\, dV)}$. Then $(\mathcal{E}, H_0^1(M))$ is a regular Dirichlet form on $L^2(M; dV)$ as the closure of a closable Markovian symmetric form (cf. [73, Theorem 3.1.1]). The above regularity can also be checked directly by using Theorem 1.1.3(d).

The Hunt process B associated with this Dirichlet form is a Brownian motion on M. That B has continuous sample paths follows from the fact that the Dirichlet form $(\mathcal{E}, H_0^1(M))$ is strongly local and Theorem 4.3.4 below.

In Example 5.1.11 below, we will present a quasi-regular but not regular Dirichlet form on $\mathbb{R}^n$ that has a Hunt process properly associated with it.

2.3. ANALYTIC POTENTIAL THEORY FOR REGULAR DIRICHLET FORMS

Let E be a locally compact separable metric space and m be a positive Radon measure on E satisfying $\text{supp}[m] = E$. Let $(\mathcal{E}, \mathcal{F})$ be a regular Dirichlet form on $L^2(E; m)$ in the sense of Definition 1.3.10. In this section, we present an analytic potential theory for $(\mathcal{E}, \mathcal{F})$ primarily due to A. Beurling and J. Deny [8] and J. Deny [39], which will be used in the sequel.

The $\mathcal{E}$-quasi-notions ($\mathcal{E}$-nest, $\mathcal{E}$-polarity, $\mathcal{E}$-quasi-continuity) are introduced in Definition 1.2.12 for a Dirichlet form $(\mathcal{E}, \mathcal{F})$ on a general topological space (E, m). These notions for the regular Dirichlet form $(\mathcal{E}, \mathcal{F})$ on $L^2(E; m)$ have been interpreted in terms of the capacity Cap_1 which is defined to be $\text{Cap}_{h,\alpha}$ with $h \equiv 1$ and $\alpha = 1$ in Section 1.2. For reader's convenience, we recall the definition of Cap_1 below. Denote by $\mathcal{O}$ the family of all open subsets of E and, for $A \in \mathcal{O}$, we put $\mathcal{L}_{A,1} = \{f \in \mathcal{F} : f \geq 1 \ m\text{-a.e. on } A\}$. The function $e_A \in \mathcal{L}_{A,1}$ minimizing $\{\mathcal{E}_1(u, u),\ u \in \mathcal{L}_{A,1}\}$ is called the (1-order) *equilibrium potential* of the open set A. (Deviating from the notation used in Section 1.2, we write it as e_A instead of $\mathbf{1}_A$ in order to distinguish it from the indicator of A). For open set A, define

$$\text{Cap}_1(A) = \begin{cases} \inf\{\mathcal{E}_1(f,f) : f \in \mathcal{L}_{A,1}\}, & A \in \mathcal{O}_0 \\ \infty & A \notin \mathcal{O}_0 \end{cases} \tag{2.3.1}$$

where $\mathcal{O}_0 = \{A \in \mathcal{O} : \mathcal{L}_{A,1} \neq \emptyset\}$. The capacity $\text{Cap}_1(B)$ for an arbitrary set $B \subset E$ is defined by

$$\text{Cap}_1(B) = \inf\{\text{Cap}_1(A) : A \in \mathcal{O},\ A \supset B\}. \tag{2.3.2}$$

Theorem 1.3.14 states that a set $N \subset E$ is $\mathcal{E}$-polar if and only if $\text{Cap}_1(N) = 0$, an increasing sequence of closed sets $\{F_k\}$ is an $\mathcal{E}$-nest if and only if $\lim_{k\to\infty} \text{Cap}_1(K \setminus F_k) = 0$ for any compact set $K \subset E$, and a function f is $\mathcal{E}$-quasi-continuous if and only if for any $\varepsilon > 0$, there is an open set O with $\text{Cap}_1(O) < \varepsilon$ such that $f|_{E\setminus O}$ is finite and continuous. Since we are dealing with a fixed regular Dirichlet form $(\mathcal{E}, \mathcal{F})$, for convenience we drop "$\mathcal{E}$-" from the terminology an "$\mathcal{E}$-nest" and "$\mathcal{E}$-quasi-continuous" and will simply call them a *nest* and *quasi continuous*, respectively. We call an increasing sequence $\{F_k\}$ of closed sets a Cap_1*-nest* if $\lim_{k\to\infty} \text{Cap}_1(E \setminus F_k) = 0$. Any

Cap_1-nest is a nest but not vice versa. However, a function f is quasi continuous if and only if there is a Cap_1-nest $\{F_k\}$ such that $f \in C(\{F_k\})$. Moreover, in view of Lemma 1.3.15, any element $f \in \mathcal{F}$ admits a quasi continuous version $\widetilde{f}$ in the restricted sense relative to a Cap_1-nest, namely, there is a Cap_1-nest $\{F_k\}$ such that $\widetilde{f} \in C_\infty(\{F_k\})$.

We also use the term *quasi everywhere* or *q.e.* to mean "except for an $\mathcal{E}$-polar set," or, equivalently, "except for a Cap_1-polar set," in view of Theorem 1.3.14.

Theorem 1.3.3 enables us to introduce the equilibrium potentials for sets which are not necessarily open. For an arbitrary set $B \subset E$. we let $\mathcal{L}_{B,1} = \{f \in \mathcal{F} : \widetilde{f} \geq 1 \text{ q.e. on } B\}$. If $\mathcal{L}_{B,1} \neq \emptyset$, then $\mathcal{L}_{B,1}$ is a closed convex subset of $(\mathcal{F}, \mathcal{E}_1)$ and there exists a unique element $e_B \in \mathcal{L}_{B,1}$ minimizing $\mathcal{E}_1(u,u)$. We call e_B the (1-order) *equilibrium potential* of B. By Theorem 1.3.7, if B is an open set, then e_B coincides with the equilibrium potential of B mentioned at the beginning of this section. e_B enjoys the following properties.

THEOREM 2.3.1. *Fix an arbitrary set $B \subset E$ with $\mathcal{L}_{B,1} \neq \emptyset$.*
(i) *It holds that*

$$\mathrm{Cap}_1(B) = \mathcal{E}_1(e_B, e_B). \tag{2.3.3}$$

(ii) $0 \leq e_B \leq 1$ $[m]$ *and* $\widetilde{e}_B = 1$ *q.e. on B.*
(iii) e_B *is the unique element of $\mathcal{F}$ satisfying the following properties: $\widetilde{e}_B = 1$ q.e. on B and $\mathcal{E}_1(e_B, f) \geq 0$ for any $f \in \mathcal{F}$ with $f \geq 0$ q.e. on B.*

Proof. In view of Theorem 1.3.3, (ii) and (iii) can be shown similarly to the proof of Theorem 1.2.5. Let us show (i).

For any $\varepsilon > 0$, there exists $A \in \mathcal{O}_0$ such that $B \subset A$ and $\mathrm{Cap}_1(B) > \mathrm{Cap}_1(A) - \varepsilon$. Since the equilibrium potential e_A of A belongs to $\mathcal{L}_{B,1}$ by Theorem 1.3.7, $\mathrm{Cap}_1(A) = \mathcal{E}_1(e_A, e_A) \geq \mathcal{E}_1(e_B, e_B)$, and so we get $\mathrm{Cap}_1(B) \geq \mathcal{E}_1(e_B, e_B)$.

To prove the converse inequality, we fix a quasi continuous version $\widetilde{e}_B$ of e_B. For any $\varepsilon > 0$, we can find an open set A_ε with $\mathrm{Cap}(A_\varepsilon) < \varepsilon$ such that $\widetilde{e}_B|_{A_\varepsilon^c}$ is continuous and $\widetilde{e}_B(x) \geq 1$ for every $x \in B \cap A_\varepsilon^c$. Then

$$G_\varepsilon = \{x \in A_\varepsilon^c : \widetilde{e}_B > 1 - \varepsilon\} \cup A_\varepsilon$$

is an open set containing B. If we denote the equilibrium potential of A_ε by e_ε, it holds that $e_B + e_\varepsilon \geq 1 - \varepsilon$ m-a.e. on G_ε. Therefore

$$\begin{aligned}\mathrm{Cap}_1(B) &\leq \mathrm{Cap}_1(G_\varepsilon) \leq (1-\varepsilon)^{-2} \|e_B + e_\varepsilon\|_{\mathcal{E}_1}^2 \\ &\leq (1-\varepsilon)^{-2} (\|e_B\|_{\mathcal{E}_1} + \|e_\varepsilon\|_{\mathcal{E}_1})^2 \leq (1-\varepsilon)^{-2} (\|e_B\|_{\mathcal{E}_1} + \sqrt{\varepsilon})^2.\end{aligned}$$

We get (2.3.3) by letting $\varepsilon \downarrow 0$. □

Next suppose that the semigroup generated by Dirichlet form $(\mathcal{E},\mathcal{F})$ is transient in the sense of Definition 2.1.1. Then the extended Dirichlet space $(\mathcal{F}_e,\mathcal{E})$ is a Hilbert space and by Theorem 2.1.5

$$\|u\|_{L^1(g\cdot m)} \leq \|u\|_{\mathcal{E}}, \qquad u \in \mathcal{F}_e$$

for some reference function g.

For $A \in \mathcal{O}$, let $\mathcal{L}^{(0)}_{A,1} = \{f \in \mathcal{F}_e : f \geq 1 \ m\text{-a.e. on } A\}$. If $\mathcal{L}^{(0)}_{A,1} \neq \emptyset$, there exists a unique element $e^{(0)}_A \in \mathcal{L}^{(0)}_{A,1}$ minimizing $\{\mathcal{E}(u,u),\ u \in \mathcal{L}^{(0)}_{A,1}\}$. $e^{(0)}_A$ is called the *0-order equilibrium potential* of A. The 0-order capacity $\mathrm{Cap}^{(0)}(A)$ of an open set A is defined by (2.3.1) with $\mathcal{E}_1$ and $\mathcal{L}_{A,1}$ being replaced by $\mathcal{E}$ and $\mathcal{L}^{(0)}_{A,1}$, respectively. $\mathrm{Cap}^{(0)}(B)$ for any set $B \subset E$ is defined analogously to (2.3.2).

An increasing sequence $\{F_k\}$ of closed sets is said to be a $\mathrm{Cap}^{(0)}$-*nest* if $\lim_{k\to\infty} \mathrm{Cap}^{(0)}(E \setminus F_k) = 0$. A set $N \subset E$ is called $\mathrm{Cap}^{(0)}$-polar if $\mathrm{Cap}^{(0)}(N) = 0$.

THEOREM 2.3.2. *Suppose $(\mathcal{E},\mathcal{F})$ is transient.*
(i) *Any* $\mathrm{Cap}^{(0)}$*-nest is a nest.*
(ii) *A subset of E is* $\mathrm{Cap}^{(0)}$*-polar if and only if it is $\mathcal{E}$-polar.*
(iii) *Any element of $\mathcal{F}_e$ admits a quasi continuous m-version in the restricted sense relative to a* $\mathrm{Cap}^{(0)}$*-nest. If a sequence of quasi continuous functions $\{f_n, n \geq 1\} \subset \mathcal{F}_e$ is $\mathcal{E}$-convergent to $f \in \mathcal{F}_e$, then there exist a subsequence $\{n_\ell\}$ and a* $\mathrm{Cap}^{(0)}$*-nest $\{F_k\}$ with $f_{n_\ell} \in C_\infty(\{F_k\})$ such that f_{n_ℓ} converges as $\ell \to \infty$ uniformly on each set F_k to a quasi continuous version of f.*

Proof. (i) For any relatively compact open set $D \subset E$, choose a non-negative function $h \in \mathcal{F} \cap C_c(E)$ such that $h \geq 1$ on $D \setminus F_1$. Define $v_k = e^{(0)}_{E\setminus F_k} \wedge h$. Then $v_k \in \mathcal{F}(= \mathcal{F}_e \cap L^2(E;m))$ and

$$\sup_{k\geq 1} \mathcal{E}_1(v_k,v_k) \leq \sup_{k\geq 1} \mathrm{Cap}^{(0)}(E \setminus F_k) + \mathcal{E}_1(h,h) < \infty.$$

Hence by Theorem A.4.1, a Cesàro mean sequence $\{w_k\}$ of a subsequence of $\{v_k\}$ is $\mathcal{E}_1$-convergent. On the other hand, by Theorem 2.1.5(i), v_k converges to 0 m-a.e. as $k \to \infty$, so is w_k. Since $w_k \in \mathcal{L}_{D\setminus F_k,1}$, this together with

$$\lim_{k\to\infty} \mathrm{Cap}_1(D \setminus F_k) \leq \lim_{k\to\infty} \mathcal{E}_1(w_k,w_k) = 0$$

yields via Theorem 1.3.14(ii) that $\{F_k\}$ is a nest.
(ii) The "Only if" part is a consequence of (i). If a set is $\mathcal{E}$-polar, then it is Cap_1-polar and consequently $\mathrm{Cap}^{(0)}$-polar.
(iii) The second assertion can be proved in the same way as that of Theorem 1.3.3 by using a 0-order counterpart of Exercise 1.3.16. The first assertion follows from the second. □

Exercise 2.3.3. Let $0 < \alpha \le 2$, $\alpha < n$. Then the rotation-invariant stable semigroup of index α on $\mathbb{R}^n$ is regular and transient. Moreover, its extended Dirichlet space is characterized as (2.2.20) in Section 2.2.2:

$$\mathcal{F}_e = \left\{u = I_{\alpha/2} * f : f \in L^2(\mathbb{R}^n)\right\} \tag{2.3.4}$$

with $\mathcal{E}(u, v) = \int_{\mathbb{R}^n} f(x)g(x)dx$ for $u = I_{\alpha/2} * f$, $v = I_{\alpha/2} * g$. Prove that every function appearing on the right hand side of (2.3.4), namely, the $\alpha/2$-order Riesz potential of an L^2-function, is quasi continuous.

The next theorem says that Theorem 2.3.2(iii) in fact holds for any (not necessarily transient) Dirichlet form $(\mathcal{E}, \mathcal{F})$.

THEOREM 2.3.4. *Any $f \in \mathcal{F}_e$ admits its quasi continuous version. If $\{f_n\} \subset \mathcal{F}$ is an approximating sequence of $f \in \mathcal{F}_e$ and f_n, $n \ge 1$, are quasi continuous, then there exists a subsequence n_ℓ such that f_{n_ℓ} converges to a quasi continuous version of f q.e. as $\ell \to \infty$.*

If $f \in \mathcal{F}_e$ is bounded on E, its approximating sequence $\{f_n\} \subset \mathcal{F}$ can be taken to be uniformly bounded on E.

Proof. Owing to Lemma 2.1.15, there exists a function g belonging to the family $\mathcal{K}_0$ defined by (2.1.27) such that $f \in \mathcal{F}_e^g$ and $\{f_n\}$ is $\mathcal{E}^g$-convergent to f, where $(\mathcal{E}^g, \mathcal{F})$ is the perturbed Dirichlet form on $L^2(E; m)$ defined by (2.1.28) and $\mathcal{F}_e^g$ is its extended Dirichlet space. Since $(\mathcal{E}^g, \mathcal{F})$ is transient and the norm $\sqrt{\mathcal{E}_1^g(u, u)}$ is equivalent to $\sqrt{\mathcal{E}_1(u, u)}$ for $u \in \mathcal{F}$, we get the desired conclusions from Theorem 2.3.2 and Exercise 2.1.13. □

A positive Radon measure μ on E is called a *measure of finite energy integral* if there exists a constant $C_\mu > 0$ such that

$$\int_E |g(x)|\mu(dx) \le C_\mu \, \|g\|_{\mathcal{E}_1} \quad \text{for every } g \in \mathcal{F} \cap C_c(E). \tag{2.3.5}$$

Let S_0 denote the family of all measures of finite energy integrals. For $\mu \in S_0$, the linear functional ℓ_μ on $\mathcal{F} \cap C_c(E)$ defined by $\ell_\mu(g) = \int_E g(x)\mu(dx)$ has the property

$$|\ell_\mu(g)| \le C_\mu \|g\|_{\mathcal{E}_1} \qquad \text{for every } g \in \mathcal{F} \cap C_c(E).$$

Since $(\mathcal{E}, \mathcal{F})$ is regular, for each $\alpha > 0$, ℓ_μ determines a unique bounded linear functional on the Hilbert space $(\mathcal{F}, \mathcal{E}_\alpha)$. By Riesz representation theorem (cf. [154]), there exists a unique element $U_\alpha \mu \in \mathcal{F}$ so that

$$\mathcal{E}_\alpha(U_\alpha \mu, g) = \int_E g(x)\mu(dx) \quad \text{for every } g \in \mathcal{F} \cap C_c(E). \tag{2.3.6}$$

$U_\alpha \mu$ is called the *α-potential* of $\mu \in S_0$.

Let $\{T_t; t > 0\}$ be the L^2-semigroup generated by the Dirichlet form $(\mathcal{E}, \mathcal{F})$. For $\alpha > 0$, the α-excessiveness of a function $f \in L^2(E; m)$ relative to $\{T_t; t > 0\}$ is defined by Definition 1.2.1.

Lemma 2.3.5. *Let $f \in \mathcal{F}$ and $\alpha > 0$. The following are equivalent.*

(i) $f = U_\alpha \mu$ for some $\mu \in S_0$.
(ii) $\mathcal{E}_\alpha(f, g) \geq 0$ for every $g \in \mathcal{F} \cap C_c(E)_+$.
(iii) $\mathcal{E}_\alpha(f, g) \geq 0$ for every $g \in \mathcal{F}$ with $g \geq 0$ $[m]$.
(iv) f is α-excessive relative to $\{T_t; t \geq 0\}$.

Proof. The equivalence of (iii) and (iv) has already been established by Lemma 1.2.4. (i) $\Rightarrow$ (ii) is trivial. Assuming (ii), take, for $g \in \mathcal{F}$ with $g \geq 0$ $[m]$, a sequence $\{g_n, n \geq 1\} \subset \mathcal{F} \cap C_c(E)$ that is $\mathcal{E}_1$-convergent to g. Since the sequence $\{g_n^+, n \geq 1\} \subset \mathcal{F} \cap C_c(E)$ is uniformly bounded in $\mathcal{E}_1$-norm and a subsequence of $\{g_n^+, n \geq 1\}$ converges m-a.e. to g, by Theorem A.4.1, there is a subsequence of $\{g_n^+, n \geq 1\}$ whose Cesàro mean sequence $\{h_n, n \geq 1\}$ is $\mathcal{E}_1$-convergent to g. We thus get (iii) from $\mathcal{E}_\alpha(f, h_n) \geq 0$ by letting $n \to \infty$. (iii) $\Rightarrow$ (ii) is trivial.

We now show (ii) $\Rightarrow$ (i). For any compact set $K \subset E$, we can find a non-negative function $g_K \in \mathcal{F} \cap C_c(E)$ with $g_K \geq 1$ on K. If we define $\ell_f(g) = \mathcal{E}_\alpha(f, g)$, $g \in \mathcal{F} \cap C_c(E)$, then, by the assumption, we have $|\ell_f(g)| \leq \|g\|_\infty \ell_f(g_K)$ for any $g \in \mathcal{F} \cap C_c(E)$ with supp$[g] \subset K$. According to Exercise 1.3.13, any $h \in C_c(E)$ is a uniform limit of a sequence $g_n \in \mathcal{F} \cap C_c(E)$ with supp$[g_n] \subset$ supp$[h]$. Therefore ℓ_f can be extended uniquely to a positive linear functional on $C_c(E)$. Therefore there is a positive Radon measure μ on E such that $\ell_f(g) = \int_E g d\mu$ for every $g \in \mathcal{F} \cap C_c(E)$. This establishes (i). □

By Remark 1.2.2, we get

Corollary 2.3.6. *If f_1, f_2 are α-potentials, then so are the functions $f_1 \wedge f_2$ and $f_1 \wedge 1$.*

Denote by $\widetilde{\mathcal{F}}$ the totality of quasi continuous functions belonging to $\mathcal{F}$.

Theorem 2.3.7. *Let $\mu \in S_0$. Then*

(i) μ charges no $\mathcal{E}$-polar set.
(ii) $\widetilde{\mathcal{F}} \subset L^1(E; \mu)$ and, for every $\alpha > 0$ and $g \in \mathcal{F}$,

$$\mathcal{E}_\alpha(U_\alpha \mu, g) = \int_E \widetilde{g}(x) \mu(dx). \tag{2.3.7}$$

Proof. (i) Define for every $n \geq 1$, $g_n := n(U_1\mu - e^{-1/n} T_{1/n}(U_1\mu))$. Note that by Lemma 2.3.5, $g_n \geq 0$ $[m]$. So it follows from (1.2.1) and (2.3.6) that, for any $h \in \mathcal{F} \cap C_c(E)$,

$$\lim_{n\to\infty} \int_E h(x)g_n(x)m(dx) = \mathcal{E}_1(U_1\mu, h) = \int_E h(x)\mu(dx).$$

Therefore, for any compact set K and any relatively compact open set G containing K, we have by Theorem 1.2.5

$$\begin{aligned}\mu(K) &\le \liminf_{n\to\infty} \int_G g_n(x)m(dx) \le \liminf_{n\to\infty}(g_n, e_G)\\ &= \mathcal{E}_1(U_1\mu, e_G) \le \|U_1\mu\|_{\mathcal{E}_1}\sqrt{\mathrm{Cap}_1(G)},\end{aligned}$$

which proves (i).
(ii) For any $g \in \mathcal{F}$, choose $g_n \in \mathcal{F} \cap C_c(E)$ so that it is $\mathcal{E}_1$-convergent to g as $n \to \infty$. By virtue of Theorem 1.3.3, a subsequence g_{n_k} converges to $\widetilde{g}$ q.e. Since (2.3.7) is valid for each g_n, it follows from (i) and Fatou's lemma that

$$\begin{aligned}\int_E |\widetilde{g}(x) - g_n(x)|\mu(dx) &= \int_E \liminf_{k\to\infty} |g_{n_k}(x) - g_n(x)|\mu(dx)\\ &\le C_\mu \cdot \liminf_{k\to\infty} \|g_{n_k} - g_n\|_{\mathcal{E}_1}.\end{aligned}$$

This shows that $\widetilde{g} \in L^1(E;\mu)$ and that g_n converges to $\widetilde{g}$ in $L^1(E;\mu)$. It now suffices to let $n \to \infty$ in the equation (2.3.7) for g_n. □

For $\mu \in S_0$, $\mathcal{E}_\alpha(\mu) = \mathcal{E}_\alpha(U_\alpha\mu, U_\alpha\mu)$ is called the *α-energy integral* of μ, which is equal to the integral $\int_E \widetilde{U_\alpha\mu}(x)\mu(dx)$ by virtue of the above theorem.

Exercise 2.3.8. Prove the following equation:

$$U_\alpha\mu - U_\beta\mu + (\alpha - \beta)G_\alpha U_\beta\mu = 0 \quad \text{for } \mu \in S_0 \text{ and } \alpha,\ \beta > 0. \tag{2.3.8}$$

LEMMA 2.3.9. (i) *Suppose, for $\mu \in S_0$ and a constant $C \ge 0$, $\widetilde{U_\alpha\mu} \le C$ μ-a.e. Then $U_\alpha\mu \le C$ $[m]$.*
(ii) *For any $\mu \in S_0$, there exists a* Cap_1*-nest $\{F_k\}$ such that*

$$\|U_1(\mathbf{1}_{F_k}\mu)\|_\infty < \infty \quad \textit{for every } k \ge 1.$$

Proof. (i) The function $f = U_\alpha\mu \wedge C$ is an α-potential by Corollary 2.3.6. By virtue of the preceding theorem

$$\mathcal{E}_\alpha(f, U_\alpha\mu) = \langle \widetilde{f}, \mu\rangle = \langle \widetilde{U_\alpha\mu}, \mu\rangle = \mathcal{E}_\alpha(U_\alpha\mu, U_\alpha\mu).$$

Since $f \le U_\alpha\mu$ $[m]$, we get by the above identity and Lemma 2.3.5 that

$$\|f - U_\alpha\mu\|^2_{\mathcal{E}_\alpha} = \mathcal{E}_\alpha(f, f - U_\alpha\mu) \le 0.$$

Hence $U_\alpha\mu = f \le C$ $[m]$.

(ii) For $\mu \in S_0$, choose a quasi continuous version of $\widetilde{U_1\mu}$ and an associated Cap_1-nest $\{F_k^0\}$. We let $F_k = \{x \in F_k^0 : \widetilde{U_1\mu}(x) \le k\}$, $k = 1, 2, \ldots$. For each k, $\widetilde{U_1(\mathbf{1}_{F_k} \cdot \mu)} \le \widetilde{U_1\mu} \le k$ q.e. on F_k, and consequently $U_1(\mathbf{1}_{F_k} \cdot \mu) \le k$ $[m]$ by (i). Furthermore, $\mathrm{Cap}_1(E \setminus F_k) \le \mathrm{Cap}_1(E \setminus F_k^0) + \mathrm{Cap}_1(\{\widetilde{U_1\mu} > k\})$, which tends to zero as $k \to \infty$ in view of Exercise 1.3.16. □

LEMMA 2.3.10. *The following conditions are mutually equivalent for $f \in \mathcal{F}$ and a closed set $F \subset E$:*

(i) $f = U_\alpha\mu$ *for some* $\mu \in S_0$ *with* $\mathrm{supp}[\mu] \subset F$.
(ii) $\mathcal{E}_\alpha(f, g) \ge 0$ *for any* $g \in \mathcal{F}$ *with* $\widetilde{g} \ge 0$ *q.e. on* F.
(iii) $\mathcal{E}_\alpha(f, g) \ge 0$ *for any* $g \in \mathcal{F} \cap C_c(E)$ *with* $g \ge 0$ *on* F.

Proof. (i) ⇒ (ii) follows from Theorem 2.3.7(ii). (ii) ⇒ (iii) is trivial. (iii) ⇒ (i) follows from the proof of Lemma 2.3.5 and Exercise 1.3.13. □

Recall the 1-order equilibrium potential e_B defined for a set $B \subset E$ with $\mathcal{L}_{B,1} \neq \emptyset$. By Theorem 2.3.1(iii) and the above lemma, there is a unique measure $\mu_B \in S_0$ such that $e_B = U_1\mu_B$ with $\mathrm{supp}[\mu_B] \subset \overline{B}$. μ_B is called *1-equilibrium measure* of the set B. For a compact set $K \subset E$, we have in particular

$$\mathrm{Cap}_1(K) = \mathcal{E}_1(e_K, e_K) = \mu_K(K) \tag{2.3.9}$$

on account of Theorem 2.3.1 and Theorem 2.3.7.

Let us define a subfamily S_{00} of S_0 by

$$S_{00} = \{\mu \in S_0 : \mu(E) < \infty,\ \|U_1\mu\|_\infty < \infty\}. \tag{2.3.10}$$

COROLLARY 2.3.11. *The following conditions for $B \in \mathcal{B}(E)$ are mutually equivalent:*

(i) $\mathrm{Cap}_1(B) = 0$.
(ii) $\mu(B) = 0$ *for every* $\mu \in S_0$.
(iii) $\mu(B) = 0$ *for every* $\mu \in S_{00}$.

Proof. The implication (i) ⇒ (ii) is given by Theorem 2.3.7(i) and the implication (ii) ⇒ (iii) is trivial. If $\mathrm{Cap}_1(B) > 0$, then there is a compact set $K \subset B$ with $\mathrm{Cap}_1(K) > 0$ by (1.2.6) with $h = \alpha = 1$ there. Since $\mu_K \in S_{00}$ has $\mu_K(K) > 0$ by (2.3.9), this establishes the implication (iii) ⇒ (i). □

Fix an $\alpha > 0$ and an arbitrary set $B \subset E$. Let $f \in \mathcal{F}$ be α-excessive relative to $\{T_t\}$ and put

$$\mathcal{L}_{f,B} = \{g \in \mathcal{F} : \widetilde{g} \ge \widetilde{f} \text{ q.e. on } B\}. \tag{2.3.11}$$

Since $\mathcal{L}_{f,B}$ is a closed convex subset of $(\mathcal{F}, \mathcal{E}_\alpha)$ by Theorem 1.3.3, there exists a unique function f_B in $\mathcal{L}_{f,B}$ minimizing $\|g\|_{\mathcal{E}_\alpha}$. Since

$$\mathcal{E}_\alpha(f_B, g) \geq 0 \quad \text{for } g \in \mathcal{F} \text{ with } \widetilde{g} \geq 0 \text{ q.e. on } B, \tag{2.3.12}$$

f_B is again α-excessive relative to $\{T_t\}$ by Lemma 2.3.5. f_B is said to be the *α-reduced function* of the α-excessive function $f \in \mathcal{F}$ with respect to B. By Lemma 2.3.10, there exist unique measures μ, $\nu \in S_0$ so that $f = U_\alpha\mu$, $f_B = U_\alpha\nu$, and $\text{supp}[\nu] \subset \overline{B}$. The correspondence $\mu \in S_0 \mapsto \nu \in S_0$ is said to be the *α-sweeping out* of μ on the set B.

LEMMA 2.3.12. *f_B is an α-reduced function of α-excessive function $f \in \mathcal{F}$ relative to a set B if and only if f_B is an element of $\mathcal{F}$ satisfying* (2.3.12) *and*

$$\widetilde{f_B} = \widetilde{f} \quad \text{q.e. on } B. \tag{2.3.13}$$

Proof. We need only show the necessity of (2.3.13). To this end, we observe that $h = f_B \wedge f$ is again α-excessive relative to $\{T_t; t > 0\}$ with $h \leq f_B \in \mathcal{F}$, and hence Lemma 1.2.3 implies that $h \in \mathcal{F}$ with $\|h\|_{\mathcal{E}_\alpha} \leq \|f_B\|_{\mathcal{E}_\alpha}$, yielding $h \in \mathcal{L}_{f,B}$ and $f_B = h \leq f$. □

This lemma leads us to another important interpretation of a reduced function. If we let

$$\mathcal{F}_{E\setminus B} = \{g \in \mathcal{F} : \widetilde{g} = 0 \text{ q.e. on } B\}, \tag{2.3.14}$$

then $\mathcal{F}_{E\setminus B}$ is a closed linear subspace of the Hilbert space $(\mathcal{F}, \mathcal{E}_\alpha)$. Denote by $\mathcal{H}_B^\alpha$ its orthogonal complement, and so $\mathcal{F} = \mathcal{F}_{E\setminus B} \oplus \mathcal{H}_B^\alpha$. Denoting by $P_{\mathcal{H}_B^\alpha}$ the orthogonal projection to the space $\mathcal{H}_B^\alpha$, we have

$$P_{\mathcal{H}_B^\alpha} f = f_B, \tag{2.3.15}$$

because $f = (f - f_B) + f_B$ gives the $\mathcal{E}_\alpha$-orthogonal decomposition of f in view of Lemma 2.3.12.

We notice that, in a particular case where B is an open set, f_B is nothing but the α-reduced function of the α-excessive function f on the set B in the sense of Section 1.2 on account of Theorem 1.2.5.

DEFINITION 2.3.13. A positive measure μ on $(E, \mathcal{B}(E))$ is called *smooth* if it satisfies the following two conditions:

(S.1) μ charges no $\mathcal{E}$-polar set.
(S.2) There exists a nest $\{F_k\}$ such that $\mu(F_k) < \infty$ for every $k \geq 1$.

The nest $\{F_k\}$ appearing in the above is said to be *attached to* the smooth measure μ. We note that $\mu(E \setminus \bigcup_k F_k) = 0$ in this case.

We denote by S the totality of smooth measures. Any positive Radon measure on E charging no $\mathcal{E}$-polar set is an element of S. In this case, for an increasing sequence $\{G_k\}$ of relatively compact open sets, we can let $F_k = \overline{G_k}$ to get a nest $\{F_k\}$ satisfying **(S.2)**. In particular, $S_{00} \subset S_0 \subset S$.

LEMMA 2.3.14. (i) *Let ν be a finite measure on $(E, \mathcal{B}(E))$. If $\nu(A) \leq C \cdot \mathrm{Cap}_1(A)$ for every $A \in \mathcal{B}(E)$ for some positive constant C, then $\nu \in S_0$.*
(ii) *Let ν be a finite measure on $(E, \mathcal{B}(E))$ charging no $\mathcal{E}$-polar set. There exists then a decreasing sequence $\{G_n\}$ of open sets such that*

$$\lim_{n\to\infty} \mathrm{Cap}_1(G_n) = 0, \qquad \lim_{n\to\infty} \nu(G_n) = 0$$

and

$$\nu(A) \leq 2^n \mathrm{Cap}_1(A) \quad \forall A \in \mathcal{B}(E),\ A \subset E \setminus G_n,\ n \geq 1.$$

Proof. (i) For any non-negative g with $g \in \mathcal{F} \cap C_c(E)$ and $\mathcal{E}_1(g, g) = 1$, we have by (1.3.2) and Exercise 1.3.16,

$$\begin{aligned} \int_E g(x)\nu(dx) &\leq \nu(E) + \sum_{k=0}^{\infty} 2^{k+1}\nu(\{x : 2^k < g(x) \leq 2^{k+1}\}) \\ &\leq \nu(E) + C\sum_{k=0}^{\infty} 2^{k+1}\mathrm{Cap}_1(\{x : g(x) > 2^k\}) \\ &\leq \nu(E) + C\sum_{k=0}^{\infty} 2^{k+1}2^{-2k} = \nu(E) + 4C, \end{aligned}$$

which means $\nu \in S_0$.
(ii) Fix $n \geq 1$ and let $\alpha := \inf\{2^n\mathrm{Cap}_1(A) - \nu(A) : A \in \mathcal{B}(E)\}$. Clearly $-\nu(E) \leq \alpha$. If $\alpha < 0$, then choose an open set B_1 with $2^n\mathrm{Cap}_1(B_1) - \nu(B_1) \leq \alpha/2$ and let $\alpha_1 = \inf\{2^n\mathrm{Cap}_1(A) - \nu(A) : A \in \mathcal{B}(E),\ A \subset E \setminus B_1\}$. Since for any $A \in \mathcal{B}(E)$ with $A \subset E \setminus B_1$ we have

$$\begin{aligned} \alpha &\leq 2^n\mathrm{Cap}_1(A \cup B_1) - \nu(A \cup B_1) \\ &\leq 2^n\mathrm{Cap}_1(A) - \nu(A) + 2^n\mathrm{Cap}_1(B_1) - \nu(B_1), \end{aligned}$$

it holds that $\alpha/2 \leq \alpha_1$. If $\alpha_1 < 0$, then choose a relatively open set $B_2 \subset E \setminus B_1$ with $2^n\mathrm{Cap}(B_2) - \nu(B_2) \leq \alpha_1/2$.

By repeating the same procedures, we find a sequence $\{B_1 \cup B_2 \cup \cdots \cup B_k\}$ of open sets such that

$$2^n\mathrm{Cap}(B_1 \cup \cdots \cup B_k) - \nu(B_1 \cup \cdots \cup B_k) \leq 0$$

$$2^n\mathrm{Cap}(A) - \nu(A) \geq 2^{-k}\alpha \quad \text{for every } A \subset E \setminus (B_1 \cup \cdots \cup B_k).$$

If we let $G_n' = \cup_{k=1}^n B_k$, then for every $A \subset E \setminus G_n'$,

$$2^n\mathrm{Cap}(A) \geq \nu(A) \quad \text{and} \quad 2^n\mathrm{Cap}(G_n') \leq \nu(G_n') \leq \nu(E) < \infty.$$

$G_n = \cup_{\ell=n}^{\infty} G'_\ell$ gives a desired sequence of decreasing open sets. In fact, $\mathrm{Cap}(G_n) \le 2^{-n+1}\nu(E) \to 0, n \to \infty$. Since ν charges no $\mathcal{E}$-polar set, $0 = \nu(\bigcap_n G_n) = \lim_{n\to\infty} \nu(G_n)$. □

THEOREM 2.3.15. *The following assertions are mutually equivalent for a positive measure μ on $(E, \mathcal{B}(E))$:*
(i) $\mu \in S$.
(ii) *There exists a nest $\{F_k\}$ satisfying*

$$\mu(F_k) < \infty,\ \mathbf{1}_{F_k}\cdot\mu \in S_0,\ \forall k \ge 1, \quad \mu\left(E \setminus \bigcup_{k=1}^{\infty} F_k\right) = 0. \qquad (2.3.16)$$

(iii) *There exists a nest $\{F_k\}$ satisfying*

$$\mathbf{1}_{F_k}\cdot\mu \in S_{00},\ \forall k \ge 1, \quad \mu\left(E \setminus \bigcup_{k=1}^{\infty} F_k\right) = 0. \qquad (2.3.17)$$

Proof. (i) ⇒ (ii): If μ is a finite measure charging no $\mathcal{E}$-polar set, then, by letting F_k be the complement of the open set G_k in Lemma 2.3.14(ii), we see that $\{F_k\}$ is a Cap_1-nest satisfying the second part of (2.3.16) as well as its first half on account of Lemma 2.3.14(i).

For a general smooth measure μ, let $\{E_\ell, \ell \ge 1\}$ be a nest associated with μ in Definition 2.3.13. For each ℓ, $\mu_\ell = \mathbf{1}_{E_\ell}\cdot\mu$ is finite and charging no $\mathcal{E}$-polar set, there exists a Cap_1-nest $\{F_k^{(\ell)}\}$ satisfying (2.3.16) for μ_ℓ. Let $F_k = \bigcup_{\ell=1}^{k}\{E_\ell \cap F_k^{(\ell)}\}$, then $\{F_k\}$ is an increasing sequence of closed sets satisfying the first half of (2.3.16).

Furthermore, for any compact set K,

$$\mathrm{Cap}_1(K \setminus F_k) \le \mathrm{Cap}_1(K \setminus E_\ell) + \mathrm{Cap}_1(E \setminus F_k^{(\ell)}), \quad \ell \le k,$$

and we conclude from Theorem 1.2.14 that $\{F_k\}$ is a nest by letting $k \to \infty$ and then $\ell \to \infty$. Since μ charges no $\mathcal{E}$-polar set, it satisfies the last part of (2.3.16) as well.

(ii) ⇒ (iii): For μ, take a nest $\{E_\ell\}$ satisfying (2.3.16) but with $\{E_\ell\}$ in place of $\{F_k\}$. For each ℓ, we associate with $\mu_\ell = \mathbf{1}_{E_\ell}\cdot\mu$ a Cap_1-nest $\{F_k^{(\ell)}\}$ satisfying condition of Lemma 2.3.9(ii) and then construct a nest $\{F_k\}$ in the similar way as in the proof of (i). Then $\mathbf{1}_{F_k}\cdot\mu \in S_{00}$ for every $k \ge 1$.

(iii) ⇒ (i): If μ satisfies (2.3.17) for some nest $\{F_k\}$, then μ charges no $\mathcal{E}$-polar set by Theorem 2.3.7 and it is smooth. □

In the rest of this section, we assume that the Dirichlet form $(\mathcal{E}, \mathcal{F})$ is transient. We will be concerned with 0-order counterparts of Theorem 2.3.7 and Lemma 2.3.12. Note that functions in the extended Dirichlet space $\mathcal{F}_e$ are not necessarily in $L^2(E; m)$ and so some modifications of the proofs are called for.

We say that a positive Radon measure μ on E is *of finite* 0*-order energy integral* ($\mu \in S_0^{(0)}$ in notation) if the inequality (2.3.5) holds with $\mathcal{E}_1$ on the right hand side being replaced by $\mathcal{E}$. Then the equation (2.3.6) with $\mathcal{E}_\alpha$ being replaced by $\mathcal{E}$ determines a unique function $U\mu \in \mathcal{F}_e$, which is called the 0*-order potential* of μ. Evidently $S_0^{(0)} \subset S_0$ and in particular any $\mu \in S_0^{(0)}$ charges no $\mathcal{E}$-polar set by Theorem 2.3.7(i).

For $\mu \in S_0^{(0)}$, we have also the 0-order counterpart of Theorem 2.3.7(ii)

$$\widetilde{\mathcal{F}}_e \subset L^1(E; \mu), \quad \mathcal{E}(U\mu, g) = \int_E \widetilde{g}(x)\mu(dx), \quad g \in \mathcal{F}_e, \tag{2.3.18}$$

where $\widetilde{\mathcal{F}}_e$ denotes the totality of quasi continuous functions belonging to $\mathcal{F}_e$, as well as the 0-order counterpart of a part of the characterization Lemma 2.3.5:

$$f = U\mu,\ \exists \mu \in S_0^{(0)} \iff \mathcal{E}(f, g) \geq 0 \quad \text{for } g \in \mathcal{F}_+ \cap C_c(E). \tag{2.3.19}$$

Furthermore, Lemma 2.3.10 is still valid with $\mathcal{F}$, $U_\alpha\mu$, S_0, and $\mathcal{E}_\alpha$ being replaced by $\mathcal{F}_e$, $U\mu$, $S_0^{(0)}$, and $\mathcal{E}$, respectively.

Lemma 2.3.16. *If $f_1, f_2 \in \mathcal{F}_e$ are* (0*-order*) *potentials, then so are $f_1 \wedge f_2$ and $f_1 \wedge 1$.*

Proof. This is the 0-order counterpart of Corollary 2.3.6. But analogous reasoning does not work because elements of $\mathcal{F}_e$ need not be in $L^2(E; m)$. Instead we first observe that equations (2.3.7) and (2.3.18) imply for $\mu \in S_0^{(0)}$

$$\|U_\alpha\mu - U_\beta\mu\|_{\mathcal{E}_\alpha}^2 \leq \langle \widetilde{U_\alpha\mu}, \mu\rangle - \langle \widetilde{U_\beta\mu}, \mu\rangle \qquad \text{for } \beta > \alpha > 0,$$

$$\|U\mu - U_\alpha\mu\|_{\mathcal{E}}^2 \leq \langle \widetilde{U\mu}, \mu\rangle - \langle \widetilde{U_\alpha\mu}, \mu\rangle \qquad \text{for } \alpha > 0,$$

which means that $\langle \widetilde{U_\alpha\mu}, \mu\rangle$ increases to a finite limit, and $U_\alpha\mu$ is $\mathcal{E}$-convergent to some $f \in \mathcal{F}_e$ as $\alpha \downarrow 0$. Moreover, since

$$\alpha(U_\alpha\mu, U_\alpha\mu) \leq \mathcal{E}_\alpha(U_\alpha\mu, U_\alpha\mu) = \langle \widetilde{U_\alpha\mu}, \mu\rangle \leq \langle \widetilde{U\mu}, \mu\rangle,$$

$\lim_{\alpha\downarrow 0} \alpha(U_\alpha\mu, g) = 0$ for every $g \in \mathcal{F}$. We then let $\alpha \downarrow 0$ in (2.3.6) to conclude that $f = U\mu$.

For $\mu, \nu \in S_0^{(0)}$, set $f_1 := U\mu$ and $f_2 := U\nu$. Then $U_\alpha\mu \wedge U_\alpha\nu$ is $\mathcal{E}$-weakly convergent to $f_1 \wedge f_2 \in \mathcal{F}_e$ as $\alpha \downarrow 0$ by the above observation and Exercise 2.1.13. Hence by letting $\alpha \downarrow 0$ in the inequality

$$\mathcal{E}_\alpha(U_\alpha\mu \wedge U_\alpha\nu, g) \geq 0 \qquad \text{for } g \in \mathcal{F}_+ \cap C_c(E),$$

we get $\mathcal{E}(f_1 \wedge f_2, g) \geq 0$. This shows that $f_1 \wedge f_2$ is a potential in view of (2.3.19). □

For an arbitrary set $B \subset E$ and a (0-order) potential $f \in \mathcal{F}_e$, the (0-order) reduced function $f_B^{(0)}$ of f relative to B can be specified as the unique element of $\mathcal{L}_B^{(0)} = \{g \in \mathcal{F}_e : \widetilde{g} \geq \widetilde{f} \text{ q.e. on } B\}$ minimizing $\|g\|_{\mathcal{E}}$. Since $f_B^{(0)}$ satisfies

$$\mathcal{E}(f_B^{(0)}, g) \geq 0, \quad \forall g \in \mathcal{F}_e,\ \widetilde{g} \geq 0 \text{ q.e. on } B, \tag{2.3.20}$$

$f_B^{(0)}$ is a potential by virtue of the 0-order version of Lemma 2.3.10.

LEMMA 2.3.17. *$f_B^{(0)}$ is a reduced function of a potential $f \in \mathcal{F}_e$ relative to a set B if and only $f_B^{(0)}$ is an element of $\mathcal{F}_e$ satisfying* (2.3.20) *and* (2.3.13).

Proof. We need to argue differently from Lemma 2.3.12. To get the necessity of (2.3.13), we observe that, together with $f_B^{(0)}$, $h = f_B^{(0)} \wedge f$ is a potential on account of Lemma 2.3.16. Thus there are $\mu, \nu \in S_0^{(0)}$ so that $f_B^{(0)} = U\mu$ and $h = U\nu$. We get from (2.3.18)

$$\begin{aligned}\mathcal{E}(h,h) &= \langle \widetilde{h}, \nu\rangle \leq \langle \widetilde{f}_B^{(0)}, \nu\rangle = \mathcal{E}(f_B^{(0)}, h)\\ &= \langle \widetilde{h}, \mu\rangle \leq \langle \widetilde{f}_B^{(0)}, \mu\rangle = \mathcal{E}(f_B^{(0)}, f_B^{(0)}).\end{aligned}$$

Consequently, $f_B^{(0)} = h \leq f$. □

If we let

$$\mathcal{F}_{e,E\setminus B} = \{g \in \mathcal{F}_e : \widetilde{g} = 0 \text{ q.e. on } B\}, \tag{2.3.21}$$

then $\mathcal{F}_{e,E\setminus B}$ is a closed linear subspace of the Hilbert space $(\mathcal{F}_e, \mathcal{E})$. Denote by $\mathcal{H}_B$ its orthogonal complement. Denoting by $P_{\mathcal{H}_B}$ the orthogonal projection to the space $\mathcal{H}_B$, we have

$$P_{\mathcal{H}_B} f = f_B^{(0)}, \tag{2.3.22}$$

because $f = (f - f_B^{(0)}) + f_B^{(0)}$ gives the orthogonal decomposition of f in view of Lemma 2.3.17.

2.4. LOCAL PROPERTIES

Let E be a Hausdorff topological space with the Borel σ-field $\mathcal{B}(E)$ being generated by the continuous functions on E and m be a σ-finite measure with $\text{supp}[m] = E$. For an m-measurable function f defined and finite m-a.e. on E, we denote by $\text{supp}[f]$ its support, namely, the support of the measure $f \cdot m$. Let $(\mathcal{E}, \mathcal{F})$ be a Dirichlet form on $L^2(E; m)$. In this section, we present some equivalent conditions for $(\mathcal{E}, \mathcal{F})$ to be local.

For a closed subset $F \subset E$, recall the subspace $\mathcal{F}_F$ of $\mathcal{F}$ defined by (1.2.2), which is a closed linear subspace of the Hilbert space $(\mathcal{F}, \mathcal{E}_1)$. Let $\mathcal{H}^1_{F^c}$ be its $\mathcal{E}_1$-orthogonal complement. The $\mathcal{E}_1$-orthogonal projection of $f \in \mathcal{F}$ to the space $\mathcal{H}^1_{F^c}$ is denoted by f_{F^c}, which is the unique element of $\mathcal{F}$ such that $f_{F^c} = f$ m-a.e. on F^c and $\mathcal{E}_1(f_{F^c}, g) = 0$ for any $g \in \mathcal{F}_F$. $f - f_{F^c}$ is the $\mathcal{E}_1$-orthogonal projection of f into $\mathcal{F}_F$. We start with the following lemma.

LEMMA 2.4.1. *For every $f \in \mathcal{F}$ and an $\mathcal{E}$-nest $\{F_k, k \geq 1\}$, $f - f_{F_k^c}$ is $\mathcal{E}_1$-convergent to f as $k \to \infty$.*

Proof. For notational convenience, let $f_k := f - f_{F_k^c} \in \mathcal{F}_{F_k}$. By the above characterizations, we see for $j > k$ that $(f_{F_j^c})_{F_k^c} = f_{F_j^c}$, $f_k = f_j - (f_j)_{F_k^c}$ and $\mathcal{E}_1(f_k, f_j) = \mathcal{E}_1(f_k, f_k)$. Therefore

$$\mathcal{E}_1(f_j - f_k, f_j - f_k) = \mathcal{E}_1(f_j, f_j) - \mathcal{E}_1(f_k, f_k).$$

It follows that $\{f_k, k \geq 1\}$ is a Cauchy sequence in $(\mathcal{F}, \mathcal{E}_1)$ and thus it is $\mathcal{E}_1$-convergent to some $g \in \mathcal{F}$. Since $f - f_k$ is $\mathcal{E}_1$-orthogonal to $\mathcal{F}_{F_k}$, we have

$$\mathcal{E}(f - g, \varphi) = \lim_{k \to \infty} \mathcal{E}(f - f_k, \varphi) = 0 \quad \text{for every } \varphi \in \bigcup_{k \geq 1} \mathcal{F}_{F_k}.$$

As $\cup_{k \geq 1} \mathcal{F}_{F_k}$ is $\mathcal{E}_1$-dense in $\mathcal{F}$, we conclude from above that $f = g$ and this completes the proof of the lemma. □

THEOREM 2.4.2. *Let $(\mathcal{E}, \mathcal{F})$ be a quasi-regular Dirichlet form on $L^2(E; m)$ and $(\mathcal{C}, \mathcal{F})$ be a symmetric form with*

$$0 \leq \mathcal{C}(u, u) \leq \mathcal{E}(u, u) \qquad \text{for every } u \in \mathcal{F}.$$

Then the following two conditions are equivalent:

(i) *For every $u, v \in \mathcal{F}$ with disjoint compact support, $\mathcal{C}(u, v) = 0$,*
(ii) *For every $u, v \in \mathcal{F}$ with $uv = 0$ m-a.e. on E, $\mathcal{C}(u, v) = 0$.*

Proof. Clearly (ii) implies (i). Since $(\mathcal{E}, \mathcal{F})$ is quasi-regular, there exist a Borel $\mathcal{E}$-polar set $N \subset E$ and a Borel standard process X on $E \setminus N$ properly associated with $(\mathcal{E}, \mathcal{F})$ in view of Theorem 1.5.2. Denote by $\{R_\alpha; \alpha > 0\}$ the resolvent kernel of X. We let $h = R_1 f$ for a strictly positive bounded m-integrable function f on E. Then h is an $\mathcal{E}$-quasi-continuous element of $\mathcal{F}$ and strictly positive on $E \setminus N$.

Suppose (i) holds. By considering u^+, u^-, v^+ and v^- separately, without loss of generality we may and do assume that $u, v \in \mathcal{F}$ are two non-negative functions with $uv = 0$ m-a.e. on E. Also we may take as u, v $\mathcal{E}$-quasi-continuous m-versions. There is an m-regular $\mathcal{E}$-nest $\{F_k, k \geq 1\}$ of compact sets so that $\{u, v, h\} \subset C(\{F_k\})$ with $N \subset E \setminus \cup_k F_k$. Then $\delta_k := \inf_{x \in F_k} h(x)$ is positive for every $k \geq 1$.

By Lemma 2.4.1, $u - u_{F_k^c}$ and $v - v_{F_k^c}$ are $\mathcal{E}_1$-convergent to u and v, respectively, as $k \to \infty$. Let $u_k := (u - u_{F_k^c})^+ \wedge u$ and $v_k := (v - v_{F_k^c})^+ \wedge v$. Then it follows from Lemma 1.1.11(ii) that u_k and v_k are $\mathcal{E}_1$-convergent to u and v, respectively. For every $\varepsilon > 0$, it is clear that

$$\text{supp}[(u_k - \varepsilon h)^+] \subset F_k \cap \{u \geq \varepsilon \delta_k\}$$

and

$$\text{supp}[(v_k - \varepsilon h)^+] \subset F_k \cap \{v \geq \varepsilon \delta_k\}.$$

Since $uv = 0$ m-a.e., $u(x)v(x) = 0$ for every $x \in F_k$ in view of Theorem 1.3.7. It follows that $(u_k - \varepsilon h)^+$ and $(v_k - \varepsilon h)^+$ have disjoint compact support. Thus by (i),

$$\mathcal{C}((u_k - \varepsilon h)^+, (v_k - \varepsilon h)^+) = 0 \qquad \text{for every } k \geq 1 \text{ and } \varepsilon > 0.$$

Letting $\varepsilon \to 0$ and then $k \to \infty$, we obtain $\mathcal{C}(u, v) = 0$. □

THEOREM 2.4.3. *Let $(\mathcal{E}, \mathcal{F})$ be a regular Dirichlet form on $L^2(E; m)$ where E is a locally compact separable metric space and m is a positive Radon measure on E with* supp$[m] = E$. *Let $(\mathcal{C}, \mathcal{F})$ be a symmetric form with*

$$0 \leq \mathcal{C}(u, u) \leq \mathcal{E}(u, u) \qquad \textit{for every } u \in \mathcal{F}.$$

Then the following are equivalent:

(i) $\mathcal{C}(u, v) = 0$ *whenever $u, v \in \mathcal{F}$ have compact support with v being constant in an open neighborhood of* supp$[u]$.
(ii) *For $u, v \in \mathcal{F}$ with v being constant in an open neighborhood of* supp$[u]$, $\mathcal{C}(u, v) = 0$.
(iii) *For $u, v \in \mathcal{F}$ with $u(v - c) = 0$ m-a.e. on E for some constant $c \in \mathbb{R}$,* $\mathcal{C}(u, v) = 0$.

Proof. Clearly the implications (iii) $\Rightarrow$ (ii) $\Rightarrow$ (i) are valid so that it suffices to show the implication (i) $\Rightarrow$ (iii). When $c = 0$, this is already shown in the preceding theorem.

Let $u, v \in \mathcal{F}$ with $u(v - c) = 0$ m-a.e. on E for $c \neq 0$. Without loss of generality, assume $c = 1$, u is bounded and $u \geq 0$ (otherwise consider $u = u^+ \wedge n$ and $u^- \wedge n$, respectively). Since $(\mathcal{E}, \mathcal{F})$ is regular, there is a sequence $\{u_k, k \geq 1\} \subset \mathcal{F} \cap C_c(E)$ such that u_k is $\mathcal{E}_1$-convergent to u. Fix $k \geq 1$. Let $h \in \mathcal{F}_+ \cap C_c(E)$ so that $h = 1$ on an open neighborhood of supp$[u_k]$. By (i), $\mathcal{C}(u_k \wedge u, h) = 0$. $\mathcal{C}(u_k^+ \wedge u, h) = 0$. On the other hand,

$$|(u_k^+ \wedge u)(v - h)| \leq |(u_k^+ \wedge u)(v - 1)| + |(u_k^+ \wedge u)(1 - h)| = 0 \quad m\text{-a.e. on } E.$$

So by the proof of Theorem 2.4.2, we have $\mathcal{C}(u_k^+ \wedge u, v - h) = 0$ and hence $\mathcal{C}(u_k^+ \wedge u, v) = 0$ for every $k \geq 1$. As $u_k^+ \wedge u$ is $\mathcal{E}_1$-convergent to u, after taking $k \to \infty$ we have $\mathcal{C}(u, v) = 0$. □

Remark 2.4.4. We may define a symmetric form $(\mathcal{E}, \mathcal{D}(\mathcal{E}))$ to be *strongly local* if $\mathcal{E}(u, v) = 0$ whenever $u, v \in \mathcal{D}(\mathcal{E})$ with $u(v - c) = 0$ m-a.e. on E for some constant c. Note that this definition is invariant under quasi-homeomorphism. So under this definition, Beurling-Deny decomposition (see Theorem 4.3.3 in Chapter 4 for its statement) carries over to quasi-regular Dirichlet forms by using the fact that it is quasi-homeomorphic to a regular Dirichlet form on a locally compact separable metric space. See Remark 4.3.5(iii) below. □

Chapter Three

SYMMETRIC HUNT PROCESSES AND REGULAR DIRICHLET FORMS

As is clearly embodied by Theorem 1.4.3, three theorems of Section 1.5, and Theorem 3.1.13 below, the study of general symmetric Markov processes can be essentially reduced to the study of a symmetric Hunt process associated with a regular Dirichlet form. So without loss of generality, we assume throughout this chapter, except for the last parts of Sections 3.1 and 3.5, that E is a locally compact separable metric space, m is a positive Radon measure on E with $\mathrm{supp}[m] = E$, and $X = (X_t, \mathbf{P}_x)$ is an m-symmetric Hunt process on $(E, \mathcal{B}(E))$ whose Dirichlet form $(\mathcal{E}, \mathcal{F})$ is regular on $L^2(E; m)$. (By redefining the process X at a properly exceptional set if needed, we may assume that X starts from every point in E.) We shall adopt without any specific notices those potential theoretic terminologies and notations that are formulated in Section 2.3 for the regular Dirichlet form $(\mathcal{E}, \mathcal{F})$.

Throughout this chapter, we adopt the convention that any numerical function on E is extended to the one-point compactification $E_\partial = E \cup \{\partial\}$ by setting its value at ∂ to be zero.

3.1. RELATIONS BETWEEN PROBABILISTIC AND ANALYTIC CONCEPTS

Denote by $\mathcal{O}_0$ the family of all open subsets of E with finite capacity Cap_1 with respect to the Dirichlet form $(\mathcal{E}, \mathcal{F})$. For $A \in \mathcal{O}_0$, let e_A be its 1-equilibrium potential and p_A^1 the 1-order hitting probability of A for the Hunt process X on E defined by

$$p_A^1(x) := \mathbf{E}_x\left[e^{-\sigma_A}\right], \quad x \in E.$$

Lemma 3.1.1. *$e_A = p_A^1$ $[m]$ for $A \in \mathcal{O}_0$.*

Proof. According to Lemma 1.1.14, the L^2-semigroup $\{T_t; t > 0\}$ generated by $(\mathcal{E}, \mathcal{F})$ is related to the transition function $\{P_t, t \geq 0\}$ of X by $T_t f = P_t f$ $[m]$ for every $t > 0$ and $f \in \mathcal{B}_+(E) \cap L^2(E; m)$. e_A is 1-excessive relative to $\{T_t\}$ in view of Theorem 1.2.5, while Lemma A.2.4 says that p_A^1 is 1-excessive relative to $\{P_t, t \geq 0\}$, and so it is relative to $\{T_t\}$. Further, $p_A^1(x) = 1$ for every $x \in A$. Therefore, it suffices to show the inequality

$$p_A^1 \leq e_A \quad [m], \tag{3.1.1}$$

because Lemma 1.2.3 then implies that $p_A^1 \in \mathcal{F}$ and $\mathcal{E}_1(p_A^1, p_A^1) \leq \mathcal{E}_1(e_A, e_A)$, and the variational characterization Theorem 1.2.5(i) of e_A yields the desired identity.

To prove (3.1.1), we take a Borel modification $\widetilde{e}_A$ of e_A with $\widetilde{e}_A(x) = 1$ for every $x \in A$ and let $Y_t(\omega) = e^{-t}\widetilde{e}_A(X_t(\omega))$, $t \geq 0$, $\omega \in \Omega$. Then, for any non-negative Borel measurable function h with $\int_E h dm = 1$, the stochastic process $(Y_t, \mathcal{F}_t^0, \mathbf{P}_{h\cdot m})_{t \geq 0}$ is a supermartingale. In fact, by the Markov property of X, we have for any $0 \leq s < t$,

$$\mathbf{E}_{h\cdot m}\left[Y_t | \mathcal{F}_s^0\right] = e^{-s}e^{-(t-s)}P_{t-s}\widetilde{e}_A(X_s) \leq e^{-s}\widetilde{e}_A(X_s), \ \mathbf{P}_{h\cdot m}\text{-a.s.}$$

The last inequality follows from $e^{-(t-s)}P_{t-s}\widetilde{e}_A \leq \widetilde{e}_A$ $[m]$.

For a finite set $\Lambda \subset (0, \infty)$, denote the minimum and maximum of Λ by a, b, respectively, and define

$$\sigma(\Lambda; A) = \min\{t \in \Lambda : X_t \in A\} \quad \text{with convention } \min\emptyset := b.$$

By the optional sampling theorem for the supermartingale,

$$\mathbf{E}_{h\cdot m}\left[e^{-\sigma(\Lambda;A)}; \sigma(\Lambda; A) < b\right] \leq \mathbf{E}_{h\cdot m}\left[Y_{\sigma(\Lambda;A)}\right] \leq \mathbf{E}_{h\cdot m}[Y_a] \leq (h, \widetilde{e}_A).$$

Letting $\Lambda \uparrow \mathbb{Q}_+ \cap (0, b)$ and then $b \uparrow \infty$, we get $(h, p_A^1) \leq (h, e_A)$, which yields (3.1.1). □

Lemma 3.1.2. *The following two conditions are equivalent for a decreasing sequence $\{A_n\}$ of open sets of finite capacity:*

$$\lim_{n\to\infty} \mathrm{Cap}_1(A_n) = 0, \tag{3.1.2}$$

$$\lim_{n\to\infty} p_{A_n}^1(x) = 0, \quad m\text{-a.e. } x \in E. \tag{3.1.3}$$

Proof. By the preceding lemma, $\mathrm{Cap}_1(A_n) = \|p_{A_n}^1\|_{\mathcal{E}_1}^2 \geq \|p_{A_n}^1\|_2^2$. Moreover, Theorem 1.2.9(iv) says that $\{p_{A_n}^1\}$ is $\mathcal{E}_1$-Cauchy. Hence we readily see the equivalence of the two stated conditions. □

Notice that the m-symmetry of X implies that the measure m is excessive with respect to $\{P_t; t \geq 0\}$ in the sense of Section A.2.2. We say a subset $N \subset E$ is *m-polar* if there is a nearly Borel set $N_1 \supset N$ such that $\mathbf{P}_m(\sigma_{N_1} < \infty) = 0$, where $\sigma_{\widetilde{N}} := \inf\{t > 0 : X_t \in N_1\}$.

THEOREM 3.1.3. *A subset of E is $\mathcal{E}$-polar if and only if it is m-polar for X.*

Proof. If $N \subset E$ is $\mathcal{E}$-polar, then there exists a decreasing sequence $\{A_n\}$ of open sets containing N satisfying (3.1.2). The set $B = \cap_{n=1}^{\infty} A_n$ then contains N and satisfies $p_B^1(x) = 0$ $[m]$ by the preceding lemma. Hence N is m-polar.

Here we make a general remark that, for a compact set $K \subset E$ and a sequence $\{D_n\}$ of relatively compact open sets such that

$$D_n \supset \overline{D}_{n+1},\ n \geq 1,\ \bigcap_{n=1}^{\infty} D_n = K, \tag{3.1.4}$$

we have

$$\lim_{n\to\infty} \sigma_{D_n} = \dot{\sigma}_K\ \mathbf{P}_x\text{-a.s. for any } x \in E \quad \text{and} \quad \dot{\sigma}_K = \sigma_K\ \mathbf{P}_m\text{-a.e.}, \tag{3.1.5}$$

where $\dot{\sigma}_K$ is the entrance time of K defined by (A.1.20). Indeed, if $\sigma = \lim_{n\to\infty} \sigma_{D_n} = \lim_{n\to\infty} \dot{\sigma}_{D_n}$, then clearly $\sigma \leq \sigma_K$. On the other hand, by the quasi-left-continuity (A.1.31) of the Hunt process X, we have $X_\sigma = \lim_{n\to\infty} X_{\sigma_{D_n}} \in \cap_n \overline{D}_n = K$, almost surely on $\{\sigma < \infty\}$, and consequently $\dot{\sigma}_K = \sigma$. If $x \notin K \setminus K^r$, then $\dot{\sigma}_K = \sigma_K$, $\mathbf{P}_x$-a.s. But, by Lemma A.2.18, the set $K \setminus K^r$ is semipolar and thus of potential zero. Hence it is m-negligible by Theorem A.2.13 and we get the second conclusion in (3.1.5).

Now, if K is a compact m-polar set, then, for a sequence $\{D_n, n \geq 1\}$ of open sets as above, we have $\lim_{n\to\infty} p_{D_n}^1 = p_K^1 = 0$ $[m]$ by (3.1.5). Since $\mathrm{Cap}_1(D_n) < \infty$, we can conclude from Lemma 3.1.2 that

$$\mathrm{Cap}_1(K) \leq \lim_{n\to\infty} \mathrm{Cap}_1(D_n) = 0.$$

If N is an arbitrary m-polar set, then there exists an m-polar Borel set $\widetilde{N}$ containing N by Theorem A.2.13. Observe that every compact subset of $\widetilde{N}$ is m-polar, and hence is $\mathcal{E}$-polar. On account of (1.2.6), $\widetilde{N}$ as well as N is $\mathcal{E}$-polar. □

Based on Theorem 3.1.3, we can and we will use the term *quasi everywhere* (q.e. in abbreviation) for the two equivalent meanings: "except for an m-polar set" or "except for an $\mathcal{E}$-polar set".

We call an increasing sequence $\{F_k\}$ of closed sets a *strong nest* if

$$\mathbf{P}_x\left(\lim_{k\to\infty} \sigma_{E\setminus F_k} < \infty\right) = 0 \qquad \text{for q.e. } x \in E. \tag{3.1.6}$$

By virtue of the second statement of the next theorem, any strong nest is indeed a nest.

THEOREM 3.1.4. *Let $\{F_k\}$ be an increasing sequence of closed subsets of E.*
(i) *$\{F_k\}$ is a* Cap_1*-nest if and only if* $\mathrm{Cap}_1(E \setminus F_n) < \infty$ *for some n and $\{F_k\}$ is a strong nest.*

(ii) $\{F_k\}$ *is a nest if and only if*

$$\mathbf{P}_x\left(\lim_{k\to\infty}\sigma_{E\setminus F_k}<\zeta\right)=0 \quad \textit{for q.e. } x\in E. \tag{3.1.7}$$

Proof. (i) Denote the complement of F_k by A_k and let

$$p(x):=\lim_{k\to\infty}p^1_{A_k}(x) \quad \text{for } x\in E.$$

By Lemma 3.1.2, the condition that $\{F_k\}$ is a Cap_1-nest is, under the finiteness of $\mathrm{Cap}_1(A_n)$ for some n, equivalent to $p=0$ $[m]$, which is also equivalent to the validity of (3.1.6) for m-a.e. $x\in E$. Hence it suffices to derive $p=0$ q.e. from $p=0$ m-a.e.

Since $\{A_k\}$ are open sets, $p^1_{A_k}(x)$ are Borel measurable functions by Theorem A.1.9 and so is $p(x)$. For any $\varepsilon>0$, consider a compact subset K of the Borel set $\{x\in E: p(x)\geq\varepsilon\}$ and let $\mathbf{H}^1_K$ be the 1-order hitting distribution of K defined by (A.2.2). From Lemma A.2.4, $\mathbf{H}^1_K p^1_{A_k}\leq p^1_{A_k}$, $k\geq 1$, and we get $\mathbf{H}^1_K p\leq p$ by letting $k\to\infty$. Hence $\varepsilon p^1_K(x)\leq p(x)$ for $x\in E$, which means that K is m-polar if $p=0$ m-a.e. Therefore, Theorem 3.1.3 and (1.2.6) imply that the set $\{p\geq\varepsilon\}$ as well as the set $\{p>0\}$ is $\mathcal{E}$-polar.

(ii) On account of Theorem 1.3.14 and Lemma 3.1.2, $\{F_k\}$ is a nest if and only if, for any relatively compact open set D, the decreasing sequence $\{A_k=D\setminus F_k,\ k\geq 1\}$ of open sets satisfies (3.1.3), which, in view of the proof of (i), is also equivalent to the validity of

$$\mathbf{P}_x\left(\lim_{k\to\infty}\sigma_{D\setminus F_k}<\infty\right)=0 \quad \text{for q.e. } x\in E. \tag{3.1.8}$$

It suffices to show the equivalence of condition (3.1.7) and property (3.1.8) holding for any relatively compact open set D. Assume (3.1.7) holds. We fix an arbitrary $x\in E$ for which (3.1.7) is true. Suppose (3.1.8) fails to be true for this fixed x and for some relatively compact open set D, namely, by setting $\sigma=\lim_{k\to\infty}\sigma_{D\setminus F_k}$, $\mathbf{P}_x(\sigma<\infty)=\delta$, we assume $\delta>0$. On account of the quasi-left-continuity (A.1.31) of the Hunt process X, we then have $\mathbf{P}_x(X_\sigma\in\overline{D}, \sigma<\infty)=\delta$, and consequently $\mathbf{P}_x(\lim_{k\to\infty}\sigma_{E\setminus F_k}\leq\sigma<\zeta)\geq\delta$, arriving at a contradiction.

Conversely, (3.1.7) can be derived from (3.1.8) by taking a sequence $\{D_\ell\}$ of relatively compact open sets increasing to E and by using the inequality $\sigma_{E\setminus F_k}\geq\sigma_{D_\ell\setminus F_k}\wedge\sigma_{E\setminus D_\ell}$. □

A nearly Borel set $A\subset E$ is called *X-invariant* for the Hunt process X if

$$\mathbf{P}_x(\sigma_{E\setminus A}\wedge\widehat{\sigma}_{E\setminus A}<\infty)=0 \quad \text{for every } x\in A,$$

where $\widehat{\sigma}_{E\setminus A}:=\inf\{t>0: X_{t-}\in E\setminus A\}$ with the convention that $\inf\emptyset=\infty$. According to Lemma A.1.27 and Remark A.1.30, the restriction X_A of X to its invariant set A is a Hunt process again.

Parallel to Definition A.2.12 of an m-inessential set for a Borel right process, we say that a set $N \subset E$ is *properly exceptional* for the Hunt process X if N is m-negligible, nearly Borel and $E \setminus N$ is X-invariant. Clearly a properly exceptional set for X is m-polar.

On account of Lemma A.1.32, the proper exceptionality of a set with respect to a Hunt process X is actually a synonym for its m-inessentiality with respect to X regarded as a Borel right process. Therefore, the next theorem is a special case of Theorem A.2.15 formulated for a Borel right process. Yet we give an alternative direct proof of it for the present symmetric Hunt process.

THEOREM 3.1.5. *Any m-polar set N is contained in a Borel properly exceptional set for X.*

Proof. If N is m-polar, then $\text{Cap}_1(N) = 0$ by Theorem 3.1.3 and hence there exists a Cap_1-nest $\{F_k\}$ such that N is contained in the set $B_0 = \cap_{k=1}^{\infty}(E \setminus F_k)$, which is a G_δ-set of zero capacity. Since $\sigma_{E\setminus F_k} \le \sigma_{B_0}$, $k \ge 1$, and also $\sigma_{E\setminus F_k} = \widehat{\sigma}_{E\setminus F_k} \le \widehat{\sigma}_{B_0}$ for $k \ge 1$, it follows from Theorem 3.1.4(i) and the property (3.1.6) that $\mathbf{P}_x(\sigma_{B_0} \wedge \widehat{\sigma}_{B_0} < \infty) = 0$ q.e. Therefore, we can find a strong nest $\{F'_k\}$ such that $B_1 = \cap_{k=1}^{\infty}(E \setminus F'_k)$ contains B_0 and the above identity holds for any $x \in E \setminus B_1$.

Repeating this procedure, we can construct an increasing sequence B_n of G_δ-sets of zero capacity satisfying

$$\mathbf{P}_x(\sigma_{B_n} \wedge \widehat{\sigma}_{B_n} < \infty) = 0 \quad \text{for every } x \in E \setminus B_{n+1}.$$

$B = \cup_{n=1}^{\infty} B_n$ is then a Borel properly exceptional set containing N. □

DEFINITION 3.1.6. A numerical function f defined q.e. on E is called *finely continuous q.e.* if there exists a properly exceptional set N such that $f\big|_{E\setminus N}$ is nearly Borel measurable and finely continuous with respect to the restricted Hunt process $X\big|_{E\setminus N}$.

For simplicity, we phrase this property as f is nearly Borel and finely continuous on $E \setminus N$ for a properly exceptional set N. We note that if f has this property, then so does it for any properly exceptional set containing N on account of Exercise A.1.31.

THEOREM 3.1.7. (i) *If a function f on E is quasi continuous, then f is finely continuous q.e. and finite q.e. Furthermore, there exists a Borel properly exceptional set N such that f is a finite Borel measurable function on $E \setminus N$ and, for any $x \in E \setminus N$,*

$$\mathbf{P}_x\left(\lim_{t'\downarrow t} f(X_{t'}) = f(X_t) \quad \textit{for every } t \ge 0\right) = 1, \tag{3.1.9}$$

$$\mathbf{P}_x\left(\lim_{t'\uparrow t} f(X_{t'}) = f(X_{t-}) \quad \textit{for every } t \in (0,\zeta)\right) = 1, \tag{3.1.10}$$

$$\begin{aligned} &\mathbf{P}_x(\lim_{t'\uparrow\zeta} f(X_{t'}) = f(X_{\zeta-}),\ \zeta < \infty,\ X_{\zeta-} \in E) \\ &= \mathbf{P}_x(\zeta < \infty,\ X_{\zeta-} \in E). \end{aligned} \tag{3.1.11}$$

If in addition $f \in \mathcal{F}$, then the time interval $(0,\zeta)$ in (3.1.10) can be strengthened to $(0,\infty)$.
(ii) *If f is finely continuous q.e. and $f \in \mathcal{F}$, then f is quasi continuous.*

Proof. (i). If f is quasi continuous, then by Theorem 1.3.14(iii) there is a Cap_1-nest $\{F_k\}$ such that $f \in C(\{F_k\})$. Denote by N_0 the m-polar set in (3.1.6) for $\{F_k\}$. We can then find a Borel properly exceptional set N containing $N_0 \cup (\cap_{k=1}^{\infty}(E \setminus F_k))$ by Theorem 3.1.5. Clearly f is finite and Borel measurable on $E \setminus N$. For $x \in E \setminus N$, $\lim_{k\to\infty} \sigma_{E\setminus F_k} = \infty$ $\mathbf{P}_x$-a.s., and thus (3.1.9), (3.1.10), and (3.1.11) are valid.

Applying Theorem A.2.9 to the Hunt process $X|_{E\setminus N}$, we conclude from (3.1.9) that f is finely continuous on $E \setminus N$.

If f is quasi continuous and $f \in \mathcal{F}$, then f equals q.e. to a quasi continuous function in the restricted sense relative to a Cap_1-nest in view of Lemma 1.3.15. This together with Theorem 3.1.4(i) yields the last assertion of (i).
(ii). Suppose f is finely continuous q.e. and $f \in \mathcal{F}$. By Lemma 1.3.15, f admits a quasi continuous m-version $\widetilde{f}$. By (i), $\widetilde{f}$ is finely continuous q.e. and $\widetilde{f} = f$ $[m]$. On account of Theorem A.2.15, there exists a Borel m-inessential set N for X as a right process such that both $\widetilde{f}$ and f are finely continuous with respect to the right process $X_{E\setminus N}$. Thus $\widetilde{f} = f$ q.e. by Theorem A.2.13(iv). Since f equals the quasi continuous function $\widetilde{f}$ except for an $\mathcal{E}$-polar set, f itself is quasi continuous. □

The condition "$f \in \mathcal{F}$" in statement (ii) of the above theorem will be removed in Theorem 3.3.3.

Remark 3.1.8. For a right process X having m as an excessive measure, it follows from Theorem A.2.9, Theorem A.2.13, Lemma A.2.14, and Theorem A.2.15 that a function f is finite q.e. and finely continuous q.e. if and only if there exists an m-inessential set N for the right process X such that f is a finite Borel measurable function on $E \setminus N$ and f satisfies (3.1.9) for every $x \in E \setminus N$.

For the present m-symmetric Hunt process X, f also satisfies (3.1.10), a reflection of the symmetry of X. This property is not necessarily true for non-symmetric right processes. For instance, the deterministic uniform motion X to the right on $\mathbb{R}$ is a right process possessing the Lebesgue measure m as an excessive measure. In this case, a real function f on $\mathbb{R}$ is finely continuous q.e. with respect to X if and only if f is right continuous as a real function and (3.1.10) is not satisfied unless f is left continuous.

If two numerical functions f, g on E satisfy $f = g\ [m]$, then f is said to be a *version* or an *m-version* of g. Let us consider the resolvent kernel $\{R_\alpha; \alpha > 0\}$ (resp. $\{G_\alpha; \alpha > 0\}$) of $\{P_t\}$ (resp. $\{T_t; t \geq 0\}$) defined by (1.1.27) (resp. (1.1.1)). According to (1.1.28), $R_\alpha f$ is a version of $G_\alpha f$ for $f \in b\mathcal{B}(E) \cap L^2(E; m)$. Hence, for a non-negative universally measurable function $f \in L^2(E; m)$, $R_\alpha f$ is a version of $G_\alpha f \in \mathcal{F}$. Since $R_\alpha f$ is α-excessive for X and finely continuous by Theorem A.2.9, $R_\alpha f$ is a quasi continuous version of $G_\alpha f$ by virtue of Theorem 3.1.7(ii), yielding (ii) of the following proposition. □

PROPOSITION 3.1.9. *Let f be a non-negative universally measurable function on E belonging to $L^2(E; m)$.*
(i) *For each $t > 0$, $P_t f$ is a quasi continuous version of $T_t f$.*
(ii) *For each $\alpha > 0$, $R_\alpha f$ is a quasi continuous version of $G_\alpha f$.*

Proof. For a non-negative universally measurable function f, there are $f_1, f_2 \in \mathcal{B}_+(E)$ so that $0 \leq f_1 \leq f \leq f_2$ and $m(\{f_2 > f_1\}) = 0$. Thus $P_t f_1 \leq P_t f \leq P_t f_2$ with $m(\{P_t f_2 > P_t f_1\}) = 0$. If $P_t f_i$, $i = 1, 2$, are quasi continuous, then so is $P_t f$ by Lemma 1.3.6 and Theorem 1.3.7. Hence it suffices to prove (i) for every Borel measurable function f.

Denote by $\mathcal{G}$ the family of all non-negative Borel measurable functions $f \in L^2(E; m)$ for which $P_t f$ is quasi continuous. $\mathcal{G}$ contains the family $C_c^+(E)$ of all non-negative continuous functions with compact support. In fact, if $f \in C_c^+(E)$, then for $x \in E$,

$$\alpha R_\alpha(P_t f)(x) = P_t(\alpha R_\alpha f)(x) \to P_t f(x) \quad \text{as } \alpha \to \infty.$$

Since this convergence also takes place strongly in $\mathcal{F}$ by virtue of Section A.4(**v**) and $R_\alpha(P_t f) \in \mathcal{G}$ by (ii), we get $f \in \mathcal{G}$ in view of Theorem 1.3.3.

If $\{f_n, n \geq 1\}$ is a sequence of functions in $\mathcal{G}$ increasing pointwise to $f \in L^2(E; m)$, then $P_t f_n$ is $\mathcal{E}_1$-convergent to $P_t f$ in the space $\mathcal{G}$ by Section A.4(**v**) again, and thus $f \in \mathcal{G}$.

We can now conclude that $\mathcal{G} = \mathcal{B}_+(E) \cap L^2(E; m)$, because, for any relatively compact open set $D \subset E$, the family $\mathcal{C} = \{A \subset D : A \in \mathcal{B}(E), \mathbf{1}_A \in \mathcal{G}\}$ becomes a Dynkin class containing all open subsets of D and so $\mathcal{C} = \mathcal{B}(D)$ by virtue of Proposition A.1.2. □

We present three important implications of Proposition 3.1.9.

THEOREM 3.1.10. *Any semipolar set of X is $\mathcal{E}$-polar.*

Proof. For any open set $D \in \mathcal{O}_0$, the 1-excessive function p_D^1 is an m-version of the 1-equilibrium potential $e_D \in \mathcal{F}$ in view of Lemma 3.1.1. By the preceding proposition, $e^{-t} P_t p_D^1$ is quasi continuous for each $t > 0$. As $t \downarrow 0$, it increases to p_D^1 pointwisely and it is also $\mathcal{E}_1$-convergent to e_D by Section A.4(**v**). Hence p_D^1 is a quasi continuous version of e_D by Theorem 1.3.3.

According to Definition A.2.6, a semipolar set is a set contained in a countable union of thin sets. Hence it is enough to show that any Borel measurable thin set is $\mathcal{E}$-polar. In view of (1.2.6), we may show this for a compact thin set K so that $K^r = \emptyset$. Choose for K relatively compact open sets $\{D_n\}$ satisfying (3.1.4). By (3.1.5), $p^1_{D_n}(x)$ decreases to $\dot{p}^1_K(x) := \mathbf{E}_x[e^{-\dot{\sigma}_K}]$ as $n \to \infty$ for each $x \in E$. Since $\mathcal{E}_1(p^1_{D_n}, p^1_{D_n}) = \text{Cap}_1(D_n)$, $p^1_{D_n}$ is $\mathcal{E}_1$-convergent as $n \to \infty$ by Theorem 1.2.9. Therefore, in view of Theorem 1.3.3, $\dot{p}^1_K$ is a quasi continuous function in $\mathcal{F}$. Since $e^{-t}P_t\dot{p}^1_K(x)$ increases to $p^1_K(x)$ as $t \downarrow 0$ for each $x \in E$, we see as above that p^1_K is also a quasi continuous version of $\dot{p}^1_K$. Consequently, $\mathcal{E}$-polar. $\dot{p}^1_K = p^1_K < 1$ $\mathcal{E}$-q.e., yielding that K is $\mathcal{E}$-polar. □

Theorem 3.1.10 is not necessarily valid for non-symmetric right processes. For instance, when X is the deterministic uniform motion to the right on $\mathbb{R}$, each one-point set $K = \{x\}$, $x \in \mathbb{R}$, has no regular point for X and hence K is semipolar, while K is neither polar nor m-polar (m is the Lebesgue measure) because it can be hit by the path starting from the left of x.

Recall the absolute continuity conditions **(AC)** and **(AC)**$'$ for the transition function $\{P_t; t \geq 0\}$ and the resolvent kernel $\{R_\alpha; \alpha > 0\}$ of X introduced in Definition A.2.16.

PROPOSITION 3.1.11. *Conditions* **(AC)** *and* **(AC)**$'$ *for X are equivalent.*

Proof. Obviously **(AC)** implies **(AC)**$'$. Conversely, assume condition **(AC)**$'$ holds. Then any m-polar set is polar by Theorem A.2.17. If $A \in \mathcal{B}(E)$ satisfies $m(A) = 0$, then $P_t(\cdot, A) = 0$ m-a.e. by the symmetry of P_t. Since $P_t(x, A)$ is quasi continuous in x by Proposition 3.1.9, we have $P_t(\cdot, A) = 0$ q.e. by Lemma 1.3.6 and Theorem 1.3.7. Hence there exists a Borel polar set N such that $P_t(x, A) = 0$ for every $x \in E \setminus N$. Therefore, for any $x \in E$, we get $P_{2t}(x, A) = \mathbf{E}_x[P_t(X_t, A); X_t \notin N] = 0$, yielding **(AC)**. □

Proposition 3.1.11 also reflects the symmetry of X. The deterministic uniform motion to the right on $\mathbb{R}$ does not satisfy **(AC)** but it does satisfy **(AC)**$'$.

As an application of Proposition 3.1.9 and Theorem 3.1.5, we formulate a uniqueness statement.

THEOREM 3.1.12. *Let $(\mathcal{E}, \mathcal{F})$ be a regular Dirichlet form on $L^2(E; m)$. Let $X^{(1)}$ and $X^{(2)}$ be m-symmetric Hunt processes on E associated with a common Dirichlet form $(\mathcal{E}, \mathcal{F})$. There exists then a Borel measurable properly exceptional set N common for $X^{(1)}$ and $X^{(2)}$ such that the transition functions of $X^{(1)}$ and $X^{(2)}$ are identical on $E \setminus N$.*

Proof. Denote the transition function of $X^{(i)}$ by $\{P_t^{(i)}; t \geq 0\}$. Let $C_1(E)$ be a countable uniformly dense subfamily of $C_c(E)$. On account of Proposition 3.1.9, Lemma 1.3.1, Lemma 1.3.6, and Theorem 1.3.7, there exists a Borel measurable $\mathcal{E}$-polar set B_0 such that

$$P_t^{(1)}f(x) = P_t^{(2)}f(x) \qquad \text{for } x \in E \setminus B_0,\ t \in \mathbb{Q}_+ \text{ and } f \in C_1(E). \tag{3.1.12}$$

By the right continuity of the sample paths of $X^{(1)}$ and $X^{(2)}$, (3.1.12) holds for all $t \geq 0$. Accordingly,

$$P_t^{(1)}(x, B) = P_t^{(2)}(x, B) \qquad \text{for } t \geq 0,\ x \in E \setminus B_0 \text{ and } B \in \mathcal{B}(E).$$

By Theorem 3.1.5, we can choose a properly exceptional Borel set N of $X^{(2)}$ that contains B_0. Then $P_t^{(1)}(x, N) = 0$ for every $x \in E \setminus N$, and the Markov processes $X^{(1)}\big|_{E\setminus N}$ and $X^{(2)}\big|_{E\setminus N}$ have the same distribution. In particular, N is also a properly exceptional set for $X^{(1)}$. □

We call two m-symmetric Hunt processes on E *equivalent* if their transition functions are the same outside some common properly exceptional set. Theorem 3.1.12 says that an m-symmetric Hunt process on E associated with a regular Dirichlet form on $L^2(E; m)$ is unique up to the equivalence.

The relationship between an m-symmetric Hunt process and an associated regular Dirichlet form studied so far in this section can be generalized to that between an m-symmetric right process and an associated quasi-regular Dirichlet form in the following fashion.

Let E be a Radon space, m be a σ-finite measure on E, and $(\mathcal{E}, \mathcal{F})$ be a quasi-regular Dirichlet form on $L^2(E; m)$ in the sense of Definition 1.3.8. Let $N_0 \in \mathcal{B}^*(E)$ be an $\mathcal{E}$-polar set and X be an m-symmetric right process on $E \setminus N_0$ whose Dirichlet form on $L^2(E; m)$ is equal to $(\mathcal{E}, \mathcal{F})$.

Denote by $\{P_t; t \geq 0\}$ and $\{T_t; t > 0\}$ the transition function of X and the corresponding semigroup on $L^2(E; m)$. The process X is said to be *properly associated with* the Dirichlet form $(\mathcal{E}, \mathcal{F})$ if $P_t f$ for any $t > 0$, $f \in \mathcal{B}^*(E) \cap L^2(E; m)$, is an $\mathcal{E}$-quasi-continuous m-version of $T_t f$. In view of the proof of Proposition 3.1.9, this condition is equivalent to the condition that the resolvent kernel $R_\alpha f$ of X is an $\mathcal{E}$-quasi-continuous m-version of $G_\alpha f$ for any $\alpha > 0$ and $f \in \mathcal{B}^*(E) \cap L^2(E; m)$.

Proposition 3.1.9 says that an m-symmetric Hunt process associated with a regular Dirichlet form is automatically properly associated with it. Hence the special Borel standard process $j^{-1}(\widehat{X})$ constructed for a quasi-regular Dirichlet form $(\mathcal{E}, \mathcal{F})$ in Theorem 1.5.2 is also properly associated with it. The next theorem gives the uniqueness of an m-symmetric right process properly associated with a quasi-regular Dirichlet form, which can be proved in a way similar to the proof of Theorem 3.1.12.

We shall use the term *a properly exceptional set* for a Borel standard process $X = (X_t, \mathbf{P}_x, \zeta)$ in a slightly weaker sense than for a Hunt process. A nearly

Borel measurable set A is said to be X-*invariant* if for every $x \in A$

$$\mathbf{P}_x (X_t \in A \text{ and } X_{t-} \in A \text{ for every } t \in [0, \zeta)) = 1.$$

A nearly Borel set N is called properly exceptional for X if $m(N) = 0$ and $E \setminus N$ is X-invariant.

THEOREM 3.1.13. *Suppose that X is an m-symmetric right process properly associated with a quasi-regular Dirichlet form $(\mathcal{E}, \mathcal{F})$ on $L^2(E; m)$. Let $\widetilde{X} := j^{-1}(\widehat{X})$ be the special Borel standard process constructed in Theorem 1.5.2 from a Hunt process $\widehat{X}$ associated with a regular Dirichlet $(\widehat{\mathcal{E}}, \widehat{\mathcal{F}})$ that is quasi-homeomorphic to $(\mathcal{E}, \mathcal{F})$ through a quasi-homeomorphism j. There exists then a Borel properly exceptional set N for $\widetilde{X}$ such that the transition functions of X and $\widetilde{X}$ are identical on $[0, \infty) \times (E \setminus N) \times \mathcal{B}(E)$. Consequently, $X|_{E\setminus N}$ and $\widetilde{X}|_{E\setminus N}$ have the same distribution. This in particular implies that $X|_{E\setminus N}$ is a special Borel standard process on $E \setminus N$. The process $X|_{E\setminus N}$ is properly associated with $(\mathcal{E}, \mathcal{F})$ and it enjoys those properties in Theorem 3.1.3, Theorem 3.1.4, and Theorem 3.1.5.*

This theorem combined with Theorem 1.5.3 yields the following:

COROLLARY 3.1.14. *Let X be an m-tight m-symmetric right process X on a Radon space E or an m-symmetric right process X on a Lusin space E.*
(i) *There exists an m-inessential Borel set $N \subset E$ such that $X|_{E\setminus N}$ is a special Borel standard process.*
(ii) *Conditions* **(AC)** *and* **(AC)**$'$ *for X are equivalent.*

We close this subsection by presenting some useful results relating a general symmetric right process to the associated Dirichlet form without assuming the quasi-regularity of the latter.

Let $(E, \mathcal{B}^*(E))$ be a Radon space and m be a σ-finite measure on it. Let $X = (X_t, \zeta, \mathbf{P}_x)$ be a right process on $(E, \mathcal{B}^*(E))$ with an m-symmetric transition function $\{P_t; t \geq 0\}$. We denote the associated Dirichlet form on $L^2(E; m)$ by $(\mathcal{E}, \mathcal{F})$. The resolvent kernel of X is denoted by $\{R_\alpha; \alpha > 0\}$.

LEMMA 3.1.15. *Let F be a nearly Borel finely closed subset of E and $\alpha > 0$. Put $C = E \setminus F$ and*

$$R^F_\alpha f(x) = \mathbf{E}_x \left[\int_0^{\sigma_C} e^{-\alpha t} f(X_t) dt \right], \quad x \in E, \ f \in \mathcal{B}(E).$$

Then

$$\begin{cases} R^F_\alpha (b\mathcal{B}(E) \cap L^2(E; m)) \subset \mathcal{F}, \\ \mathcal{E}_\alpha (R^F_\alpha f, R^F_\alpha g) = (R^F_\alpha f, g), \quad f, g \in b\mathcal{B}(E) \cap L^2(E; m). \end{cases} \tag{3.1.13}$$

Proof. We put for $\alpha > 0$, $\gamma > 0$

$$R_\alpha^{\gamma C} f(x) = \mathbf{E}_x\left[\int_0^\infty e^{-\alpha t - \gamma \int_0^t \mathbf{1}_C(X_s)ds} f(X_t)dt\right].$$

Since C is finely open, $\sigma_C = \inf\{t > 0 : \int_0^t \mathbf{1}_C(X_s)ds > 0\}$ and $R_\alpha^F f(x) = \lim_{\gamma\to\infty} R_\alpha^{\gamma C} f(x)$. Therefore, we have

$$\mathcal{E}(R_\alpha^F f, R_\alpha^F f) = \lim_{\beta\to\infty}\lim_{\gamma\to\infty} \beta(R_\alpha^{\gamma C} f - \beta R_\beta R_\alpha^{\gamma C} f, R_\alpha^{\gamma C} f) \tag{3.1.14}$$

in the sense that, if the right hand side is finite, then $R_\alpha^F f \in \mathcal{F}$ and the limit coincides with the left hand side.

We compute the right hand side by using a generalized resolvent equation

$$R_\beta f - R_\alpha^{\gamma C} f + (\beta - \alpha) R_\beta R_\alpha^{\gamma C} f - \gamma R_\beta(\mathbf{1}_C \cdot R_\alpha^{\gamma C} f) = 0. \tag{3.1.15}$$

This equation for $\beta = \alpha$ implies

$$R_\alpha^{\gamma C} f \in \mathcal{F},\ \mathcal{E}_\alpha(R_\alpha^{\gamma C} f, v) + \gamma(R_\alpha^{\gamma C} f, v)_{\mathbf{1}_C \cdot m} = (f, v).$$

In particular, $R_\alpha^{\gamma C}$ is m-symmetric. Keeping this in mind, we can rewrite the right hand side of (3.1.14) as

$$\begin{aligned}
&\lim_{\beta\to\infty}\lim_{\gamma\to\infty} \beta(R_\beta f - \alpha R_\beta R_\alpha^{\gamma C} f - \gamma R_\beta(\mathbf{1}_C \cdot R_\alpha^{\gamma C} f), R_\alpha^F f) \\
&= (f, R_\alpha^F f) - \alpha(R_\alpha^F f, R_\alpha^F f) - \lim_{\beta\to\infty}\lim_{\gamma\to\infty} \gamma(f, R_\alpha^{\gamma C}(\mathbf{1}_C \cdot \beta R_\beta R_\alpha^F f)).
\end{aligned}$$

If we put $\tau_t = \inf\{s > 0 : \int_0^s \mathbf{1}_C(X_v)dv > t\}$, then we have from Lemma A.3.7

$$\gamma R_\alpha^{\gamma C}(\mathbf{1}_C \cdot g)(x) = \gamma \mathbf{E}_x\left[\int_0^\infty e^{-\alpha\tau_t - \gamma t} g(X_{\tau_t})dt\right],$$

which converges to $\mathbf{E}_x[e^{-\alpha\tau_0} g(X_{\tau_0})]$ as $\gamma \to \infty$ provided that g is bounded and finely continuous. Since $X_{\tau_0} = X_{\sigma_C} \in C^r$ by (A.2.4), and $R_\alpha^F f(x) = 0,\ x \in C^r$, we arrive at the desired identity

$$\lim_{\beta\to\infty}\lim_{\gamma\to\infty} \gamma(f, R_\alpha^{\gamma C}(\mathbf{1}_C \cdot \beta R_\beta R_\alpha^F f)) = \mathbf{E}_{f\cdot m}\left[e^{-\alpha\tau_0} R_\alpha^F f(X_{\tau_0})\right] = 0.$$

□

Exercise 3.1.16. Derive equation (3.1.15) from (4.1.7) of chapter 4. Prove further that (3.1.13) holds for any open set D in place of the finely closed set F when X is an m-symmetric Hunt process on a locally compact separable metric space E where m is a positive Radon measure on E with full support.

LEMMA 3.1.17. *Let $\{F_n\}$ be an increasing sequence of closed subsets of E. If*

$$\mathbf{P}_m(\lim_{n\to\infty} \sigma_{E\setminus F_n} < \zeta) = 0, \tag{3.1.16}$$

then $\{F_n\}$ is an $\mathcal{E}$-nest.

Proof. Take any increasing sequence $\{F_k\}$ of closed sets satisfying (3.1.16). Then for $f \in b\mathcal{B}(E) \cap L^2(E; m)$ the function $R_1^{F_k} f(x)$ increases to $R_1 f(x)$ as $k \to \infty$ for m-a.e. $x \in E$. Since (3.1.13) implies $\|R_1 f - R_1^{F_k} f\|_{\mathcal{E}_1}^2 = (R_1 f - R_1^{F_k} f, f)$, $R_1^{F_k} f \in \mathcal{F}_{F_k}$ is $\mathcal{E}_1$-convergent to $R_1 f$. $\{R_1 f\}$ being $\mathcal{E}_1$-dense in $\mathcal{F}$, $\cup_{k=1}^{\infty} \mathcal{F}_{F_k}$ is $\mathcal{E}_1$-dense in $\mathcal{F}$, namely, $\{F_k\}$ is an $\mathcal{E}$-nest. □

3.2. HITTING DISTRIBUTIONS AND PROJECTIONS I

From this section until the first half of Section 3.5, we shall again work with a fixed m-symmetric Hunt process X and an associated regular Dirichlet form $(\mathcal{E}, \mathcal{F})$ on $L^2(E; m)$ as is formulated in the beginning of Chapter 3.

For $\alpha > 0$, the α-order hitting distribution $\mathbf{H}_B^\alpha$ of X for a nearly Borel measurable set B is defined as

$$\mathbf{H}_B^\alpha g(x) := \mathbf{E}_x \left[e^{-\alpha \sigma_B} g(X_{\sigma_B}) \right].$$

According to Theorem A.1.22, we have, for any non-negative universally measurable function f, the identity

$$\mathbf{H}_B^\alpha (R_\alpha f)(x) = \mathbf{E}_x \left[\int_{\sigma_B}^{\infty} e^{-\alpha t} f(X_t) dt \right], \quad x \in E. \tag{3.2.1}$$

Our aim in this section is to identify in Theorem 3.2.2 the function $\mathbf{H}_B^\alpha \tilde{g}$ for $g \in \mathcal{F}$ with the projection of g on the orthogonal complement $\mathcal{H}_B^\alpha$ of the closed subspace $\mathcal{F}_{E\setminus B}$ of $(\mathcal{F}, \mathcal{E}_\alpha)$.

For a nearly Borel measurable set $B \subset E$, recall from (2.3.14) the space

$$\mathcal{F}_{E\setminus B} = \{f \in \mathcal{F} : \tilde{f} = 0 \quad \text{q.e. on } B\}. \tag{3.2.2}$$

$\mathcal{F}_{E\setminus B}$ is a closed subspace of the Hilbert space $(\mathcal{F}, \mathcal{E}_\alpha)$ for every $\alpha > 0$. Its orthogonal complement is denoted by $\mathcal{H}_B^\alpha$ and the projection on $\mathcal{H}_B^\alpha$ by $P_{\mathcal{H}_B^\alpha}$.

LEMMA 3.2.1. *If f is α-excessive with respect to X and $f \in \mathcal{F}$, then $P_{\mathcal{H}_B^\alpha} f = \mathbf{H}_B^\alpha f$.*

Proof. If $f \in \mathcal{F}$ is α-excessive with respect to X, then so it is with respect to $\{T_t\}$ and, by (2.3.15), $P_{\mathcal{H}_B^\alpha} f$ coincides with the α-reduced function f_B of f

relative to B. We need to prove $\mathbf{H}_B^\alpha f = f_B$ but it suffices to show the inequality

$$\mathbf{H}_B^\alpha f \leq f_B \quad [m]. \tag{3.2.3}$$

Indeed, since $f \in \mathcal{F}$ and f is finely continuous, f is quasi continuous by Theorem 3.1.7(ii). Hence the space defined by (2.3.11) can be rewritten as $\mathcal{L}_{f,B} = \{g \in \mathcal{F} : \widetilde{g} \geq f \text{ q.e. on } B\}$ and f_B is the unique element of $\mathcal{L}_{f,B}$ minimizing the norm $\|g\|_{\mathcal{E}_\alpha}$. Since $\mathbf{H}_B^\alpha f$ is α-excessive, (3.2.3) combined with Lemma 1.2.3 implies $\mathbf{H}_B^\alpha f \in \mathcal{F}$, $\|\mathbf{H}_B^\alpha f\|_{\mathcal{E}_\alpha} \leq \|f_B\|_{\mathcal{E}_\alpha}$. Furthermore, $\mathbf{H}_B^\alpha f = f$ q.e. on B because $B \setminus B^r$ is m-polar by virtue of Lemma A.2.18, Theorem 3.1.3, and Theorem 3.1.10. Hence $\mathbf{H}_B^\alpha f \in \mathcal{L}_{f,B}$ and the inequality (3.2.3) turns out to be an equality.

The proof of the inequality (3.2.3) is essentially analogous to that of Lemma 3.1.1. We fix a non-negative Borel quasi continuous version $\widetilde{f_B}$ of f_B. There exists then a Borel measurable properly exceptional set N such that

$$e^{-\alpha t} P_t \widetilde{f_B}(x) \leq \widetilde{f_B}(x) \quad \text{for every } t > 0 \text{ and } x \in E \setminus N, \tag{3.2.4}$$

$$\widetilde{f_B}(x) = f(x) \quad \text{for every } x \in B \setminus N, \tag{3.2.5}$$

$$\mathbf{P}_x \left(\lim_{s \downarrow t} \widetilde{f_B}(X_s) = \widetilde{f_B}(X_t) \text{ for every } t \geq 0 \right) = 1 \quad \text{for } x \in E \setminus N. \tag{3.2.6}$$

This can be verified in the following way. Since f_B is α-excessive relative to $\{T_t\}$ and $P_t \widetilde{f_B}$ is quasi continuous by Proposition 3.1.9, the inequality in (3.2.4) holds q.e. for each $t > 0$ in view of Lemma 1.3.6 and Theorem 1.3.7. Further, taking (2.3.13) and Theorem 3.1.7 into account, we can use Theorem 3.1.5 to find a Borel measurable properly exceptional set N such that (3.2.5) and (3.2.6) are satisfied and (3.2.4) is valid for all $t \in \mathbb{Q}_+$. By (3.2.6) and Fatou's lemma, (3.2.4) is then valid for all $t > 0$.

We fix $x \in E \setminus N$. Owing to (3.2.4) and (3.2.6), the stochastic process $(e^{-\alpha t} \widetilde{f_B}(X_t), \mathcal{F}_t, \mathbf{P}_x)_{t \geq 0}$ is a non-negative right continuous supermartingale and consequently, by the optional sampling theorem, we have

$$\mathbf{H}_B^\alpha \widetilde{f_B}(x) = \mathbf{E}_x \left[e^{-\alpha \sigma_B} \widetilde{f_B}(X_{\sigma_B}) \right] \leq \widetilde{f_B}(x).$$

On the other hand, we can see from (3.2.5), (3.2.6), and the right continuity of f along the sample path that $\mathbf{H}_B^\alpha \widetilde{f_B}(x) = \mathbf{H}_B^\alpha f(x)$, arriving at the desired inequality (3.2.3). □

THEOREM 3.2.2. *Let $B \subset E$ be a nearly Borel measurable set. For any $f \in \mathcal{F}$, $\mathbf{H}_B^\alpha |\widetilde{f}|(x) < \infty$ q.e. and $\mathbf{H}_B^\alpha \widetilde{f}$ is a quasi continuous version of $P_{\mathcal{H}_B^\alpha} f$.*

Proof. First we assume that $f \in \mathcal{F}$ is bounded and we fix a bounded Borel measurable quasi continuous version $\widetilde{f}$ of f. For each $\beta > 0$, $R_\beta \widetilde{f}$ is a difference of bounded α-excessive functions belonging to $\mathcal{F}$ and hence $\mathbf{H}_B^\alpha(R_\beta \widetilde{f})$ is a quasi continuous version of $P_{\mathcal{H}_B^\alpha}(G_\beta f)$ by virtue of the preceding lemma. Since $\beta G_\beta f$ is $\mathcal{E}_1$-convergent to f as $\beta \to \infty$, so is $\beta P_{\mathcal{H}_B^\alpha}(G_\beta f)$ to $P_{\mathcal{H}_B^\alpha} f$. On the other hand, we can find a properly exceptional set N such that $\widetilde{f}$ satisfies (3.2.6). If $x \in E \setminus N$, then $\beta \mathbf{H}_B^\alpha(R_\beta \widetilde{f})(x) \to \mathbf{H}_B^\alpha \widetilde{f}(x)$, $\beta \to \infty$, and accordingly we can conclude that $\mathbf{H}_B^\alpha \widetilde{f}$ is a quasi continuous version of $P_{\mathcal{H}_B^\alpha} f$ on account of Theorem 1.3.3.

For a general $f \in \mathcal{F}$, functions $f_\ell = ((-\ell) \vee f) \wedge \ell \in \mathcal{F}$ are $\mathcal{E}_1$-convergent to f as $\ell \to \infty$ and so are the functions $P_{\mathcal{H}_B^\alpha} f_\ell$ to $P_{\mathcal{H}_B^\alpha} f$. Hence we get the desired conclusion by repeating the same argument as above. □

We now state two important corollaries of Theorem 3.2.2. The first one extends Lemma 3.1.1 from open sets to Borel sets. The 1-equilibrium potential e_B for a Borel set B with $\mathcal{L}_{B,1} \neq \emptyset$ is determined in the paragraph preceding Theorem 2.3.1. p_B^1 denotes the 1-order hitting probability of B for X defined by (A.2.3).

COROLLARY 3.2.3. *For any Borel set B with $\mathcal{L}_{B,1} \neq \emptyset$, p_B^1 is a quasi continuous version of the* 1*-equilibrium potential e_B.*

Proof. On account of Theorem 2.3.1(iii) and (2.3.15), we have $e_B = P_{\mathcal{H}_B^1} e_B$. By the above theorem, $\mathbf{H}_B^1 \widetilde{e}_B$ is a quasi continuous version of e_B. Since $\widetilde{e}_B(x) = 1$ for q.e. $x \in B$, we conclude from Theorem 3.1.5 and Theorem 3.1.7 that, for q.e. $x \in E$,

$$\mathbf{H}_B^1 \widetilde{e}_B(x) = \mathbf{E}_x \left[e^{-\sigma_B} \widetilde{e}_B(X_{\sigma_B}) \right] = \mathbf{E}_x \left[e^{-\sigma_B} \right] = p_B^1(x).$$

□

For a nearly Borel measurable subset B of E, its complement $E \setminus B$ will be denoted by D for simplicity. $\mathcal{F}_D$ is the subspace of $\mathcal{F}$ defined by the right hand side of (3.2.2). (Sometimes we also denote it by $\mathcal{F}^D$.) For $f \in \mathcal{B}_+(E)$, we put

$$R_\alpha^D f(x) = \mathbf{E}_x \left[\int_0^{\sigma_B} e^{-\alpha t} f(X_t) dt \right], \quad x \in E. \tag{3.2.7}$$

By (3.2.1), we then have

$$R_\alpha f(x) = R_\alpha^D f(x) + \mathbf{H}_B^\alpha (R_\alpha f)(x), \quad x \in E, \tag{3.2.8}$$

which still make sense for q.e. $x \in E$ when $f \in \mathcal{B}(E) \cap L^2(E; m)$. Further, we see from Lemma A.2.18 and Theorem 3.1.10 that $R_\alpha^D f \in \mathcal{F}_D$ in this case. Therefore, we obtain the following corollary immediately from Theorem 3.2.2.

COROLLARY 3.2.4. (i) *For $\alpha > 0$, $f \in \mathcal{B}(E) \cap L^2(E; m)$,* (3.2.8) *represents the orthogonal decomposition of $R_\alpha f$ as a sum of elements of $\mathcal{F}_D$ and $\mathcal{H}_B^\alpha$ in* *the Hilbert space* $(\mathcal{F}, \mathcal{E}_\alpha)$. $R^D_\alpha f \in \mathcal{F}_D$ *is quasi continuous and satisfies*

$$\mathcal{E}_\alpha(R^D_\alpha f, v) = \int_D f(x)v(x)m(dx) \quad \text{for every } v \in \mathcal{F}_D. \tag{3.2.9}$$

(ii) $\{R^D_\alpha; \alpha > 0\}$ *is m-symmetric in the sense that, for any* $f, g \in \mathcal{B}(E) \cap L^2(E; m)$,

$$\int_D f(x)R^D_\alpha g(x)m(dx) = \int_D R^D_\alpha f(x)g(x)m(dx). \tag{3.2.10}$$

3.3. QUASI PROPERTIES, FINE PROPERTIES, AND PART PROCESSES

Corollary 3.2.4 enables us to study the quasi properties for the regular Dirichlet form $(\mathcal{E}, \mathcal{F})$ more thoroughly in relation to the fine topology for the Hunt process X.

DEFINITION 3.3.1. (i) For $B_i \subset E$, $i = 1, 2$, we write as $B_1 \subset B_2$ q.e. if $B_1 \setminus B_2$ is $\mathcal{E}$-polar. If $B_1 = B_2$ q.e., then we say that B_1, B_2 are *q.e. equivalent.*
(ii) A set $D \subset E$ is called *quasi open* if there exists a nest $\{F_k\}$ such that $D \cap F_k$ is open subset of F_k with respect to the relative topology for each $k \geq 1$. The complement of a quasi open set is called *quasi closed.*
(iii) A set $D \subset E$ is called *q.e. finely open* if there exists a properly exceptional set N such that $D \setminus N$ is a nearly Borel measurable finely open set of $X\big|_{E \setminus N}$.

For simplicity, we phrase the last property in the above definition as $D \setminus N$ is nearly Borel finely open on $E \setminus N$ for a properly exceptional set N. Note that if a set D has this property, then so does it for any properly exceptional set containing N on account of Exercise A.1.31.

We further notice that, as in the proof of Theorem 1.3.14(iii), we can replace a nest $\{F_k\}$ with a Cap_1-nest in the above definition of a quasi open set.

Exercise 3.3.2. Prove the following:

(i) A set q.e. equivalent to a quasi open set is again quasi open.
(ii) A function f defined q.e. on E is quasi continuous if and only if f is finite q.e. and $f^{-1}(I)$ is quasi open for any open set $I \subset \mathbb{R}$.
(iii) Let f be a quasi continuous function on E and D be a quasi open subset of E. Using Lemma 1.3.6, prove that if $f \geq 0$ m-a.e. on D, then $f \geq 0$ q.e. on D.

The second assertion in the next theorem removes the condition "$f \in \mathcal{F}$" from Theorem 3.1.7 (ii).

THEOREM 3.3.3. (i) $D \subset E$ *is quasi open if and only if it is q.e. finely open.* (ii) *The following conditions are mutually equivalent for a numerical function f defined q.e. on E.*

(a) *f is quasi continuous.*
(b) *f is finite q.e. and finely continuous q.e.*
(c) *There exists a Borel properly exceptional set N such that f is finite, Borel measurable and finely continuous on $E \setminus N$.*

(iii) *$B \subset E$ is quasi closed if and only if there exists a non-negative quasi continuous function g in $\mathcal{F}$ with $B = g^{-1}(\{0\})$ q.e.*

Proof. (i) Let D be a quasi open set and $\{F_k\}$ be a nest such that $D \cap F_k$ is relatively open in F_k for each $k \geq 1$. By Theorem 3.1.3, Theorem 3.1.4, and Theorem 3.1.5, we can find a Borel properly exceptional set N containing $\cap_k(E \setminus F_k)$ such that (3.1.7) holds for any $x \in E \setminus N$. Then $D \setminus N$ is Borel finely open on $E \setminus N$.

Conversely, assume that $D_1 = D \setminus N$ is nearly Borel finely open on $E \setminus N$ for some properly exceptional set N. We let $B_1 = E \setminus D_1$ and we consider the function $g = R_1^{D_1} f$ defined by (3.2.7) for $\alpha = 1, B = B_1$ and $f \in b\mathcal{B}(E) \cap L^2(E; m)$ strictly positive on E. Then $\{x \in E : g(x) > 0\} = D_1 \cup (B_1 \setminus B_1^r)$. Since g is quasi continuous by Corollary 3.2.4 and $B_1 \setminus B_1^r$ is $\mathcal{E}$-polar by Theorem 3.1.3 and Theorem 3.1.10, D_1 and hence D are quasi open by Exercise 3.3.2.

(ii) The implication $(a) \Rightarrow (c)$ was shown in Theorem 3.1.7(i). $(c) \Rightarrow (b)$ is trivial. If f is finite q.e. and finely continuous q.e., then, for any open set $I \subset \mathbb{R}, f^{-1}(I)$ is q.e. finely open and hence quasi open by (i). Hence f is quasi continuous by Exercise 3.3.2 (ii).

(iii) The "if" part is obvious. If B is quasi closed, then, by (i), $B_1 = B \setminus N$ is nearly Borel finely closed on $E \setminus N$ for some properly exceptional set N. Then the function g appearing in the proof of (i) has the desired property. □

The third statement of the above theorem will now be applied to obtain a useful characterization of the quasi support of a measure.

DEFINITION 3.3.4. Let μ be a positive Borel measure on E charging no $\mathcal{E}$-polar set. A set $F \subset E$ is called the *quasi support of* μ if the following two conditions are satisfied:

(a) F is quasi closed and $\mu(E \setminus F) = 0$.
(b) If $\widetilde{F}$ is another set with property (a), then $F \subset \widetilde{F}$ q.e.

A quasi support of a positive Borel measure charging no $\mathcal{E}$-polar set is unique up to q.e. equivalence.

THEOREM 3.3.5. *For a positive Borel measure μ on E charging no $\mathcal{E}$-polar set and a quasi closed set $F \subset E$, the following conditions are equivalent.*

(a) *F is a quasi support of μ,*
(b) *$u = 0$ μ-a.e. if and only if $u = 0$ q.e. on F for any quasi continuous function $u \in \mathcal{F}$.*
(c) *$u = 0$ μ-a.e. if and only if $u = 0$ q.e. on F for any quasi continuous function u on E.*

Proof. $(b) \Rightarrow (a)$: We put

$$\mathcal{N}_\mu = \left\{ u \in \mathcal{F} : \int_E |\widetilde{u}| d\mu = 0 \right\}$$

and we assume that $\mathcal{N}_\mu = \mathcal{F}_{E \setminus F}$ where $\mathcal{F}_{E \setminus F}$ is the space defined by the right hand side of (3.2.2) for $B = F$. By Theorem 3.3.3(iii), there exists a non-negative quasi continuous function $g \in \mathcal{F}$ with $F = g^{-1}(\{0\})$ q.e. Then $g \in \mathcal{F}_{E \setminus F}$ and hence $g \in \mathcal{N}_\mu$, which means $\mu(E \setminus F) = 0$. Consider another quasi closed set F_1 with $\mu(E \setminus F_1) = 0$ and choose a corresponding function g_1 for F_1 by Theorem 3.3.3(iii). Then $g_1 \in \mathcal{N}_\mu$ and hence $g_1 \in \mathcal{F}_{E \setminus F}$, which implies that $F \subset F_1$ q.e., yielding that F is a quasi support of μ.
$(a) \Rightarrow (c)$: Suppose F is a quasi support of μ. Then the "if" part of condition (b) is obviously satisfied for any Borel function u. If u is quasi continuous and $u = 0$ μ-a.e., then the set $F_1 = \{u = 0\}$ satisfies condition (a) in Definition 3.3.4 and consequently $F \subset F_1$ q.e. and $u = 0$ q.e. on F.
The implication $(c) \Rightarrow (b)$ is trivial. □

COROLLARY 3.3.6. *If $\mu = f \cdot m$ for a measurable function f strictly positive m-a.e., then μ has the full quasi support E.*

We shall prove in Theorem 5.2.1 that any smooth measure μ in the sense of Definition 2.3.13 admits its quasi support, which actually coincides q.e. with the support of a PCAF with Revuz measure μ.

So much for the study of $\mathcal{E}$-quasi notions with a specific use of the function $R^D_\alpha f$ in Corollary 3.2.4. We come back again to the general setting of Corollary 3.2.4 where B is nearly Borel measurable subset of E and $D = E \setminus B$ and investigate its implications in some details.

The space $L^2(D; m)$ consisting of all m-square integrable real functions on D can be identified with the subspace $\{f \in L^2(E; m) : f = 0 \ m\text{-a.e. on } B\}$ of $L^2(E; m)$. $\mathcal{F}_D$ can be viewed as a linear subspace of $L^2(D; m)$ under this identification but it is not necessarily a dense subspace. However,

$$\mathcal{D}(\mathcal{E}^D) = \mathcal{F}_D \quad \text{and} \quad \mathcal{E}^D(f, g) = \mathcal{E}(f, g) \ \text{ for } f, g \in \mathcal{F}_D \tag{3.3.1}$$

defines a bilinear form $\mathcal{E}^D$ on $L^2(D; m)$ satisfying all other requirements as a Dirichlet form on $L^2(D; m)$. $\mathcal{E}^D$ is called the *part of the Dirichlet form* $(\mathcal{E}, \mathcal{F})$ on the set D. We will see that $\mathcal{E}^D$ becomes a genuine Dirichlet form on $L^2(D; m)$ if D is finely open.

Let us now assume that D is a nearly Borel measurable and finely open with respect to the Hunt process $X = (\Omega, \mathcal{M}, X_t, \zeta, \mathbf{P}_x)$. We let for $\omega \in \Omega$

$$X_t^D(\omega) := \begin{cases} X_t(\omega), & 0 \le t < \tau_D(\omega) \\ \partial, & t \ge \tau_D \end{cases} \quad \text{and} \quad \zeta^D(\omega) := \tau_D(\omega).$$

Here $\tau_D := \inf\{t > 0 : X_t \notin D\}$ is the first time the process X exits D. Then $X^D = (\Omega, \mathcal{M}, X_t^D, \zeta^D, \mathbf{P}_x)$ can be verified to be a standard process on the Radon space $(D, \mathcal{B}^*(D))$ in the sense of Definition A.1.39.

Exercise 3.3.7. (i) Prove the last statement using Theorem A.1.37.
(ii) Prove that X^D is a Hunt process on $(D, \mathcal{B}(D))$ if D is an open set.

We call X^D the *part process* of X on D. The transition function $\{P_t^D; t \ge 0\}$ and the resolvent kernel $\{R_\alpha^D; \alpha > 0\}$ of X^D have the following expressions, respectively: for $f \in \mathcal{B}_+(D)$, $x \in D$,

$$P_t^D f(x) = \mathbf{E}_x\left[f(X_t); t < \tau_D\right], \quad R_\alpha^D f(x) = \mathbf{E}_x\left[\int_0^{\tau_D} e^{-\alpha t} f(X_t)dt\right]. \tag{3.3.2}$$

In particular, $\{P_t^D; t \ge 0\}$ is a transition function on $(D, \mathcal{B}^*(D))$ in the sense of Definition 1.1.13.

If we denote the restriction of $f \in \mathcal{B}_+(E)$ to D by f_D, then we have the relation $R_\alpha^D f(x) = R_\alpha^D f_D(x)$ for $x \in D$. Thus (3.2.9) and (3.2.10) can be rewritten as follows: for $f \in \mathcal{B}(D) \cap L^2(D; m)$, $R_\alpha^D f \in \mathcal{D}(\mathcal{E}^D)$ and

$$\mathcal{E}_\alpha^D(R_\alpha^D f, v) = \int_D f(x)v(x)m(dx) \quad \text{for } v \in \mathcal{D}(\mathcal{E}^D). \tag{3.3.3}$$

Moreover,

$$\int_D f(x)R_\alpha^D g(x)m(dx) = \int_D R_\alpha^D f(x)g(x)m(dx) \tag{3.3.4}$$

for every $g \in \mathcal{B}(D) \cap L^2(D; m)$.

The last identity means the m-symmetry of $\{R_\alpha^D; \alpha > 0\}$ on the set D. Consequently, P_t^D is also symmetric on D with respect to m. By Lemma 1.1.14, the transition function $\{P_t^D; t \ge 0\}$ determines uniquely a strongly continuous contraction semigroup $\{T_t^D; t > 0\}$ of Markovian symmetric operators on $L^2(D; m)$. Further, equation (3.3.3) confirms that $\mathcal{E}^D$ is just the Dirichlet form on $L^2(D; m)$ generated by $\{T_t^D; t \ge 0\}$.

THEOREM 3.3.8. *Let $D = E \setminus B$ be a nearly Borel measurable finely open set with respect to X, X^D be the part process of X on D, and $\mathcal{E}^D$ be the part of $\mathcal{E}$ on D.*
(i) *If an increasing sequence $\{F_k\}$ of closed subsets of E is a nest, then $\{F_k \cap D\}$ is an $\mathcal{E}^D$-nest.*

(ii) $(\mathcal{E}^D, \mathcal{F}_D)$ *is a quasi-regular Dirichlet form on* $L^2(D;m)$ *and* X^D *is a standard process properly associated with it.*
(iii) $N \subset D$ *is* $\mathcal{E}^D$*-polar if and only if it is* $\mathcal{E}$*-polar.*
(iv) *For* $u \in \mathcal{F}_D$, u *is* $\mathcal{E}^D$*-quasi-continuous if and only if* u *is the restriction to* D *of a quasi continuous function on* E.

Proof. (i) Denote by σ_A^D the hitting time of the part process X^D for a set $A \subset D$. For any nest $\{F_k\}$, we then have

$$\mathbf{P}_x\left(\lim_{k\to\infty} \sigma^D_{D\setminus D\cap F_k} < \zeta^D\right) \leq \mathbf{P}_x\left(\lim_{k\to\infty} \sigma_{E\setminus F_k} < \zeta\right), \qquad x \in D,$$

which equals zero by Theorem 3.1.4. Hence $\{F_k \cap D\}$ is an $\mathcal{E}^D$-nest by virtue of Lemma 3.1.17.

(ii) By Exercise 3.3.7, X^D is a standard process. (Moreover, by [13, IV:(4.33)] X^D is special.) Consider the 1-order hitting probability $p_B^1(x) := \mathbf{E}_x\left[e^{-\sigma_B}\right]$, $x \in E$, for the set $B = E \setminus D$. Then $\{p_B^1 > 0\} = D \cup N$ where $N = B \setminus B^r$. Since N is $\mathcal{E}$-polar by Theorem 3.1.3 and Theorem 3.1.10 and further p_B^1 is quasi continuous by Corollary 3.2.3, there is a compact nest $\{K_k\}$ such that $\cup_k K_k \subset E \setminus N$ and $p_B^1 \in C(\{K_k\})$. We let $D_k := \left\{p_B^1 \geq \frac{1}{k}\right\}$ and $K_k^D := K_k \cap D_k$ for $k \geq 1$. Obviously each K_k^D is a compact subset of D.

Let $\sigma = \lim_{k\to\infty} \sigma_{D\setminus D_k}$. By the quasi-left-continuity of X and the left continuity (3.1.10) of the quasi continuous function p_B^1, we have for q.e. $x \in D$

$$\mathbf{P}_x(\sigma < \sigma_B \wedge \zeta) = \mathbf{P}_x\left(p_B^1(X_\sigma) = 0,\ \sigma < \sigma_B \wedge \zeta\right) = 0.$$

Since $\sigma^D_{D\setminus K_k^D} = \sigma^D_{D\setminus K_k} \wedge \sigma^D_{D\setminus D_k}$, it holds that, for q.e. $x \in D$,

$$\mathbf{P}_x\left(\lim_{k\to\infty} \sigma^D_{D\setminus K_k^D} < \zeta^D\right) \leq \mathbf{P}_x\left(\lim_{k\to\infty} \sigma_{D\setminus K_k} < \zeta^D\right) + \mathbf{P}_x(\sigma < \zeta^D)$$
$$= 0,$$

and consequently $\{K_k^D\}$ is a compact $\mathcal{E}^D$-nest by Lemma 3.1.17 again.

We have shown that $(\mathcal{E}^D, \mathcal{F}_D)$ satisfies condition (a) in Definition 1.3.8. Condition (b) is clear from (i) because any quasi continuous m-version of $f \in \mathcal{F}_D$ is $\mathcal{E}^D$-quasi-continuous. Condition (c) and the proper association of the right process X_D with $(\mathcal{E}^D, \mathcal{F}_D)$ follows from the fact that the resolvent $R^D_\alpha f$ is quasi continuous and hence $\mathcal{E}^D$-quasi-continuous for any $f \in \mathcal{B}(D) \cap L^2(D;m)$.

(iii) (i) implies that, if $N \subset D$ is $\mathcal{E}$-polar, then it is $\mathcal{E}^D$-polar. The converse implication can be proved by a comparison of the 1-capacity. For a relatively open set $A \subset D$, its 1-capacity relative $(\mathcal{E}^D, \mathcal{F}_D)$ is defined by

$$\mathrm{Cap}_1^D(A) = \inf_{u \in \mathcal{L}^D_{A,1}} \mathcal{E}_1(u,u)$$

with the convention of $\inf \emptyset = \infty$, where

$$\mathcal{L}_{A,1}^D := \{u \in \mathcal{F}_D,\ u \geq 1 \ m\text{-a.e. on } A\}.$$

By Theorem 1.2.14, a set $N \subset D$ is $\mathcal{E}^D$-polar if and only if $\mathrm{Cap}_1^D(N) = 0$.

On the other hand, owing to Theorem 2.3.1, the 1-capacity $\mathrm{Cap}_1(A)$ relative to $(\mathcal{E}, \mathcal{F})$ can be evaluated as

$$\mathrm{Cap}_1(A) = \inf_{u \in \mathcal{L}_{A,1}} \mathcal{E}_1(u,u) \quad \text{with } \mathcal{L}_{A,1} = \{u \in \mathcal{F} : \widetilde{u} \geq 1 \text{ q.e. on } A\}.$$

Since A is quasi open by Theorem 3.3.3, we can invoke Exercise 3.3.2(iii) to identify the space $\mathcal{L}_{A,1}$ with $\{u \in \mathcal{F} : u \geq 1 \ m\text{-a.e. on } A\}$. Therefore, we get the inequality $\mathrm{Cap}_1^D(A) \geq \mathrm{Cap}_1(A)$ for any relatively open $A \subset D$, and consequently for any $B \subset D$, as was to be proved.

(iv) The "if" part is implied by (i). Suppose $u \in \mathcal{F}_D$ is $\mathcal{E}^D$-quasi-continuous. Extend u to a function u_1 on E by setting $u_1(x) = 0,\ x \in E \setminus D$. By the definition of $\mathcal{E}^D$, there is a quasi continuous function $v \in \mathcal{F}$ vanishing q.e. on $E \setminus D$ such that $u = v$ m-a.e. on D. By Exercise 3.3.2(iii), $u = v$ $\mathcal{E}^D$-q.e. on D. Accordingly, $u = v$ q.e. on E by (iii) and u_1 is a quasi continuous function on E. □

When D is open, the part $\mathcal{E}^D$ of $\mathcal{E}$ on D can be seen to enjoy much simpler properties:

THEOREM 3.3.9. *Let $D \subset E$ be an open set. Then the following hold.*
(i) *$(\mathcal{E}^D, \mathcal{F}_D)$ is a regular Dirichlet form on $L^2(D; m)$.*
(ii) *For any special standard core $\mathcal{C}$* (cf. Definition 1.3.17) *of $(\mathcal{E}, \mathcal{F})$, the space*

$$\mathcal{C}_D = \{f \in \mathcal{C} : \mathrm{supp}[f] \subset D\} \tag{3.3.5}$$

is a core of $(\mathcal{E}^D, \mathcal{F}_D)$.

Proof. (i) By Exercise 1.3.13, we see that $C_c(D) \cap \mathcal{F}$ is uniformly dense in $C_c(D)$. Let $\mathcal{G}$ be the $\mathcal{E}_1$-closure of $C_c(D) \cap \mathcal{F}$. By Theorem 1.3.3, $\mathcal{G} \subset \mathcal{F}_D$. So it remains to show that $\mathcal{F}_D \subset \mathcal{G}$.

We know from Theorem 3.3.8 that $(\mathcal{E}^D, \mathcal{F}_D)$ is the Dirichlet form of the part process X^D of X killed upon leaving D, where X is the symmetric Hunt process associated with $(\mathcal{E}, \mathcal{F})$. Let $\{D_k, k \geq 1\}$ be an increasing sequence of relatively compact open subsets of D so that $\overline{D}_k \subset D_{k+1}$ for every $k \geq 1$ and $\cup_{k \geq 1} D_k = D$. Define $F_k := \overline{D}_k$. As $\mathbf{P}_x\left(\lim_{k\to\infty} \tau_{F_k} = \tau_D\right) = 1$ for every $x \in D$ by virtue of the quasi-left-continuity of X, we have by Lemma 3.1.17 applied to subprocess X^D that $\{F_k, k \geq 1\}$ is an $\mathcal{E}^D$-nest. Here $\tau_A := \inf\{t > 0 : X_t \notin A\}$.

Observe that in view of (3.2.2) and Theorem 1.3.7, we have

$$(\mathcal{F}_D)_{F_k} = \{g \in \mathcal{F}_D : g = 0 \ m\text{-a.e. on } D \setminus F_k\}$$

$$= \{g \in \mathcal{F} : \widetilde{g} = 0 \text{ q.e. on } E \setminus F_k\} = \mathcal{F}_{F_k}.$$

Therefore, $\cup_{k\geq 1}\mathcal{F}_{F_k}$ is $\mathcal{E}_1$-dense in $\mathcal{F}_D$. Let $k \geq 1$ and $u \in b\mathcal{F}_{F_k}$. Since $(\mathcal{E}, \mathcal{F})$ is a regular Dirichlet form on $L^2(E; m)$, there is a sequence $\{u_j, j \geq 1\} \subset C_c(E) \cap \mathcal{F}$ that is $\mathcal{E}_1$-convergent to u. On account of Lemma 1.1.11(ii), replacing u_j by $((-\|u\|_\infty) \vee u_j) \wedge \|u\|_\infty$ if necessary, we may and do assume that $\sup_{j\geq 1} \|u_j\|_\infty < \infty$. Since $C_c(E) \cap \mathcal{F}$ is uniformly dense in $C_c(E)$, there is some $\phi \in C_c(D) \cap \mathcal{F}$ so that $\phi = 1$ on F_k. Clearly $u_j\phi \in C_c(D) \cap \mathcal{F}$ and by Exercise 1.1.10, $\sup_{j\geq 1} \mathcal{E}_1(u_j\phi, u_j\phi) < \infty$. As $u_j\phi$ is L^2-convergent to $u\phi = u$, we have by Banach-Saks Theorem (Theorem A.4.1), the Cesàro mean sequence of a subsequence of $\{u_j\phi\}$ is $\mathcal{E}_1$-convergent to u. This proves that $u \in \mathcal{G}$ and so $b\mathcal{F}_{F_k} \subset \mathcal{G}$ for every $k \geq 1$. Consequently, $\cup_{k\geq 1}\mathcal{F}_{F_k} \subset \mathcal{G}$. We then have $\mathcal{F}_D \subset \mathcal{G}$ as $\cup_{k\geq 1}\mathcal{F}_{F_k}$ is $\mathcal{E}_1$-dense in $\mathcal{F}_D$.
(ii) For any $v \in \mathcal{F} \cap C_c(E)$ with $\text{supp}[v] \subset D$, there exists $\{v_n\} \subset \mathcal{C}_D$ which is $\mathcal{E}_1$-convergent to v. To see this, it suffices to assume $0 \leq v \leq 1$. Take $w \in \mathcal{C}_D$ with $w = 1$ on $\text{supp}[v]$ and choose $w_n \in \mathcal{C}$ to be $\mathcal{E}_1$-convergent to v and let, for a fixed $\varepsilon \in (0, 1)$, $v_n = \varphi_\varepsilon(w_n) \cdot w$ for φ_ε appearing in Definition 1.3.17. Then $v_n \in \mathcal{C}_D$, v_n converges to v as $n \to \infty$, and $\mathcal{E}_1(v_n, v_n)$ can be easily verified to be uniformly bounded in n. Hence, by Theorem A.4.1, the Cesàro mean sequence of a subsequence of $\{v_n\}$ is $\mathcal{E}_1$-convergent to v. □

The concept of quasi continuity can be localized as follow. Let D be an open set. We call a function u defined q.e. on D *quasi continuous on D* if there exists a decreasing sequence $\{G_n, n \geq 1\}$ of open subsets of D such that $\lim_{n\to\infty} \text{Cap}_1(G_n) = 0$ and $u\big|_{D\setminus G_n}$ is continuous for each n. Of course, the restriction to D of a quasi continuous function on E is quasi continuous on D. We notice that, if u is quasi continuous on an open set D, then u is $\mathcal{E}^D$-quasi-continuous, because, for $\{G_n\}$ as above, $\{E \setminus G_n\}$ is an $\mathcal{E}$-nest so $\{D \setminus G_n\}$ is an $\mathcal{E}^D$-nest by Theorem 3.3.8(i).

The part process X^D of X on an open set D is a Hunt process on D (Exercise 3.3.7). All the results obtained so far are directly applicable to X^D and its associated regular Dirichlet form $(\mathcal{E}^D, \mathcal{F}_D)$. In particular, we have the following.

PROPOSITION 3.3.10. *Let μ be a measure on E charging no $\mathcal{E}$-polar set and F be its quasi support. Consider an open set $D \subset E$ and a quasi continuous function u on D. If $u = 0$ μ-a.e. on D, then $u = 0$ $\mathcal{E}$-q.e. on $D \cap F$.*

Proof. Let $\nu = \mu\big|_D$. Then $\nu(D \setminus (D \cap F)) \leq \mu(E \setminus F) = 0$. Suppose $v \in \mathcal{F}_D$ is $\mathcal{E}^D$-quasi-continuous and $v = 0$ ν-a.e. on D. Then v is the restriction to D of an $\mathcal{E}$-quasi-continuous function w on E that vanishes $\mathcal{E}$-q.e. on $E \setminus D$. It follows then that $w = 0$ μ-a.e. on E and so we have by Theorem 3.3.5 $w = 0$ $\mathcal{E}$-q.e. on F. Accordingly, $v = 0$ $\mathcal{E}^D$-q.e. on $D \cap F$. This means that $D \cap F$ is a quasi support of ν relative to the Dirichlet form $(\mathcal{E}^D, \mathcal{F}_D)$ by virtue of Theorem 3.3.5 applied to the regular Dirichlet form $(\mathcal{E}^D, \mathcal{F}_D)$.

If u satisfies the stated condition, then u is $\mathcal{E}^D$-quasi-continuous on D. By Theorem 3.3.5(c) applied to $(\mathcal{E}^D, \mathcal{F}_D)$, we have $u = 0$ $\mathcal{E}^D$-q.e. on $D \cap F$, and consequently $\mathcal{E}$-q.e. on $D \cap F$. □

3.4. HITTING DISTRIBUTIONS AND PROJECTIONS II

We keep the setting in the preceding section. The purpose of this section is to establish 0-order counterparts of Theorem 3.2.2.

Let $B \subset E$ be a nearly Borel measurable set. By Theorem 2.3.4, any function u in the extended Dirichlet space $\mathcal{F}_e$ admits its quasi continuous version $\widetilde{u}$. We consider a linear subspace of $\mathcal{F}_e$ defined by

$$\mathcal{F}_{e, E\setminus B} = \{u \in \mathcal{F}_e : \widetilde{u} = 0 \text{ q.e. on } B\}. \tag{3.4.1}$$

The hitting distribution $\mathbf{H}_B$ of X for the set B is defined by (A.2.2):

$$\mathbf{H}_B g(x) = \mathbf{E}_x \left[g(X_{\sigma_B}); \sigma_B < \infty\right] \quad \text{for } x \in E.$$

PROPOSITION 3.4.1. *If $u \in b\mathcal{F}_e$, then $\mathbf{H}_B\widetilde{u}$ is a quasi continuous element of $\mathcal{F}_e$ and*

$$\mathcal{E}(\mathbf{H}_B\widetilde{u}, v) = 0, \qquad \forall v \in \mathcal{F}_{e, E\setminus B}. \tag{3.4.2}$$

Proof. Suppose $u \in \mathcal{F}_e$ and $|u| \le M$ for some $M > 0$. Take an approximating sequence $\{u_n\} \subset \mathcal{F}$ of u. By Theorem 2.3.4, we may assume that $|u_n| \le M$, $n \ge 1$, and $\widetilde{u}_n \to \widetilde{u}$ q.e. as $n \to \infty$.

Choose $\alpha_n \downarrow 0$ such that $\alpha_n(u_n, u_n) \le 1$. By virtue of Theorem 3.2.2,

$$\|\mathbf{H}_B^{\alpha_n}\widetilde{u}_n\|_{\mathcal{E}_{\alpha n}}^2 \le \|u_n\|_{\mathcal{E}_{\alpha n}}^2 \le \|u_n\|_{\mathcal{E}}^2 + 1.$$

Thus both $\|\mathbf{H}_B^{\alpha_n}\widetilde{u}_n\|_{\mathcal{E}}^2$ and $\alpha_n \|\mathbf{H}_B^{\alpha_n}\widetilde{u}_n\|_2^2$ are uniformly bounded in n. In particular, by Theorem A.4.1, we can find a subsequence of $\{\mathbf{H}_B^{\alpha_n}\widetilde{u}_n\}$ such that its Cesàro mean sequence, denoted by $\{f_n\}$, is $\mathcal{E}$-Cauchy.

Since $\lim_{n\to\infty} \mathbf{H}_B^{\alpha_n}\widetilde{u}_n = \mathbf{H}_B\widetilde{u}$ q.e. on E and $\mathbf{H}_B^{\alpha_n}\widetilde{u}_n$ is quasi continuous by Theorem 3.2.2, $\{f_n\}$ is an approximating sequence of $\mathbf{H}_B\widetilde{u}$, which is therefore a quasi continuous element of $\mathcal{F}_e$ in view of Theorem 2.3.4.

To prove the identity (3.4.2) for $v \in \mathcal{F}_{E\setminus B}$, we observe from Theorem 3.2.2

$$\mathcal{E}_{\alpha_n}\left(\mathbf{H}_B^{\alpha_n}\widetilde{u}_n, v\right) = 0, \quad v \in \mathcal{F}_{E\setminus B}.$$

Since

$$\lim_{n\to\infty} \left|\alpha_n \left(\mathbf{H}_B^{\alpha_n}\widetilde{u}_n, v\right)\right| \leq \lim_{n\to\infty} \alpha_n \|\mathbf{H}_B^{\alpha_n}\widetilde{u}_n\|_2 \|v\|_2 = 0,$$

we have $\lim_{n\to\infty} \mathcal{E}(\mathbf{H}_B^{\alpha_n}\widetilde{u}_n, v) = 0$ and so

$$\mathcal{E}(\mathbf{H}_B\widetilde{u}, v) = \lim_{n\to\infty} \mathcal{E}(f_n, v) = 0 \qquad \text{for } v \in \mathcal{F}_{E\setminus B}.$$

We next show (3.4.2) for bounded $v \in \mathcal{F}_{e,E\setminus B}$. By applying the same argument as above to v instead of u, we can find $v_n,\ g_n \in \mathcal{F},\ n \geq 1$, such that $v_n - g_n \in \mathcal{F}_{E\setminus B}$ and $v_n - g_n$ is $\mathcal{E}$-convergent to $v - \mathbf{H}_B\widetilde{v} = v$. Hence we get (3.4.2) for v from that for $v_n - g_n$ by letting $n \to \infty$.

Finally, for any $v \in \mathcal{F}_{e,E\setminus B}$, we put $v_\ell = \varphi^\ell(v)$ by the contraction φ^ℓ of (1.1.19). When $(\mathcal{E}, \mathcal{F})$ is transient, v_ℓ is $\mathcal{E}$-convergent to v as $\ell \to \infty$ by Exercise 2.1.13 and we obtain (3.4.2) for v from that for v_ℓ. In general, we can use Lemma 2.1.15 to find $g \in \mathcal{K}_0$ with $v \in \mathcal{F}_e^g$ so that v_ℓ converges to v in $\mathcal{E}^g$ and hence in $\mathcal{E}$. □

If the Dirichlet form $(\mathcal{E}, \mathcal{F})$ is transient, then by the $\mathcal{E}$ (0-order) version of Theorem 1.3.3, the space $\mathcal{F}_{e,E\setminus B}$ defined by (3.4.1) is a closed subspace of the Hilbert space $(\mathcal{F}_e, \mathcal{E})$. Denote its orthogonal complement by $\mathcal{H}_B$ and the orthogonal projection on $\mathcal{H}_B$ by $P_{\mathcal{H}_B}$.

THEOREM 3.4.2. *If $(\mathcal{E}, \mathcal{F})$ is transient, then, for any $u \in \mathcal{F}_e$, $\mathbf{H}_B|\widetilde{u}|(x) < \infty$ q.e. and $\mathbf{H}_B\widetilde{u}$ is a quasi continuous version of $P_{\mathcal{H}_B}u$.*

Proof. If $u \in \mathcal{F}_e$ is bounded, this is contained in Proposition 3.4.1. It suffices to prove this for a non-negative $u \in \mathcal{F}_e$. We then put $u_\ell = u \wedge \ell$. By Proposition 3.4.1, $\mathbf{H}_B\widetilde{u}_\ell$ is a quasi continuous element of $\mathcal{H}_B$ and $\|\mathbf{H}_B\widetilde{u}_\ell\|_\mathcal{E} \leq \|u_\ell\|_\mathcal{E} \leq \|u\|_\mathcal{E}$. In particular, by Theorem A.4.1, the Cesàro mean sequence of a subsequence of $\{\mathbf{H}_B\widetilde{u}_\ell\}$ is $\mathcal{E}$-convergent as $\ell \to \infty$. Since $\mathbf{H}_B\widetilde{u}_\ell$ increases pointwise to $\mathbf{H}_B\widetilde{u}$ as $\ell \to \infty$, $H_B\widetilde{u}$ is a quasi continuous element of $\mathcal{H}_B$ and $u = (u - \mathbf{H}_B\widetilde{u}) + \mathbf{H}_B\widetilde{u}$ give the $\mathcal{E}$-orthogonal decomposition of u. □

When $(\mathcal{E}, \mathcal{F})$ is transient, we have the 0-order version of Theorem 2.3.1: for any set $B \subset E$ with the space $\mathcal{L}_B^{(0)} = \{f \in \mathcal{F}_e : \widetilde{f} \geq 1 \text{ q.e. on } B\}$ being non-empty, there exists a unique function $e_B^{(0)} \in \mathcal{L}_B^{(0)}$ minimizing $\mathcal{E}(f,f)$ and

$$\mathrm{Cap}^{(0)}(B) = \|e_B^{(0)}\|_\mathcal{E}^2, \tag{3.4.3}$$

where $\mathrm{Cap}^{(0)}$ denotes the 0-order capacity defined in (2.3.1) with $\mathcal{E}_1$ and $\mathcal{L}_{A,1}$ being replaced by $\mathcal{E}$, $\mathcal{L}_{A,1}^{(0)}$ respectively.

The hitting probability p_B of X for a nearly Borel set B is defined by (A.2.3):

$$p_B(x) = \mathbf{P}_x(\sigma_B < \infty),\ x \in E.$$

Exactly in the same way as Corollary 3.2.3 is derived from Theorem 3.2.2 combined with Theorem 2.3.1 and (2.3.15), we can get the following corollary from Theorem 3.4.2 combined with the 0-order version of Theorem 2.3.1 and (2.3.22).

COROLLARY 3.4.3. *Assume that $(\mathcal{E}, \mathcal{F})$ is transient. For any Borel set B with $\mathcal{L}_B^{(0)} \neq \emptyset$, p_B is a quasi continuous version of the equilibrium potential $e_B^{(0)}$.*

COROLLARY 3.4.4. *Assume that $(\mathcal{E}, \mathcal{F})$ is transient. An increasing sequence $\{F_k\}$ of closed subsets of E is a nest if and only if $\cup_{k=1}^{\infty}(\mathcal{F}_e)_{F_k}$ is $\mathcal{E}$-dense in $\mathcal{F}_e$, where*

$$(\mathcal{F}_e)_{F_k} = \{u \in \mathcal{F}_e : u = 0 \ m\text{-a.e. on } E \setminus F_k\}. \tag{3.4.4}$$

Proof. Recall that a nest is an abbreviation of an $\mathcal{E}$-nest in the sense of Definition 1.2.12. Hence, the "only if" part is trivial.

In order to prove the "if" part, we assume that an increasing sequence $\{F_k\}$ of closed sets satisfies the stated denseness property. We shall derive the stochastic property (3.1.7) of $\{F_k\}$ to deduce that it is a nest.

Let $\lim_{k\to\infty} \sigma_{F_k} = \sigma$ and, for a strictly positive $f \in b\mathcal{B}(E)$ with $\int_E f \cdot Rf\, dm < \infty$,

$$u(x) = \lim_{k\to\infty} \mathbf{H}_{E\setminus F_k}(Rf)(x) = \mathbf{E}_x\left[\int_{\sigma\wedge\zeta}^{\zeta} f(X_t)dt\right], \quad x \in E.$$

Since $Rf \in \mathcal{F}_e$ by Theorem 2.1.12 and Rf is excessive and hence quasi continuous by Theorem 3.3.3, we see from Theorem 3.4.2 that for $k \leq \ell$,

$$\|\mathbf{H}_{E\setminus F_k}(Rf) - \mathbf{H}_{E\setminus F_\ell}(Rf)\|_{\mathcal{E}}^2 = (\mathbf{H}_{E\setminus F_k}(Rf) - \mathbf{H}_{E\setminus F_\ell}(Rf), f),$$

and hence $\{\mathbf{H}_{E\setminus F_k}(Rf)\}$ is $\mathcal{E}$-Cauchy and $\mathcal{E}$-convergent to $u \in \mathcal{F}_e$. Consequently, u is $\mathcal{E}$-orthogonal to the space $\cup_{k=1}^{\infty}(\mathcal{F}_e)_{F_k}$, and quasi continuous. Accordingly, u vanishes q.e. and, by the above expression of u, we get $\mathbf{P}_x(\sigma < \zeta) = 0$ q.e., namely, (3.1.7). □

In a general (not necessarily transient) case, we can still formulate a theorem which is obtained from Proposition 3.4.1 just by removing the boundedness requirement for u. To this end, we need to ensure the finiteness of $\mathbf{H}_B\widetilde{u}$ for $u \in \mathcal{F}_e$ in advance.

Let us fix a nearly Borel measurable set $B \subset E$. We put

$$D = \{x \in E : p_B(x) > 0\}. \tag{3.4.5}$$

Since p_B is excessive, D is nearly Borel measurable and finely open.

PROPOSITION 3.4.5. (i) $\mathbf{P}_x(\sigma_D < \infty) = 0$ *for* $x \in D^c$.
(ii) *There exists a Borel properly exceptional set* $N \subset D$ *such that both* $D \setminus N$ *and* D^c *are* X*-invariant.*

Proof. (i) For $n \geq 1$, define $D_n := \{x \in E : p_B(x) \geq 1/n\}$, which is finely closed. Then, for $x \in D^c$,

$$\begin{aligned} p_{D_n}(x) &= \mathbf{P}_x(\sigma_B = \infty,\ \sigma_{D_n} < \infty) \\ &= \mathbf{E}_x\left[\mathbf{P}_{X_{\sigma_{D_n}}}(\sigma_B = \infty);\ \sigma_{D_n} < \infty\right] \\ &\leq \left(1 - \frac{1}{n}\right) p_{D_n}(x). \end{aligned}$$

It follows that $p_{D_n}(x) = 0$. We then get (i) by letting $n \to \infty$.
(ii) Note that (i) and Lemma A.1.32 imply that the set D^c is X-invariant. Further, (i) implies that D is $\{T_t\}$-invariant in view of Proposition 2.1.6 and that D^c is finely open. Thus D is both finely open and finely closed. Consequently, the function $\mathbf{1}_D$ is finely continuous and therefore quasi continuous by Theorem 3.3.3. By the $\{T_t\}$-invariance of D and Proposition 3.1.9, we have $P_t(\mathbf{1}_D u) = \mathbf{1}_D \cdot P_t u$ q.e. for any $t > 0$ and $u \in C_c(E)$. Taking a sequence of non-negative functions $\{u_n, n \geq 1\} \subset C_c(E)$ that increases to 1, we obtain $P_t \mathbf{1}_D = \mathbf{1}_D \cdot P_t 1$ q.e.; namely, for every $t > 0$, there is an m-polar set N_t so that $\mathbf{P}_x(X_t \in D) = \mathbf{P}_x(t < \zeta)$ for $x \in D \setminus N_t$. Choose a properly exceptional set N_1 containing $\cup_{t \in \mathbb{Q}_+} N_t$ and let $N := D \cap N_1$. As $\mathbf{1}_D$ is finely continuous, $\mathbf{P}_x(X_t \in D \setminus N) = \mathbf{P}_x(t < \zeta)$ for any $t \geq 0$ and any $x \in D \setminus N$, namely, $\mathbf{P}_x(\sigma_{E \setminus (D \setminus N)} < \infty) = 0$ for any $x \in D \setminus N$. This combined with Lemma A.1.32 yields the X-invariance of the set $D \setminus N$. □

We denote by $\widetilde{D}$ the X-invariant set $D \setminus N$ in the above proposition and by D_1 the fine interior of $\widetilde{D} \setminus B$:

$$D_1 := \{x \in \widetilde{D} \setminus B : \mathbf{P}_x(\sigma_B > 0) = 1\}. \tag{3.4.6}$$

D_1 is then nearly Borel measurable and finely open. By virtue of Theorem 3.3.8, the part $(\mathcal{E}^{D_1}, \mathcal{F}_{D_1})$ of $(\mathcal{E}, \mathcal{F})$ on D_1 is a Dirichlet form on $L^2(D_1; m)$ which is associated with the part process X^{D_1} of X on D_1.

LEMMA 3.4.6. *The part* $\mathcal{E}^{D_1}$ *of* $\mathcal{E}$ *on the set* D_1 *is a transient Dirichlet form on* $L^2(D_1; m)$.

Proof. If we denote the resolvent kernel of the part process X^{D_1} by $\{R_\alpha^{D_1}; \alpha > 0\}$, then

$$R_1^{D_1} 1(x) = \mathbf{E}_x\left[\int_0^{\sigma_B \wedge \zeta} e^{-t} dt\right] \leq 1 - \mathbf{E}_x\left[e^{-\sigma_B}\right] < 1,\ x \in D_1,$$

and we get the transience of $\mathcal{E}^{D_1}$ on account of Proposition 2.1.10. □

LEMMA 3.4.7. *For $u \in \mathcal{F}_e$, $\mathbf{H}_B|\widetilde{u}|(x)$ is finite m-a.e.*

Proof. We may assume that $u \in \mathcal{F}_e$ is non-negative. Since $\mathbf{H}_B\widetilde{u}(x) = \widetilde{u}(x) < \infty$ q.e. $x \in B \cup B^r$ by Theorem 3.1.10 and $\mathbf{H}_B\widetilde{u}(x) = 0$ for every $x \in D^c$, it suffices to show that

$$\mathbf{H}_B\widetilde{u}(x) < \infty \quad m\text{-a.e. on } D_1. \tag{3.4.7}$$

Since $\mathcal{E}^D$ is transient, it admits a reference function g on D_1: g is a strictly positive bounded m-integrable function on D_1 such that $\int_{D_1} g(x)R^{D_1}g(x)m(dx) \leq 1$. By replacing g with $g/(u \vee 1)$ if necessary, we may assume that $\int_{D_1} gudm < \infty$. Recall that the domain of $\mathcal{E}^{D_1}$ is defined by

$$\mathcal{F}_{D_1} = \{v \in \mathcal{F} : \widetilde{v} = 0 \text{ q.e. on } E \setminus D_1\}.$$

We put $u_n = u \wedge n$ for a fixed n. By Theorem 2.3.4, we can find a sequence $\{v_k\} \subset \mathcal{F}$ such that $0 \leq v_k \leq n$, $\{v_k\}$ is $\mathcal{E}$-Cauchy and $\lim_{k\to\infty} \widetilde{v}_k = \widetilde{u}_n$ q.e. In view of Theorem 3.2.2 and Proposition 2.1.6, we see that $\mathbf{1}_{\widetilde{D}}(v_k - \mathbf{H}_B^\alpha\widetilde{v}_k) \in \mathcal{F}_{D_1}$ and

$$\begin{aligned}\int_{D_1} g|v_k - \mathbf{H}_B^\alpha\widetilde{v}_k|dm &= \mathcal{E}_\alpha\left(R_\alpha^{D_1}g, \mathbf{1}_{\widetilde{D}}|v_k - \mathbf{H}_B^\alpha\widetilde{v}_k|\right)\\ &\leq \left(\int_{D_1} gR_\alpha^{D_1}gdm\right)^{1/2} \|v_k - \mathbf{H}_B^\alpha\widetilde{v}_k\|_{\mathcal{E}_\alpha} \leq \|v_k\|_{\mathcal{E}_\alpha}.\end{aligned}$$

By letting $\alpha \downarrow 0$ and $k \to \infty$, we have

$$\int_{D_1} g|u_n - \mathbf{H}_B\widetilde{u}_n|dm \leq \|u_n\|_{\mathcal{E}} \leq \|u\|_{\mathcal{E}}.$$

Hence $\int_{D_1} g \cdot \mathbf{H}_B\widetilde{u}_n dm \leq \int_{D_1} g \cdot u_n dm + \|u\|_{\mathcal{E}}$, and, by letting $n \to \infty$, we obtain $\int_{D_1} g \cdot \mathbf{H}_B\widetilde{u}dm \leq \int_G g \cdot udm + \|u\|_{\mathcal{E}} < \infty$, proving (3.4.7). □

THEOREM 3.4.8. *For any $u \in \mathcal{F}_e$ and for any nearly Borel measurable set $B \subset E$, $\mathbf{H}_B|\widetilde{u}|$ is finite q.e. and $\mathbf{H}_B\widetilde{u}$ is a quasi continuous element of $\mathcal{F}_e$ satisfying* (3.4.2).

Proof. We may assume that $u \in \mathcal{F}_e$ is non-negative. We put $u_n = u \wedge n$. By Proposition 3.4.1, u_n then satisfies the desired properties and in particular $\|\mathbf{H}_B\widetilde{u}_n\|_{\mathcal{E}} \leq \|u_n\|_{\mathcal{E}} \leq \|u\|_{\mathcal{E}}$. Since $\mathbf{H}_B\widetilde{u}_n$ increases pointwise to $\mathbf{H}_B\widetilde{u}$ as $n \to \infty$ and $\mathbf{H}_B\widetilde{u}$ is finite m-a.e. by Lemma 3.4.7, in view of Theorem A.4.1, we can conclude that $\mathbf{H}_B\widetilde{u}$ is an element of $\mathcal{F}_e$ admitting a Cesàro mean of a subsequence of $\{\mathbf{H}_B\widetilde{u}_n\}$ as its approximating sequence. Equation (3.4.2) for $\mathbf{H}_B\widetilde{u}$ then follows from that for $\mathbf{H}_B\widetilde{u}_n$. The quasi continuity of $\mathbf{H}_B\widetilde{u}$ follows from that of $\mathbf{H}_B\widetilde{u}_n$ and Theorem 2.3.4. □

THEOREM 3.4.9. *Suppose that D is a nearly Borel measurable finely open subset of E. Then the extended Dirichlet space $\mathcal{F}_{D,e}$ of $(\mathcal{E}^D, \mathcal{F}_D)$ can be characterized as*

$$\mathcal{F}_{D,e} = \{u \in \mathcal{F}_e : \widetilde{u} = 0 \text{ q.e. on } D^c\}. \tag{3.4.8}$$

Proof. We denote the right hand side of (3.4.8) by $\mathcal{F}_{e,D}$. Clearly the extended Dirichlet space $\mathcal{F}_{D,e}$ of $(\mathcal{E}^D, \mathcal{F}_D)$ is a subspace of $\mathcal{F}_e$ and it follows from Theorem 2.3.4 that $\mathcal{F}_{D,e} \subset \mathcal{F}_{e,D}$.

Let us prove the inverse inclusion. Assume first that $(\mathcal{E}, \mathcal{F})$ is transient. By Theorem 2.3.2, $\mathcal{F}_{e,D}$ is a closed linear subspace of the Hilbert space $\mathcal{F}_e$ with inner product $\mathcal{E}$. Let R and R^D be the 0-order resolvent kernel for the Hunt process X associated with $(\mathcal{E}, \mathcal{F})$ and its subprocess X^D, respectively. Any $f \in L^2(D; m)$ is extended to E by setting $f(x) = 0,\ x \in D^c$. For $f \in L^2(D; m)$ with $\int_D f(x) R(\mathbf{1}_D f)(x) m(dx) < \infty$, we know by Theorem 2.1.12 that $Rf \in \mathcal{F}_e$ and $R^D f \in \mathcal{F}_{D,e}$. As $\mathbf{H}_{D^c} Rf = Rf - R^D f$, we have by Theorem 3.4.8

$$\mathcal{E}(R^D f, v) = \mathcal{E}(Rf, v) = (f, v) \quad \text{for every } v \in \mathcal{F}_{e,D}.$$

This implies that the subspace of $\mathcal{F}_{D,e}$ defined by

$$\{R^D f : f \in L^2(D; m) \text{ with } \int_D f(x) R(\mathbf{1}_D f)(x) m(dx) < \infty\}$$

is $\mathcal{E}$-dense in $\mathcal{F}_{e,D}$. This proves that $\mathcal{F}_{D,e} = \mathcal{F}_{e,D}$ when $(\mathcal{E}, \mathcal{F})$ is transient.

In the general case, for $u \in \mathcal{F}_{e,D}$, let $g \in \mathcal{K}_0$ be the function in Lemma 2.1.15 so that $u \in \mathcal{F}_e^g$. Note that $(\mathcal{E}^g, \mathcal{F})$ is a transient Dirichlet form with $\mathcal{E}^g(u, u) \geq \mathcal{E}(u, u)$ for $u \in \mathcal{F}$ and shares the same quasi notions with $(\mathcal{E}, \mathcal{F})$. We conclude from the above transient case that $u \in \mathcal{F}_{e,D}^g = \mathcal{F}_{D,e}^g \subset \mathcal{F}_{D,e}$. This completes the proof that $\mathcal{F}_{D,e} = \mathcal{F}_{e,D}$. □

In view of Theorem 3.4.9, there is no ambiguity to use notation $\mathcal{F}_e^D$ to denote either of $\mathcal{F}_{D,e}$ and $\mathcal{F}_{e,D}$.

3.5. TRANSIENCE, RECURRENCE, AND PATH BEHAVIOR

We say that the m-symmetric Hunt process X on E is *transient*, *recurrent*, and *irreducible* if so is its Dirichlet form $(\mathcal{E}, \mathcal{F})$ on $L^2(E; m)$ in the sense of Section 2.1, respectively. In this section, we present basic stochastic features of transience, recurrence, and irreducibility of X.

In dealing with the transience and its applications, we continue to work under the present setting that X is an m-symmetric Hunt process on a locally compact separable metric space E whose Dirichlet form $(\mathcal{E}, \mathcal{F})$ on $L^2(E; m)$ is regular. However, the recurrence and irreducibility will be discussed for a more general symmetric right process without assuming the regularity of the associated Dirichlet form in the second part of this section.

We first state an application of Corollary 3.4.3.

LEMMA 3.5.1. *Assume that X is transient. If, for a decreasing sequence $\{A_n\}$ of open sets, $\{E \setminus A_n\}$ is a $\mathrm{Cap}^{(0)}$-nest in the sense that $\lim_{n\to\infty} \mathrm{Cap}^{(0)}(A_n) = 0$, then*

$$\mathbf{P}_x\left(\bigcup_{n=1}^{\infty}\{\sigma_{A_n} = \infty\}\right) = 1, \quad q.e.\ x \in E. \tag{3.5.1}$$

In particular, $\{E \setminus A_n\}$ is a strong nest in the sense of (3.1.6).

Proof. By Corollary 3.4.3 and (3.4.3), p_{A_n} is quasi continuous and $\mathrm{Cap}^{(0)}(A_n) = \|p_{A_n}\|_{\mathcal{E}}^2$ for an open set A_n with $\mathrm{Cap}^{(0)}(A_n) < \infty$. If this quantity tends to zero as $n \to \infty$, then $\lim_{n\to\infty} p_{A_n}(x) = 0$ q.e. $x \in E$, which in turn implies (3.5.1). □

Obviously any Cap_1-nest $\{E \setminus A_n\}$ is a $\mathrm{Cap}^{(0)}$-nest and hence $\{A_n\}$ enjoys the property (3.5.1) in transient case. Property (3.5.1) is stronger than (3.1.6) and it reflects a specific property of a transient process X that escapes to infinity ∂ of E when time goes to infinity, as will be formulated below.

THEOREM 3.5.2. *Assume that X is transient. Then the path wanders out to infinity whenever its lifetime is infinite:*

$$\mathbf{P}_x\left(\zeta = \infty \text{ and } \lim_{t\to\infty} X_t = \partial\right) = \mathbf{P}_x(\zeta = \infty) \quad \text{for } q.e.\ x \in E. \tag{3.5.2}$$

Proof. Let R be the 0-order resolvent kernel of X defined by (2.1.3) and f be a strictly positive Borel measurable function on $L^1(E;m)$. Then Rf is excessive and the set $N = \{x \in E : Rf(x) = \infty\}$ is m-negligible in view of Proposition 2.1.3. Consequently, N is m-polar by Theorem A.2.13(v). On account of Theorem 3.1.5, we can find a Borel properly exceptional set N_0 such that

$$Rf(x) > 0 \quad \text{for } x \in E \qquad \text{and} \qquad Rf(x) < \infty \quad \text{for } x \in E \setminus N_0.$$

By virtue of Theorem 3.3.3, Rf is quasi continuous and hence by Theorem 1.3.14(iii) there is a Cap_1-nest $\{E \setminus A_n\}$ such that the restriction of Rf to each set $E \setminus A_n$ is continuous. We may assume $N_0 \subset A_n$ for every $n \geq 1$. Furthermore, on account of Lemma 3.5.1, the identity (3.5.1) holds for $x \in E \setminus N_1$, N_1 being some Borel properly exceptional set containing N_0.

Take an increasing sequence of compact sets $\{K_n, n \geq 1\}$ that increases to E, and let $F_n = K_n \setminus A_n = K_n \cap (E \setminus A_n)$. Since F_n is compact and Rf is continuous on F_n, $c_n = \inf_{x\in F_n} Rf(x)$ is positive and

$$c_n\, p_{F_n}(x) \le \mathbf{E}_x\left[(Rf)(X_{\sigma_{F_n}})\right] \le Rf(x) < \infty \quad \text{for } x \in E\setminus N_1.$$

Therefore, for $x \in E \setminus N_1$, $P_t p_{F_n}(x) \le c_n^{-1}\, P_t Rf(x) \downarrow 0$ as $t \uparrow \infty$, and consequently,

$$\mathbf{P}_x\left(\bigcap_{j=1}^{\infty} \Lambda_\ell\right) = \lim_{j\to\infty} \mathbf{P}_x\left(\Lambda_j\right) = 0 \text{ for } \Lambda_j := \{\sigma_{F_n}\circ\theta_j < \infty\} \cap \{j < \zeta\}.$$

Since

$$\bigcap_{j=1}^{\infty} \Lambda_j = \bigcap_{j=1}^{\infty}\{\sigma_{F_n}\circ\theta_j < \infty\} \bigcap \{\zeta = \infty\},$$

we get

$$\mathbf{P}_x\left(\bigcup_{j=1}^{\infty}\{\sigma_{F_n}\circ\theta_j = \infty\}\bigcap\{\zeta = \infty\}\right) = \mathbf{P}_x(\zeta = \infty) \quad \text{for } x \in E\setminus N_1.$$

This holds for every $n \ge 1$ and so for $x \in E \setminus N_1$,

$$\mathbf{P}_x\left(\{\zeta = \infty\}\bigcap\left(\bigcap_{n=1}^{\infty}\bigcup_{j=1}^{\infty}\{X(j,\infty) \subset E\setminus F_n\}\right)\right) = \mathbf{P}_x(\zeta = \infty).$$

On account of (3.5.1), we can replace F_n with K_n in the above to get the desired identity (3.5.2). □

Corollary 3.5.3. *Suppose X is transient. Then any quasi continuous function $f \in \mathcal{F}_e$ satisfies*

$$\mathbf{P}_x\left(\lim_{t\to\zeta} f(X_t) = 0;\, \Lambda^c\right) = \mathbf{P}_x(\Lambda^c) \quad \textit{for q.e. } x \in E, \tag{3.5.3}$$

where $\Lambda = \{\zeta < \infty,\ X_{\zeta-} \in E\}$.

Proof. By Theorem 3.5.2, $\mathbf{P}_x(X_{\zeta-} = \partial; \Lambda^c) = \mathbf{P}_x(\Lambda^c)$ for q.e. $x \in E$. By the 0-order counterpart of Lemma 1.3.15, any $f \in \mathcal{F}_e$ admits a quasi continuous version in the restricted sense with respect to $\mathrm{Cap}^{(0)}$: there exists a decreasing open sets $\{A_k\}$ with $\lim_{k\to\infty}\mathrm{Cap}^{(0)}(A_k) = 0$ such that $f \in C_\infty(\{E\setminus A_k\})$. If $f \in \mathcal{F}_e$ is quasi continuous, then f differs from a function of this property by a set of zero capacity and f itself enjoys this property. Accordingly, (3.5.3) follows from Lemma 3.5.1. □

We can now combine Lemma 3.5.1 with Theorem 2.3.2 and Lemma 2.1.15 to deduce an important statement stronger than a part of Theorem 2.3.4 for a general (not necessarily transient) m-symmetric Hunt process $X = (X_t, \mathbf{P}_x)$ whose Dirichlet form $(\mathcal{E}, \mathcal{F})$ on $L^2(E; m)$ is regular. By Theorem 2.3.4, any element of the extended Dirichlet space $\mathcal{F}_e$ admits a quasi continuous m-version.

THEOREM 3.5.4. *Assume that $\{f_n, n \geq 1\} \subset \mathcal{F}_e$ are both m-a.e. convergent and $\mathcal{E}$-convergent to $f \in \mathcal{F}_e$ as $n \to \infty$. Let $\widetilde{f}_n$, $n \geq 1$, and $\widetilde{f}$ be quasi continuous versions of f_n, $n \geq 1$, and f, respectively. There exist then a subsequence $\{n_\ell\}$ and a strong nest $\{F_k\}$ such that $\{\widetilde{f}, \widetilde{f}_{n_\ell}, \ell \geq 1\} \subset C_\infty(\{F_k\})$ and $\widetilde{f}_{n_\ell}$ converges to $\widetilde{f}$ uniformly on each set F_k as $\ell \to \infty$. In particular, it holds for every $T > 0$ that*

$$\mathbf{P}_x\left(\lim_{\ell\to\infty} \sup_{0\leq t\leq T} \left|\widetilde{f}_{n_\ell}(X_t) - \widetilde{f}(X_t)\right| > 0\right) = 0 \quad \textit{for q.e. } x \in E. \tag{3.5.4}$$

Proof. Property (3.5.4) is a consequence of the first assertion in view of Definition 3.1.6 of a strong nest. Furthermore, the first assertion follows from Theorem 2.3.2 and Lemma 3.5.1 when X is transient.

In a general (not necessarily transient) case, we shall prove the first assertion by a reduction to a transient case. By Lemma 2.1.15, we can find $g \in \mathcal{K}_0$ such that $\{f, f_n, n \geq 1\} \subset \mathcal{F}_e^g$ and f_n is $\mathcal{E}^g$-convergent to f as $n \to \infty$, where $(\mathcal{E}^g, \mathcal{F})$ is the perturbed Dirichlet form on $L^2(E; m)$ defined by (2.1.28). It is a transient regular Dirichlet form and $\widetilde{f}_n$, $n \geq 1$, $\widetilde{f}$ are $\mathcal{E}^g$-quasi-continuous.

Let $\widehat{X} = (\widehat{X}_t, \widehat{\mathbf{P}}_x)$ be the canonical subprocess of the Hunt process $X = (X_t, \mathbf{P}_x)$ with respect to its multiplicative functional e^{-A_t} for

$$A_t = \int_0^t g(X_s)ds, \quad t \geq 0,$$

as is defined by (A.3.28) and (A.3.29). Since $e^{-A_t} \in \mathcal{F}_t^0$, the transition function $\widehat{P}_t(x, B) = \mathbf{E}_x[e^{-A_t}\mathbf{1}_B(X_t)]$ of $\widehat{X}$ is $\mathcal{B}(E)$-measurable in $x \in E$ for $B \in \mathcal{B}(E)$. Hence $\widehat{X}$ is a Hunt process on $(E, \mathcal{B}(E))$ in view of Theorem A.3.13. Moreover, $\widehat{X}$ is m-symmetric and its Dirichlet form on $L^2(E; m)$ coincides with $(\mathcal{E}^g, \mathcal{F})$ by virtue of Theorem 5.1.3.

Therefore, the first assertion of the theorem holds true for some increasing sequence $\{F_k\}$ of closed sets that is a strong nest with respect to $\widehat{X}$, namely,

$$\widehat{\mathbf{P}}_x(\sigma < \infty) = 0, \quad \sigma = \lim_{k\to\infty} \sigma_{E\setminus F_k} \quad \text{for q.e. } x \in E.$$

But, for each $T > 0$, we have from (A.3.29)

$$\widehat{\mathbf{P}}_x(\sigma < T) = \mathbf{E}_x\left[e^{-A_t}; \sigma < T\right],$$

and consequently

$$\mathbf{P}_x(\sigma < T) \leq e^{T\|g\|_\infty}\widehat{\mathbf{P}}_x(\sigma < T) = 0 \quad \text{for q.e. } x \in E,$$

which means that $\{F_k\}$ is a strong nest with respect to X, as was to be proved. $\square$

We now turn to the probabilistic descriptions of irreducibility and recurrence under a more general setting. Let $(E, \mathcal{B}(E))$ be a Lusin space and m be a σ-finite measure on it. Let $X = (X_t, \mathbf{P}_x)$ be a right process on $(E, \mathcal{B}(E))$ whose transition function $\{P_t; t \geq 0\}$ is m-symmetric. We do not consider the associated Dirichlet form. The term "q.e." will mean "except for an m-polar set".

Lemma 3.5.5. (i) *If X is recurrent, then any bounded excessive function f is $\{P_t\}$-invariant: for q.e. $x \in E$,*

$$P_t f(x) = f(x) \qquad \text{for } t > 0. \tag{3.5.5}$$

(ii) *If X is irreducible recurrent, then any excessive function is constant q.e.*

Proof. (i) Assume that X is recurrent and take a strictly positive function $g \in L^1(E; m)$. Then for any bounded excessive function f, the resolvent equation gives

$$\begin{aligned}(R_\beta g, f - \alpha R_\alpha f) &= (g, R_\beta f - \alpha R_\beta R_\alpha f)\\ &= (g, R_\alpha f - \beta R_\alpha R_\beta f) \leq (g, R_\alpha f) < \infty.\end{aligned}$$

By letting $\beta \to 0$, we get from Proposition 2.1.3 that $\alpha R_\alpha f = f$ m-a.e. and hence q.e. Since $P_t f(x)$ is right continuous in $t > 0$ owing to the fine continuity of f, we obtain (3.5.5).

(ii) If an excessive function f is not constant q.e., then it is not constant m-a.e. and there are constant $0 < a < b$ such that two finely open sets $A = \{x \in E : f(x) < a\}$ and $B = \{x \in E : f(x) > b\}$ are both of positive m-measures. Since $f \wedge b$ is bounded and excessive by Exercise A.2.3, we have from (i)

$$b = f(x) \wedge b = \mathbf{E}_x[f(X_t) \wedge b], \quad t > 0, \ x \in B \setminus N,$$

for some m-inessential set N. This implies for $x \in B \setminus N$ that $\mathbf{P}_x(X_t \in A) = 0$ and consequently $\mathbf{P}_x(X_t \in A, \ \exists t > 0) = 0$ on account of the right continuity of $f(X_t)$ in $t > 0$. Hence $R\mathbf{1}_A(x) = 0, x \in B \setminus N$, while $R\mathbf{1}_A(x) > 0$, $x \in A$, arriving at a contradiction to the irreducible recurrence criterion Proposition 2.1.3(iv). $\square$

Theorem 3.5.6. (i) *If X is irreducible, then for any non-m-polar nearly Borel measurable set B*

$$\mathbf{P}_x(\sigma_B < \infty) > 0, \quad q.e.\ x \in E. \tag{3.5.6}$$

(ii) *If X is irreducible and recurrent, then for any non-m-polar nearly Borel measurable set B*

$$\mathbf{P}_x(\sigma_B \circ \theta_n < \infty \text{ for every } n \geq 1) = 1 \quad \text{for q.e. } x \in E. \tag{3.5.7}$$

Proof. (i) The set D defined by (3.4.5) is of positive m-measure because of the non-m-polarity of B. By virtue of Proposition 3.4.5, D is $\{T_t\}$-invariant and hence $m(D^c) = 0$ by the irreducibility assumption. Since D^c is finely open by Proposition 3.4.5(i), it is m-polar on account of Theorem A.2.13, yielding (3.5.6).

(ii) In view of Lemma 3.5.5, $p_B(x) = c$, $x \in E \setminus N$, for some positive constant c and a Borel m-inessential set N. We have then, for any $t > 0$, $x \in E \setminus N$,

$$\begin{aligned} c &= \mathbf{P}_x(\sigma_B < \infty) = \mathbf{P}_x(\sigma_B \leq t) + \mathbf{P}_x(t < \sigma_B,\ \sigma_B < \infty) \\ &= \mathbf{P}_x(\sigma_B \leq t) + \mathbf{E}_x\left[\mathbf{P}_{X_t}(\sigma_B < \infty); X_t \in E \setminus N, t < \sigma_B\right] \\ &= \mathbf{P}_x(\sigma_B \leq t) + c\mathbf{P}_x(t < \sigma_B). \end{aligned}$$

By letting $t \to \infty$, we get $c(1 - c) = 0$ and hence $c = 1$. We are then readily led to (3.5.7) by the Markov property of X. □

At the end of this section, we present two examples of symmetric diffusions and study their transience, recurrence, and other properties. As is stated in Theorem 1.5.1, there exists for any regular Dirichlet form $(\mathcal{E}, \mathcal{F})$ on $L^2(E; m)$ an m-symmetric Hunt process X on E, which is unique up to $\mathcal{E}$-polar sets in view of Theorem 3.1.12. If furthermore $\mathcal{E}$ is strongly local, then the Hunt process X can be taken to be a diffusion with no killing inside E, as will be proved in Theorem 4.3.4 at the end of the next chapter. We shall utilize these facts in the considerations of the following examples.

Example 3.5.7. (One-dimensional diffusion) We consider the one-dimensional local Dirichlet form studied in Section 2.2.3. Let $I = (r_1, r_2)$ be a one-dimensional open interval and (s, m) be a pair of a canonical scale and a canonical measure on I. Defining the Dirichlet integral $\mathcal{E}^{(s)}$ based on s and the associated space $\mathcal{F}^{(s)}$ by (2.2.29) and (2.2.28), respectively, we are concerned with the Dirichlet form

$$(\mathcal{E}, \mathcal{F}) = (\mathcal{E}^{(s)}, \mathcal{F}^{(s)} \cap L^2(I; m)) \tag{3.5.8}$$

on $L^2(I; m)$.

The boundary r_1 (resp. r_2) of I is said to be approachable if $s(r_1) > -\infty$ (resp. $s(r_2) < \infty$), and it is called regular if it is approachable and $m(r_1, c) < \infty$ (resp. $m(c, r_2) < \infty$) for $c \in I$. We denote by I^* the interval obtained from I by adjoining its boundary r_i whenever r_i is regular ($i = 1, 2$). m is extended to I^* by $m(\cdot) = m(\cdot \cap I)$.

According to Theorem 2.2.11, (3.5.8) is a regular, strongly local and irreducible Dirichlet form on $L^2(I^*; m)(= L^2(I; m))$. The Dirichlet form in (3.5.8) is transient if and only if either r_1 or r_2 is approachable but non-regular. Otherwise (3.5.8) is recurrent. Furthermore, In view of (2.2.40), the Dirichlet form (3.5.8) has a quite simple feature that every point of I^* is not $\mathcal{E}$-polar.

Therefore, by the remark made in advance, there exists a unique m-symmetric Hunt process $X = (X_t, \zeta, \mathbf{P}_x)$ on I^* with continuous sample path and no killing inside I^* associated with (3.5.8). On account of Theorem 3.5.6(i),

$$\mathbf{P}_x(\sigma_y < \infty) > 0 \quad \text{for every } x, y \in I^*.$$

If (3.5.8) is recurrent, then X is *point recurrent* in the sense that

$$\mathbf{P}_x(\sigma_y < \infty) = 1 \quad \text{for every } x, y \in I^*,$$

by Theorem 3.5.6(ii). When $I^* \setminus I \neq \emptyset$, X is called the *reflecting diffusion* on I^* associated with the pair (s, m) reflected at $I^* \setminus I$.

Let $X^0 = (X_t^0, \zeta^0, \mathbf{P}_x^0)$ be the part process of X on I. X^0 is obtained from X by killing upon the hitting time of $I^* \setminus I$. In view of Theorem 3.3.8, X^0 is m-symmetric and its Dirichlet form on $L^2(I; m)$ equals the part $(\mathcal{E}^I, \mathcal{F}_I)$ of the Dirichlet form (3.5.8) on $L^2(I)$ defined by (3.2.2) and (3.3.1).

Accordingly,

$$\begin{cases} \mathcal{F}_I = \left\{u \in \mathcal{F}^{(s)} \cap L^2(I; m) : u(r_i) = 0 \text{ if } \ r_i \ \text{ is regular}\right\}, \\ \mathcal{E}^I = \mathcal{E}^{(s)}\big|_{\mathcal{F}_I \times \mathcal{F}_I}. \end{cases} \tag{3.5.9}$$

By Theorem 3.3.9, $(\mathcal{E}^I, \mathcal{F}_I)$ is a regular Dirichlet form on $L^2(I; m)$. Furthermore, by Theorem 3.4.9 combined with (2.2.39), its extended Dirichlet space $(\mathcal{F}_{I,e}, \mathcal{E}^I)$ can be described as

$$\begin{cases} \mathcal{F}_{I,e} = \left\{u \in \mathcal{F}^{(s)} : u(r_i) = 0 \text{ if } r_i \ \text{ is approachable}\right\}, \\ \mathcal{E}^I = \mathcal{E}^{(s)}\big|_{\mathcal{F}_{I,e} \times \mathcal{F}_{I,e}}. \end{cases} \tag{3.5.10}$$

Suppose $u \in \mathcal{F}_{I,e}$ and $\mathcal{E}^{(s)}(u, u) = 0$. Then u is a constant, which vanishes if and only if either r_1 or r_2 is approachable. Hence $(\mathcal{E}^I, \mathcal{F}_I)$ is transient if either r_1 or r_2 is approachable and it is otherwise recurrent on account of Theorem 2.1.8 and Theorem 2.1.9.

X^0 enjoys the following properties:

(d.1) X^0 is a Hunt process on I.

(d.2) X^0 is a diffusion process: X_t^0 is continuous in $t \in (0, \zeta^0)$ almost surely.

(d.3) X^0 is irreducible: $\mathbf{P}^0_x(\sigma_y < \infty) > 0$ for any $x, y \in I$.
(d.4) X^0 admits no killing inside I: $\mathbf{P}^0_x(\zeta^0 < \infty, X^0_{\zeta^0-} \in I) = 0,\ x \in I$.

It follows from **(d.1)** and **(d.4)** that

$$\mathbf{P}^0_x(X^0_{\zeta^0-} \in \{r_1, r_2\},\ \zeta^0 < \infty) = \mathbf{P}^0_x(\zeta^0 < \infty) \quad \text{for every } x \in I. \tag{3.5.11}$$

Let us call a process X^0 satisfying properties **(d.1)**, **(d.2)**, and **(d.3)** a *minimal diffusion* or an absorbing diffusion on an open interval $I = (r_1, r_2)$. We notice that under **(d.1)** and **(d.2)**, the condition **(d.3)** is equivalent to the requirement for each point $a \in I$ to be regular in the sense that $\mathbf{E}_a[e^{-\sigma_{a+}}] = \mathbf{E}_a[e^{-\sigma_{a-}}] = 1$ where $E_a[e^{-\sigma_{a\pm}}] = \lim_{b\to\pm a} E_a[e^{-\sigma_b}]$.

We have started with a pair (s, m) of a canonical scale and a canonical measure on I and constructed an m-symmetric minimal diffusion X^0 on I with no killing inside whose Dirichlet form and extended Dirichlet space are given by (3.5.9) and (3.5.10), respectively.

Conversely, suppose we are given a minimal diffusion $X^0 = (X^0_t, \zeta^0, \mathbf{P}^0_x)$ on I with no killing inside. It is well known [95, 97] that there is an associated pair (s, m) of a canonical scale and a canonical measure such that, for each $J = (a, b)$, $r_1 < a < b < r_2$, they coincide with

$$s_J(x) = \mathbf{P}^0_x(\sigma_a > \sigma_b), \quad m_J(x) = -\frac{d\mathbf{E}^0_x[\sigma_a \wedge \sigma_b]}{ds_J(x)},\ x \in J,$$

respectively, up to a linear transformation. The pair associated with X^0 is unique up to a multiplicative constant $c > 0$ in the sense that, for another associated pair $(\widetilde{s}, \widetilde{m})$, $d\widetilde{s} = c\,ds$, $d\widetilde{m} = c^{-1}dm$.

The minimal diffusion X^0 is known to be symmetric with respect to an associated canonical measure m [95, 97]. We refer the reader to [69] for a proof of the following theorem.

THEOREM 3.5.8. *Let X^0 be a minimal diffusion on I with no killing inside. Then its Dirichlet form on $L^2(I; m)$ and its extended Dirichlet space are given by* (3.5.9) *and* (3.5.10), *respectively, in terms of an associated pair* (s, m).

In Example **(5°)** of Section 5.3, we shall identify the Dirichlet form and the extended Dirichlet space of a general minimal diffusion allowing killings inside. Thus any minimal diffusion on an open interval can be studied entirely in the framework of the Dirichlet form. See also a remark stated at the end of that example.

Feller's classification of the boundaries stated in K. Itô [95] and K. Itô-H. McKean [97] particularly is concerned with the approachability of X^0 to r_1, r_2 at a finite lifetime ζ^0. For $i = 1, 2$, we call r_i *approachable in finite time* if

$$\mathbf{P}^0_x(X^0_{\zeta^0-} = r_i,\ \zeta^0 < \infty) > 0, \qquad \text{for any } x \in I. \tag{3.5.12}$$

In view of (3.5.11), we see that X^0 is conservative, namely, $\mathbf{P}_x^0(\zeta^0 = \infty) = 1$ for every $x \in I$, if and only if neither r_1 nor r_2 is approachable in finite time.

In Theorem 5.15.2 of [95], it is proved that r_1 (resp. r_2) is approachable in finite time if and only if, for $c \in I$,

$$\int_{r_1}^{c} m((x,c))s(dx) < \infty \quad \left(\text{resp. } \int_{c}^{r_2} m((c,x))s(dx) < \infty\right). \tag{3.5.13}$$

When the condition (3.5.13) is fulfilled, r_i is called *exit or regular* (resp. *exit*) in the terminology of [95] (resp. [97]).

Suppose now r_1 is non-approachable, while r_2 is approachable but non-regular. Then $I^* = I$, $X = X^0$, and the Dirichlet form (3.5.8) is transient by Theorem 2.2.11. We can then show that

$$\mathbf{P}_x\left(\lim_{t\to\zeta} X_t = r_2\right) = 1 \quad \text{for } x \in I. \tag{3.5.14}$$

In fact, combining (3.5.11) with Theorem 3.5.2, we have

$$\mathbf{P}_x\left(\lim_{t\to\zeta} X_t \in \{r_1, r_2\}\right) = 1 \quad \text{for every } x \in I.$$

On the other hand, the extended Dirichlet space $\mathcal{F}_e$ is equal to $\mathcal{F}_e = \{u \in \mathcal{F}^{(s)} : u(r_2) = 0\}$ in the present case by Theorem 2.2.11(ii). In particular, $\mathcal{F}_e$ contains a function which is non-zero constant near r_1. If (3.5.14) were not true, we would encounter a contradiction to Corollary 3.5.3 which states that $f(X_t)$ converges to zero as $t \to \zeta$ $\mathbf{P}_x$-a.s. for any $f \in \mathcal{F}_e$. Property (3.5.14) legitimates our usage of the term "approachable".

For instance, if $I = (0,\infty)$, $ds = x^{1-n}dx$, $dm = 2x^{n-1}dx$, $n \geq 2$, then 0 is non-approachable, while ∞ is non-approachable when $n = 2$ and approachable, non-regular when $n \geq 3$. Hence the associated Dirichlet form $\mathcal{E}(u,v) = \int_I u'v'x^{n-1}dx$ on $L^2(I; 2x^{n-1}dx)$ is recurrent when $n = 2$ and transient when $n \geq 3$. Furthermore, both integrals in (3.5.13) for $r_1 = 0$, $r_2 = \infty$ diverge for any $n \geq 2$ and hence the associated diffusion $X = (X_t, \mathbf{P}_x)$ on $(0,\infty)$ is always conservative. When $n \geq 3$, we have from (3.5.14)

$$\mathbf{P}_x\left(\zeta = \infty,\ \lim_{t\to\infty} X_t = \infty\right) = 1, \quad \text{for any } x \in (0,\infty).$$

X is called the *Bessel process*. The corresponding generator $\frac{d}{dm}\frac{du}{ds}$ equals $\frac{1}{2}(u'' + \frac{n-1}{x}u')$, which is nothing but the radial part of half of the n-dimensional Laplace operator Δ. Accordingly, X is equivalent in law to the radial part $|B_t|$ of the n-dimensional Brownian motion and the above-mentioned properties of X are obvious from those of B_t.

As is directly derived at the end of Section 2.2.3 and will be used in the next example, we have

$$2\int |u(x)|g(x)x^{n-1}dx \le \left(\int_0^\infty (u')^2 x^{n-1}dx\right)^{1/2}, \quad u \in \mathcal{G}, \tag{3.5.15}$$

for some strictly positive bounded $x^{n-1}dx$-integrable function g, which expresses a transience inequality (2.1.11) for the Bessel process with $n \ge 3$. Here $\mathcal{G}$ denotes the space $\mathcal{F}$ defined by (3.5.8) for the present choice of I, s, m. □

Example 3.5.9. (Brownian motion and related diffusions) For $n \ge 1$, the n-dimensional *standard Brownian motion* (BM in abbreviation) X is a diffusion process on $\mathbb{R}^n$ with transition function (2.2.13). It is symmetric with respect to the Lebesgue measure and the associated Dirichlet form on $L^2(\mathbb{R}^n)$ is $(\frac{1}{2}\mathbf{D}, H^1(\mathbb{R}^n))$ given by (2.2.15).

Consider a domain D of $\mathbb{R}^n$. The part process X^D of X on D is called the *absorbing Brownian motion* on D. X^D is obtained from X by killing upon its hitting time of $\mathbb{R}^n \setminus D$. By Theorem 3.3.8, X^D is symmetric with respect to the Lebesgue measure and its Dirichlet form on $L^2(D)$ equals the part

$$\left(\tfrac{1}{2}\mathbf{D}, H_0^1(D)\right) \tag{3.5.16}$$

on D of the Dirichlet form (2.2.15):

$$H_0^1(D) = \{u \in H^1(\mathbb{R}^n) : \widetilde{u} = 0 \text{ q.e. on } \mathbb{R}^n \setminus D\}, \ (H_0^1(\mathbb{R}^n) = H^1(\mathbb{R}^n)). \tag{3.5.17}$$

The pair in (3.5.16) is a regular Dirichlet form on $L^2(D)$ and actually it has $C_c^\infty(D)$ as a core by virtue of Theorem 3.3.9(ii). Accordingly, it can be identified with the ($\mathcal{E}_1$-)closure of $C_c^\infty(D)$ in the Sobolev space $H^1(D)$ of (2.2.45), which is a customary way to introduce the space $H_0^1(D)$.

Let us denote by $H_{0,e}^1(D)$ the extended Dirichlet spaces of the Dirichlet form (3.5.16). On account of Theorem 3.4.9 and (3.5.17), we then have

$$H_{0,e}^1(D) = \{u \in H_e^1(\mathbb{R}^n) : \widetilde{u} = 0 \text{ q.e. on } \mathbb{R}^n \setminus D\}, \ (H_{0,e}^1(\mathbb{R}^n) = H_e^1(\mathbb{R}^n)). \tag{3.5.18}$$

PROPOSITION 3.5.10. (i) *When $n = 1$ or 2, X^D is transient if and only if* $\mathrm{Cap}_1(\mathbb{R}^n \setminus D) > 0$. *In this case, the lifetime of X^D is finite a.s.*
(ii) *When $n \ge 3$, X^D is always transient.*

Proof. (i) When $n = 1$, $\mathrm{Cap}_1(\mathbb{R}^1 \setminus D) > 0$ if and only if $\mathbb{R}^1 \setminus D \ne \emptyset$, which is in turn equivalent to the transience of X^D, as seen in the preceding example. Assume that $n = 2$. In view of (3.5.18) and Theorem 2.2.13, the extended

Dirichlet space of (3.5.16) equals

$$H^1_{0.e}(D) = \{u \in \mathrm{BL}(\mathbb{R}^2) : \widetilde{u} = 0 \text{ q.e. on } \mathbb{R}^2 \setminus D\}, \ \mathcal{E} = \frac{1}{2}\mathbf{D}.$$

Hence, if $u \in H^1_{0.e}(D)$ satisfies $\mathcal{E}(u,u) = 0$, then u is constant a.e. on $\mathbb{R}^2$ and hence $\widetilde{u}$ is constant q.e. on $\mathbb{R}^2$. If $\mathbb{R}^2 \setminus D$ is of positive capacity, $\widetilde{u}$ must be zero q.e. on $\mathbb{R}^2$, yielding the transience of the form by virtue of Theorem 2.1.9. If $\mathrm{Cap}_1(\mathbb{R}^2 \setminus D) = 0$, then $H^1_{0,e}(D) = H^1_e(\mathbb{R}^2) = \mathrm{BL}(\mathbb{R}^2)$, the recurrent extended Dirichlet space. The second assertion follows from the irreducible recurrence of X and Theorem 3.5.6.

(ii) When $n \geq 3$, the transience property (2.2.47) is fulfilled by $H^1_e(\mathbb{R}^n)$. Hence X^D is transient by (3.5.18) and Theorem 2.1.9. □

In what follows, we only consider a domain D possessing a continuous boundary in the sense of Section 2.2.4. Then the Dirichlet form $(\frac{1}{2}\mathbf{D}, H^1(D))$ becomes a regular strongly local Dirichlet form on $L^2(\overline{D})$ $(= L^2(\overline{D}; \mathbf{1}_D \cdot dx))$. Accordingly, there exists an associated diffusion process X^r on $\overline{D}$ uniquely up to the equivalence in the sense of Theorem 3.1.12. X^r is called a *reflecting Brownian motion* on $\overline{D}$. X^r is irreducible because its transition function dominates that of the absorbing Brownian motion on D and the latter is known to possesses a strictly positive density function. If either $n \leq 2$ or D is of finite Lebesgue measure, $X^r = (X_t, \mathbf{P}_x)$ is recurrent by virtue of Theorem 2.2.13 and Corollary 2.2.15. In particular, by Proposition 2.1.10, X^r is conservative in the sense that its lifetime is infinity $\mathbf{P}_x$-a.s. for q.e. $x \in \overline{D}$. Further, for any non-$\mathcal{E}$-polar nearly Borel set $B \subset \overline{D}$, it follows from Theorem 3.5.6 that

$$\mathbf{P}_x(\sigma_B \circ \theta_n < \infty, \ \forall n) = 1, \quad q.e. \ x \in \overline{D}. \tag{3.5.19}$$

Suppose that $n \geq 3$ and D is unbounded. Owing to Takeda's test (cf. [146], [73, Example 5.7.1]), we can verify that X^r is still conservative. But the transience of X^r depends on the geometric shape of the unbounded domain D. For instance, X^r is transient if $D = \mathbb{R}^3$, while X^r is recurrent if $D = \{x \in \mathbb{R}^3 : x_1 \in \mathbb{R}, \ x_2^2 + x_3^2 \leq 1\}$ the infinite tube. In the latter case, X^r is a direct product of the one-dimensional Brownian motion and the reflecting Brownian motion on the closed unit disk so that the sample path of X^r cannot escape to infinity as time goes to infinity. We notice that if $D_1 \subset D_2$ and the reflecting Brownian motion on $\overline{D}_1$ is transient, then so is the reflecting Brownian motion on $\overline{D}_2$ because of the recurrence criteria in Theorem 2.1.8.

We show that for a domain $D \subset \mathbb{R}^n$ with $n \geq 3$, the reflecting Brownian motion X^r on $\overline{D}$ is transient if D contains an infinite cone $C_A = \{r\xi \in \mathbb{R}^n : r > 0, \xi \in A\}$ where A is a non-void open subset of the unit sphere $\Sigma_1 = \{x \in \mathbb{R}^n : |x| = 1\}$. By the above observation, it suffices to derive the transience inequality (2.1.11) of the Dirichlet form $(\frac{1}{2}\mathbf{D}, \mathbf{H}^1(D))$ for $D = C_A$ by assuming that C_A has a continuous boundary.

Let $\mathcal{G}$ denote the space that appeared at the end of Example 3.5.7. Take any $u \in C_c^\infty(\overline{C}_A)$. Then $u(x) = u(r\xi)$ belongs to $\mathcal{G}$ as a function of r for each $\xi \in A$ and the inequality $u_r(r\xi)^2 \leq |\nabla u(x)|^2$ with $x = r\xi$ is readily verified using the polar coordinate. Hence, denoting the surface measure on Σ_1 by $\sigma(d\xi)$, it follows from (3.5.15) that

$$\int_{C_A} |u(x)| g(|x|) dx = \int_A \int_0^\infty |u(r\xi)| g(r) r^{n-1} dr \sigma(d\xi)$$

$$\leq \frac{\sqrt{\sigma(\Sigma_1)}}{2} \left(\int_A \int_0^\infty u_r(r\xi)^2 r^{n-1} dr d\sigma(\xi) \right)^{1/2} \leq \frac{\sqrt{\sigma(\Sigma_1)}}{2} \cdot \mathbf{D}(u,u)^{1/2},$$

which readily extends to $u \in H^1(C_A)$.

When the reflecting Brownian motion $X^r = (X_t, \mathbf{P}_x)$ on $\overline{D}$ is transient, X_t approaches to the point at infinity ∂ of $\overline{D}$ as $t \to \infty$ $\mathbf{P}_x$-a.s. for q.e. $x \in \overline{D}$ by virtue of Theorem 3.5.2. Further, any quasi continuous function in the extended Sobolev space $H_e^1(D)$ goes to zero along X_t as $t \to \infty$ by Corollary 3.5.3. In the transient case, $H_e^1(D)$ does not contain non-zero constant functions, while the larger space $\mathrm{BL}(D)$ always does.

When D is a bounded domain and a Lipschitz domain in the sense that the function F in the condition (2.2.55) can be taken to be Lipschitz continuous, we can choose a representative X^r of the equivalence class of reflecting Brownian motions such that its transition function $\{P_t^*; t \geq 0\}$ is strong Feller in the sense that $P_t^*(b\mathcal{B}(\overline{D})) \subset C(\overline{D})$ for every $t > 0$.[1] Since X^r then satisfies the absolute continuity condition **(AC)**, we see from Remark 4.3.5 that X^r is a conservative diffusion starting at every point of $\overline{D}$ and satisfies (3.5.19) for every $x \in \overline{D}$. □

[1] [15].

Chapter Four

ADDITIVE FUNCTIONALS OF SYMMETRIC MARKOV PROCESSES

This chapter is devoted to the study of additive functionals of symmetric Markov processes under the same setting as in the preceding chapter, namely, we let E be a locally compact separable metric space, $\mathcal{B}(E)$ be the family of all Borel sets of E, and m be a positive Radon measure on E with $\text{supp}[m] = E$, and we consider an m-symmetric Hunt process $X = (\Omega, \mathcal{M}, X_t, \zeta, \mathbf{P}_x)$ on $(E, \mathcal{B}(E))$ whose Dirichlet form $(\mathcal{E}, \mathcal{F})$ on $L^2(E; m)$ is regular on $L^2(E; m)$. The transition function and the resolvent of X are denoted by $\{P_t; t \geq 0\}$, $\{R_\alpha, \alpha > 0\}$, respectively. $\mathcal{B}^*(E)$ will denote the family of all universally measurable subsets of E. Any numerical function f defined on E will be always extended to E_∂ by setting $f(\partial) = 0$.

4.1. POSITIVE CONTINUOUS ADDITIVE FUNCTIONALS AND SMOOTH MEASURES

We shall adopt the notions and results in Section A.3.1 of Appendix A, where, under a more general setting that X is a Borel right process on a Lusin space E with an excessive measure m, the notion of an *additive functional* (*admitting an exceptional set*) of X, the concept of the *m-equivalence* of two additive functionals as well as the notion of a *positive continuous additive functional* (*PCAF* in abbreviation) are introduced. It is further proved in Theorem A.3.5 that for any PCAF A of X, a measure μ_A on $(E, \mathcal{B}(E))$ called the *Revuz measure* of A is uniquely determined.

Under the present more special setting, m is of course excessive with respect to the m-symmetric Hunt process X, and accordingly we can use those notions and results in Section A.3.1 without any change except for one necessary modification mentioned below.

Recall that a set $N \subset E$ is said to be an m-inessential set (resp. a properly exceptional set) for a Borel right process (resp. for a Hunt process) X if N is an m-negligible nearly Borel set such that $E \setminus N$ is X-invariant. See Section A.1.3 for the X-invariance. By Remark A.1.30, the restriction of a Borel right process (resp. a Hunt process) to its invariant set is again a Borel right process (resp. a Hunt process). According to Lemma A.1.32, N is properly exceptional for a Hunt process X if and only if it is m-inessential for X regarded as a Borel right process. Therefore, as the exceptional set N appearing in Definition A.3.1 of an additive functional A, an *m-inessential set* for a Borel right process X can now be replaced by a *properly exceptional set* for the Hunt process X. In what follows, we shall use the notion of an additive functional for the present Hunt process X with this replacement.

Let us denote by $\mathbf{A}_c^+$ the totality of positive continuous additive functionals of the m-symmetric Hunt process X. A, $B \in \mathbf{A}_c^+$ are called *m-equivalent* if $\mathbf{P}_m(A_t \neq B_t) = 0$ for every $t > 0$. We write $A \sim B$ in this case. By virtue of Lemma A.3.2 and the above remark, we see that $A \sim B$ if and only if there are a common defining set Λ and a common Borel exceptional set N such that

$$A_t(\omega) = B_t(\omega) \quad \text{for every } t \geq 0 \text{ and } \omega \in \Lambda.$$

We are concerned with an analytic characterization of the family $\mathbf{A}_c^+/\!\sim$ of all equivalence classes. The notion of a *smooth measure* with respect to a regular Dirichlet form $(\mathcal{E}, \mathcal{F})$ is introduced in Definition 2.3.13. The family of all smooth measures is denoted by S.

A purpose of this section is to prove the next theorem on the one-to-one correspondence of $\mathbf{A}_c^+/\!\sim$ and S formulated in terms of the notion of the Revuz measure. For $A \in \mathbf{A}_c^+$, its Revuz measure will be denoted by μ_A. According to Theorem A.3.5, μ_A is characterized by the following formula: for any $f \in \mathcal{B}_+(E)$,

$$\begin{aligned} \langle \mu_A, f \rangle &= \lim_{t \downarrow 0} \frac{1}{t} \mathbf{E}_m \left[\int_0^t f(X_s) dA_s \right] \\ &= \lim_{\alpha \to \infty} \alpha \mathbf{E}_m \left[\int_0^\infty e^{-\alpha t} f(X_s) dA_s \right]. \end{aligned} \tag{4.1.1}$$

THEOREM 4.1.1. (i) *For any $A \in \mathbf{A}_c^+$, $\mu_A \in S$.*
(ii) *For any $\mu \in S$, there exists $A \in \mathbf{A}_c^+$ satisfying $\mu_A = \mu$ uniquely up to the m-equivalence.*
(iii) *For $A \in \mathbf{A}_c^+$ and $\mu \in S$, the following three conditions are mutually equivalent.*

(a) $\mu_A = \mu$.

(b) *For any $f, h \in \mathcal{B}_+(E)$ and $t > 0$,*

$$\mathbf{E}_{h \cdot m} \left[\int_0^t f(X_s) dA_s \right] = \int_0^t \langle P_s h, f \cdot \mu \rangle ds. \tag{4.1.2}$$

(c) *For any $f, h \in \mathcal{B}_+(E)$ and $\alpha > 0$,*

$$\mathbf{E}_{h \cdot m} \left[\int_0^\infty e^{-\alpha s} f(X_s) dA_s \right] = \langle R_\alpha h, f \cdot \mu \rangle. \tag{4.1.3}$$

We shall use the following notations in the sequel: for $A \in \mathbf{A}_c^+$, $f \in b\mathcal{B}(E)$,

$$U_A^\alpha f(x) = \mathbf{E}_x\left[\int_0^\infty e^{-\alpha t} f(X_t) dA_t\right], \quad \alpha > 0,\ x \in E \setminus N, \tag{4.1.4}$$

$$R_\alpha^A f(x) = \mathbf{E}_x\left[\int_0^\infty e^{-\alpha t} e^{-A_t} f(X_t) dt\right], \quad \alpha > 0,\ x \in E \setminus N, \tag{4.1.5}$$

where N is an exceptional set of A.

Exercise 4.1.2. Derive the following identities using $e^{-\beta t} - e^{-\alpha t} = (\alpha - \beta)e^{-\beta t} \int_0^t e^{-(\alpha-\beta)s} ds$ and $e^{A_t} - 1 = \int_0^t e^{A_s} dA_s$, respectively. For $\alpha,\ \beta > 0$, $f \in b\mathcal{B}(E)$,

$$U_A^\alpha f(x) - U_A^\beta f(x) + (\alpha - \beta) R_\alpha U_A^\beta f(x) = 0, \quad x \in E \setminus N, \tag{4.1.6}$$

$$R_\alpha^A f(x) - R_\alpha f(x) + U_A^\alpha R_\alpha^A f(x) = 0, \quad x \in E \setminus N. \tag{4.1.7}$$

LEMMA 4.1.3. *Let $A \in \mathbf{A}_c^+$, $f \in \mathcal{B}_+(E)$. If $U_A^1 1 \le R_1 f$, $[m]$, then $\mu_A(E) \le \langle m, f\rangle$.*

Proof. By Lemma A.2.11, $\beta m R_\beta \le m$. From (4.1.6) and the resolvent equation (1.1.25), we have for $\beta > 1$,

$$\begin{aligned}\beta\langle m, R_\beta f - U_A^\beta 1\rangle &= \beta\langle m, R_1 f - U_A^1 1\rangle - (\beta - 1)\beta\langle m, R_\beta(R_1 f - U_A^1 1)\rangle \\ &\ge \beta\langle m, R_1 f - U_A^1 1\rangle - (\beta - 1)\langle m, R_1 f - U_A^1 1\rangle \\ &= \langle m, R_1 f - U_A^1 1\rangle \ge 0.\end{aligned}$$

By (4.1.1), $\mu_A(E) = \lim_{\beta\to\infty} \beta\langle m, U_A^\beta 1\rangle \le \lim_{\beta\to\infty} \beta\langle m R_\beta, f\rangle = \langle m, f\rangle$. □

Proof of Theorem 4.1.1(i). For $A \in \mathbf{A}_c^+$, with a properly exceptional set N and a strictly positive bounded function $f \in \mathcal{B}(E) \cap L^1(E; m)$, we let φ be the function $R_1^A f$ defined by (4.1.5). Then $\varphi(x) > 0$ for every $x \in E \setminus N$. By (4.1.7), φ is a difference of finite 1-excessive functions $R_1 f$, $U_A^1(R_1^A f)$ relative to the Hunt process $X_{E\setminus N}$ and hence finely continuous q.e. relative to X, and accordingly quasi continuous by virtue of Theorem 3.3.3.

Since N is $\mathcal{E}$-polar by Theorem 3.1.3, by Theorem 1.3.14(iii), there is a Cap_1-nest $\{E_n\}$ so that $N \subset \cap_n(E \setminus E_n)$ and $\varphi|_{E_n}$ is continuous for each n. If we put

$$F_n = \left\{x \in E_n : \varphi(x) \ge \frac{1}{n}\right\},$$

then the increasing sequence $\{F_n\}$ of closed sets becomes a nest. To see this, put $B_n = \{x \in E \setminus N : \varphi(x) \le \frac{1}{n}\}$, $\sigma_n = \sigma_{B_n}$, $\sigma = \lim_{n\to\infty} \sigma_n$. Since φ is finely continuous relative to $X_{E\setminus N}$, we have for $x \in E \setminus N$

$$\mathbf{E}_x\left[\int_{\sigma_n}^{\zeta} e^{-t} f(X_t) e^{-A_t} dt\right] = \mathbf{E}_x\left[e^{-\sigma_n} e^{-A_{\sigma_n}} \varphi(X_{\sigma_n})\right] \le \frac{1}{n}.$$

By letting $n \to \infty$, we get $\mathbf{P}_x(\sigma < \zeta) = 0$, $x \in E \setminus N$. Hence, by the inclusion $E \setminus F_n \subset (E \setminus E_n) \cup B_n$ and Theorem 3.1.4, we can conclude that $\{F_n\}$ is a nest.

On account of Theorem A.3.5, the Revuz measure μ_A of A charges no m-polar set and hence it charges no $\mathcal{E}$-polar set. By the same theorem, we see that for each n, the Revuz measure of the positive continuous additive functional $A_n = \mathbf{1}_{F_n} \cdot A$ equals $\mathbf{1}_{F_n} \cdot \mu_A$. Since

$$U^1_{A_n} 1 = U^1_A \mathbf{1}_{F_n} \le n U^1_A \varphi \le n R_1 f,$$

we get $\mu_A(F_n) \le n \int_E f dm < \infty$ by Lemma 4.1.3. Thus $\mu_A \in S$. □

We next prove Theorem 4.1.1(iii). We prepare a lemma to this end.

LEMMA 4.1.4. *Suppose $A \in \mathbf{A}_c^+$, $\mu_A(E) < \infty$. Then*

$$\langle R_\alpha h, \mu_A \rangle \le \langle h \cdot m, U^\alpha_A 1 \rangle \quad \textit{for any } \alpha > 0 \textit{ and } h \in \mathcal{B}_+(E). \tag{4.1.8}$$

Proof. We may assume that $h \in b\mathcal{B}_+(E) \cap L^1(E; m)$. By virtue of Theorem A.3.5(iv),

$$\langle R_\alpha h, \mu_A \rangle = \lim_{\beta \to \infty} \beta \mathbf{E}_m \left[\int_0^\infty e^{-\beta s} R_\alpha h(X_s) dA_s\right]. \tag{4.1.9}$$

Since the discontinuous points of the sample path $s \mapsto X_s$ are at most countable and A is continuous, we have

$$\mathbf{E}_m \left[\int_0^\infty e^{-\beta s} R_\alpha h(X_s) dA_s\right] = \mathbf{E}_m \left[\int_0^\infty e^{-\beta s} R_\alpha h(X_{s-}) dA_s\right]. \tag{4.1.10}$$

We notice that if we put $Z_n := \sum_{k \ge 0} e^{-\beta \frac{k}{n}} R_\alpha h(X_{\frac{k}{n}-})(A_{\frac{k+1}{n}} - A_{\frac{k}{n}})$ and $Y := \sum_{k \ge 0} e^{-\beta k}(A_{k+1} - A_k)$, then $Z_n \le \frac{\|h\|_\infty}{\alpha} Y$ for $n \ge 1$ and

$$\begin{aligned} \mathbf{E}_m[Y] &= \sum_{k \ge 0} e^{-\beta k} \mathbf{E}_m\left[\mathbf{E}_{X_k}(A_1)\right] \le \sum_{k \ge 0} e^{-\beta k} \mathbf{E}_m[A_1] \\ &\le \frac{1}{1 - e^{-\beta}} \mu_A(E) < \infty. \end{aligned}$$

Since $R_\alpha h$ is an α-excessive function belonging to $\mathcal{F}$, $R_\alpha h(X_{s-})$ is left continuous in $s \in (0, \zeta)$ $\mathbf{P}_m$-a.e. on account of Theorem 3.1.7. Moreover, the measure dA_s concentrates on $(0, \zeta)$, $R_\alpha h$ is uniformly bounded, and $e^{-\beta s} dA_s$

is integrable $\mathbf{P}_m$-a.e. Consequently, Z_n converges to $\int_0^\infty e^{-\beta s} R_\alpha h(X_{s-})dA_s$ as $n \to \infty$ $\mathbf{P}_m$-a.e. Combining this with the above notice, we can see that $\lim_{n\to\infty} \mathbf{E}_m[Z_n]$ coincides with the right hand side of (4.1.10), and hence $\lim_{\beta\to\infty} \lim_{n\to\infty} \beta \mathbf{E}_m[Z_n]$ is identical with the left hand side of (4.1.9).

Because of the quasi-left-continuity of X, we have $X_{(k/n)-} = X_{k/n}$ $\mathbf{P}_m$-a.e., and so we get by the Markov property

$$\begin{aligned}
\langle R_\alpha h, \mu_A \rangle &= \lim_{\beta\to\infty} \lim_{n\to\infty} \beta \mathbf{E}_m[Z_n] \\
&= \lim_{\beta\to\infty} \lim_{n\to\infty} \beta \sum_{k\geq 0} \mathbf{E}_m\left[e^{-\beta k/n} R_\alpha h(X_{k/n}) \mathbf{E}_{X_{k/n}}[A_{1/n}]\right] \\
&\leq \lim_{\beta\to\infty} \lim_{n\to\infty} \beta \sum_{k\geq 0} e^{-\beta k/n} \int_E R_\alpha h(y) \mathbf{E}_y[A_{1/n}] m(dy) \\
&= \lim_{\beta\to\infty} \lim_{n\to\infty} \frac{\beta}{1 - e^{-\beta/n}} (h, R_\alpha(\mathbf{E}_\cdot[A_{1/n}])).
\end{aligned}$$

Since $\frac{1}{n}\frac{\beta}{1-e^{-\beta/n}} \to 1$, $n \to \infty$, for each $\beta > 0$, we are led to

$$\begin{aligned}
\langle R_\alpha h, \mu_A \rangle &\leq \lim_{n\to\infty} n \int_E h(y) R_\alpha(\mathbf{E}_\cdot[A_{1/n}])(y) m(dy) \\
&= \lim_{n\to\infty} \int_0^\infty n e^{-\alpha t} \mathbf{E}_{h\cdot m}\left[\mathbf{E}_{X_t}[A_{1/n}]\right] dt \\
&= \lim_{n\to\infty} \int_0^\infty n e^{-\alpha t} \mathbf{E}_{h\cdot m}[A_{t+(1/n)} - A_t] dt \\
&= \lim_{n\to\infty} \left(n(e^{\alpha/n} - 1) \int_{1/n}^\infty e^{-\alpha s} \mathbf{E}_{h\cdot m}[A_s] ds - n \int_0^{1/n} e^{-\alpha s} \mathbf{E}_{h\cdot m}[A_s] ds \right) \\
&= \alpha \int_0^\infty e^{-\alpha s} \mathbf{E}_{h\cdot m}[A_s] ds = \langle h \cdot m, U_A^\alpha 1 \rangle.
\end{aligned}$$

The last equality is obtained by a computation similar to (A.3.10). □

Proof of Theorem 4.1.1(iii). Since the Laplace transform of both sides of (4.1.2) gives (4.1.3), the implication **(c)** ⇒ **(b)** follows from the uniqueness of the Laplace transform. Substituting $h = 1$ in the equality (4.1.2), we get the first identity in (4.1.1) with μ_A being replaced by μ. This means the implication **(b)** ⇒ **(a)**. Hence it suffices to show the implication **(a)** ⇒ **(c)**. To this end, we only need to prove the identity

$$\langle R_\alpha h, \mu_A \rangle = \langle h \cdot m, U_A^\alpha 1 \rangle \quad \text{for any } \alpha > 0 \quad \text{and } h \in \mathcal{B}_+(E), \tag{4.1.11}$$

holding for any $A \in \mathbf{A}_c^+$, because, if (4.1.11) is established, then we know the validity of the same identity with μ_A, 1 being replaced by $f \cdot \mu_A$, f ($f \in \mathcal{B}_+(E)$), respectively, owing to Theorem A.3.5(iii).

We have already shown in Lemma 4.1.4 that when $A \in \mathbf{A}_c^+$ and $\mu_A(E) < \infty$, (4.1.11) holds with the equality being replaced by the inequality $\leq$. Using this, let us first derive the equality (4.1.11) when $A \in \mathbf{A}_c^+$ satisfies

$$\mu_A(E) < \infty, \qquad \mu_A \in S_0. \tag{4.1.12}$$

For $\mu \in S_0$, its α-potential $U_\alpha\mu \in \mathcal{F}$ is defined by (2.3.6). By (1.1.28), (2.3.7), and Proposition 3.1.9, we have for any $\alpha > 0$ and $h \in b\mathcal{B}_+(E) \cap L^1(E; m)$

$$\langle R_\alpha h, \mu\rangle = \mathcal{E}_\alpha(R_\alpha h, U_\alpha\mu) = \langle h \cdot m, U_\alpha\mu\rangle. \tag{4.1.13}$$

Therefore, by the preceding lemma, we get the inequality $U_\alpha\mu_A \leq U_A^\alpha 1$ $[m]$ holding for any $\alpha > 0$.

Let us put $g_\alpha = U_A^\alpha 1 - U_\alpha\mu_A$, $\alpha > 0$. From equations (2.3.8) and (4.1.6),

$$\langle m, g_\alpha\rangle = \beta\langle m, R_\alpha g_{\beta+\alpha}\rangle + \langle m, g_{\beta+\alpha}\rangle \leq \left(\frac{1}{\alpha} + \frac{1}{\beta}\right)\beta\langle m, g_{\beta+\alpha}\rangle.$$

If we let $\beta \to \infty$, then $\beta\langle m, U_A^{\alpha+\beta} 1\rangle \to \mu_A(E)$ by Theorem A.3.5(iv) and $\beta\langle m, U_{\alpha+\beta}\mu_A\rangle = \beta\langle R_{\alpha+\beta} 1, \mu_A\rangle \to \mu_A(E)$ by (4.1.13). Hence we get $g_\alpha = 0$ $[m]$, which combined with (4.1.13) again implies (4.1.11).

For any $A \in \mathbf{A}_c^+$, $\mu_A \in S$ by the already shown statement (i) of Theorem 4.1.1 and hence, in accordance with Theorem 2.3.15, there exists a nest $\{F_k\}$ such that the measure $\mathbf{1}_{F_k} \cdot \mu_A$ satisfies the condition (4.1.12) for each k. If we put $A_k = \mathbf{1}_{F_k} \cdot A$, then $\mu_{A_k} = \mathbf{1}_{F_k} \cdot \mu_A$ by Theorem A.3.5(iii) and we have the identity (4.1.11) for A_k. By letting $k \to \infty$ and using the stochastic characterization Theorem 3.1.4, we arrive at (4.1.11) for A. □

Theorem 4.1.1(ii) follows from two propositions presented below.

LEMMA 4.1.5. *Consider a measure μ on E satisfying*

$$\mu \in S_0, \quad \mu(E) < \infty. \tag{4.1.14}$$

A necessary and sufficient condition for $A \in \mathbf{A}_c^+$ to satisfy $\mu_A = \mu$ is

$$U_A^1 1 = U_1\mu \quad [m]. \tag{4.1.15}$$

In this case, it also holds that

$$U_A^\alpha 1 = U_\alpha\mu \quad [m] \quad \textit{for any } \alpha > 0,$$

where $U_\alpha\mu$ denotes the α-potential of μ.

Proof. Suppose μ satisfies (4.1.14). If $A \in \mathbf{A}_c^+$ satisfies $\mu_A = \mu$, then, by substituting this into the left hand side of (4.1.11), we get by (4.1.13) that $(h, U_1\mu) = (h, U_A^1 1)$ for every $h \in b\mathcal{B}_+(E) \cap L^1(E; m)$, from which follows (4.1.15).

Conversely, if the equality (4.1.15) holds, then, by equations (2.3.8) and (4.1.6), we have $U_A^\alpha 1 = U_\alpha \mu$ $[m]$ for any $\alpha > 0$. By substituting this into the right hand side of (4.1.11), we obtain $\alpha\langle R_\alpha h, \mu_A\rangle = \alpha\langle R_\alpha h, \mu\rangle$ for every $h \in \mathcal{B}_+(E)$. We put $h = 1$ and let $\alpha \to \infty$ to get $\mu_A(E) = \mu(E) < \infty$. Then the same procedure for $h \in bC_+(E)$ gives $\langle h, \mu_A\rangle = \langle h, \mu\rangle$, yielding $\mu_A = \mu$. □

PROPOSITION 4.1.6. *For any $\mu \in S_{00}$, there exists $A \in \mathbf{A}_c^+$ such that $\mu_A = \mu$.*

Proof. Since $\mu \in S_{00}$ satisfies (4.1.14), it suffices to construct $A \in \mathbf{A}_c^+$ satisfying (4.1.15) by virtue of Lemma 4.1.5. The 1-potential $U_1\mu$ of μ is in $\mathcal{F}$ and 1-excessive relative to $\{T_t\}$, and hence we can find a finite Borel measurable quasi continuous version f of $U_1\mu$ and a Borel properly exceptional set N such that

$$\begin{cases} nR_{n+1}f(x) \uparrow f(x),\ n \to \infty & \text{for any } x \in E \setminus N \\ f(x) = 0, & \text{for any } x \in N. \end{cases} \tag{4.1.16}$$

We put

$$g_n(x) = \begin{cases} n(f(x) - nR_{n+1}f(x)), & x \in E \setminus N \\ 0, & x \in N. \end{cases}$$

Then $R_1 g_n(x) \uparrow f(x),\ x \in E \setminus N,\ n \to \infty$ and further $R_1 g_n$ is $\mathcal{E}_1$-convergent to f because $\|R_1 g_n - f\|_{\mathcal{E}_1}^2 \le -(g_n, f) + \|f\|_{\mathcal{E}_1}^2 \to 0,\ n \to \infty$.

For each n, we define the functional $\widetilde{A}_n$ by

$$\widetilde{A}_n(t, \omega) = \int_0^t e^{-s} g_n(X_s(\omega))ds, \quad t \ge 0,\ \omega \in \Omega.$$

Let us show for any $\nu \in S_{00}$ that

$$\mathbf{E}_\nu\left[(\widetilde{A}_n(\infty) - \widetilde{A}_\ell(\infty))^2\right] \le 2M_\nu\sqrt{\mathcal{E}_1(\mu)}\,\|R_1 g_n - R_1 g_\ell\|_{\mathcal{E}_1} \tag{4.1.17}$$

with $M_\nu = \|U_2\nu\|_\infty$. Without loss of generality, we assume that $\nu \in S_{00}$ is a probability measure on E.

If we put $g_{n,\ell} = g_n - g_\ell$, $n > \ell$, then the left hand side of (4.1.17) equals

$$\begin{aligned}
&2\mathbf{E}_\nu \left[\int_0^\infty e^{-s} g_{n,\ell}(X_s) ds \int_s^\infty e^{-u} g_{n,\ell}(X_u) du \right] \\
&= 2\mathbf{E}_\nu \left[\int_0^\infty e^{-2s} (g_{n,\ell} \cdot R_1 g_{n,\ell})(X_s) ds \right] = 2\langle \nu, R_2(g_{n,\ell} R_1 g_{n,\ell}) \rangle \\
&= 2\langle U_2 \nu, g_{n,\ell} R_1 g_{n,\ell} \rangle \le 2\langle U_2 \nu, g_n R_1 g_{n,\ell} \rangle \\
&\le 2M_\nu (g_n, R_1 g_n - R_1 g_\ell) = 2M_\nu \mathcal{E}_1(R_1 g_n, R_1 g_n - R_1 g_\ell).
\end{aligned}$$

Then the Schwarz inequality and the bound

$$\|R_1 g_n\|_{\mathcal{E}_1}^2 = (g_n, R_1 g_n) \le (g_n, f) = \langle \mu, R_1 g_n \rangle \le \langle \mu, f \rangle = \mathcal{E}_1(\mu)$$

lead us to (4.1.17).

On the other hand, we have

$$\mathbf{E}_\nu \left[\widetilde{A}_n(\infty) | \mathcal{F}_t \right] = \widetilde{A}_n(t) + e^{-t} E_{X_t}[\widetilde{A}_n(\infty)] = \widetilde{A}_n(t) + e^{-t} R_1 g_n(X_t).$$

Accordingly,

$$M_n(t) = \widetilde{A}_n(t) + e^{-t} R_1 g_n(X_t), \quad 0 \le t \le \infty \tag{4.1.18}$$

is a martingale relative to $(\{\mathcal{F}_t\}, \mathbf{P}_\nu)$, $\nu \in S_{00}$, and, by Doob's inequality, it holds for any $\varepsilon > 0$ that

$$\mathbf{P}_\nu \left(\sup_{0 \le t \le \infty} |M_n(t) - M_\ell(t)| > \varepsilon \right) \le \frac{1}{\varepsilon^2} \mathbf{E}_\nu \left[(\widetilde{A}_n(\infty) - \widetilde{A}_\ell(\infty))^2 \right]. \tag{4.1.19}$$

Choose a subsequence $\{n_k\}$ satisfying $\|R_1 g_{n_{k+1}} - R_1 g_{n_k}\|_{\mathcal{E}_1} \le 2^{-3k}$ and put $\Lambda_k = \{\sup_{0 \le t \le \infty} |M_{n_k}(t) - M_{n_{k+1}}(t)| > 2^{-k}\}$. It follows from (4.1.17) and (4.1.19) that $\mathbf{P}_\nu(\Lambda_k) \le 2M_\nu \sqrt{\mathcal{E}_1(\mu)} 2^{-k}$. By the Borel-Cantelli lemma, we are led to $\mathbf{P}_\nu(\limsup_{k \to \infty} \Lambda_k) = 0$ for every $\nu \in S_{00}$, which further implies that $\mathbf{P}_x(\limsup_{k \to \infty} \Lambda_k) = 0$ for q.e. $x \in E$, on account of Corollary 2.3.11.

Combining this with the expression (4.1.18) and Theorem 3.5.4, we arrive at the following conclusion: there exist a subsequence $\{n_k\}$ and a properly exceptional set $\widetilde{N} \supset N$ such that $\mathbf{P}_x(\Lambda) = 1$ for every $x \in E \setminus \widetilde{N}$, where

$$\begin{aligned}
\Lambda = \{\omega \in \widetilde{\Omega} : \widetilde{A}_{n_k}(\infty, \omega) < \infty, \ \widetilde{A}_{n_k}(t, \omega) \text{ is uniformly} \\
\text{convergent on each finite subinterval of } [0, \infty)\},
\end{aligned} \tag{4.1.20}$$

and $\widetilde{\Omega} = \{\omega \in \Omega : \sigma_{\widetilde{N}} = \infty, \ \widehat{\sigma}_{\widetilde{N}} = \infty\}$.

For $\omega \in \Lambda$, we put $\widetilde{A}(t, \omega) = \lim_{k \to \infty} \widetilde{A}_{n_k}(t, \omega)$, while, for $\omega \notin \Lambda$ we let $\widetilde{A}(t, \omega) = 0$. Finally, we define $A(t, \omega) = \int_0^t e^s d\widetilde{A}(s, \omega)$, $t \in [0, \infty]$. Then A is a positive continuous additive functional of X with defining set Λ and exceptional set $\widetilde{N}$.

In order to complete the proof of Proposition 4.1.6, it suffices to show $\mathbf{E}_\nu[\widetilde{A}(\infty)] = \langle \nu, f\rangle$ for every $\nu \in S_{00}$, on account of Corollary 2.3.11. Since $M_n(\infty) = \widetilde{A}_n(\infty)$ is $L^2(\mathbf{P}_\nu)$-convergent, so is $M_n(t)$. Therefore,

$$\mathbf{E}_\nu[\widetilde{A}(t)] + e^{-t}\langle \nu, P_t f\rangle = \lim_{n\to\infty} \mathbf{E}_\nu[M_n(t)]$$
$$= \lim_{n\to\infty} \mathbf{E}_\nu[\widetilde{A}_n(\infty)] = \lim_{n\to\infty} \langle \nu, R_1 g_n\rangle = \langle \nu, f\rangle.$$

Since $\mu \in S_{00}$, $\|f\|_\infty < \infty$ and $\langle \nu, P_t f\rangle \le \nu(E)\|f\|_\infty < \infty$. We get the desired equality by letting $t \to \infty$. □

Exercise 4.1.7. Show that for $A,\ B \in \mathbf{A}_c^+$, the identity

$$\mathbf{E}_x\left[\int_0^\infty e^{-t} dA_t \cdot \int_0^\infty e^{-s} dB_s\right] = U_A^2(U_B^1 1)(x) + U_B^2(U_A^1 1)(x) \quad (4.1.21)$$

holds for $x \in E \setminus (N_A \cup N_B)$, N_A, N_B being exceptional sets of A, B, respectively.

PROPOSITION 4.1.8. *For a given $\mu \in S_{00}$, $A \in \mathbf{A}_c^+$ satisfying $\mu_A = \mu$ is unique up to the m-equivalence.*

Proof. Given $\mu \in S_{00}$, suppose $A^{(1)}$, $A^{(2)} \in \mathbf{A}_c^+$ satisfy $\mu_{A^{(1)}} = \mu_{A^{(2)}} = \mu$. Denoting by f a Borel measurable quasi continuous version of $U_1\mu$, we see from Lemma 4.1.5 that $U^1_{A^{(1)}} 1 = U^1_{A^{(2)}} 1 = f$ q.e. We use the notation $\widetilde{A}_t = \int_0^t e^{-s} dA_s$, $t \in [0,\infty]$ for $A \in \mathbf{A}_c^+$. If we put

$$g_{ij}(x) = \mathbf{E}_x\left[\widetilde{A}^{(i)}_\infty \cdot \widetilde{A}^{(j)}_\infty\right], \quad i,j = 1 \text{ or } 2,$$

then, by Exercise 4.1.7, $g_{ij} = U^2_{A^{(i)}} f + U^2_{A^{(j)}} f$ q.e.

For any strictly positive bounded m-integrable function h on E, we get from (4.1.3)

$$\langle hm, g_{ij}\rangle = \langle R_2 h, f\cdot \mu_{A^{(i)}}\rangle + \langle R_2 h, f\cdot \mu_{A^{(j)}}\rangle$$
$$= 2\langle R_2 h, f\mu\rangle \le \|h\|_\infty \|f\|_\infty \mu(E) < \infty.$$

Consequently,

$$\mathbf{E}_{h\cdot m}\left[(\widetilde{A}^{(1)}_\infty - \widetilde{A}^{(2)}_\infty)^2\right] = \langle hm, g_{11} - 2g_{12} + g_{22}\rangle = 0.$$

Since $\widetilde{A}^{(i)}_\infty = \widetilde{A}^{(i)}_t + e^{-t}\widetilde{A}^{(i)}_\infty \circ \theta_t$ and

$$\mathbf{E}_{h\cdot m}\left[(\widetilde{A}^{(1)}_\infty - \widetilde{A}^{(2)}_\infty)^2 \circ \theta_t\right] = \mathbf{E}_{P_t h\cdot m}\left[(\widetilde{A}^{(1)}_\infty - \widetilde{A}^{(2)}_\infty)^2\right] = 0,$$

it follows that $\mathbf{E}_{h\cdot m}[(\widetilde{A}^{(1)}_t - \widetilde{A}^{(2)}_t)^2] = 0$ for every $t \ge 0$, yielding the m-equivalence of $A^{(1)}$, $A^{(2)}$. □

Proof of Theorem 4.1.1(ii). For any $\mu \in S$, there exists a nest $\{F_k\}$ such that $\mu^{(k)} = \mathbf{1}_{F_k} \cdot \mu \in S_{00}$ for each k by virtue of Theorem 2.3.15.

By Proposition 4.1.6, there exists $A^{(k)} \in \mathbf{A}_c^+$ satisfying $\mu_{A^{(k)}} = \mu^{(k)}$ for each k. According to Theorem A.3.5, the Revuz measure of $\mathbf{1}_{F_k} \cdot A^{(k+1)}$ equals $\mathbf{1}_{F_k} \cdot \mu_{A^{(k+1)}} = \mathbf{1}_{F_k} \cdot \mu^{(k+1)} = \mu^{(k)}$, and consequently, $\mathbf{1}_{F_k} \cdot A^{(k+1)}$ and $A^{(k)}$ are m-equivalent, $k = 1, 2, \ldots$, on account of Proposition 4.1.8.

By Lemma A.3.2, we can take a common defining set $\Lambda \subset \Omega$ and an exceptional set $N \subset E$ for $\{A^{(k)}\}$ such that

$$(\mathbf{1}_{F_k} \cdot A^{(k+1)})_t(\omega) = A_t^{(k)}(\omega) \quad \text{for any } t \geq 0,\ \omega \in \Lambda \text{ and } k \geq 1.$$

Since $\{F_k\}$ is a nest, we may assume that

$$\sigma(\omega) = \lim_{k \to \infty} \sigma_{E \setminus F_k}(\omega) \geq \zeta(\omega) \quad \text{for any } \omega \in \Lambda$$

by redefining Λ, N if necessary.

Putting $F_0 = \emptyset$, we now let

$$A_t(\omega) = \begin{cases} A_t^{(k)}(\omega), & \sigma_{E \setminus F_{k-1}} \leq t < \sigma_{E \setminus F_k},\ k = 1, 2, \ldots, \\ A_{\sigma(\omega)-}(\omega), & t \geq \sigma(\omega). \end{cases}$$

A is then a positive continuous additive functional of X possessing Λ, N as its defining set and exceptional set. When $k < \ell$, $A^{(k)} = \mathbf{1}_{F_k} \cdot A^{(\ell)}$. By letting $\ell \to \infty$, we obtain $A^{(k)} = \mathbf{1}_{F_k} \cdot A$, which means that $\mathbf{1}_{F_k} \cdot \mu_A = \mathbf{1}_{F_k} \cdot \mu$ for every $k \geq 1$, namely, $\mu_A = \mu$.

Finally we assume, for $\mu \in S$, A, $B \in \mathbf{A}_c^+$ satisfy $\mu_A = \mu_B = \mu$. Take a nest $\{F_k\}$ satisfying $\mu^{(k)} = \mathbf{1}_{F_k} \cdot \mu \in S_{00}$ as above. Since $\mu_{\mathbf{1}_{F_k} \cdot A} = \mu_{\mathbf{1}_{F_k} \cdot B} = \mu^{(k)}$, Proposition 4.1.8 applies in concluding that $\mathbf{1}_{F_k} \cdot A$ and $\mathbf{1}_{F_k} \cdot B$ are m-equivalent. k being arbitrary, A, B are m-equivalent. □

The proof of Theorem 4.1.1 is now complete.

Let $D \subset E$ be a nearly Borel, finely open set and X^D be the part process of X on D. X^D is an $m|_D$-symmetric standard process on D. The transition function $\{P_t^D; t \geq 0\}$ and the resolvent $\{R_\alpha^D; \alpha > 0\}$ of X^D are given by (3.3.2).

Exercise 4.1.9. (i) Define the shift operator θ_t^0 on Ω by

$$\theta_t^0 \omega = \begin{cases} \theta_t \omega & \text{for } t < \tau_D(\omega), \\ \omega_\partial & \text{for } t \geq \tau_D(\omega), \end{cases} \tag{4.1.22}$$

where ω_∂ denotes a specific element of Ω with $X_t(\omega_\partial) = \partial$ for every $t \geq 0$. Using the relation $t + \tau_D(\theta_t \omega) = \tau_D(\omega)$ whenever $t < \tau_D(\omega)$, show that $X_t^D(\theta_s^0 \omega) = X_{s+t}^D(\omega)$, $s, t \geq 0$.

(ii) Denote the minimum admissible filtration $\sigma\{X_s^D : s \leq t\}$ of X^D by $\mathcal{G}_t^0$ and, for $\mu \in \mathcal{P}(D_\partial)$, let $\mathcal{G}_t^\mu = \sigma\{\mathcal{G}_t^0, \mathcal{N}\}$ for each $t \geq 0$, where $\mathcal{N}$ is the family of all $\mathbf{P}_\mu$-null sets. Further let $\mathcal{G}_t = \cap_{\mu \in \mathcal{P}(D_\partial)} \mathcal{G}_t^\mu$. Define the σ-field of events strictly prior to time $\tau_D \wedge t$ by $\mathcal{F}^{\mu}_{(\tau_D\wedge t)-} = \sigma\{\Lambda \cap \{s < \tau_D \wedge t\} : \Lambda \in \mathcal{F}^{\mu}_s, s \geq 0\}$. Prove the inclusion

$$\mathcal{F}^{\mu}_{(\tau_D\wedge t)-} \subset \mathcal{G}^{\mu}_t \subset \mathcal{F}^{\mu}_{\tau_D\wedge t}. \tag{4.1.23}$$

Show that, for any $F \in b\mathcal{G}_\infty$ and $\mu \in \mathcal{P}(D_\partial)$,

$$F \circ \theta^0_t \in b\mathcal{G}_\infty, \ \mathbf{E}_\mu[F \circ \theta^0_t \mid \mathcal{F}_t] = \mathbf{E}_{X^D_t}[F]\ \mathbf{P}_\mu\text{-a.s.} \tag{4.1.24}$$

(iii) For an additive functional A of X, let $B_t = A_{t\wedge\tau_D}$, $t \geq 0$. Show that $\{B_t\}_{t\geq 0}$ satisfies the additivity

$$B_{s+t} = B_s + B_t \circ \theta^0_s, \qquad s, t \geq 0,$$

with respect to the shift operator θ^0_s defined by (4.1.22).
Show that if A is continuous, then B is a $\{\mathcal{G}_t\}$-adapted continuous additive functional of X^D.

Consider $A \in \mathbf{A}^+_c$ and $\mu \in S$ such that μ is the Revuz measure of A.

PROPOSITION 4.1.10. *$B_t := A_{t\wedge\tau_D}$, $t \geq 0$, is a PCAF of X^D with Revuz measure $\mu_D := \mu|_D$. More specifically, it holds for any $f, h \in \mathcal{B}_+(D)$ that*

$$\mathbf{E}_{h\cdot m}\left[\int_0^{t\wedge\tau_D} f(X_s)dA_s\right] = \int_0^t \langle P^D_s h, f\cdot\mu_D\rangle ds, \quad t > 0, \tag{4.1.25}$$

$$\mathbf{E}_{h\cdot m}\left[\int_0^{\tau_D} e^{-\alpha s} f(X_s)dA_s\right] = \langle R^D_\alpha h, f\cdot\mu_D\rangle, \quad \alpha > 0. \tag{4.1.26}$$

Proof. By Exercise 4.1.9, $\{B_t\}$ is a PCAF of X^D. It suffices to show (4.1.26) for a bounded f and $\mu \in S_{00}$ in view of Theorem 2.3.15. Let $U^{0,\alpha}_A f(x) = \mathbf{E}_x\left[\int_0^{\tau_D} e^{-\alpha s} f(X_s)dA_s\right]$. Then $U^{0,\alpha}_A f = U^\alpha_A f - \mathbf{H}^\alpha_{E\setminus D} U^\alpha_A f$ and $U^\alpha_A f$ is a quasi continuous version of the α-potential $U_\alpha(f\cdot\mu)$ by Lemma 4.1.5. On account of Theorem 3.2.2, the left hand side of (4.1.26) is equal for a bounded $h \in L^2(E; m)$ to

$$\begin{aligned}(h, U^{0,\alpha}_A f) &= \mathcal{E}_\alpha(R_\alpha h, U^\alpha_A f - \mathbf{H}^\alpha_{E\setminus D} U^\alpha_A f) = \mathcal{E}_\alpha(R_\alpha h - \mathbf{H}^\alpha_{E\setminus D} R_\alpha h, U^\alpha_A f) \\ &= \mathcal{E}_\alpha(R^D_\alpha h, U_\alpha(f\cdot\mu)) = \langle R^D_\alpha h, f\cdot\mu_D\rangle.\end{aligned}$$

□

We close this section by showing that under the absolute continuity condition for X,

(AC) the transition function $P_t(x, \cdot)$ of X is absolutely continuous with respect to m for every $t > 0$, $x \in E$, positive continuous additive functionals in the strict sense of X can also be characterized. If the exceptional set of $A \in \mathbf{A}_c^+$ can be taken to be empty, namely, if its defining set Λ can be chosen in such a way that $\mathbf{P}_x(\Lambda) = 1$ for every $x \in E$, A is said to be a *positive continuous additive functional in the strict sense*. Let us denote by $\mathbf{A}_{c1}^+$ the family of all positive continuous additive functionals of X in the strict sense. A, $B \in \mathbf{A}_{c1}^+$ are called *equivalent in the strict sense* if

$$\mathbf{P}_x(A_t = B_t) = 1, \quad \text{for any } t > 0 \quad \text{and } x \in E. \tag{4.1.27}$$

In this case, we can find a common defining set Λ such that $\mathbf{P}_x(\Lambda) = 1$ for every $x \in E$, and $A_t(\omega) = B_t(\omega)$ for every $t \geq 0$ and $\omega \in \Lambda$.

A measure μ on $(E, \mathcal{B}(E))$ is called *smooth in the strict sense* if there exists an increasing sequence $\{E_n\}$ of Borel measurable finely open sets such that $\bigcup_{n=1}^{\infty} E_n = E$ and $\mathbf{1}_{E_n} \cdot \mu \in S_{00}$ for every $n \geq 1$. We denote by S_1 the totality of smooth measures in the strict sense.

THEOREM 4.1.11. *Suppose X satisfies the condition* **(AC)**.
(i) *For any* $A \in \mathbf{A}_{c1}^+$, $\mu_A \in S_1$.
(ii) *For any* $\mu \in S_1$, *there exists* $A \in \mathbf{A}_{c1}^+$ *satisfying* $\mu_A = \mu$ *uniquely up to the equivalence in the strict sense.*

Proof of Theorem 4.1.11(i). Suppose $A \in \mathbf{A}_{c1}^+$ is given. For a strictly positive bounded Borel m-integrable function f on E, $\varphi(x) = R_\alpha^A f(x)$ is well defined for all $x \in E$ by (4.1.5). Since equation (4.1.7) holds for all $x \in E$, φ is a difference of the 1-excessive functions $R_\alpha f$, $U_A^\alpha R_\alpha^A f$, and is finely continuous. Further, φ is Borel measurable on account of Theorem A.2.17 and the assumption **(AC)**. If we put

$$E_n = \{x \in E : \varphi(x) > 1/n\}, \quad n = 1, 2, \ldots, \tag{4.1.28}$$

then $\{E_n\}$ is a sequence of Borel finely open sets increasing to E.

For each n, the Revuz measure of $A_n = \mathbf{1}_{E_n} \cdot A \in \mathbf{A}_{c1}^+$ equals $\mathbf{1}_{E_n} \cdot \mu_A$ by virtue of Theorem A.3.5. Since

$$U_{A_n}^1 1(x) \leq n U_A^1 \varphi(x) \leq n R_1 f(x), \quad x \in E, \tag{4.1.29}$$

we get $\mu_{A_n}(E) < \infty$ in view of Lemma 4.1.3. Moreover, by the above inequality, Lemma 2.3.5, and Lemma 1.2.3, there exists $\mu \in S_0$ such that $U_{A_n}^1 1 = U_1 \mu$ $[m]$. In the same way as in the proof of Lemma 4.1.5, the last identity implies that $\mu(E) = \mu_{A_n}(E) < \infty$ and furthermore $\mu = \mu_{A_n}$. Hence $\mathbf{1}_{E_n} \cdot \mu_A \in S_{00}$ □

The proof of Theorem 4.1.11(ii) can be reduced to the next proposition. Under the condition **(AC)**, the resolvent kernel $R_\alpha(x, \cdot)$ is absolutely continuous with respect to m. It is known (cf. [73, Lemma 4.2.4]) that the density function $r_\alpha(x, y)$ can be taken in such a way that it is symmetric in x and y, α-excessive in each variable, and satisfies the resolvent equation. In this case, if we let for $\mu \in S_0$

$$R_\alpha\mu(x) = \int_E r_\alpha(x, y)\mu(dy), \quad x \in E,$$

then $R_\alpha\mu$ can be verified to be a quasi continuous and α-excessive version of the potential $U_\alpha\mu$. In particular, a necessary and sufficient condition for μ to be in S_{00} is that $\mu(E) < \infty$ and $R_1\mu$ is bounded.

Proposition 4.1.12. *Suppose X satisfies* **(AC)**. *For any $\mu \in S_{00}$, there exists $A \in \mathbf{A}^+_{c1}$ satisfying*

$$U_A^1 1(x) = R_1\mu(x) \quad \textit{for every } x \in E, \tag{4.1.30}$$

uniquely up to the equivalence in the strict sense.

Proof. Such an $A \in \mathbf{A}^+_{c1}$ can be constructed by making use of Proposition 4.1.6. Indeed, according to that proposition, we can construct an $A \in \mathbf{A}^+_c$ with a suitable defining set Λ and exceptional set N satisfying (4.1.30) for m-a.e. $x \in E$. Since both sides of (4.1.30) are 1-excessive relative to $X_{E\setminus N}$, this identity holds for all $x \in E \setminus N$ under the condition **(AC)**.

For this A, Λ, we put $\Lambda_0 = \bigcap_n \theta_{1/n}^{-1}\Lambda$ and we let, for $\omega \in \Lambda_0$,

$$\widetilde{A}_t(\omega) = \begin{cases} \lim_{n\to\infty} A_{t-(1/n)}(\theta_{1/n}\omega) & \text{if the limit exists,} \\ 0 & \text{otherwise.} \end{cases}$$

We further define

$$\widetilde{\Lambda} = \left\{\omega \in \Lambda_0 : \widetilde{A}_t(\omega) < \infty \text{ for every } t > 0 \text{ and } \widetilde{A}_{0+}(\omega) = 0\right\}.$$

Then $\mathbf{P}_x(\widetilde{\Lambda}) = 1$ for every $x \in E$, $\widetilde{A}$ becomes a positive continuous additive functional in the strict sense with a defining set $\widetilde{\Lambda}$ and $\widetilde{A}$ can be verified to satisfy (4.1.30). We refer to [73, Theorem 5.1.6] for the details.

The uniqueness can be easily shown. If A, $B \in \mathbf{A}^+_{c1}$ satisfy the identity (4.1.30), then A, B are m-equivalent by Lemma 4.1.5 and Proposition 4.1.8, and consequently (4.1.27) holds for q.e. $x \in E$ by Lemma A.3.2. Hence, by **(AC)**, we have $\mathbf{P}_x(A_t - A_{1/n} = B_t - B_{1/n}) = \mathbf{E}_x[\mathbf{P}_{X_{1/n}}(A_{t-(1/n)} = B_{t-(1/n)})] = 0$ for any $x \in E$. It suffices to let $n \to \infty$. □

Proof of Theorem 4.1.11(ii). For $\mu \in S_1$, there is a sequence $\{E_n\}$ of Borel finely open sets increasing to E such that $\mathbf{1}_{E_n} \cdot \mu \in S_{00}$ for every $n \geq 1$. Then, due to the quasi-left-continuity of the Hunt process X, it holds that

$$\mathbf{P}_x\left(\lim_{n\to\infty}\sigma_{E\setminus E_n}\geq\zeta\right)=1,\quad\text{for any } x\in E.$$

Noting this, we can make use of Proposition 4.1.12 to prove the existence and uniqueness of the desired positive continuous additive functional in the strict sense A just as in the proof of Theorem 4.1.1(ii). □

4.2. DECOMPOSITIONS OF ADDITIVE FUNCTIONALS OF FINITE ENERGY

The notion of an additive functional $A_t(\omega)$ accompanying a defining set Λ and a (properly) exceptional set N of the m-symmetric Hunt process X is formulated in the beginning of the preceding section. In the preceding section, we dealt exclusively with a positive continuous additive functional. In the present and the next sections, we are concerned with an additive functional which is not necessarily positive but finite càdlàg. By this we mean that for each $\omega\in\Lambda$, $A_t(\omega)$ is finite, right continuous in t on $[0,\infty)$, and has left limits in t on $(0,\infty)$. Thus, in Sections 4.2 and 4.3, we always require an additive functional of X to be finite càdlàg without a specific notice.

We define the *energy* $\mathbf{e}(A)$ of an additive functional A_t of X by

$$\mathbf{e}(A)=\lim_{t\to 0}\frac{1}{2t}\mathbf{E}_m[A_t^2],\tag{4.2.1}$$

whenever the limit on the right hand side exists. For two additive functionals A_t, B_t of X, their *mutual energy* $\mathbf{e}(A,B)$ is defined by

$$\mathbf{e}(A,B)=\lim_{t\to 0}\frac{1}{2t}\mathbf{E}_m[A_tB_t],\tag{4.2.2}$$

whenever the limit exists.

We shall consider three types of additive functionals mentioned and studied below.

(1°) Additive functionals generated by functions in $\mathcal{F}_e$

For $u\in\mathcal{F}_e$, we let

$$A_t^{[u]}=\widetilde{u}(X_t)-\widetilde{u}(X_0),\quad t\geq 0.\tag{4.2.3}$$

Since the quasi continuous modification $\widetilde{u}$ is decided uniquely up to q.e., $A^{[u]}$ is uniquely determined up to the m-equivalence. But we always let $\widetilde{u}(\partial)=0$ by convention. In view of Theorem 3.5.4, $\widetilde{u}$ is then an element of $C_\infty(\{F_k\})$ for some strong nest $\{F_k\}$ and $\widetilde{u}|_{F_k\cup\{\partial\}}$ is continuous and bounded for each k.

Therefore, we can assume $A^{[u]}$ to be finite càdlàg on $[0,\infty)$ by choosing its exceptional set appropriately.

In order to compute the energy of the additive functional $A^{[u]}$, we need first to draw from $(\mathcal{E},\mathcal{F})$ a positive Radon measure κ which will be called its killing measure.

For $u \in \mathcal{F}$, since

$$0 \le \frac{1}{2t}\mathbf{E}_m\left[(A_t^{[u]})^2\right] = \frac{1}{t}(u - P_t u, u) - \frac{1}{2t}(1 - P_t 1, u^2), \tag{4.2.4}$$

we have

$$\frac{1}{2t}\mathbf{E}_m\left[(A_t^{[u]})^2\right] \le \mathcal{E}(u,u), \quad \frac{1}{2t}(1 - P_t 1, u^2) \le \mathcal{E}(u,u). \tag{4.2.5}$$

Hence it holds for any compact set $K \subset E$ that

$$\sup_{0<t<\infty} \frac{1}{t}\int_K (1 - P_t 1)(x)dm(x) < \infty,$$

namely, the family $\{(1/t)(1 - P_t 1)\cdot m\}_{t>0}$ of measures on E is uniformly bounded on K. Therefore, there exists a subsequence $t_n \downarrow 0$ such that the sequence of measures $(1/t_n)(1 - P_{t_n}1)\cdot m$ converges vaguely to a certain positive Radon measure κ on E; in other words, we have

$$\int_E v(x)^2\kappa(dx) = \lim_{n\to\infty}\frac{1}{t_n}(1 - P_{t_n}1, v^2) \le 2\mathcal{E}(v,v) \tag{4.2.6}$$

for any $v \in \mathcal{F}\cap C_c(E)$.

In particular, κ satisfies for each compact set K

$$\begin{aligned}\int_E |v(x)|\mathbf{1}_K(x)\kappa(dx) &\le \kappa(K)^{1/2}\left(\int_E v(x)^2\kappa(dx)\right)^{1/2}\\ &\le (2\kappa(K))^{1/2}\mathcal{E}(v,v)^{1/2},\end{aligned}$$

which means that $\mathbf{1}_K\cdot\kappa \in S_0$ and consequently κ is smooth ($\kappa \in \mathcal{S}$). We remark further that the inequality between the first and third terms of (4.2.6) remains valid for any quasi continuous function $\widetilde{v} \in \mathcal{F}_e$ by Theorem 2.3.4 and Fatou's lemma. Similarly, two inequalities of (4.2.5) extend to $u \in \mathcal{F}_e$.

We next take, for any quasi continuous function $\widetilde{u} \in \mathcal{F}_e$, a sequence $v_l \in \mathcal{F}\cap C_c(X)$ which is $\mathcal{E}$-convergent as well as q.e. convergent to $\widetilde{u}$. By the above remark,

$$\left|\left(\int_E \widetilde{u}(x)^2\kappa(dx)\right)^{1/2} - \left(\frac{1}{t_n}(1-P_{t_n}1,u^2)\right)^{1/2}\right|$$
$$\leq 2\sqrt{2}\mathcal{E}(u-v_l,u-v_l)^{1/2}$$
$$+\left|\left(\int_E v_l(x)^2\kappa(dx)\right)^{1/2} - \left(\frac{1}{t_n}(1-P_{t_n}1,v_l^2)\right)^{1/2}\right|,$$

which can be made arbitrarily small by (4.2.6). We have shown that

$$\int_E \widetilde{u}(x)^2\kappa(dx) = \lim_{n\to\infty}\frac{1}{t_n}(1-P_{t_n}1,u^2), \quad \forall u\in\mathcal{F}_e. \tag{4.2.7}$$

We now recall the family $\mathbf{A}_c^+$ of all positive continuous additive functionals of the Hunt process X studied in the preceding section. As $\kappa\in S$, there exists $C\in\mathbf{A}_c^+$ with Revuz measure κ uniquely up to the m-equivalence by virtue of Theorem 4.1.1. Actually C can be expressed in terms of the Lévy system (N,H) of the Hunt process X introduced in Section A.3.4.

THEOREM 4.2.1. (i) *It holds that*

$$\kappa(dx) = N(x,\{\partial\})\mu_H(dx), \qquad C_t = \int_0^t N(X_s,\{\partial\})dH_s,$$

where μ_H denotes the Revuz measure of the PCAF H.
(ii) *For any $u\in\mathcal{F}_e$,*

$$\int_E \widetilde{u}(x)^2\kappa(dx) = \lim_{t\downarrow 0}\frac{1}{t}(1-P_t1,u^2) < \infty, \tag{4.2.8}$$

$$\mathbf{e}(A^{[u]}) = \mathcal{E}(u,u) - \frac{1}{2}\int_E \widetilde{u}(x)^2\kappa(dx). \tag{4.2.9}$$

Proof. It follows from the identity (A.3.33) with $T=\infty$ and $g(s)=e^{-\alpha s}$ that, for any $f\in C_c(E)$ and any $x\in E$,

$$\mathbf{E}_x\left[e^{-\alpha\zeta}f(X_{\zeta-})\right] = \mathbf{E}_x\left[\sum_{s<\infty} e^{-\alpha s}f(X_{s-})1_{\{\partial\}}(X_s)\right]$$
$$= \mathbf{E}_x\left[\int_0^\infty e^{-\alpha t}f(X_t)N(X_t,\{\partial\})dH_t\right]. \tag{4.2.10}$$

Combining (4.2.10) with the following lemma, we get Theorem 4.2.1(i) immediately from Theorem 4.1.1.

Theorem 4.2.1(i) particularly implies that the measure κ is determined uniquely by the Hunt process X independently of the choice of the subsequence $\{t_n\}$. Accordingly, we get (4.2.8) from (4.2.7). The expression (4.2.9) of the energy of $A^{[u]}$ is obtained from (4.2.4) for $u \in \mathcal{F}$. The expression is then extended to $u \in \mathcal{F}_e$ using the estimates (4.2.5) holding for any $u \in \mathcal{F}_e$ and $t > 0$. □

LEMMA 4.2.2. *For $f \in C_c(E)$, $h = R_\alpha g$ with $g \in b\mathcal{B}(E)$ and $\alpha > 0$,*

$$\mathbf{E}_{h\cdot m}\left[e^{-\alpha\zeta} f(X_{\zeta-})\right] = \langle R_\alpha h, f \cdot \kappa \rangle.$$

Proof. It suffices to show this for $f \in C_c(E) \cap \mathcal{F}$ and $g \in b\mathcal{B}(E) \cap L^1(E; m)$. For a sequence $\{t_n\}$ decreasing to 0 and satisfying (4.2.6), we have

$$\begin{aligned}
\mathbf{E}_{h\cdot m}\left[e^{-\alpha\zeta} f(X_{\zeta-})\right] &= \lim_{n\to\infty} \mathbf{E}_{h\cdot m}\left[\sum_{k=1}^{\infty} e^{-\alpha k t_n} f(X_{(k-1)t_n}) : (k-1)t_n < \zeta \le k t_n\right] \\
&= \lim_{n\to\infty} \sum_{k=1}^{\infty} \mathbf{E}_{h\cdot m}\left[f(X_{(k-1)t_n}) e^{-\alpha k t_n}(1 - P_{t_n}1)(X_{(k-1)t_n})\right] \\
&= \lim_{n\to\infty} \sum_{k=1}^{\infty} \left(e^{-\alpha k t_n} f \cdot P_{(k-1)t_n} h,\ 1 - P_{t_n}1\right).
\end{aligned}$$

Since we get from (4.2.7)

$$\lim_{n\to\infty} \frac{1}{t_n}(f \cdot R_\alpha h,\ 1 - P_{t_n}1) = \int_E f(x) R_\alpha h(x) \kappa(dx),$$

the last expression in the above display equals

$$\lim_{n\to\infty} I_n + \int_E f(x) R_\alpha h(x) \kappa(dx),$$

where

$$I_n = \left(\sum_{k=1}^{\infty} f \cdot \left(e^{-\alpha k t_n} t_n P_{(k-1)t_n} h - \int_{(k-1)t_n}^{k t_n} e^{-\alpha s} P_s h ds\right),\ \frac{1}{t_n}(1 - P_{t_n}1)\right).$$

On the other hand, we obtain from $|R_\alpha 1(x) - P_s R_\alpha 1(x)| \le 2s$ for $s > 0$, the estimates

$$|P_{(k-1)t_n} h - P_s h| \le 2t_n \|g\|_\infty \quad \text{for } (k-1)t_n \le s < k t_n.$$

It follows

$$\begin{aligned}
&\left| e^{-\alpha k t_n} P_{(k-1)t_n} h \cdot t_n - \int_{(k-1)t_n}^{kt_n} e^{-\alpha s} P_s h ds \right| \\
&= \left| \int_{(k-1)t_n}^{kt_n} \left((e^{-\alpha k t_n} - e^{-\alpha s}) P_{(k-1)t_n} h + e^{-\alpha s} \left(P_{(k-1)t_n} h - P_s h \right) \right) ds \right| \\
&\leq \alpha e^{-\alpha(k-1)t_n} \|h\|_\infty t_n^2 + 2 e^{-\alpha(k-1)t_n} \|g\|_\infty t_n^2 \\
&\leq 3 \|g\|_\infty \cdot e^{-\alpha(k-1)t_n} t_n^2.
\end{aligned}$$

Therefore, $|I_n| \leq 3\|g\|_\infty \frac{t_n}{1-e^{-\alpha t_n}} (|f|, 1 - P_{t_n} 1)$, which tends to 0 as $n \to \infty$. □

The identity (4.2.8) particularly implies that the probability of X continuously escaping to ∂ is small compared to that of jumping directly to ∂.

Proposition 4.2.3. *Let* $r_t(x) := \mathbf{P}_x(X_{\zeta-} = \partial$ *and* $\zeta \leq t)$ *for* $x \in E$ *and* $t > 0$. *Then*

$$\lim_{t \to 0} \frac{1}{t} (u^2, r_t) = 0 \qquad \text{for every } u \in \mathcal{F}_e. \tag{4.2.11}$$

Proof. Since $1 - P_t 1(x) = \mathbf{P}_x(X_{\zeta-} \in E$ and $\zeta \leq t) + r_t(x)$ for $x \in E$, we have $\frac{1}{t}(1 - P_t 1, u^2) = I_t + \frac{1}{t}(u^2, r_t)$, where

$$I_t = \frac{1}{t} \mathbf{E}_{u^2 \cdot m} \left[\sum_{s \leq t} 1_E(X_{s-}) 1_{\{\partial\}}(X_s) \right].$$

By (A.3.31) and (4.1.2),

$$I_t \frac{1}{t} \mathbf{E}_{u^2 \cdot m} \left[\int_0^t N(X_s, \{\partial\}) dH_s \right] = \frac{1}{t} \int_0^t \langle P_s \widetilde{u}^2, \kappa \rangle ds.$$

Therefore, (4.2.8) and Fatou's lemma yield

$$\langle \widetilde{u}^2, \kappa \rangle \leq \liminf_{t \downarrow 0} I_t \leq \limsup_{t \downarrow 0} I_t \leq \langle \widetilde{u}^2, \kappa \rangle,$$

which then implies (4.2.11). □

(2°) Martingale additive functionals

An additive functional M_t of X in the sense of Section 4.1 is called a *martingale additive functional* (*MAF* in abbreviation) if, for each $t > 0$,

$$\mathbf{E}_x[M_t^2] < \infty, \qquad \mathbf{E}_x[M_t] = 0, \qquad \text{q.e. } x \in E. \tag{4.2.12}$$

Then from the additivity of M and the Markov property of X,

$$\mathbf{E}_x[M_{t+s} | \mathcal{F}_s] = \mathbf{E}_x[M_s + M_t \circ \theta_s | \mathcal{F}_s] = M_s + \mathbf{E}_{X_s}[M_t] = M_s, \ \mathbf{P}_x\text{-a.s.},$$

namely, M_t is a $\mathbf{P}_x$-martingale for q.e. $x \in E$. Denote by $\mathcal{M}$ the totality of martingale additive functionals of X.

For any $M \in \mathcal{M}$, there exists $\langle M \rangle \in \mathbf{A}_c^+$ satisfying

$$\mathbf{E}_x[M_t^2] = \mathbf{E}_x[\langle M \rangle_t], \quad \forall t \geq 0, \quad \text{q.e. } x \in E,$$

which is unique up to the m-equivalence. In fact, if we let N be an exceptional set of M, then M can be regarded as the martingale additive functional in the strict sense of the restricted Hunt process $X_{E \setminus N}$, and Section A.3.3(**1**) yields $\langle M \rangle$ as above. $\langle M \rangle$ is called the *predictable quadratic variation* of M. The above identity implies that the energy $\mathbf{e}(M)$ of M can be expressed using the Revuz measure $\mu_{\langle M \rangle}$ of $\langle M \rangle$ as

$$\mathbf{e}(M) = \lim_{t \downarrow 0} \frac{1}{2t} \mathbf{E}_m[M_t^2] = \lim_{t \downarrow 0} \frac{1}{2t} \mathbf{E}_m[\langle M \rangle_t] = \frac{1}{2} \mu_{\langle M \rangle}(E). \tag{4.2.13}$$

In what follows, we denote $\overset{\circ}{\mathcal{M}} = \{M \in \mathcal{M} : \mathbf{e}(M) < \infty\}$.

Lemma 4.2.4. *For $A \in \mathbf{A}_c^+$ with Revuz measure $\mu \in S$ and for $\nu \in S_{00}$,*

$$\mathbf{E}_\nu[A_t] \leq (1 + t) \, \|U_1 \nu\|_\infty \, \mu(E), \quad t \geq 0. \tag{4.2.14}$$

Proof. On account of Theorem 2.3.15 and Theorem A.3.5, it suffices to derive (4.2.14) for $\mu \in S_{00}$. In this case, $c_t(x) = \mathbf{E}_x[A_t] \in \mathcal{F}$. Indeed, we have from (4.1.2)

$$\mathbf{E}_{fm}[A_t] = \int_0^t \langle \mu, P_s f \rangle ds, \quad t \geq 0,$$

and, for $s < t$,

$$\begin{aligned} \frac{1}{s}(c_t - P_s c_t, c_t) &= \frac{1}{s} \left\langle \mu, \int_0^t (P_r c_t - P_{s+r} c_t) dr \right\rangle \\ &= \frac{1}{s} \left\langle \mu, \int_0^s P_r c_t dr \right\rangle - \frac{1}{s} \left\langle \mu, \int_t^{t+s} P_r c_t dr \right\rangle. \end{aligned}$$

Since, by Lemma 4.1.5,

$$\langle \mu, \mathbf{E}_\cdot[A_t] \rangle \leq e^t \langle \mu, U_A^1 1 \rangle = e^t \langle \mu, U_1 \mu \rangle < \infty, \tag{4.2.15}$$

the limit $\langle \mu, c_t - P_t c_t \rangle$ is finite and we obtain $c_t \in \mathcal{F}$ together with $\mathcal{E}(c_t, c_t) = \langle \mu, c_t - P_t c_t \rangle$ in view of (1.1.4) and (1.1.5). Similarly, we obtain

$$\mathcal{E}(c_t, u) = \langle \mu, \widetilde{u} \rangle - \langle \mu, P_t u \rangle, \quad u \in \mathcal{F}.$$

In particular, it holds for $\nu \in S_{00}$ that

$$\begin{aligned} \mathbf{E}_\nu[A_t] &= \langle \nu, c_t \rangle = \mathcal{E}_1(c_t, U_1 \nu) = \langle \mu, U_1 \nu - P_t U_1 \nu \rangle + (c_t, U_1 \nu) \\ &\leq \|U_1 \nu\|_\infty (\mu(E) + \langle m, c_t \rangle), \end{aligned}$$

which leads us to (4.2.14) because $\langle m, c_t \rangle \leq t \mu(E)$. □

Theorem 4.2.5. *For any* **e**-*Cauchy sequence* $\{M_t^{(n)}\}$ *in* $\overset{\circ}{\mathcal{M}}$, *there exists a unique* $M \in \overset{\circ}{\mathcal{M}}$ *such that* $\lim_{n\to\infty} \mathbf{e}(M^{(n)} - M) = 0$. *Further, there exists a subsequence* $\{n_k\}$ *such that* $M_t^{(n_k)}$ *converges to* M_t *uniformly in* t *on each compact subinterval of* $[0, \infty)$ $\mathbf{P}_x$-*a.s. for q.e.* $x \in E$.

Proof. By (4.2.13), the total mass $\mu_{\langle M\rangle}(E)$ of the Revuz measure of $\langle M\rangle \in \mathbf{A}_c^+$ equals $2\mathbf{e}(M)$. Hence, by the martingale inequality and Lemma 4.2.4, we are led to the inequality

$$\mathbf{P}_\nu\left(\sup_{0\le s\le T} |M_s| > \lambda\right) \le \frac{1}{\lambda^2}\mathbf{E}_\nu[M_T^2] \le \frac{2}{\lambda^2}(1+T)\|U_1\nu\|_\infty \mathbf{e}(M) \qquad (4.2.16)$$

holding for any $\nu \in S_{00}$ and $T > 0$. Choosing a subsequence $\{n_k\}$ such that $\mathbf{e}(M^{(n_{k+1})} - M^{(n_k)}) < 2^{-3k}$ and taking $\lambda = 2^{-k}$, $M_t = M_t^{(n_{k+1})} - M_t^{(n_k)}$ in (4.2.16), we obtain

$$\mathbf{P}_\nu\left(\sup_{0\le s\le T} |M_s^{(n_{k+1})} - M_s^{(n_k)}| > 2^{-k}\right) \le 2(1+T)\|U_1\nu\|_\infty 2^{-k}.$$

By the Borel-Cantelli lemma, we have $\mathbf{P}_\nu(\Lambda) = 0$ for

$$\Lambda = \bigcap_{n=1}^\infty \bigcup_{k=n}^\infty \left\{\sup_{0\le s\le T} |M_s^{(n_{k+1})} - M_s^{(n_k)}| > 2^{-k}\right\}.$$

Since the above holds for every $\nu \in S_{00}$, we conclude by Corollary 2.3.11 that $\mathbf{P}_x(\Lambda) = 0$ for q.e. $x \in E$. For $\omega \notin \Lambda$, $\lim_{k\to\infty} M_s^{(n_k)}(\omega) = M_s(\omega)$ exists as a uniform convergent limit on $[0, T]$. By the inequality (4.2.16), this convergence also takes place in $L^2(\mathbf{P}_\nu)$ so that $\mathbf{E}_\nu(M_t^2) < \infty$ and $\mathbf{E}_\nu[M_t] = 0$. Consequently, $\mathbf{E}_x(M_t^2) < \infty$, $\mathbf{E}_x[M_t] = 0$ for q.e. $x \in E$ and $M \in \mathcal{M}$.

Furthermore, by Fatou's lemma,

$$\begin{aligned}\frac{1}{2t}\mathbf{E}_m\left[(M_t^{(n)} - M_t)^2\right] &\le \liminf_{k\to\infty} \frac{1}{2t}\mathbf{E}_m\left[(M_t^{(n)} - M_t^{(n_k)})^2\right]\\ &\le \liminf_{k\to\infty} \mathbf{e}(M^{(n)} - M^{(n_k)}),\end{aligned}$$

and accordingly, $\mathbf{e}(M^{(n)} - M) \le \liminf_{k\to\infty} \mathbf{e}(M^{(n)} - M^{(n_k)})$. In particular, $M \in \overset{\circ}{\mathcal{M}}$ and $M^{(n)}$ is **e**-convergent to M as $n \to \infty$. □

(3°) Continuous additive functionals of zero energy

A continuous additive functional N of X is called a *continuous additive functional of zero energy* if

$$\mathbf{E}_x(|N_t|) < \infty \text{ q.e. } x \in E \text{ for each } t > 0, \quad \mathbf{e}(N) = 0. \qquad (4.2.17)$$

$\mathcal{N}_c$ will denote the totality of continuous additive functionals of zero energy.

A typical example of $N \in \mathcal{N}_c$ is given by $N_t = \int_0^t f(X_s)ds$, $t \geq 0$ for a nearly Borel function $f \in L^2(E;m)$. In fact, the right hand side of $\mathbf{E}_x[\int_0^t |f(X_s)|ds] \leq e^t R_1|f|(x)$ is finite for q.e. $x \in E$ by Proposition 3.1.9, so that N is a continuous additive functional. Moreover,

$$\begin{aligned}\mathbf{E}_m[N_t^2] &= 2\mathbf{E}_m\left[\int_0^t f(X_s)\int_s^t f(X_v)dvds\right] \\ &= 2\mathbf{E}_m\left[\int_0^t\int_0^{t-s} f(X_s)P_v f(X_s)dvds\right] \\ &= 2\int_0^t\int_0^{t-s}(P_s 1, f\cdot P_v f)dvds \\ &= 2\int_0^t (P_s 1, f\, S_{t-s}f)ds,\end{aligned}$$

where $S_s f(x) := \int_0^s P_r f(x)dr$, and so we have

$$\mathbf{E}_m[N_t^2] \leq t^2\cdot(f,f).$$

We are now in a position to prove the following decomposition theorem due to the second-named author by making use of Theorem 3.5.4, Theorem 4.2.5, and the above example.

Theorem 4.2.6. *For any $u \in \mathcal{F}_e$, there exist $M^{[u]} \in \overset{\circ}{\mathcal{M}}$ and $N^{[u]} \in \mathcal{N}_c$ uniquely such that*

$$A^{[u]} = M^{[u]} + N^{[u]}, \qquad \mathbf{P}_x\text{-}a.s. \text{ for } q.e.\ x \in E. \tag{4.2.18}$$

Proof. Uniqueness: If $M \in \overset{\circ}{\mathcal{M}} \cap \mathcal{N}_c$, then (4.2.13) implies that $\mu_{\langle M\rangle}(E) = 0$ and $\langle M\rangle = 0$, $\mathbf{P}_x$-a.s. for q.e. $x \in E$. Therefore, $M = 0$.

Existence: For any nearly Borel function $f \in L^2(E;m)$ and $u = R_1 f \in \mathcal{F}$, define $N_t^{[u]}$ by

$$N_t^{[u]} = \int_0^t (u(X_s) - f(X_s))ds$$

and let $M_t^{[u]} = A_t^{[u]} - N_t^{[u]}$. By the above example, $N^{[u]} \in \mathcal{N}_c$. It is easy to verify that $M^{[u]} \in \mathcal{M}$. Moreover, $M^{[u]} \in \overset{\circ}{\mathcal{M}}$ because $\mathbf{e}(A^{[u]}) < \infty$.

For a general $u \in \mathcal{F}_e$, choose a sequence of functions $u_n = R_1 f_n$, $f_n \in L^2(E;m)$ such that $\mathcal{E}(u_n - u, u_n - u) \to 0$ and $u_n \to u$ m-a.e. as $n \to \infty$. Let $A^{[u_n]} = M^{[u_n]} + N^{[u_n]}$ be the decomposition given above. By virtue of Theorem 3.5.4, there exists a subsequence $\{n_k\}$ such that, as $k \to \infty$, $A_t^{[u_{n_k}]}$ converges to $A_t^{[u]}$ uniformly on each compact time interval $\mathbf{P}_x$-a.s. for q.e.

$x \in E$. Since we have from (4.2.9)

$$\mathbf{e}(M^{[u_m]} - M^{[u_n]}) = \mathbf{e}(A^{[u_m]} - A^{[u_n]}) \leq \mathcal{E}(u_m - u_n, u_m - u_n),$$

we find a suitable subsequence of $\{n_k\}$ (denoted by $\{n_k\}$ again) such that $M^{[u_{n_k}]}$ converges as $k \to \infty$ to some $M^{[u]} \in \overset{\circ}{\mathcal{M}}$ in $\mathbf{e}$-metric as well as uniformly on each compact time interval $\mathbf{P}_x$-a.s. for q.e. $x \in E$ by virtue of Theorem 4.2.5.

As a result, $N_t^{[u_{n_k}]}$ converges to a continuous additive functional $N_t^{[u]}$ and it holds that $A_t^{[u]} = M_t^{[u]} + N_t^{[u]}$. $N^{[u]}$ is of zero energy because

$$\begin{aligned}\overline{\lim_{t\to 0}} \frac{1}{2t}\mathbf{E}_m\left[(N_t^{[u]})^2\right] &\leq \overline{\lim_{t\to 0}} \frac{3}{2t}\mathbf{E}_m\Big[\left(A_t^{[u]} - A_t^{[u_n]}\right)^2 \\ &\quad + \left(M_t^{[u]} - M_t^{[u_n]}\right)^2 + \left(N_t^{[u_n]}\right)^2\Big] \\ &\leq 6\,\mathcal{E}(u - u_n, u - u_n) \to 0 \quad \text{as } n \to \infty. \qquad \square\end{aligned}$$

4.3. PROBABILISTIC DERIVATION OF BEURLING-DENY FORMULA

Any regular Dirichlet form $(\mathcal{E}, \mathcal{F})$ admits the following representation. This was first announced in the seminal paper [8] of A. Beurling and J. Deny published in 1959 and its analytic proof was given more than a decade later. For $u,\ v \in \mathcal{F}$,

$$\begin{aligned}\mathcal{E}(u, v) = \mathcal{E}^{(c)}(u, v) &+ \frac{1}{2}\int_{E\times E\setminus d} (\widetilde{u}(x) - \widetilde{u}(y))(\widetilde{v}(x) - \widetilde{v}(y))J(dxdy) \\ &+ \int_E \widetilde{u}(x)\widetilde{v}(x)\kappa(dx),\end{aligned} \tag{4.3.1}$$

where $\widetilde{u}$ denotes a quasi continuous version of $u \in \mathcal{F}$. Here J is a symmetric Radon measure on $E \times E \setminus d$, where d denotes the diagonal set, and κ is a Radon measure on E. $\mathcal{E}^{(c)}$ is a symmetric form possessing the strongly local property in the sense of Definition 1.3.17(v).

In this section, we maintain the setting of the preceding two sections on $E,\ m,\ X = (X_t, \mathbf{P}_x)$, and the Dirichlet form $(\mathcal{E}, \mathcal{F})$. For $u \in \mathcal{F}_e$, we have drawn in Theorem 4.2.6 of the preceding section the martingale part $M^{[u]}$ from the additive functional $A_t^{[u]} = \widetilde{u}(X_t) - \widetilde{u}(X_0)$. In this section, we shall derive the Beurling-Deny formula (4.3.1) holding for $u, v \in \mathcal{F}_e$ by means of a further decomposition of the MAF $M^{[u]}$. As a result, we can give more specific probabilistic expressions of the measures J, k in terms of the Lévy system of the Hunt process X, which describes the jumping and killing behaviors of X. They will also readily imply the sample path characterizations of the local property and the strongly local property of the form.

To this end, we first make a general observation. As stated at the beginning of Section 4.1, an additive functional A of X is accompanied by a properly exceptional set N called an exceptional set of A so that A can be regarded as an additive functional in the strict sense (an AF with a null exceptional set) with respect to the restricted Hunt process $X|_{E\setminus N}$.

The MAF $M \in \mathcal{M}$ of X already studied in the preceding section is, denoting its exceptional set by N, an element of the family $\mathbb{M}$ of the martingale additive functionals in the strict sense with respect to the Hunt process $X_{E\setminus N}$ in the sense of Section A.3.3. Therefore, by invoking the statements **(1), (2), (4)** of Section A.3.3, we can produce uniquely its continuous part M^c, purely discontinuous part M^d, predictable quadratic variation $\langle M\rangle$, and quadratic variation $[M]$ as additive functionals in the strict sense of $X_{E\setminus N}$. By regarding them as additive functionals (admitting exceptional sets) of X again, they are determined by M in the following ways uniquely up to the m-equivalence:

We introduce subfamilies of the space $\mathcal{M}$ of the MAFs of X by

$$\mathcal{M}_c = \{M \in \mathcal{M};\ \mathbf{P}_x(M_t \text{ is continuous in } t) = 1,\ \text{q.e. } x\}$$

$$\mathcal{M}_d = \{M \in \mathcal{M};\ \mathbf{P}_x(\langle M, L\rangle = 0) = 1 \text{ for q.e. } x \text{ and for any } L \in \mathcal{M}_c\}.$$

Then $M \in \mathcal{M}$ admits a unique decomposition $M = M^c + M^d$, $M^c \in \mathcal{M}_c$, $M^d \in \mathcal{M}_d$. The predictable quadratic variation $\langle M\rangle$ is a positive continuous additive functional ($\langle M\rangle \in \mathbf{A}_c^+$) characterized by the following property:

$$\mathbf{E}_x[M_t^2] = \mathbf{E}_x[\langle M\rangle_t], \quad \text{q.e. } x \in E, \quad \text{for any } t \geq 0. \tag{4.3.2}$$

Further,

$$[M]_t = \langle M^c\rangle_t + \sum_{s\leq t}(\Delta M_s)^2, \quad [M]_t^p = \langle M\rangle_t, \quad \text{for any } t \geq 0. \tag{4.3.3}$$

Here $\Delta M_s = M_s - M_{s-}$ and A^p denotes the additive functional obtained as the dual predictable projection of the integrable additive functional A of bounded variation.

For $u \in \mathcal{F}_e$, the decomposition

$$A^{[u]} = M^{[u]} + N^{[u]}, \quad M^{[u]} \in \mathring{\mathcal{M}}(\subset \mathcal{M}),\ N^{[u]} \in \mathcal{N}_c$$

holds by Theorem 4.2.6. Let us denote the continuous part and purely discontinuous part of $M^{[u]}$ by $M^{[u],c}$ and $M^{[u],d}$, respectively.

As $N_t^{[u]}$ is a finite continuous additive functional, we see from the above decomposition that the jumps of $M^{[u]}$ coincide with those of $A^{[u]}$. In particular, we have from Theorem 3.1.7(i)

$$\Delta M_\zeta^{[u]} = \widetilde{u}(X_\zeta) - \widetilde{u}(X_{\zeta-}) = -\widetilde{u}(X_{\zeta-}). \tag{4.3.4}$$

If we let $K_t = -\widetilde{u}(X_{\zeta-})1_{\{\zeta\leq t\}}$, $t \geq 0$, then K_t is an additive functional of X and it changes only by a jump when X jumps from E to ∂ at its lifetime ζ.

By (4.3.4) and Section A.3.3(**3**), K has its dual predictable projection K^p with $K - K^p \in \mathcal{M}_d$. Due to the quasi-left-continuity, X_t does not jump at any predictable stopping time. Hence K^p is a continuous additive functional by virtue of Section A.3.3(**3**).

We now let

$$M_t^{[u],k} = -\widetilde{u}(X_{\zeta-})\mathbf{1}_{\{\zeta \le t\}} - (-\widetilde{u}(X_{\zeta-})\mathbf{1}_{\{\zeta \le t\}})^p, \quad t \ge 0. \tag{4.3.5}$$

We further let $M^{[u],j} = M^{[u],d} - M^{[u],k}$. Then $M^{[u],k}$, $M^{[u],j} \in \mathcal{M}^d$, $M^{[u],k}$ jumps only at $t = \zeta$ and $\Delta M_\zeta^{[u],j} = 0$ in view of Section A.3.3(**5**), and $M^{[u]} \in \overset{\circ}{\mathcal{M}}$ can be decomposed as

$$M^{[u]} = M^{[u],c} + M^{[u],j} + M^{[u],k}. \tag{4.3.6}$$

We next define for $M,\ N \in \overset{\circ}{\mathcal{M}}$,

$$[\,M,N\,] = \frac{1}{2}([\,M+N\,] - [\,M\,] - [\,N\,]).$$

It then follows from the first identity of (4.3.3) that

$$[\,M^{[u],j}, M^{[u],k}\,]_t = \sum_{0<s\le t} \Delta M_s^{[u],j} \cdot \Delta M_s^{[u],k} = 0.$$

The second identity of (4.3.3) then implies $\langle M^{[u],j}, M^{[u],k} \rangle = 0$. Accordingly, we conclude that (4.3.6) is actually an orthogonal decomposition with respect to $\langle\ ,\ \rangle$.

Consequently, if we denote the Revuz measure of the predictable quadratic variation $\langle M \rangle \in \mathbf{A}_c^+$ of $M \in \mathcal{M}$ by $\mu_{\langle M \rangle}$, we obtain from the decomposition (4.3.6) that

$$\mu_{\langle M^{[u]} \rangle} = \mu_{\langle M^{[u],c} \rangle} + \mu_{\langle M^{[u],j} \rangle} + \mu_{\langle M^{[u],k} \rangle}. \tag{4.3.7}$$

In what follows, we write for $u, v \in \mathcal{F}_e$

$$\mu_{\langle u,v \rangle} = \mu_{\langle M^{[u]}, M^{[v]} \rangle}, \quad \mu^i_{\langle u,v \rangle} = \mu_{\langle M^{[u],i}, M^{[v],i} \rangle} \text{ with } i = c,\, j,\, k, \tag{4.3.8}$$

When $u = v$, we simply write $\mu_{\langle u \rangle}$ for $\mu_{\langle u,u \rangle}$ and $\mu^i_{\langle u \rangle}$ for $\mu^i_{\langle u,u \rangle}$ with $i = c,\, j,\, k$. Observe that $\mu_{\langle u,v \rangle} = \frac{1}{4}(\mu_{\langle u+v \rangle} - \mu_{\langle u-v \rangle})$ and similar relations hold for $\mu^i_{\langle u,v \rangle}$ with $i = c,\, j,\, k$.

Now consider the Lévy system $(N(x,dy), H)$ of the Hunt process X introduced in Section A.3.4. For $u \in \mathcal{F}_e$, by the definitions of $M^{[u],j}$, $M^{[u],k}$

combined with (4.3.3) and (A.3.34), we have

$$\langle M^{[u],j}\rangle_t = [\, M^{[u],j}\,]_t^p = \left(\sum_{0<s\le t<\zeta} (\Delta M_s^{[u]})^2\right)^p$$

$$= \left(\sum_{0<s\le t} \left(\widetilde{u}(X_s) - \widetilde{u}(X_{s-})\right)^2 1_{\{X_s\in E\}}\right)^p$$

$$= \int_0^t \int_E (\widetilde{u}(X_s) - \widetilde{u}(y))^2 N(X_s, dy)dH_s, \tag{4.3.9}$$

$$\langle M^{[u],k}\rangle_t = \left((\Delta M_\zeta^{[u]})^2 1_{\{\zeta\le t\}}\right)^p = \left((\widetilde{u}(X_{\zeta-})^2 1_{\{\zeta\le t\}}\right)^p$$

$$= \left(\sum_{0<s\le t} 1_{\{\partial\}}(X_s)\widetilde{u}(X_{s-})^2 1_{\{X_s\ne X_{s-}\}}\right)^p$$

$$= \int_0^t \widetilde{u}(X_s)^2 N(X_s, \{\partial\})dH_s. \tag{4.3.10}$$

Denoting the Revuz measure of $H \in \mathbf{A}_c^+$ by μ_H, we let

$$J(dx, dy) = N(x, dy)\mu_H(dx), \quad \kappa(dx) = N(x, \{\partial\})\mu_H(dx). \tag{4.3.11}$$

We are then led from (4.3.9), (4.3.10), and Theorem A.3.5 to the expression

$$\begin{cases} \mu^j_{\langle u\rangle}(dx) = \int_E (\widetilde{u}(x) - \widetilde{u}(y))^2 J(dx, dy), \\ \mu^k_{\langle u\rangle}(dx) = \widetilde{u}(x)^2\kappa(dx). \end{cases} \tag{4.3.12}$$

For $u \in \mathcal{F}_e$, the total mass of each measure appearing on the right hand side of (4.3.7) is finite. Hence $M^{[u],i} \in \overset{\circ}{\mathcal{M}}$, $i = c, j, k$, in view of (4.2.13). Using their mutual energies, we introduce three bilinear forms on $\mathcal{F}_e$ by

$$\begin{cases} \mathcal{E}^{(c)}(u, v) = \mathbf{e}(M^{[u],c}, M^{[v],c}), \\ \mathcal{E}^{(j)}(u, v) = \mathbf{e}(M^{[u],j}, M^{[v],j}), \\ \mathcal{E}^{(k)}(u, v) = 2\mathbf{e}(M^{[u],k}, M^{[v],k}). \end{cases} \tag{4.3.13}$$

By (4.2.13) and (4.3.12) it then holds for $u \in \mathcal{F}_e$ that

$$\begin{cases} \mathcal{E}^{(j)}(u, u) = \dfrac{1}{2}\displaystyle\int_{E\times E\setminus d} (\widetilde{u}(x) - \widetilde{u}(y))^2 J(dx, dy), \\ \mathcal{E}^{(k)}(u, u) = \displaystyle\int_E \widetilde{u}(x)^2\kappa(dx). \end{cases} \tag{4.3.14}$$

Since $\widetilde{u}(x) - \widetilde{u}(y)$ vanishes along the diagonal d of $E \times E$, the integral $\int_{E\times E\setminus d}(\widetilde{u}(x) - \widetilde{u}(y))^2 J(dx, dy)$ will also be denoted as $\int_{E\times E}(\widetilde{u}(x) - \widetilde{u}(y))^2 J(dx, dy)$ in this book. This remark also applies to other integrals involving $J(dx, dy)$ as long as the integrand vanishes along the diagonal.

On the other hand, by virtue of Theorem 4.2.1(ii), Theorem 4.2.6, and (4.2.13), we have

$$\begin{aligned} \mathcal{E}(u,u) &= \mathbf{e}(M^{[u]}) + \frac{1}{2}\int_E \widetilde{u}(x)^2 \kappa(dx) \\ &= \frac{1}{2}\mu_{\langle u\rangle}(E) + \frac{1}{2}\int_E \widetilde{u}(x)^2\kappa(dx), \qquad u \in \mathcal{F}_e, \end{aligned} \tag{4.3.15}$$

and consequently we arrive at the decomposition

$$\mathcal{E}(u,v) = \mathcal{E}^{(c)}(u,v) + \mathcal{E}^{(j)}(u,v) + \mathcal{E}^{(k)}(u,v), \quad u, v \in \mathcal{F}_e. \tag{4.3.16}$$

PROPOSITION 4.3.1. (i) *For $u \in \mathcal{F}_e$ and a finely open set $D \subset E$, suppose $\widetilde{u}$ is constant q.e. on D. Then $\mu^c_{\langle u\rangle}(D) = 0$.*
(ii) *For $u \in \mathcal{F}_e$ and a quasi open set $D \subset E$, suppose $\widetilde{u}$ is constant q.e. on D. Then $\mu^c_{\langle u\rangle}(D) = 0$.*
(iii) *$\mathcal{E}^{(c)}$ is strongly local in the sense of Definition 1.3.17(v).*

Proof. (i) By Theorem 4.2.6,

$$\widetilde{u}(X_t) - \widetilde{u}(X_0) = M_t^{[u]} + N_t^{[u]}, \quad t \geq 0, \quad \mathbf{P}_x\text{-a.s. for q.e. } x \in E, \tag{4.3.17}$$

where $M^{[u]} \in \overset{\circ}{\mathcal{M}}$ and $N^{[u]} \in \mathcal{N}_c$. Define

$$C_t^{(n)} := \sum_{k=1}^{[2^n t]} \left(N^{[u]}_{k2^{-n}} - N^{[u]}_{(k-1)2^{-n}}\right)^2, \quad t \geq 0.$$

Here for $x \in \mathbb{R}$, $[x]$ denotes the largest integer that is no larger than x. Then for every integer $j \geq 1$,

$$\begin{aligned} \lim_{n\to\infty} \mathbf{E}_m\left[C_j^{(n)}\right] &= \lim_{n\to\infty}\sum_{k=1}^{j2^n} \mathbf{E}_m\left[\mathbf{E}_{X_{(k-1)2^{-n}}}\left[\left(N^{[u]}_{2^{-n}}\right)^2\right]\right] \\ &\leq \lim_{n\to\infty} j2^n \mathbf{E}_m\left[\left(N^{[u]}_{2^{-n}}\right)^2\right] = 0. \end{aligned}$$

It follows that for every integer $j \geq 1$ and $\varepsilon > 0$,

$$\lim_{n\to\infty}\mathbf{P}_m\left(\sup_{t\in[0,j]} C_t^{(n)} > \varepsilon\right) = \lim_{n\to\infty}\mathbf{P}_m\left(C_j^{(n)} > \varepsilon\right) = 0. \tag{4.3.18}$$

Define

$$B_t^{(n)} := \sum_{k=1}^{[2^n t]} \left(\widetilde{u}(X_{k2^{-n}}) - \widetilde{u}(X_{(k-1)2^{-n}})\right)^2, \quad t \geq 0.$$

Then $B_t^{(n)} = 0$ for $t < \tau_D$ where $\tau_D = \sigma_{E \setminus D} \wedge \zeta$. On account of (4.3.17)–(4.3.18) and Section A.3.3(**4**), it holds that

$$\begin{aligned}\lim_{n\to\infty} B_t^{(n)} &= \lim_{n\to\infty} \sum_{k=1}^{[2^n t]} \left(M_{k2^{-n}}^{[u]} - M_{(k-1)2^{-n}}^{[u]}\right)^2 = [M^{[u]}]_t \\ &= \langle M^{[u],c} \rangle_t + \sum_{0<s\leq t} (\widetilde{u}(X_s) - \widetilde{u}(X_{s-}))^2, \qquad (4.3.19)\end{aligned}$$

where the convergence is in measure with respect to $\mathbf{P}_m$ uniformly on every compact time interval. Therefore,

$$\langle M^{[u],c} \rangle_t = 0, \quad t < \tau_D, \quad \mathbf{P}_m\text{-a.e.} \qquad (4.3.20)$$

It then follows from (4.1.25) that $\mu^c_{\langle u \rangle}(D) = 0$.
(ii) follows from (i), Theorem 3.3.3, and the fact that $\mu^c_{\langle u \rangle}$ is a smooth measure so it charges no $\mathcal{E}$-polar set.
(iii) Suppose, for $u, v \in \mathcal{F}_e$, $K = \text{supp}[u \cdot m]$ is compact and v is constant m-a.e. on a neighborhood D of K. Since a quasi continuous version $\widetilde{v}$ of v is constant q.e. on D, we have $\mu^c_{\langle v \rangle}(D) = 0$ by (i). In the same way, $\mu^c_{\langle u \rangle}(E \setminus K) = 0$. Since $\mu^c_{\langle u,v \rangle}(B)^2 \leq \mu^c_{\langle u \rangle}(B)\mu^c_{\langle v \rangle}(B)$, $B \in \mathcal{B}(E)$, we get $\mu^c_{\langle u,v \rangle}(E) = 0$ and hence $\mathcal{E}^{(c)}(u, v) = 0$. □

We need the next proposition to ensure the symmetry of the measure J defined by (4.3.11).

PROPOSITION 4.3.2. *For any* $u, v \in \mathcal{F} \cap C_c^+(E)$ *with* $\text{supp}\,[u] \cap \text{supp}\,[v] = \emptyset$,

$$\int_{E\times E} u(x)v(y)J(dx, dy) = -\mathcal{E}(u, v).$$

Proof. We consider a relatively compact open set $D \subset E$ satisfying $\text{supp}[u] \subset D \subset \overline{D} \subset (\text{supp}[v])^c$. Let $(\mathcal{E}^D, \mathcal{F}_D)$ be the part of the Dirichlet form $(\mathcal{E}, \mathcal{F})$ on the open set D: $\mathcal{F}_D$ is defined by (3.2.2) and $\mathcal{E}^D$ is the restriction of $\mathcal{E}$ on $\mathcal{F}_D \times \mathcal{F}_D$. According to Theorem 3.3.9, $\mathcal{E}^D$ is a regular Dirichlet form on $L^2(D; m)$.

Denoting $\int_E f(y)N(x,dy)$ by $Nf(x)$, we have for $f \in \mathcal{F} \cap C_c(D)$

$$\begin{aligned}\mathcal{E}^{(j)}(|f|, v) &= \int_{E\times E} (|f(x)| - |f(y)|)(v(x) - v(y))N(x,dy)\mu_H(dx) \\ &= -\left(\int_E v \cdot N|f| d\mu_H + \int_E |f| \cdot Nv d\mu_H\right),\end{aligned}$$

which implies that

$$\int_E |f| \cdot Nv d\mu_H \leq \mathcal{E}^{(j)}(|f|,|f|)^{1/2} \cdot \mathcal{E}^{(j)}(v,v)^{1/2} \leq C\sqrt{\mathcal{E}(f,f)},$$

namely, $\mathbf{1}_D \cdot Nv \cdot \mu_H$ is a measure of finite energy integral with respect to the regular Dirichlet form $\mathcal{E}^D$.

On the other hand, by noting that

$$e^{-\sigma_{E\setminus D}} v(X_{\sigma_{E\setminus D}}) = \sum_{0<s\leq\sigma_{E\setminus D}} e^{-\alpha s}(v(X_s) - v(X_{s-})),$$

we can get from the identity (A.3.33) with $T = \sigma_{E\setminus D}$, $g(s) = e^{-\alpha s}$, and $f(x,y) = v(y) - v(x)$ that

$$\begin{aligned}H^{\alpha}_{E\setminus D}v(x) &= \mathbf{E}_x\left[\int_0^{\sigma_{E\setminus D}} e^{-\alpha s}\int_E (v(y) - v(X_s))N(X_s,dy)dH_s\right] \\ &= \mathbf{E}_x\left[\int_0^{\sigma_{E\setminus D}} e^{-\alpha s} Nv(X_s)dH_s\right].\end{aligned}$$

Therefore, we can deduce from (4.1.26) that $\mathbf{H}^{\alpha}_{E\setminus D}v(x)$ coincides m-a.e. on D with the α-potential $U^D_\alpha(\mathbf{1}_D Nv \cdot \mu_H)$ of the measure $\mathbf{1}_D Nv \cdot \mu_H$ with respect to $\mathcal{E}^D$.

Furthermore, $\mathbf{H}^{\alpha}_{E\setminus D}v(x) = v(x)$ on $E \setminus D$ and hence

$$\mathbf{H}^{\alpha}_{E\setminus D}v = U^D_\alpha(\mathbf{1}_D Nv \cdot \mu_H) + v.$$

Since $\mathcal{E}_\alpha(\mathbf{H}^{\alpha}_{E\setminus D}v, u) = 0$ by Theorem 3.2.2, we finally get

$$\int_{E\times E} u(x)v(y)J(dx,dy) = \mathcal{E}_\alpha(U^D_\alpha(\mathbf{1}_D Nv \cdot \mu_H), u) = -\mathcal{E}(u,v). \qquad \square$$

Summing up what has been studied, we obtain the following theorem on the Beurling-Deny representation for the regular Dirichlet form $(\mathcal{E}, \mathcal{F})$.

THEOREM 4.3.3. (i) *The form $\mathcal{E}$ on $\mathcal{F}_e \times \mathcal{F}_e$ admits a unique representation* (4.3.1) *by means of a strongly local symmetric form $\mathcal{E}^{(c)}$, a symmetric Radon measure J on $E \times E \setminus d$, and a Radon measure κ on E. Moreover, these parts are related to martingale additive functionals by* (4.3.13).

(ii) *The measures* J, κ *are expressed as* (4.3.11) *by means of the Lévy system* (N, H) *of* X.
(iii) *The strongly local part admits the expression* $\mathcal{E}^{(c)}(u,v) = \frac{1}{2}\mu^c_{\langle u,v\rangle}(E)$ *for every* $u, v \in \mathcal{F}_e$ *in terms of the measure* μ^c *defined by* (4.3.8).

Proof. We defined J, κ by (4.3.11) and derived the representation (4.3.13)–(4.3.16) of $\mathcal{E}$. It follows from (4.3.13) and (4.2.13) that $\mathcal{E}^{(c)}(u,v) = \frac{1}{2}\mu^c_{\langle u,v\rangle}(E)$ for every $u, v \in \mathcal{F}_e$. By Proposition 4.3.1, $\mathcal{E}^{(c)}$ is strongly local and, by Proposition 4.3.2, J is symmetric. The uniqueness of the expression (4.3.1) is then immediate. □

The measures J, κ defined by (4.3.11) in terms of the Lévy system of X are called the *jumping measure, killing measure* of X, respectively.

From Theorem 4.3.3, we can deduce probabilistic characterizations of the notions of the local property and the strongly local property of the Dirichlet form introduced in Definition 1.3.17.

THEOREM 4.3.4. (i) $\mathcal{E}$ *is local if and only if* X *is equivalent to a diffusion, namely, for some Borel properly exceptional set* N

$$\mathbf{P}_x(X_t \text{ is continuous in } t \in [0,\zeta)) = 1, \quad x \in E \setminus N. \tag{4.3.21}$$

(ii) $\mathcal{E}$ *is strongly local if and only if* X *is equivalent to a diffusion with no killing inside* E, *namely, for some Borel properly exceptional set* N, (4.3.21) *is valid for* $x \in E \setminus N$ *and furthermore*

$$\mathbf{P}_x(X_{\zeta-} \in E,\ \zeta < \infty) = 0, \quad x \in E \setminus N. \tag{4.3.22}$$

Proof. Theorem 4.3.3(i) implies that $\mathcal{E}$ is local (resp. strongly local) if and only if $J = 0$ (resp. $J = 0$ and $\kappa = 0$).

On account of Theorem 4.3.3(ii), we can combine (A.3.31) with Theorem 4.1.1 to get

$$\mathbf{E}_{h\cdot m}\left[\sum_{s\le t} f(X_{s-}, X_s)\mathbf{1}_E(X_s)\right] = \int_0^t ds \int_{E\times E} P_s h(x) f(x,y) J(dx,dy), \tag{4.3.23}$$

for any $h \in \mathcal{B}_+(E)$ and $f \in \mathcal{B}_+(E \times E)$ with $f(x,x) = 0$ for every $x \in E$, and further

$$\mathbf{E}_{h\cdot m}\left[\sum_{s\le t} g(X_{s-})\mathbf{1}_{\{\partial\}}(X_s)\right] = \int_0^t ds \int_{E\times E} P_s h(x) g(x)\kappa(dx), \tag{4.3.24}$$

for any $h \in \mathcal{B}_+(E)$ and $g \in \mathcal{B}_+(E)$.

Therefore, $J = 0$ if and only if the left hand side of (4.3.23) vanishes always, or, equivalently, by taking $f(x, y) = d(x, y)$ where d is the metric on E, (4.3.21) holds for m-a.e. $x \in E$. Since the function $g(x) = \mathbf{P}_x(X_t$ is continuous in $t \in (0, \zeta))$ of $x \in E$ is excessive with respect to X, we can conclude from Theorem 3.3.3 or Theorem A.2.13 that (4.3.21) is valid for q.e. $x \in E$ and we obtain (i) by using Theorem 3.1.5.

In the same way, we can obtain (ii) by making additional use of (4.3.24). □

Remark 4.3.5. (i) The proof of the above theorem in fact showed the following: the Hunt process X admits no killing inside E in the sense that (4.3.22) holds for some Borel properly exceptional set N if and only if $\kappa = 0$.
(ii) Assume that the Hunt process X on E additionally satisfies the absolute continuity condition **(AC)** in Definition A.2.16. We then see by Theorem A.2.17(ii) that the above theorem can be strengthened as follows: $\mathcal{E}$ is local (resp. strongly local) if and only if (4.3.21) holds (resp. (4.3.21) and (4.3.22) hold) for every $x \in E$. Furthermore, $\kappa = 0$ if and only if (4.3.22) holds for every $x \in E$.
(iii) The Beurling-Deny decomposition can be established for quasi-regular Dirichlet form by using quasi-homeomorphism of Theorem 1.4.3 and an equivalent formulation on the strongly local part $\mathcal{E}^{(c)}$ in Theorem 2.4.3(iii) that is invariant under quasi-homeomorphism. See Proposition 6.4.1 for another equivalent formulation on strongly local part $\mathcal{E}^{(c)}$. □

Let us study more detailed properties of the measure μ^c expressing the strongly local part $\mathcal{E}^{(c)}$ in Theorem 4.3.3. The next *derivation property* of μ^c was discovered by Y. LeJan [117] by an analytic method. The proof given below is probabilistic.

LEMMA 4.3.6. *It holds for any $u, v, w \in b\mathcal{F}_e$ that*

$$d\mu^c_{\langle uv,w\rangle} = \widetilde{u}d\mu^c_{\langle v,w\rangle} + \widetilde{v}d\mu^c_{\langle u,w\rangle}. \tag{4.3.25}$$

Proof. For $u = R_\alpha f$, $v = R_\alpha g$ with $f, g \in C_c(E)$, we let $Z^1_t := u(X_t)$ and $Z^2_t := v(X_t)$. Then $Z^1_t = u(X_0) + M^{[u]}_t + N^{[u]}_t$ and $Z^2_t = v(X_0) + M^{[v]}_t + N^{[v]}_t$ are two semimartingales. In fact,

$$N^{[u]}_t = \int_0^t (\alpha u - f)(X_s)ds \quad \text{and} \quad N^{[v]}_t = \int_0^t (\alpha v - g)(X_s)ds.$$

By the uniqueness of Fukushima's decomposition for

$$u(X_t)v(X_t) - u(X_0)v(X_0) = M^{[uv]}_t + N^{[uv]}_t, \qquad t \geq 0,$$

we draw the martingale parts from product rule (A.3.40) for $u(X_t)v(X_t)$ to conclude that

$$M_t^{[uv]} = \int_0^t \widetilde{u}(X_{s-})dM_s^{[v]} + \int_0^t \widetilde{v}(X_{s-})dM_s^{[u]} + \left([M^{[u]}, M^{[v]}]_t - \langle M^{[u]}, M^{[v]}\rangle_t\right). \tag{4.3.26}$$

Now we can observe that

$$M_t^{[uv],c} = \int_0^t \widetilde{u}(X_{s-})dM_s^{[v],c} + \int_0^t \widetilde{v}(X_{s-})dM_s^{[u],c}, \tag{4.3.27}$$

$$M_t^{[uv],d} = \int_0^t \widetilde{u}(X_{s-})dM_s^{[v],d} + \int_0^t \widetilde{v}(X_{s-})dM_s^{[u],d} + \left([M^{[u]}, M^{[v]}]_t - \langle M^{[u]}, M^{[v]}\rangle_t\right) \tag{4.3.28}$$

are the continuous part and the purely discontinuous part of the MAF $M^{[uv]}$, respectively. Indeed, we know that the latter belongs to the space $\mathbb{M}^d$ from **(4)** and **(5)** of Section A.3.3, while the former belongs to $\mathbb{M}^c$ because the stochastic integral based on a continuous martingale is continuous.

Therefore, we obtain

$$\langle M^{[uv],c}, M^{[w],c}\rangle_t = \int_0^t \widetilde{u}(X_s)d\langle M^{[v],c}, M^{[w],c}\rangle_s + \int_0^t \widetilde{v}(X_s)d\langle M^{[u],c}, M^{[w],c}\rangle_s$$

and we arrive at (4.3.25) by taking the Revuz measures of both sides. For general $u, v \in b\mathcal{F}_e$, it suffices to note that $G_1(L^2(E;m))$ is dense in $(\mathcal{F}, \mathcal{E}_1^{1/2})$, which in turn is dense in $(\mathcal{F}_e, \mathcal{E}^{1/2})$, and that for $w \in \mathcal{F}_e$ and any bounded Borel function f on E,

$$\left| \int_E f d\mu^c_{\langle u,w\rangle} - \int_E f d\mu^c_{\langle v,w\rangle} \right| \le 2\|f\|_\infty \sqrt{\mathcal{E}(w,w)}\sqrt{\mathcal{E}(u-v,u-v)}. \qquad \square$$

As we see from (4.3.28), the measure $\mu^d = \mu^j + \mu^k$ corresponding to the purely discontinuous part does not satisfy the derivation property in general.

The derivation property of μ^c can be interpreted in a more general fashion. For $\Phi \in C^1(\mathbb{R}^m)$ satisfying $\Phi(0) = 0$ and for $u_1, \ldots, u_m \in b\mathcal{F}$, we have

$$|\Phi(\mathbf{u}(x))| \le \sum_{i=1}^m \|\Phi_{x_i}\|_{L^\infty(V)} |u_i(x)|$$

$$|\Phi(\mathbf{u}(x)) - \Phi(\mathbf{u}(y))| \le \sum_{i=1}^m \|\Phi_{x_i}\|_{L^\infty(V)} |u_i(x) - u_i(y)|,$$

where $V \subset \mathbb{R}^m$ denotes the range of $\mathbf{u}$. Hence, by noting that the approximating form $\mathcal{E}^{(t)}$ admits the expression

$$\mathcal{E}^{(t)}(u,u) = \frac{1}{2t}\int_E\int_E (u(x)-u(y))^2 P_t(x,dy)m(dx) + \frac{1}{t}\int_E u(x)^2(1-P_t1)(x)m(dx),$$

we can conclude that $\Phi(\mathbf{u}) = \Phi(u_1,\ldots,u_m) \in \mathcal{F}$ and

$$\mathcal{E}_\alpha(\Phi(\mathbf{u}),\Phi(\mathbf{u}))^{1/2} \le \sum_{i=1}^m \|\Phi_{x_i}\|_{L^\infty(V)}\mathcal{E}_\alpha(u_i,u_i)^{1/2}. \tag{4.3.29}$$

THEOREM 4.3.7. *It holds for any $\Phi \in C^1(\mathbb{R}^m)$ satisfying $\Phi(0) = 0$ and any $u_1,\ldots,u_m \in b\mathcal{F}$ that $\Phi(\mathbf{u}) \in \mathcal{F}$ and*

$$d\mu^c_{\langle\Phi(\mathbf{u}),v\rangle} = \sum_{i=1}^m \Phi_{x_i}(\mathbf{u})d\mu^c_{\langle u_i,v\rangle} \quad \textit{for every } v \in b\mathcal{F}. \tag{4.3.30}$$

Proof. Denote by $\mathcal{A}$ the totality of functions Φ satisfying the properties of the above statement. By Lemma 4.3.6, it holds for any $\Phi, \Psi \in \mathcal{A}$ that

$$\begin{aligned} d\mu^c_{\langle\Phi(\mathbf{u})\Psi(\mathbf{u}),v\rangle} &= \Phi(\mathbf{u})d\mu^c_{\langle\Psi(\mathbf{u}),v\rangle} + \Psi(\mathbf{u})d\mu^c_{\langle\Phi(\mathbf{u}),v\rangle} \\ &= \sum_{i=1}^m (\Phi\Psi)_{x_i}(\mathbf{u})d\mu^c_{\langle u_i,v\rangle} \end{aligned}$$

and consequently $\Phi\Psi \in \mathcal{A}$. Since $\mathcal{A}$ contains coordinate functions, it contains all polynomials vanishing at the origin.

For $\Phi \in C^1(\mathbb{R}^m)$ vanishing at the origin and for a cube V containing the range of $\mathbf{u}$, we can find a sequence $\Phi^{(k)}$ of polynomials such that $\Phi^{(k)}(0)=0$, $\|\Phi^{(k)} - \Phi\|_{L^\infty(V)} \to 0$ and $\|\Phi^{(k)}_{x_i} - \Phi_{x_i}\|_{L^\infty(V)} \to 0$. Inequality (4.3.29) then implies that $\{\Phi^{(k)}(\mathbf{u})\}$ is $\mathcal{E}_\alpha$-convergent to $\Phi(\mathbf{u})$. Hence (4.3.30) can be obtained as a limit of that for $\Phi^{(k)}$. □

The following result shows that the push forward measure $\mu^c_{\langle u\rangle} \circ u^{-1}$ of $\mu^c_{\langle u\rangle}$ under map $u \in b\mathcal{F}$ is absolutely continuous with respect to the Lebesgue measure on $\mathbb{R}$.

THEOREM 4.3.8. *Suppose that $u \in b\mathcal{F}$. Then the push forward measure ν of $\mu^c_{\langle u\rangle}$ under map u defined by*

$$\nu(A) := \mu^c_{\langle u\rangle}(u^{-1}(A)), \quad A \in \mathcal{B}(\mathbb{R}),$$

is absolutely continuous with respect to the Lebesgue measure on $\mathbb{R}$.

Proof. It suffices to show that for any compact set $K \subset \mathbb{R}$ having zero Lebesgue measure, $\nu(K) = 0$. Let K be a compact set having zero Lebesgue measure. There exists a sequence $\{\varphi_k, k \geq 1\}$ of continuous functions having compact support in $\mathbb{R}$ such that $|\varphi_k| \leq 1$, $\lim_{k\to\infty} \varphi_k(r) = \mathbf{1}_K(r)$ on $\mathbb{R}$, and

$$\int_0^\infty \varphi_k(r)dr = \int_{-\infty}^0 \varphi_k(r)dr = 0 \quad \text{for } k \geq 1.$$

The last display implies that each $\Phi_k(x) := \int_0^x \varphi_k(r)dr$ is a C^1 function with compact support, $\Phi_k(0) = 0$ and $|\Phi_k'(x)| \leq 1$. Hence $\Phi_k(u)$ is a normal contraction of u and so $\Phi_k(u) \in \mathcal{F}$ with $\mathcal{E}(\Phi_k(u), \Phi_k(u)) \leq \mathcal{E}(u,u)$. Since $\lim_{k\to\infty} \Phi_k(r) = 0$ on $\mathbb{R}$, by dominated convergence theorem, $\Phi_k(u) \to 0$ in $L^2(E;m)$. Thus by Banach-Saks Theorem (Theorem A.4.1), taking the Cesàro mean sequence of a suitable subsequence of $\{\varphi_k, k \geq 1\}$, and then redefining them as $\{\varphi_k, k \geq 1\}$ if necessary, we may and do assume that $\Phi_k(u)$ is $\mathcal{E}_1$-convergent to $0 \in \mathcal{F}$. Now by Fatou's lemma and Theorems 4.3.3(iii) and 4.3.7, we have

$$\begin{aligned}
\nu(K) &\leq \lim_{k\to\infty} \int_{\mathbb{R}} \varphi_k(r)^2 \nu(dr) = \lim_{k\to\infty} \int_E \varphi_k(u(x))^2 \mu^c_{\langle u\rangle}(dx) \\
&= \lim_{k\to\infty} 2\mathcal{E}^c(\Phi_k(u), \Phi_k(u)) \leq 2 \lim_{k\to\infty} \mathcal{E}(\Phi_k(u), \Phi_k(u)) = 0.
\end{aligned}$$

This completes the proof. □

The next two results will be used later in Chapter 6.

Lemma 4.3.9. *Suppose that $u \in \mathcal{F}$ and φ is a normal contraction on $\mathbb{R}$. Then $\mu^c_{\langle\varphi(u)\rangle} \leq \mu^c_{\langle u\rangle}$ on E.*

Proof. Let $v := \varphi(u)$. Since v is a normal contraction of u, we have by (4.3.19) that $\mathbf{P}_m$ -a.e.

$$\langle M^{[v],c}\rangle_t - \langle M^{[v],c}\rangle_s \leq \langle M^{[u],c}\rangle_t - \langle M^{[u],c}\rangle_s + \sum_{s<r\leq t} (\widetilde{u}(X_r) - \widetilde{u}(X_{r-}))^2$$

for every $t > s \geq 0$. For every $\varepsilon > 0$, let $T_0^\varepsilon := 0$ and define for $k \geq 1$

$$T_k^\varepsilon := \inf\left\{t > T_{k-1}^\varepsilon : |\widetilde{u}(X_t) - \widetilde{u}(X_{t-})| > \varepsilon\right\}.$$

Note that T_k^ε is a sequence of stopping times that increases to infinity. We have from above that $\mathbf{P}_m$-a.e.,

$$\langle M^{[v],c}\rangle_t - \langle M^{[v],c}\rangle_{T_{k-1}^\varepsilon} \leq \langle M^{[u],c}\rangle_t - \langle M^{[u],c}\rangle_{T_{k-1}^\varepsilon} + \sum_{T_{k-1}^\varepsilon<r\leq t} (\widetilde{u}(X_r) - \widetilde{u}(X_{r-}))^2$$

for every $k \geq 1$ and every $t \in (T^{\varepsilon}_{k-1}, T^{\varepsilon}_{k})$. Consequently, we have $\mathbf{P}_m$-a.e.,

$$\begin{aligned}\langle M^{[v],c}\rangle_t - \langle M^{[v],c}\rangle_s \leq \langle M^{[u],c}\rangle_t - \langle M^{[u],c}\rangle_s \\ + \sum_{s<r\leq t} \mathbf{1}_{\{|\widetilde{u}(X_r)-\widetilde{u}(X_{r-})|\leq\varepsilon\}} (\widetilde{u}(X_r) - \widetilde{u}(X_{r-}))^2\end{aligned}$$

for every $t > s \geq 0$. Letting $\varepsilon \downarrow 0$, we conclude that $\mathbf{P}_m$-a.e.

$$\langle M^{[v],c}\rangle_t - \langle M^{[v],c}\rangle_s \leq \langle M^{[u],c}\rangle_t - \langle M^{[u],c}\rangle_s \quad \text{for every } t > s \geq 0.$$

It now follows from (4.1.1) that $\mu^c_{\langle v\rangle} \leq \mu^c_{\langle u\rangle}$ on E. □

Define the following version of a local Dirichlet space:

$$\begin{aligned}\overset{\circ}{\mathcal{F}}_{\text{loc}} = \Big\{ u : \text{ there is an increasing sequence of quasi open sets } \{D_n\} \\ \text{with } \bigcup_{n=1}^{\infty} D_n = E \text{ q.e. and a sequence } \{u_n\} \subset \mathcal{F} \\ \text{such that } u = u_n \ m\text{-a.e. on } D_n \Big\}.\end{aligned} \tag{4.3.31}$$

Clearly, every $u \in \overset{\circ}{\mathcal{F}}_{\text{loc}}$ admits an m-version that is quasi continuous on E.

THEOREM 4.3.10. (i) *Suppose D is a quasi open set in E. Then $\mu^c_{\langle u\rangle} = \mu^c_{\langle v\rangle}$ on D for every $u, v \in \mathcal{F}$ such that $\widetilde{u}$ and $\widetilde{v}$ differ by a constant q.e. on D.*
(ii) *For $u \in \overset{\circ}{\mathcal{F}}_{\text{loc}}$, let $\{u_n, n \geq 1\} \subset \mathcal{F}$ and an increasing sequence of quasi open sets $\{D_n, n \geq 1\}$ with $\cup_{n=1}^{\infty} D_n = E$ q.e. be such that $u = u_n$ m-a.e. on D_n. Then there is a unique smooth measure $\mu^c_{\langle u\rangle}$ on E so that for every $n \geq 1$, $\mu^c_{\langle u\rangle} = \mu^c_{\langle u_n\rangle}$ on D_n.*
(iii) *Suppose that φ is a normal contraction on $\mathbb{R}$. Then for every $u \in \overset{\circ}{\mathcal{F}}_{\text{loc}}$, $v := \varphi(u) \in \overset{\circ}{\mathcal{F}}_{\text{loc}}$ and $\mu^c_{\langle v\rangle} \leq \mu^c_{\langle u\rangle}$ on E.*

Proof. (i) Since $\widetilde{u} - \widetilde{v}$ is a constant q.e. on D by (4.3.20), $\langle M^{[u],c} - M^{[v],c}\rangle_t = \langle M^{[u-v],c}\rangle_t = 0$ for every $t < \tau_D$ $\mathbf{P}_m$-a.s. Therefore, $\mathbf{P}_m$-a.e. $\langle M^{[u],c}\rangle_t = \langle M^{[v],c}\rangle_t$ for every $t < \tau_D$. It then follows from (4.1.25) that $\mu^c_{\langle u\rangle} = \mu^c_{\langle v\rangle}$ on D.

(ii) As $\widetilde{u}_j = \widetilde{u}_k$ q.e. on D_k for every $j > k$, we have by (i) $\mu^c_{\langle u_j\rangle} = \mu^c_{\langle u_k\rangle}$ on D_k for every $j > k$. Consequently,

$$\mu^c_{\langle u\rangle}(A) := \lim_{j\to\infty} \mu^c_{\langle u_j\rangle}(A \cap D_j)$$

defines the unique smooth measure on E that coincides with $\mu^c_{\langle u_k\rangle}$ when restricted to D_k for every $k \geq 1$.

(iii) By the definition of $\overset{\circ}{\mathcal{F}}_{\rm loc}$ and because every normal contraction operates on the Dirichlet form $(\mathcal{E}, \mathcal{F})$, it is clear that $v \in \overset{\circ}{\mathcal{F}}_{\rm loc}$. That $\mu^c_{\langle v\rangle} \leq \mu^c_{\langle u\rangle}$ on E follows directly from (ii) and Lemma 4.3.9. □

We end this section by giving an analytic characterization of the measure $\mu_{\langle u,v\rangle}$ for $u, v \in \mathcal{F}_e$ introduced by (4.3.8).

Theorem 4.3.11. *It holds for $u \in b\mathcal{F}_e$ that*

$$\int_E \widetilde{f}(x)\mu_{\langle u\rangle}(dx) = 2\mathcal{E}(u, uf) - \mathcal{E}(u^2, f), \quad f \in b\mathcal{F}_e. \tag{4.3.32}$$

Proof. First, we assume that $u, f \in b\mathcal{F}$. We see from Theorem 4.1.1, (4.3.2), and Theorem 4.2.6 that

$$\int_E \widetilde{f} d\mu_{\langle u\rangle} = \lim_{t\to 0}\frac{1}{t}\mathbf{E}_{f\cdot m}[\langle M^{[u]}\rangle_t] = \lim_{t\to 0}\frac{1}{t}\mathbf{E}_{f\cdot m}[(\widetilde{u}(X_t) - \widetilde{u}(X_0))^2].$$

Since

$$\begin{aligned}\mathbf{E}_{f\cdot m}[(\widetilde{u}(X_t) - \widetilde{u}(X_0))^2] &= (f, P_t u^2 - 2uP_t u + u^2) \\ &= 2(fu, u - P_t u) - (f, u^2 - P_t u^2),\end{aligned}$$

we get (4.3.32) by (1.1.4) and (1.1.5).

For general $u, f \in b\mathcal{F}_e$, we can choose their approximating sequences $\{u_n\}$, $\{f_n\} \subset \mathcal{F}$ with $|u_n| \leq M$, $|f_n| \leq M$ for some $M > 0$ and $\lim_{n\to\infty}\widetilde{f}_n = f$ q.e. in view of Theorem 2.3.4.

By Lemma 2.1.15, there exists $g \in \mathcal{K}_0$ such that u, $f \in \mathcal{F}^g_e$ and u_n (resp. f_n) is $\mathcal{E}^g$-convergent to u (resp. f) as $n \to \infty$ in the transient Dirichlet space $\mathcal{F}^g_e$. Since u_n^2 and u^2 are normal contractions of $2Mu_n$ and $2Mu$, respectively, we see from Exercise 2.1.13(i) that $\{u_n^2\}$ is weakly $\mathcal{E}^g$-convergent and hence weakly $\mathcal{E}$-convergent to u^2. Similarly, $u_n f_n = \frac{1}{4}\{(u_n + f_n)^2 - (u_n - f_n)^2\}$ is weakly $\mathcal{E}$-convergent to uf as $n \to \infty$.

If we let $\|v\| = \left(\int_E g d\mu_{\langle v\rangle}\right)^{1/2}$ for $g \in b\mathcal{B}_+(E)$ and $v \in b\mathcal{F}_e$, then the triangle inequality $\|v_1 + v_2\| \leq \|v_1\| + \|v_2\|$ holds for $v_1, v_2 \in b\mathcal{F}_e$. Moreover, by (4.3.15), $\|v\|^2 \leq 2\|g\|_\infty \mathcal{E}(v, v)$ for $v \in b\mathcal{F}_e$. Therefore, the identity (4.3.32) for u, f follows from that for u_n, f_n by letting $n \to \infty$. □

The measure $\mu_{\langle u\rangle}$ was introduced by M. L. Silverstein [138] by means of equation (4.3.32) and is called the *energy measure* of $u \in \mathcal{F}_e$. Theorem 4.3.11 shows that it coincides with the Revuz measure of $\langle M^{[u]}\rangle \in \mathbf{A}_c^+$. Identity (4.3.32) implies

$$\int_E \widetilde{f} d\mu_{\langle u,v\rangle} = \mathcal{E}(u, vf) + \mathcal{E}(v, uf) - \mathcal{E}(uv, f), \quad u, v \in b\mathcal{F}_e. \tag{4.3.33}$$

When $\mu_{\langle u,v\rangle}$ is absolutely continuous with respect to the measure m, its density $\Gamma(u,v)$ can be informally expressed as

$$\Gamma(u,v) = \mathcal{A}(uv) - \mathcal{A}u \cdot v - u \cdot \mathcal{A}v,$$

which coincides with the so-called *square field operator*, or *carré du champ*. Here $\mathcal{A}$ denotes the generator of L^2-semigroup.

Exercise 4.3.12. Use Lemma 4.3.6 to derive the equation

$$\int_E \widetilde{f}(x) d\mu^c_{\langle u\rangle}(x) = 2\mathcal{E}^{(c)}(u, uf) - \mathcal{E}^{(c)}(u^2, f) \quad \text{for } u, f \in b\mathcal{F}_e. \tag{4.3.34}$$

Chapter Five

TIME CHANGES OF SYMMETRIC MARKOV PROCESSES

Time change is one of the most basic and very useful transformations for Markov processes, which has been studied by many authors. The following is a prototype of the problem that will be studied in this chapter. Suppose X is a Lévy process in $\mathbb{R}^n$ that is the sum of a Brownian motion in $\mathbb{R}^n$ and an independent rotationally symmetric α-stable process in $\mathbb{R}^n$, where $n \geq 1$ and $\alpha \in (0, 2)$. Denote by $B(x, r)$ the open ball in $\mathbb{R}^n$ centered at $x \in \mathbb{R}^n$ with radius r. Its Euclidean closure is denoted by $\overline{B(x, r)}$. Let $F = \overline{B(0, 1)} \cup \partial B(x_0, 1)$, where $x_0 \in \mathbb{R}^n$ with $|x_0| = 3$. What is the trace process of X on the closed set F? More precisely, let $\mu(dx) := \mathbf{1}_{\overline{B(0,1)}}(x)dx + \sigma_{\partial B(x_0,1)}(dx)$, where $\sigma_{\partial B(x_0,1)}$ denotes the Lebesgue surface measure of $\partial B(x_0, 1)$. It is easy to see that μ is a smooth measure of X and it uniquely determines a positive continuous additive functional $A^\mu = \{A_t^\mu, t \geq 0\}$ of X having μ as its Revuz measure. Define its inverse

$$\tau_t := \inf\{s > 0 : A_s^\mu > t\} \qquad \text{for } t \geq 0.$$

Then the time-changed process $Y_t := X_{\tau_t}$ is a μ-symmetric Markov process on F, which can be regarded as the trace process of X on F. So the more precise question is

Question: Can we characterize the time-changed process Y?

We could assume that $(\mathcal{E}, \mathcal{F})$ is a quasi-regular symmetric Dirichlet form on $L^2(E; m)$, where E is a Hausdorff space and m is a σ-finite measure with $\text{supp}[m] = E$ and that X is an m-symmetric right process associated with $(\mathcal{E}, \mathcal{F})$. However, in view of the quasi-homeomorphism method in Theorem 1.4.3 and Theorem 1.5.1, without loss of generality, we may and do assume throughout this chapter that E is a locally compact separable metric space, m is a positive Radon measure on E with $\text{supp}[m] = E$, $(\mathcal{E}, \mathcal{F})$ is a regular symmetric Dirichlet form in $L^2(E; m)$, and $X = (X_t, \zeta, \mathbf{P}_x)$ is an m-symmetric Hunt process associated with $(\mathcal{E}, \mathcal{F})$. Recall the convention that any function takes value 0 at the cemetery point ∂ added to E as a one-point compactification.

We will use the same notations as before. The transition function and the resolvent of X are denoted by $\{P_t; t \geq 0\}$, $\{R_\alpha; \alpha > 0\}$, respectively. $\mathcal{B}^*(E)$ will denote the family of all universally measurable subsets of E. $\mathbf{A}_c^+$ denotes the family of all positive continuous additive functionals (PCAFs in abbreviation) of the Hunt process X and S denotes the family of all smooth measures on E with respect to the regular Dirichlet form $(\mathcal{E}, \mathcal{F})$. The equivalence classes of $\mathbf{A}_c^+$ and S are in one-to-one correspondence according to Theorem 4.1.1.

Recall that the Dirichlet space $\mathcal{F}$ consists of functions defined m-a.e. on E and $\mathcal{F}$ is regarded as the family of equivalence classes in the sense of m-a.e. coincidence. But each element of $\mathcal{F}$ admits its quasi continuous version and two quasi continuous functions coincide q.e. if and only if they coincide m-a.e. The same is true for the extended Dirichlet space $\mathcal{F}_e$ of $(\mathcal{E}, \mathcal{F})$ in view of Theorem 2.3.4.

For notational convenience, in what follows, We shall use the symbol $\mathcal{F}$ (resp. $\mathcal{F}_e$) to denote the totality of quasi continuous functions belonging to it and we regard $\mathcal{F}$ (resp. $\mathcal{F}_e$) as the family of equivalence classes in the sense of q.e. coincidence. We see from Theorem 2.3.4 that $u \in \mathcal{F}_e$ if and only if

(ED) u is defined q.e. on E and there exists an $\mathcal{E}$-Cauchy sequence $\{u_n\} \subset \mathcal{F}$ such that $\lim_{n\to\infty} u_n(x) = u(x)$ q.e. $x \in E$.

In Section 5.1, we relate the perturbation of the Dirichlet form to a Feynman-Kac transform of X. Section 5.2 deals with characterization of the Dirichlet form $(\check{\mathcal{E}}, \check{\mathcal{F}})$ of a time-changed process $\check{X}$ of X in terms of the trace of $(\mathcal{E}, \mathcal{F})$ on the quasi support F of the measure $\mu \in S$ used in the time change. In particular, the extended Dirichlet space $(\check{\mathcal{F}}_e, \check{\mathcal{E}})$ of $(\check{\mathcal{E}}, \check{\mathcal{F}})$ will be shown to be dependent on F but independent of the choice of μ quasi-supported by F.

Feller measures on F relative to the part process X^0 of X on the quasi open set $E_0 = E \setminus F$ are introduced in Section 5.5. Using detailed stochastic computations made in Section 5.5 by means of additive functionals and martingales, the Beurling-Deny decomposition of the extended Dirichlet space $(\check{\mathcal{F}}_e, \check{\mathcal{E}})$ living on F is derived in Section 5.6 in terms of the due restriction of $\mathcal{E}$ to F with additional contributions by Feller measures.

In Section 5.7, Feller measures are described probabilistically as the joint distributions of starting and end points of the excursions of the process X away from the set F using an associated exit system.

Various examples related to Brownian motions and reflecting Brownian motions are collected in two sections, Sections 5.3 and 5.8.

Section 5.4 is a short introduction to the concept of the energy functional of a general symmetric transient right process, which will be utilized in later sections as well as later chapters.

5.1. SUBPROCESSES AND PERTURBED DIRICHLET FORMS

We fix a smooth measure $\mu \in S$ and, under the convention mentioned above, we define

$$\mathcal{F}^{\mu} = \mathcal{F} \cap L^2(E;\mu), \tag{5.1.1}$$

$$\mathcal{E}^{\mu}(u,v) = \mathcal{E}(u,v) + (u,v)_{\mu}, \quad u,v \in \mathcal{F}^{\mu}, \tag{5.1.2}$$

$$\mathcal{E}^{\mu}_{\alpha}(u,v) = \mathcal{E}^{\mu}(u,v) + \alpha(u,v), \quad u,v \in \mathcal{F}^{\mu},\ \alpha > 0,$$

where $(\,\cdot,\cdot\,)_{\mu}$ is the inner product in $L^2(E;\mu)$. Since μ charges no set of zero capacity, the above definition makes sense.

LEMMA 5.1.1. *$(\mathcal{E}^{\mu},\mathcal{F}^{\mu})$ is a Dirichlet form on $L^2(E;m)$.*

Proof. The Markov property of $(\mathcal{E}^{\mu},\mathcal{F}^{\mu})$ is clear and it suffices to prove its closedness. Let $\{u_n\}$ be an $\mathcal{E}^{\mu}_1$-Cauchy sequence in $\mathcal{F}^{\mu}$. Then it is $\mathcal{E}_1$-convergent to some $u \in \mathcal{F}$. Further, it converges to some $v \in L^2(E;\mu)$ in $L^2(E;\mu)$ and μ-a.e. as well by taking a subsequence if necessary. It remains to show that $u = v$ μ-a.e.

Let F be a closed set such that $\mathbf{1}_F \cdot \mu \in S_0$. By (2.3.7) and Fatou's lemma,

$$\begin{aligned}\int_F |u(x) - v(x)|\mu(dx) &\le \liminf_{n\to\infty} \int_E |u(x) - u_n(x)|\mathbf{1}_F(x)\mu(dx) \\ &\le \lim_{n\to\infty} \sqrt{\mathcal{E}_1(\mathbf{1}_F \cdot \mu)}\ \|u - u_n\|_{\mathcal{E}_1} = 0.\end{aligned}$$

Hence $u = v$ μ-a.e. on each closed set F and accordingly on E on account of Theorem 2.3.15. □

Let us consider a PCAF $A \in \mathbf{A}^+_c$ with $\mu_A = \mu$. Let N be the exceptional set of A and put

$$P^A_t f(x) = \mathbf{E}_x\left[e^{-A_t} f(X_t)\right], \quad t \ge 0,\ x \in E \setminus N,\ f \in \mathcal{B}_+(E). \tag{5.1.3}$$

Exercise 5.1.2. Show that (5.1.3) gives a transition function $\{P^A_t; t \ge 0\}$ on $(E \setminus N, \mathcal{B}^*(E \setminus N))$ in the sense of Definition 1.1.13.

$\{P^A_t; t \ge 0\}$ is called a *Feynman-Kac semigroup* generated by $A \in \mathbf{A}^+_c$. Its resolvent kernel $\{R^A_{\alpha}; \alpha > 0\}$ is given by (4.1.5):

$$R^A_{\alpha} f(x) = \mathbf{E}_x\left[\int_0^{\infty} e^{-\alpha t} e^{-A_t} f(X_t) dt\right], \quad x \in E \setminus N.$$

Since the set N is m-negligible, we can identify $L^2(E \setminus N; m)$ with $L^2(E;m)$.

THEOREM 5.1.3. (i) *The transition function $\{P^A_t; t \ge 0\}$ is m-symmetric and its Dirichlet form on $L^2(E;m)$ is $(\mathcal{E}^{\mu},\mathcal{F}^{\mu})$.*

(ii) *For any* $f \in \mathcal{B}(E) \cap L^2(E;m)$,

$$R_\alpha^A f \in \mathcal{F}^\mu \quad \text{and} \quad \mathcal{E}_\alpha^\mu(R_\alpha^A f, v) = (f, v) \text{ for every } v \in \mathcal{F}^\mu. \tag{5.1.4}$$

Proof. (i) follows from (ii). We shall show (ii) for $f \in b\mathcal{B}_+(E) \cap L^2(E;m)$. (ii) then holds for any $f \in \mathcal{B}_+(E) \cap L^2(E;m)$ on account of the bound $\|R_\alpha^A f\|_2 \leq \|R_\alpha f\|_2 \leq \frac{1}{\alpha}\|f\|_2$.

First suppose $\mu \in S_{00}$. By Exercise 4.1.2,

$$R_\alpha^A f(x) = R_\alpha f(x) - U_A^\alpha R_\alpha^A f(x), \quad x \in E \setminus N.$$

as a difference of two α-excessive functions of $X_{E\setminus N}$, $R_\alpha^A f$ is q.e. finely continuous and hence quasi continuous by Theorem 3.3.3. Since $R_\alpha^A f$ is non-negative and bounded, the measure $\mu' = R_\alpha^A f \cdot \mu$ is in S_{00} and $U_A^\alpha R_\alpha^A f$ is the α-potential of μ' in $\mathcal{F}$ by Lemma 4.1.5 and its proof. Hence $R_\alpha^A f \in \mathcal{F} \cap L^2(E;\mu) = \mathcal{F}^\mu$ because μ is finite. Moreover, for $v \in \mathcal{F}^\mu$,

$$\begin{aligned}\mathcal{E}_\alpha(R_\alpha^A f, v) &= \mathcal{E}_\alpha(R_\alpha f, v) - \mathcal{E}_\alpha(U_A^\alpha R_\alpha^A f, v)\\ &= (f, v) - (R_\alpha^A f, v)_\mu.\end{aligned}$$

This proves (5.1.4).

For a general $\mu \in S$, we may choose a nest $\{F_n\}$ such that $\mu_n = \mathbf{1}_{F_n} \cdot \mu \in S_{00}$ by virtue of Theorem 2.3.15. Let $A_n = \mathbf{1}_{F_n} \cdot A$. Then $\mu_{A_n} = \mu_n$ and

$$R_\alpha^{A_n} f \in \mathcal{F}^{\mu_n}, \quad \mathcal{E}_\alpha^{\mu_n}(R_\alpha^{A_n} f, v) = (f, v), \quad v \in \mathcal{F}^{\mu_n}, \tag{5.1.5}$$

which implies

$$\|R_\alpha^{A_n} f - R_\alpha^{A_\ell} f\|^2_{\mathcal{E}_\alpha^{\mu_n}} \leq (f, R_\alpha^{A_n} f - R_\alpha^{A_\ell} f) \quad \text{for } n \leq \ell. \tag{5.1.6}$$

On account of Theorem 3.1.4, $R_\alpha^{A_n} f(x)$ decreases to $R_\alpha^A f(x)$ for q.e. $x \in E$ as $n \to \infty$. Consequently, $R_\alpha^{A_n} f \in \mathcal{F}$ is $\mathcal{E}_1$-convergent and we conclude that $R_\alpha^A f \in \mathcal{F}$. Property (5.1.5) further implies

$$\|R_\alpha^{A_n} f\|^2_{L^2(E;\mu_n)} \leq (f, R_\alpha^{A_n} f) \leq \frac{1}{\alpha}\|f\|^2_{L^2(E;m)},$$

and we get by Fatou's lemma

$$\|R_\alpha^A f\|_{L^2(E;\mu)} \leq \frac{1}{\sqrt{\alpha}}\|f\|_{L^2(E;m)},$$

proving that $R_\alpha^A f \in \mathcal{F}^\mu$.

Finally we let $n \to \infty$ in (5.1.5) by observing the inequality: for $v \in L^2(E;\mu)$,

$$|(R_\alpha^{A_n} f, v)_{\mu_n} - (R_\alpha^A f, v)_\mu| \leq \|R_\alpha^{A_n} f - R_\alpha^A f\|_{L^2(E;\mu_n)} \cdot \|v\|_{L^2(E;\mu)} + (R_\alpha^A f, v)_{\mu - \mu_n}.$$

The second term on the right hand side tends to zero as $n \to \infty$. The first term also tends to zero because we have from (5.1.6)

$$\|R_\alpha^{A_n} f - R_\alpha^{A_\ell} f\|^2_{L^2(E;\mu_n)} \le (f, R_\alpha^{A_n} f - R_\alpha^{A_\ell} f),$$

and it is enough to let $\ell \to \infty$ first and then $n \to \infty$. The proof of (5.1.4) is complete. □

Recall Definition 1.2.12 on an $\mathcal{E}$-nest for a general Dirichlet form $(\mathcal{E}, \mathcal{F})$. For the present fixed regular Dirichlet form $(\mathcal{E}, \mathcal{F})$, an $\mathcal{E}$-nest will be simply called a nest. The next theorem says that $(\mathcal{E}^\mu, \mathcal{F}^\mu)$ has the same set of quasi notions as $(\mathcal{E}, \mathcal{F})$.

THEOREM 5.1.4. *An increasing sequence $\{F_k\}$ of closed subsets of E is a nest if and only if it is an $\mathcal{E}^\mu$-nest.*

Proof. Suppose an increasing sequence $\{F_k\}$ of closed sets is an $\mathcal{E}^\mu$-nest. As $\mathcal{E}$ is dominated by $\mathcal{E}^\mu$, $\bigcup_{k=1}^\infty (\mathcal{F}^\mu)_{F_k}$ is $\mathcal{E}_1$-dense in $\mathcal{F}^\mu$. Moreover, we can easily verify that $\mathcal{F}^\mu$ is $\mathcal{E}_1$-dense in $\mathcal{F}$. Indeed, by replacing μ, A with $p\mu, pA$, respectively, we have from (5.1.4) that for $f \in \mathcal{B}(E) \cap L^2(E; m)$ and $v \in \mathcal{F}^\mu$,

$$R_1^{pA} f \in \mathcal{F}^\mu \qquad \text{and} \qquad \mathcal{E}_1^{p\mu}(R_1^{pA} f, v) = (f, v).$$

Hence, for $p' < p$,

$$\|R_1^{pA} f - R_1^{p'A} f\|^2_{\mathcal{E}_1} \le \|R_1^{pA} f - R_1^{p'A} f\|^2_{\mathcal{E}_1^{p'\mu}} \le (f, R_1^{p'A} f) - (f, R_1^{pA} f).$$

Since $(f, R_1^{pA} f)$ is dominated by $\|f\|_2^2$ and $R_1^{pA} f(x)$ converges to $R_1 f(x)$ for $x \in E \setminus N$ as $p \downarrow 0$, $R_1^{pA} f$ is $\mathcal{E}_1$-convergent to $R_1 f$ as $p \downarrow 0$. Therefore $\{F_k\}$ is a nest.

Conversely, suppose an increasing sequence $\{F_k\}$ of closed subsets of E is a nest. We let $C_k = E \setminus F_k$, $k \ge 1$, and $\sigma = \lim_{k\to\infty} \sigma_{C_k}$. By Theorem 3.1.4,

$$\mathbf{P}_x(\sigma < \zeta) = 0 \qquad \text{for q.e. } x \in E. \tag{5.1.7}$$

Let $\widehat{X} = (\widehat{\Omega}, \widehat{X}_t, \widehat{\zeta}, \widehat{\mathbf{P}}_x)$ be the canonical subprocess of $X|_{E\setminus N}$ with respect to its multiplicative functional e^{-A_t} introduced at the end of Section A.3.2. By Lemma A.3.12, the transition function of $\widehat{X}$ equals $\{P_t^A; t \ge 0\}$ of (5.1.3). Denote by $\widehat{\sigma}_{C_k}$ the hitting time of $\widehat{X}$ for C_k and put $\widehat{\sigma} = \lim_{k\to\infty} \widehat{\sigma}_{C_k}$. For $\widehat{\omega} = (\omega, \lambda) \in \widehat{\Omega}$, we have from (A.3.28)

$$\left\{\widehat{\omega} : \widehat{\sigma}(\widehat{\omega}) < \widehat{\zeta}(\widehat{\omega})\right\} = \left\{\widehat{\omega} : \sigma(\omega) < \zeta(\omega),\ \sigma(\omega) < \lambda\right\}.$$

Therefore, we get from (A.3.29)

$$\widehat{\mathbf{P}}_x(\widehat{\sigma} < \widehat{\zeta}) = \mathbf{E}_x\left[e^{-A_\sigma}; \sigma < \zeta\right], \quad x \in E \setminus N. \tag{5.1.8}$$

In view of Theorem 5.1.3, the transition function $\{P_t^A; t \geq 0\}$ of $\widehat{X}$ is m-symmetric and the associated Dirichlet form on $L^2(E;m)$ is $(\mathcal{E}^\mu, \mathcal{F}^\mu)$. $(\mathcal{E}^\mu, \mathcal{F}^\mu)$ may not be a regular Dirichlet form, but we have $\widehat{\mathbf{P}}_x(\widehat{\sigma} < \widehat{\zeta}) = 0$ for m-a.e. $x \in E$ from (5.1.7) and (5.1.8). Hence we can use Lemma 3.1.17 to conclude that $\{F_k\}$ is an $\mathcal{E}^\mu$-nest. □

THEOREM 5.1.5. *$(\mathcal{E}^\mu, \mathcal{F}^\mu)$ is a quasi-regular Dirichlet form on $L^2(E;m)$. The canonical subprocess $\widehat{X}$ of $X|_{E\setminus N}$ with respect to e^{-A_t} is a Hunt process properly associated with it.*

Proof. Among the three conditions in Definition 1.3.8 of quasi-regularity, (a) and (b) are trivially satisfied by $(\mathcal{E}^\mu, \mathcal{F}^\mu)$ on account of the preceding theorem. Let $C_1(E)$ be a uniformly dense countable subfamily of $C_c(E)$. Then $\{R_\alpha^A f : \alpha \in \mathbb{Q}_+, f \in C_1(E)\}$ consists of quasi continuous and hence $\mathcal{E}^\mu$-quasi-continuous functions in $\mathcal{F}^\mu$ separating points of $E \setminus N$ yielding condition (c).

$\widehat{X}$ is a Hunt process on $(E \setminus N, \mathcal{B}^*(E \setminus N))$ by virtue of Theorem A.3.13. Since its resolvent $R_\alpha^A f$ for $f \in \mathcal{B}(E) \cap L^2(E;m)$ is $\mathcal{E}^\mu$-quasi-continuous, $\widehat{X}$ is properly associated with $(\mathcal{E}^\mu, \mathcal{F}^\mu)$. □

A measure $\mu \in S$ is not necessarily finite on each compact subset of E and so the Dirichlet form $(\mathcal{E}^\mu, \mathcal{F}^\mu)$ is in general not regular on $L^2(E;m)$. But if μ is a positive Radon measure on E charging no $\mathcal{E}$-polar set, then μ is still an element of S and we have the following theorem. The notion of a special standard core $\mathcal{C}$ of the regular Dirichlet form $(\mathcal{E}, \mathcal{F})$ is introduced in Definition 1.3.17.

THEOREM 5.1.6. *If μ is a positive Radon measure on E charging no $\mathcal{E}$-polar set, then $(\mathcal{E}^\mu, \mathcal{F}^\mu)$ is a regular Dirichlet form on $L^2(E;m)$. More specifically any special standard core of $(\mathcal{E}, \mathcal{F})$ remains to be a special standard core of $(\mathcal{E}^\mu, \mathcal{F}^\mu)$.*

Proof. First suppose $\mu(E) < \infty$ and μ charges no set of zero capacity. Take a bounded $u \in \mathcal{F}$. Due to the regularity of $(\mathcal{E}, \mathcal{F})$, there exist $u_n \in \mathcal{F} \cap C_c(E)$, $n \geq 1$, which is $\mathcal{E}_1$-convergent to u. We may assume that $\{u_n\}$ converges to u q.e. as $n \to \infty$. We may further assume that $\{u_n\}$ is uniformly bounded by Lemma 1.1.11. Then $\{u_n\}$ is convergent to u in $L^2(E;\mu)$ and consequently in metric $\mathcal{E}_1^\mu$ as well.

Next assume that μ is a positive Radon measure on E charging no set of zero capacity. Let $\{D_k\}$ be a sequence of relatively compact open subsets of E increasing to E such that $\overline{D}_k \subset D_{k+1}$, $k \geq 1$. Put $F_k = \overline{D}_k$, $k \geq 1$, then $\{F_k\}$ is an $\mathcal{E}$-nest and hence an $\mathcal{E}^\mu$-nest by Theorem 5.1.4, which means that any $u \in \mathcal{F}^\mu$ can be $\mathcal{E}_1^\mu$-approximated by a sequence of functions in $\bigcup_{k=1}^\infty (\mathcal{F}^\mu)_{F_k}$ where

$$(\mathcal{F}^\mu)_{F_k} = \{v \in \mathcal{F}^\mu : v = 0 \; m\text{-a.e. on } E \setminus F_k\}.$$

Keeping in mind our present convention that each element of $\mathcal{F}$ is quasi continuous, we have, for each $k \geq 2$,

$$
\begin{aligned}
(\mathcal{F}^{\mu})_{F_{k-1}} &= \{v \in \mathcal{F} : v = 0 \text{ q.e. on } E \setminus F_{k-1}\} \cap L^2(E;\mu) \\
&\subset \mathcal{F}_{D_k} \cap L^2(E;\mu),
\end{aligned}
$$

where $(\mathcal{E}, \mathcal{F}_{D_k})$ is the part on the open set D_k of $(\mathcal{E}, \mathcal{F})$ introduced in (3.2.2). In view of Theorem 3.3.9, $(\mathcal{E}, \mathcal{F}_{D_k})$ is a regular Dirichlet form on $L^2(D_k; m)$, and more specifically any function $v \in \mathcal{F}_{D_k}$ can be $\mathcal{E}_1$-approximated by functions v_n belonging to the space $\mathcal{C}_{D_k} = \{v \in \mathcal{C} : \text{supp}[v] \subset D_k\}$ for a special standard core $\mathcal{C}$ of $\mathcal{E}$. Since $\mu(D_k) < \infty$, we see as above that v can be $\mathcal{E}_1^{\mu}$-approximated by functions in $\mathcal{C}_{D_k}$. Accordingly, any function in $\mathcal{F}^{\mu}$ can be $\mathcal{E}_1^{\mu}$-approximated by functions in $\mathcal{C}$. □

We next introduce a kernel $U^{\alpha}_{p,A}$ on $(E \setminus N, \mathcal{B}^*(E \setminus N))$ for $\alpha \geq 0$, $p \geq 0$, by

$$
U^{\alpha}_{p,A} f(x) = \mathbf{E}_x \left[\int_0^{\infty} e^{-\alpha t - pA_t} f(X_t) dA_t \right], \quad x \in E \setminus N,\ f \in \mathcal{B}_+(E). \tag{5.1.9}
$$

$U^{\alpha}_{0,A}$ equals U^{α}_A defined by (4.1.4). Analogously to (4.1.7), we have for $f \in b\mathcal{B}(E)$

$$
U^{\alpha}_A f(x) - U^{\alpha}_{p,A} f(x) - p U^{\alpha}_A U^{\alpha}_{p,A} f(x) = 0, \quad x \in E \setminus N. \tag{5.1.10}
$$

Lemma 5.1.7. *For any α, $p > 0$ and $f \in \mathcal{B}(E) \cap L^2(E;\mu)$,*

$$
U^{\alpha}_{p,A} f \in \mathcal{F}^{\mu} \quad \textit{and} \quad \mathcal{E}^{p\mu}_{\alpha}(U^{\alpha}_{p,A} f, v) = (f, v)_{\mu} \quad \textit{for } v \in \mathcal{F}^{\mu}. \tag{5.1.11}
$$

Proof. First suppose that $\mu \in S_{00}$ and f is a non-negative bounded Borel function. As in the proof of Theorem 5.1.3, we then see that $U^{\alpha}_A f$ (resp. $U^{\alpha}_A U^{\alpha}_{p,A} f$) is a bounded quasi continuous version of the α-potential of the measure $f \cdot \mu \in S_{00}$ (resp. $U^{\alpha}_{p,A} f \cdot \mu \in S_{00}$). Therefore, we conclude from (5.1.10) that $U^{\alpha}_{p,A} f \in \mathcal{F}^{\mu}$ and

$$
\begin{aligned}
\mathcal{E}_{\alpha}(U^{\alpha}_{p,A} f, v) &= \mathcal{E}_{\alpha}(U_{\alpha}(f \cdot \mu) - p U_{\alpha}(U^{\alpha}_{p,A} f \cdot \mu), v) \\
&= (f, v)_{\mu} - p(U^{\alpha}_{p,A} f, v)_{\mu}, \quad v \in \mathcal{F}^{\mu},
\end{aligned}
$$

obtaining (5.1.11).

For a general $\mu \in S$, we choose a nest $\{F_n\}$ such that $\mu_n = \mathbf{1}_{F_n} \cdot \mu \in S_{00}$. Let us assume that f is a bounded Borel function vanishing on $E \setminus F_{n_0}$ for some n_0. For $A_n = \mathbf{1}_{F_n} \cdot A$, we then have

$$
U^{\alpha}_{p,A_n} f \in \mathcal{F}^{\mu_n} \quad \text{and} \quad \mathcal{E}^{p\mu_n}_{\alpha}(U^{\alpha}_{p,A_n} f, v) = (f, v)_{\mu_{n_0}} \tag{5.1.12}
$$

for $v \in \mathcal{F}^{\mu_n}$ and $n \geq n_0$. By the expression (5.1.9), $U^{\alpha}_{p,A_n} f$ converges to $U^{\alpha}_{p,A} f$ pointwise and boundedly as $n \to \infty$. Since μ_{n_0} is finite, we see from (5.1.12) that the convergence is in $\mathcal{E}_1$ as well and hence $U^{\alpha}_{p,A} f \in \mathcal{F}$. Just as in the proof of Theorem 5.1.3, we can also derive from (5.1.12) that $U^{\alpha}_{p,A}f \in L^2(E;\mu)$ and the equation in (5.1.11) is valid.

We have proved (5.1.11) for any bounded Borel function vanishing off F_{n_0} for some n_0. For such a function f, we have a bound $\|U^{\alpha}_{p,A}f\|_{L^2(E;\mu)} \leq \frac{1}{p}\|f\|_{L^2(E;\mu)}$ from (5.1.11). Since any non-negative Borel function in $L^2(E;\mu)$ is an increasing limit of non-negative functions of this type, we readily get (5.1.11) for $f \in \mathcal{B}(E) \cap L^2(E;\mu)$. □

COROLLARY 5.1.8. *For any $\alpha \geq 0$, $p \geq 0$ and $f, g \in \mathcal{B}_+(E)$,*

$$(U^{\alpha}_{p,A}f, g)_{\mu} = (f, U^{\alpha}_{p,A}g)_{\mu}, \tag{5.1.13}$$

$$(R^{pA}_{\alpha}f, g)_{\mu} = (f, U^{\alpha}_{p,A}g)_{m}. \tag{5.1.14}$$

Proof. (5.1.13) is a consequence of Lemma 5.1.7, while (5.1.14) follows from (5.1.4) and (5.1.11) □

We next let

$$(\mathcal{F}_e)^{\mu} = \mathcal{F}_e \cap L^2(E;\mu), \quad \mathcal{E}^{\mu}(u,v) = \mathcal{E}(u,v) + (u,v)_{\mu}, \quad u, v \in (\mathcal{F}_e)^{\mu}. \tag{5.1.15}$$

Recall the conventions made at the beginning of this chapter. We denote by $((\mathcal{F}^{\mu})_e, \mathcal{E}^{\mu})$ the extended Dirichlet space of the Dirichlet form $(\mathcal{E}^{\mu}, \mathcal{F}^{\mu})$ on $L^2(E;m)$. The next proposition asserts that the procedures of taking an extended Dirichlet space and making a perturbation by $\mu \in S$ are exchangeable.

PROPOSITION 5.1.9. $((\mathcal{F}^{\mu})_e, \mathcal{E}^{\mu}) = ((\mathcal{F}_e)^{\mu}, \mathcal{E}^{\mu})$.

Proof. Let u be a member of an m-equivalence class belonging to $(\mathcal{F}^{\mu})_e$ and $\{u_n\} \subset \mathcal{F}^{\mu} \subset \mathcal{F}$ be its $\mathcal{E}^{\mu}$-approximating sequence. Then $\{u_n\}$ is an $\mathcal{E}$-approximating sequence of u and, in view of Theorem 2.3.4, its suitable subsequence converges $\mathcal{E}$-q.e. To a quasi continuous version of u. Hence we can take u to be a quasi continuous member in $\mathcal{F}_e$. Since $\{u_n\}$ is Cauchy in $L^2(E;\mu)$, u then belongs to $L^2(E;\mu)$ so that $u \in (\mathcal{F}_e)^{\mu}$.

Conversely, suppose $u \in (\mathcal{F}_e)^{\mu}$. We may assume that u is non-negative. By Theorem 2.3.4, we can choose $\{u_n\}$ in $\mathcal{F}$ which is $\mathcal{E}$-Cauchy and convergent to u q.e. If we let $v_n = 0 \vee u_n \wedge u (\in \mathcal{F})$, $n \geq 1$, then v_n is convergent to u q.e. and $\|v_n\|^2_{\mathcal{E}} \leq \|u_n\|^2_{\mathcal{E}} + \|u\|^2_{\mathcal{E}}$ which is uniformly bounded in n. Therefore, by Theorem A.4.1, the Cesàro mean sequence $\{w_n\}$ of a suitable subsequence of $\{v_n\}$ is $\mathcal{E}$-Cauchy. Since $\{w_n\}$ is non-negative, uniformly bounded by $u \in L^2(E;\mu)$, and convergent to u q.e., it is an $\mathcal{E}^{\mu}$-approximating sequence in $\mathcal{F}^{\mu}$ for u, yielding that $u \in (\mathcal{F}^{\mu})_e$. □

PROPOSITION 5.1.10. *For any $p > 0$ and $f \in \mathcal{B}(E) \cap L^2(E;\mu)$,*

$$U^0_{p,A}f \in (\mathcal{F}_e)^{\mu} \quad \textit{and} \quad \mathcal{E}^{p\mu}(U^0_{p,A}f, v) = (f, v)_{\mu} \quad \textit{for } v \in (\mathcal{F}_e)^{\mu}. \tag{5.1.16}$$

Proof. First consider a non-negative bounded Borel function $f \in L^2(E;\mu)$. From (5.1.11), we have for $\beta > \alpha > 0$

$$\|U^\alpha_{p,A} f - U^\beta_{p,A} f\|^2_{\mathcal{E}^{p\mu}} \leq \|U^\alpha_{p,A} f\|^2_{\mathcal{E}^{p\mu}_\alpha} - 2\mathcal{E}^{p\mu}_\alpha(U^\alpha_{p,A} f, U^\beta_{p,A} f) + \|U^\beta_{p,A} f\|^2_{\mathcal{E}^{p\mu}_\beta}$$

$$= (f, U^\alpha_{p,A} f)_\mu - (f, U^\beta_{p,A} f)_\mu,$$

which tends to zero as $\alpha, \beta \to 0$, because $U^\alpha_{p,A} f$ increases pointwise to a bounded function $U^0_{p,A} f$ as $\alpha \downarrow 0$. Hence $\{U^\alpha_{p,A} f\}$ is $\mathcal{E}^{p\mu}$-Cauchy and $U^0_{p,A} f \in (\mathcal{F}^\mu)_e = (\mathcal{F}_e)^\mu$. We then get (5.1.16) for $v \in \mathcal{F}^\mu$ by letting $\alpha \downarrow 0$ in (5.1.11).

Property (5.1.16) readily extend to $v \in (\mathcal{F}_e)^\mu$. They also extend to any $f \in \mathcal{B}(E) \cap L^2(E;\mu)$ by the same reasoning as in the last part of the proof of Lemma 5.1.7. □

We close this section by presenting an example of a finite dimensional quasi-regular but not regular Dirichlet form due to Albeverio and Ma [3].

Example 5.1.11. Let $E = \mathbb{R}^n$ and $(\mathcal{E}, \mathcal{F})$ be the regular Dirichlet form of Brownian motion on $L^2(\mathbb{R}^n; dx)$, that is, $\mathcal{F} = H^1(\mathbb{R}^n)$ and

$$\mathcal{E}(u, v) = \frac{1}{2} \int_{\mathbb{R}^n} \nabla u(x) \cdot \nabla v(x)\, dx \quad \text{for } u, v \in \mathcal{F}.$$

Let $\{x_k, k \geq 1\}$ be a dense countable subset of $\mathbb{R}^n$ and $\{\alpha_k, k \geq 1\}$ a sequence of negative numbers so that each $\alpha_k \leq -n$. Consider a function q on $\mathbb{R}^n$ defined by

$$q(x) = \sum_{k=1}^\infty c_k |x - x_k|^{\alpha_k},$$

where $c_k > 0$ for every $k \geq 1$. It is shown in [3, Proposition 1.3] that there exists a sequence of strictly positive real numbers $\{c_k, k \geq 1\}$, so that $m(dx) := q(x)dx$ is a smooth measure on $\mathbb{R}^n$ with respect to the regular Dirichlet form $(\mathcal{E}, \mathcal{F})$ on $L^2(\mathbb{R}^n; dx)$. By Theorem 5.1.5, $(\mathcal{E}^\mu, \mathcal{F}^\mu)$ is a quasi-regular Dirichlet form on $L^2(\mathbb{R}^n; dx)$ and there is a Hunt process (Brownian motion on $\mathbb{R}^n$ killed at rate $q(x)$) properly associated with it. However, $(\mathcal{E}^\mu, \mathcal{F}^\mu)$ is not a regular Dirichlet form on $L^2(\mathbb{R}^n; dx)$ because $\mu(D) = \infty$ for any non-empty open subset $D \subset \mathbb{R}^n$. □

5.2. TIME CHANGES AND TRACE DIRICHLET FORMS

Let us consider a PCAF $A \in \mathbf{A}^+_c$ of the Hunt process $X = (X_t, \zeta, \mathbf{P}_x)$ and a smooth measure $\mu \in S$ such that μ equals the Revuz measure μ_A of A.

Let N be an exceptional set of A. We may regard A as a PCAF in the strict sense of the restricted Hunt process $X|_{E\setminus N}$. In accordance with (A.3.11), the *support* F of A is defined by

$$F = \{x \in E \setminus N : \mathbf{P}_x(R = 0) = 1\}, \quad R(\omega) = \inf\{t > 0 : A_t(\omega) > 0\}. \quad (5.2.1)$$

As is observed in the paragraph below (A.3.11), F is a nearly Borel finely closed set with respect to the Hunt process $X|_{E\setminus N}$. Hence F is quasi closed by virtue of Theorem 3.3.3. Since F is nearly Borel measurable, $F \in \mathcal{B}^*(E)$. We denote $F \cup \{\partial\}$ by F_∂ regarding it as a topological subspace of E_∂. Recall that $X_\infty(\omega)$ is defined to be ∂.

The right continuous inverse τ_t of the PCAF A is defined by

$$\tau_t(\omega) = \begin{cases} \inf\{s : A_s(\omega) > t\} & \text{if } t < A_{\zeta(\omega)-}(\omega), \\ \infty & \text{if } t \geq A_{\zeta(\omega)-}(\omega). \end{cases}$$

We let

$$\check{X}_t(\omega) = X_{\tau_t(\omega)}(\omega), \ t \geq 0, \quad \check{\zeta}(\omega) = A_{\zeta(\omega)-}(\omega).$$

By virtue of Theorem A.3.9, $\check{X} = (\check{X}_t, \check{\zeta}, \{\mathbf{P}_x\}_{x\in F_\partial})$ is then a right process with state space $(F_\partial, \mathcal{B}^*(F_\partial))$. $\check{X}$ is called the *time-changed process* of the Hunt process $X|_{E\setminus N}$ by the PCAF A.

According to Definition 3.3.4, the *quasi support* of a Borel measure is the smallest (up to q.e. equivalence) quasi closed set outside which the measure vanishes. We have already observed that the support F of the PCAF A is quasi closed.

THEOREM 5.2.1. (i) *F is a quasi support of μ.*
(ii) *The transition function $\{\check{P}_t; t \geq 0\}$ of the time-changed process $\check{X}$ is μ-symmetric.*

Proof. (i) Since $\mu = \mu_A$, μ does not charge the set $E \setminus F$ in view of (A.3.14) of Proposition A.3.6 and Theorem A.3.5. Therefore, it suffices to verify the condition (c) in Theorem 3.3.5: if u is quasi continuous and vanishes μ-a.e. on E, then $u = 0$ q.e. on F.

We let $v_n(x) = |u(x)| \wedge n$, $x \in E$, and $v_n(\partial) = 0$. By Theorem A.3.5, the PCAF $v_n \cdot A$ vanishes identically. From the time-change formula Lemma A.3.7, we get, for q.e. $x \in E$,

$$\beta \mathbf{E}_x \left[\int_0^\infty e^{-\beta t} v_n(X_{\tau_t}) dt \right] = \beta \mathbf{E}_x \left[\int_0^\infty e^{-\beta A_t} v_n(X_t) dA_t \right] = 0.$$

Since v_n is finely continuous q.e. by Theorem 3.3.3, we let $\beta \to \infty$ to see that the left hand side converges to $\mathbf{E}_x[v_n(X_R)] = v_n(x)$ for q.e. $x \in F$, and hence $u = 0$ for q.e. $x \in F$.

(ii) According to the usual convention, any numerical function on F will be extended to F_∂ by defining its value at ∂ to be zero. Since $\check{X}$ is a right process, its transition function $\{\check{P}_t; t \geq 0\}$ is a transition function on $(F, \mathcal{B}^*(F))$ in the sense of Definition 1.1.13. Denote its resolvent kernel by $\{\check{R}_p; p > 0\}$. We have then from $\check{P}_t f(x) = \mathbf{E}_x\left[f(X_{\tau_t})\right]$ and Lemma A.3.7

$$\check{R}_p f(x) = U^0_{p,A} f(x) \quad \text{for } x \in F,\ p > 0 \text{ and } f \in \mathcal{B}_+(F), \tag{5.2.2}$$

where $U^0_{p,A}$ is defined by (5.1.9). The μ-symmetry of $\check{R}_p$ then follows from (5.1.13) by noting that μ charges on F. Hence $\{\check{P}_t; t \geq 0\}$ is μ-symmetric. □

By Theorem 5.2.1 and Lemma 1.1.14, the transition function of the time-changed process $\check{X}$ determines uniquely a strongly continuous contraction semigroup $\{\check{T}_t; t > 0\}$ of Markovian symmetric operators on $L^2(F;\mu)$. The associated Dirichlet form on $L^2(F;\mu)$ will be called the *Dirichlet form* of the time-changed process $\check{X}$. A primary purpose of this section is to identify the Dirichlet form of $\check{X}$ and its extended Dirichlet space.

We denote by $\mathcal{Q}$ the collection of all non-$\mathcal{E}$-polar, quasi closed subsets of E. Sets F_1, $F_2 \in \mathcal{Q}$ are said to be q.e. equivalent if $F_1 \Delta F_2$ is $\mathcal{E}$-polar. The equivalence class to which $F \in \mathcal{Q}$ belongs is designated by $\dot{F}$. For $F \in \mathcal{Q}$, we let

$$S_F = \{\mu \in S: \text{ the quasi support of } \mu = F\}. \tag{5.2.3}$$

S_F is uniquely determined for an equivalence class so that it may also be written as $S_{\dot{F}}$. Notice that by virtue of Theorem 3.3.3, any equivalence class $\dot{F}$ admits a nearly Borel measurable finely closed representative. Recall also the conventions made for $\mathcal{F}$ and $\mathcal{F}_e$ at the beginning of this chapter.

Suppose $F \in \mathcal{Q}$ is nearly Borel measurable and finely closed. The hitting distribution $\mathbf{H}_F$ of X for F is defined by (A.2.2):

$$\mathbf{H}_F g(x) = \mathbf{E}_x\left[g(X_{\sigma_F}); \sigma_F < \infty\right], \quad x \in E,\ g \in \mathcal{B}_+(E).$$

Since F is finely closed, $\mathbf{H}_F(x, F^c) = 0$, $x \in E$, by (A.2.5), and the function $\mathbf{H}_F g$ is uniquely determined by the restriction of g to the set F. On the other hand, by virtue of Theorem 3.4.8, it holds for any $u \in \mathcal{F}_e$ that $\mathbf{H}_F|u|(x) < \infty$ q.e. and $\mathbf{H}_F u \in \mathcal{F}_e$. Therefore, the following definition of the function space $(\check{\mathcal{F}}_e, \check{\mathcal{E}})$ makes sense:

$$\check{\mathcal{F}}_e = \mathcal{F}_e\big|_F \quad \text{and} \quad \check{\mathcal{E}}(u|_F, v|_F) = \mathcal{E}(\mathbf{H}_F u, \mathbf{H}_F v) \quad \text{for } u, v \in \mathcal{F}_e. \tag{5.2.4}$$

Two elements of $\check{\mathcal{F}}_e$ can be identified if they coincide q.e. on F.

For any $F \in \mathcal{Q}$, we take a nearly Borel finely closed representative $\widetilde{F}$ of $\dot{F}$ and define the space $(\check{\mathcal{F}}_e, \check{\mathcal{E}})$ by (5.2.4) for $\widetilde{F}$. This space is uniquely determined up to q.e. coincidence of functions independently of the choice of $\widetilde{F}$ and depending only on $\dot{F}$. So, in defining this space, we may well assume F to be nearly Borel finely closed from the beginning to avoid the use of notation like $\widetilde{F}$.

Now take any $\mu \in S_F$. On account of Theorem 3.3.5, two elements of $\check{\mathcal{F}}_e$ coincide q.e. on F if and only if they coincide μ-a.e. Accordingly,

$$\check{\mathcal{F}} = \check{\mathcal{F}}_e \cap L^2(F;\mu) \tag{5.2.5}$$

defines a symmetric form $(\check{\mathcal{E}}, \check{\mathcal{F}})$ on $L^2(F;\mu)$, where we denote the restriction of $\check{\mathcal{E}}$ to $\check{\mathcal{F}} \times \check{\mathcal{F}}$ by the same notation again.

THEOREM 5.2.2. *Let $A \in \mathbf{A}_c^+$ and $\mu \in S$ are related by $\mu = \mu_A$. Let F be the support of A and $\check{X}$ be the time-changed process of X by A. Then the Dirichlet form of $\check{X}$ on $L^2(F;\mu)$ is $(\check{\mathcal{E}}, \check{\mathcal{F}})$ of* (5.2.4)–(5.2.5).

To prove Theorem 5.2.2, we first need a lemma. Consider the perturbed Dirichlet form $(\mathcal{E}^\mu, \mathcal{F}^\mu)$ of Lemma 5.1.1 and the Feynman-Kac semigroup $\{P_t^A, t \geq 0\}$ of (5.1.3). By Theorem 5.1.3, the latter is m-symmetric and the former is the Dirichlet form on $L^2(E;m)$ of the latter. The semigroup on $L^2(E;m)$ determined by the latter is denoted by $\{T_t^\mu; t > 0\}$, while $\{T_t; t > 0\}$ will designate the L^2-semigroup determined by the transition function of X. Recall the identity $(\mathcal{F}^\mu)_e = (\mathcal{F}_e)^\mu$ in Proposition 5.1.9 and the current convention that any function in $\mathcal{F}_e$ is assumed to be quasi continuous.

LEMMA 5.2.3. *Let $A \in \mathbf{A}_c^+$ and define*

$$D = \{x \in E \setminus N : \mathbf{P}_x(A_\infty > 0) > 0\}, \tag{5.2.6}$$

where N is the properly exceptional set associated with A. Then
(i) *D is both $\{T_t\}$-invariant and $\{T_t^\mu\}$-invariant.*
(ii) *If $u \in (\mathcal{F}_e)^\mu$, then $\mathbf{1}_D \cdot u \in (\mathcal{F}_e)^\mu$. If $\mathcal{E}^\mu(\mathbf{1}_D \cdot u, \mathbf{1}_D \cdot u) = 0$ for $u \in (\mathcal{F}_e)^\mu$, then $\mathbf{1}_D \cdot u = 0$. Furthermore, if $\{\mathbf{1}_D \cdot u_n\}$ is an $\mathcal{E}^\mu$-Cauchy sequence for $u_n \in (\mathcal{F}_e)^\mu$, then there exists $u \in (\mathcal{F}_e)^\mu$ such that $\mathbf{1}_D \cdot u_n$ is $\mathcal{E}^\mu$-convergent to $\mathbf{1}_D \cdot u$.*

Proof. (i) In view of (A.3.12),

$$D = \{x \in E \setminus N : \mathbf{P}_x(\sigma_F < \infty) > 0\} \tag{5.2.7}$$

and hence $\mathbf{P}_x(\sigma_D < \infty) = 0$ for every $x \in D^c \setminus N$ by virtue of Proposition 3.4.5 applied to $X|_{E\setminus N}$. By Proposition 2.1.6, D is $\{T_t\}$-invariant and consequently $\{T_t^\mu\}$-invariant as well.
(ii) In view of the proof of Proposition 3.4.5, $\mathbf{1}_D$ is quasi continuous and so is $\mathbf{1}_D \cdot u$ for $u \in \mathcal{F}_e$. Therefore, $\mathbf{1}_D \cdot u \in \mathcal{F}_e$ for $u \in \mathcal{F}_e$ by virtue of (i) and Proposition 2.1.6. The first assertion of (ii) follows from this.

In accordance with the paragraph below Proposition 2.1.6, the restriction $(\mathcal{E}^{\mu,D}, \mathcal{F}^{\mu,D})$ of $(\mathcal{E}^{\mu}, \mathcal{F}^{\mu})$ to the $\{T_t^{\mu}\}$-invariant set D is well defined by

$$\mathcal{F}^{\mu,D} = \{u|_D : u \in \mathcal{F}^{\mu}\}, \quad \mathcal{E}^{\mu,D}(u|_D, v|_D) = \mathcal{E}^{\mu}(\mathbf{1}_D u, \mathbf{1}_D v), \quad u, v \in \mathcal{F}^{\mu}, \tag{5.2.8}$$

where $u|_D$ denotes the restriction of u to D. This is the Dirichlet form on $L^2(E; m_D)$ associated with the semigroup $T_t^{\mu,D}(f|_D) = T_t^{\mu}(\mathbf{1}_D f)|_D$. From (5.1.3), we have, for $x \in D$,

$$P_t^A \mathbf{1}_D(x) \leq \mathbf{E}_x \left[e^{-A_t}; t < \zeta \right],$$

which is strictly less than 1 for some $t > 0$. So we get the transience of $\{T_t^{\mu,D}; t > 0\}$ by virtue of Proposition 2.1.10.

Moreover, the extended Dirichlet space $(\mathcal{F}_e^{\mu,D}, \mathcal{E}^{\mu,D})$ of $(\mathcal{E}^{\mu,D}, \mathcal{F}^{\mu,D})$ is obtained from (5.2.8) with $(\mathcal{F}_e)^{\mu}$ in place of $\mathcal{F}^{\mu}$. The second and third assertions of (ii) follow from this and Theorem 2.1.9. □

Proof of Theorem 5.2.2. We first show that for any $f \in \mathcal{B}(F) \cap L^2(F; \mu)$,

$$\check{R}_p f \in \check{\mathcal{F}}, \quad \check{\mathcal{E}}_p(\check{R}_p f, g) = (f, g)_{\mu}, \quad g \in \check{\mathcal{F}},\ p > 0. \tag{5.2.9}$$

Since $A_{\sigma_F} = 0$ by (A.3.12), we deduce from (5.2.2), (A.3.12), and the strong Markov property that

$$U_{p,A}^0 f(x) = \check{R}_p f(x), \quad x \in F, \tag{5.2.10}$$

$$U_{p,A}^0 f(x) = \mathbf{E}_x \left[\int_{\sigma_F}^{\infty} e^{-pA_s} f(X_s) dA_s \right] = \mathbf{H}_F(\check{R}_p f)(x), \quad x \in E \setminus N. \tag{5.2.11}$$

From (5.1.16) we get $\check{R}_p f \in (\mathcal{F}_e)^{\mu}|_F = \check{\mathcal{F}}$ and, for $g = v|_F$ with $v \in (\mathcal{F}_e)^{\mu}$,

$$\check{\mathcal{E}}_p(\check{R}_p f, g) = \mathcal{E}^{p\mu}(U_{p,A}^0 f, \mathbf{H}_F v) = (f, \mathbf{H}_F v)_{\mu} = (f, g)_{\mu},$$

arriving at (5.2.9).

We next prove the closedness of the form $(\check{\mathcal{E}}, \check{\mathcal{F}})$ on $L^2(F; \mu)$. We note that for $f \in \check{\mathcal{F}}$ with $f = u|_F$ and $u \in \mathcal{F}_e$, $w = \mathbf{H}_F u$ is a function belonging to $(\mathcal{F}_e)^{\mu}$ and $w = \mathbf{1}_D \cdot w$ on account of (5.2.7). Further, we have from (5.2.4) and (5.2.5)

$$\check{\mathcal{E}}_1(f, f) = \mathcal{E}^{\mu}(w, w).$$

Let $\{f_n\} \subset \check{\mathcal{F}}$ be an $\check{\mathcal{E}}_1$-Cauchy sequence with $f_n = u_n|_F$ and $u_n \in \mathcal{F}_e$. Then, for $w_n = \mathbf{H}_F u_n$, $\{w_n = \mathbf{1}_D \cdot w_n; n \geq 1\}$ is an $\mathcal{E}^{\mu}$-Cauchy sequence in $(\mathcal{F}_e)^{\mu}$, and hence is $\mathcal{E}^{\mu}$-convergent as $n \to \infty$ to $\mathbf{1}_D \cdot w_D$ for some $w_D \in (\mathcal{F}_e)^{\mu}$ on account of Lemma 5.2.3. Denote $\mathbf{1}_D \cdot w_D$ by w which is a member of $(\mathcal{F}_e)^{\mu}$.

For any $v \in \mathcal{F}_e$ vanishing q.e. on the set F, $\mathcal{E}(w_n, v) = 0$ by virtue of Theorem 3.4.8. By letting $n \to \infty$, we have $\mathcal{E}(w, v) = 0$. By the same theorem, we also have $\mathcal{E}(\mathbf{H}_F w, v) = 0$. Since $w - \mathbf{H}_F w$ vanishes q.e. on F, we get $\mathcal{E}^\mu(w - \mathbf{H}_F w, w - \mathbf{H}_F w) = \mathcal{E}(w - \mathbf{H}_F w, w - \mathbf{H}_F w) = 0$. Since $w - \mathbf{H}_F w = \mathbf{1}_D \cdot (w - \mathbf{H}_F w)$, we conclude that $w = \mathbf{H}_F w$ from Lemma 5.2.3.

Let $f = w|_F$. Then $f \in \check{\mathcal{F}}$, $w = \mathbf{H}_F w$, and

$$\lim_{n\to\infty} \|f_n - f\|_{\check{\mathcal{E}}_1} = \lim_{n\to\infty} \|w_n - w\|_{\mathcal{E}^\mu} = 0,$$

establishing the completeness of $\check{\mathcal{F}}$ with metric $\check{\mathcal{E}}_1$. □

The quadratic form $(\check{\mathcal{E}}, \check{\mathcal{F}})$ defined by (5.2.4) and (5.2.5) is called the *trace Dirichlet form* of $(\mathcal{E}, \mathcal{F})$ on $L^2(F; \mu)$.

LEMMA 5.2.4. *Denote by $((\check{\mathcal{F}})_e, \mathcal{A})$ the extended Dirichlet space of the trace Dirichlet form $(\check{\mathcal{E}}, \check{\mathcal{F}})$ of $(\mathcal{E}, \mathcal{F})$ on $L^2(F; \mu)$. Then*

$$\check{\mathcal{F}}_e \subset (\check{\mathcal{F}})_e \quad \text{and} \quad \mathcal{A}(f, g) = \check{\mathcal{E}}(f, g) \quad \text{for } f, g \in \check{\mathcal{F}}_e. \tag{5.2.12}$$

Proof. Let $\{F_k\}$ be a nest associated with the smooth measure μ. In particular, $\mu(F_k) < \infty$ for every $k \geq 1$. By Theorem 1.3.7, the space $\mathcal{G} = \bigcup_{k=1}^\infty \mathcal{F}_{F_k}$ is $\mathcal{E}_1$-dense in $\mathcal{F}$, where

$$\mathcal{F}_{F_k} = \{u \in \mathcal{F} : u = 0 \quad \text{q.e. on } E \setminus F_k\}.$$

For any bounded function $u \in \mathcal{F}_e$, we can therefore find a uniformly bounded sequence $\{u_n\} \subset \mathcal{G}$ which is $\mathcal{E}$-Cauchy and convergent to u q.e. as $n \to \infty$ in view of Theorem 2.3.4. Let $f = u|_F$, $f_n = u_n|_F$, $n \geq 1$. Then $f \in \check{\mathcal{F}}_e$, $f_n \in \check{\mathcal{F}} = \check{\mathcal{F}}_e \cap L^2(F; \mu)$, and

$$\|f_n - f\|_{\check{\mathcal{E}}} = \|\mathbf{H}_F u_n - \mathbf{H}_F u\|_{\mathcal{E}} \leq \|u_n - u\|_{\mathcal{E}} \to 0, \quad n, m \to \infty,$$

as was to be proved. □

We shall prove in Theorem 5.2.15 that actually $(\check{\mathcal{F}})_e = \check{\mathcal{F}}_e$ for any $\mu \in S_F$.

The time-changed $\check{X}$ is called *transient, recurrent, irreducible* if so is the trace Dirichlet form $(\check{\mathcal{E}}, \check{\mathcal{F}})$ on $L^2(F; \mu)$, respectively. The next result says that time change does not alter the transience or recurrence of the processes (or Dirichlet forms).

THEOREM 5.2.5. (i) *If X is transient, then $\check{X}$ is transient.*
(ii) *If X is recurrent, then $\check{X}$ is recurrent.*

Proof. For any $f \in \mathcal{B}_+(F) \cap L^1(F; \mu)$, $u \in \mathcal{B}_+(E) \cap L^1(E; m)$, we have from (5.1.14)

$$(u, U^0_{0,A} f)_m = (R_0 u, f)_\mu. \tag{5.2.13}$$

Suppose X is transient. Then, for any strictly positive function $u \in \mathcal{B}_+(E) \cap L^1(E; m)$, $R_0 u$ is finite m-a.e. and accordingly q.e. on E by Theorem A.2.13(v) and Theorem 3.1.3. Take a strictly positive function $g \in \mathcal{B}_+(F) \cap L^1(F;\mu)$ and put

$$f(x) := \frac{g(x)}{1 \vee R_0 u(x)}, \qquad x \in F.$$

f is then in $L^1(F;\mu)$ and strictly positive q.e. on F and hence μ-a.e. on F. Furthermore, the right hand side of (5.2.13) is finite so that $U^0_{0,A} f(x) < \infty$ for m-a.e. $x \in E$ and consequently q.e. $x \in E$. In view of (5.2.10), $\check{R}_0 f(x) < \infty$ for μ-a.e. $x \in F$, yielding the transience of $\check{X}$.

Suppose next X is recurrent. We take an arbitrary $f \in \mathcal{B}_+(F) \cap L^1(F;\mu)$. Since, for any $u \in \mathcal{B}_+(E) \cap L^1(E;m)$, the set $\{x \in E : 0 < R_0 u < \infty\}$ is finely open and m-negligible, it is m-polar by Theorem A.2.13 and thus is μ-negligible. Hence the right hand side of (5.2.13) is either 0 or ∞ and so is its left hand side for any $u \in \mathcal{B}_+(E) \cap L^1(E;m)$. Therefore, the set $B = \{x \in E : 0 < U^0_{0,A} f < \infty\}$ is m-negligible, and by the same reason as above, $\mu(B) = 0$. By (5.2.10), the set $\{x \in F : 0 < \check{R}_0 f(x) < \infty\}$ equals $B \cap F$, which is μ-negligible, yielding the recurrence of $\check{X}$. □

Let us now state a property of the $\check{\mathcal{E}}$-nest for the Dirichlet form $(\check{\mathcal{E}}, \check{\mathcal{F}})$ in the sense of Definition 1.2.12. The state space F of the time-changed process $\check{X}$ is endowed with the relative topology as a subspace of E. In particular, a closed subset of F is of a form $F_1 \cap F$ for a closed subset F_1 of E.

THEOREM 5.2.6. *If an increasing sequence $\{F_k\}$ of closed subsets of E is a nest, then its trace $\{F_k \cap F\}$ on F is an $\check{\mathcal{E}}$-nest. The restriction to F of any quasi continuous function on E is $\check{\mathcal{E}}$-quasi-continuous. The trace on F of any $\mathcal{E}$-polar set is $\check{\mathcal{E}}$-polar.*

Proof. Suppose an increasing sequence $\{F_k\}$ of closed sets is a nest. By Theorem 5.1.4, it is also an $\mathcal{E}^\mu$-nest. In particular, for any $v \in (\mathcal{F}^\mu)_e$, there exist functions $\{v_n\}$ such that

$$v_n \in \bigcup_{k=1}^{\infty} (\mathcal{F}^\mu)_{F_k}, \qquad \|v_n - v\|_{\mathcal{E}^\mu} \to 0,\ n \to \infty. \tag{5.2.14}$$

Take any function $f \in \check{\mathcal{F}}$. There is then $v \in (\mathcal{F}_e)^\mu$ such that $f = v|_F$ and $\|f\|_{\check{\mathcal{E}}_1} = \|\mathbf{H}_F v\|_{\mathcal{E}^\mu}$. By virtue of Proposition 5.1.9, $v \in (\mathcal{F}^\mu)_e$. Let $\{v_n\}$ be functions satisfying (5.2.14) and set $f_n = v_n|_F$, $n \geq 1$. Then $f_n \in \check{\mathcal{F}}$ and

$$\|f_n - f\|_{\check{\mathcal{E}}_1} = \|\mathbf{H}_F v_n - \mathbf{H}_F v\|_{\mathcal{E}^\mu} \leq \|v_n - v\|_{\mathcal{E}^\mu} \to 0,\ n \to \infty.$$

Since $f_n = 0$ q.e. on $F \setminus F_k$ for some k,

$$f_n \in \bigcup_{k=1}^{\infty} \{g \in \check{\mathcal{F}} : g = 0\ \mu\text{-a.e. on } F \setminus F_k\}.$$

Therefore, $\{F_k \cap F\}$ is an $\check{\mathcal{E}}$-nest. The second assertion follows from the first. The last assertion follows from Definition 1.2.12. □

THEOREM 5.2.7. *$(\check{\mathcal{E}}, \check{\mathcal{F}})$ is a quasi-regular Dirichlet form on $L^2(F;\mu)$. The time-changed process $\check{X}$ is a right process properly associated with it.*

Proof. Let $\varphi_A(x) = \mathbf{E}_x[e^{-\sigma_F}]$, $x \in E \setminus N$. Then $F = \{x \in E \setminus N : \varphi_A(x) = 1\}$ on account of Proposition A.3.6. Since φ_A is a 1-excessive function relative to $X|_{E\setminus N}$, it is quasi continuous by virtue of Theorem 3.3.3. Therefore, there is a compact nest $\{K_k\}$ such that $\cup_k K_k \subset E \setminus N$ and $\varphi_A \in C(\{K_k\})$. Clearly $\check{K}_k = K_k \cap F = K_k \cap \{\varphi_A = 1\}$ is compact for each $k \geq 1$ and $\{\check{K}_k\}$ is an $\check{\mathcal{E}}$-nest by the preceding theorem. Condition (a) in Definition 1.2.12 of the quasi-regularity is just verified for $(\check{\mathcal{E}}, \check{\mathcal{F}})$. Other conditions (b) and (c) can be also verified using the preceding theorem as in the proof of Theorem 5.1.5.

The time-changed process $\check{X}$ is a right process on the Radon space $(F, \mathcal{B}^*(F))$ by virtue of Theorem A.3.11. Its resolvent $\check{R}_p f$ for $f \in \mathcal{B}(F) \cap L^2(F;\mu)$ is $\check{\mathcal{E}}$-quasi-continuous by (5.2.2) and the preceding theorem, and consequently $\check{X}$ is properly associated with $(\check{\mathcal{E}}, \check{\mathcal{F}})$. □

Denote by $\overset{\circ}{S}$ the totality of positive Radon measures on E charging no $\mathcal{E}$-polar set. For $F \in \mathcal{Q}$, we let

$$\overset{\circ}{S}_F = \{\mu \in \overset{\circ}{S} : \ \text{the quasi support of } \mu \text{ is } F\}, \tag{5.2.15}$$

which is a subfamily of S_F. For $\mu \in \overset{\circ}{S}_F$, $C_c(E) \subset L^2(E;\mu)$ and we may discuss the regularity of the trace Dirichlet form $(\check{\mathcal{E}}, \check{\mathcal{F}})$.

THEOREM 5.2.8. *Let $\mu \in \overset{\circ}{S}_F$.*
(i) *For any special standard core $\mathcal{C}$ of $(\mathcal{E}, \mathcal{F})$, $\mathcal{C}|_F$ is $\check{\mathcal{E}}_1$-dense in $\check{\mathcal{F}}$.*
(ii) *A set $N \subset F$ is $\check{\mathcal{E}}$-polar if and only if N is $\mathcal{E}$-polar as a subset of E.*

Proof. (i) By virtue of Theorem 5.1.6, $\mathcal{C}$ becomes a special standard core of the Dirichlet form $(\mathcal{E}^\mu, \mathcal{F}^\mu)$ on $L^2(E;m)$. Take any $f \in \check{\mathcal{F}}$ and $v \in \mathcal{F}_e$ such that $v|_F = f$. Since $v \in (\mathcal{F}_e)^\mu = (\mathcal{F}^\mu)_e$, we can find $v_n \in \mathcal{C}(\subset \mathcal{F}^\mu)$ which is $\mathcal{E}^\mu$-convergent to v. Then $f_n = v_n|_F \in \check{\mathcal{F}}$ and

$$\|f_n - f\|_{\check{\mathcal{E}}_1} = \|\mathbf{H}_F v_n - \mathbf{H}_F v\|_{\mathcal{E}^\mu} \leq \|v_n - v\|_{\mathcal{E}^\mu} \to 0, \ n \to \infty,$$

as was to be proved.
(ii) The "if" part was shown in Theorem 5.2.6. The "only if" part will be shown by means of capacitary comparison. Let D be the $\{T_t\}$-invariant set defined by (5.2.7) and $(\mathcal{E}^\nu, \mathcal{F}^\nu)$ be the perturbed Dirichlet form on $L^2(E;m)$ for the measure $\nu = \mu + \mathbf{1}_{D^c} \cdot m$. Let A be the PCAF of X with Revuz measure

μ. Observe that the transition function $\{q_t; t \geq 0\}$ associated with $(\mathcal{E}^\nu, \mathcal{F}^\nu)$ is given by

$$q_t f(x) = \mathbf{1}_D(x)\mathbf{E}_x[e^{-A_t}(\mathbf{1}_D f)(X_t)] + \mathbf{1}_{D^c}(x)e^{-t}\mathbf{E}_x[(\mathbf{1}_{D^c} f)(X_t)].$$

So, for $x \in E \setminus N$, there is some $t > 0$ such that $q_t 1(x) < 1$. This implies by Proposition 2.1.10 that $(\mathcal{E}^\nu, \mathcal{F}^\nu)$ is transient. The Dirichlet form $(\mathcal{E}^\nu, \mathcal{F}^\nu)$ is regular on $L^2(E; m)$ in view of Theorem 5.1.6. Denote by $\mathrm{Cap}^{(0),\nu}$ the 0-order $\mathcal{E}^\nu$-capacity on E as is introduced in Section 2.3. Any $\mathrm{Cap}^{(0),\nu}$-polar set is $\mathcal{E}^\nu$-polar by Theorem 2.3.2 and consequently $\mathcal{E}$-polar by Theorem 5.1.4.

Define the capacity $\check{\mathrm{Cap}}_1$ on F as the (h, α)-capacity for the Dirichlet form $(\check{\mathcal{E}}, \check{\mathcal{F}})$ on $L^2(F; \mu)$ with $h = 1$ and $\alpha = 1$. By Theorem 1.2.14, a set $N \subset F$ is $\check{\mathcal{E}}$-polar if and only if $\check{\mathrm{Cap}}_1(N) = 0$. Let $A \subset E$ be a relatively compact open subset and $B = F \cap A$. Then $\check{\mathrm{Cap}}_1(B) = \check{\mathcal{E}}_1(e_B, e_B)$ where $e_B \in \check{\mathcal{F}}$ is the 1-reduced function of 1 on the relatively open subset B of F. As an element of $\check{\mathcal{F}}$, e_B is the restriction to F of a function in $\mathcal{F}_e$ so that $\mathbf{H}_F e_B$ is a quasi continuous function in $\mathcal{F}_e$, $\mathbf{H}_F e_B|_F = e_B$ q.e. on F, and $e_B = 1$ μ-a.e. on B.

Since F is a quasi support of μ and $\mathbf{H}_F e_B = 1$ μ-a.e. on A, $\mathbf{H}_F e_B = 1$ q.e. on B by Proposition 3.3.10. Further, $\mathbf{H}_F e_B = 0$ q.e. on D^c in view of the definition of D and $\mathbf{H}_F$. In particular, $\mathbf{H}_F e_B$ is a member of $(\mathcal{F}_e)^\nu = (\mathcal{F}^\nu)_e$. It follows from the 0-order counterpart of Theorem 2.3.1 that

$$\begin{aligned}\mathrm{Cap}^{(0),\nu}(B) &\leq \mathcal{E}^\nu(\mathbf{H}_F e_B, \mathbf{H}_F e_B) = \mathcal{E}^\mu(\mathbf{H}_F e_B, \mathbf{H}_F e_B)\\ &= \check{\mathcal{E}}_1(e_B, e_B) = \check{\mathrm{Cap}}_1(B),\end{aligned}$$

yielding the validity of the "only if" part. □

For $\mu \in \overset{\circ}{S}_F$, we denote by F^*_μ the topological support of μ. Since any closed set is quasi closed, $F \subset F^*_\mu$ q.e., but the set $F^*_\mu \setminus F$ is not necessarily $\mathcal{E}$-polar. See [73, Example 5.1.2] for a counterexample.

LEMMA 5.2.9. (i) *$\overset{\circ}{S}_F \neq \emptyset$ for any $F \in \mathcal{Q}$.*

(ii) *Suppose that F is non-$\mathcal{E}$-polar, nearly Borel finely closed, $\nu \in \overset{\circ}{S}$, $\nu(E \setminus F) = 0$ and that the* 1*-order hitting distribution $H^1_F(x, \cdot)$ of X for F is absolutely continuous with respect to ν for m-a.e. $x \in E$. Then $\nu \in \overset{\circ}{S}_F$.*

(iii) *Suppose F is a closed set satisfying*

$$\text{there exists } \nu \in \overset{\circ}{S}_F \text{ such that } F^*_\nu = F. \tag{5.2.16}$$

Then $F^*_\mu = F$ for any $\mu \in \overset{\circ}{S}_F$.

Proof. (i) We may assume that F is nearly Borel and finely closed by choosing from $\dot{F}$ such a representative if necessary. For a strictly positive function

$g \in L^1(E;m)$ we let

$$\mu(B) = \mathbf{P}_{g\cdot m}(X_{\sigma_F} \in B,\ \sigma_F < \infty), \quad B \in \mathcal{B}(E).$$

μ is a non-trivial finite measure on $(E, \mathcal{B}(E))$ vanishing off F and charging no m-polar set. If f is a quasi continuous function in $\mathcal{F}_e$ vanishing μ-a.e. on E, then

$$\int_E g(x)\mathbf{H}_F|f|(x)m(dx) = \int_E |f(x)|\mu(dx) = 0.$$

Since $\mathbf{H}_F f$ is quasi continuous by Theorem 3.4.8, it vanishes m-a.e. on E and consequently, $f = 0$ q.e. on F, yielding that F is the quasi support of μ by Theorem 3.3.5.

(ii) The measure $\mu \in \mathring{S}_F$ constructed in (i) is then absolutely continuous with respect to ν.

(iii) Take any $\mu \in \mathring{S}_F$. Then $F_\mu^* \subset F$ because μ vanishes off the closed set F. Since F_μ^* includes q.e. the quasi support of μ which equals F, the relatively open subset $O = F \setminus F_\mu^*$ of F is $\mathcal{E}$-polar so that ν does not charge on it. Hence the condition (5.2.16) forces O to be empty. □

As an immediate consequence of Theorem 5.2.8(i) combined with Lemma 5.2.9(iii), we have

COROLLARY 5.2.10. *If F is a closed set satisfying condition* (5.2.16), *then, for any $\mu \in \mathring{S}_F$, $(\check{\mathcal{E}}, \check{\mathcal{F}})$ is a regular Dirichlet form on $L^2(F;\mu)$. Further, for any special standard core $\mathcal{C}$ of $(\mathcal{E}, \mathcal{F})$, $\mathcal{C}|_F$ is a core of $(\check{\mathcal{E}}, \check{\mathcal{F}})$.*

The whole space E satisfies condition (5.2.16) because $m \in \mathring{S}_E$ and the topological support of m equals E. Consider any measure $\mu \in \mathring{S}_E$. Then the definition (5.2.4), (5.2.5) read simply as follows: the trace Dirichlet form on $L^2(E;\mu)$ is

$$(\check{\mathcal{E}}, \check{\mathcal{F}}) = (\mathcal{E}, \mathcal{F}_e \cap L^2(E;\mu)). \tag{5.2.17}$$

By Corollary 5.2.10, (5.2.17) is regular and any special standard core of $(\mathcal{E}, \mathcal{F})$ is a core of (5.2.17).

The next theorem says that the Dirichlet form (5.2.17) has the same set of quasi notions as the original Dirichlet form.

THEOREM 5.2.11. *For $\mu \in \mathring{S}_E$, an increasing sequence $\{F_k\}$ of closed subsets of E is a nest for the regular Dirichlet form* (5.2.17) *on $L^2(E;\mu)$ if and only if it is a nest for the original Dirichlet form $(\mathcal{E}, \mathcal{F})$ on $L^2(E;m)$.*

Proof. The "if" part was shown in Theorem 5.2.6.

Take any $\mu \in \overset{\circ}{S}_E$. Since μ has a full quasi support, Proposition 3.3.10 implies that if a quasi continuous function vanishes μ-a.e. on an open set $U \subset E$, then it vanishes q.e. on U.

Suppose an increasing sequence $\{F_k\}$ is an $\check{\mathcal{E}}$-nest. Take any $f \in (\mathcal{F}^\mu)_e$. Since $f \in \check{\mathcal{F}}$ by Proposition 5.1.9, it can be $\check{\mathcal{E}}_1(= \mathcal{E}^\mu)$-approximated by functions f_n in $\check{\mathcal{F}} = (\mathcal{F}^\mu)_e$ vanishing μ-a.e. on $E \setminus F_k$ for some k. By the above remark, each f_n vanishes q.e. and consequently m-a.e. on $E \setminus F_k$ for some k.

On account of Theorem 5.1.6, $(\mathcal{E}^\mu, \mathcal{F}^\mu)$ is a regular Dirichlet form on $L^2(E; m)$, which is also transient in view of Lemma 5.2.3 because the set D in the lemma equals E q.e. Therefore, Corollary 3.4.4 applies in concluding that $\{F_k\}$ is an $\mathcal{E}^\mu$-nest in the sense that $\bigcup_{k=1}^{\infty} (\mathcal{F}^\mu)_{F_k}$ is $\mathcal{E}_1^\mu$-dense in $\mathcal{F}^\mu$. By virtue of Theorem 5.1.4, $\{F_k\}$ is then an $\mathcal{E}$-nest. □

COROLLARY 5.2.12. *For any $\mu \in \overset{\circ}{S}_E$, denote by $((\check{\mathcal{F}})_e, \mathcal{A})$ the extended Dirichlet space of the Dirichlet form $(\check{\mathcal{E}}, \check{\mathcal{F}})$ on $L^2(E; \mu)$ defined by* (5.2.17). *Then*

$$((\check{\mathcal{F}})_e, \mathcal{A}) = (\mathcal{F}_e, \mathcal{E}). \tag{5.2.18}$$

Proof. Two Dirichlet forms $\check{\mathcal{E}}$ and $\mathcal{E}$ coincide on a common core $\mathcal{F} \cap C_c(E)$. They share the set of quasi notions by Theorem 5.2.11. Hence (5.2.18) follows from Theorem 2.3.4. □

Identity (5.2.18) will be extended to any $\mu \in S_F$ and any $F \in \mathcal{Q}$ in Theorem 5.2.15.

We now return to a general quasi closed set $F \in \mathcal{Q}$ and a measure $\mu \in \overset{\circ}{S}_F$. Let $F^* = F^*_\mu$ be the topological support of μ, which is a closed set. As noted earlier, $F \subset F^*$ q.e. Recall that F has been taken to be the support of a PCAF $A \in \mathbf{A}_c^+$ with Revuz measure μ. By enlarging the exceptional set N of A if necessary, we may assume from the beginning that

$$F \subset F^*. \tag{5.2.19}$$

Since $\mu(F^* \setminus F) = 0$, it holds that $L^2(F^*; \mu) = L^2(F; \mu)$. So the trace Dirichlet form $(\check{\mathcal{E}}, \check{\mathcal{F}})$ can be regarded as a Dirichlet form on $L^2(F^*; \mu)$. For emphasis, we denote $(\check{\mathcal{E}}, \check{\mathcal{F}})$ as $(\mathcal{E}^*, \mathcal{F}^*)$ when viewed as a Dirichlet form on $L^2(F^*; \mu)$.

THEOREM 5.2.13. (i) *$(\mathcal{E}^*, \mathcal{F}^*)$ is a regular Dirichlet form on $L^2(F^*; \mu)$. The set $F^* \setminus F$ is $\mathcal{E}^*$-polar.*

(ii) *There exists an $\mathcal{E}^*$-polar subset N of F such that $F \setminus N$ is Borel measurable and the restriction to $F \setminus N$ of the time-changed process $\check{X}$ is a Hunt process properly associated with $(\mathcal{E}^*, \mathcal{F}^*)$.*
(iii) *Let A be a PCAF of X with Revuz measure μ. There exists a Borel properly exceptional set $N \subset E$ containing an exceptional set of A such that the following holds. Let F be the support of A being considered as a PCAF in the strict sense of the restricted Hunt process $X_{E\setminus N}$ and $\check{X}$ be the time-changed process on F of $X_{E\setminus N}$ by means of this PCAF. Then $\check{X}$ is a Hunt process on F properly associated with $(\mathcal{E}^*, \mathcal{F}^*)$.*

Proof. (i) For a function f on E, $f|_{F^*}$ differs from $f|_F$ only on a μ-negligible set. Hence for a special standard core $\mathcal{C}$ of $(\mathcal{E}, \mathcal{F})$, $\mathcal{C}|_{F^*}$ is $\mathcal{E}^*_1$-dense in $\mathcal{F}^*$ by virtue of Theorem 5.2.8. Since $\mathcal{C}|_{F^*}$ is uniformly dense in $C_c(F^*)$, $(\mathcal{E}^*, \mathcal{F}^*)$ is a regular Dirichlet form on $L^2(F^*; \mu)$.

As $(\check{\mathcal{E}}, \check{\mathcal{F}})$ is quasi-regular on $L^2(F; \mu)$ by Theorem 5.2.7, there exists an $\check{\mathcal{E}}$-nest $\{K_n\}$ of increasing compact subsets of F. $\{K_n\}$ is then an increasing sequence of closed subsets of F^*, which is clearly an $\mathcal{E}^*$-nest. Therefore, the set $F^* \setminus F \subset (\cup_n K_n)^c$ is $\mathcal{E}^*$-polar by definition.
(ii) By taking intersections with an $\check{\mathcal{E}}$-nest $\{K_n\}$ as above, every $\check{\mathcal{E}}$-nest becomes an $\mathcal{E}^*$-nest and vice versa. Therefore, a function on F^* is $\mathcal{E}^*$-quasi-continuous if and only if its restriction to F is $\check{\mathcal{E}}$-quasi-continuous and a subset of F is $\mathcal{E}^*$-polar if and only if it is $\check{\mathcal{E}}$-polar. Consequently, the time-changed process $\check{X}$ of X on F is a right process properly associated with $(\mathcal{E}^*, \mathcal{F}^*)$ by virtue of Theorem 5.2.7. Further, the restriction to F^* of any quasi continuous function on E is $\mathcal{E}^*$-quasi-continuous and the trace on F^* of any $\mathcal{E}$-polar set is $\mathcal{E}^*$-polar in view of Theorem 5.2.6.

As defined at the beginning of this section, the sample path $X_{\tau_t(\omega)}(\omega)$ of the time-changed process $\check{X}$ of X by the right continuous inverse of $A_t(\omega)$ takes its value in $F_\partial = F \cup \{\partial\}$ where F_∂ is viewed as a topological subset of the one-point compactification E_∂ of E. We also regard $F^*_\partial = F^* \cup \{\partial\}$ as the topological subset of E_∂, which is closed and hence compact. Since $X_t(\omega)$ admits the left limit $X_{t-}(\omega) \in E_\partial$ for every $t \geq 0$ and F^*_∂ is a compact set containing F_∂, $\check{X}_t(\omega)$ admits the left limit $\check{X}_{t-}(\omega) \in F^*_\partial$ for every $t \geq 0$. It could happen that $\check{X}_{t-} \in F^* \setminus F$ and the quasi-left-continuity of $\check{X}$ may be violated. However, by Proposition 5.2.14 below, there is an $\mathcal{E}^*$-polar set $N_0 \subset F$ so that $\check{X}|_{F\setminus N_0}$ is a quasi-left-continuous Borel right process and hence a Hunt process.
(iii) The above $\mathcal{E}^*$-polar set N_0 is $\mathcal{E}$-polar on account of Theorem 5.2.8(ii) so that we can find a properly exceptional Borel set N for X containing N_0 and the exceptional set of A. This set N has the properties stated in (iii). □

PROPOSITION 5.2.14. *Let $(\mathcal{E}, \mathcal{F})$ be a regular Dirichlet form on $L^2(E; m)$ and $X = (X_t, \mathbf{P}_x)$ be a right process on a subset $E_1 \subset E$ such that $E \setminus E_1$ is*

$\mathcal{E}$-polar. Assume that X is properly associated with $(\mathcal{E}, \mathcal{F})$ and that the left limits of X_t exist in E_∂ for every $t > 0$. Then there exists a Borel set $E_2 \subset E_1$ such that $E \setminus E_2$ is $\mathcal{E}$-polar, E_2 is X-invariant in the sense that, for every $x \in E_2$,

$$\mathbf{P}_x\,(X_t \in (E_2)_\partial \textit{ for every } t \geq 0 \textit{ and } X_{t-} \in (E_2)_\partial \textit{ for every } t > 0) = 1,$$

and the restriction $X|_{E_2}$ of X to E_2 is a Hunt process properly associated with $(\mathcal{E}, \mathcal{F})$.

Proof. By virtue of Theorem 3.1.13, there exists a Borel set $E_1' \subset E_1$ such that $E \setminus E_1'$ is $\mathcal{E}$-polar and $X|_{E_1'}$ is a special Borel standard process enjoying the properties of Theorem 3.1.3 and Theorem 3.1.4. In particular, any Cap_1-nest is a strong nest for $X|_{E_1'}$. Let $u \in \mathcal{F}$ be $\mathcal{E}$-quasi-continuous. Then u is q.e. equivalent to a quasi continuous function in the restricted sense relative to a Cap_1-nest by virtue of Lemma 1.3.15. Since X_t admits the left limits on E_∂, the above-mentioned strong nest property yields

$$\mathbf{P}_x\left(\lim_{t'\uparrow t} u(X_{t'}) = u(X_{t-}) \text{ for } t > 0\right) = 1 \quad \text{for q.e. } x \in E_1'. \tag{5.2.20}$$

Choose a countable subfamily $C_1 \subset \mathcal{F} \cap C_c(E)$ that is uniformly dense in $C_c(E)$. Since $P_t f$ is an $\mathcal{E}$-quasi-continuous function in $\mathcal{F}$ for $f \in C_1$, we can find an $\mathcal{E}$-polar set $N \subset E$ such that (5.2.20) holds for each $u = P_s f$, $s \in \mathbb{Q}_+$, $f \in C_1$ and for all $x \in E_1' \setminus N$.

Using again the strong nest property of any Cap_1-nest and the existence of the left limit of X_t, we can construct, in exactly the same manner as the proof of Theorem 3.1.5, a Borel set $E_2 \subset E_1' \setminus N$ such that $E \setminus E_2$ is $\mathcal{E}$-polar and E_2 is X-invariant. Since $X|_{E_2}$ is a Borel right process on E_2 and (5.2.20) is valid for each $u = P_{sf}$, $s \in \mathbb{Q}_+$, $f \in C_1$ and for every $x \in E_2$, precisely the same argument as in [13, p. 50] leads us to the quasi-left-continuity on $(0, \infty)$ of $X|_{E_2}$. □

Theorem 5.2.13 enables us to extend Corollary 5.2.12 to any measure $\mu \in S_F$ for any set $F \in \mathcal{Q}$:

THEOREM 5.2.15. *For any $F \in \mathcal{Q}$ and for any $\mu \in S_F$, denote by $((\check{\mathcal{F}})_e, \mathcal{A})$ the extended Dirichlet space of the trace Dirichlet form $(\check{\mathcal{E}}, \check{\mathcal{F}})$ of $(\mathcal{E}, \mathcal{F})$ on $L^2(F; \mu)$ defined by* (5.2.4), (5.2.5). *Then*

$$((\check{\mathcal{F}})_e, \mathcal{A}) = (\check{\mathcal{F}}_e, \check{\mathcal{E}}), \tag{5.2.21}$$

which is dependent on $F \in \mathcal{Q}$ but independent of $\mu \in S_F$.

Proof. On account of Lemma 5.2.4, it suffices to prove

$$(\check{\mathcal{F}})_e \subset \check{\mathcal{F}}_e.$$

First we assume that $\mu \in \overset{\circ}{S}_F$ and take any $f \in (\check{\mathcal{F}})_e$. Without loss of generality, we assume that f is bounded. By Theorem 5.2.13, $(\check{\mathcal{E}}, \check{\mathcal{F}})$ can be regarded as a regular Dirichlet form on $L^2(F^*; \mu)$. Therefore, by virtue of Theorem 2.3.4, we can find a sequence $\{f_n\}$ of uniformly bounded functions in $\check{\mathcal{F}} = \check{\mathcal{F}}_e \cap L^2(F^*; \mu)$ such that $\{f_n\}$ is $\check{\mathcal{E}}$-Cauchy and $\lim_{n\to\infty} f_n(\xi) = f(\xi)$ for $x \in F \setminus N$, where N is an $\check{\mathcal{E}}$-polar subset of F.

Let $u_n = \mathbf{H}_F f_n$, $n \geq 1$. Then $\{u_n\} \subset \mathcal{F}_e$ is $\mathcal{E}$-Cauchy and $\lim_{n\to\infty} u_n(x) = \mathbf{H}_F\widetilde{f}(x)$ for q.e. $x \in E$, because N is $\mathcal{E}$-polar by virtue of Theorem 5.2.8. Therefore, $\mathbf{H}_F\widetilde{f} \in \mathcal{F}_e$ and $\widetilde{f} = \mathbf{H}_F\widetilde{f}$ q.e. on F, as was to be proved.

Next consider any $\mu \in S_F$ and denote $\check{\mathcal{F}} = \check{\mathcal{F}}_e \cap L^2(F; \mu)$ by $\check{\mathcal{F}}^\mu$ to indicate its dependence on μ. Choose a function $f \in L^1(F; \mu)$ with $0 < f \leq 1$ μ-a.e. Then $\check{\mathcal{F}}^\mu \subset \check{\mathcal{F}}^{f\cdot\mu}$ so that $(\check{\mathcal{F}}^\mu)_e \subset (\check{\mathcal{F}}^{f\cdot\mu})_e$. Since $f \cdot \mu$ is finite, $f \cdot \mu \in \overset{\circ}{S}_F$ and we get from the above $(\check{\mathcal{F}}^{f\cdot\mu})_e \subset \check{\mathcal{F}}_e$. Therefore, the desired inclusion $(\check{\mathcal{F}})_e \subset \check{\mathcal{F}}_e$ is obtained. □

In the remainder of this section, we shall present two important applications of the invariance properties embodied by Theorem 5.2.11 and Corollary 5.2.12 for fully supported Radon measures $\overset{\circ}{S}_E$.

First we prove that Theorem 2.1.11 can be generalized as follows.

THEOREM 5.2.16. *Let $(\mathcal{E}, \mathcal{F})$ be a regular Dirichlet form on $L^2(E; m)$.*
(i) *Suppose $(\mathcal{E}, \mathcal{F})$ is recurrent. Then $(\mathcal{E}, \mathcal{F})$ is irreducible if and only if the following condition is satisfied.*

$$u \in \mathcal{F}_e \text{ with } \mathcal{E}(u, u) = 0 \implies u \text{ is constant.} \tag{5.2.22}$$

(ii) *If $(\mathcal{E}, \mathcal{F})$ is irreducible, then* (5.2.22) *holds.*

Proof. (i) This was shown in Theorem 2.1.11 under the additional condition that $m(E) < \infty$. If $m(E) = \infty$, we make a time change relative to a finite measure μ which is mutually absolutely continuous with respect to m. Obviously, $\mu \in \overset{\circ}{S}_E$ and the extended Dirichlet space is preserved under the time change due to Corollary 5.2.12. Since the irreducibility is characterized entirely in terms of the extended Dirichlet space in view of Proposition 2.1.6, it is invariant under such a transformation. Condition (5.2.22) is also invariant. The recurrence is invariant under the time change by Theorem 5.2.5. It follows that (i) is true.
(ii) If $(\mathcal{E}, \mathcal{F})$ is irreducible, then it is either transient or recurrent. In the latter case, (5.2.22) is valid by (i), while it is trivially true in the former case. □

Our second application is on resurrection of a Dirichlet form. Let $(\mathcal{E}, \mathcal{F})$ be a regular Dirichlet form on $L^2(E; m)$. Its extended Dirichlet space $(\mathcal{F}_e, \mathcal{E})$ will be regarded as a collection of equivalence classes under $\mathcal{E}$-q.e. coincidence of $\mathcal{E}$-quasi-continuous functions. Recall from Theorem 4.3.3 that for $u, v \in \mathcal{F}_e$, $\mathcal{E}(u, v)$ admits the following unique Beurling-Deny representation:

$$\mathcal{E}(u, v) = \mathcal{E}^{(c)}(u, v) + \frac{1}{2} \int_{E \times E \setminus d} (u(x) - u(y))(v(x) - v(y)) J(dxdy) + \int_E u(x)v(x)\kappa(dx). \tag{5.2.23}$$

For $u, v \in \mathcal{F}_e$, define

$$\mathcal{E}^{\text{res}}(u, v) = \mathcal{E}^{(c)}(u, v) + \frac{1}{2} \int_{E \times E \setminus d} (u(x) - u(y))(v(x) - v(y)) J(dxdy)$$

and $m_\kappa = m + \kappa$. Then

$$\begin{cases} \mathcal{F} = \mathcal{F}_e \cap L^2(E; m_\kappa), \\ \mathcal{E}^{\text{res}}(u, v) + (u, v)_{L^2(E; m_\kappa)} = \mathcal{E}_1(u, v), \quad u, v \in \mathcal{F}, \end{cases} \tag{5.2.24}$$

which in particular implies that $(\mathcal{E}^{\text{res}}, \mathcal{F})$ is a regular Dirichlet form on $L^2(E; m_\kappa)$. Moreover, these two regular Dirichlet forms have the same set of quasi notions as $\{F_k, k \geq 1\}$ is an $\mathcal{E}$-nest if and only if it is an $\mathcal{E}^{\text{res}}$-nest.

Let $(\mathcal{F}_e^{\text{res}}, \mathcal{E}^{\text{res}})$ be the extended Dirichlet space of $(\mathcal{E}^{\text{res}}, \mathcal{F})$. In view of Theorem 2.3.4 and the preceding remark, $u \in \mathcal{F}_e^{\text{res}}$ if and only if there exists an $\mathcal{E}^{\text{res}}$-Cauchy sequence $\{u_n\} \subset \mathcal{F}$ such that $u_n \to u$ $\mathcal{E}$-q.e. on E. Therefore, $\mathcal{F}_e^{\text{res}} \supset \mathcal{F}_e$. $(\mathcal{F}_e^{\text{res}}, \mathcal{E}^{\text{res}})$ is called the *extended resurrected Dirichlet space* of $(\mathcal{E}^{\text{res}}, \mathcal{F})$.

THEOREM 5.2.17. *Let $(\mathcal{E}, \mathcal{F})$ be a regular Dirichlet form on $L^2(E; m)$ and X be its associated Hunt process. Then*

$$(\mathcal{E}^{\text{res}}, \mathcal{F}_e^{\text{res}} \cap L^2(E; m)) \quad \text{on } L^2(E; m) \tag{5.2.25}$$

is a regular Dirichet form sharing the same set of quasi notions with the regular Dirichlet form $(\mathcal{E}, \mathcal{F})$ on $L^2(E; m)$. Any special standard core of $(\mathcal{E}, \mathcal{F})$ is a core of (5.2.25). *Let X^{res} be a Hunt process on E associated with* (5.2.25) *and A be the PCAF of X^{res} with Revuz measure κ. Then the subprocess of X^{res} relative to the multiplicative functional e^{-A_t} has the same law as X. Furthermore,*

$$\mathcal{F}_e = \mathcal{F}_e^{\text{res}} \cap L^2(E; \kappa). \tag{5.2.26}$$

We call (5.2.25) and X^{res} in the above theorem a *resurrected Dirichlet form* of $(\mathcal{E}, \mathcal{F})$ and a *resurrected Hunt process* of X, respectively.

Proof. Let Y be a Hunt process associated with the regular Dirichlet form $(\mathcal{E}^{\rm res}, \mathcal{F})$ on $L^2(E; m_\kappa)$. The measure m is a member of $\overset{\circ}{S}_E$ for $(\mathcal{E}^{\rm res}, \mathcal{F})$ because it is so for $(\mathcal{E}, \mathcal{F})$. Let $\check{Y}$ be a time-changed process of Y by its PCAF with Revuz measure m. Then the Dirichlet form of $\check{Y}$ on $L^2(E; m)$ is given by (5.2.25) in view of (5.2.17). By Corollary 5.2.10, (5.2.25) is regular and any special standard core of $(\mathcal{E}, \mathcal{F})$ is a core of (5.2.25). $\check{Y}$ can be taken to be a Hunt process by Theorem 5.2.13 and it can be regarded as a special realization of a Hunt process $X^{\rm res}$ associated with (5.2.25).

By Theorem 5.2.11, the Dirichlet form (5.2.25) shares the same set of quasi notions with $(\mathcal{E}^{\rm res}, \mathcal{F})$ on $L^2(E; m_\kappa)$, and thus with the original Dirichlet form $(\mathcal{E}, \mathcal{F})$. In particular, the measure κ is smooth with respect to (5.2.25). Let $(X^{\rm res})^\kappa$ be the subprocess of $X^{\rm res}$ in the statement of the theorem. By Theorem 5.1.3, its Dirichlet form on $L^2(E; m)$ is given by

$$(\mathcal{E},\ \mathcal{F}_e^{\rm res} \cap L^2(E; m + \kappa)), \tag{5.2.27}$$

which is regular in view of Theorem 5.1.6. Since (5.2.27) is identical with $(\mathcal{E}, \mathcal{F})$ on a common core $\mathcal{F} \cap C_c(E)$, they must be the same. This implies that $(X^{\rm res})^\kappa$ and X are the same in the sense of Theorem 3.1.12. On account of Proposition 5.1.9 and Corollary 5.2.12, we can get (5.2.26). □

Theorem 5.2.17 will be utilized in the last example of Section 5.3.

We now consider a resurrection of the part process of the original Hunt process X on an open set. Let D be an open subset of E. Consider the part $(\mathcal{E}^D, \mathcal{F}_D)$ (resp. X^D) of the Dirichlet form $(\mathcal{E}, \mathcal{F})$ (resp. the associated Hunt process X) on D. In view of Theorem 3.3.9, $(\mathcal{E}^D, \mathcal{F}_D)$ is a regular Dirichlet form on $L^2(D; m)$ and, for any special standard core $\mathcal{C}$ of the Dirichlet form $(\mathcal{E}, \mathcal{F})$ on $L^2(E; m)$, $\mathcal{C}_D = \{f \in \mathcal{C} : {\rm supp}[f] \subset D\}$ is a core of $(\mathcal{E}^D, \mathcal{F}_D)$. By Theorem 3.4.9, its extended Dirichlet space $(\mathcal{F}_{D,e}, \mathcal{E}^D)$ can be described as follows:

$$\mathcal{F}_{e,D} = \{u \in \mathcal{F}_e : u = 0 \text{ q.e. on } D^c\}, \tag{5.2.28}$$

and $\mathcal{E}^D$ is the restriction of $\mathcal{E}$ on $\mathcal{F}_{e,D} \times \mathcal{F}_{e,D}$. We see from (5.2.23) that for any $u, v \in \mathcal{F}_{e,D}$,

$$\begin{aligned}\mathcal{E}(u, v) = \mathcal{E}^{(c)}(u, v) &+ \frac{1}{2}\int_{D\times D\setminus d} (u(x) - u(y))(v(x) - v(y))J(dxdy)\\ &+ \int_E u(x)v(x)\kappa_D(dx),\end{aligned} \tag{5.2.29}$$

where $\kappa_D(B) := \kappa(B) + J(B \times E \setminus d)$ for $B \in \mathcal{B}(D)$. Hence the killing measure of $(\mathcal{E}^D, \mathcal{F}_D)$ is κ_D, which may not vanish even when $\kappa = 0$.

Let

$$(\mathcal{E}^{D,\mathrm{res}}, \mathcal{F}_{D,\mathrm{res}}) \quad \text{on } L^2(D; m) \tag{5.2.30}$$

be the resurrected Dirichlet form of $(\mathcal{E}^D, \mathcal{F}_D)$. On account of Theorem 5.2.17, $\mathcal{C}_D := \{f \in \mathcal{C} : \mathrm{supp}[f] \subset D\}$ is a core of (5.2.30) for any special standard core $\mathcal{C}$ of $(\mathcal{E}, \mathcal{F})$. Moreover, for $u, v \in \mathcal{C}_D$,

$$\mathcal{E}^{D,\mathrm{res}}(u, v) = \mathcal{E}^{(c)}(u, v) + \frac{1}{2}\int_{D\times D\setminus d} (u(x) - u(y))(v(x) - v(y))J(dxdy), \tag{5.2.31}$$

and the space $\mathcal{F}_{D,\mathrm{res}}$ is the $\mathcal{E}_1^{D,\mathrm{res}}$-closure of $\mathcal{C}_D$.

Accordingly, the associated resurrected Hunt process $X^{D,\mathrm{res}}$ on D admits no killing inside D, while the part process X^D of X on D may have killings inside caused by jumps of X from D to $E \setminus D$. When X is a rotationally symmetric α-stable process on $\mathbb{R}^n$, $X^{D,\mathrm{res}}$ will be called in Section 6.5(**6°**) a censored stable process on an open set $D \subset \mathbb{R}^n$.

5.3. EXAMPLES

The Brownian motion on $\mathbb{R}^n$ and absorbing and reflecting Brownian motions for a domain $D \subset \mathbb{R}^n$ were considered in Example 3.5.9. The first four examples of this section concern their time changes.

(1°) Measures of full support and time change

Let X be the standard Brownian motion on $\mathbb{R}^n$ with $n \geq 1$. It is symmetric with respect to the Lebesgue measure dx, its Dirichlet form on $L^2(\mathbb{R}^n)$ is $(\frac{1}{2}\mathbf{D}, H^1(\mathbb{R}^n))$ defined by (2.2.15), and its extended Dirichlet space is denoted by $H^1_e(\mathbb{R}^n)$ in Section 2.2.4. In this and the next two examples, contrary to the convention being made in this chapter, we specifically use $\widetilde{H}^1_e(\mathbb{R}^n)$ to denote the collection of all quasi continuous versions of functions in $H^1_e(\mathbb{R}^n)$. When $n \geq 3$, for instance, we see by Exercise 2.3.3 that the space $\widetilde{H}^1_e(\mathbb{R}^n)$ admits a more direct description

$$\widetilde{H}^1_e(\mathbb{R}^n) = \{I_1 * f : f \in L^2(\mathbb{R}^n)\}$$

using the 1-order Riesz kernel $I_1(x) = \gamma_{n,1}|x|^{-(n-1)}$, $x \in \mathbb{R}^n$.

Let m be a positive Radon measure on $\mathbb{R}^n$ charging no $\mathcal{E}$-polar set with full quasi support ($m \in \overset{\circ}{S}_{\mathbb{R}^n}$ in the notation of this section) and κ be a positive Radon measure on $\mathbb{R}^n$ charging no $\mathcal{E}$-polar set (i.e. $\kappa \in \overset{\circ}{S}$). For instance, in view of Corollary 3.3.6, m can be $f(x)dx$ for a function $f \in L^1_{\mathrm{loc}}(\mathbb{R}^n)$ that is strictly positive a.e. on $\mathbb{R}^n$. Denote by A and B the positive continuous additive functionals of X with Revuz measure m and κ, respectively. Consider the right process Y obtained from the Brownian motion X first by a time change using the inverse of the PCAF A and then by killing using the PCAF B.

Combining (5.2.17) with Theorem 5.1.6 and Proposition 5.1.9, we conclude as follows: Y is m-symmetric and its Dirichlet form $(\mathcal{E}, \mathcal{F})$ on $L^2(\mathbb{R}^n; m)$ admits the expression

$$\begin{cases} \mathcal{F} = \widetilde{H}^1_e(\mathbb{R}^n) \cap L^2(\mathbb{R}^n; m) \cap L^2(\mathbb{R}^n; \kappa), \\ \mathcal{E}(u, v) = \frac{1}{2}\mathbf{D}(u, v) + (u, v)_\kappa, \quad u, v \in \mathcal{F}. \end{cases} \tag{5.3.1}$$

Furthermore, its extended Dirichlet space $(\mathcal{F}_e, \mathcal{E})$ is expressed as

$$\begin{cases} \mathcal{F}_e = \widetilde{H}^1_e(\mathbb{R}^n) \cap L^2(\mathbb{R}^n; \kappa), \\ \mathcal{E}(u, v) = \frac{1}{2}\mathbf{D}(u, v) + (u, v)_\kappa, \quad u, v \in \mathcal{F}_e, \end{cases} \tag{5.3.2}$$

independently on $m \in \overset{\circ}{S}_{\mathbb{R}^n}$ by Corollary 5.2.12 and Proposition 5.1.9.

When $n \geq 3$, X is transient and so is Y by Theorem 5.2.5. If $m(\mathbb{R}^n) < \infty$ additionally, then $1 \in L^1(\mathbb{R}^n; m)$ and $R^Y 1 < \infty$ $[m]$ for the 0-order resolvent R^Y of $Y = (Y_t, \mathbf{P}^Y_x, \zeta^Y)$ by Proposition 2.1.3. Hence the lifetime ζ^Y is finite $\mathbf{P}_x$-a.s. for q.e. $x \in \mathbb{R}^n$.

Let D be a domain of $\mathbb{R}^n$ and X^D be the absorbing Brownian motion on D. Let Y^D be the right process obtained from X^D by time change and killing with respect to positive Radon measures $m \in \overset{\circ}{S}_D$ and $k \in \overset{\circ}{S}$. Then similar to the above, Y^D is m-symmetric and its Dirichlet form on $L^2(D; m)$ and its extended Dirichlet space can be described as (5.3.1) and (5.3.2), respectively, with $\mathbb{R}^n$ and $\widetilde{H}^1_e(\mathbb{R}^n)$ being replaced by D and $\widetilde{H}^1_{0,e}(D)$.

When $n = 2$, the absorbing Brownian motion $X^D = (X^D_t, \mathbf{P}_x)$ has the following conformal invariant property. Recall that X^D is defined from the two-dimensional Brownian motion $X = (X_t, \mathbf{P}_x)$ by $X^D_t = X_t$ if $t < \tau_D$ and $X^D_t = \partial$ otherwise. Any function g on D is extended to $D \cup \{\partial\}$ by setting $g(\partial) = 0$.

THEOREM 5.3.1. *Suppose that* $n = 2$ *and* φ *is a conformal map from a domain* $D \subset \mathbb{R}^2$ *onto another planar domain* U. *Then up to a time change,* $\phi(X^D)$ *is an absorbing Brownian motion in* U. *More precisely, let* $A_t = \int_0^t |\phi'(X^D_s)|^2\, ds$ *and* $\tau_t := \inf\{s > 0 : A_s > t\}$ *be the right continuous inverse of* A. *Then* $\{Y_t := \phi(X^D_{\tau_t}),\ t \geq 0\}$ *with lifetime* A_∞ *is an absorbing Brownian motion on* U. *This in particular implies that*

(i) *if* U^c *is non-polar, then* $A_\infty < \infty$ *a.s. and* $\lim_{t \uparrow \tau_D} \phi(X_t)$ *exists and takes values in* U^c *a.s.;*

(ii) *if* U^c *is polar, then* $A_\infty = \infty$ *a.s.*

Proof. Using the Cauchy-Riemann equations, an application of Itô's formula to $\phi(X_t) - \phi(X_0) =: (Z^1_t, Z^2_t)$ for $t < \tau_D$ yields that Z^1 and Z^2 are continuous local martingales on the random time interval $[0, \tau_D)$ with $\langle Z^k \rangle_t = A_t$ and $\langle Z^1, Z^2 \rangle_t = 0$ for $t < \tau_D$ and $k = 1, 2$. According to Theorem 7.3′ of Chapter II in Ikeda-Watanabe [91], there exists an extension (in the sense of Definition 7.1 there) of the original filtered probability space and a two-dimensional Brownian motion $\overline{Y}$ on it starting at $\phi(X_0)$ such that (i) X is a two-dimensional Brownian motion with respect to this enlarged filtered probability space; and (ii) a.s.

$$\overline{Y}_t = \phi(X^D_{\tau_t}) \quad \text{for } t \in [0, A_\infty).$$

Let σ be the exit time of $\overline{Y}$ from U, then almost surely $A_\infty \leq \sigma$ and $X^D_{\tau_t} = \phi^{-1}(\overline{Y}_t)$ for $t \in [0, A_\infty)$. Since $\{\phi^{-1}(\overline{Y}_t) : t \in [0, \sigma)\}$ is a.s. a continuous curve in D and $\lim_{t \uparrow A_\infty} X^D_{\tau_t} \in \partial D$ a.s. on $A_\infty < \infty$, we must have $A_\infty = \sigma$ a.s.

If U^c is non-polar, by the recurrence of planar Brownian motion, the lifetime of Y has to be finite a.s. and the left limit of Y at that time takes values in U^c, which proves (i). If U^c is polar, then Y is Brownian motion on $\mathbb{R}^2$ that has infinite lifetime and so $A_\infty = \infty$ a.s., that is, (ii) holds. This completes the proof of the theorem. □

When U^c is non-polar, Theorem 5.3.1(i) indicates in some sense that although the conformal map $\phi : D \to U$ may not be extended continuously to $\overline{D}$, it can be extend along the sample path of Brownian motion up to the boundary of D.

Finally, let D be a domain of $\mathbb{R}^n$ possessing a continuous boundary and X^r be the reflecting Brownian motion on $\overline{D}$. Let Y^r be the right process obtained from X^r by time change and killing with respect to positive Radon measures $m \in \overset{\circ}{S}_{\overline{D}}$ and $k \in \overset{\circ}{S}$. Then similarly to the above again, Y^r is m-symmetric and its Dirichlet form on $L^2(\overline{D}; m)$ and its extended Dirichlet space can be described as (5.3.1) and (5.3.2), respectively, but with $\overline{D}$ and $\widetilde{H}^1_e(D)$ in place of $\mathbb{R}^n$ and $\widetilde{H}^1_e(\mathbb{R}^n)$.

(2°) Conformal invariance of RBM

Let $\mathbb{D}$ be the unit disk and D be a Jordan domain in $\mathbb{R}^2$. By Riemann mapping theorem, there is a conformal map ϕ that maps $\mathbb{D}$ onto D, which can be extended to a topological homeomorphism from $\overline{\mathbb{D}}$ onto $\overline{D}$. We identify $\mathbb{R}^2$ with the complex plane $\mathbb{C}$ and denote by $\lambda(dz)$ the Lebesgue measure on it.

Let $Z = (Z_t, \mathbf{P}_z)$ be a reflecting Brownian motion (RBM in abbreviation) on $\overline{\mathbb{D}}$, whose Dirichlet form on $L^2(\overline{\mathbb{D}}; \mathbf{1}_{\mathbb{D}}(z)\lambda(dz))$ is $(\frac{1}{2}\mathbf{D}, H^1(\mathbb{D}))$. By noting that the transition function $P_t(z, \cdot)$ of Z is absolutely continuous with respect to λ for each $t > 0$ and $z \in \overline{\mathbb{D}}$, we can deduce the following two properties of Z holding $\mathbf{P}_z$-a.s. for every $z \in \overline{\mathbb{D}}$: first, $\int_0^\infty \mathbf{1}_{\partial\mathbb{D}}(Z_s)ds = 0$, and second, Z is Harris recurrent in the sense that for any $B \in \mathcal{B}(\overline{\mathbb{D}})$ with $\lambda(B) > 0$, $\lim_{t\to\infty} \int_0^t \mathbf{1}_B(Z_s)ds = \infty$ because Z is irreducible and recurrent by Theorem 2.2.13 (see, e.g., Lemma 4.8.1 in 2nd Ed. of [73]).

We let

$$m(dz) = |\phi'(z)|^2 \mathbf{1}_{\mathbb{D}}(z)\lambda(dz), \quad A_t = \int_0^t |\phi'(Z_s)|^2 \mathbf{1}_{\mathbb{D}}(Z_s)ds. \tag{5.3.3}$$

Since $\phi'(z) \neq 0$ for $z \in \mathbb{D}$, we conclude from the above observation that for every $z \in \overline{\mathbb{D}}$, $\mathbf{P}_z$-a.s. A_t is strictly increasing to ∞ as $t \to \infty$, and so is its inverse τ_t. The time-changed process $X = (Z_{\tau_t}, \mathbf{P}_z)_{z \in \overline{\mathbb{D}}}$ is therefore a conservative diffusion process on $\overline{\mathbb{D}}$. In fact, X is recurrent in view of Theorem 5.2.5.

The process X is symmetric with respect to m and, by virtue of (5.2.17), the Dirichlet form $(\mathcal{E}^X, \mathcal{F}^X)$ of X on $L^2(\mathbb{D}; m)$ is given by

$$\mathcal{F}^X = H_e^1(\mathbb{D}) \cap L^2(\mathbb{D}; m), \quad \mathcal{E}^X(f, g) = \frac{1}{2}\mathbf{D}(f, g), \quad f, g \in \mathcal{F}^X.$$

We define a process $Y = (Y_t, \mathbf{P}_w^Y)_{w \in \overline{D}}$ on $\overline{D}$ by

$$Y_t = \phi(Z_{\tau_t}), \quad \mathbf{P}_w^Y = \mathbf{P}_{\phi^{-1}(w)}, \quad w \in \overline{D}. \tag{5.3.4}$$

Y is then a conservative diffusion process on $\overline{D}$. We claim that Y is actually a RBM on $\overline{D}$.

Denote by $\{P_t^X, t > 0\}$ and $\{P_t^Y, t > 0\}$ the transition function of X and Y, respectively. It then hold that $P_t^Y f(\phi(z)) = P_t^X(f \circ \phi)(z)$, $z \in \mathbb{D}$. By the change-of-variables formula

$$\int_{\mathbb{D}} u(\phi(z))m(dx) = \int_D u(w)\lambda(dw),$$

we immediately obtain $\|P_t^Y f\|_{L^2(D;\lambda)} = \|P_t^X(f \circ \phi)\|_{L^2(\mathbb{D};m)}$ and

$$(P_t^Y f, g)_{L^2(D;\lambda)} = (P_t^X(f \circ \phi), g \circ \phi)_{L^2(\mathbb{D};m)},$$

from which λ-symmetry of Y follows. Let $(\mathcal{E}^Y, \mathcal{F}^Y)$ be the Dirichlet form of Y on $L^2(D; \lambda)$. In order to identify Y with the RBM on $\overline{D}$, it suffices to show that

$$\mathcal{F}^Y = H^1(D), \tag{5.3.5}$$

$$\mathcal{E}^Y(f,f) = \frac{1}{2}\int_D |\nabla f|^2(w)\lambda(dw), \quad f \in \mathcal{F}^Y. \tag{5.3.6}$$

For $f \in L^2(D; \lambda)$, we let $t \downarrow 0$ in the equality

$$t^{-1}(f - P_t^Y f, f)_{L^2(D;\lambda)} = t^{-1}(f \circ \phi - P_t^X(f \circ \phi), f \circ \phi)_{L^2(\mathbb{D};m)}$$

to see that $f \in \mathcal{F}^Y$ if and only if $f \circ \phi \in \mathcal{F}^X$, and in this case,

$$\mathcal{E}^Y(f,f) = \frac{1}{2}\int_{\mathbb{D}} |\nabla(f \circ \phi)|^2(z)\lambda(dz)$$
$$= \frac{1}{2}\int_{\mathbb{D}} |\nabla f|^2(\phi(z))m(dz) = \frac{1}{2}\int_{D} |\nabla f|^2(w)\lambda(dw).$$

Thus (5.3.6) holds true. If $f \in \mathcal{F}^Y$, then $f \in L^2(D;\lambda)$ and we have also $f \in \mathrm{BL}(D)$ from the above so that $f \in H^1(D)$, getting the inclusion $\subset$ in (5.3.5). Conversely, if $f \in H^1(D)$, then $f \circ \phi \in \mathrm{BL}(\mathbb{D}) \cap L^2(\mathbb{D}; m)$ from the above, and consequently $f \circ \phi \in \mathcal{F}^X$ because $\mathrm{BL}(\mathbb{D}) = H^1_e(\mathbb{D})$ in view of (2.2.54). Therefore, $f \in \mathcal{F}^Y$, as was to be proved.

Property (5.3.4) gives a construction of a RBM Y on the closure of a Jordan domain D which can start at every point $z \in \overline{D}$. By this construction and Proposition 3.1.11, we know that its transition function $P^Y(z;\cdot)$ is absolutely continuous with respect to the Lebesgue measure λ for every $z \in \overline{D}$.

Conformal invariance of RBM holds not only for simply connected planar domains. As we already saw in Section 2.2.4, if D is a domain in $\mathbb{R}^2$ having a continuous boundary, then the Dirichlet form $(\frac{1}{2}\mathbf{D}, H^1(D))$ is regular on $L^2(\overline{D}; \mathbf{1}_D dx)$. Its associated continuous Hunt process X is a RBM on $\overline{D}$. Suppose U is another domain in $\mathbb{R}^2$ and ϕ is a conformal map from D onto U that can be extended to be a homeomorphism from $\overline{D}$ onto $\overline{U}$. Then the same argument above in this example shows that $\phi(X)$ is a RBM on $\overline{U}$ up to a time change.

(3°) Trace of RBM on the unit disk

Let us consider the planar unit disk $\mathbb{D} = \{x \in \mathbb{R}^2 : |x| < 1\}$. Let T denote the Euclidean boundary of $\mathbb{D}$, which will be parameterized as $T = \{\theta : 0 \le \theta < 2\pi\}$. The reflecting Brownian motion (RBM in abbreviation) X on $\overline{\mathbb{D}}$ is a recurrent diffusion associated with the regular Dirichlet form $(\mathcal{E},\mathcal{F}) = (\frac{1}{2}\mathbf{D}, H^1(\mathbb{D}))$ on $L^2(\overline{\mathbb{D}}; \mathbf{1}_{\mathbb{D}}(x)dx)$. By (2.2.57), $H^1(\mathbb{D}) = H^1_e(\mathbb{D}) = \mathrm{BL}(\mathbb{D})$. We denote by $\widetilde{H}^1(\mathbb{D})$ the totality of functions in $H^1(\mathbb{D})$ that are quasi continuous on $\overline{\mathbb{D}}$. Let σ be the uniform measure on T. By Theorem 3.4.8, for every $u \in H^1(\mathbb{D})$, $\mathbf{H}\tilde{u}$ is a function in $\widetilde{H}^1(\mathbb{D})$ that is $\mathbf{D}$-orthogonal to $H^1_0(\mathbb{D})$ and so, in particular, it is harmonic on $\mathbb{D}$. We can deduce from this that for every $x \in \mathbb{D}$, the hitting distribution $\mathbf{H}(x, d\theta)$ for T by X starting from x is $K(x,\theta)\sigma(d\theta)$, where K is the Poisson kernel (2.2.61).

We deduce from the identity $u(e^{i\theta}) = \int_0^1 \frac{d}{dr}\left(u(re^{i\theta})r^2\right)dr$ that there is a constant $C > 0$ so that

$$\int_T u(e^{i\theta})^2\sigma(d\theta) \le C\,\|u\|^2_{\mathcal{E}_1} \tag{5.3.7}$$

for every $u \in C^1(\overline{\mathbb{D}})$ and hence for every $u \in \widetilde{H}^1(\mathbb{D})$, by Cauchy-Schwarz inequality, the above in particular implies that

$$\int_T |u(e^{i\theta})|\sigma(d\theta) \leq \sqrt{C}\, \|u\|_{\mathcal{E}_1} \quad \text{for every } u \in \widetilde{H}^1(D);$$

that is, σ is of finite energy integral and so it charges no $\mathcal{E}$-polar set. Since the hitting distribution of X for T is absolutely continuous with respect to σ, we conclude from Lemma 5.2.9 (ii) that $\sigma \in \overset{\circ}{S}_T$.

Let L_t be the positive continuous additive functional of X with Revuz measure σ and Y be the time-changed process of X with respect to L_t. Then Y is a σ-symmetric right process on T and, by virtue of Theorem 5.2.2, Corollary 5.2.10, and (5.3.7), the Dirichlet form $(\check{\mathcal{E}}, \check{\mathcal{F}})$ of Y on $L^2(T;\sigma)$ is regular and is given by

$$\begin{cases} \check{\mathcal{F}} = \widetilde{H}^1(\mathbb{D})\big|_T \cap L^2(T;\sigma) = \widetilde{H}^1(\mathbb{D})\big|_T, \\ \check{\mathcal{E}}(f,g) = \frac{1}{2}\mathbf{D}(\mathbf{H}f, \mathbf{H}g) \qquad \text{for } f,g \in \check{\mathcal{F}}. \end{cases} \tag{5.3.8}$$

Observe that $u = \mathbf{H}\widetilde{u}$ for $u \in H^1(\mathbb{D})$ if and only if u is the Poisson integral of some function $\varphi \in L^2(T;\sigma)$ with finite Dirichlet integral, which is in turn, as is shown by (2.2.60), equivalent to $\varphi \in \mathcal{G}$ for the space $\mathcal{G}$ defined by (2.2.58):

$$\begin{cases} \mathcal{G} = \left\{\varphi \in L^2(T;\sigma) : \mathbf{C}(\varphi,\varphi) < \infty\right\}, \\ \mathbf{C}(\varphi,\varphi) = \dfrac{1}{2}\displaystyle\int_{T\times T} (\varphi(\theta) - \varphi(\theta'))^2 U(\theta - \theta')\sigma(d\theta)\sigma(d\theta'), \end{cases}$$

where U is the kernel on T defined by (2.2.59).

In this case, $\frac{1}{2}\mathbf{D}(u,u) = \mathbf{C}(\varphi,\varphi)$ and furthermore $\widetilde{u}|_T$ is a $\mathbf{C}$-quasi-continuous version of $\varphi \in \mathcal{G}$ on account of Theorem 5.2.6. Denote by $\widetilde{\mathcal{G}}$ the totality of $\mathbf{C}$-quasi-continuous functions in $\mathcal{G}$. Then we have the identification of the trace Dirichlet form

$$\check{\mathcal{F}} = \widetilde{\mathcal{G}}, \quad \check{\mathcal{E}}(\varphi,\psi) = \mathbf{C}(\varphi,\psi), \quad \varphi,\psi \in \widetilde{\mathcal{G}}. \tag{5.3.9}$$

On the other hand, we obtained in Section 2.2.4 the characterization

$$\mathcal{G} = \left\{\varphi \in L^2(T;\sigma) : \mathbf{C}(\varphi,\varphi) = \sum_{k=-\infty}^{\infty} |k|\,|c_k|^2 < \infty\right\}$$

in terms of the Fourier coefficient c_k of φ: $c_k = \frac{1}{2\pi}\int_0^{2\pi} e^{-ik\xi}\varphi(\xi)d\xi$. Thus $(\check{\mathcal{E}}, \check{\mathcal{F}})$ is the Dirichlet form of the symmetric Cauchy process on the circle introduced at the end of Section 2.2.2. According to [65], the Fourier partial sum $S_n(\varphi;\xi) = \sum_{k=-n}^{n} c_k e^{ik\xi}$ of $\varphi \in \mathcal{G}$ converges for $\mathbf{C}$-q.e. $\xi \in T$ to a $\mathbf{C}$-quasi-continuous version of φ as $n \to \infty$. Thus we have the following direct description of the space $\widetilde{\mathcal{G}}$:

$$\widetilde{\mathcal{G}} = \left\{ \lim_{n\to\infty} S_n(\varphi;\xi) : \ \xi \in T, \ \varphi \in \mathcal{G} \right\}. \tag{5.3.10}$$

Furthermore, the above description of the form $\mathbf{C}$ in terms of the Fourier coefficients immediately implies

$$\int_T (\varphi - c_0)^2 d\sigma \le \mathbf{C}(\varphi,\varphi) \quad \text{for } \varphi \in \mathcal{G},$$

where $c_0 = \int_T \varphi(x)\sigma(dx)$. From this we can conclude that the extended Dirichlet space of the trace Dirichlet form (5.3.9) on $L^2(T;\sigma)$ is again the space $\mathcal{G}$.

We finally take any $\mu \in \overset{\circ}{S}_T$ in place of σ and consider the time-changed process Z on T obtained from X by means of the positive continuous additive functional with Revuz measure μ. By virtue of Corollary 5.2.10, the Dirichlet form $(\check{\mathcal{E}},\check{\mathcal{F}})$ of Z on $L^2(T;\mu)$ is regular and, by combining Theorem 5.2.2 and Theorem 5.2.15 with the above observation, we see that it is describable as

$$\check{\mathcal{F}} = \widetilde{\mathcal{G}} \cap L^2(T;\mu), \quad \check{\mathcal{E}}(\varphi,\psi) = \mathbf{C}(\varphi,\psi), \quad \varphi,\psi \in \check{\mathcal{F}}, \tag{5.3.11}$$

where $\widetilde{\mathcal{G}}$ admits a direct expression as the right hand side of (5.3.10).

(4°) Time change of RBM on a half space

We consider the upper half space $D = \{x = (x', x_n) : x' \in \mathbb{R}^{n-1}, x_n > 0\} \subset \mathbb{R}^n, n \ge 1$. The reflecting Brownian motion on $\overline{D}$ is by definition a diffusion process on $\overline{D}$ associated with the regular Dirichlet form $(\frac{1}{2}\mathbf{D}, H^1(D))$ on $L^2(\overline{D})$. The RBM $\widehat{X}$ can be obtained from the n-dimensional Brownian motion $X = (X_t = (X_t', X_t^{(n)}), \mathbf{P}_x)$ by reflection:

$$\widehat{X}_t = (X_t', |X_t^{(n)}|). \tag{5.3.12}$$

To see this, for a function u on $\overline{D}$, denote by $\widetilde{u}$ its extension to $\mathbb{R}^n$ by reflection; that is, $\widetilde{u}((x',x_n)) = u((x',-x_n))$ when $x_n \le 0$. Denote by $\{\widehat{R}_\alpha;\alpha>0\}$ and $\{R_\alpha;\alpha>0\}$ the resolvent kernels of $\widehat{X}$ and X, respectively. Then, for $f \in L^2(D)$, $\widehat{R}_\alpha f = R_\alpha \widetilde{f}|_D$ so that $\widehat{R}_\alpha f \in H^1(D)$. Further, for any $v \in H^1(D)$, $\widetilde{v} \in H^1(\mathbb{R}^n)$ and

$$\frac{1}{2}\int_{\mathbb{R}^n} \nabla R_\alpha \widetilde{f} \cdot \nabla \widetilde{v} dx + \alpha(R_\alpha \widetilde{f}, \widetilde{v})_{L^2(\mathbb{R}^n)} = (\widetilde{f}, \widetilde{v})_{L^2(\mathbb{R}^n)}.$$

Since $R_\alpha \widetilde{f}$ coincides with the extension of $\widehat{R}_\alpha f$ by reflection, we obtain the desired equation

$$\frac{1}{2}\mathbf{D}(\widehat{R}_\alpha f, v) + \alpha(\widehat{R}_\alpha f, v)_{L^2(D)} = (f,v)_{L^2(D)}, \quad v \in H^1(D).$$

In particular, the RBM $\widehat{W}$ on the one-dimensional half line $I = [0, \infty)$ is obtained from the one-dimensional Brownian motion $W = (W_t, \mathbf{P}_x)$ by $\widehat{W}_t = |W_t|$. The associated Dirichlet form is $(\frac{1}{2}\mathbf{D}, H^1(I))$, which is regular on $L^2([0,\infty))$. As is studied in Example 3.5.7, each one-point set of $[0, \infty)$ is non-polar. Hence the unit mass δ_0 concentrated at the one point $\{0\}$ is a smooth measure and admits a unique positive continuous additive functional L_t of $\widehat{W}$ with Revuz measure δ_0. L_t is called the *local time* of $\widehat{W}$ at the origin.

We put $\widehat{r}_\alpha(x,y) = \int_0^\infty e^{-\alpha t}\widehat{p}_t(x,y)dt$ for the transition density $\widehat{p}_t(x,y)$ of $\widehat{W}$. By virtue of Lemma 4.1.5, it holds then that

$$\mathbf{E}_x\left[\int_0^\infty e^{-\alpha t}dL_t\right] = r_\alpha(x,0), \quad x \in [0,\infty). \tag{5.3.13}$$

By letting $x = 0$, we get

$$\mathbf{E}_0\left[\int_0^\infty e^{-\alpha t}dL_t\right] = \sqrt{\frac{2}{\alpha}}. \tag{5.3.14}$$

Let τ_t be the right continuous inverse of L_t. Then the function $f(t) = \mathbf{E}_0[e^{-\alpha\tau_t}]$ can be seen to enjoy the property $f(s+t) = f(s)f(t)$ by virtue of Proposition A.3.8(ii), (iv) and the strong Markov property of $\widehat{W}$. Hence $f(t) = e^{-kt}$ for some $k \geq 0$. Since the left hand side of (5.3.14) equals $\int_0^\infty f(t)dt$ by Lemma A.3.7, we arrive at the following identification of the distribution of τ_t:

$$\mathbf{E}_0\left[e^{-\alpha\tau_t}\right] = \exp\left(-\sqrt{\alpha/2}\, t\right), \qquad t \geq 0. \tag{5.3.15}$$

We now return to the RBM $\widehat{X} = ((X'_t, |X_t^{(n)}|), \mathbf{P}_x)$ specified by (5.3.12) on the half-space $\overline{D} \subset \mathbb{R}^n$ but for $n \geq 2$. Denote by σ the Lebesgue measure on the hyperplane $F = \{(x', 0) : x' \in \mathbb{R}^{n-1}\}$: $\sigma(dx) = dx'\delta_0(dx^{(n)})$. In the same way as in the preceding example, we can see that $\mathbf{1}_{\{|x'| \leq R\}} \cdot \sigma(dx)$ is of finite energy integral for each $R > 0$ so that σ is a smooth measure for $(\frac{1}{2}\mathbf{D}, H^1(D))$. Moreover, we can similarly verify that the hitting distribution of $\widehat{X}$ for F is absolutely continuous with respect to σ with density being the Poisson kernel. Hence we conclude from Lemma 5.2.9(ii) that $\sigma \in \mathring{S}_F$.

Let L_t be the local time at the origin of the nth component process $|X_t^{(n)}|$, which is the RBM on $[0, \infty)$. Then L_t can be viewed as the positive continuous additive functional of $\widehat{X}$ with Revuz measure σ, namely, the relation (4.1.2)

$$\int_D h(x)\mathbf{E}_x\left[\int_0^t f(X'_s, |X_s^{(n)}|)dL_s\right]dx = \int_0^t \int_F (\widehat{P}_s h)(y', 0) f(y', 0)dy'ds$$

holds for $f, h \in \mathcal{B}_+(D)$ and the transition function $\widehat{P}_t$ of $\widehat{X}$. Indeed, (5.3.13) implies that $\mathbf{E}_{x^{(n)}}\left[\int_0^\infty g(s)dL_s\right] = \int_0^\infty g(s)\widehat{p}_s^{(n)}(x^{(n)}, 0)ds$ for the transition density $\widehat{p}_t^{(n)}$ of $|X_t^{(n)}|$ and for $g \in b\mathcal{B}([0,\infty))$ with compact support, from which the above identity is readily derived.

Let τ_t be the right continuous inverse of L_t and $Y = ((X'_{\tau_t}, 0), \mathbf{P}_{(x',0)})$ be the time-changed process of the RBM $\widehat{X}$. Since $(X'_t, \mathbf{P}_{(x',0)})$ is the $(n-1)$-dimensional Brownian motion independent of τ_t, it follows from (5.3.15) that

$$\mathbf{E}_0\left[\exp(i\langle\xi, X'_{\tau_t}\rangle)\right] = \int_0^\infty \mathbf{E}_0[\exp(i\langle\xi, X'_s\rangle)]\mathbf{P}_0(\tau_t \in ds)$$

$$= \int_0^\infty \exp\left(-\frac{|\xi|^2}{2} s\right)\mathbf{P}_0(\tau_t \in ds) = e^{-\frac{|\xi|}{2}t}, \quad \xi \in F,$$

and accordingly the transition function of Y is determined as (2.2.10) by a symmetric stable convolution semigroup $\{\nu_t\}$ on F of index 1. We call Y a *symmetric Cauchy process* on the hyperplane F.

According to (2.2.9), (2.2.16), and the identity[1]

$$\frac{\Gamma(n/2)}{2\pi^{n/2}} \int_{\mathbb{R}^{n-1}} (1 - \cos\langle\xi, \eta\rangle)|\eta|^{-n} d\eta = \frac{1}{2}|\xi|, \quad \xi \in \mathbb{R}^{n-1},$$

we can see that the Dirichlet form of Y on $L^2(F;\sigma)$ is given by $(\mathbf{C}, \mathcal{D})$, which is expressed by using a kernel

$$U(\eta) = \frac{\Gamma(n/2)}{2\pi^{n/2}} |\eta|^{-n} \tag{5.3.16}$$

on F as

$$\begin{cases} \mathbf{C}(\varphi, \varphi) = \dfrac{1}{2}\displaystyle\int_{F\times F} (\varphi(\xi+\eta) - \varphi(\xi))^2 U(\eta) d\xi d\eta, \\ \mathcal{D} = \{\varphi \in L^2(F;\sigma) : \mathbf{C}(\varphi, \varphi) < \infty\}. \end{cases} \tag{5.3.17}$$

Let $\mathcal{D}_e$ be the extended Dirichlet space of the Dirichlet form $(\mathbf{C}, \mathcal{D})$ on $L^2(F;\sigma)$ and $\widetilde{\mathcal{D}}_e$ be the collection of all $\mathbf{C}$-quasi-continuous functions in $\mathcal{D}_e$. Then we see from Theorem 5.2.15 and Theorem 5.2.6 that

$$\widetilde{\mathcal{D}}_e = \widetilde{H}^1_e(D)\big|_F.$$

In place of σ, we can consider a general $\mu \in \overset{\circ}{S}_F$ and the time-changed process Z of the RBM $\widehat{X}$ by means of the PCAF with Revuz measure μ. By Corollary 5.2.10 and Theorem 5.2.15, the Dirichlet form $(\check{\mathcal{E}}, \check{\mathcal{F}})$ of Z on $L^2(F;\mu)$ is regular and admits the expression

$$\check{\mathcal{F}} = \widetilde{\mathcal{D}}_e \cap L^2(F;\mu), \quad \check{\mathcal{E}}(\varphi, \varphi) = \mathbf{C}(\varphi, \varphi), \quad \varphi \in \check{\mathcal{F}}. \tag{5.3.18}$$

---

[1] Cf. [132, solutions to E6.12, E18.9].

We note that in view of Exercise 2.3.3, the space $\widetilde{\mathcal{D}}_e$ is expressible directly as $\widetilde{\mathcal{D}}_e = \{\varphi = I_{1/2} * \psi : \psi \in L^2(\mathbb{R}^{n-1})\}$ in terms of the $1/2$-order Riesz convolution kernel $I_{1/2}(x) = \gamma_{n-1,1/2}|x|^{-n+3/2}$, $x \in \mathbb{R}^{n-1}$, when $n \geq 3$. When $n = 2$, the Dirichlet form (5.3.17) is recurrent. By using (6.5.4) below, we can conclude in this case that

$$\check{\mathcal{F}} = \{\varphi \in L^2(F;\mu) : \mathbf{C}\text{-quasi-continuous}, \mathbf{C}(\varphi,\varphi) < \infty\}, \qquad (5.3.19)$$

analogously to (5.3.11).

Sections 5.5, 5.6, and 7.2 will be devoted to the investigation of the generality of the expression of the trace Dirichlet forms as appeared in the above two examples.

(5°) General one-dimensional minimal diffusion

For an open interval $I = (r_1, r_2) \subset \mathbb{R}$, we consider a general minimal diffusion $X^0 = (X_t,{}^0, \zeta^0, \mathbf{P}_x^0)$ on I. We only require that X^0 satisfies conditions **(d.1)**, **(d.2)**, and **(d.3)** in Example 3.5.7, where the Dirichlet form of X^0 is identified under the additional condition **(d.4)** that X^0 admits no killings inside I. We now remove the condition **(d.4)** by an application of Theorem 5.2.17 on the resurrection.

We denote by $\{R_\alpha; \alpha > 0\}$ the resolvent kernel of X^0. Since I consists of regular points for X^0, we have from [97, §3.6] that $R_\alpha(\mathcal{B}_b(I)) \subset C_b(I)$ and the generator $\mathcal{G}$ of X^0 is well defined by

$$\begin{cases} \mathcal{D}(\mathcal{G}) = R_\alpha(C_b(I)), \\ (\mathcal{G}u)(x) = \alpha u(x) - f(x) \quad \text{for } u = R_\alpha f,\ f \in C_b(I),\ x \in I, \end{cases}$$

independently of $\alpha > 0$. On account of [97, §4.3], there exist a canonical scale s, a canonical measure m, and a positive Radon measure k called a *killing measure* on I such that, for any $u \in \mathcal{D}(\mathcal{G})$,

$$\mathcal{G}u \cdot dm = dD_s u - u dk \qquad \text{on } I. \qquad (5.3.20)$$

The triplet (s, m, k) is unique up to a multiplicative constant in the sense that, for another such triplet $(\widetilde{s}, \widetilde{m}, \widetilde{k})$, there exists a constant $c > 0$ so that $d\widetilde{s} = c ds$, $d\widetilde{m} = c^{-1} dm$, and $d\widetilde{k} = c^{-1} dk$.

Let $J = (j_1, j_2)$ with $r_1 < j_1 < j_2 < r_2$ and $\{R_\alpha^J, \alpha > 0\}$ be the resolvent kernel of the part process X^J of X^0 on J.

Lemma 5.3.2. *Let $u = R_\alpha^J f$ for $f \in C_b(I)$. Then $u \in C_c(I)$,*

$$-dD_s u + u dk + \alpha u dm = f dm \qquad \text{on } J, \qquad (5.3.21)$$

and

$$u(j_1+) = u(j_2-) = 0. \tag{5.3.22}$$

Proof. For $i = 1, 2$ and $x \in J$, define $\varphi_i(x) = \mathbf{E}_x^0[e^{-\alpha\tau_J}; X_{\tau_J-} = j_i]$. If $g \in C_b(I)$ vanishes on J, then $R_\alpha g(x) = \sum_{k=1}^2 R_\alpha g(j_k)\varphi_k(x)$, $x \in J$. For $g_1 \in C_b(I)$ vanishing on (r_1, j_2) and positive on (j_2, r_2), $R_\alpha g_1$ is strictly increasing on J. A similar choice of $g_2 \in C_b(I)$ gives $R_\alpha g_2$ strictly decreasing on J so that $R_\alpha g_1(j_1)R_\alpha g_2(j_2) - R_\alpha g_2(j_1)R_\alpha g_1(j_2) < 0$ and φ_i becomes a linear combination of $R_\alpha g_k$, $k = 1, 2$, on J. Thus (5.3.20) implies $-dD_s\varphi_i + \varphi_i dk + \alpha\varphi_i dm = 0$ on J. Further $\varphi_1(j_1+) = 1$, $\varphi_1(j_2-) = 0$, $\varphi_2(j_1+) = 0$, $\varphi_2(j_2-) = 1$, as in [69, Lemma 2.1].

Equations (5.3.21) and (5.3.22) then follow from (5.3.20) and the identity

$$u(x) = R_\alpha f(x) - \varphi_1(x)R_\alpha f(j_1) - \varphi_2(x)R_\alpha f(j_2), \quad x \in J.$$

Since each point of J is regular with respect to X^J, $u \in C_b(J)$ and consequently $u \in C_c(I)$. □

PROPOSITION 5.3.3. *The diffusion X^0 is symmetric with respect to m and the Dirichlet form $(\mathcal{E}, \mathcal{F})$ of X^0 on $L^2(I; m)$ is regular. For any interval $J = (j_1, j_2)$ with $r_1 < j_1 < j_2 < r_2$, $R_\alpha^J(C_c(I)) \subset \mathcal{F}$ and*

$$\mathcal{E}(u, v) = \mathcal{E}^{(s)}(u, v) + \int_I u \cdot v dk, \quad u, v \in R_\alpha^J(C_c(I)), \tag{5.3.23}$$

where $\mathcal{E}^{(s)}$ is defined by (2.2.29).

Proof. For $u = R_\alpha^J f$, $v = R_\alpha^J g$, $f, g \in C_c(I)$, we get from (5.3.21)

$$-\int_J vdD_s u + \int_J uvdk + \alpha\int_J uvdm = \int_J vfdm.$$

Since v satisfies (5.3.22), $v(j_1+)D_s u(j_1+) - v(j_2-)D_s u(j_2-) = 0$ and an integration by parts yields

$$\mathcal{E}_\alpha^{(s)}(u, v) + \int_J uvdk = \int_J vfdm. \tag{5.3.24}$$

The left hand side also equals $\int_J ugdm$ so that $\int_J fR_\alpha^J gdm = \int_J R_\alpha^J fgdm$. By letting $J \uparrow I$, we obtain $\int_I fR_\alpha gdm = \int_I R_\alpha f \cdot gdm$ for $f, g \in C_c(I)$. This proves the symmetry of the resolvent kernel of X^0 with respect to the measure m.

On the other hand, in view of Exercise 3.1.16, we have for any $f \in C_c(I)$, $u = R_\alpha^J f \in \mathcal{F}$ and $\mathcal{E}_\alpha(u, u) = \int_I fudm$, which combined with (5.3.24) yields (5.3.23). The above equation also implies that for any $f \in C_c(I)$, $R_\alpha f \in \mathcal{F}$ is $\mathcal{E}_\alpha$-approximated by $R_\alpha^J f \in \mathcal{F} \cap C_c(I)$ as $J \uparrow I$, and we can conclude that $(\mathcal{E}, \mathcal{F})$ is regular. □

Denote by $(\mathcal{F}_e, \mathcal{E})$ the extended Dirichlet space of the Dirichlet form $(\mathcal{E}, \mathcal{F})$ of the minimal diffusion X^0 on $L^2(I; m)$. For $i = 1, 2$, recall that the boundary r_i of I is called approachable if $|s(r_i)| < \infty$.

THEOREM 5.3.4.

$$\begin{cases} \mathcal{F}_e = \{u \in \mathcal{F}^{(s)} : u(r_i) = 0 \textit{ if } r_i \textit{ is approachable}\} \cap L^2(I; k), \\ \mathcal{E}(u, v) = \mathcal{E}^{(s)}(u, v) + \int_I u \cdot v dk, \quad u, v \in \mathcal{F}_e. \end{cases} \tag{5.3.25}$$

$$\mathcal{F} = \mathcal{F}_e \cap L^2(I; m), \tag{5.3.26}$$

where the space $\mathcal{F}^{(s)}$ is defined by (2.2.28).

Proof. By Proposition 5.3.3, X^0 is m-symmetric and the Dirichlet form $(\mathcal{E}, \mathcal{F})$ of X^0 on $L^2(I; m)$ is regular. Let $(\mathcal{E}^{\text{res}}, \mathcal{F}_e^{\text{res}} \cap L^2(I; m))$ and $X^{0,\text{res}}$ be the resurrected Dirichlet form of $(\mathcal{E}, \mathcal{F})$ and the resurrected Hunt process of X^0, respectively. By virtue of Theorem 5.2.17, $(\mathcal{E}^{\text{res}}, \mathcal{F}_e^{\text{res}} \cap L^2(I; m))$ and $(\mathcal{E}, \mathcal{F})$ share the same set of quasi notions. By condition **(d.3)**, each point of I is not m-polar for X^0 and consequently not for $X^{0,\text{res}}$.

Since X^0 is a subprocess of $X^{0,\text{res}}$ in view of Theorem 5.2.17, the transition function of $X^{0,\text{res}}$ dominates that of X^0 and hence the irreducibility of $(\mathcal{E}^{\text{res}}, \mathcal{F}_e^{\text{res}} \cap L^2(I; m))$ follows from the property **(d.3)** for X^0. Therefore, $X^{0,\text{res}}$ also satisfies **(d.3)** owing to Theorem 3.5.6 and the observed fact that the one-point set is not m-polar for $X^{0,\text{res}}$. Consequently, $X^{0,\text{res}}$ is a minimal diffusion on I admitting no killing inside.

Let $(\widetilde{s}, \widetilde{m})$ be a canonical scale and a canonical measure associated with $X^{0,\text{res}}$. $X^{0,\text{res}}$ is $\widetilde{m}$-symmetric. However, the symmetrizing measure of $X^{0,\text{res}}$ is unique up to a multiplicative constant by Ying-Zhao [153] and the strong irreducibility **(d.3)** of $X^{0,\text{res}}$. Hence we can take m as $\widetilde{m}$. By virtue of Theorem 3.5.8, the extended Dirichlet space $(\mathcal{F}_e^{\text{res}}, \mathcal{E}^{\text{res}})$ of $(\mathcal{E}^{\text{res}}, \mathcal{F}_e^{\text{res}} \cap L^2(I; m))$ is expressed as

$$\begin{cases} \mathcal{F}_e^{\text{res}} = \{u \in \mathcal{F}^{(\widetilde{s})} : u(r_i) = 0 \quad \text{if } r_i \text{ is approachable}\}, \\ \mathcal{E}^{\text{res}}(u, v) = \mathcal{E}^{(\widetilde{s})}(u, v), \quad u, v \in \mathcal{F}_e^{\text{res}}. \end{cases}$$

Therefore, we can use Theorem 5.2.17 to conclude that the extended Dirichlet space and the Dirichlet form of X^0 admit the representations (5.3.25) and (5.3.26) with $\widetilde{s}$ and the killing measure κ for X^0 in place of s and k, respectively. But by Proposition 5.3.3, the form $\mathcal{E}$ admits the expression (5.3.23) in terms of s and k, so that $\widetilde{s} = s$ and $\kappa = k$. □

Theorem 5.15.1 in Itô's book [95] provided us with an explicit construction of a symmetric density of the resolvent kernel with respect to the canonical measure for a minimal diffusion process on I under the additional condition

(d.4), namely, by assuming that $k = 0$. Readers are invited to extend this theorem by removing condition **(d.4)**.

5.4. ENERGY FUNCTIONALS FOR TRANSIENT PROCESSES

In this section, we introduce the concept of the energy functional for a general transient symmetric right process X and study its basic properties that will be utilized in later sections and later chapters as well.

Let E be a Hausdorff topological space whose Borel field is generated by continuous functions, m be σ-finite measure on E with $\mathrm{supp}[m] = E$, $(\mathcal{E}, \mathcal{F})$ be a quasi-regular Dirichlet form on $L^2(E; m)$, and $X = (X_t, \mathbf{P}_x, \zeta)$ be an m-symmetric special Borel standard process on E that is properly associated with the form $(\mathcal{E}, \mathcal{F})$. In dealing with a general symmetric right process, we lose no generality to study under this setting on account of Theorem 1.5.3 and Theorem 3.1.13. Our basic additional assumption in this section is the transience of $(\mathcal{E}, \mathcal{F})$, or, equivalently that of X. $\{P_t; t \geq 0\}$ and $\{R_\alpha; \alpha > 0\}$ will denote the transition function and resolvent kernel of X, respectively. The 0-order resolvent kernel R_0 will be denoted by R. The extended Dirichlet space of $(\mathcal{E}, \mathcal{F})$ will be designated by $\mathcal{F}_e$. We make a convention that any function in $\mathcal{F}_e$ is $\mathcal{E}$-quasi continuous already.

Potential theory was formulated for a transient regular Dirichlet form in Section 2.3. For the present transient quasi-regular Dirichlet form $(\mathcal{E}, \mathcal{F})$ on $L^2(E; m)$, we can well introduce notions of measures of finite 0-order energy integrals and its potentials. Let $S(E)$ be the space of smooth measures on E for $(\mathcal{E}, \mathcal{F})$: $\nu \in S(E)$ if and only if ν is a Borel measure on E charging no $\mathcal{E}$-polar set and there is an $\mathcal{E}$-nest $\{F_k, k \geq 1\}$ such that $\nu(F_k) < \infty$ for each k. We write as $\nu \in S_0^{(0)}(E)$ if $\nu \in S(E)$ and, for some constant $C > 0$.

$$\int_E |v(x)|\nu(dx) \leq C\|v\|_{\mathcal{E}}, \quad \forall v \in \mathcal{F}_e,$$

or, equivalently, there exists a function $U\nu \in \mathcal{F}_e$ such that

$$\mathcal{E}(U\nu, v) = \int_E v(x)\nu(dx) \qquad \text{for every } v \in \mathcal{F}_e. \tag{5.4.1}$$

$\nu \in S_0^{(0)}(E)$ is called a measure of finite 0-*order energy integral* and $U\nu$ is called the 0-*order potential* of $\nu \in S_0^{(0)}(E)$. Notice that as a function v on the right hand side above, only an $\mathcal{E}$-quasi continuous function is allowed to appear.

Further, for $\alpha > 0$, the α-order potential $U_\alpha \nu$ of $\nu \in S_0^{(0)}(E)$ is also well defined as an element of $\mathcal{F}$ satisfying the above equation with $\mathcal{E}_\alpha$, $\mathcal{F}$ in place of $\mathcal{E}$, $\mathcal{F}_e$, respectively.

We now introduce the notion of an energy functional for the m-symmetric Borel standard process X properly associated with $(\mathcal{E}, \mathcal{F})$. We denote by $\langle \nu, f\rangle$ or $\langle f, \nu\rangle$ the integral $\int_E f d\nu$ for a measure ν and a function f, and by (f, g) the integral $\int_E fgdm$ for functions f, g on E whenever they make sense.

The concept of the excessive function relative to X is defined by Definition A.1.35. A universally measurable function f defined q.e. on E will be called q.e. *excessive* relative to X if it is finite excessive q.e. in the sense that, for some $\mathcal{E}$-polar set $N \subset E$ and every $x \in E \setminus N$,

$$0 \le f(x) < \infty, \qquad P_t f(x) \uparrow f(x) \text{ as } t \downarrow 0. \tag{5.4.2}$$

By virtue of Theorem 3.1.13, we can take as N a properly exceptional set in the above definition. A q.e. excessive function f is called q.e. *purely excessive* if

$$\lim_{t\to\infty} P_t f(x) = 0, \quad \forall x \in E \setminus N, \tag{5.4.3}$$

for some $\mathcal{E}$-polar set N. We denote by $\mathcal{S}(E)$ the totality of q.e. excessive functions on E relative to X. We also define

$$\mathcal{S}^{\rm pur}(E) := \{f \in \mathcal{S}(E) : f \text{ is q.e. purely excessive}\}.$$

For any $f \in \mathcal{S}^{\rm pur}(E)$ and $g \in \mathcal{S}(E)$, we let

$$L(f, g) := \uparrow \lim_{t\downarrow 0} \frac{1}{t}(f - P_t f, g) \tag{5.4.4}$$

and call it the *energy functional*. Here $\uparrow \lim_{t\downarrow 0}$ means that it is an increasing limit as $t \downarrow 0$. In fact, if we denote $(f - P_t f, g)$ by $e(t)$, then for $s, t > 0$,

$$\begin{aligned} e(t+s) &= e(t) + (P_t f - P_{t+s} f, g) = e(t) + (P_t(f - P_s f), g) \\ &= e(t) + (f - P_s f, P_t g). \end{aligned} \tag{5.4.5}$$

Exercise 5.4.1. Using (5.4.5) and following the method of the proof of Lemma A.3.4, show that (5.4.4) is an increasing limit.

It also holds that

$$L(f, g) = \uparrow \lim_{\alpha\uparrow\infty} \alpha(f - \alpha R_\alpha f, g), \tag{5.4.6}$$

where $\uparrow \lim_{\alpha\uparrow\infty}$ means that it is an increasing limit as $\alpha \uparrow \infty$. This is because

$$\alpha(f - \alpha R_\alpha f, g) = \int_0^\infty e^{-t}(t/\alpha)^{-1}(f - P_{t/\alpha} f, g)\, t dt.$$

The notion L is a special case of the energy functional of a purely excessive measure and an excessive function in the context of a general right process in R. Getoor [81].

LEMMA 5.4.2. (i) *For any $f \in \mathcal{S}^{\rm pur}(E)$, let $N \subset E$ be an X-properly exceptional set outside which* (5.4.2) *and* (5.4.3) *hold. Put*

$$f_t(x) = \frac{1}{t}(f(x) - P_t f(x)), \quad x \in E \setminus N. \tag{5.4.7}$$

Then, for each $x \in E \setminus N$,

$$Rf_t(x) = \frac{1}{t}\int_0^t P_s f(x)ds \uparrow f(x) \quad t \downarrow 0.$$

(ii) *For any $g \in \mathcal{S}(E)$, there exists an increasing sequence of non-negative universally measurable functions $\{g_n, n \geq 1\}$ such that for q.e. $x \in E$,*

$$Rg_n(x) \uparrow g(x) \quad \textit{as } n \uparrow \infty.$$

Proof. (i) For each $T > 0$, we have for every $t \in (0, T)$

$$\int_0^T P_s f_t(x)ds = \frac{1}{t}\int_0^T P_s f(x)ds - \frac{1}{t}\int_t^{T+t} P_s f(x)ds$$

$$= \frac{1}{t}\int_0^t P_s f(x)ds - \frac{1}{t}\int_T^{T+t} P_s f(x)ds.$$

Since the last term is bounded by $P_T f(x)$, we get by letting $T \to \infty$ the identity $Rf_t(x) = \frac{1}{t}\int_0^t P_s f(x)ds$, which in turn increases to $f(x)$ as $t \downarrow 0$.
(ii) Take a strictly positive m-integrable Borel function h on E. Rh is X-excessive and strictly positive on E. We shall show that $Rh \in \mathcal{S}^{\rm pur}(E)$. By Proposition 2.1.3, $Rh < \infty$ m-a.e. and consequently q.e. by Theorem A.2.13 and Theorem 3.1.13. Further, Rh is q.e. purely excessive because $P_t Rh(x) = \int_t^\infty P_s h(x)ds \leq Rh(x) < \infty$ q.e.

We now let $v_n = g \wedge (nRh)$. Then $v_n \in \mathcal{S}^{\rm pur}(E)$ and v_n increases to g as $n \to \infty$. By (i), $R(v_n)_t$ increases to v_n as $t \downarrow 0$. Since $R(v_n)_t(x) = \frac{1}{t}\int_0^t P_s v_n(x)ds$ increases as $n \uparrow$ and $t \downarrow$ and admits g as the supremum in two variables (n, t), we see that $Rg_n \uparrow g$ q.e. as $n \to \infty$ for $g_n = (v_n)_{1/n}$. □

THEOREM 5.4.3. *The energy functional L defined by* (5.4.4) *enjoys the following properties:*
(i) *If $f = Rh$ for some $h \in \mathcal{B}_+^*(E)$ and $f < \infty$ q.e. on E, then $f \in \mathcal{S}^{\rm pur}(E)$ and $L(f, g) = (h, g)$ for every $g \in \mathcal{S}(E)$.*
(ii) *If $f_1, f_2 \in \mathcal{S}^{\rm pur}(E)$ satisfy $f_1 \leq f_2$, m-a.e. on E, then*

$$L(f_1, g) \leq L(f_2, g) \quad \textit{for every } g \in \mathcal{S}(E).$$

(iii) *If* $f_n, f \in \mathcal{S}^{\rm pur}(E)$ *with* $f_n \uparrow f$ *as* $n \to \infty$, *then*

$$L(f_n, g) \uparrow L(f, g) \quad \text{as } n \uparrow \infty \text{ for every } g \in \mathcal{S}(E).$$

(iv) *For* $f \in \mathcal{S}^{\rm pur}(E)$, $g \in \mathcal{S}(E)$,

$$L(f, g) = \sup\{(h, g) : Rh \le f,\ h \in \mathcal{B}^*_+(E)\}.$$

If $f \in \mathcal{S}^{\rm pur}(E)$ *and* f *equals the* 0*-order potential* $U\nu$ *of some measure* $\nu \in S_0^{(0)}(E)$, *then*

$$L(f, g) = \langle \nu, g \rangle \quad \text{for every } g \in \mathcal{S}(E).$$

(v) *If* $f, g \in \mathcal{S}^{\rm pur}(E)$, *then* $L(f, g) = L(g, f)$.

Proof. (i) In this case, we see in the same way as in the proof of Lemma 5.4.2(ii) that $f \in \mathcal{S}^{\rm pur}(E)$. Since, for any $g \in \mathcal{S}(E)$,

$$(f_t, g) = \frac{1}{t}\left(\int_0^t P_s h ds,\ g\right) \uparrow (h, g) \quad \text{as } t \downarrow 0,$$

we get (i).

(ii) and (iii) For any $g \in \mathcal{S}(E)$, choose a sequence $\{g_n\}$ as in Lemma 5.4.2(ii). By symmetry of R and Lemma 5.4.2, we interchange the order of taking increasing limits to get for any $f \in \mathcal{S}^{\rm pur}(E)$

$$L(f, g) = \lim_{t\downarrow 0}\lim_{k\to\infty}(f_t, Rg_k) = \lim_{k\to\infty}\lim_{t\downarrow 0}(Rf_t, g_k) = \lim_{k\to\infty}(f, g_k),$$

from which follow (ii) and (iii) immediately.

(iv) Let $f \in \mathcal{S}^{\rm pur}(E)$, $g \in \mathcal{S}(E)$. If $Rh \le f$ for $h \in \mathcal{B}^*_+(E)$, then (i) and (ii) imply $(h, g) = L(Rh, g) \le L(f, g)$. Since $f_t \ge 0$ and $Rf_t \uparrow f$ as $t \downarrow 0$, we obtain from (i) and (iii) that $(f_t, g) = L(Rf_t, g) \uparrow L(f, g)$ as $t \downarrow 0$, yielding (iv).

(v) Let $h_\alpha = \alpha(U\nu - \alpha R_\alpha U\nu)$, $\alpha > 0$. We can then see from (2.3.8) and the proof of Lemma 2.3.16 that $h_\alpha = \alpha U_\alpha \nu$. Take a strictly positive function $w \in \mathcal{B}^*_+(E) \cap L^2(E; m)$ and put $g_k = g \wedge (kw)$. Then we have

$$(h_\alpha, g_k) = \alpha\mathcal{E}_\alpha(U_\alpha\nu, R_\alpha g_k) = \alpha\langle \nu, R_\alpha g_k\rangle.$$

The second identity in the above holds because $R_\alpha g_k$ is an $\mathcal{E}$-quasi-continuous element of $\mathcal{F}$ due to the proper association of X with $(\mathcal{E}, \mathcal{F})$. We let $k \to \infty$ to get $(h_\alpha, g) = \alpha\langle \nu, R_\alpha g\rangle$. It follows that

$$(h_\alpha, g) \uparrow \langle \nu, g\rangle \quad \text{as } \alpha \uparrow \infty,$$

which yields (v) in view of (5.4.6).

(vi) Define $f_s = \frac{1}{s}(f - P_{sf})$, $g_t = \frac{1}{t}(g - P_t g)$ and look at the identity

$$(Rf_s, g_t) = (f_s, Rg_t).$$

By virtue of Lemma 5.4.2 and (5.4.4), the right hand side increases to (f_s, g) as $t \downarrow 0$, which then increases to $L(f, g)$ as $s \downarrow 0$. Changing the order of the increasing limits, the left hand side converges to $L(g,f)$. □

5.5. TRACE DIRICHLET FORMS AND FELLER MEASURES

We return to the setting and convention made at the beginning of this chapter: we consider an m-symmetric Hunt process $X = (X_t, \mathbf{P}_x)$ whose Dirichlet form $(\mathcal{E}, \mathcal{F})$ on $L^2(E; m)$ is regular and every element in $\mathcal{F}_e$ is represented by its quasi continuous version already. But throughout this and next section, we make an additional assumption that $(\mathcal{E}, \mathcal{F})$ is irreducible. The transition function and resolvent kernel of X are denoted by $\{P_t; t \geq 0\}$ and $\{R_\alpha; \alpha > 0\}$, respectively.

Denote by Cap_1 the 1-capacity defined by the form $(\mathcal{E}, \mathcal{F})$. Let F be a quasi closed subset F of E such that

$$\mathrm{Cap}_1(F) > 0. \tag{5.5.1}$$

We remark that the notion of being quasi closed is invariant under the quasi-homeomorphism of Dirichlet forms. Furthermore, by Theorem 3.3.3(i), a quasi closed set is q.e. finely closed in the sense that there is a properly exceptional set N such that $F \setminus N$ is nearly Borel measurable and finely closed with respect to $X|_{E \setminus N}$. Since we are only concerned with assertions holding q.e., we may and do make a convention that the quasi closed set F is nearly Borel and finely closed already.

Let $E_0 = E \setminus F$. Under the present convention, E_0 is nearly Borel and finely open with respect to X. The part process of X on E_0 killed upon leaving E_0 will be denoted by X^0. To be more precise, we let

$$\tau_0 := \tau_{E_0} = \inf\{t \in [0, \zeta] : X_t \notin E_0\}, \tag{5.5.2}$$

so that $\tau_0 = \sigma_F \wedge \zeta$ $\mathbf{P}_x$-a.s. for $x \in E_0$, where $\sigma_F := \inf\{t > 0 : X_t \in F\}$. The part process X^0 is then defined by $X^0 = (X_t^0, \zeta^0, \mathbf{P}_x)_{x \in E_0}$, where

$$\zeta^0(\omega) =: \tau_0(\omega) \quad \text{and} \quad X_t^0(\omega) = \begin{cases} X_t(\omega) & \text{for } t < \zeta^0(\omega); \\ \partial & \text{for } t \geq \zeta^0(\omega). \end{cases} \tag{5.5.3}$$

The process X^0 is an m-symmetric standard process on E_0 by Exercise 3.3.7.

By Theorem 3.3.8 the Dirichlet form $(\mathcal{E}^0, \mathcal{F}^0)$ of the m-symmetric standard process X^0 on $L^2(E_0; m)$ can be identified with the part of $(\mathcal{E}, \mathcal{F})$ on E_0:

$$\mathcal{F}^0 = \{u \in \mathcal{F} : u = 0 \text{ q.e. on } F\} \quad \text{and} \quad \mathcal{E}^0 = \mathcal{E}|_{\mathcal{F}^0 \times \mathcal{F}^0}. \tag{5.5.4}$$

On account of Theorem 3.3.8, $(\mathcal{E}^0, \mathcal{F}^0)$ is a quasi-regular Dirichlet form on $L^2(E_0; m)$ and the standard process X^0 is properly associated with it. Therefore, in view of Theorem 3.1.13, X^0 can be considered as a special Borel standard process by restricting it to the complement of its suitable properly exceptional set if necessary. Theorem 3.3.8 further states that a set $N \subset E_0$ is $\mathcal{E}^0$-polar if and only if it is $\mathcal{E}$-polar, and moreover, under the present convention that every element of $\mathcal{F}$ is assumed to be quasi continuous, the space $\mathcal{F}^0$ defined above can be viewed as the collection of all $\mathcal{E}^0$-quasi-continuous functions belonging to this space.

Since $(\mathcal{E}, \mathcal{F})$ is assumed to be irreducible, the assumption (5.5.1) combined with Theorem 3.5.6(i) and Proposition 2.1.10 immediately implies that $(\mathcal{E}^0, \mathcal{F}^0)$ is a transient Dirichlet form on $L^2(E_0; m)$. By Theorem 3.4.9, its extended Dirichlet space is given by

$$\mathcal{F}_e^0 = \{u \in \mathcal{F}_e : u = 0 \quad \text{q.e. on } F\}. \tag{5.5.5}$$

Every element in $\mathcal{F}_e^0$ is represented by its quasi continuous and hence $\mathcal{E}^0$-quasi-continuous version. $S(E_0)$, $S_0^{(0)}(E_0)$ denote the associated spaces of smooth measures and measures of finite 0-order energy integrals on E_0, respectively. The 0-order potential of $\nu \in S_0^{(0)}(E_0)$ is designated by $U^0\nu$.

The transition function, the resolvent, and the 0-order resolvent of X^0 will be denoted by $\{P_t^0; t \geq 0\}$, $\{R_\alpha^0; \alpha > 0\}$, and R^0, respectively. $\mathcal{S}(E_0)$ (resp. $\mathcal{S}^{\rm pur}(E_0)$) will denote the space of X^0-q.e. excessive functions (resp. X^0-q.e. purely excessive functions) on E_0. Finally we define the energy functional $L^0(f, g), f \in \mathcal{S}^{\rm pur}(E_0)$, $g \in \mathcal{S}(E_0)$, for X^0 by (5.4.4).

For $\alpha \geq 0$, let $\mathbf{H}^\alpha$ denote the α-order hitting measure of F; that is,

$$\mathbf{H}^\alpha(x, B) = \mathbf{E}_x\left[e^{-\alpha\tau_0}\mathbf{1}_B(X_{\tau_0}); \ \tau_0 < \infty\right] \quad \text{for } x \in E_0 \text{ and } B \in \mathcal{B}(F).$$

When $\alpha = 0$, $\mathbf{H}^0$ will simply be denoted by $\mathbf{H}$. Since F is a finely closed set, $\mathbf{H}^\alpha(x, \cdot)$ is carried by F. For $f \in \mathcal{B}_+^*(F)$, define

$$\mathbf{H}^\alpha f(x) := \mathbf{E}_x[e^{-\alpha\tau_0} f(X_{\tau_0}); \ \tau_0 < \infty] \qquad \text{for } x \in E.$$

Exercise 5.5.1. Use the Markov property of X^0 to show the following: for any $\alpha, \beta \geq 0$,

$$\mathbf{H}^\alpha f(x) - \mathbf{H}^\beta f(x) + (\alpha - \beta) R_\alpha^0 \mathbf{H}^\beta f(x) = 0, \quad x \in E_0, \ f \in b\mathcal{B}^*(F).$$

For any $f \in \mathcal{B}_+^*(F)$ and $\alpha \geq 0$, $\mathbf{H}^\alpha f$ is α-excessive with respect to the part process X^0 because, for each $x \in E_0$,

$$e^{-\alpha t} P_t^0 \mathbf{H}^\alpha f(x) = \mathbf{E}_x[e^{-\alpha\tau_0} f(X_{\tau_0}); \ t < \tau_0 < \infty] \uparrow \mathbf{H}^\alpha f(x), \ t \downarrow 0.$$

Moreover,

$$P_t^0 \mathbf{H} f(x) = \mathbf{E}_x \left[f(X_{\tau_0}); t < \tau_0 < \infty \right] \leq \mathbf{H} f(x), \quad x \in E_0. \tag{5.5.6}$$

Therefore, $\mathbf{H}f \in \mathcal{S}^{\mathrm{pur}}(E_0)$ whenever $f \in b\mathcal{B}_+^*(F)$. Furthermore, by Theorem 3.4.8, it holds for $u \in \mathcal{F}_e$ that $\mathbf{H}|u| < \infty$ q.e. on E and $\mathbf{H}u \in \mathcal{F}_e$. In particular, $\mathbf{H}u$ is a member of $\mathcal{S}^{\mathrm{pur}}(E_0)$ for any $u \in (\mathcal{F}_e)_+$.

Define $q(x) := 1 - \mathbf{H}1(x) = \mathbf{P}_x(\tau_0 \geq \zeta)$. For $f, g \in b\mathcal{B}_+(F)$, define

$$U(f \otimes g) := L^0(\mathbf{H}f, \mathbf{H}g) \quad \text{and} \quad V(f) := L^0(\mathbf{H}f, q). \tag{5.5.7}$$

By Theorem 5.4.3, U is a symmetric bimeasure on $F \times F$ and V is a measure on F. U will be called the *Feller measure* for F and V will be called the *supplementary Feller measure* for F. Notice that $q \in \mathcal{S}(E_0)$ but q is not necessarily a member of $\mathcal{S}^{\mathrm{pur}}(E_0)$.

Recall that for $u \in \mathcal{F}_e$, the following Fukushima's decomposition holds uniquely:

$$u(X_t) - u(X_0) = M_t^{[u]} + N_t^{[u]} \qquad \text{for } t \geq 0,$$

where $M^{[u]}$ is a martingale additive functional of X having finite energy and $N^{[u]}$ is a continuous additive functional of X having zero energy. In the sequel, we will use $\mu_{\langle u \rangle}$ to denote the Revuz measure for the predictable quadratic variation $\langle M^{[u]} \rangle$ for the square integrable martingale $M^{[u]}$.

Let $(N(x, dy), H)$ denote a Lévy system for the m-symmetric Hunt process X on E. The Revuz measure of the PCAF H of X will be denoted as μ_H. As before, we define

$$J(dx, dy) := N(x, dy)\mu_H(dx) \quad \text{and} \quad \kappa(dx) := N(x, \{\partial\})\mu_H(dx)$$

as the jumping measure and the killing measure of X (or, equivalently, of $(\mathcal{E}, \mathcal{F})$).

LEMMA 5.5.2. (i) *For any $u \in b\mathcal{F}_e$, let $w = \mathbf{H}(u^2) - (\mathbf{H}u)^2$. Then $w \in b\mathcal{F}_e^0 \cap \mathcal{S}^{\mathrm{pur}}(E_0)$ and $w = U^0 \nu$ with $\nu = \mu_{\langle \mathbf{H}u \rangle}|_{E_0} \in S_0^{(0)}(E_0)$.*
(ii) *Assume that $m(E_0) < \infty$. Then*

$$\mu_{\langle \mathbf{H}u \rangle}(E_0) + \int_{E_0} (\mathbf{H}u)^2(x)\kappa(dx)$$

$$= \lim_{\alpha \to \infty} \alpha(\mathbf{H}^\alpha 1, w)_{E_0} + \int_F u(x)^2 V(dx). \tag{5.5.8}$$

Proof. Since $b\mathcal{F}_e$ is an algebra and $w = 0$ q.e. on F, we have $w \in b\mathcal{F}_e^0$. For $v \in b\mathcal{F}_e^0$, $v\mathbf{H}u \in \mathcal{F}_e^0$ and so by Proposition 3.4.1 and Theorem 4.3.11

$$\mathcal{E}(w, v) = -\mathcal{E}((\mathbf{H}u)^2, v) = \int_E v(x)\mu_{\langle \mathbf{H}u\rangle}(dx) - 2\mathcal{E}(v\mathbf{H}u, \mathbf{H}u)$$
$$= \int_E v(x)\mu_{\langle \mathbf{H}u\rangle}(dx) = \int_{E_0} v(x)\nu(dx).$$

This proves that $w = U^0\nu$.

To prove that $w \in \mathcal{S}^{\mathrm{pur}}(E_0)$, observe that w together with $\mathbf{H}u^2$ and $(\mathbf{H}u)^2$ are bounded and finely continuous relative to X^0 and moreover

$$\lim_{t\to\infty} P_t^0 w(x) = 0 \quad \text{for } x \in E_0.$$

So it suffices to show that for $\alpha > 0$ and $g \in \mathcal{B}_+^*(E_0) \cap L^1(E_0; m)$,

$$\alpha(w, R_\alpha^0 g)_{E_0} \le (w, g)_{E_0}.$$

But $w = U^0\nu$ with $\nu \in S_0(E_0)$. For each $\alpha > 0$, ν has its α-order potential $U_\alpha^0\nu$, which is an element of $\mathcal{F}^0(\subset L^2(E_0; m))$ and α-excessive in the weak sense of Definition 1.2.1 by Lemma 2.3.5. In the same way as in the proof of Lemma 2.3.16 and by using equation (2.3.8), we can see that $U_{\alpha_n}^0\nu$ increases to $w = U^0\nu$ m-a.e. on E_0 as $\alpha_n \downarrow 0$. Hence the desired inequality follows from $(\alpha - \alpha_n)(U_{\alpha_n}^0\nu, R_\alpha^0 g)_{E_0} \le (U_{\alpha_n}^0\nu, g)_{E_0}$ by letting $n \to \infty$.

We next show the identity (5.5.8). We have by Theorem 5.4.3(v), (vi) that

$$\nu(E_0) = L^0(w, 1) = L^0(w, \mathbf{H}1) + L^0(w, q) = L^0(\mathbf{H}1, w) + L^0(w, q). \quad (5.5.9)$$

Thus by (5.4.6) and Exercise 5.5.1,

$$\mu_{\langle \mathbf{H}u\rangle}(E_0) = \lim_{\alpha\to\infty} \alpha(\mathbf{H}^\alpha 1, w)_{E_0} + L^0(w, q). \quad (5.5.10)$$

On the other hand, owing to the assumption that $m(E_0) < \infty$, the boundedness of u, and the symmetry of $\{P_t^0\}$, we have

$$L^0(w, q) = L^0(\mathbf{H}(u^2), q) - \lim_{t\downarrow 0} \frac{1}{t}((\mathbf{H}u)^2, q - P_t^0 q)_{E_0}. \quad (5.5.11)$$

As $q(x) = 1 - \mathbf{H}1(x) = \mathbf{P}_x(\tau_0 = \zeta)$, we have

$$q(x) - P_t^0 q(x) = \mathbf{P}_x(\tau_0 = \zeta) - \mathbf{P}_x(\tau_0 = \zeta,\ t < \tau_0)$$
$$= \mathbf{P}_x(t \ge \tau_0 = \zeta) = q_1(t, x) + q_2(t, x),$$

where

$$\begin{cases} q_1(t, x) = \mathbf{P}_x(t \ge \tau_0 = \zeta,\ X_{\zeta-} \in E_0), \\ q_2(t, x) = \mathbf{P}_x(t \ge \tau_0 = \zeta,\ X_{\zeta-} = \partial). \end{cases}$$

Since $\mathbf{H}u \in \mathcal{F}_e$ and $q_2(t,x) \leq \mathbf{P}_x(t \geq \zeta, X_{\zeta-} = \partial)$, it follows from Proposition 4.2.3 that $\langle (\mathbf{H}u)^2, q_2(t,\cdot)\rangle_{E_0} = o(t)$, while we get from (A.3.33) with $T = t \wedge \tau_0$, $h = 1$, and $f(x,y) = \mathbf{1}_{E_0}(x)\mathbf{1}_{\{\partial\}}(y)$ that

$$q_1(t,x) = \mathbf{E}_x\left[\sum_{s \leq t\wedge\tau_0} \mathbf{1}_{E_0}(X_{s-})\mathbf{1}_{\{\partial\}}(X_s)\right]$$

$$= \mathbf{E}_x\left[\int_0^{t\wedge\tau_0} \mathbf{1}_{E_0}(X_s)N(X_s,\{\partial\})dH_s\right].$$

By noting that $A_t := \int_0^t \mathbf{1}_{E_0}(X_s)N(X_s,\{\partial\})dH_s$ is the PCAF of X with Revuz measure $\mathbf{1}_{E_0}(x)\kappa(dx)$, we get from Proposition 4.1.10

$$\lim_{t\to 0}\frac{1}{t}((\mathbf{H}u)^2, q_1(t,\cdot))_{E_0} = \lim_{t\to 0}\frac{1}{t}\int_{E_0}(\mathbf{H}u)^2(x)\mathbf{E}_x[A_{t\wedge\tau_0}]m(dx)$$

$$= \lim_{t\to 0}\int_{E_0}\left(\frac{1}{t}\int_0^t P_s^0\left((\mathbf{H}u)^2\right)(x)ds\right)\kappa(dx).$$

Since $(\mathbf{H}u)^2$ is dominated by the X^0-excessive function $\mathbf{H}u^2$, $\frac{1}{t}\int_0^t P_s^0((\mathbf{H}u)^2)(x)ds$ is dominated by $\mathbf{H}u^2(x)$, which we claim to be integrable with respect to the killing measure κ. Hence by the dominated convergence theorem, we obtain from the above display and (5.5.11),

$$L^0(w,q) = \int_F u(x)^2 V(dx) - \int_{E_0}(\mathbf{H}u)^2(x)\kappa(dx). \tag{5.5.12}$$

This combined with (5.5.10) proves identity (5.5.8) provided that

$$\int_{E_0}\mathbf{H}u^2(x)\kappa(dx) < \infty.$$

We now show that the last claim is true. Note that as $\mathbf{H}u^2 = (\mathbf{H}u)^2 + U^0\nu$ and $\mathbf{H}u \in \mathcal{F}_e$, it suffices to show that $U^0\nu$ is integrable on E_0 with respect to κ. As κ is a smooth measure of X, by Theorem 2.3.15 there is an $\mathcal{E}$-nest $\{F_k\}$ with $\mathbf{1}_{F_k}\cdot\kappa \in S_0$, $k \geq 1$ and so that $\mathbf{1}_{F_k^0}\cdot\kappa \in S_0(E_0)$ for $F_k^0 = F_k \cap E_0$, $k \geq 0$. The relation (4.1.26) then implies that $x \mapsto \mathbf{E}_x[\int_0^{\tau_0} e^{-\alpha t}\mathbf{1}_{F_k}(X_t)dA_t]$ is an $\mathcal{E}^0$-quasi-continuous version of the potential $U_\alpha^0(\mathbf{1}_{F_k^0}\cdot k) \in \mathcal{F}^0$ for each $\alpha > 0$. Since $\nu = \mu_{\langle\mathbf{H}u\rangle}|_{E_0} \in S_0^{(0)}(E_0)$ and $\mathbf{E}_x[A_{\tau_0}] = q_1(\infty,x) \leq 1$ for q.e. $x \in E_0$, we have

$$\begin{aligned}\int_{E_0} U_\alpha^0\nu(x)\kappa(dx) &= \lim_{k\to\infty}\int_{E_0} U_\alpha^0\nu(x)(\mathbf{1}_{F_k^0}\cdot\kappa)(dx)\\ &= \lim_{k\to\infty}\mathcal{E}_\alpha^0(U_\alpha^0\nu,\ U_\alpha^0(\mathbf{1}_{F_k}\kappa))\\ &= \lim_{k\to\infty}\int_{E_0}\mathbf{E}_x\left[\int_0^{\tau_0} e^{-\alpha t}\mathbf{1}_{F_k}(X_t)dA_t\right]\nu(dx)\\ &\le \nu(E_0) = \mu_{\langle \mathbf{H}u\rangle}(E_0) \le 2\mathcal{E}(u,u) < \infty.\end{aligned}$$

As $\alpha\downarrow 0$, $U_\alpha^0\nu$ increases to $U^0\nu$ q.e. on E_0 on account of the proof of Lemma 2.3.16. By the monotone convergence theorem, we conclude that $\int_{E_0} U^0\nu(x)\kappa(dx) < \infty$. □

For $\alpha > 0$, define the α-order Feller measure U_α on $F\times F$ by

$$U_\alpha(f\otimes g) := \alpha(\mathbf{H}^\alpha f, \mathbf{H}g)_{E_0} \qquad \text{for } f,\ g\in b\mathcal{B}_+(F). \tag{5.5.13}$$

By Exercise 5.5.1, it is easy to see that U_α is symmetric in $f, g \in b\mathcal{B}_+(F)$. It follows from $\mathbf{H}^\alpha g = \mathbf{H}g - \alpha R_\alpha^0\mathbf{H}g$, $g\in b\mathcal{B}_+(F)$, and (5.4.6) that

$$\uparrow \lim_{\alpha\uparrow\infty} U_\alpha(f\otimes g) = U(f\otimes g) \qquad \text{for } f,\ g\in b\mathcal{B}_+(F). \tag{5.5.14}$$

Both U_α and U are bimeasures on $F\times F$, which can be extended to measures on $\mathcal{B}(F\times F)$ in the following way. Choose a sequence $\{D_n\}$ of Borel subsets of E_0 increasing to E_0 with $m_0(D_n) < \infty$ for every $n\ge 1$. For functions u, v on D_n, denote by $(u,v)_n$ the integral $\int_{D_n} u(x)v(x)m_0(dx)$. Then $U_n^\alpha(f,g) := \alpha(\mathbf{H}^\alpha f, \mathbf{H}g)_n$, $f, g\in b\mathcal{B}_+(F)$, is a finite symmetric bimeasure on $F\times F$ which can be extended uniquely to a finite symmetric measure U_n^α on $\mathcal{B}(F\times F)$.[2] The extended measures are increasing in n on $\mathcal{B}(F\times F)$. Then the measure defined by

$$U_\alpha(B) = \uparrow \lim_{n\to\infty} U_n^\alpha(B), \quad B\in\mathcal{B}(F\times F),$$

extends the bimeasure U_α. The constructed measure U_α is easily seen to be increasing in α on $\mathcal{B}(F\times F)$ so that $U(B) = \uparrow \lim_{\alpha\uparrow\infty} U_\alpha(B)$, $B\in\mathcal{B}(F\times F)$, gives a measure on $\mathcal{B}(F\times F)$ extending the symmetric bimeasure U.

The Feller measure U satisfies a property that

$$\text{if a A Borel set } N\subset F \text{ is } \mathcal{E}\text{-polar, then } U(N\times F) = 0, \tag{5.5.15}$$

because if N is $\mathcal{E}$-polar, then it is m-polar with respect to X by Theorem 3.1.3 and $\mathbf{H}\mathbf{1}_N(x) = \mathbf{P}_x(\sigma_N < \infty) = 0$ for m-a.e. $x\in E_0$.

[2] Cf. [37, III:74].

Lemma 5.5.3. *For $\alpha > 0$ and for $u \in b\mathcal{B}^*(F)$, let $w = \mathbf{H}(u^2) - (\mathbf{H}u)^2$. Then*

$$\alpha(\mathbf{H}^\alpha 1, w)_{E_0} + \alpha \int_{E_0 \times F} (\mathbf{H}u(x) - u(\xi))^2 \mathbf{H}^\alpha(x, d\xi) m(dx)$$

$$= \int_{F \times F} (u(\xi) - u(\eta))^2 U_\alpha(d\xi, d\eta) + \alpha(\mathbf{H}^\alpha(u^2), q)_{E_0}.$$

Proof. For $\{D_n\}$ as above,

$$\alpha(\mathbf{H}^\alpha 1,\ \mathbf{H}u^2 - (\mathbf{H}u)^2)_n + \alpha \int_{D_n \times F} (\mathbf{H}u(x) - u(\xi))^2 \mathbf{H}^\alpha(x, d\xi) m(dx)$$

$$= U_n^\alpha(1,\ u^2) - 2U_n^\alpha(u,\ u) + \alpha(\mathbf{H}^\alpha u^2,\ 1)_n$$

$$= \int_{F \times F} (u(\xi) - u(\eta))^2 U_n^\alpha(d\xi, d\eta) + \alpha(\mathbf{H}^\alpha u^2,\ q)_n.$$

It then suffices to let $n \to \infty$. □

The next three results are needed for the proof of Theorem 5.5.8.

Lemma 5.5.4. *Suppose $v \in b\mathcal{F}_e$. Then*

$$\limsup_{t \to 0} \frac{1}{t} \mathbf{E}_{m_0} \left[(v(X_t) - v(X_0))^2;\ t < \tau_0 \right]$$

$$\le \mu^c_{\langle v \rangle}(E_0) + \int_{E_0 \times E_0} (v(x) - v(y))^2 J(dx, dy).$$

Proof. First note that

$$\mathbf{E}_x \left[(M^{[v]}_{t \wedge \tau_0})^2 \right] = \mathbf{E}_x \left[\langle M^{[v]} \rangle_{t \wedge \tau_0} \right].$$

By Proposition 4.1.10, $t \to \langle M^{[v]} \rangle_{t \wedge \tau_0}$ is a PCAF of X^0 and

$$\lim_{t \to 0} \frac{1}{t} \mathbf{E}_{m_0} \left[\langle M^{[v]} \rangle_{t \wedge \tau_0} \right] = \mu_{\langle v \rangle}(E_0)$$

$$= \mu^c_{\langle v \rangle}(E_0) + \int_{E_0 \times E_\partial} (v(x) - v(y))^2 N(x, dy) \mu_H(dx). \tag{5.5.16}$$

Define, for $t \ge 0$, $A_t := (v(X_{\tau_0}) - v(X_{\tau_0 -}))\mathbf{1}_{\{t \ge \tau_0 > 0\}}$, and let A^p be its dual predictable projection. Since A is a process of bounded variation, A^p can be expressed as

$$A^p_t = \int_0^{t \wedge \tau_0} \int_{F_\partial} (v(y) - v(X_s)) N(X_s, dy) dH_s, \tag{5.5.17}$$

on account of the Lévy system formula (A.3.33). By (A.3.38), $M := A - A^p$ is a purely discontinuous square integrable martingale that is orthogonal to

$M^{[v]}_{\cdot\wedge\tau_0} - M$ in the sense that $[M, M^{[v]}_{\cdot\wedge\tau_0} - M] = 0$. We claim that

$$\lim_{t\to 0}\frac{1}{t}\mathbf{E}_{m_0}\left[(A^p_t)^2\right] = 0. \tag{5.5.18}$$

To prove it, for $k \geq 1$, define

$$A^k_t := (v(X_{\tau_0}) - v(X_{\tau_0-}))\mathbf{1}_{\{|v(X_{\tau_0})-v(X_{\tau_0-})|>1/k\}}\,\mathbf{1}_{\{t\geq\tau_0>0\}}$$

and

$$A^{k,p}_t := \int_0^{t\wedge\tau_0}\int_{F_\partial}(v(y) - v(X_s))\mathbf{1}_{\{|v(y)-v(X_s)|>1/k\}}\,N(X_s, dy)dH_s.$$

Then $M^k := A^k - A^{k,p}$ is a purely discontinuous square integrable martingale and $[M - M^k]_t = (A_t - A^k_t)^2$ in view of Section A.3.3(**4**). Therefore, by the Lévy system formula mentioned above,

$$\begin{aligned}
&\limsup_{t\to 0}\frac{1}{t}\mathbf{E}_{m_0}\left[(A^p_t - A^{k,p}_t)^2\right]\\
&\quad\leq \limsup_{t\to 0}\frac{2}{t}\mathbf{E}_{m_0}\left[(M_t - M^k_t)^2\right] + \limsup_{t\to 0}\frac{2}{t}\mathbf{E}_{m_0}\left[(A_t - A^k_t)^2\right]\\
&\quad\leq 4\int_{E_0\times F_\partial}(v(x) - v(y))^2\mathbf{1}_{\{|v(x)-v(y)|\leq 1/k\}}J(dx, dy),
\end{aligned} \tag{5.5.19}$$

which tends to 0 as $k\to\infty$. Now define

$$B^k_t := |v(X_{\tau_0}) - v(X_{\tau_0-})|\mathbf{1}_{\{|v(X_{\tau_0})-v(X_{\tau_0-})|>1/k\}}\,\mathbf{1}_{\{t\geq\tau_0>0\}}$$

and

$$B^{k,p}_t := \int_0^{t\wedge\tau_0}\int_{F_\partial}|v(y) - v(X_s)|\mathbf{1}_{\{|v(y)-v(X_s)|>1/k\}}\,N(X_s, dy)dH_s.$$

Then

$$\mathbf{E}_x[B^{k,p}_t] = \mathbf{E}_x[B^k_t] \leq 2\|v\|_\infty\mathbf{P}_x(t\geq\tau_0) \qquad \text{for } x\in E_0 \tag{5.5.20}$$

and $B^{k,p}$ is a PCAF of X^0 having Revuz measure μ_k with

$$\begin{aligned}
\mu_k(E_0) &= \int_{E_0\times F_\partial}|v(x) - v(y)|\mathbf{1}_{\{|v(x)-v(y)|>1/k\}}N(x, dy)\mu_H(dx)\\
&\leq k\int_{E_0\times F_\partial}(v(x) - v(y))^2N(x, dy)\mu_H(dx) < \infty.
\end{aligned}$$

By the Markov property of X^0, (5.5.20), and Theorem 4.1.1,

$$\begin{aligned}
\mathbf{E}_{m_0}\left[(A_t^{k,p})^2\right] &\leq \mathbf{E}_{m_0}\left[(B_t^{k,p})^2\right] = 2\mathbf{E}_{m_0}\left[\int_0^t \left(\int_s^t dB_r^{k,p}\right) dB_s^{k,p}\right] \\
&= 2\mathbf{E}_{m_0}\left[\int_0^t \mathbf{E}_{X_s^0}\left[B_{t-s}^{k,p}\right] dB_s^{k,p}\right] \\
&\leq 4\|v\|_\infty \mathbf{E}_{m_0}\left[\int_0^t (1 - P_{t-s}^0 1(X_s^0)) dB_s^{k,p}\right] \\
&\leq 4\|v\|_\infty \int_0^t \left(\langle P_s^0 1, \mu_k\rangle - \langle P_s^0 1, P_t^0 1 \cdot \mu_k\rangle\right) ds.
\end{aligned}$$

It then follows from the dominated convergence theorem that

$$\limsup_{t\to 0} \frac{1}{t}\mathbf{E}_{m_0}\left[(A_t^{k,p})^2\right] \leq 4\|v\|_\infty(\mu_k(E_0) - \mu_k(E_0)) = 0.$$

This together with (5.5.19) establishes the claim (5.5.18).

Next by Fukushima's decomposition, (5.5.18), the stated martingale orthogonality between M and $M^{[v]}_{\cdot\wedge\tau_0} - M$, the identity $[M]_t = A_t^2$, and finally by (5.5.16) and the Lévy system formula (A.3.33), we have

$$\begin{aligned}
&\limsup_{t\to 0} \frac{1}{t}\, \mathbf{E}_{m_0}\left[(v(X_t) - v(X_0))^2;\, t < \tau_0\right] \\
&= \limsup_{t\to 0} \frac{1}{t}\, \mathbf{E}_{m_0}\left[(M^{[v]}_{t\wedge\tau_0})^2;\, t < \tau_0\right] \\
&= \limsup_{t\to 0} \frac{1}{t}\, \mathbf{E}_{m_0}\left[(M^{[v]}_{t\wedge\tau_0} - M_t - A_t^p)^2;\, t < \tau_0\right] \\
&\leq \limsup_{t\to 0} \frac{1}{t}\, \mathbf{E}_{m_0}\left[(M^{[v]}_{t\wedge\tau_0} - M_t)^2\right] \\
&= \limsup_{t\to 0} \frac{1}{t}\, \mathbf{E}_{m_0}[(M^{[v]}_{t\wedge\tau_0})^2] - \lim_{t\to 0}\frac{1}{t}\mathbf{E}_{m_0}[A_t^2] \\
&= \mu^c_{\langle v\rangle}(E_0) + \int_{E_0\times E_0} (v(x) - v(y))^2 N(x, dy)\mu_H(dx).
\end{aligned}$$

This completes the proof of the lemma. □

LEMMA 5.5.5. *For $v = \mathbf{H}u$ with $u \in \mathcal{F}_e$, we have*

$$v(X_t) - v(X_0) = M_t^{[v]} \qquad \textit{for } t \in [0, \tau_0].$$

Proof. Denote by F^r the set of all regular points of F. Since $F \setminus F^r$ is $\mathcal{E}$-polar by Theorem 3.1.10, we can choose a properly exceptional set $N \supset F \setminus F^r$. It then holds $\mathbf{P}_x$-a.s. for $x \in E \setminus N$ that $X_{\tau_0} \in F^r \cup \{\partial\}$ and $\tau_0 \circ \theta_{\tau_0}(\omega) = 0$. This means that $v(X_{t\wedge\tau_0}) - v(X_0) = \mathbf{E}_x\left[u(X_{\tau_0})\middle|\mathcal{F}_{t\wedge\tau_0}\right] - v(X_0)$ $\mathbf{P}_x$-a.s. $x \in E \setminus N$; namely, $v(X_{t\wedge\tau_0}) - v(X_0)$ is a martingale relative to $\{\mathcal{F}_{t\wedge\tau_0}\}$ under $\mathbf{P}_x$ for each $x \in E \setminus N$.

Thus if we let $C_t = v(X_{t\wedge\tau_0}) - v(X_0) - M^{[v]}_{t\wedge\tau_0}$, then $C_t = N^{[v]}_{t\wedge\tau_0}$, $t \geq 0$, and $\{C_t\}_{\{t\geq 0\}}$ is a continuous $\mathbf{P}_x$-martingale relative to the filtration $\{\mathcal{F}_{t\wedge\tau_0}\}$ for q.e. $x \in E$. Since $N^{[v]}$ has zero energy, we have for each fixed $t > 0$,

$$\mathbf{E}_{\mathbf{1}_{E_0}\cdot m}\left[\langle C\rangle_t;\ t < \tau_0\right] = \mathbf{E}_{\mathbf{1}_{E_0}\cdot m}\left[\lim_{n\to\infty}\sum_{k=1}^{n}\left(N^{[v]}_{kt/n} - N^{[v]}_{(k-1)t/n}\right)^2;\ t < \tau_0\right]$$

$$\leq \lim_{n\to\infty}\mathbf{E}_m\left[\sum_{k=1}^{n}\left(N^{[v]}_{kt/n} - N^{[v]}_{(k-1)t/n}\right)^2\right] = 0.$$

Hence, for every $t > 0$, $\langle C\rangle_t = 0$ $\mathbf{P}_{\mathbf{1}_{E_0}\cdot m}$-a.e. on $\{t < \tau_0\}$. By the continuity of $\langle C\rangle$, we have $\langle C\rangle_{\tau_0} = 0$ $\mathbf{P}_{\mathbf{1}_{E_0}\cdot m}$-a.e. Thus $\mathbf{P}_{\mathbf{1}_{E_0}\cdot m}$-a.e., $C_t = 0$, namely, $v(X_{t\wedge\tau_0}) - v(X_0) = M^{[v]}_{t\wedge\tau_0}$ for every $t \geq 0$. □

THEOREM 5.5.6. *For $v = \mathbf{H}u$ with $u \in b\mathcal{F}_e$,*

$$\lim_{t\to 0}\frac{1}{t}\mathbf{E}_{m_0}\left[(v(X_t) - v(X_0))^2; t < \tau_0\right]$$

$$= \mu^c_{\langle v\rangle}(E_0) + \int_{E_0\times E_0}(v(x) - v(y))^2 J(dx, dy). \tag{5.5.21}$$

Proof. For $f \in \mathcal{F}^0 \subset \mathcal{F}$, let the Fukushima decomposition of $f(X^0_t) - f(X_0)$ be denoted as $M^{0,[f]}_t + N^{0,[f]}_t$, while the Fukushima decomposition for $f(X_t) - f(X_0)$ is denoted by $M^{[f]}_t + N^{[f]}_t$. Since $f(X_{t\wedge\tau_0}) - f(X_0) = f(X^0_t) - f(X^0_0)$, we have

$$M^{[f]}_{t\wedge\tau_0} - M^{0,[f]}_t = N^{0,[f]}_t - N^{[f]}_{t\wedge\tau_0}, \qquad t \geq 0.$$

In view of Exercise 4.1.9, $M^{0,[f]}_t$ is a square-integrable martingale with respect to the filtration $\{\mathcal{F}_{t\wedge\tau_0}, t \geq 0\}$ and so is $M^{[f]}_{t\wedge\tau_0} - M^{0,[f]}_t$. On the other hand, by the same argument as that in the proof of Lemma 5.5.5, we have

$$\mathbf{E}_{\mathbf{1}_{E_0}\cdot m}\left[\langle M^{[f]}_{\cdot\wedge\tau_0} - M^{0,[f]}\rangle_t; t < \tau_0\right]$$

$$= \mathbf{E}_{\mathbf{1}_{E_0}\cdot m}\left[\langle N^{[f]}_{\cdot\wedge\tau_0} - N^{0,[f]}\rangle_t; t < \tau_0\right] = 0.$$

By the continuity of $\langle M^{[f]}_{\cdot\wedge\tau_0} - M^{0,[f]}\rangle_t$, we conclude that $\langle M^{[f]}_{\cdot\wedge\tau_0} - M^{0,[f]}\rangle_{\tau_0} = 0$ and therefore $M^{[f]}_{t\wedge\tau_0} = M^{0,[f]}_t$. Consequently, $N^{[f]}_{t\wedge\tau_0} = N^{0,[f]}_t$.

Now let $f = \alpha R^0_\alpha \mathbf{1}_{E_0\cap K} \in \mathcal{F}^0$ for a fixed compact set $K \subset E$. Note that $0 \le f \le 1$. By Fukushima's decomposition and Proposition 4.1.10,

$$\begin{aligned}
&\lim_{t\to 0}\frac{1}{t}\mathbf{E}_{m_0}\left[(v(X_t)-v(X_0))^2; t<\tau_0\right] \\
&= \lim_{t\to 0}\frac{1}{t}\mathbf{E}_{m_0}\left[(M^{[v]}_{t\wedge\tau_0})^2; t<\tau_0\right] \\
&\ge \lim_{t\to 0}\frac{1}{t}\mathbf{E}_{m_0}\left[(M^{[v]}_{t\wedge\tau_0})^2 f(X^0_t)\right] \\
&= \lim_{t\to 0}\frac{1}{t}\mathbf{E}_{f\cdot m_0}\left[(M^{[v]}_{t\wedge\tau_0})^2\right] + \lim_{t\to 0}\frac{1}{t}\mathbf{E}_{m_0}\left[(M^{[v]}_{t\wedge\tau_0})^2(f(X^0_t)-f(X^0_0))\right] \\
&= \lim_{t\to 0}\frac{1}{t}\mathbf{E}_{f\cdot m_0}\left[(M^{[v]}_{t\wedge\tau_0})^2\right] + \lim_{t\to 0}\frac{1}{t}\mathbf{E}_{m_0}\left[(M^{[v]}_{t\wedge\tau_0})^2(f(X_{t\wedge\tau_0})-f(X_0))\right] \\
&= \int_{E_0} f(x)\mu_{\langle v\rangle}(dx) + \lim_{t\to 0}\frac{1}{t}\mathbf{E}_{m_0}\left[(M^{[v]}_{t\wedge\tau_0})^2 M^{[f]}_{t\wedge\tau_0}\right] \\
&=: \int_{E_0} f(x)\mu_{\langle v\rangle}(dx) + I. \qquad (5.5.22)
\end{aligned}$$

In the second to the last equality, we used the fact that

$$N^{[f]}_{t\wedge\tau_0} = N^{0,[f]}_t = \int_0^{t\wedge\tau_0} \alpha(f - \mathbf{1}_{E_0\cap K})(X_s)ds,$$

whose absolute value is bounded by αt. By Itô's formula (A.3.39),

$$\begin{aligned}
I &= \lim_{t\to 0}\frac{1}{t}\mathbf{E}_{m_0}\Bigg[\int_0^{t\wedge\tau_0} M^v_{s-}d\langle M^{[v],c} M^{[f],c}\rangle_s \\
&\quad + \sum_{s\le t\wedge\tau_0}((M^{[v]}_s)^2 - (M^{[v]}_{s-})^2)(M^{[f]}_s - M^{[f]}_{s-})\Bigg] \\
&= \lim_{t\to 0}\frac{1}{t}\mathbf{E}_{m_0}\Bigg[\int_0^{t\wedge\tau_0} M^v_{s-}d\langle M^{[v],c} M^{[f],c}\rangle_s \\
&\quad + \sum_{s\le t\wedge\tau_0} 2M^{[v]}_{s-}(M^{[v]}_s - M^{[v]}_{s-})(M^{[f]}_s - M^{[f]}_{s-}) \\
&\quad + \sum_{s\le t\wedge\tau_0} (M^{[v]}_s - M^{[v]}_{s-})^2(M^{[f]}_s - M^{[f]}_{s-})\Bigg].
\end{aligned}$$

Since $M^{[v]}_t = v(X_t) - v(X_0)$ for $t \le \tau_0$ by Lemma 5.5.5, we then have by the Revuz formula in Proposition 4.1.10, the Lévy system formula (A.3.33), and the symmetry of $J(dx, dy)$,

$$
\begin{aligned}
I = & \lim_{t\to 0} \frac{1}{t} \mathbf{E}_{m_0} \Bigg[\int_0^{t\wedge\tau_0} (v(X_s) - v(X_0)) d\langle M^{[v],c} M^{[f],c}\rangle_s \\
& + \sum_{s\le t\wedge\tau_0} 2(v(X_{s-}) - v(X_0))(v(X_s) - v(X_{s-}))(f(X_s) - f(X_{s-})) \\
& + \sum_{s\le t\wedge\tau_0} (v(X_s) - v(X_{s-}))^2 (f(X_s) - f(X_{s-})) \Bigg] \\
= & \, 0 + \lim_{t\to 0} \frac{1}{t} \mathbf{E}_{m_0} \left[2\int_0^{t\wedge\tau_0} v(X_s) \int_{E_\partial} (v(X_s) - v(y))(f(X_s) - f(y)) N(X_s, dy) dH_s \right] \\
& - \lim_{t\to 0} \frac{1}{t} \mathbf{E}_{v\cdot m_0} \left[2\int_0^{t\wedge\tau_0} \int_{E_\partial} (v(X_s) - v(y))(f(X_s) - f(y)) N(X_s, dy) dH_s \right] \\
& + \lim_{t\to 0} \frac{1}{t} \mathbf{E}_{m_0} \left[\int_0^{t\wedge\tau_0} \int_{E_\partial} (v(y) - v(X_s))^2 (f(y) - f(X_s)) N(X_s, dy) dH_s \right] \\
= & \int_{E_0\times E_\partial} (v(y) - v(x))^2 (f(y) - f(x)) N(x, dy) d\mu_H(dx) \\
= & - \int_{E_0\times F_\partial} f(x)(v(x) - v(y))^2 N(x, dy) d\mu_H(dx).
\end{aligned}
$$

Thus we have, by (5.5.22),

$$
\begin{aligned}
& \lim_{t\to 0} \frac{1}{t} \mathbf{E}_{m_0} \left[(v(X_t) - v(X_0))^2; t < \tau_0 \right] \\
& \quad \ge \int_{E_0} f(x) \mu_{\langle v\rangle}(dx) - \int_{E_0\times F_\partial} f(x)(v(x) - v(y))^2 N(x, dy) d\mu_H(dx) \\
& \quad = \int_{E_0} f(x) \mu^c_{\langle v\rangle}(dx) + \int_{E_0\times E_0} f(x)(v(x) - v(y))^2 N(x, dy) d\mu_H(dx).
\end{aligned}
$$

Since this is true for all $f = \alpha R^0_\alpha \mathbf{1}_{E_0\cap K}$ where $\alpha > 0$ and K is a compact subset of E, we conclude by first letting $K \uparrow E$ and then $\alpha \uparrow \infty$ that

$$
\begin{aligned}
& \lim_{t\to 0} \frac{1}{t} \mathbf{E}_{m_0} \left[(v(X_t) - v(X_0))^2; t < \tau_0 \right] \\
& \qquad \ge \mu^c_{\langle v\rangle}(E_0) + \int_{E_0\times E_0} (v(x) - v(y))^2 N(x, dy) d\mu_H(dx).
\end{aligned}
$$

This together with Lemma 5.5.4 completes the proof of the theorem. □

The proof of Theorem 5.5.6 can be refined to show the following.

Exercise 5.5.7. Show that (5.5.21) holds for every $v \in \mathcal{F}_e$.

The next theorem relates Feller measures to the jumping measure J and the killing measure κ of $(\mathcal{E}, \mathcal{F})$.

THEOREM 5.5.8. *Assume that $m(E_0) < \infty$. For any $u \in \mathcal{F}_e$,*

$$\mu_{\langle \mathbf{H}u \rangle}(E_0) + \int_{E_0 \times F} (\mathbf{H}u(x) - u(\xi))^2 J(dx, d\xi) + \int_{E_0} (\mathbf{H}u)^2(x)\kappa(dx)$$

$$= \int_{F \times F} (u(\xi) - u(\eta))^2 U(d\xi, d\eta) + 2 \int_F u(\xi)^2 V(d\xi).$$

Proof. Without loss of generality, we may assume that $u \in b\mathcal{F}_e$ since otherwise we consider $u_n = ((-n) \vee u) \wedge n$ and then pass $n \to \infty$. For $\alpha > 0$, by Lemma 5.5.3,

$$\int_{F \times F} (u(\xi) - u(\eta))^2 U_\alpha(d\xi, d\eta) + \alpha(\mathbf{H}^\alpha(u^2), q)_{E_0}$$

$$= \alpha(\mathbf{H}^\alpha 1, w)_{E_0} + \alpha \int_{E_0 \times F} (\mathbf{H}u(x) - u(\xi))^2 \mathbf{H}^\alpha(x, d\xi) m(dx), \qquad (5.5.23)$$

where $w = \mathbf{H}(u^2) - (\mathbf{H}u)^2$ and $q = 1 - \mathbf{H}1$.

It follows from (5.5.14) that

$$\lim_{\alpha \to \infty} \int_{F \times F} (u(\xi) - u(\eta))^2 U_\alpha(d\xi, d\eta) = \int_{F \times F} (u(\xi) - u(\eta))^2 U(d\xi, d\eta). \qquad (5.5.24)$$

By definition (5.5.7) and (5.4.6), we have

$$\lim_{\alpha \to \infty} \alpha(\mathbf{H}^\alpha(u^2), q)_{E_0} = \int_F u(\xi)^2 V(d\xi). \qquad (5.5.25)$$

The limit in α of the first term of the right hand side of (5.5.23) has the expression as is exhibited in (5.5.8) under the assumption $m(E_0) < \infty$. Moreover, the last term in (5.5.23) can be rewritten as

$$I_\alpha := \alpha \mathbf{E}_m \left[e^{-\alpha \tau_0} (\mathbf{H}u(X_0) - u(X_{\tau_0}))^2 \mathbf{1}_{\{\tau_0 < \zeta\}} \right].$$

Hence it only remains to prove that

$$\lim_{\alpha \to \infty} I_\alpha = \int_{E_0 \times F} (\mathbf{H}u(x) - u(\xi))^2 J(dx, d\xi). \qquad (5.5.26)$$

Let $v := \mathbf{H}u$, which is a bounded function in $\mathcal{F}_e$. By the argument given after (5.5.11), we have

$$\lim_{t \to 0} \frac{1}{t} \int_{E_0} v(x)^2 \mathbf{P}_x(t \geq \tau_0; \tau_0 = \zeta) m_0(dx) = \int_{E_0} v(x)^2 \kappa(dx). \qquad (5.5.27)$$

Note that $u(X_{\tau_0}) = \mathbf{H}u(X_{\tau_0})$ $\mathbf{P}_{m_0}$-a.s. By a change of variable $r = \alpha s$,

$$\begin{aligned}
&\lim_{\alpha\to\infty} \alpha \mathbf{E}_{m_0}\left[e^{-\alpha\tau_0}(u(X_{\tau_0}) - \mathbf{H}u(X_0))^2;\ \tau_0 < \zeta\right] \\
&\quad = \lim_{\alpha\to\infty} \alpha \mathbf{E}_{m_0}\left[\int_0^\infty \alpha e^{-\alpha s}(v(X_{\tau_0}) - v(X_0))^2 1_{\{s\geq\tau_0;\tau_0<\zeta\}} ds\right] \\
&\quad = \lim_{\alpha\to\infty} \alpha \mathbf{E}_{m_0}\left[\int_0^\infty e^{-r}(v(X_{\tau_0}) - v(X_0))^2 1_{\{r/\alpha\geq\tau_0;\tau_0<\zeta\}} dt\right] \\
&\quad = \lim_{\alpha\to\infty} \int_0^\infty re^{-r}\,(\alpha/r)\mathbf{E}_{m_0}\left[(v(X_{\tau_0}) - v(X_0))^2 1_{\{r/\alpha\geq\tau_0;\tau_0<\zeta\}}\right] dt.
\end{aligned} \tag{5.5.28}$$

By Lemma 5.5.5, $v(X_{t\wedge\tau_0}) - v(X_0) = M^{[v]}_{t\wedge\tau_0}$. Then by (5.5.16), (5.5.27), and Theorem 5.5.6, we have

$$\begin{aligned}
&\lim_{t\to0}\frac{1}{t}\mathbf{E}_{m_0}\left[(v(X_{\tau_0}) - v(X_0))^2 1_{\{t\geq\tau_0;\tau_0<\zeta\}}\right] \\
&= \lim_{t\to0}\frac{1}{t}\left(\mathbf{E}_{m_0}\left[(v(X_{t\wedge\tau_0}) - v(X_0))^2\, 1_{\{t\geq\tau_0\}}\right] - \mathbf{P}_{v^2\cdot m_0}(t\geq\tau_0;\tau_0=\zeta)\right) \\
&= \lim_{t\to0}\frac{1}{t}\left(\mathbf{E}_{m_0}\left[(M^{[v]}_{t\wedge\tau_0})^2\right] - \mathbf{E}_{m_0}\left[(v(X_t) - v(X_0))^2 1_{\{t<\tau_0\}}\right]\right) \\
&\quad - \int_{E_0} v(x)^2\kappa(dx) = \int_{E_0\times F}(v(x) - v(y))^2 J(dx,dy).
\end{aligned}$$

This together with (5.5.28) and the dominated convergence theorem establish the claim (5.5.26) and hence the theorem. □

THEOREM 5.5.9. *For any $u \in \mathcal{F}_e$,*

$$\begin{aligned}
\mathcal{E}(\mathbf{H}u,\mathbf{H}u) = \frac{1}{2}\mu^c_{\langle\mathbf{H}u\rangle}(F) + \frac{1}{2}\int_{F\times F}(u(x) - u(y))^2(U(dx,dy)J(dx,dy)) \\
+ \int_F u(x)^2(V(dx) + \kappa(dx)).
\end{aligned}$$

Proof. By (4.3.7), (4.3.8), (4.3.12), and (4.3.15),

$$\begin{aligned}
\mathcal{E}(\mathbf{H}u,\mathbf{H}u) &= \tfrac{1}{2}\mu_{\langle\mathbf{H}u\rangle}(E) + \tfrac{1}{2}\mu^k_{\langle\mathbf{H}u\rangle}(E) \\
&= \tfrac{1}{2}\mu_{\langle\mathbf{H}u\rangle}(E_0) + \tfrac{1}{2}\mu_{\langle\mathbf{H}u\rangle}(F) + \tfrac{1}{2}\mu^k_{\langle\mathbf{H}u\rangle}(E) \\
&= \tfrac{1}{2}\mu_{\langle\mathbf{H}u\rangle}(E_0) + \tfrac{1}{2}\mu^k_{\langle\mathbf{H}u\rangle}(E_0) + \tfrac{1}{2}\mu^c_{\langle\mathbf{H}u\rangle}(F) + \tfrac{1}{2}\mu^j_{\langle\mathbf{H}u\rangle}(F) + \mu^k_{\langle\mathbf{H}u\rangle}(F),
\end{aligned}$$

where

$$\begin{cases} \mu^j_{\langle \mathbf{H}u \rangle}(dx) = \int_E (\mathbf{H}u(x) - \mathbf{H}u(y))^2 \, J(dx, dy), \\ \mu^k_{\langle \mathbf{H}u \rangle}(dx) = (\mathbf{H}u)^2(x)\kappa(dx). \end{cases}$$

The assertion of the theorem now follows from Theorem 5.5.8 under the condition $m(E_0) < \infty$.

The only place where the condition $m(E_0) < \infty$ is used is to ensure that the symmetry can be applied in (5.5.11). By using a time change method, however, the assumption $m(E_0) < \infty$ can be dropped as follows.

Choose a strictly positive Borel function φ on E_0 such that $\int_{E_0} \varphi(x)m(dx) < \infty$. Let

$$\nu(dx) := \big(\mathbf{1}_{E_0}(x)\varphi(x) + \mathbf{1}_F(x)\big)\, m(dx).$$

Then $\nu \in \overset{\circ}{S}_E$, namely, the quasi support of ν equals the whole space E by Corollary 3.3.6. Let A^ν be the PCAF of X with Revuz measure ν and $Z = (X_{\sigma_t}, \mathbf{P}_x)$ be the time-changed process of X on E by means of the right continuous inverse σ_t of A^ν. Then Z is a ν-symmetric right process on E whose Dirichlet form $(\mathcal{E}^Z, \mathcal{F}^Z)$ on $L^2(E;\nu)$ is regular, and its extended Dirichlet space $(\mathcal{E}^Z, \mathcal{F}^Z_e)$ is identical with $(\mathcal{E}, \mathcal{F}_e)$ in view of (5.2.17) and (5.2.18). Furthermore, $(\mathcal{E}^Z, \mathcal{F}^Z)$ has the same set of quasi notions as $(\mathcal{E}, \mathcal{F})$ by virtue of Theorem 5.2.11.

We can readily deduce from these identifications the following conclusions: $(\mathcal{E}^Z, \mathcal{F}^Z_e)$ and $(\mathcal{E}, \mathcal{F})$ admit the same Beurling-Deny decomposition (4.3.1) holding for u, v in their common extended Dirichlet space with each component in (4.3.1) being identical. In view of (4.3.34), the local mutual energy measures $\mu^c_{\langle u,v \rangle}$ are also the same for them. Moreover, for $u \in \mathcal{F}_e (= \mathcal{F}^Z_e)$, two functions $\mathbf{H}u, \mathbf{H}^Z u \in \mathcal{F}_e$ possess the same values on F and $\mathcal{E}$-orthogonal to the common subspace $\mathcal{F}^0_e$ defined by (5.5.5) so that $\mathbf{H}u = \mathbf{H}^Z u$ on account of Theorem 5.2.16.

Since $\nu(E_0) < \infty$, Theorem 5.5.9 holds true for the Dirichlet form $(\mathcal{E}^Z, \mathcal{F}^Z)$ on $L^2(E;\nu)$. But due to the preceding observations, nothing changes in the above theorem if we replace $(\mathcal{E}^Z, \mathcal{F}^Z)$ with $(\mathcal{E}, \mathcal{F})$ except for the Feller measures U and V.

We now show that Feller measures are also invariant under the current time change. For this, let U^X, U^Z, V^X, V^Z, and L^X, L^Z denote the Feller measures and energy functionals for the part processes of X and Z on E_0, respectively. We will use $R^{X,0}$ and $R^{Z,0}$ to denote the 0-order resolvent of the part process X^0 and Z^0 in E_0, respectively. Clearly for a Borel measurable function $f \geq 0$ on E_0 and $x \in E_0$,

$$\begin{aligned} R^{Z,0}f(x) &= \mathbf{E}_x \left[\int_0^\infty f(X^0_{\sigma_t})dt \right] = \mathbf{E}_x \left[\int_0^\infty f(X^0_t)\varphi(X^0_t)dt \right] \\ &= R^{X,0}(\varphi f)(x). \end{aligned}$$

By (5.5.7) above and Theorem 5.4.3(iv), we have for $u, v \in b\mathcal{B}_+(F)$,

$$\begin{aligned} U^X(u \otimes v) &= L^X(\mathbf{H}u, \mathbf{H}v) \\ &= \sup\left\{\int_{E_0} \mathbf{H}v(x)g(x)m(dx) : \ g \in \mathcal{B}^*_+(E_0), \ R^{X,0}g \leq \mathbf{H}u \ [m] \text{ on } E_0\right\}. \end{aligned}$$

Hence

$$\begin{aligned} U^Z(u \otimes v) &= \sup\left\{\int_{E_0} \mathbf{H}v(x)g(x)\nu(dx) : \ g \in \mathcal{B}^*_+(E_0), R^{Z,0}g \leq \mathbf{H}u[\nu]\text{on}E_0\right\} \\ &= \sup\left\{\int_{E_0} \mathbf{H}v(x)g(x)\varphi(x)m(dx) : \ g \in \mathcal{B}^*_+(E_0), \ R^{X,0}(g\varphi) \leq \mathbf{H}u \ [m] \text{ on } E_0\right\} \\ &= U^X(u \otimes v). \end{aligned}$$

By replacing $\mathbf{H}v$ with $q = 1 - \mathbf{H}1$ above, we also get $V^X = V^Z$. □

5.6. CHARACTERIZATION OF TIME-CHANGED PROCESSES

We now study the trace of the process X on the quasi closed set F satisfying (5.5.1). Recall the space S_F of smooth measures having quasi support F as is defined by (5.2.3). Fix a measure $\mu \in S_F$, and let A^μ be the PCAF of X with Revuz measure μ. Let N be a properly exceptional set in the definition of A^μ. The support of A^μ is defined as

$$\left\{x \in E \setminus N : \ \mathbf{P}_x(\inf\{t > 0 : A^\mu_t > 0\} = 0) = 1\right\}.$$

By Theorem 5.2.1, the support of A^μ is nearly Borel, finely closed, and equal to F q.e. Therefore, we may and shall assume that the support of A^μ is just F in accordance with the preceding convention. Note that $\mathbf{H}u$ with $u \in \mathcal{F}_e$ remains the same q.e. if F is replaced by another quasi closed set that equals F q.e.

Let Y be the time-changed process of the Hunt process X by the right continuous inverse of A^μ. Then by Theorem A.3.9 and Theorem 5.2.1, Y is a μ-symmetric right process on F. By Theorem 5.2.2 and Theorem 5.2.15, the Dirichlet form $(\check{\mathcal{E}}, \check{\mathcal{F}})$ of Y on $L^2(F; \mu)$ and its extended Dirichlet space $(\check{\mathcal{F}}_e, \check{\mathcal{E}})$ admit the expression

$$\begin{cases} \check{\mathcal{F}}_e = \mathcal{F}_e|_F, \\ \check{\mathcal{F}} = \mathcal{F}_e\big|_F \cap L^2(F; \mu), \\ \check{\mathcal{E}}(u, v) = \mathcal{E}(\mathbf{H}u, \mathbf{H}v) \qquad \text{for } u, v \in \check{\mathcal{F}}_e. \end{cases} \tag{5.6.1}$$

According to Theorem 5.2.7, the Dirichlet form $(\check{\mathcal{E}}, \check{\mathcal{F}})$ is quasi-regular on $L^2(F; \mu)$ with which Y is properly associated. Recall our convention that functions in $\mathcal{F}_e$ are always represented by their quasi continuous versions. Expression (5.6.1) particularly states that the extended Dirichlet space $\check{\mathcal{F}}_e$ of $(\check{\mathcal{F}}, \check{\mathcal{E}})$ coincides with $\mathcal{F}_e|_F$, independently of the choice of $\mu \in S_F$.

Thus Theorem 5.5.9 can be restated as follows:

COROLLARY 5.6.1. *For* $u \in \check{\mathcal{F}}_e$,

$$\check{\mathcal{E}}(u,u) = \frac{1}{2}\mu^c_{\langle \mathbf{H}u \rangle}(F) + \frac{1}{2}\int_{F\times F}(u(x)-u(y))^2(U(dx,dy)+J(dx,dy)) + \int_F u(x)^2(V(dx)+\kappa(dx)). \tag{5.6.2}$$

We now choose $\mu \in \overset{\circ}{S}_F$ ($\subset S_F$), namely, μ is a positive Radon measure on E charging no $\mathcal{E}$-polar set whose quasi support is F. Such a measure μ exists in view of Lemma 5.2.9. Let F^* be the topological support of μ. F^* is then a closed set, $\mu(F^*\Delta F) = 0$, and $L^2(F;\mu) = L^2(F^*;\mu)$. Denote by $(\mathcal{E}^*, \mathcal{F}^*)$ the form $(\check{\mathcal{E}}, \check{\mathcal{F}})$ being considered as a Dirichlet form on $L^2(F^*;\mu)$ in place of $L^2(F;\mu)$. According to Theorem 5.2.13, $(\mathcal{E}^*, \mathcal{F}^*)$ becomes a regular Dirichlet form on $L^2(F^*;\mu)$ and the set $F^* \setminus F$ is polar with respect to this form. Moreover, by a suitable choice of a properly exceptional set N of A^μ and by redefining F to be the support of A^μ, it holds that $F \subset F^*$ and the time-changed process Y living on F becomes just a Hunt process properly associated with the regular Dirichlet form $(\mathcal{E}^*, \mathcal{F}^*)$.

Let us show that the decomposition of $(\check{\mathcal{E}}, \check{\mathcal{F}})$ in (5.6.2) is the Beurling-Deny decomposition for $(\mathcal{E}^*, \mathcal{F}^*)$.

THEOREM 5.6.2. *The bilinear form* $(u,v) \mapsto \mu^c_{\langle \mathbf{H}u, \mathbf{H}v\rangle}(F)$ *has the strongly local property on* $\mathcal{F}^*$*; that is, if* $u, v \in \check{\mathcal{F}} \cap C_c(F^*)$ *and* u *is constant in a neighborhood of* supp$[v]$, *then* $\mu^c_{\langle \mathbf{H}u, \mathbf{H}v\rangle}(F) = 0$. *In other words,*

$$\mathcal{E}^{*c}(u,v) = \tfrac{1}{2}\mu^c_{\langle \mathbf{H}u, \mathbf{H}v\rangle}(F) \quad \textit{for } u, v \in \mathcal{F}^*_e.$$

Proof. Let $u \in b\check{\mathcal{F}} \cap C_c(F^*)$ such that $u = c$ for some $c \in \mathbb{R}$ on a relative open subset I of F^*. $I = D \cap F^*$ for some open set $D \subset E$. Take a relatively compact open set D_1 with $\overline{D}_1 \subset D$ and let $I_1 = D_1 \cap F^*$. Since $(\mathcal{E}, \mathcal{F})$ is a regular Dirichlet space on $L^2(E;m)$, there is a $\varphi \in \mathcal{F} \cap C_c(E)$ such that $\varphi = 1$ on D_1 and $\varphi = 0$ on D^c. Let $v = c\varphi + (1-\varphi)\mathbf{H}u$. Then $v \in b\mathcal{F}_e$ and v is constant on D_1. Hence $\mu^c_{\langle v\rangle}(D_1) = 0$ by Proposition 4.3.1 and thus we conclude

$$\mu^c_{\langle v\rangle}(I_1) = 0. \tag{5.6.3}$$

Since $v = u$ q.e. on F, we have $\mathbf{H}v = \mathbf{H}u$ q.e. Define $v_0 = v - \mathbf{H}v$, which is in $b\mathcal{F}^0_e$. Since F is quasi closed, there exists an $\mathcal{E}$-nest $\{K_n, n \geq 1\}$ so that $F \cap K_n$ is a closed set. Let $(\mathcal{E}, \mathcal{F}^{E\setminus(F\cap K_n)})$ be the Dirichlet space for the part process of X killed upon leaving $E \setminus (F \cap K_n)$. Clearly $v_0 \in b\mathcal{F}_e^0 \subset b\mathcal{F}_e^{E\setminus(F\cap K_n)}$. Since $(\mathcal{E}, \mathcal{F}^{E\setminus(F\cap K_n)})$ is regular on $L^2(E \setminus (F \cap K_n); m)$ by Theorem 3.3.9 and $\mu^c_{\langle\psi\rangle}(F \cap K_n) = 0$ for any $\psi \in \mathcal{F} \cap C_c(E \setminus (F \cap K_n))$ by Proposition 4.3.1, we have $\mu^c_{\langle v_0\rangle}(F \cap K_n) = 0$, and passing n to infinity we get $\mu^c_{\langle v_0\rangle}(F) = 0$. In particular, $\mu^c_{\langle v_0\rangle}(I \cap F) = 0$.

On the other hand,

$$\begin{aligned}\mu^c_{\langle v\rangle}(I_1 \cap F) &= \mu^c_{\langle \mathbf{H}v+v_0\rangle}(I_1 \cap F)\\ &= \mu^c_{\langle \mathbf{H}v\rangle}(I_1 \cap F) + 2\mu^c_{\langle \mathbf{H}v,v_0\rangle}(I_1 \cap F) + \mu^c_{\langle v_0\rangle}(I_1 \cap F)\\ &= \mu^c_{\langle \mathbf{H}v\rangle}(I_1 \cap F) = \mu^c_{\langle \mathbf{H}u\rangle}(I_1 \cap F).\end{aligned}$$

Thus by (5.6.3), $\mu^c_{\langle \mathbf{H}u\rangle}(I_1 \cap F) = 0$. By letting $D_1 \uparrow D$, we get

$$\mu^c_{\langle \mathbf{H}u\rangle}(I \cap F) = 0. \tag{5.6.4}$$

Now for $u, v \in \check{\mathcal{F}} \cap C_c(F^*)$ such that u is constant in a neighborhood of supp$[v]$, we let $F_1 = \text{supp}[v]$ and $F_2 = F^* \setminus \text{supp}[v]$. By (5.6.4),

$$\mu^c_{\langle \mathbf{H}u\rangle}(F_1 \cap F) = 0 \quad \text{and} \quad \mu^c_{\langle \mathbf{H}v\rangle}(F_2 \cap F) = 0.$$

Since $F \subset F^*$ q.e. and $\mu^c_{\langle \mathbf{H}u,\mathbf{H}v\rangle}$ does not charge on $\mathcal{E}$-polar sets, it follows then that

$$\begin{aligned}\left|\mu^c_{\langle \mathbf{H}u,\mathbf{H}v\rangle}(F)\right| &= \left|\mu^c_{\langle \mathbf{H}u,\mathbf{H}v\rangle}(F_1 \cap F) + \mu^c_{\langle \mathbf{H}u,\mathbf{H}v\rangle}(F_2 \cap F)\right|\\ &\le \sqrt{\mu^c_{\langle \mathbf{H}u\rangle}(F_1 \cap F)\,\mu^c_{\langle \mathbf{H}v\rangle}(F_1 \cap F)}\\ &\quad + \sqrt{\mu^c_{\langle \mathbf{H}u\rangle}(F_2 \cap F)\,\mu^c_{\langle \mathbf{H}v\rangle}(F_2 \cap F)}\\ &= 0.\end{aligned}$$

This proves the theorem. □

As is described above, the time-changed process Y of X with respect to a PCAF A^μ (under a suitable choice of an exceptional set of A^μ) becomes a μ-symmetric Hunt process on $F(\subset F^*)$ which is properly associated with the regular Dirichlet form $(\mathcal{E}^*, \mathcal{F}^*)$.

Let $(\check{N}, \check{H})$ be a Lévy system of the Hunt process Y. Then the jumping measure $\check{J}$ and the killing measure $\check{\kappa}$ of Y are defined by

$$\check{J}(dx, dy) = \check{N}(x, dy)\check{\mu}_{\check{H}}(dx), \quad \check{\kappa}(dx) = \check{N}(x, \{\partial\})\check{\mu}_{\check{H}}(dx),$$

where $\check{\mu}_{\check{H}}$ denotes the Revuz measure on F^* of the PCAF $\check{H}$ of Y relative to μ.

By virtue of Theorem 4.3.3, $\check{J}$ and $\check{\kappa}$ are nothing but the measures appearing in the second and third terms of the Beurling-Deny decomposition (4.3.1) of the regular Dirichlet form $\mathcal{E}^*$. On the other hand, Theorem 5.6.2 assures that the formula (5.6.2) represents the Beurling-Deny decomposition of $(\mathcal{E}^*, \mathcal{F}^*)$. Thus we arrive at the following identification of the jumping and killing measures of the time-changed process Y.

THEOREM 5.6.3. *It holds that*

$$\begin{cases} \check{J}(dx, dy) = U(dx, dy) + J(dx, dy)|_{F\times F}, \\ \check{\kappa}(dx) = V(dx) + \kappa(dx)|_F. \end{cases} \tag{5.6.5}$$

The jumping measure $\check{J}$ of Y is carried on $F \times F$ and F is a subset of F^* as is stated above. On account of the Beurling-Deny decomposition of the regular Dirichlet form $\mathcal{E}^*$, it holds that if $\varphi, \psi \in \check{\mathcal{F}}_+ \cap C_c(F^*)$ are of disjoint support, then

$$\int_{F^*\times F^*} \varphi(\xi)\psi(\eta)\check{J}(d\xi, d\eta) = -\check{\mathcal{E}}(\varphi, \psi) < \infty,$$

which means that $\check{J}$ is a σ-finite measure on $F \times F \setminus d$. By combining this with (5.6.5) and (5.5.15), we can draw the following properties of the Feller measure U:

[**(U.1)**] $U|_{F\times F\setminus d}$ is σ-finite .
[**(U.2)**] For $B \in \mathcal{B}(F \times F \setminus d)$, $U(B) = 0$ whenever its projection on the factor F is $\mathcal{E}$-polar.

For $\varphi \in \mathcal{B}(F)$ that is finite q.e. on F, define

$$\mathbf{C}(\varphi, \varphi) = \frac{1}{2}\int_{F\times F} (\varphi(\xi) - \varphi(\eta))^2 U(d\xi, d\eta) + \int_F \varphi(\xi)^2 V(d\xi)(\leq \infty). \tag{5.6.6}$$

$\mathbf{C}(\varphi, \varphi)$ is well defined in view of **(U.1), (U.2)** and, if such functions φ_1, φ_2 coincide q.e. on F, then $\mathbf{C}(\varphi_1, \varphi_1) = \mathbf{C}(\varphi_2, \varphi_2)$. We call $\mathbf{C}(\varphi, \varphi)$ the *Douglas integral* of φ in accordance with the integral appearing at the end of Section 2.2.4 and in Section 5.3(**3**°) when X is the reflecting Brownian motion on the unit disk and F is the unit circle on the plane.

Thus we have established the following representation of the extended trace Dirichlet space $(\check{\mathcal{F}}_e, \check{\mathcal{E}})$: For any $\varphi \in \check{\mathcal{F}}_e$, φ is of finite Douglas integral and

$$\check{\mathcal{E}}(\varphi, \varphi) = \frac{1}{2}\mu^c_{\langle \mathbf{H}\varphi\rangle}(F) + \mathbf{C}(\varphi, \varphi) + \frac{1}{2}\int_{F\times F} (\varphi(\xi) - \varphi(\eta))^2 J(d\xi, d\eta)$$
$$+ \int_F \varphi(\xi)^2 \kappa(d\xi). \tag{5.6.7}$$

Moreover, the first term $\frac{1}{2}\mu^c_{\langle \mathbf{H}\varphi\rangle}(F)$ on the right hand side of (5.6.7) is the strongly local part of $\check{\mathcal{E}}(\varphi, \varphi)$ being regarded as a regular Dirichlet form on $L^2(F^*; \mu)$.

5.7. EXCURSIONS, EXIT SYSTEM, AND FELLER MEASURES

In the preceding two sections, we have seen that Feller measures play a key role in characterizing trace Dirichlet forms on the quasi closed set F. They were defined by (5.5.7) analytically. An objective of this section is to give probabilistic descriptions of Feller measures in terms of excursions of X away from F. Moreover, the related notion of the entrance law on $E_0 = E \setminus F$ will also play an important role in Sections 7.5 and 7.6 on one-point extensions.

We maintain the setting made in Section 5.5. We call a family of σ-finite measures $\{\nu_t,\ t > 0\}$ on E_0 an *X^0-entrance law* if

$$\nu_t P_s^0 = \nu_{t+s} \qquad \text{for } t, s > 0. \tag{5.7.1}$$

Lemma 5.7.1. (i) *For any $f \in b\mathcal{B}_+(E)$, there exists a unique X^0-entrance law $\{\mu_t^f, t > 0\}$ such that*

$$\mathbf{H}f \cdot m_0 = \int_0^\infty \mu_t^f \, dt. \tag{5.7.2}$$

(ii) *Denote by $\{\bar{\mu}_\alpha^f, \alpha > 0\}$ the Laplace transform of $\{\mu_t^f, t > 0\}$. Then, for any $v \in \mathcal{B}_+(E_0)$,*

$$\int_t^\infty \langle \mu_s^f, v\rangle ds = (P_t^0 \mathbf{H}f, v) \quad \textit{for every } t > 0, \tag{5.7.3}$$

$$\langle \bar{\mu}_\alpha^f, v\rangle = (\mathbf{H}_\alpha f, v) \quad \textit{for } \alpha > 0, \tag{5.7.4}$$

and

$$\int_0^t \langle \mu_s^f, v\rangle ds = \mathbf{E}_{v\cdot m_0}\left[f(X_{\sigma_F}); \sigma_F \le t\right] \quad \textit{for every } t > 0. \tag{5.7.5}$$

(iii) *For any $f \in b\mathcal{B}_+(E)$ and X^0-excessive function v on E_0, $\langle \mu_t^f, v\rangle$ is a right continuous decreasing function in $t > 0$ and*

$$L^{(0)}(\mathbf{H}f, v) = \uparrow \lim_{t\downarrow 0} \langle \mu_t^f, v\rangle.$$

In particular,

$$U(f\otimes g) = \uparrow \lim_{t\downarrow 0} \langle \mu_t^f, \mathbf{H}g\rangle \quad \textit{and} \quad V(f) = \uparrow \lim_{t\downarrow 0} \langle \mu_t^f, q\rangle. \tag{5.7.6}$$

Proof. (i) Since $\mathbf{H}f \in \mathcal{S}^{\rm pur}(E_0)$, $\mathbf{H}f \cdot m_0$ is an excessive measure of X^0 which is pure in the sense that

$$\lim_{t\to\infty} \langle (\mathbf{H}f \cdot m_0)P_t^0,\ v\rangle = \lim_{t\to\infty} (P_t^0 \mathbf{H}f, v) = 0 \qquad \text{for } v \in L^1(E_0, m_0).$$

Since X^0 is transient, the assertion follows from a theorem due to P. J. Fitzsimmons which we refer the readers to [81, Theorem 5.21].

(ii) For $v \in L^1(E_0; m_0)$, we have

$$\int_t^\infty \langle \mu_s^f, v\rangle ds = \int_0^\infty \langle \mu_{t+s}^f, v\rangle dt = \int_0^\infty \langle \mu_s^f, P_t^0 v\rangle ds$$
$$= \langle \mathbf{H}f \cdot m_0, P_t^0 v\rangle = (P_t^0 \mathbf{H}f, v)$$

and

$$\langle \mu_t^f, v\rangle = -\frac{d}{dt}(P_t^0 \mathbf{H}f, v) \qquad \text{for a.e. } t > 0.$$

Hence

$$\langle \bar{\mu}_\alpha^f, v\rangle = -\int_0^\infty e^{-\alpha t}\frac{d}{dt}(P_t^0 \mathbf{H}f, v)dt$$
$$= -e^{-\alpha t}(P_t^0 \mathbf{H}f, v)\Big|_0^\infty - \alpha\int_0^\infty e^{-\alpha t}(P_t^0 \mathbf{H}f, v)dt$$
$$= (\mathbf{H}f - \alpha G_\alpha^0 \mathbf{H}f, v) = (\mathbf{H}_\alpha f, v).$$

Equation (5.7.5) follows from (5.7.2), (5.7.3), and (5.5.6).

(iii) If v is X^0-excessive, then $\langle \mu_{t+s}, v\rangle = \{\mu_t, P_s^0 v\rangle \uparrow \langle \mu_t, v\rangle$ as $s \downarrow 0$. For $v \in L^1(E_0; m_0)$, we get from (i) and (ii) that

$$(\mathbf{H}f - P_t^0 \mathbf{H}f, v) = \int_0^t \langle \mu_s^f, v\rangle ds,$$

which extends to any X^0-excessive v as in the proof of Lemma 5.4.2(ii). Thus we get the desired identity. □

For a set $B \subset E$, define $\widehat{\sigma}_B = \inf\{t > 0 : X_{t-} \in B\}$. From [13, p. 59] or [73, Theorem A.2.3], we have then

Lemma 5.7.2. *$\sigma_F \leq \widehat{\sigma}_F$ a.s., or, equivalently,*

$$\inf\{t > 0 : X_t \in F\} = \inf\{t > 0 : X_t \in F \text{ or } X_{t-} \in F\} \quad a.s.$$

We will be concerned with a random subset $M(\omega)$ of $[0, \zeta(\omega))$ defined by

$$M(\omega) = \{t \in [0, \zeta(\omega)) : X_t(\omega) \in F \text{ or } X_{t-}(\omega) \in F\}, \qquad \omega \in \Omega, \tag{5.7.7}$$

with the convention that $X_{0-}(\omega) = X_0(\omega)$. $M(\omega)$ is *homogeneous* on $(0, \infty)$ in the sense that

$$(M - s) \cap (0, \infty) = (M \circ \theta_s) \cap (0, \infty), \qquad s > 0. \tag{5.7.8}$$

LEMMA 5.7.3. *$M(\omega)$ is a relatively closed subset of $[0, \zeta(\omega))$ $\mathbf{P}_x$-a.s. for $x \in E \setminus N_0$, where N_0 is some Borel properly exceptional set.*

Proof. Let $f(x) = \mathbf{E}_x[e^{-\sigma_F}]$, $x \in E$. f is 1-excessive by Lemma A.2.4 and so it is quasi continuous by Theorem 3.3.3. By virtue of Theorem 3.1.7, f is right continuous and left continuous as well along the sample path X_t in the sense that (3.1.9), (3.1.10), and (3.1.11) hold true $\mathbf{P}_x$-a.s. for every $x \in E \setminus N$ for some $\mathcal{E}$-polar set N. Denote by F^r the set of all regular points of F. $F^r \subset F$ and $F \setminus F^r$ is $\mathcal{E}$-polar by Theorem 3.1.10 and Lemma A.2.18. Choose a Borel properly exceptional set N_0 containing $N \cup (F \setminus F^r)$. Then

$$F \setminus N_0 = \{x \in E \setminus N_0 : f(x) = 1\}. \tag{5.7.9}$$

$X|_{E \setminus N_0}$ is a Hunt process and there is $\widetilde{\Omega} \in \mathcal{F}_\infty$ with $\mathbf{P}_x(\widetilde{\Omega}) = 1$ for every $x \in E \setminus N_0$, such that, for every $\omega \in \widetilde{\Omega}$, $t \in M(\omega)$ if and only if either $f(X_t(\omega)) = 1$ or $f(X_{t-}(\omega)) = 1$. Furthermore, if $t_n \in M(\omega)$ decreases to $t \geq 0$, then $f(X_t(\omega)) = 1$, and if $t_n \in M(\omega)$ increases to $t < \zeta(\omega)$, then $f(X_{t-}(\omega)) = 1$. This means that $M(\omega)$ is a relatively closed subset of $[0, \zeta(\omega))$. □

By the above lemma, the set $[0, \zeta(\omega)) \setminus M(\omega)$ is a disjoint union of relatively open subintervals of $[0, \zeta(\omega))$. Each subinterval is called an *excursion interval.* Each segment of the sample path $X_\cdot(\omega)$ corresponding to an excursion (time) interval is said to be an *excursion* away from the set F. Denote by $I(\omega)$ the collection of all left endpoints of excursion intervals. Since $M(\omega)$ is homogeneous on $(0, \infty)$ in the sense of (5.7.8), so is $I(\omega)$.

Define

$$R = R(\omega) = \inf\{s > 0 : s \in M(\omega)\}.$$

We have then from Lemma 5.7.2

$$R = \sigma_F, \tag{5.7.10}$$

and from (5.7.9)

$$F \setminus N_0 = \{x \in E \setminus N_0 : \mathbf{P}_x(R = 0) = 1\}.$$

DEFINITION 5.7.4. A system $(\mathbf{P}_x^*, L + J)$ is called an *exit system* relative to the homogeneous random set M, or, for (X, F) if

[(1)] $\mathbf{P}^*$ is a kernel from $E \setminus N_0$ to Ω such that $\mathbf{P}_x^* = \mathbf{P}_x$ for $x \in E_0 \setminus N_0$ and, for any $x \in E \setminus N_0$,

$$\mathbf{P}_x^*(X_0 \neq x) = 0, \quad \mathbf{P}_x^*(R = 0) = 0, \quad \textit{and} \quad \mathbf{E}_x^*\left[1 - e^{-R}\right] \leq 1.$$

[(2)] L is a PCAF in the strict sense of $X|_{E \setminus N_0}$ carried on $F \setminus N_0$ and $dJ_t = \sum_{s \in I : X_s \in E \setminus F} \varepsilon_s(dt)$, where $\varepsilon_s(dt)$ is the unit atomic measure on $\mathbb{R}$ concentrated at the point $\{s\}$.

[(**3**)] For any positive optional process $\{Z_s, s \geq 0\}$ and non-negative random variable Γ,

$$\mathbf{E}_x\left[\sum_{s\in I} Z_s \cdot (\Gamma \circ \theta_s)\right] = \mathbf{E}_x\left[\int_0^\infty Z_s \cdot \mathbf{E}^*_{X_s}(\Gamma)d(L_s + J_s)\right] \qquad (5.7.11)$$

for $x \in E \setminus N_0$, where $\mathbf{E}^*_x$ stands for the integration with respect to the measure $\mathbf{P}^*_x$.

We refer the readers to B. Maisonneuve [120] (see also [38]) for a proof of the existence of an exit system relative to the present homogeneous random set M. Since $\mathbf{P}^*_x(R > t) \leq (1 - e^{-t})^{-1}\mathbf{E}^*_x[1 - e^{-R}]$, $t > 0$, we see from (**1**) that $\mathbf{P}^*_x$ is σ-finite for $x \in F \setminus N_0$. We also notice that the following Markov property can be deduced for an exit system (cf. [120, §5]): for any non-negative $\Gamma \in \mathcal{F}^0_s$ and non-negative $\Lambda \in \mathcal{F}^0_\infty$,

$$\mathbf{E}^*_x[\Gamma \cdot \Lambda \circ \theta_s] = \mathbf{E}_x\left[\Gamma\, \mathbf{E}^*_{X_s}[\Lambda]\right], \quad s > 0,\ x \in E \setminus N_0. \qquad (5.7.12)$$

Define the *entrance law* $\{Q^*_t(x, \cdot),\ t > 0,\ x \in E \setminus N_0\}$ *for the exit system* $(\mathbf{P}^*_x, L + J)$ by

$$Q^*_t g(x) := \mathbf{E}^*_x\left[g(X_t); t < R\right], \quad g \in \mathcal{B}_+(E). \qquad (5.7.13)$$

Note that $Q^*_t g(x) = P^0_t g(x)$ for $x \in E_0 \setminus N_0$. By the Markov property, we see that for each $x \in F \setminus N_0$, $\nu_t := Q^*_t(x, \cdot)$ enjoys the X^0-entrance law property $\nu_t P^0_s = \nu_{t+s}$.

Since X is a Hunt process, it has a Lévy system (N, H). According to Section A.3.4, $N(x, dy)$ is a kernel on $(E_\partial, \mathcal{B}(E_\partial))$ and H is a PCAF in the strict sense of X such that the identity (A.3.31) holds for any non-negative Borel function f on $E \times E_\partial$ that vanishes on the diagonal and is extended to be zero elsewhere. The Revuz measure of H (with respect to the excessive measure m) will be denoted as μ_H.

THEOREM 5.7.5. *For any Borel subset* $B \subset E_0$ *and* $f \in \mathcal{B}_b(F)$,

$$\mu^f_t(B) = \int_F f(x)Q^*_t(x, B)\mu_L(dx) + \int_{F\times(E\setminus F)} f(x)P^0_t \mathbf{1}_B(y)N(x, dy)\mu_H(dx), \qquad (5.7.14)$$

where μ_L *is the Revuz measure of the PCAF* L *with respect to the measure* m.

Proof. Put $\mathbf{Q}^*_\alpha g(x) = \int_0^\infty e^{-\alpha t} Q^*_t g(x)dt$, $g \in \mathcal{B}_b(E)$. By virtue of the identity (5.7.11), we have for any $v \in \mathcal{B}_b(E)$ vanishing on F and for every $x \in E \setminus N_0$,

$$\mathbf{H}_\alpha R_\alpha v(x) = R_\alpha v(x) - R^0_\alpha v(x) = \mathbf{E}_x\left[\int_R^\infty e^{-\alpha t} v(X_t) 1_{M^c}(t)dt\right]$$
$$= \mathbf{E}_x\left[\sum_{s\in I}\int_s^{s+R\circ\theta_s} e^{-\alpha t} v(X_t)dt\right] = \mathbf{E}_x\left[\sum_{s\in I} e^{-\alpha s}\int_0^R e^{-\alpha t} v(X_t)dt \circ \theta_s\right]$$
$$= \mathbf{E}_x\left(\int_0^\infty e^{-\alpha s}\mathbf{E}^*_{X_s}\left[\int_0^R e^{-\alpha t} v(X_t)dt\right] d(L_s + J_s)\right).$$

Therefore, we have for any $f \in \mathcal{B}^+_b(E)$,

$$(f, \mathbf{H}_\alpha R_\alpha v) = \mathbf{E}_{f\cdot m}\left[\int_0^\infty e^{-\alpha s}\mathbf{Q}^*_\alpha v(X_s)d(L_s + J_s)\right]. \tag{5.7.15}$$

On the other hand, owing to the formula (4.1.3), we obtain

$$\mathbf{E}_{f\cdot m}\left[\int_0^\infty e^{-\alpha s}\mathbf{Q}^*_\alpha v(X_s)dL_s\right] = \langle R_\alpha f, \mathbf{Q}^*_\alpha v \cdot \mu_L\rangle. \tag{5.7.16}$$

Furthermore, since $\mathbf{Q}^*_\alpha(x, \cdot) = R^0_\alpha(x, \cdot)$ for $x \in E_0$, we have

$$\mathbf{E}_{f\cdot m}\left[\int_0^\infty e^{-\alpha s}\mathbf{Q}^*_\alpha v(X_s)dJ_s\right]$$
$$= \mathbf{E}_{f\cdot m}\left[\sum_{s\in I, X_s\in E\setminus F} e^{-\alpha s} R^0_\alpha v(X_s)\right]$$
$$= \mathbf{E}_{f\cdot m}\left[\sum_s e^{-\alpha s}\mathbf{1}_F(X_{s-})\mathbf{1}_{E\setminus F}(X_s) R^0_\alpha v(X_s)\right]$$
$$= \mathbf{E}_{f\cdot m}\left[\int_0^\infty e^{-\alpha s}\mathbf{1}_F(X_s)\int_{E\setminus F} N(X_s, dz) R^0_\alpha v(z)dH_s\right]$$
$$= \left\langle R_\alpha f, \mathbf{1}_F(\cdot)\int_{E\setminus F} N(\cdot, dz)R^0_\alpha v(z)\cdot \mu_H\right\rangle. \tag{5.7.17}$$

Here in the last equality we used (4.1.3) again.

We get from (5.7.15), (5.7.16), and (5.7.17)

$$(\mathbf{H}_\alpha R_\alpha f, v) = \left\langle R_\alpha f, \mathbf{Q}^*_\alpha v \cdot \mu_L + \mathbf{1}_F(\cdot) \int_{E\setminus F} N(\cdot, dz) R^0_\alpha v(z) \cdot \mu_H \right\rangle.$$

Since this identity holds for an arbitrary $f \in \mathcal{B}_b(E)$, we obtain for any $\beta > 0$, $f \in bC_+(E) \cap \mathcal{F}$, and $v \in L^1_+(E_0; m)$,

$$(\mathbf{H}_\alpha R_\beta f, v) = \left\langle R_\beta f, \mathbf{Q}^*_\alpha v \cdot \mu_L + \mathbf{1}_F(\cdot) \int_{E\setminus F} N(\cdot, dz) R^0_\alpha v(z) \cdot \mu_H \right\rangle. \tag{5.7.18}$$

Multiplying β on both side of (5.7.18) with $f = 1$ and then letting $\beta \to \infty$, we have, by monotone convergence theorem,

$$\left\langle 1, \mathbf{Q}^*_\alpha v \cdot \mu_L + \mathbf{1}_F(\cdot) \int_{E\setminus F} N(\cdot, dz) R^0_\alpha v(z) \cdot \mu_H \right\rangle$$

$$= \langle \mathbf{H}_\alpha 1, v \rangle \le \int_{E_0} v(x) m(dx) < \infty.$$

Now multiplying both side of (5.7.18) by β and letting $\beta \to \infty$ in the above equation, we have, by bounded convergence theorem,

$$(\mathbf{H}_\alpha f,\, v) = \left\langle f, \mathbf{Q}^*_\alpha v \cdot \mu_L + \mathbf{1}_F(\cdot) \int_{E\setminus F} N(\cdot, dz) R^0_\alpha v(z) \cdot \mu_H \right\rangle.$$

This combined with (5.7.4) proves the desired identity (5.7.14) since the above display is nothing but the Laplace transform of (5.7.14). □

The following theorem describes Feller measures as joint distributions of the starting and end points of excursions of X away from the quasi closed set F.

THEOREM 5.7.6. *The Feller measures are expressed by the exit system as*

$$U(dx, dy) = \mu_L(dx) \mathbf{P}^*_x (X_{\sigma_F} \in dy)$$

$$+ \mu_H(dx)\big|_F \int_{E\setminus F} N(x, dz) \mathbf{P}_z (X_{\sigma_F} \in dy) \tag{5.7.19}$$

and

$$V(dx) = \mu_L(dx) \mathbf{P}^*_x (\zeta > 0,\ \sigma_F = \infty)$$

$$+ \mu_H(dx)\big|_F \int_{E\setminus F} N(x, dz) \mathbf{P}_z (\sigma_F = \infty). \tag{5.7.20}$$

Proof. It follows from (5.7.6), (5.7.14), and the definition of $Q_t^*(x, dy)$ that for any $f, g \in \mathcal{B}_+(E)$,

$$\int_{F\times F} f(x)g(y)U(dx, dy) = \lim_{t\downarrow 0}\langle \mu_t^f, \mathbf{H}g\rangle$$

$$= \lim_{t\downarrow 0}\left[\int_{F\times E_0} f(x)\mathbf{H}g(y)Q_t^*(x, dy)\mu_L(dx) + \int_{F\times E_0} f(x)P_t^0\mathbf{H}g(y)N(x, dy)\mu_H(dx)\right]$$

$$= \lim_{t\downarrow 0}\int_F f(x)\mathbf{E}_x^*[\mathbf{H}g(X_t); t < R]\mu_L(dx) + \int_{F\times E_0} f(x)\mathbf{H}g(y)N(x, dy)\mu_H(dx)$$

$$= \lim_{t\downarrow 0}\int_F f(x)\mathbf{E}_x^*[g(X_{\sigma_F})\circ\theta_t; t < R]\mu_L(dx) + \int_{F\times E_0} f(x)\mathbf{H}g(y)N(x, dy)\mu_H(dx)$$

$$= \lim_{t\downarrow 0}\int_F f(x)\mathbf{E}_x^*[g(X_{\sigma_F}); t < R]\mu_L(dx) + \int_{F\times E_0} f(x)\mathbf{H}g(y)N(x, dy)\mu_H(dx)$$

$$= \int_F f(x)\mathbf{E}_x^*[g(X_{\sigma_F})]\mu_L(dx) + \int_{F\times E_0} f(x)\mathbf{E}_y[g(X_{\sigma_F})]N(x, dy)\mu_H(dx).$$

In the third to the last equality, we used the Markov property (5.7.12). Identity (5.7.19) now follows.

The proof for (5.7.20) is similar to that for (5.7.19). Note that for $x \in E_0$, $q(x) = 1 - \mathbf{H}1(x) = \mathbf{P}_x(\sigma_F = \infty)$. Thus by (5.7.6),

$$\int_F f(x)V(dx) = \lim_{t\downarrow 0}\langle \mu_t^f, q\rangle$$

$$= \lim_{t\downarrow 0}\left[\int_{F\times E_0} f(x)q(y)Q_t^*(x, dy)\mu_L(dx) + \int_{F\times E_0} f(x)P_t^0 q(y)N(x, dy)\mu_H(dx)\right]$$

$$= \lim_{t\downarrow 0}\int_F f(x)\mathbf{E}_x^*[q(X_t); t < \zeta\wedge R]\mu_L(dx) + \int_{F\times E_0} f(x)q(y)N(x, dy)\mu_H(dx)$$

$$= \lim_{t\downarrow 0}\int_F f(x)\mathbf{E}_x^*[1_{\{\sigma_F=\infty\}}\circ\theta_t; t < \zeta\wedge R]\mu_L(dx) + \int_{F\times E_0} f(x)q(y)N(x, dy)\mu_H(dx)$$

$$= \int_F f(x)\mathbf{P}_x^*(\zeta > 0,\ \sigma_F = \infty)\mu_L(dx) + \int_{F\times E_0} f(x)\mathbf{P}_y(\sigma_F = \infty)N(x, dy)\mu_H(dx).$$

for any $f \in \mathcal{B}_+(E)$. This establishes (5.7.20). □

5.8. MORE EXAMPLES

The first four examples of Section 5.3 concern the Dirichlet forms associated with time-changed processes of Brownian motion and reflecting Brownian motion. In this section, we give more examples of them with emphasis on the corresponding Feller measures.

(1°) Trace of RBM on the boundary

Let $D \subset \mathbb{R}^n$ be a bounded domain with C^3-boundary ∂D for $n \geq 2$. The surface measure on ∂D will be denoted by σ. Let $X = (X_t, \mathbf{P}_x)$ be the standard Brownian motion on $\mathbb{R}^n$ and X^D be the absorbing Brownian motion on D as has been considered in Example 3.5.9. X^D is the part process of X on D and it admits the transition density $p_t^0(x, y)$ expressed as

$$p_t^0(x, y) = g_t(x - y) - \mathbf{E}_x[g_{t-\tau}(X_\tau - y); \tau < t],$$

where g_t is given by (2.2.13) and τ is the exit time of X from D (the hitting time of $\mathbb{R}^n \setminus D$). We let

$$h(s, x, \xi) = \frac{1}{2} \frac{\partial p_s^0(x, \xi)}{\partial \mathbf{n}_\xi}, \qquad x \in D, \ \xi \in \partial D,$$

where $\mathbf{n}_\xi$ denotes the inward normal vector at ξ.

We show that h is a density function of the joint distribution of (τ, X_τ) under $\mathbf{P}_x$ for $x \in D$:

$$\mathbf{P}_x(\tau \in ds, X_\tau \in d\xi) = h(s, x, \xi) ds\sigma(d\xi). \tag{5.8.1}$$

To see this, let u be the harmonic function on D with a smooth boundary function f on ∂D, namely, $u(x) = \mathbf{E}_x[f(X_\tau); \tau < \infty]$, $x \in D$. Since $p_t^0(x, y)$ is symmetric in x, y, and a fundamental solution of the heat equation $\frac{\partial}{\partial t} u(t, x) = \frac{1}{2}\Delta_x u(t, x)$ with the Dirichlet boundary condition on ∂D, Green's formula yields

$$\begin{aligned} &\frac{1}{2}\int_0^t ds \int_{\partial D} h(s, x, \xi) f(\xi) d\sigma(\xi) \\ &\quad = \frac{1}{2}\int_0^t ds \int_D (\Delta_y p_s^0(x, y)) u(y) dy \\ &\quad = u(x) - P_t^0 u(x) = \mathbf{E}_x[f(X_\tau); \tau \leq t], \end{aligned}$$

holding for $x \in D$, $t > 0$, proving (5.8.1).

Therefore, the hitting distribution $H(x, d\xi)$ and the α-order hitting distribution $H_\alpha(x, d\xi)$ admit density functions given for $x \in D$, $\xi \in \partial D$ by

$$K(x, \xi) = \int_0^\infty h(s, x, \xi) ds, \quad K_\alpha(x, \xi) = \int_0^\infty e^{-\alpha s} h(s, x, \xi) ds.$$

K is called the Poisson kernel for D. Define

$$U_\alpha(\xi,\eta) = \alpha \int_D K_\alpha(x,\xi)K(x,\eta)dx, \quad \xi,\ \eta \in \partial D,$$

which can be easily rewritten as

$$U_\alpha(\xi,\eta) = \frac{1}{4}\int_0^\infty (1-e^{-\alpha t})\frac{\partial^2 p_t^0(\xi,\eta)}{\partial \mathbf{n}_\xi \partial \mathbf{n}_\eta}dt.$$

Let $U(\xi,\eta) = \lim_{\alpha\to\infty} U_\alpha(\xi,\eta)$. Then

$$U(\xi,\eta) = \frac{1}{2}\frac{\partial K(\xi,\eta)}{\partial \mathbf{n}_\xi}, \quad \xi,\ \eta \in \partial D. \tag{5.8.2}$$

This function $U(\xi,\eta)$ is the density function of the Feller measure U with respect to $\sigma\times\sigma$. Indeed we get from (5.5.14)

$$\begin{aligned} U(f\otimes g) &= \lim_{\alpha\to\infty} \alpha(H^\alpha f, Hg)_D \\ &= \lim_{\alpha\to\infty}\int_{\partial D\times\partial D} U_\alpha(\xi,\eta)f(\xi)g(\eta)\sigma(d\xi)\sigma(d\eta) \\ &= \int_{\partial D\times\partial D} U(\xi,\eta)f(\xi)g(\eta)\sigma(d\xi)\sigma(d\eta), \quad f,g\in\mathcal{B}_+(\partial D). \end{aligned}$$

We call $U(\xi,\eta)$ defined by (5.8.2) the *Feller kernel* for the domain D.

Consider the special case where D equals $B_R = \{x\in\mathbb{R}^n : |x| < R\}$ and σ is the uniform spherical measure on $\Sigma_R = \{x\in\mathbb{R}^n : |x| = R\}$. Expression (5.8.2) is then reduced to

$$U(\xi,\eta) = \frac{\Gamma(n/2)}{2\pi^{n/2}}\frac{1}{|\xi-\eta|^n}, \quad \xi,\eta\in\Sigma_R,\ \xi\neq\eta, \tag{5.8.3}$$

because the Poisson kernel for B_R equals $\frac{1}{\Omega_n R}\frac{R^2-|x|^2}{|x-\eta|^n}$, $x\in B_R$, $\eta\in\Sigma_R$, with $\Omega_n = 2\pi^{n/2}\Gamma(n/2)^{-1}$ being the area of Σ_1. When $n=2$, (5.8.3) can be rewritten as

$$U(\xi,\eta) = [4\pi R^2(1-\cos(\theta-\theta'))]^{-1}, \quad \xi = Re^{i\theta},\ \eta = Re^{i\theta'},$$

which coincides with the kernel (2.2.59) for $R=1$. The expression of $U(\xi,\eta)$ in (5.8.3) depends only on the distance of ξ and η, and it even coincides with the corresponding kernel (5.3.16) for the unbounded upper half-space $\mathbb{R}^n$ for $n\geq 2$.

Let us consider the reflecting Brownian motion $X^r = (X_t^r, \mathbf{P}_x^r)$ on $\overline{D} = D\cup\partial D$, as has been considered in Example 3.5.9. X^r is associated with the regular Dirichlet form $(\frac{1}{2}\mathbf{D}, H^1(D))$ on $L^2(\overline{D})$. The part process of X^r on D is identical with the absorbing Brownian motion X^D so that X^r and the Brownian motion X have the same hitting distribution H, H^α from D to ∂D as well as the Feller measure U and supplementally Feller measure V for the domain D.

In particular, X^r admits (5.8.2) as its Feller kernel. Since X^r is irreducible recurrent, we see from (3.5.19) that $\mathbf{H}1(x) = 1,\ x \in D$, and hence $q(x) = 1 - \mathbf{H}1(x) = 0,\ x \in D$, and V is vanishing. Further, the energy measure for $(\frac{1}{2}\mathbf{D}, H^1(D))$ is expressed as

$$\mu_{\langle u \rangle}(B) = \int_B |\nabla u(x)|^2 dx, \quad u \in bH^1(D),\ B \in \mathcal{B}(\overline{D}),$$

which does not charge on ∂D. Moreover, the jumping and killing measures of X^r are vanishing. Therefore, we get from Theorem 5.5.9

$$\frac{1}{2}\int_D |\nabla \mathbf{H} f(x)|^2 dx = \frac{1}{2}\int_{\partial D \times \partial D} (f(\xi) - f(\eta))^2 U(\xi,\eta)\sigma(d\xi)\sigma(d\eta) \tag{5.8.4}$$

for any $f \in \widetilde{H}_e^1(D)|_{\partial D}$, where $U(\xi,\eta)$ is given by (5.8.2) and $\widetilde{H}_e^1(D)$ denotes the collection of quasi continuous functions in $H_e^1(D)$. This particularly recovers the expression of $\check{\mathcal{E}}$ in (5.3.9) when D is the planer unit disk.

(2°) Trace of Brownian motion on a hypersurface

As in the preceding example, we consider the standard Brownian motion $X = (X_t, \mathbf{P}_x)$ on $\mathbb{R}^n$ with $n \geq 2$. Let S be a C^3 compact hypersurface so that $E_0 = \mathbb{R}^n \setminus S$ is the union of the interior domain D_i and exterior domain D_e. We denote by σ the surface measure on S. Further, $\partial/\partial \mathbf{n}_\xi^i$ and $\partial/\partial \mathbf{n}_\xi^e$ will denote the inward normal and outward normal derivative at $\xi \in S$ from the view of D_i, respectively. We consider the Poisson kernel $K(x,\xi),\ x \in E_0,\ \xi \in S$, and the escape probability of X from S defined by

$$q(x) = 1 - \mathbf{H}1(x) = \mathbf{P}_x(\sigma_S = \infty), \qquad x \in E_0.$$

Note that $q(x) = 0$ if $n = 2$ because X is then irreducible recurrent and (3.5.19) holds. If $n \geq 3$, then $q(x)$ is positive but only for $x \in D_e$.

We shall show that the Feller measure U and the supplementary Feller measure V with respect to X and S have densities $U(\xi,\eta)$ and $v(\xi)$ relative to $\sigma \times \sigma$ and σ, respectively, expressed as

$$U(\xi,\eta) = U^i(\xi,\eta) + U^e(\xi,\eta), \qquad \xi,\eta \in S,\ \xi \neq \eta$$

$$U^i(\xi,\eta) = \frac{1}{2}\frac{\partial K(\xi,\eta)}{\partial \mathbf{n}_\xi^i}, \qquad U^e(\xi,\eta) = \frac{1}{2}\frac{\partial K(\xi,\eta)}{\partial \mathbf{n}_\xi^e}, \tag{5.8.5}$$

$$v(\xi) = \frac{1}{2}\frac{\partial q(\xi)}{\partial \mathbf{n}_\xi^e}, \qquad \xi \in S. \tag{5.8.6}$$

In the special case of $S = \Sigma_R$, (5.8.5) is reduced to

$$U(\xi,\eta) = \frac{\Gamma(n/2)}{\pi^{n/2}}\frac{1}{|\xi - \eta|^n}, \quad \xi,\eta \in \Sigma_R,\ \xi \neq \eta, \tag{5.8.7}$$

because $U^i(\xi,\eta)$ equals (5.8.3), while the Poisson kernel for the exterior domain $\{|x|>R\}$ equals $\frac{1}{\Omega_n R}\frac{|x|^2-R^2}{|x-\eta|^n}$. In this special case with $n\geq 3$, (5.8.6) is also reduced to a constant function

$$v(\xi)=\frac{n-2}{2r}, \qquad \xi\in\Sigma_R, \tag{5.8.8}$$

because $q(x)=1-\frac{R^{n-2}}{|x|^{n-2}}$ for $|x|>R$.

For the proof of (5.8.5) and (5.8.6), it suffices to show that

$$\frac{1}{2}\int_{D_e}|\nabla \mathbf{H}f(x)|^2dx=\frac{1}{2}\int_{S\times S}(f(\xi)-f(\eta))^2U^e(\xi,\eta)\sigma(d\xi)\sigma(d\eta) + \int_S f(\xi)^2 v(\xi)\sigma(d\xi) \tag{5.8.9}$$

holds for any $f\in\widetilde{H}^1_e(\mathbb{R}^n)$ on account of (5.8.4), where $\widetilde{H}^1_e(\mathbb{R}^n)$ denotes the collection of quasi continuous functions in $H^1_e(\mathbb{R}^n)$.

To this end, take a sufficiently large N such that the ball $B_N=\{x\in\mathbb{R}^n : |x|<N\}$ contains the surface S. The bounded domain $D_{e,N}=D_e\cap B_N$ has the boundary consisting of two disjoint set S and Σ_N. We denote by $\mathbf{D}^e(u,v)$ and $\mathbf{D}^{e,N}(u,v)$ the integrals of $\nabla u(x)\cdot\nabla v(x)$ with respect to dx over D_e and $D_{e,N}$, respectively. By virtue of (5.8.4), we have, for $f,g\in bC^1(\mathbb{R}^n)$,

$$\frac{1}{2}\mathbf{D}^{e,N}(\mathbf{H}^Nf,\mathbf{H}^Ng) =\frac{1}{2}\int_{\partial D_{e,N}\times\partial D_{e,N}}(f(\xi)-f(\eta))(g(\xi)-g(\eta))U^{e,N}(\xi,\eta)\sigma(d\xi)\sigma(d\eta), \tag{5.8.10}$$

where $\mathbf{H}^N$ denotes the hitting distribution of X from $D_{e,N}$ to $\partial D_{e,N}$ and $U^{e,N}$ denotes the Feller kernel (5.8.2) for the domain $D_{e,N}$.

As will be observed in the next example (**3°**), we have $\widetilde{H}^1_e(\mathbb{R}^n)|_{D_{e,N}}\subset\widetilde{H}^1_e(D_{e,N})$. So we can derive from (5.8.10) that for any $f\in\widetilde{H}^1_e(\mathbb{R}^n)$ vanishing on $\mathbb{R}^n\setminus B_N$,

$$\begin{aligned}\frac{1}{2}\mathbf{D}^e(f,f)&\geq\frac{1}{2}\mathbf{D}^{e,N}(f,f)\geq\frac{1}{2}\mathbf{D}^{e,N}(\mathbf{H}^Nf,\mathbf{H}^Nf)\\ &=\frac{1}{2}\int_{S\times S}(f(\xi)-f(\eta))^2U^{e,N}(\xi,\eta)\sigma(d\xi)\sigma(d\eta)\\ &\quad+\int_S f(\xi)^2v^N(\xi)\sigma(d\xi),\end{aligned} \tag{5.8.11}$$

where

$$v^N(\xi)=\int_{\Sigma_N}U^{e,N}(\xi,\eta)\sigma(d\eta), \quad \xi\in S.$$

If we let $q^N(x) = \mathbf{P}_x(\sigma_S > \sigma_{\Sigma_N})$, then we get from (5.8.2) the expression

$$v^N(\xi) = \frac{1}{2}\frac{\partial q^N(\xi)}{\partial \mathbf{n}^e_\xi} \qquad \text{for } \xi \in S. \tag{5.8.12}$$

We next prove that for $\xi, \eta \in S$ with $\xi \neq \eta$, as $N \uparrow \infty$,

$$U^{e,N}(\xi, \eta) \uparrow U^e(\xi, \eta), \tag{5.8.13}$$

$$v^N(\xi) \downarrow v(\xi). \tag{5.8.14}$$

For $\xi, \eta \in S$ with $\xi \neq \eta$, choose a ball B centered at ξ with $\eta \notin B$ and $B \subset B_N$. Let $C = \partial B \cap D_e$ and $h(x, y)$, $x \in B \cap D_e$, $y \in C$, be the density function of the distribution $\mathbf{P}_x(X_{\sigma_C} \in dy, \sigma_C < \sigma_S)$ with respect to the surface measure $\sigma(dy)$ on C. For the Poisson kernel $K^N(x, \eta)$ of the domain $D_{e,N}$, it then holds that

$$K^N(x, \eta) = \int_C h(x, y)K^N(y, \eta)\sigma(dy), \quad x \in B \cap D_e, \ \eta \in S.$$

Similarly, we have

$$K(x, \eta) = \int_C h(x, y)K(y, \eta)\sigma(dy), \quad x \in B \cap D_e, \ \eta \in S.$$

By taking the outward normal derivatives in x at $\xi \in S$ in the above two equations, we get (5.8.13) from (5.8.2) and (5.8.5) because $K^N(y, \eta)$ increases to $K(y, \eta)$ as $N \to \infty$ for $y \in D_e$, $\eta \in S$.

To prove (5.8.14), we take N_1 with $S \subset B_{N_1}$ and denote by σ_1 the surface measure on Σ_{N_1}. Then we have

$$q^N(x) = \int_{\Sigma_{N_1}} K^{N_1}(x, \eta)q^N(\eta)\sigma_1(d\eta) \qquad \text{for } N > N_1$$

and $q(x) = \int_{\Sigma_{N_1}} K^{N_1}(x, \eta)q(\eta)\sigma_1(d\eta)$. By taking the outward normal derivative in x at S on both sides of each of the above two equations, we arrive at (5.8.14) because $q^N(\eta)$ decreases to $q(\eta)$ as $N \to \infty$ for each $\eta \in \Sigma_{N_1}$.

Finally, denote by $\check{\mathbf{D}}(f,f)$ the right hand side of (5.8.9). We can conclude from (5.8.11), (5.8.13), and (5.8.14) the followings. First, $\frac{1}{2}\mathbf{D}^e(f,f) \geq \check{\mathbf{D}}(f,f)$ for any $f \in \widetilde{H}^1_e(\mathbb{R}^n)$ and, by substituting $\mathbf{H}f$ in place of f,

$$\frac{1}{2}\mathbf{D}^e(\mathbf{H}f, \mathbf{H}f) \geq \check{\mathbf{D}}(f,f), \quad f \in \widetilde{H}^1_e(\mathbb{R}^n). \tag{5.8.15}$$

Second,

$$\lim_{N\to\infty} \frac{1}{2}\mathbf{D}^{e,N}(\mathbf{H}^N f, \mathbf{H}^N f) = \check{\mathbf{D}}(f,f) < \infty, \quad f \in \widetilde{H}^1_e(\mathbb{R}^n). \tag{5.8.16}$$

Since for every $x \in D_e$,

$$\mathbf{H}^N f(x) = \mathbf{E}_x\left[f(X_{\sigma_S}); \sigma_S < \sigma_{\Sigma_N}\right] \uparrow \mathbf{H}f(x) \quad \text{as } N \uparrow \infty,$$

and

$$\mathbf{D}_1^{e,N_0}(\mathbf{H}^N f, \mathbf{H}^N f) = \mathbf{D}^{e,N_0}(\mathbf{H}^N f, \mathbf{H}^N f) + \int_{D_{e,N_0}} (\mathbf{H}^N f(x))^2 dx$$

are uniformly bounded in N for each N_0 in view of (5.8.16), the Cesàro mean of a subsequence of $\{\mathbf{H}^N f\}$ is $\mathcal{D}_1^{e,N_0}$-strongly convergent to $\mathbf{H}f$ by Theorem A.4.1. Hence

$$\frac{1}{2}\mathbf{D}^{e,N_0}(\mathbf{H}f, \mathbf{H}f) \leq \check{\mathbf{D}}(f,f)$$

and, by letting $N_0 \to \infty$, we arrive at the converse inequality to (5.8.15). The proof of (5.8.9) is now complete.

We remark that, in the trace Dirichlet form (5.3.17) of the reflecting Brownian motion X on the upper half-space of $\mathbb{R}^n$, only the Feller kernel U defined by (5.3.16) is involved and the supplementary Feller measure V does not appear even when $n \geq 3$ and X is transient. The reason is in that the vertical component of X is the one-dimensional reflecting Brownian motion so that the hitting probability $\mathbf{H}1(x)$ by X of the hyperplane always equals 1. □

(3°) Trace of Brownian motion on a closed region

We continue to consider the standard Brownian motion $X = (X_t, \mathbf{P}_x)$ on $\mathbb{R}^n$ $(n \geq 2)$ and a bounded domain $D \subset \mathbb{R}^n$ with C^3-boundary ∂D. The Dirichlet form $(\mathcal{E}, \mathcal{F})$ of X on $L^2(\mathbb{R}^n)$ is given by

$$(\mathcal{E}, \mathcal{F}) = \left(\frac{1}{2}\mathbf{D}, H^1(\mathbb{R}^n)\right). \tag{5.8.17}$$

We denote by $\widetilde{H}_e^1(\mathbb{R}^n)$ the family of all $\mathcal{E}$-quasi-continuous functions in the extended Sobolev space $H_e^1(\mathbb{R}^n)$. Let us identify the trace Dirichlet form of (5.8.17) on the closed set $\overline{D}$ by (5.8.23), (5.8.24), and (5.8.25) below.

To this end, we first compare X with the reflecting Brownian motion X^r on $\overline{D}$ that appeared in Example 3.5.9. The Dirichlet form $(\mathcal{E}^r, \mathcal{F}^r)$ of X^r on $L^2(\overline{D})(= L^2(D))$ is given by

$$(\mathcal{E}^r, \mathcal{F}^r) = \left(\frac{1}{2}\mathbf{D}, H^1(D)\right). \tag{5.8.18}$$

By Corollary 2.2.15, the extended Sobolev space $H_e^1(D)$ coincides with the space $\mathrm{BL}(D)$. We shall denote by $\widetilde{\mathrm{BL}}(D)$ the family of all $\mathcal{E}^r$-quasi-continuous functions in $\mathrm{BL}(D)$. The capacity Cap_1 for (5.8.18) will be designated by Cap_1^r.

As $\mathrm{Cap}_1^r(A \cap \overline{D}) \le \mathrm{Cap}_1(A)$ for any open set $A \subset \mathbb{R}^n$,

$$f \text{ on } \mathbb{R}^n \text{ is } \mathcal{E}\text{-quasi-continuous} \implies f|_{\overline{D}} \text{ is } \mathcal{E}^r\text{-quasi-continuous.} \tag{5.8.19}$$

Moreover,

$$N \subset \overline{D} \text{ is } \mathcal{E}^r\text{-polar} \implies N \text{ is } \mathcal{E}\text{-polar.} \tag{5.8.20}$$

Indeed, for any compact set $K \subset \overline{D}$, it follows from [73, Lemma 2.2.7] that

$$\mathrm{Cap}_1^r(K) = \inf\{\mathcal{E}_1^r(u,u) : u \in C^\infty(\overline{D}),\ u \ge 1 \text{ on } K\},$$

$$\mathrm{Cap}_1(K) = \inf\{\mathcal{E}_1(u,u) : u \in C^\infty(\mathbb{R}^n) \cap H^1(\mathbb{R}^n),\ u \ge 1 \text{ on } K\}.$$

On the other hand, by virtue of [142, VI:3], there exists a linear operator T called an *extension operator* from $H^1(D)$ to $H^1(\mathbb{R}^n)$ such that

$$Tf|_D = f, \quad \mathcal{E}_1(Tf, Tf) \le M\mathcal{E}_1^r(f,f), \quad f \in H^1(D),$$

for some constant $M > 0$ independent of $f \in H^1(D)$. Moreover, T sends $C^\infty(\overline{D})$ into $C^\infty(\mathbb{R}^n) \cap H^1(\mathbb{R}^n)$. Therefore, $\mathrm{Cap}_1(K) \le M\mathrm{Cap}_1^r(K)$ for any compact set $K \subset \overline{D}$, which in particular means (5.8.20).

The above considerations lead us to

$$\widetilde{H}_e^1(\mathbb{R}^n)|_{\overline{D}} = \widetilde{\mathrm{BL}}(D). \tag{5.8.21}$$

The inclusion $\subset$ is immediate from (5.8.19). Conversely, any $u \in \mathrm{BL}(D)$ admits $v \in \mathrm{BL}(\mathbb{R}^n)$ with $v|_D = u$ (cf. [24, (2.21)]). Then $v \in H_e^1(\mathbb{R}^n)$ by Theorem 2.2.13 if $n = 2$, while $v = v_0 + c$ for some $v_0 \in H_e^1(\mathbb{R}^n)$ and $c \in \mathbb{R}$, by Theorem 2.2.12 if $n \ge 3$. In the latter case, replace c by a function $w \in C_c^\infty(\mathbb{R}^n)$ with $w = c$ on D so that $v \in H_e^1(\mathbb{R}^n)$ with $v|_D = u$. Take any $\mathcal{E}^r$-(resp. $\mathcal{E}$-)quasi-continuous version $\widetilde{u}$ (resp. $\widetilde{v}$) of u (resp. v). Then $\widetilde{v}|_{\overline{D}}$ is $\mathcal{E}^r$-quasi-continuous on $\overline{D}$ by (5.8.19) so that $\widetilde{u} = \widetilde{v}|_{\overline{D}}$ $\mathcal{E}^r$-q.e. on $\overline{D}$, and consequently $\mathcal{E}$-q.e. on $\overline{D}$, proving the inclusion $\supset$ in (5.8.21).

We now consider a measure μ on $\mathbb{R}^n$ defined by $\mu(dx) = g(x)dx$ for any bounded measurable function g which is strictly positive a.e. on D and vanishing on $\mathbb{R}^n \setminus D$. μ charges no $\mathcal{E}$-polar set and further

$$\mu \in \mathring{S}_{\overline{D}}, \tag{5.8.22}$$

namely, $\overline{D}$ is an $\mathcal{E}$-quasi-support of μ. To see this, take any $f \in \widetilde{H}_e^1(\mathbb{R}^n)$ and assume that $f = 0$ μ-a.e. $f|_{\overline{D}}$ is then $\mathcal{E}^r$-quasi-continuous by (5.8.19) and vanishing a.e. on D. Therefore, $f|_{\overline{D}} = 0$ $\mathcal{E}^r$-q.e. on $\overline{D}$, and so $f = 0$ $\mathcal{E}$-q.e. on $\overline{D}$, which proves (5.8.22) in view of Theorem 3.3.5.

Let Y be the time-changed process of X by means of its PCAF $A_t = \int_0^t g(X_s)ds$ with Revuz measure μ. By virtue of Theorem 5.2.13, Y can be realized as a μ-symmetric Hunt process on $\overline{D}$ whose Dirichlet form $(\check{\mathcal{E}}, \check{\mathcal{F}})$ on $L^2(\overline{D}; \mu)(= L^2(D; \mu))$ is regular. Denote by $(\check{\mathcal{F}}_e, \check{\mathcal{E}})$ the extended Dirichlet space of $(\check{\mathcal{E}}, \check{\mathcal{F}})$. It then holds that

$$\check{\mathcal{F}} = \check{\mathcal{F}}_e \cap L^2(\overline{D}; \mu) \tag{5.8.23}$$

on account of Theorem 5.2.15. We claim that

$$\check{\mathcal{F}}_e = \widetilde{\mathrm{BL}}(D) \tag{5.8.24}$$

and, for $f \in \widetilde{\mathrm{BL}}(D)$,

$$\begin{aligned}\check{\mathcal{E}}(f,f) = &\frac{1}{2}\int_D |\nabla f|^2(x)dx \\ &+\frac{1}{2}\int_{\partial D\times\partial D}(f(\xi)-f(\eta))^2 U^e(\xi,\eta)\sigma(d\xi)\sigma(d\eta) \\ &+\int_{\partial D} f(\xi)^2 v(\xi)\sigma(d\xi).\end{aligned} \tag{5.8.25}$$

Identity (5.8.24) follows from Theorem 5.2.15 and (5.8.21). Since $\mathbf{H}_{\overline{D}}f = f\cdot \mathbf{1}_D + \mathbf{H}_{\partial D}f \cdot \mathbf{1}_{\mathbb{R}^n\setminus D}$, we get (5.8.25) from (5.2.4) and the identity (5.8.9) in the preceding example holding for any $f \in \widetilde{H}^1_e(\mathbb{R}^n)|_{\partial D} = \widetilde{\mathrm{BL}}(D)|_{\partial D}$. □

Chapter Six

REFLECTED DIRICHLET SPACES

Reflected Dirichlet space was introduced for a regular Dirichlet form by M. L. Silverstein in 1974 in [138, 139] and further investigated by the first-named author in [16]. Reflected Dirichlet space plays an important role for the boundary theory of symmetric Markov processes. While it is possible to introduce reflected Dirichlet form for a quasi-regular Dirichlet form directly, we choose to do it first in the regular Dirichlet form setting as this allows us to define the reflected Dirichlet space via terminal random variables and harmonic functions of finite energy. This approach sheds more insight into the probabilistic meaning and the structure of the reflected Dirichlet form. As illuminated by the relationship between the Dirichlet spaces for the absorbing Brownian motion and reflecting Brownian motion, given a Dirichlet form $(\mathcal{E}, \mathcal{F})$, heuristically, there are two ways to define its reflected Dirichlet space: (i) the linear span of $\mathcal{F}$ and all harmonic functions of finite "$\mathcal{E}$-energy"; (ii) all functions that are "locally" in $\mathcal{F}$ and have finite "$\mathcal{E}$-energy." In this chapter, we will develop these two approaches simultaneously and show that they give the same object.

In Section 6.1, we introduce the notion of terminal random variables and harmonic functions of finite energy for a Hunt process associated with a transient regular Dirichlet form. In Section 6.2, we will present and establish several equivalent notions of reflected Dirichlet space $(\mathcal{E}^{\rm ref}, \mathcal{F}^{\rm ref})$ for a regular transient Dirichlet form $(\mathcal{E}, \mathcal{F})$. One of these equivalent notions can be used to define reflected Dirichlet space for a regular recurrent Dirichlet form in Section 6.3, where we will show that in this case $\mathcal{F}^{\rm ref} = \mathcal{F}_e$. In Section 6.4 we give yet another equivalent definition of reflected Dirichlet space that is invariant under quasi-homeomorphism of Dirichlet forms, which allows us to define reflected Dirichlet space for any general quasi-regular Dirichlet forms.

Various concrete examples of reflected Dirichlet spaces will be exhibited in Section 6.5 for regular Dirichlet forms including most of those presented in Section 2.2.

In Section 6.6, we first define a Silverstein extension of a quasi-regular Dirichlet form $(\mathcal{E}, \mathcal{F})$ on $L^2(E; m)$. At the probabilistic level, a symmetric Dirichlet form $(\widetilde{\mathcal{E}}, \widetilde{\mathcal{F}})$ is a Silverstein extension of $(\mathcal{E}, \mathcal{F})$ if the symmetric Markov process $\widetilde{X}$ associated with $(\widetilde{\mathcal{E}}, \widetilde{\mathcal{F}})$ extends the symmetric Markov process X associated with $(\mathcal{E}, \mathcal{F})$ to some state space $\widetilde{E}$, which spends zero sojourn time at the "boundary" $\widetilde{E} \setminus E$. We then show that the active reflected space $(\mathcal{E}^{\mathrm{ref}}, \mathcal{F}_a^{\mathrm{ref}})$ is the maximum among the Silverstein extensions of $(\mathcal{E}, \mathcal{F})$ with respect to a partial ordering defined among symmetric Dirichlet forms on $L^2(E; m)$. The maximality of the reflected Dirichlet space is essentially due to Silverstein [138].

A probabilistic notion of a harmonic function for a symmetric Hunt process plays a role in the first two sections. In Section 6.7, the equivalence of analytic and probabilistic concepts of harmonicity is investigated.

6.1. TERMINAL RANDOM VARIABLES AND HARMONIC FUNCTIONS

Let $(\mathcal{E}, \mathcal{F})$ be a transient regular Dirichlet form on $L^2(E; m)$, and let $X = \{X_t, \{\mathcal{F}_t\}_{t\geq 0}, \mathbf{P}_x, x \in E\}$ be the m-symmetric Hunt process on E associated with it. The transition function and the resolvent kernel of X are denoted by $\{P_t; t \geq 0\}$ and $\{R_\alpha; \alpha \geq 0\}$, respectively. We adopt the conventions that any element of the extended Dirichlet space $\mathcal{F}_e$ is quasi continuous already and that any numerical function f on E is extended to E_∂ by setting $f(\partial) = 0$.

Recall the class S of smooth measures on E and the class $S_0^{(0)}$ of positive Radon measures on E of finite 0-order energy integrals defined in Section 2.3. Any $\mu \in S_0^{(0)}$ admits a unique 0-order potential $U\mu \in \mathcal{F}_e$. We denote by $S_{00}^{(0)}$ the 0-order counterpart of the class S_{00} defined by (2.3.10) so that $\mu \in S_{00}^{(0)}$ if and only if $\mu \in S_0^{(0)}$ and $\mu(E) < \infty$, $\|U\mu\|_\infty < \infty$.

LEMMA 6.1.1. *For $\mu \in S$, let A^μ be the PCAF of X with Revuz measure μ. Then*

$$\mathbf{E}_\nu[A_\zeta^\mu] = \int_E U\nu(x)\mu(dx) \leq \|U\nu\|_\infty\, \mu(E) \quad \textit{for every } \nu \in S_{00}^{(0)}.$$

In particular, A_ζ^μ is $\mathbf{P}_x$-integrable for q.e. $x \in E$ if $\mu(E) < \infty$.

Proof. If $\mu \in S_0^{(0)}$, then we have from (4.1.3) that for any $h \in \mathcal{B}_+(E)$ with $(h, R_0 h) < \infty$,

$$\mathbf{E}_{h\cdot m}[A_\zeta^\mu] = \langle R_0 h, \mu\rangle = \mathcal{E}(R_0 h, U\mu) = (h, U\mu).$$

Consequently, $\mathbf{E}_x[A_\zeta^\mu]$ as a function of x is a quasi continuous version of $U\mu$ and for $\nu \in S_{00}^{(0)}$,

$$\mathbf{E}_\nu[A_\zeta^\mu] = \langle \nu, U\mu\rangle = \mathcal{E}(U\nu, U\mu) = \langle U\nu, \mu\rangle \leq \|U\nu\|_\infty \mu(E).$$

For a general $\mu \in S$, there exists an $\mathcal{E}$-nest $\{F_k, k \geq 1\}$ such that $\mathbf{1}_{F_k}\mu \in S_0^{(0)}$ for every $k \geq 1$ owing to the 0-order counterpart of Theorem 2.3.15. By replacing

A^μ and μ in the above with $\mathbf{1}_{F_k} \cdot A^\mu$ and $\mathbf{1}_{F_k} \cdot \mu$, respectively, and by letting $k \to \infty$, we get the desired relation. The second statement of the lemma follows from it and the 0-order counterpart of Corollary 2.3.11. □

Let M be a MAF of X, namely, M is an additive functional of X satisfying (4.2.12). Denote by $\langle M \rangle$ and $\mathbf{e}(M)$ the predictable quadratic variation process and the energy of M, respectively. $\langle M \rangle$ is an element of $\mathbf{A}_c^+$ and we denote by $\mu_{\langle M \rangle}$ its Revuz measure. Then $\mathbf{e}(M) = \frac{1}{2}\mu_{\langle M \rangle}(E)$. $\mathcal{M}$ denotes the totality of MAFs of X and we let $\overset{\circ}{\mathcal{M}} = \{M \in \mathcal{M} : \mathbf{e}(M) < \infty\}$.

Theorem 6.1.2. *Let $M \in \overset{\circ}{\mathcal{M}}$ and define $M^* := \sup_{t \geq 0} |M_t|$. Then*

$$\mathbf{E}_\nu[(M^*)^2] \leq 8\|U\nu\|_\infty \, \mathbf{e}(M) < \infty \quad \textit{for every } \nu \in S_{00}^{(0)}.$$

In particular, M_t converges as $t \to \infty$ to a random variable M_∞ in $L^2(\mathbf{P}_x)$ as well as $\mathbf{P}_x$-a.s. for q.e. $x \in E$, and

$$\mathbf{E}_\nu[(M_\infty)^2] \leq 8\|U\nu\|_\infty \, \mathbf{e}(M) < \infty \quad \textit{for every } \nu \in S_{00}^{(0)}. \tag{6.1.1}$$

Proof. By Doob's maximal inequality (A.3.36), for $T > 0$,

$$\begin{aligned}\mathbf{E}_\nu\left[\sup_{0 \leq t \leq T}(M_t)^2\right] &\leq 4\mathbf{E}_\nu[\langle M \rangle_T] \leq 4\mathbf{E}_\nu[\langle M \rangle_\zeta] \\ &\leq 8\|U\nu\|_\infty \, \mathbf{e}(M) < \infty,\end{aligned}$$

where the last inequality follows from Lemma 6.1.1. Letting $T \to \infty$, we get the desired estimate of M^* and hence $\{M_t : t \geq 0\}$ is a $\mathbf{P}_\nu$-uniformly integrable martingale for each $\nu \in S_{00}^{(0)}$. We then get the desired conclusions with the help of the 0-order counterpart of Corollary 2.3.11. □

Definition 6.1.3. A measurable function h is said to be *harmonic* on E if it is specified and finite up to quasi equivalence and if for every relatively compact open subset $D \subset E$, $\mathbf{E}_x[|h(X_{\tau_D})|] < \infty$ and $h(x) = \mathbf{E}_x[h(X_{\tau_D})]$ for q.e. $x \in E$. Here we use the convention that $X_\infty = \partial$ and $h(\partial) := 0$.

Definition 6.1.4. A *terminal random variable* φ is a random variable on $(\Omega, \mathcal{F}_\infty)$ that is $\mathbf{P}_x$-integrable for q.e. $x \in E$ and that for q.e. $x \in E$, $\mathbf{P}_x$-a.s. $\{\varphi = 0\} \supset \{X_{\zeta-} \in E, \zeta < \infty\} \cup \{\zeta = 0\}$ and $\varphi \circ \theta_t = \varphi$ for every $t < \zeta$.

Here we notice that, due to the transience of X,

$$\lim_{t \to \infty} X_t = \partial \quad \mathbf{P}_x\text{-a.s. on } \{\zeta = \infty\} \text{ for q.e. } x \in E, \tag{6.1.2}$$

by Theorem 3.5.2.

LEMMA 6.1.5. *If φ is a terminal random variable, then $h(x) := \mathbf{E}_x[\varphi]$ is harmonic. Conversely, a harmonic function h on E can be represented by $h = \mathbf{E}_x[\varphi]$ for some terminal random variable φ if and only if for q.e. $x \in E$, $\{h(X_{\tau_{D_k}}), \mathcal{F}_{\tau_{D_k}}\}$ is a $\mathbf{P}_x$-uniformly integrable martingale, where $\{D_k, k \geq 1\}$ is an increasing sequence of relatively compact open subsets with $\cup_{k\geq 1} D_k = E$. In this case,*

$$\lim_{k\to\infty} h(X_{\tau_{D_k}}) = \varphi, \quad \mathbf{P}_x - a.s. \text{ for } q.e.\ x \in E. \tag{6.1.3}$$

Proof. Let φ be a terminal random variable. Note that for any relatively compact open subset $D \subset E$, $\{\tau_D = \zeta\} \subset \{X_{\zeta-} \in E,\ \zeta < \infty\}$ because of (6.1.2). Hence $\varphi = 0$ on $\{\tau_D = \zeta\}$, while $\varphi \circ \theta_{\tau_D} = \varphi$ on $\{\tau_D < \zeta\}$. Define $h(x) := \mathbf{E}_x[\varphi]$, which is finite q.e. on E. It follows that for q.e. $x \in E$,

$$h(x) = \mathbf{E}_x[\varphi \circ \theta_{\tau_D};\ \tau_D < \zeta] = \mathbf{E}_x[h(X_{\tau_D});\ \tau_D < \zeta] = \mathbf{E}_x[h(X_{\tau_D})].$$

So h is harmonic on E. Moreover, if $\{D_k, k \geq 1\}$ is an increasing sequence of relatively compact open subsets with $\cup_{k\geq 1} D_k = E$, then, $\mathbf{P}_x$-a.s. for q.e. $x \in E$,

$$\begin{aligned} h(X_{\tau_{D_k}}) &= \mathbf{1}_{\{\tau_{D_k}<\zeta\}}\, \mathbf{E}_{X_{\tau_{D_k}}}[\varphi] = \mathbf{1}_{\{\tau_{D_k}<\zeta\}}\, \mathbf{E}_x[\varphi \circ \theta_{\tau_{D_k}} \mid \mathcal{F}_{\tau_{D_k}}] \\ &= \mathbf{E}_x[\varphi \circ \theta_{\tau_{D_k}} \cdot \mathbf{1}_{\{\tau_{D_k}<\zeta\}} \mid \mathcal{F}_{\tau_{D_k}}] = \mathbf{E}_x[\varphi\, \mathbf{1}_{\{\tau_{D_k}<\zeta\}} \mid \mathcal{F}_{\tau_{D_k}}] \\ &= \mathbf{E}_x[\varphi \mid \mathcal{F}_{\tau_{D_k}}]. \end{aligned}$$

Therefore, $\{h(X_{\tau_{D_k}}), \mathcal{F}_{\tau_{D_k}}\}$ is a $\mathbf{P}_x$-uniformly integrable martingale for q.e. $x \in E$ and further (6.1.3) holds true due to the fact that filtration $\{\mathcal{F}_t, t \geq 0\}$ is quasi-left-continuous by Theorem A.1.40 and so $\bigvee_{k\geq 1} \mathcal{F}_{\tau_{D_k}} = \mathcal{F}_\infty$. Here for an increasing sequence $\{\mathcal{G}_k, k \geq 1\}$ of σ-fields, $\bigvee_{k\geq 1} \mathcal{G}_k$ denotes the smallest σ-field generated by the sets in $\bigcup_{k\geq 1} \mathcal{G}_k$.

Conversely, suppose that h is a harmonic function so that $\{h(X_{\tau_{D_k}}), \mathcal{F}_{\tau_{D_k}}\}$ is a $\mathbf{P}_x$-uniformly integrable martingale for q.e. $x \in E$. Define $\varphi := \lim_{k\to\infty} h(X_{\tau_{D_k}})$. Then clearly $\mathbf{E}_x[|\varphi|] < \infty$, $h(x) = \mathbf{E}_x[\varphi]$ for q.e. $x \in E$, and $\varphi \circ \theta_t = \varphi$ on $\{\zeta > t\}$ because $\lim_{k\to\infty} \tau_{D_k} = \zeta$ by the quasi-left-continuity of X. On $\{X_{\zeta-} \neq \partial,\ \zeta < \infty\}$, $\tau_{D_k} = \zeta$ for some (random) $k \geq 1$ and so $\varphi = 0$. This proves that φ is a terminal random variable. □

THEOREM 6.1.6. *If φ is a terminal random variable and $h(x) := \mathbf{E}_x[\varphi]$, then h is quasi continuous and for q.e. $x \in E$,*

$$\mathbf{P}_x\Big(h(X_t) \text{ is right continuous on } [0,\infty) \\ \text{and } \lim_{s\uparrow t} h(X_s) = h(X_{t-}) \text{ on } (0,\zeta)\Big) = 1. \tag{6.1.4}$$

Moreover, it holds that, $\mathbf{P}_x$*- a.s. for q.e.* $x \in E$,

$$\lim_{t \uparrow \zeta} h(X_t) = h(X_{\zeta-})\mathbf{1}_{\{X_{\zeta-} \neq \partial,\ \zeta < \infty\}} + \varphi \mathbf{1}_{\{X_{\zeta-} = \partial\}}. \tag{6.1.5}$$

Proof. Without loss of generality, we may and do assume that $\varphi \geq 0$. Let N_0 be a properly exceptional set so that φ is $\mathbf{P}_x$-integrable for every $x \in E \setminus N_0$. Then h is an excessive function with respect to the Hunt process X restricted to the state space $E \setminus N_0$ because, by the Markov property of X, for $x \in E \setminus N_0$,

$$P_t h(x) = \mathbf{E}_x[\varphi \circ \theta_t; t < \zeta] = \mathbf{E}_x[\varphi; t < \zeta] \uparrow h(x) \quad \text{as } t \downarrow 0.$$

By Theorems A.2.2 and A.2.9, h is finely continuous q.e. on E and so by Theorem 3.3.3(ii), h is quasi continuous on E. Now (6.1.4) follows from Theorem 3.1.7, which asserts moreover that $\lim_{s \uparrow \zeta} h(X_s) = h(X_{\zeta-})$ on $\{\zeta < \infty, X_{\zeta-} \in E\}$.

Since h is excessive, $h(X_t)$ is a non-negative supermartingale and admits the left limit in t on $(0, \infty]$ (cf. [37, VI:1]). Take $\{D_k\}$ as in the preceding lemma. As $\{X_{\zeta-} = \partial\} = \{\tau_{D_k} < \zeta,\ \forall k,\ \lim_{k \to \infty} \tau_{D_k} = \zeta\}$ by virtue of the quasi-left-continuity of X and (6.1.2), we get $\lim_{t \uparrow \zeta} h(X_t) = \varphi$ on $\{X_{\zeta-} = \partial\}$ on account of (6.1.3). This establishes (6.1.5). □

Let φ be a terminal random variable and define

$$h(x) := \mathbf{E}_x[\varphi], \qquad x \in E, \tag{6.1.6}$$

$$M_t^h := h(X_t)\mathbf{1}_{\{t < \zeta\}} + \varphi \mathbf{1}_{\{t \geq \zeta\}} - h(X_0), \qquad t \geq 0. \tag{6.1.7}$$

M^h is then a $\mathbf{P}_x$-uniformly integrable martingale for q.e. $x \in E$ with $M_0^h = 0$. This is due to the fact that $\varphi = \varphi \circ \theta_t$ on $\{t < \zeta\}$ and $\varphi \in \mathcal{F}_\infty = \mathcal{F}_\zeta$ and so, for q.e. $x \in E$,

$$M_t^h = \mathbf{E}_x[\varphi \mid \mathcal{F}_t]\, \mathbf{1}_{\{t < \zeta\}} + \varphi \mathbf{1}_{\{t \geq \zeta\}} - h(X_0) = \mathbf{E}_x[\varphi \mid \mathcal{F}_t] - h(X_0). \tag{6.1.8}$$

Lemma 6.1.7. *Let* φ *be a terminal random variable of* X *with*

$$\mathbf{E}_x\left[\varphi^2\right] < \infty \quad \text{for q.e. } x \in E. \tag{6.1.9}$$

Consider the function h *and the martingale* M^h *defined by* (6.1.6) *and* (6.1.7), *respectively. Let*

$$g(x) := \mathbf{E}_x\left[\varphi^2\right] - h(x)^2, \qquad x \in E. \tag{6.1.10}$$

Then $M^h = \{M_t^h, t \geq 0\}$ *is a martingale additive functional of* X, $\{(M_t^h)^2; t \geq 0\}$ *is* $\mathbf{P}_x$*-uniformly integrable, and further*

$$g(x) = P_t g(x) + \mathbf{E}_x\left[(M_t^h)^2\right], \qquad t \geq 0, \tag{6.1.11}$$

for q.e. $x \in E$.

Proof. M^h is a $\mathbf{P}_x$-martingale by (6.1.8), which combined with (6.1.9) means that $\{(M_t^h)^2; t \geq 0\}$ is uniformly integrable. The additivity $M_{s+t}^h = M_s^h + M_t^h \circ \theta_s$ of M^h is clear when $s+t < \zeta$. If $s < \zeta$, $s+t \geq \zeta$, then $t \geq \zeta \circ \theta_s$ and so $M_t^h \circ \theta_s = \varphi - h(X_s)$ and hence the additivity follows. If $s \geq \zeta$, then $\zeta \circ \theta_s = 0$ so that $\varphi \circ \theta_s = 0$, $h(X_0 \circ \theta_s) = h(\partial) = 0$ and $M^h \circ \theta_s = 0$, yielding the additivity $M_{s+t}^h = M_s^h = \varphi - h(X_0)$. Moreover,

$$
\begin{aligned}
&\mathbf{E}_x[(M_t^h)^2] \\
&\quad = P_t h^2(x) + \mathbf{E}_x[\varphi^2 \mathbf{1}_{\{t \geq \zeta\}}] + h(x)^2 - 2h(x)\mathbf{E}_x[h(X_t)\mathbf{1}_{t<\zeta} + \varphi \mathbf{1}_{\{t \geq \zeta\}}] \\
&\quad = P_t h^2(x) + \mathbf{E}_x[\varphi^2] - P_t(\mathbf{E}_\cdot[\varphi^2])(x) - h(x)^2 = g(x) - P_t g(x).
\end{aligned}
$$

□

Since X is an m-symmetric transient Hunt process associated with a regular Dirichlet form $(\mathcal{E}, \mathcal{F})$, the concept of the energy functional $L(f, g)$, $f \in \mathcal{S}^{\text{pur}}(E)$, $\mathcal{S}(E)$, introduced in Section 5.4 by (5.4.4) is well defined for X. For simplicity, we denote the space of X-q.e. excessive functions (resp. X-q.e. purely excessive functions) on E by $\mathcal{S}$ (resp. $\mathcal{S}^{\text{pur}}$) with E being dropped.

THEOREM 6.1.8. *Let φ be a terminal random variable satisfying* (6.1.9) *and h, M^h, g be defined by* (6.1.6), (6.1.7), (6.1.10), *respectively.*
(i) *$g \in \mathcal{S}^{\text{pur}}$ and*

$$
\frac{1}{2}L(g, 1) = \mathbf{e}(M^h). \tag{6.1.12}
$$

In particular, $M^h \in \overset{\circ}{\mathcal{M}}$ if and only if $L(g, 1) < \infty$.
(ii) *Denote by $M^{h,c}$ the continuous part of M^h. Then*

$$
\begin{aligned}
\mathbf{e}(M^h) = &\frac{1}{2}\mu_{\langle M^{h,c}\rangle}(E) + \frac{1}{2}\int_{E\times E}(h(x) - h(y))^2 J(dx, dy) \\
&+ \frac{1}{2}\int_E h(x)^2 \kappa(dx) \leq \infty.
\end{aligned} \tag{6.1.13}
$$

Proof. (i) We have $\lim_{t\to\infty} M_t^h = \varphi - h(X_0)$ in view of (6.1.7) and (6.1.5). Since $\{(M_t^h)^2; t \geq 0\}$ is uniformly integrable by (6.1.8), we have for q.e. $x \in E$

$$
\lim_{t\to\infty} \mathbf{E}_x[(M_t^h)^2] = \mathbf{E}_x[(\varphi - h(X_0))^2] = g(x).
$$

Formula (6.1.11) now implies that $g \in \mathcal{S}^{\text{pur}}$ together with the validity of (6.1.12).

(ii) On account of **(4)** of Section A.3.3, Theorem 6.1.6, (6.1.5), and (6.1.7), we obtain, using the identity (A.3.31) for the Lévy system (N, H) of X,

$$\begin{aligned}
\mathbf{e}(M^h) &= \lim_{t\to 0}\frac{1}{2t}\mathbf{E}_m\left[[M^h]_t\right] \\
&= \lim_{t\to 0}\frac{1}{2t}\mathbf{E}_m\left[\langle M^{h,c}\rangle_t + \sum_{0<s\le t}(M^h_s - M^h_{s-})^2\right] \\
&= \mathbf{e}(M^{h,c}) + \lim_{t\to 0}\frac{1}{2t}\mathbf{E}_m\left[\sum_{0<s\le t}(h(X_s)-h(X_{s-}))^2\mathbf{1}_{\{X_s\neq X_{s-}\}}\right] \\
&= \mathbf{e}(M^{h,c}) + \lim_{t\to 0}\frac{1}{2t}\mathbf{E}_m\left[\int_0^t\left(\int_{E_\partial}(h(y)-h(X_s))^2N(X_s,dy)\right)dH_s\right] \\
&= \mathbf{e}(M^{h,c}) + \frac{1}{2}\int_E\left(\int_{E_\partial}(h(y)-h(x))^2N(x,dy)\right)\mu_H(dx) \\
&= \frac{1}{2}\mu_{\langle M^{h,c}\rangle}(E) + \frac{1}{2}\int_{E\times E}(h(x)-h(y))^2J(dx,dy) + \frac{1}{2}\int_E h(x)^2\kappa(dx).
\end{aligned} \tag{6.1.14}$$

In the last equality, the expressions $J(dx, dy) := N(x, dy)\mu_H(dx)$, $\kappa(dx) := N(x, \{\partial\})\mu_H(dx)$, and the convention $h(\partial) = 0$ are used. □

6.2. REFLECTED DIRICHLET SPACES: TRANSIENT CASE

We continue to work under the setting of the preceding section. Let

$$\begin{aligned}
\mathbf{N} := \{\varphi :\ &\text{terminal random variable with } \mathbf{E}_x[\varphi^2] < \infty \text{ for q.e. } x \in E, \\
&\text{and } L(g, 1) < \infty\},
\end{aligned} \tag{6.2.1}$$

where g is related to φ via (6.1.10), and

$$\mathbf{HN} := \{h = \mathbf{E}_\cdot[\varphi] : \varphi \in \mathbf{N}\}. \tag{6.2.2}$$

The space $\mathbf{N}$ is sometimes called the *universal Dirichlet space* and $\mathbf{HN}$ the *space of harmonic functions of finite energy*.

DEFINITION 6.2.1. The *reflected Dirichlet space* $(\mathcal{F}^{\text{ref}}, \mathcal{E}^{\text{ref}})$ of a transient regular Dirichlet form $(\mathcal{E}, \mathcal{F})$ is defined to be

$$\mathcal{F}^{\text{ref}} := \mathcal{F}_e + \mathbf{HN} \tag{6.2.3}$$

$$\mathcal{E}^{\text{ref}}(f, f) := \mathcal{E}(f_0, f_0) + \mathbf{e}(M^h) + \frac{1}{2}\int_E h(x)^2\kappa(dx) \tag{6.2.4}$$

for $f = f_0 + h$ with $f_0 \in \mathcal{F}_e$ and $h = \mathbf{E}_x[\varphi] \in \mathbf{HN}$. M^h is the MAF of finite energy as is defined by (6.1.7). Here κ is the killing measure for Dirichlet form $(\mathcal{E}, \mathcal{F})$.

We let $\mathcal{F}_a^{\text{ref}} = \mathcal{F}^{\text{ref}} \cap L^2(E; m)$ and call $(\mathcal{F}_a^{\text{ref}}, \mathcal{E}^{\text{ref}})$ the *active reflected Dirichlet space*.

Remark 6.2.2. (i) Let $\{D_k\}$ be a sequence of relatively compact open sets increasing to E. Then $\lim_{k\to\infty} f_0(X_{\tau_{D_k}}) = 0$ for any $f_0 \in \mathcal{F}_e$ by Corollary 3.5.3, while $\lim_{k\to\infty} h(X_{\tau_{D_k}}) = \varphi$ for $h = \mathbf{E}_x[\varphi] \in \mathbf{HN}$ by (6.1.3). Thus $\mathcal{F}_e \cap \mathbf{HN} = \{0\}$ and so the definition of $\mathcal{E}^{\text{ref}}$ given by (6.2.4) is well defined on $\mathcal{F}^{\text{ref}}$.

For $f \in \mathcal{F}^{\text{ref}}$ with $f = f_0 + h$, $f_0 \in \mathcal{F}_e$, $h \in \mathbf{HN}$, the components f_0, h are uniquely determined by f as $h = \mathbf{E}_\cdot[\varphi]$ with $\varphi = \lim_{k\to\infty} f(X_{\tau_{D_k}})$ and $f_0 = f - h$. In particular, if f is bounded, then both f_0 and h are bounded.

(ii) For $f, g \in \mathcal{F}^{\text{ref}}$, we define

$$\mathcal{E}^{\text{ref}}(f, g) := \frac{1}{4}\left(\mathcal{E}^{\text{ref}}(f + g, f + g) - \mathcal{E}^{\text{ref}}(f - g, f - g)\right).$$

On account of the expression (6.1.13) of $\mathbf{e}(M^h)$, we see that $\mathcal{E}^{\text{ref}}(f, g)$ is a nonnegative definite symmetric bilinear form on $\mathcal{F}^{\text{ref}}$. Further, $\mathcal{E}^{\text{ref}}(f, h) = 0$ for $f \in \mathcal{F}_e$ and $h \in \mathbf{HN}$. □

The main result of this section is that the active reflected Dirichlet space $(\mathcal{E}^{\text{ref}}, \mathcal{F}_a^{\text{ref}})$ is a Dirichlet form on $L^2(E; m)$. Along the way, we will present several useful equivalent characterizations of the reflected Dirichlet space $\mathcal{F}^{\text{ref}}$. We begin with

LEMMA 6.2.3. *Suppose that $\{h_k, k \geq 1\} \subset \mathbf{HN}$ is $\mathcal{E}^{\text{ref}}$-Cauchy and that $h_k(x)$ converges to a finite limit for m-a.e. $x \in E$. Then h_k is $\mathcal{E}^{\text{ref}}$-convergent to a harmonic function $h \in \mathbf{HN}$. Moreover, there is a subsequence h_{k_j} that converges to h q.e. on E.*

Proof. Let $\varphi_k \in \mathbf{N}$ so that $h_k = \mathbf{E}_\cdot[\varphi_k]$ and let M^{h_k} be the MAF of X defined by (6.1.7) with h_k and φ_k in place of h and φ there. Then $M^{(k)} := M^{h_k}$ is an $\mathbf{e}$-Cauchy sequence in $\mathring{\mathcal{M}}$ in view of (6.2.4). By virtue of Theorem 4.2.5, there exist a unique $M \in \mathring{\mathcal{M}}$, a subsequence $\{k_j, j \geq 1\}$, and a Borel properly exceptional set N_0 such that $\{M^{(k)}, k \geq 1\}$ is $\mathbf{e}$-convergent to M and $\{M_t^{(k_j)}, j \geq 1\}$ is uniformly convergent to M_t on every compact time interval of $[0, \infty)$ $\mathbf{P}_x$-a.s. for every $x \in E \setminus N_0$.

Let $E_1 := \{x \in E \setminus N_0 : h_{k_j}(x) \text{ converges as } j \to \infty\}$. By assumption, $m(E \setminus E_1) = 0$. Suppose $\text{Cap}^{(0)}(E \setminus E_1) > 0$. Then $\text{Cap}^{(0)}(K) > 0$ for some compact set $K \subset E \setminus E_1$ by the 0-order version of (1.2.6). On account of Corollary 3.4.3, the hitting probability $p_K(x) = \mathbf{P}_x(\sigma_K < \zeta)$ is strictly positive for x in a set of positive m-measure, and especially for some point $x_0 \in E_1$. If we consider the event $\Lambda = \{\sigma_K < \zeta,\ M^{(k_j)}_{\sigma_K} = h_{k_j}(X_{\sigma_K}) - h_{k_j}(x)\}$, then $\mathbf{P}_{x_0}(\Lambda) = p_K(x_0) > 0$. But $M^{(k_j)}_{\sigma_K}$ diverges on Λ as $j \to \infty$ because $X_{\sigma_K} \in K \subset E \setminus E_1$. This contradiction shows that $\mathrm{Cap}^{(0)}(E \setminus E_1) = 0$ and so $E \setminus E_1$ is $\mathcal{E}$-polar by Theorem 2.3.2.

We define $h(x) = \lim_{j\to\infty} h_{k_j}(x)$ for $x \in E_1$. In view of (6.1.5) and (6.1.7), it holds that $M^{(k)}_\infty = \varphi_k - h_k(X_0)$. By making use of the estimate (6.1.1), we can find, as in the proof of Theorem 4.2.5, a subsequence of $\{k_j\}$ (denoted as $\{k_j\}$ again) and a properly exceptional set N_1 such that $M^{(k_j)}_\infty$ converges to M_∞ in $L^2(\mathbf{P}_x)$ as well as $\mathbf{P}_x$-a.s., as $j \to \infty$, for every $x \in E \setminus N_1$. Consider the event $\Lambda = \{\lim_{j\to\infty} \varphi_{k_j} \text{ exists}\}$. Then $\mathbf{P}_x(\Lambda) = 1$ for every $x \in E_1 \setminus N_1$. If we let $\varphi = \lim_{j\to\infty} \varphi_{k_j}$ on Λ, then $\varphi \in \mathbf{N}$ and

$$M^{(k_j)}_\infty = \varphi_{k_j} - h_{k_j}(x) \to \varphi - h(x),\ j \to \infty,\ \mathbf{P}_x\text{-a.s. } x \in E_1 \setminus N_1.$$

Since $M^{(k_j)}_\infty$ is $L^2(\mathbf{P}_x)$-convergent to M_∞, we have $\varphi - h(x) = M_\infty$ and

$$h(x) = \mathbf{E}_x[\varphi], \qquad M_t = M^h_t, \quad t \geq 0.$$

We can now readily conclude that $h \in \mathbf{HN}$ and h_k is $\mathcal{E}^{\mathrm{ref}}$-convergent to h. □

For $u \in \mathcal{F}^{\mathrm{ref}}_a$, define $\mathcal{E}^{\mathrm{ref}}_1(u,u) := \mathcal{E}^{\mathrm{ref}}(u,u) + \|u\|^2_{L^2(E;m)}$.

THEOREM 6.2.4. *$(\mathcal{F}^{\mathrm{ref}}_a, \mathcal{E}^{\mathrm{ref}}_1)$ is a Hilbert space.*

Proof. By Remark 6.2.2, $(\mathcal{F}^{\mathrm{ref}}_a, \mathcal{E}^{\mathrm{ref}}_1)$ is a pre-Hilbert space. Suppose $\{u_n, n \geq 1\} \subset \mathcal{F}^{\mathrm{ref}}_a$ is $\mathcal{E}^{\mathrm{ref}}_1$-Cauchy. Then u_n converges to some u in $L^2(E;m)$ and there are $f_n \in \mathcal{F}_e$ and $h_n \in \mathbf{HN}$ so that $u_n = f_n + h_n$. Since $\{f_n, n \geq 1\}$ is $\mathcal{E}$-Cauchy and $(\mathcal{F}_e, \mathcal{E})$ is a Hilbert space, there is $f \in \mathcal{F}_e$ so that f_n is $\mathcal{E}$-convergent to f. By Theorem 2.1.5, there is a subsequence $\{n_k, k \geq 1\}$ so that f_{n_k} and u_{n_k} converge to f and u m-a.e. on E, respectively. It follows then that h_{n_k} converges to function $h := u - f$ m-a.e. on E. As $\{h_n, n \geq 1\} \subset \mathbf{HN}$ is $\mathcal{E}^{\mathrm{ref}}$-Cauchy, by Lemma 6.2.3, $h \in \mathbf{HN}$ and h_{n_k} (and hence h_n itself) is $\mathcal{E}^{\mathrm{ref}}$-convergent to h. This proves that u_n is $\mathcal{E}^{\mathrm{ref}}_1$-convergent to $u = f + h \in \mathcal{F}^{\mathrm{ref}}_a$. □

Before proving that every normal contraction operates on $(\mathcal{F}^{\mathrm{ref}}_a, \mathcal{E}^{\mathrm{ref}})$, we shall establish the following characterization and a related one for the space $(\mathcal{F}^{\mathrm{ref}}, \mathcal{E}^{\mathrm{ref}})$.

Let $\mathcal{F}_{\mathrm{loc}}$ denote the space of functions u on E so that for every relatively compact open subset $D \subset E$, there is a function $v \in \mathcal{F}$ with $u = v$ m-a.e. on D. Clearly $\mathcal{F}_{\mathrm{loc}}$ is a subspace of $\mathring{\mathcal{F}}_{\mathrm{loc}}$ defined by (4.3.31). For $f \in \mathcal{F}_{\mathrm{loc}}$, let $\mu^c_{\langle f\rangle}$ be the Revuz measure of the PCAF $\langle M^{[f],c}\rangle$, where $M^{[f],c}$ is the continuous part of the martingale $M^{[f]}$. Note that by Theorem 4.3.10(ii) the energy measure $\mu^c_{\langle f\rangle}$ is well-defined for every $f \in \mathcal{F}_{\rm loc}$. Moreover, by Theorem 4.3.10(iii), if $g := \phi(f)$ is a normal contraction of $f \in \mathcal{F}_{\rm loc}$, then

$$\mu^c_{\langle g\rangle} \leq \mu^c_{\langle f\rangle} \qquad \text{on } E. \tag{6.2.5}$$

For $f \in \mathcal{F}_{\rm loc}$, define

$$\widehat{\mathcal{E}}(f,f) := \frac{1}{2}\mu^c_{\langle f\rangle}(E) + \frac{1}{2}\int_{E\times E}(f(x)-f(y))^2 J(dx,dy) + \int_E f(x)^2\kappa(dx). \tag{6.2.6}$$

THEOREM 6.2.5. *The reflected Dirichlet space $\mathcal{F}^{\rm ref}$ consists of all the functions f on E finite m-a.e. that can be approximated m-a.e. on E by an $\widehat{\mathcal{E}}$-Cauchy sequence $\{f_n, n\geq 1\}\subset \mathcal{F}_{\rm loc}$. For each such $f\in\mathcal{F}^{\rm ref}$,*

$$\widehat{\mathcal{E}}(f,f) := \lim_{n\to\infty}\widehat{\mathcal{E}}(f_n,f_n)$$

is well-defined (that is, independently of the approximating functions chosen) and $\widehat{\mathcal{E}}(f,f) = \mathcal{E}^{\rm ref}(f,f)$ for $f\in\mathcal{F}^{\rm ref}$.

Remark 6.2.6. (i) Using a Cesàro means of a subsequence, one can conclude from Theorem 6.2.5 that $\mathcal{F}^{\rm ref}$ consists of all the functions f on E finite m-a.e. that can be approximated m-a.e. on E by an $\widehat{\mathcal{E}}$-bounded sequence $\{f_n, n\geq 1\}\subset \mathcal{F}_{\rm loc}$. In this case, $\mathcal{E}^{\rm ref}(f,f) = \widehat{\mathcal{E}}(f,f) \leq \sup_{n\geq 1}\widehat{\mathcal{E}}(f_n,f_n)$.

(ii) In fact, $\widehat{\mathcal{E}}(f,f)$ of (6.2.6) can be defined for every $f \in \mathring{\mathcal{F}}_{\rm loc}$ and the reflected Dirichlet space $\mathcal{F}^{\rm ref}$ can be characterized to be all the functions f in $\mathring{\mathcal{F}}_{\rm loc}$ so that $\widehat{\mathcal{E}}(f,f) < \infty$; see Theorem 6.4.2. One of the reasons that we choose to characterize $\mathcal{F}^{\rm ref}$ via $\mathcal{F}_{\rm loc}$ by Theorem 6.2.5 in this section is that in concrete examples, $\mathcal{F}_{\rm loc}$ is easier to be identified than $\mathring{\mathcal{F}}_{\rm loc}$, as we shall see in Section 6.5. □

To prove Theorem 6.2.5, we need some preparations.

LEMMA 6.2.7. *Bounded harmonic functions are in $\mathcal{F}_{\rm loc}$. A bounded harmonic function h is in* **HN** *if and only if $\widehat{\mathcal{E}}(h,h)<\infty$. In this case, $\mu_{\langle M^{h,c}\rangle} = \mu^c_{\langle h\rangle}$ and $\mathcal{E}^{\rm ref}(h,h) = \widehat{\mathcal{E}}(h,h)$.*

Proof. Let φ be the bounded terminal random variable that corresponds to h. To show that $h\in\mathcal{F}_{\rm loc}$, by considering $\mathbf{E}_x[\varphi^+]$ and $\mathbf{E}_x[\varphi^-]$ separately, we may assume that φ and $h(x) = \mathbf{E}_x[\varphi]$ are both non-negative. As we see from the proof of Theorem 6.1.6, h is excessive with respect to X. For any relatively compact open subset D, $f(x) := \mathbf{E}_x[e^{-\sigma_D}]$ is the 1-equilibrium potential of D. As $g := h\wedge(\|h\|_\infty f)$ is 1-excessive and $f\in\mathcal{F}$, we have $g\in\mathcal{F}$. Note that $h = g$ on D. This proves that $h\in\mathcal{F}_{\rm loc}$.

For a bounded harmonic function h, let φ be its corresponding terminal random variable. The martingale M^h defined by (6.1.7) is bounded and hence square integrable under $\mathbf{P}_x$ and so $\langle M^h \rangle$ exists. By Theorem 6.1.8(i), $h \in \mathbf{HN}$ if and only if M^h is a MAF of finite energy. In this case, by (6.2.4) and Theorem 6.1.8(ii),

$$\begin{aligned}\mathcal{E}^{\text{ref}}(h,h) &= \mathbf{e}(M^h) + \frac{1}{2}\int_E h(x)^2 \kappa(dx)\\ &= \frac{1}{2}\mu_{\langle M^{h,c}\rangle}(E) + \frac{1}{2}\int_{E\times E}(h(x)-h(y))^2 J(dx,dy) + \int_E h(x)^2\kappa(dx).\end{aligned}$$

The above is equal to $\widehat{\mathcal{E}}(h,h)$ if we can show that $\mu_{\langle M^{h,c}\rangle} = \mu^c_{\langle h\rangle}$, which we are going to do in the following.

As h is assumed to be in $b\mathcal{F}_{\text{loc}}$, for any relatively compact open subset $D \subset E$, there is a function $f \in b\mathcal{F}$ such that $h = f$ on D. By Fukushima's decomposition, $M_t^h = h(X_t) - h(X_0) = M_t^{[f]} + N_t^{[f]}$ on $\{t < \tau_D\}$ and so

$$M_t^{h,c} - M_t^{[f],c} = M_t^{[f],d} - M_t^{h,d} + N_t^{[f]} \qquad \text{for } t < \tau_D, \tag{6.2.7}$$

where $M^{[f],d}$ and $M^{h,d}$ are the purely discontinuous martingale parts of the square integrable martingales $M^{[f]}$ and M^h, respectively. Define

$$\begin{aligned}A_t &:= \Delta(M^{[f]}_{\tau_D} - M^h_{\tau_D})\mathbf{1}_{\{t\ge\tau_D\}}\\ &= ((f-h)(X_{\tau_D}) - (f-h)(X_{\tau_D-}) - \varphi\mathbf{1}_{\{\tau_D=\zeta\}})\mathbf{1}_{[\tau_D,\infty)}(t),\end{aligned}$$

which is a process of bounded variation. Then by **(3)** and **(5)** of Section A.3.3, $M := A - A^p$ is a purely discontinuous MAF of X that has jumps only at time τ_D and A^p is a continuous additive functional of X of finite variation. Hence the process $M_t^{[f],d} - M_t^{h,d} - M_t - A_t^p$ is continuous at τ_D and equal to $M_t^{[f],d} - M_t^{h,d}$ for $t < \tau_D$. So in particular we have by (6.2.7) that

$$M_t^{h,c} - M_t^{[f],c} = M_t^{[f],d} - M_t^{h,d} - M_t + (N_t^{[f]} - A_t^p) \qquad \text{for } t \le \tau_D.$$

It follows that the purely discontinuous square integrable martingale $t \mapsto M^{[f],d}_{t\wedge\tau_D} - M^{h,d}_{t\wedge\tau_D} - M_{t\wedge\tau_D}$ is continuous in $t \ge 0$. Consequently, $M_t^{[f],d} - M_t^{h,d} - M_t = 0$ for every $t \in [0, \tau_D]$ and so

$$M_t^{h,c} - M_t^{[f],c} = N_t^{[f]} - A_t^p \qquad \text{for } t \le \tau_D. \tag{6.2.8}$$

Let w be a strictly positive bounded Borel function on E so that $\int_E w(x)m(dx) = 1$. For a stochastic process $\{B_t\}$, define

$$V_B(t,\tau_D,n) = \sum_{k=1}^n \left(B_{(\frac{k}{n}t)\wedge\tau_D} - B_{(\frac{k-1}{n}t)\wedge\tau_D}\right)^2.$$

We have by (A.3.37) that

$$\langle M^{h,c} - M^{[f],c}\rangle_{t\wedge\tau_D} = [M^{h,c} - M^{[f],c}]_{t\wedge\tau_D} = \lim_{n\to\infty} V_{M^{h,c}-M^{[f],c}}(t,\tau_D,n)$$

in probability with respect to $\mathbf{P}_{w\cdot m}$. Since A^p is continuous and of bounded variation, $\lim_{n\to\infty} V_{A^p}(t, \tau_D, n) = 0$ $\mathbf{P}_{w\cdot m}$-a.s. and hence in probability with respect to $\mathbf{P}_{w\cdot m}$. Since $N^{[f]}$ is a continuous additive functional of X of zero energy,

$$\begin{aligned}\lim_{n\to\infty} \mathbf{E}_m \left[V_{N^{[f]}}(t, \tau_D, n)\right] &\le \lim_{n\to\infty} \mathbf{E}_m \left[\sum_{k=1}^{n} \left(N^{[f]}_{kt/n} - N^{[f]}_{(k-1)t/n}\right)^2\right] \\ &= \lim_{n\to\infty} \sum_{k=1}^{n} \mathbf{E}_m \left[\mathbf{E}_{X_{(n-1)t/n}}\left[\left(N^{[f]}_{1/n}\right)^2\right]\right] \\ &\le \lim_{n\to\infty} n\mathbf{E}_m \left[\left(N^{[f]}_{t/n}\right)^2\right] = 0.\end{aligned}$$

Consequently, $V_{N^{[f]}}(t, \tau_D, n)$ and so $V_{N^{[f]}-A^p}(t, \tau_D, n)$ converges to 0 in probability with respect to $\mathbf{P}_{w\cdot m}$. In view of (6.2.8), we have shown that $\langle M^{h,c} - M^{[f],c}\rangle_{t\wedge\tau_D} = 0$, $\mathbf{P}_{w\cdot m}$-a.s. This together with (4.1.25) yields $\mu_{\langle M^{h,c}-M^{[f],c}\rangle}(D) = 0$. Hence $\mu_{\langle M^{h,c}\rangle} = \mu_{\langle M^{[f],c}\rangle} = \mu^c_{\langle h\rangle}$ on D. Since this holds for every relatively compact open subset D of E, we have $\mu_{\langle M^{h,c}\rangle} = \mu^c_{\langle h\rangle}$ on E, which completes the proof of the lemma. □

Lemma 6.2.8. *Suppose that $\varphi \in \mathbf{N}$ and $h(x) := \mathbf{E}_x[\varphi]$. For $n \ge 1$, define $\varphi_n := ((-n) \vee \varphi) \wedge n$ and $h_n(x) := \mathbf{E}_x[\varphi_n]$. Then $\varphi_n \in \mathbf{N}$ and h_n converges to h both q.e. on E and in $\mathcal{E}^{\text{ref}}$-norm.*

Proof. Clearly $\lim_{n\to\infty} h_n = h$ q.e. on E. Let Var_x and Cov_x denote variance and covariance under the probability measure $\mathbf{P}_x$. Since

$$\begin{aligned}&\mathrm{Cov}_x(\varphi - \varphi_n, \varphi_n) \\ &\quad= \mathbf{E}_x[(\varphi - \varphi_n)\varphi_n] - \mathbf{E}_x[\varphi - \varphi_n]\mathbf{E}_x[\varphi_n] \\ &\quad\ge \mathbf{E}_x[n(\varphi - n); \varphi > n] + \mathbf{E}_x[(-n)(\varphi + n); \varphi < -n] - n\mathbf{E}_x[|\varphi - \varphi_n|] \\ &\quad= 0,\end{aligned}$$

we have

$$\begin{aligned}\mathrm{Var}_x(\varphi) &= \mathrm{Var}_x(\varphi - \varphi_n + \varphi_n) \\ &= \mathrm{Var}_x(\varphi - \varphi_n) + \mathrm{Var}_x(\varphi_n) + 2\mathrm{Cov}_x(\varphi - \varphi_n, \varphi_n) \\ &\ge \mathrm{Var}_x(\varphi - \varphi_n) + \mathrm{Var}_x(\varphi_n). \qquad (6.2.9)\end{aligned}$$

Let $g(x) := \mathbf{E}_x[\varphi^2] - (h(x))^2$ and $g_n(x) := \mathbf{E}_x[\varphi_n^2] - (h_n(x))^2$. Then $g \ge g_n$ as $g(x) = \mathrm{Var}_x(\varphi)$ and $g_n(x) = \mathrm{Var}_x(\varphi_n)$. So by Theorem 5.4.3(ii) applied to the present energy functional L of X, we have $L(g_n, 1) \le L(g, 1)$ for every $n \ge 1$. Consequently, $\varphi_n \in \mathbf{N}$ for every $n \ge 1$ and $\mathbf{e}(M^{h_n}) \le \mathbf{e}(M^h)$ by (6.1.12).

Similarly, for $j > n$, with φ_j in place of φ, we have $\mathrm{Var}_x(\varphi_j) \geq \mathrm{Var}_x(\varphi_j - \varphi_n) + \mathrm{Var}_x(\varphi_n)$ and so

$$\mathbf{e}(M^{h_j}) = \frac{1}{2}L(g_j, 1) \geq \mathbf{e}(M^{h_j} - M^{h_n}) + \mathbf{e}(M^{h_n}).$$

As $\lim_{n\to\infty} \mathbf{e}(M^{h_n})$ exists and is finite, $\mathbf{e}(M^{h_j} - M^{h_n}) \to 0$ as $j, n \to \infty$. Since by (6.1.13) and (6.2.4) $\mathcal{E}^{\mathrm{ref}}(h_j - h_k, h_j - h_k) \leq 2\mathbf{e}(M^{h_j} - M^{h_n})$, $\{h_j, j \geq 1\}$ is an $\mathcal{E}^{\mathrm{ref}}$-Cauchy sequence. As $h_j \to h$ q.e. on E, we have by Lemma 6.2.3 that h_j is $\mathcal{E}^{\mathrm{ref}}$-convergent to h. □

Remark 6.2.9. By a similar argument as that leads to (6.2.9), we have for any random variable φ, $\mathrm{Var}_x(\varphi) \geq \mathrm{Var}_x(\varphi^+) + \mathrm{Var}_x(\varphi - \varphi^+)$. Thus $\varphi^+ \in \mathbf{N}$ for any $\varphi \in \mathbf{N}$ and $\mathbf{e}(M^{h_1}) \leq \mathbf{e}(M^h)$, where $h_1(x) := \mathbf{E}_x[\varphi^+]$. □

The next lemma holds for any regular Dirichlet form $(\mathcal{E}, \mathcal{F})$ on $L^2(E; m)$. For an open set $D \subset E$, let

$$\mathcal{F}_{e,D} = \{u \in \mathcal{F}_e : u = 0 \quad q.e. \ on\ D^c\}.$$

Lemma 6.2.10. *Let f be a bounded function in $\mathcal{F}_{\mathrm{loc}}$ and D a relatively compact open subset of E. Then $f - \mathbf{H}_{D^c}f \in \mathcal{F}_{e,D}$ and*

$$\begin{aligned}
&\mathcal{E}(f - \mathbf{H}_{D^c}f, f - \mathbf{H}_{D^c}f) \\
&\quad \leq \frac{1}{2}\mu^c_{\langle f\rangle}(\overline{D}) + \frac{1}{2}\int_{D\times D}(f(x) - f(y))^2 J(dx, dy) \\
&\qquad + \int_{D\times D^c}(f(x) - f(y))^2 J(dx, dy) + \int_D f(x)^2 \kappa(dx). \qquad (6.2.10)
\end{aligned}$$

Proof. Clearly by definition of $\mathbf{H}_{D^c}f(x) := \mathbf{E}_x[f(X_{\tau_D})]$, $\mathbf{H}_{D^c}f = f$ q.e. on D^c. If $f \in \mathcal{F}_e$, we know from Theorem 3.4.8 that $f - \mathbf{H}_{D^c}f \in \mathcal{F}_{e,D}$ and that

$$\mathcal{E}(f, f) = \mathcal{E}(f - \mathbf{H}_{D^c}f, f - \mathbf{H}_{D^c}f) + \mathcal{E}(\mathbf{H}_{D^c}f, \mathbf{H}_{D^c}f).$$

The Beurling-Deny formula (Theorem 4.3.3) tells us that

$$\mathcal{E}(f, f) = \frac{1}{2}\mu^c_{\langle f\rangle}(E) + \frac{1}{2}\int_{E\times E}(f(x) - f(y))^2 J(dx, dy) + \int_E f(x)^2\kappa(dx)$$

and

$$
\begin{aligned}
&\mathcal{E}(\mathbf{H}_{D^c}f, \mathbf{H}_{D^c}f) \\
&\quad = \frac{1}{2}\mu^c_{\langle \mathbf{H}_{D^c}f\rangle}(E) + \frac{1}{2}\int_{E\times E}(\mathbf{H}_{D^c}f(x) - \mathbf{H}_{D^c}f(y))^2 J(dx, dy) \\
&\qquad + \int_E \mathbf{H}_{D^c}f(x)^2\kappa(dx) \\
&\quad \geq \frac{1}{2}\mu^c_{\langle \mathbf{H}_{D^c}f\rangle}(E\setminus \overline{D}) + \frac{1}{2}\int_{D^c\times D^c}(f(x) - f(y))^2 J(dx, dy) \\
&\qquad + \int_{D^c} f(x)^2\kappa(dx) \\
&\quad = \frac{1}{2}\mu^c_{\langle f\rangle}(E\setminus \overline{D}) + \frac{1}{2}\int_{D^c\times D^c}(f(x) - f(y))^2 J(dx, dy) + \int_{D^c} f(x)^2\kappa(dx).
\end{aligned}
$$

So (6.2.10) holds when $f \in \mathcal{F}_e$.

For a general bounded $f \in \mathcal{F}_{\text{loc}}$, take an increasing sequence of relatively compact open subsets $\{D_k, k \geq 1\}$ so that $\cup_{k\geq 1} D_k = E$ and $\overline{D} \subset D_k$. For each $k \geq 1$, there is a bounded $g_k \in \mathcal{F}$ so that $f = g_k$ m-a.e. on D_k and $\|g_k\|_\infty \leq \|f\|_\infty$. Note that $\mu^c_{\langle f\rangle} = \mu^c_{\langle g_k\rangle}$ on D_1. From (6.2.10) for g_k,

$$
\begin{aligned}
&\mathcal{E}(g_k - \mathbf{H}_{D^c}g_k, g_k - \mathbf{H}_{D^c}g_k) \\
&\quad \leq \frac{1}{2}\mu^c_{\langle f\rangle}(\overline{D}) + \frac{1}{2}\int_{D\times D}(f(x) - f(y))^2 J(dx, dy) \\
&\qquad + \int_{D\times(D^c\cap D_1)}(f(x) - f(y))^2 J(dx, dy) \\
&\qquad + \int_{D\times(D^c\cap D_1^c)}(f(x) - g_k(y))^2 J(dx, dy) + \int_D f(x)^2\kappa(dx).
\end{aligned}
$$

As $J(D, D_1^c) < \infty$ by the regularity of $(\mathcal{E}, \mathcal{F})$, we have by the bounded convergence theorem

$$
\begin{aligned}
&\limsup_{k\to\infty} \mathcal{E}(g_k - \mathbf{H}_{D^c}g_k, g_k - \mathbf{H}_{D^c}g_k) \\
&\quad \leq \frac{1}{2}\mu^c_{\langle f\rangle}(\overline{D}) + \frac{1}{2}\int_{D\times D}(f(x) - f(y))^2 J(dx, dy) \\
&\qquad + \int_{D\times D^c}(f(x) - f(y))^2 J(dx, dy) + \int_D f(x)^2\kappa(dx) < \infty.
\end{aligned}
$$

The above argument in particular implies that the right hand side of (6.2.10) is finite. By Theorem A.4.1, there is a subsequence $\{g_{k_j} - \mathbf{H}_{D^c}g_{k_j}, j \geq 1\}$ so that its Cesàro means $\{\psi_j := (1/j)\sum_{l=1}^{j}(g_{k_l} - \mathbf{H}_{D^c} g_{k_l}), j \geq 1\}$ is $\mathcal{E}$-Cauchy. Since ψ_j converges to $f - \mathbf{H}_{D^c} f$ on E, $f - \mathbf{H}_{D^c} f \in \mathcal{F}_{e,D}$ and

$$\mathcal{E}(f - \mathbf{H}_{D^c} f, f - \mathbf{H}_{D^c} f) = \lim_{j\to\infty} \mathcal{E}(\psi_j, \psi_j)$$

$$\leq \limsup_{k\to\infty} \mathcal{E}(g_k - \mathbf{H}_{D^c} g_k, g_k - \mathbf{H}_{D^c} g_k).$$

Thus (6.2.10) follows. □

The next lemma says that bounded functions in $\mathcal{F}_{\text{loc}}$ having finite $\widehat{\mathcal{E}}$-norm are functions in $\mathcal{F}^{\text{ref}}$ and the $\widehat{\mathcal{E}}$- and $\mathcal{E}^{\text{ref}}$-norms are equivalent on $\mathcal{F}^{\text{ref}}$.

Lemma 6.2.11. *For every bounded $u \in \mathcal{F}_{\text{loc}}$ with $\widehat{\mathcal{E}}(u,u) < \infty$, there is a unique function $f \in \mathcal{F}_e$ and a unique $h \in \mathbf{HN}$ so that $u = f + h$. Moreover, both f and h are bounded,*

$$\mathcal{E}(f,f) \leq \widehat{\mathcal{E}}(u,u), \quad \mathcal{E}^{\text{ref}}(h,h) \leq 4\widehat{\mathcal{E}}(u,u), \quad \text{and} \quad \widehat{\mathcal{E}}(u,u) \leq 2\mathcal{E}^{\text{ref}}(u,u).$$

Proof. Let $\{D_k, k \geq 1\}$ be an increasing sequence of relatively compact open sets with $\cup_{k\geq 1} D_k = E$. By Lemma 6.2.10 $u - \mathbf{H}_{D_k^c} u \in \mathcal{F}_e$ with $\sup_{k\geq 1} \mathcal{E}(u - \mathbf{H}_{D_k^c} u, u - \mathbf{H}_{D_k^c} u) \leq \widehat{\mathcal{E}}(u,u) < \infty$. Therefore, by Theorem A.4.1, there is a subsequence $\{u - \mathbf{H}_{D_{k_j}^c} u, j \geq 1\}$ so that its Cesàro means $\{\psi_j := (1/j) \sum_{l=1}^{j}(u - \mathbf{H}_{D_{k_l}^c} u), j \geq 1\}$ is $\mathcal{E}$-convergent to some $f \in \mathcal{F}_e$. Clearly f is bounded and

$$\mathcal{E}(f,f) = \lim_{j\to\infty} \mathcal{E}(\psi_j, \psi_j) \leq \limsup_{k\to\infty} \mathcal{E}(u - \mathbf{H}_{D_k^c} u, u - \mathbf{H}_{D_k^c} u) \leq \widehat{\mathcal{E}}(u,u).$$

Define $h := u - f$. For any relatively compact open subset D of E, there is some $k \geq 1$ so that $\overline{D} \subset D_k$. Note that for $j > k$,

$$\left(u - \mathbf{H}_{D_j^c} u\right) - \mathbf{H}_{D^c}\left(u - \mathbf{H}_{D_j^c} u\right) = u - \mathbf{H}_{D^c} u.$$

Since $\{\psi_j\}$ is uniformly bounded and convergent to f q.e. by taking a subsequence (denoted by $\{\psi_j\}$ again) in view of Theorem 2.3.2, we have $\lim_{j\to\infty}(\psi_j - \mathbf{H}_{D^c}\psi_j) = f - \mathbf{H}_{D^c} f$ and further $f - \mathbf{H}_{D^c} f = u - \mathbf{H}_{D^c} u$ by the above observation. In other words, $h = \mathbf{H}_{D^c} h$. Hence h is a bounded harmonic function on E. By Lemma 6.2.7, $h \in b\mathcal{F}_{\text{loc}}$ and so by the triangle inequality

$$\widehat{\mathcal{E}}(h,h) = \widehat{\mathcal{E}}(u - f, u - f) \leq 2\widehat{\mathcal{E}}(u,u) + 2\widehat{\mathcal{E}}(f,f) \leq 4\widehat{\mathcal{E}}(u,u).$$

By Lemma 6.2.7 again, $h \in \mathbf{HN}$ with $\mathcal{E}^{\text{ref}}(h,h) = \widehat{\mathcal{E}}(h,h)$. So $u = f + h \in \mathcal{F}^{\text{ref}}$. On the other hand,

$$\widehat{\mathcal{E}}(u,u) \leq 2\widehat{\mathcal{E}}(f,f) + 2\widehat{\mathcal{E}}(h,h) = 2\mathcal{E}(f,f) + 2\mathcal{E}^{\text{ref}}(h,h) = 2\mathcal{E}^{\text{ref}}(u,u).$$

The uniqueness of f and h follows from the unique decomposition of $u \in \mathcal{F}^{\text{ref}}$ into a function in $\mathcal{F}_e$ and a function in $\mathbf{HN}$. □

We now remove the boundedness assumption on u from the preceding lemma.

LEMMA 6.2.12. *For every $u \in \mathcal{F}_{\rm loc}$ with $\widehat{\mathcal{E}}(u,u) < \infty$, there is a unique function $f \in \mathcal{F}_e$ and a unique $h \in \mathbf{HN}$ so that $u = f + h$. Moreover,*

$$\mathcal{E}(f,f) \le \widehat{\mathcal{E}}(u,u), \quad \mathcal{E}^{\rm ref}(h,h) \le 4\widehat{\mathcal{E}}(u,u), \quad \text{and } \widehat{\mathcal{E}}(u,u) \le 2\mathcal{E}^{\rm ref}(u,u). \tag{6.2.11}$$

Proof. Define $u_n := ((-n) \vee u) \wedge n$. Note that u_n is a normal contraction of u and so $u_n \in \mathcal{F}_{\rm loc}$ with $\sup_{n\ge 1} \widehat{\mathcal{E}}(u_n,u_n) \le \widehat{\mathcal{E}}(u,u)$. By Lemma 6.2.11, there are bounded $f_n \in \mathcal{F}_e$ and $h_n \in \mathbf{HN}$ so that $u_n = f_n + h_n$. Since $\sup_{n\ge 1} \mathcal{E}(f_n,f_n) \le \sup_{n\ge 1} \widehat{\mathcal{E}}(u_n,u_n) \le \widehat{\mathcal{E}}(u,u)$, by Theorem A.4.1, there is a subsequence $\{f_{n_k}, k \ge 1\}$ so that the sequence of its Cesàro means $\{g_k := (1/k)\sum_{j=1}^k f_{n_j}, k \ge 1\}$ is $\mathcal{E}$-convergent to some $f \in \mathcal{F}_e$. Clearly, $\mathcal{E}(f,f) \le \sup_{n\ge 1} \mathcal{E}(f_n,f_n) \le \widehat{\mathcal{E}}(u,u)$. As in the proof of the preceding lemma, we may assume without loss of generality that g_k converges to f q.e. on E by taking a subsequence if necessary.

Define $h := u - f$ and $\psi_k := (1/k)\sum_{j=1}^k h_{n_j}$. Then ψ_k converges to h q.e. on E and, $\psi_k \in \mathbf{HN}$ with

$$\begin{aligned}
\mathcal{E}^{\rm ref}(\psi_k,\psi_k)^{1/2} &\le \frac{1}{k}\sum_{j=1}^k \mathcal{E}^{\rm ref}(h_{n_j},h_{n_j})^{1/2} \le \frac{1}{k}\sum_{j=1}^k 2\widehat{\mathcal{E}}(u_{n_j},u_{n_j})^{1/2} \\
&\le 2\widehat{\mathcal{E}}(u,u)^{1/2}.
\end{aligned}$$

By using the Cesàro means of a subsequence of $\{\psi_k, k \ge 1\}$, we conclude from Theorem A.4.1 and Lemma 6.2.3 that $h \in \mathbf{HN}$ with

$$\mathcal{E}^{\rm ref}(h,h) \le \sup_{k\ge 1} \mathcal{E}^{\rm ref}(\psi_k,\psi_k) \le 4\widehat{\mathcal{E}}(u,u).$$

This proves the first two inequalities in (6.2.11).

Note that $u - u_n = \mathbf{1}_{\{|u|\ge n\}} u$. In view of Proposition 4.3.1 and dominated convergence theorem, we have $\lim_{n\to\infty} \widehat{\mathcal{E}}(u_n - u, u_n - u) = 0$ and so $v_k := (1/k)\sum_{j=1}^k u_{n_j} \in b\mathcal{F}_{\rm loc}$ is $\widehat{\mathcal{E}}$-convergent to u. For $j > k$, since $v_j - v_k = (g_j - g_k) + (\psi_j - \psi_k)$ is the unique decomposition of $v_j - v_k \in b\mathcal{F}_{\rm loc}$ into $g_j - g_k \in \mathcal{F}_e$ and $\psi_j - \psi_k \in \mathbf{HN}$, we conclude from the second inequality of (6.2.11) that $\{\psi_k, k \ge 1\}$ is $\mathcal{E}^{\rm ref}$-Cauchy. Thus by Lemmas 6.2.3 and 6.2.11,

$$\begin{aligned}
\widehat{\mathcal{E}}(u,u) &= \lim_{k\to\infty} \widehat{\mathcal{E}}(v_k,v_k) \le \limsup_{k\to\infty} 2\mathcal{E}^{\rm ref}(v_k,v_k) \\
&= 2\limsup_{k\to\infty} \left(\mathcal{E}(g_k,g_k) + \mathcal{E}^{\rm ref}(\psi_k,\psi_k)\right) \\
&= 2\mathcal{E}(f,f) + 2\mathcal{E}^{\rm ref}(h,h) = 2\mathcal{E}^{\rm ref}(u,u).
\end{aligned}$$

The lemma is established. □

We are now ready to give the

Proof of Theorem 6.2.5. Let $\mathcal{G}$ denote the linear space of functions u for which there is an $\widehat{\mathcal{E}}$-Cauchy sequence of $\{u_n, n \geq 1\} \subset \mathcal{F}_{\text{loc}}$ that converges to u m-a.e. on E. For $u \in \mathcal{G}$ and its approximating sequence $\{u_n, n \geq 1\}$, by Lemma 6.2.12, there are $\mathcal{E}$-Cauchy sequence $\{f_n \geq 1\} \subset \mathcal{F}_e$ and $\mathcal{E}^{\text{ref}}$-Cauchy sequence $\{h_n, n \geq 1\} \subset \mathbf{HN}$ such that $u_n = f_n + h_n$ for every $n \geq 1$. Therefore, f_n is $\mathcal{E}$-convergent to some $f \in \mathcal{F}_e$. By taking a subsequence if necessary, we may assume that f_n converges to f q.e. on E. Consequently, h_n converges m-a.e. to $h := u - f$. By Lemma 6.2.3 and (6.2.11), $h \in \mathbf{HN}$ and $\mathcal{E}^{\text{ref}}(h, h) = \lim_{n\to\infty} \mathcal{E}^{\text{ref}}(h_n, h_n) \leq 4\widehat{\mathcal{E}}(u, u)$. It follows then $u = f + h \in \mathcal{F}^{\text{ref}}$; in other words, we have shown $\mathcal{G} \subset \mathcal{F}^{\text{ref}}$. On the other hand, evidently $\mathcal{F}_e \subset \mathcal{G}$ and $\mathbf{HN} \subset \mathcal{G}$ by Lemmas 6.2.7 and 6.2.8. It follows $\mathcal{F}^{\text{ref}} \subset \mathcal{G}$ and so $\mathcal{G} = \mathcal{F}^{\text{ref}}$.

We now show that $\widehat{\mathcal{E}}(u, u) := \lim_{n\to\infty} \widehat{\mathcal{E}}(u_n, u_n)$ is well-defined on $\mathcal{G}$. Suppose that $\{u'_n, n \geq 1\} \subset \mathcal{F}_{\text{loc}}$ is another $\widehat{\mathcal{E}}$-Cauchy sequence that converges to u m-a.e. on E. Let $f'_n \in \mathcal{F}_e$ and $h'_n \in \mathbf{HN}$ be such that $u'_n = f'_n + h'_n$. By the above reasoning, f'_n is $\mathcal{E}$-convergent to some $f' \in \mathcal{F}_e$, h'_n is $\mathcal{E}^{\text{ref}}$-convergent to some $h' \in \mathbf{HN}$, and $u = f' + h'$. By the uniqueness of the decomposition for $u \in \mathcal{F}^{\text{ref}}$ (see Remark 6.2.2), $f' = f$ and $h' = h$. It follows that

$$\begin{aligned}
&\lim_{n\to\infty} \mathcal{E}^{\text{ref}}(u_n - u'_n, u_n - u'_n) \\
&\quad = \lim_{n\to\infty} \mathcal{E}(f_n - f'_n, f_n - f'_n) + \lim_{n\to\infty} \mathcal{E}^{\text{ref}}(h_n - h'_n, h_n - h'_n) = 0.
\end{aligned}$$

Thus by (6.2.11),

$$\lim_{n\to\infty} \widehat{\mathcal{E}}(u_n - u'_n, u_n - u'_n) \leq \lim_{n\to\infty} 2\mathcal{E}^{\text{ref}}(u_n - u'_n, u_n - u'_n) = 0$$

and so $\lim_{n\to\infty} \widehat{\mathcal{E}}(u_n, u_n) = \lim_{n\to\infty} \widehat{\mathcal{E}}(u'_n, u'_n)$.

We finally prove that $\widehat{\mathcal{E}} = \mathcal{E}^{\text{ref}}$ on $\mathcal{F}^{\text{ref}}$. For $f, g \in \mathcal{F}_{\text{loc}}$ with $\widehat{\mathcal{E}}(f, f) < \infty$, $\widehat{\mathcal{E}}(g, g) < \infty$, we define $\widehat{\mathcal{E}}(f, g) = \frac{1}{4}\widehat{\mathcal{E}}(f + g, f + g) - \frac{1}{4}\widehat{\mathcal{E}}(f - g, f - g)$ so that

$$\begin{aligned}
\widehat{\mathcal{E}}(f, g) = {}& \frac{1}{2}\mu^c_{\langle f,g\rangle}(E) + \frac{1}{2}\int_{E\times E} ((f(x) - f(y))(g(x) - g(y))J(dx, dy) \\
&+ \int_E f(x)g(x)\kappa(dx),
\end{aligned} \tag{6.2.12}$$

where $\mu^c_{\langle f,g\rangle} = \frac{1}{4}\mu^c_{\langle f+g\rangle} - \frac{1}{4}\mu^c_{\langle f-g\rangle}$. As $\widehat{\mathcal{E}} = \mathcal{E}$ on $\mathcal{F}_e$ and, by Lemmas 6.2.7 and 6.2.8, $\widehat{\mathcal{E}} = \mathcal{E}^{\text{ref}}$ on $\mathbf{HN}$, it suffices to show that $\widehat{\mathcal{E}}(f, h) = 0$ for $f \in \mathcal{F}_e$ and $h \in \mathbf{HN}$. In fact, one needs only to show that $\widehat{\mathcal{E}}(f, h) = 0$ for non-negative functions $f \in \mathcal{F} \cap C_c(E)$ and for $h(x) = \mathbf{E}_x[\varphi]$ with non-negative bounded functions $\varphi \in \mathbf{N}$, since the linear spans of such functions are $\widehat{\mathcal{E}}$-dense in $\mathcal{F}_e$ and in $\mathbf{HN}$, respectively.

In this case, it holds that

$$\widehat{\mathcal{E}}(f,h)=\mathbf{e}(M^{[f]},M^h)+\frac{1}{2}\int_E f(x)h(x)\kappa(dx)$$
$$=\lim_{t\to 0}\frac{1}{2t}\mathbf{E}_m\big[M_t^{[f]}M_t^h\big]+\frac{1}{2}\int_E f(x)h(x)\kappa(dx), \qquad (6.2.13)$$

where $M^{[f]}$ and $M^h \in \overset{\circ}{\mathcal{M}}$ are martingale additive functionals of finite energy specified by Theorem 4.2.6 and (6.1.8), respectively, and $\mathbf{e}$ denotes the mutual energy defined by (4.2.2). In fact, by using the polarized expression of the quadratic variation in **(4)** of Section A.3.3, we can derive just as in the proof of Theorem 6.1.8 the equality

$$\mathbf{e}(M^{[f]},M^h)=\frac{1}{2}\mu_{\langle M^{[f],c},M^{h,c}\rangle}(E)$$
$$+\frac{1}{2}\int_{E\times E}(f(x)-f(y))(h(x)-h(y))J(dx,dy)+\frac{1}{2}\int_E f(x)h(x)\kappa(dx).$$

But, in view of the proof of Lemma 6.2.7, we have $\mu_{\langle M^{h,c}\rangle}=\mu^c_{\langle h\rangle}$ and so $\mu_{\langle M^{[f],c},M^{h,c}\rangle}=\mu^c_{\langle h,f\rangle}$. This combined with the above equality and (6.2.12) leads us to (6.2.13).

From (6.2.13), we obtain

$$\widehat{\mathcal{E}}(f,h)=\lim_{t\to 0}\frac{1}{2t}\mathbf{E}_m\left[(f(X_t)-f(X_0))M_t^h\right]+\frac{1}{2}\int_E f(x)h(x)\kappa(dx)$$
$$=\lim_{t\to 0}\frac{1}{2t}\mathbf{E}_m\left[f(X_t)M_t^h\right]+\frac{1}{2}\int_E f(x)h(x)\kappa(dx)$$
$$=\lim_{t\to 0}\frac{1}{2t}\mathbf{E}_m\left[f(X_t)(h(X_t)-h(X_0));t<\zeta\right]$$
$$+\frac{1}{2}\int_E f(x)h(x)\kappa(dx), \qquad (6.2.14)$$

where in the first equality we used Fukushima's decomposition $M_t^{[f]}=f(X_t)-f(X_0)-N_t^{[f]}$ and the fact that $N_t^{[f]}$ has zero energy. Due to the symmetry of X and the expression (6.1.7) of the MAF M^h,

$$\mathbf{E}_m\left[f(X_t)(h(X_t)-h(X_0));t<\zeta\right]$$
$$=\mathbf{E}_m\left[f(X_0)(h(X_0)-h(X_t));\ t<\zeta\right]$$
$$=\mathbf{E}_{f\cdot m}\left[\varphi\mathbf{1}_{\{t\geq\zeta\}}-h(X_0)\mathbf{1}_{\{t\geq\zeta\}}\right]$$
$$=\mathbf{E}_{f\cdot m}\left[\varphi\mathbf{1}_{\{t\geq\zeta,X_{\zeta-}=\partial\}}\right]-\mathbf{P}_{fh\cdot m}(t\geq\zeta). \qquad (6.2.15)$$

Since $f \in \mathcal{F} \cap C_c(E)$, there exists some $g \in \mathcal{F} \cap C_c(E)$ with $f \le g^2$ on E. Then $\mathbf{E}_{f\cdot m}\big[\varphi \mathbf{1}_{\{t\ge\zeta,\, X_{\zeta-}=\partial\}}\big]$ is dominated by

$$||\varphi||_\infty \cdot \mathbf{P}_{g^2\cdot m}(t > \zeta,\ X_{\zeta-} = \partial),$$

which is $o(t)$ on account of Proposition 4.2.3. Furthermore, by virtue of (4.2.8),

$$\lim_{t\to\infty} \frac{1}{2t} \mathbf{P}_{fh\cdot m}(t \ge \zeta) = \frac{1}{2} \int_E f(x)h(x)\kappa(dx).$$

Therefore, (6.2.14) and (6.2.15) yield $\widehat{\mathcal{E}}(f, h) = 0$. This completes the proof of the theorem. □

For $k \ge 1$, define truncation operator τ_k by $\tau_k u := ((-k) \vee u) \wedge k$.

THEOREM 6.2.13. *It holds that*

$$\mathcal{F}^{\text{ref}} = \left\{u : |u| < \infty\ [m],\ \tau_k u \in \mathcal{F}_{\text{loc}},\ \forall k \ge 1,\ \sup_{k\ge1} \widehat{\mathcal{E}}(\tau_k u, \tau_k u) < \infty\right\}$$

and

$$\mathcal{E}^{\text{ref}}(u, u) = \lim_{k\to\infty} \widehat{\mathcal{E}}(\tau_k u, \tau_k u) \qquad \text{for } u \in \mathcal{F}^{\text{ref}}.$$

Proof. By Remark 6.2.6,

$$\mathcal{F}^{\text{ref}} \supset \left\{u : |u| < \infty\ [m],\ \tau_k u \in \mathcal{F}_{\text{loc}},\ \forall k \ge 1,\ \sup_{k\ge1} \widehat{\mathcal{E}}(\tau_k u, \tau_k u) < \infty\right\}.$$

For $u \in \mathcal{F}^{\text{ref}}$, by Theorem 6.2.5, there is an $\widehat{\mathcal{E}}$-Cauchy sequence $\{u_n, n \ge 1\}$ in $\mathcal{F}_{\text{loc}}$ that converges to u m-a.e. on E. By (6.2.5) and (6.2.6), $\tau_k u_n \in \mathcal{F}^{\text{ref}}$ and for each $k \ge 1$,

$$\limsup_{n\to\infty} \widehat{\mathcal{E}}(\tau_k u_n, \tau_k u_n) \le \lim_{n\to\infty} \widehat{\mathcal{E}}(u_n, u_n) = \mathcal{E}^{\text{ref}}(u, u) < \infty.$$

So by Remark 6.2.6, $\tau_k u \in \mathcal{F}^{\text{ref}}$ and $\mathcal{E}^{\text{ref}}(\tau_k u, \tau_k u) \le \mathcal{E}^{\text{ref}}(u, u)$. As $\tau_k u$ is a bounded function in $\mathcal{F}^{\text{ref}}$, in view of Remark 6.2.2, there are bounded $f \in \mathcal{F}_e$ and bounded $h \in \mathbf{HN}$ so that $\tau_k u = f + h$. It follows from Lemma 6.2.7 that $\tau_k u \in \mathcal{F}_{\text{loc}}$. This shows that

$$\mathcal{F}^{\text{ref}} \subset \left\{u : |u| < \infty\ [m],\ \tau_k u \in \mathcal{F}_{\text{loc}},\ \forall k \ge 1,\ \sup_{k\ge1} \widehat{\mathcal{E}}(\tau_k u, \tau_k u) < \infty\right\}$$

and so these two function spaces are the same.

Note that for $u \in \mathcal{F}^{\text{ref}}$, by (6.2.5) and (6.2.6), $\widehat{\mathcal{E}}(\tau_k u, \tau_k u)$ is increasing in $k \ge 1$ and bounded by $\mathcal{E}(u, u)$. Since $\tau_k u$ is convergent pointwise to u as $k \to \infty$, we have from Remark 6.2.6

$$\mathcal{E}^{\text{ref}}(u, u) = \widehat{\mathcal{E}}(u, u) \le \lim_{k\to\infty} \widehat{\mathcal{E}}(\tau_k u, \tau_k u) \le \widehat{\mathcal{E}}(u, u).$$

□

We are now ready to present the main theorem of this section.

THEOREM 6.2.14. *Define* $\mathcal{F}_a^{\rm ref} := \mathcal{F}^{\rm ref} \cap L^2(E;m)$. *Then* $(\mathcal{E}^{\rm ref}, \mathcal{F}_a^{\rm ref})$ *is a Dirichlet form on* $L^2(E;m)$.

Proof. For $u \in \mathcal{F}_a^{\rm ref}$, there is an $\widehat{\mathcal{E}}$-Cauchy sequence $\{u_n, n \geq 1\} \subset \mathcal{F}_{\rm loc}$ that converges to u m-a.e. on E. Let $g_n := (0 \vee u_n) \wedge 1$. Then $g_n \in \mathcal{F}_{\rm loc}$, $g_n \to g := (0 \vee u) \wedge 1$ m-a.e. on E and $\widehat{\mathcal{E}}(g_n, g_n) \leq \widehat{\mathcal{E}}(u_n, u_n)$ by (6.2.5) and (6.2.6). Thus by Remark 6.2.6, $g \in \mathcal{F}_a^{\rm ref}$ with

$$\mathcal{E}^{\rm ref}(g,g) \leq \limsup_{n\to\infty} \widehat{\mathcal{E}}(g_n, g_n) \leq \limsup_{n\to\infty} \widehat{\mathcal{E}}(u_n, u_n) = \mathcal{E}^{\rm ref}(u,u).$$

Since $(\mathcal{F}_a^{\rm ref}, \mathcal{E}_1^{\rm ref})$ is a Hilbert space by Theorem 6.2.4, it follows that $(\mathcal{E}^{\rm ref}, \mathcal{F}_a^{\rm ref})$ is a Dirichlet form on $L^2(E;m)$. □

The relation between $\mathcal{F}^{\rm ref}$ and the extended Dirichlet space $\left(\mathcal{F}_a^{\rm ref}\right)_e$ of $(\mathcal{E}^{\rm ref}, \mathcal{F}_a^{\rm ref})$ will be given in Theorem 6.6.10.

On account of Theorem 6.2.13, $(\mathcal{E}_\alpha^{\rm ref}, \mathcal{F}_a^{\rm ref})$ can be viewed as the reflected Dirichlet space of $(\mathcal{E}_\alpha, \mathcal{F})$ for any $\alpha > 0$. For $u \in \mathcal{F}^{\rm ref}$, there are unique $f \in \mathcal{F}_e$ and $\varphi \in \mathbf{N}$ so that $u = f + h$ where $h(x) := \mathbf{E}_x[\varphi]$. In the following theorem, we write φ_u for φ.

THEOREM 6.2.15. *Let* $\alpha > 0$ *and* $u \in \mathcal{F}_a^{\rm ref}$. *Then* $\varphi_u = 0$ *on* $\{\zeta = \infty\}$. *Moreover,* $u - h_\alpha \in \mathcal{F}$, *where* $h_\alpha(x) := \mathbf{E}_x[e^{-\alpha\zeta}\varphi_u]$.

Proof. Let Y be the subprocess of X killed at an independent exponential random time T of rate $\alpha > 0$; that is,

$$Y_t(\omega) = \begin{cases} X_t(\omega) & \text{if } t < T(\omega), \\ \partial & \text{otherwise.} \end{cases}$$

Since the transition function of Y equals $\{e^{-\alpha t}P_t; t \geq 0\}$, the Dirichlet space on $L^2(E;m)$ and the extended Dirichlet space of Y are both $(\mathcal{E}_\alpha, \mathcal{F})$. Let $\mathbf{N}^Y$ and $\mathbf{HN}_Y$ be the universal Dirichlet space and the harmonic function space defined by (6.2.1) and (6.2.2), respectively, but with Y in place of X. As u is in the reflected Dirichlet space of Y, there is a unique $\psi \in \mathbf{N}^Y$ so that $u - \widehat{h}_\alpha \in \mathcal{F}$, where $\widehat{h}_\alpha(x) := \mathbf{E}_x^Y[\psi]$. Let $\{D_k, k \geq 1\}$ be an increasing sequence of relatively compact open sets with $\cup_{k\geq 1} D_k = E$. By noting that φ_u is a terminal random variable of X, it follows from Remark 6.2.2 that

$$\psi = \lim_{k\to\infty} u(Y_{\tau_{D_k}^Y}) = \lim_{k\to\infty} u(X_{\tau_{D_k}})\mathbf{1}_{\{\tau_{D_k} < \zeta\wedge T\}} = \varphi_u \mathbf{1}_{\{\zeta \leq T\}} \tag{6.2.16}$$

and so $\widehat{h}_\alpha(x) = \mathbf{E}_x[\varphi_u \mathbf{1}_{\{\zeta\leq T\}}] = \mathbf{E}_x[e^{-\alpha\zeta}\varphi_u] = h_\alpha(x)$.

Since $h - h_\alpha = (u - h_\alpha) - (u - h) \in \mathcal{F}_e$, we have from (6.1.3) and Corollary 3.5.3

$$\varphi_u = \lim_{k\to\infty} h(X_{\tau_{D_k}}) = \lim_{k\to\infty} h_\alpha(X_{\tau_{D_k}})$$

$\mathbf{P}_x$-a.s. for q.e. $x \in E$. As φ_u is a terminal random variable of X, we see in a similar manner to the proof of (6.1.3) that for q.e. $x \in E$, $\mathbf{P}_x$-a.s.

$$h_\alpha(X_{\tau_{D_k}}) = \mathbf{E}_x[\eta_{\alpha,k} | \mathcal{F}_{\tau_{D_k}}] \qquad \text{for } k \geq 1,$$

where $\eta_{\alpha,k} = e^{-\alpha(\zeta - \tau_{D_k})}\varphi_u \mathbf{1}_{\{\zeta<\infty\}}$. Hence

$$\begin{aligned}
&\lim_{k\to\infty} \mathbf{E}_x\big[\big|h_\alpha(X_{\tau_{D_k}}) - \varphi_u \mathbf{1}_{\{\zeta<\infty\}}\big|\big] \\
&\quad \leq \lim_{k\to\infty} \mathbf{E}_x\big[\big|\mathbf{E}_x[\eta_{\alpha,k} - \varphi_u \mathbf{1}_{\{\zeta<\infty\}} | \mathcal{F}_{\tau_{D_k}}]\big|\big] \\
&\qquad + \lim_{k\to\infty} \mathbf{E}_x\big[\big|E_x[\varphi_u \mathbf{1}_{\{\zeta<\infty\}} | \mathcal{F}_{\tau_{D_k}}] - \varphi_u \mathbf{1}_{\{\zeta<\infty\}}\big|\big] \\
&\quad \leq \lim_{k\to\infty} \mathbf{E}_x\big[\big|\eta_{\alpha,k} - \varphi_u \mathbf{1}_{\{\zeta<\infty\}}\big|\big] = 0.
\end{aligned}$$

Consequently, $\varphi_u = \varphi_u \cdot \mathbf{1}_{\{\zeta<\infty\}}$. □

COROLLARY 6.2.16. *If X is conservative, then $(\mathcal{E}^{\mathrm{ref}}, \mathcal{F}^{\mathrm{ref}}_a) = (\mathcal{E}, \mathcal{F})$.*

Proof. By above theorem, $\mathcal{F}^{\mathrm{ref}}_a = \mathcal{F}_e \cap L^2(E; m) = \mathcal{F}$. □

Remark 6.2.17. When X is conservative, **HN** can be non-trivial although $\mathcal{F}^{\mathrm{ref}}_a = \mathcal{F}$. For example, suppose $n \geq 3$ and $D := \{x = (x_1, \ldots, x_{n-1}, x_n) \in \mathbb{R}^n : |x_n| > (x_1^2 + \cdots + x_{n-1}^2)^{1/2}\} \cup \{x \in \mathbb{R}^n : |x| < 1\}$. Let X be the symmetric reflecting Brownian motion on $E := \overline{D}$. We know that X is conservative and transient by Example 3.5.9. Its associated Dirichlet form on $L^2(E; dx)$ is $(\mathcal{E}, W^{1,2}(D))$, where $\mathcal{E}(f, g) = \frac{1}{2}\int_D \nabla f(x) \cdot \nabla g(x) dx$. Then **HN** is non-trivial. In fact it can be shown (cf. [24]) that **HN** is a linear space of dimension 2. □

6.3. RECURRENT CASE

In this section, $(\mathcal{E}, \mathcal{F})$ is a recurrent regular Dirichlet form on $L^2(E; m)$. In this case, it follows from Lemma 3.5.5 and the arguments in Section 6.1 that every terminal random variable is a constant on each irreducible component. So it is not suitable to define its reflected Dirichlet space $(\mathcal{E}^{\mathrm{ref}}, \mathcal{F}^{\mathrm{ref}})$ by Definition 6.2.1 as in the transient case. Instead, we use its equivalent version given by Theorem 6.2.13. That is, we define

$$\mathcal{F}^{\text{ref}} := \left\{ u : |u| < \infty \ m\text{-a.e.}, \ \tau_k u \in \mathcal{F}_{\text{loc}}, \ \sup_{k \geq 1} \widehat{\mathcal{E}}(\tau_k u, \tau_k u) < \infty \right\}, \qquad (6.3.1)$$

where $\widehat{\mathcal{E}}$ is defined by (6.2.6), and

$$\mathcal{E}^{\text{ref}}(u, u) := \sup_{k \geq 1} \widehat{\mathcal{E}}(\tau_k u, \tau_k u) \qquad \text{for } u \in \mathcal{F}^{\text{ref}}. \qquad (6.3.2)$$

Note that by the normal contraction property, $\widehat{\mathcal{E}}(\tau_k u, \tau_k u)$ is non-decreasing in k and so

$$\mathcal{E}^{\text{ref}}(u, u) = \lim_{k \to \infty} \widehat{\mathcal{E}}(\tau_k u, \tau_k u) \qquad \text{for } u \in \mathcal{F}^{\text{ref}}.$$

LEMMA 6.3.1. *Suppose that* $\{f_k, k \geq 1\}$ *is a sequence in* $\mathcal{F}_e$ *that is uniformly bounded on* E *and has* $\sup_{k \geq 1} \mathcal{E}(f_k, f_k) < \infty$. *Then there is a subsequence* $\{f_{k_j}, j \geq 1\}$ *so that a suitable subsequence of the Cesàro means* $\{\frac{1}{n}\sum_{j=1}^n f_{k_j}, n \geq 1\}$ *converges q.e. on* E *to some* $f \in \mathcal{F}_e$.

Proof. Take a strictly positive $w \in C_b(E) \cap L^1(E; m)$ and define $\mu(dx) = w(x)m(dx)$. Let $(\mathcal{E}^Z, \mathcal{F}^Z)$ be the Dirichlet form on $L^2(E; \mu)$ associated with the time-changed process Z of X by the PCAF $A_t = \int_0^t w(X_s)ds$, $t \geq 0$. Here $X = (X_t, \mathbf{P}_x)$ is the symmetric Hunt process associated with $(\mathcal{E}, \mathcal{F})$ and $Z = (Z_t, \mathbf{P}_x)$ is given by $Z_t = X_{\tau_t}$ with $\tau_t := \inf\{s : A_s > t\}$. By Corollary 5.2.12, $(\mathcal{E}^Z, \mathcal{F}_e^Z) = (\mathcal{E}, \mathcal{F}_e)$. As $\sup_{k \geq 1} \mathcal{E}_1^Z(f_k, f_k) < \infty$, by Theorem A.4.1, there is a subsequence $\{f_{k_j}, j \geq 1\}$ so that the sequence of its Cesàro means $\{g_n, n \geq 1\} = \{\frac{1}{n}\sum_{j=1}^n f_{k_j}, n \geq 1\}$ is $\mathcal{E}_1^Z$-convergent to some $f \in \mathcal{F}^Z \subset \mathcal{F}_e$ and a suitable subsequence of $\{g_n, n \geq 1\}$ converges to f $\mathcal{E}^Z$-q.e. on E. By Theorem 5.2.11, a set is $\mathcal{E}^Z$-polar if and only if it is $\mathcal{E}$-polar. This completes the proof of the lemma. □

THEOREM 6.3.2. $(\mathcal{E}^{\text{ref}}, \mathcal{F}^{\text{ref}}) = (\mathcal{E}, \mathcal{F}_e)$.

Proof. Clearly from the definition, $\mathcal{F}_e \subset \mathcal{F}^{\text{ref}}$. To show the other direction, let $\{D_k, k \geq 1\}$ be an increasing sequence of relatively compact open subsets increasing to E. For bounded $u \in \mathcal{F}^{\text{ref}}$, by Lemma 6.2.10, $u - \mathbf{H}_{D_k^c} u \in \mathcal{F}_{e, D_k}$ with $\mathcal{E}(u - \mathbf{H}_{D_k^c} u, u_n - \mathbf{H}_{D_k^c} u) \leq \widehat{\mathcal{E}}(u, u) < \infty$. Since u is bounded, by Lemma 6.3.1 there is a subsequence $\{k_j, j \geq 1\}$ so that, if we let $g_n = \frac{1}{n}\sum_{j=1}^n (u - \mathbf{H}_{D_{k_j}^c} u)$, then g_n converges q.e. on E to some bounded function $f \in \mathcal{F}_e$. Note that $u - g_n$ converges boundedly q.e. on E to $h := u - f$. By bounded convergence theorem, for every relatively compact open subset $D \subset E$, $h(x) = \mathbf{H}_{D^c} h(x) = \mathbf{E}_x[h(X_{\tau_D})]$ q.e. on E and so h is a bounded harmonic function on E. In particular, $\{h(X_{\tau_{D_k}})\}$ is a bounded $\{\mathcal{F}_{\tau_{D_k}}\}$-martingale under $\mathbf{P}_x$ for q.e. $x \in E$ and so

$$h(x) = \mathbf{E}_x[\varphi] \qquad \text{for q.e. } x \in E, \text{ where } \varphi := \lim_{k\to\infty} h(X_{\tau_{D_k}}).$$

Note that $\lim_{k\to\infty} \tau_{D_k} = \zeta = \infty$ $\mathbf{P}_x$-a.s. for q.e. $x \in E$, because X is recurrent and hence conservative.

Let Z be the time-changed process of X as appeared in the proof of the preceding lemma and P_t^Z be the transition function of Z. It follows from Theorem 5.2.5 that Z is recurrent. This in particular implies that the lifetime A_∞ of Z is infinity, and so $\tau_t < \infty$ for each $t \geq 0$. Therefore, $\varphi \circ \theta_{\tau_t} = \varphi$ and we have, for q.e. $x \in E$,

$$P_t^Z h(x) = \mathbf{E}_x\left[h(X_{\tau_t})\right] = \mathbf{E}_x\left[\mathbf{E}_{X_{\tau_t}}[\varphi]\right] = \mathbf{E}_x[\varphi \circ \theta_{\tau_t}] = h(x).$$

Since μ is finite and h is bounded, we conclude from (1.1.4) and (1.1.5) that $h \in \mathcal{F}^Z \subset \mathcal{F}^Z_e = \mathcal{F}_e$ and $\mathcal{E}(h,h) = \mathcal{E}^Z(h,h) = 0$. This shows that $u = f + h \in \mathcal{F}_e$ with $\mathcal{E}^{\text{ref}}(u,u) = \mathcal{E}(u,u)$. Now for general $u \in \mathcal{F}^{\text{ref}}$, clearly $\tau_k u \in \mathcal{F}^{\text{ref}} \subset \mathcal{F}_e$ and $\mathcal{E}(\tau_k u, \tau_k u) = \widehat{\mathcal{E}}(\tau_k u, \tau_k u)$ for every $k \geq 1$. We conclude by Lemma 1.1.12 that $u \in \mathcal{F}_e$. Moreover, $\mathcal{E}^{\text{ref}}(u,u) = \lim_{k\to\infty} \mathcal{E}(\tau_k u, \tau_k u) = \mathcal{E}(u,u)$. This proves that $(\mathcal{E}^{\text{ref}}, \mathcal{F}^{\text{ref}}) = (\mathcal{E}, \mathcal{F}_e)$. □

6.4. TOWARD QUASI-REGULAR CASES

We will extend the notion of a reflected Dirichlet form $(\mathcal{E}^{\text{ref}}, \mathcal{F}^{\text{ref}})$ to any quasi-regular Dirichlet form $(\mathcal{E}, \mathcal{F})$. Recall that by Theorem 1.4.3 a quasi-regular Dirichlet form is quasi-homeomorphic to a regular Dirichlet form on a separable locally compact metric space. So we will first find an equivalent definition of $(\mathcal{E}^{\text{ref}}, \mathcal{F}^{\text{ref}})$ for a regular Dirichlet form $(\mathcal{E}, \mathcal{F})$ on $L^2(E;m)$ that is invariant under quasi-homeomorphism. Along the way, we shall also see that the reflected Dirichlet space of a regular Dirichlet form is invariant under the full support time change of the associated Hunt process.

For a non-negative Borel measure μ changing no $\mathcal{E}$-polar sets on E, denote by $\mathcal{E}$-supp$[\mu]$ the quasi support of μ as defined in Definition 3.3.4. For an m-a.e.-defined function u on E, define $\mathcal{E}$-supp$[u]$ as $\mathcal{E}$-supp$[u(x)m(dx)]$. An $\mathcal{E}$-quasi-open set containing a set B is called an *$\mathcal{E}$-neighborhood* of B. Note that by Theorem 4.3.10(i)

$$\mu^c_{\langle u,v\rangle} := \tfrac{1}{4}(\mu^c_{\langle u+v\rangle} - \mu^c_{\langle u-v\rangle}) = 0$$

for any bounded u, v in $\mathcal{F}$ such that $\widetilde{u}$ is constant in an $\mathcal{E}$-neighborhood of $\mathcal{E}$-supp$[v]$. This is because $\mu^c_{\langle u-v\rangle} = \mu^c_{\langle v-u\rangle}$ on E and, for the above u and v, $u+v$ and $v-u$ differ by a constant in an $\mathcal{E}$-neighborhood of $\mathcal{E}$-supp$[v]$, and $u+v = u-v$ on the complement of $\mathcal{E}$-supp$[v]$. So it follows from

Theorem 4.3.10(i) that $\mu^c_{\langle u+v\rangle} = \mu^c_{\langle u-v\rangle}$ on E. Now by the Beurling-Deny decomposition for regular Dirichlet forms in Theorem 4.3.3 together with the quasi-homeomorphism Theorem 1.4.3, we readily get the following Beurling-Deny formula for quasi-regular Dirichlet forms.

PROPOSITION 6.4.1. *Any symmetric quasi-regular Dirichlet form $(\mathcal{E}, \mathcal{F})$ on $L^2(E; m)$ can be uniquely decomposed into*

$$\mathcal{E}(u, v) = \mathcal{E}^{(c)}(u, v) + \frac{1}{2}\int_{E\times E} (\widetilde{u}(x) - \widetilde{u}(y))(\widetilde{v}(x) - \widetilde{v}(y))J(dx, dy)$$

$$+ \int_E \widetilde{u}(x)\widetilde{v}(x)\kappa(dx)$$

for $u, v \in \mathcal{F}$, with $\mathcal{E}^{(c)}$, J and K satisfying the following conditions:
(i) *$(\mathcal{E}^{(c)}, \mathcal{F})$ is a positive definite symmetric bilinear form with strongly local property: $\mathcal{E}(u, v) = 0$ for any u, v in $\mathcal{F}$ such that $\widetilde{u}$ is constant in an $\mathcal{E}$-neighborhood of $\mathcal{E}$-*supp$[v]$.
(ii) *J is a σ-finite symmetric positive measure on $E \times E \setminus d$ such that J does not change any subset of $E \times E \setminus d$ whose marginal projection is $\mathcal{E}$-polar, where d denotes the diagonal of $E \times E$.*
(iii) *κ is a σ-finite positive measure on E changing no $\mathcal{E}$-polar set.*
In fact, $\mathcal{E}^{(c)}(u, v) = \frac{1}{2}\mu^c_{\langle u,v\rangle}(E)$. Furthermore, every normal contraction operates on the form $\mathcal{E}^{(c)}$.

Assume that $(\mathcal{E}, \mathcal{F})$ is a regular Dirichlet space on $L^2(E; m)$ that is either transient or recurrent. Recall from (4.3.31) another version of the local Dirichlet space $\overset{\circ}{\mathcal{F}}_{\rm loc}$ defined in terms of quasi open sets. Clearly $\mathcal{F}_{\rm loc} \subset \overset{\circ}{\mathcal{F}}_{\rm loc}$ and each $u \in \overset{\circ}{\mathcal{F}}_{\rm loc}$ has a quasi continuous version $\widetilde{u}$ on E. In the following, every $u \in \overset{\circ}{\mathcal{F}}_{\rm loc}$ is always represented by such a quasi continuous version $\widetilde{u}$. In view of Theorem 4.3.10, the energy measure $\mu^c_{\langle u\rangle}$ is well-defined for each $u \in \overset{\circ}{\mathcal{F}}_{\rm loc}$ by letting $\mu^c_{\langle u\rangle} = \mu^c_{\langle u_n\rangle}$ on D_n.

For $u \in \overset{\circ}{\mathcal{F}}_{\rm loc}$, define

$$\widehat{\mathcal{E}}(u, u) = \frac{1}{2}\mu^c_{\langle u\rangle}(E) + \frac{1}{2}\int_{E\times E} (u(x) - u(y))^2 J(dx, dy)$$

$$+ \int_E u(x)^2\kappa(dx). \tag{6.4.1}$$

We know from Theorem 6.2.13 and from Section 6.3 that the reflected Dirichlet space $(\mathcal{E}^{\mathrm{ref}}, \mathcal{F}^{\mathrm{ref}})$ can be defined as

$$\mathcal{F}^{\mathrm{ref}} = \left\{ u : \text{finite } m\text{-a.e., } \tau_n u \in \mathcal{F}_{\mathrm{loc}}, \ \forall n \geq 1, \ \sup_{n\geq 1} \widehat{\mathcal{E}}(\tau_n u, \tau_n u) < \infty \right\},$$

$$\mathcal{E}^{\mathrm{ref}}(u,u) = \lim_{n\to\infty} \widehat{\mathcal{E}}(\tau_n u, \tau_n u) \qquad \text{for } u \in \mathcal{F}^{\mathrm{ref}}.$$

Moreover, for each $u \in \mathcal{F}^{\mathrm{ref}}$, by Theorem 6.1.6, Definition 6.2.1, and Theorem 6.3.2, u has a quasi continuous version $\widetilde{u}$ and we always take its quasi continuous version. Note that $\{|u| < n\}$ is quasi open and $\bigcup_{n=1}^{\infty}\{|u| < n\} = E$ $\mathcal{E}$-q.e. We denote $\tau_n u$ by u_n. Since $u = u_n$ m-a.e. on $\{|u| < n\}$ and $u_n \in \mathcal{F}_{\mathrm{loc}} \subset \mathring{\mathcal{F}}_{\mathrm{loc}}$, we have $u \in \mathring{\mathcal{F}}_{\mathrm{loc}}$ and $\mu^c_{\langle u\rangle} = \mu^c_{\langle u_n\rangle}$ on $\{|u| < n\}$. Therefore,

$$\widehat{\mathcal{E}}(u,u) = \lim_{n\to\infty} \widehat{\mathcal{E}}(u_n, u_n) = \mathcal{E}^{\mathrm{ref}}(u,u) < \infty.$$

THEOREM 6.4.2. *Suppose that $(\mathcal{E}, \mathcal{F})$ is a regular Dirichlet form on $L^2(E;m)$ that is either transient or recurrent. Then $\mathcal{F}^{\mathrm{ref}} = \{u \in \mathring{\mathcal{F}}_{\mathrm{loc}} : \ \widehat{\mathcal{E}}(u,u) < \infty\}$ and $\mathcal{E}^{\mathrm{ref}}(u,u) = \widehat{\mathcal{E}}(u,u)$ for $u \in \mathcal{F}^{\mathrm{ref}}$.*

Proof. It suffices to show that if u is a function in $\mathring{\mathcal{F}}_{\mathrm{loc}}$ with $\widehat{\mathcal{E}}(u,u) < \infty$, then $u_n = -n \vee (u \wedge n) \in \mathcal{F}_{\mathrm{loc}}$ and therefore $u \in \mathcal{F}^{\mathrm{ref}}$.

For fixed n, let $f = u_n$. For any relatively compact open subset D of E, let $\varphi \in C_c(E) \cap \mathcal{F}$ such that $0 \leq \varphi \leq 1$ on E and $\varphi = 1$ on D. Since $f \in \mathring{\mathcal{F}}_{\mathrm{loc}}$, there is a sequence of quasi open sets G_k with $\bigcup_{k=1}^{\infty} G_k = E$ q.e. and $f_k \in \mathcal{F}$ such that $f = f_k$ m-a.e. on G_k. By taking $(-n) \vee (f_k \wedge n)$ if necessary, we may assume that $\|f_k\|_\infty \leq n$. Put $\varphi_k(x) = \varphi(x) - \mathbf{E}_x[\varphi(X_{\tau_{G_k}})]$. Then $\varphi_k \in \mathcal{F}$ and $\varphi_k = 0$ q.e on G_k^c. Note that $\|\varphi_k\|_\infty \leq 1$ and $f\varphi_k = f_k\varphi_k \in \mathcal{F}$.

$$\begin{aligned}\mathcal{E}(f\varphi_k, f\varphi_k) = \mu^c_{\langle f\varphi_k\rangle}(E) &+ \frac{1}{2}\int_{E\times E} (f(x)\varphi_k(x) - f(y)\varphi_k(y))^2 J(dx,dy) \\ &+ \int_E f(x)^2 \varphi_k(x)^2\,\kappa(dx).\end{aligned}$$

By Lemma 4.3.6,

$$\begin{aligned}\mu^c_{\langle f\varphi_k\rangle}(E) &= \mu^c_{\langle f_k\varphi_k\rangle}(E) \\ &= \int_E f_k(x)^2\,\mu^c_{\langle\varphi_k\rangle}(dx) + 2\int_E f_k(x)\varphi_k(x)\,\mu^c_{\langle f_k,\varphi_k\rangle}(dx) + \int_E \varphi_k(x)^2\,\mu^c_{\langle f_k\rangle}(dx) \\ &\leq 2\int_E f_k(x)^2\,\mu^c_{\langle\varphi_k\rangle}(dx) + 2\int_E \varphi_k(x)^2\,\mu^c_{\langle f_k\rangle}(dx) \\ &\leq 2\|f_k\|^2_\infty \mu^c_{\langle\varphi_k\rangle}(E) + 2\|\varphi_k\|^2_\infty \mu^c_{\langle f_k\rangle}(G_k).\end{aligned}$$

Thus

$$\begin{aligned}
&\mathcal{E}(f\varphi_k,f\varphi_k)\\
&\quad\le 2n^2\mu^c_{\langle\varphi_k\rangle}(E)+2\|\varphi_k\|^2_\infty\mu^c_{\langle f\rangle}(G_k)+\|f\|^2_\infty\int(\varphi_k(x)-\varphi_k(y))^2J(dxdy)\\
&\qquad+\|\varphi_k\|^2_\infty\iint(f(x)-f(y))^2J(dxdy)+\|\varphi_k\|^2_\infty\int f(x)^2\kappa(dx)\\
&\quad\le 2n^2\mathcal{E}(\varphi_k,\varphi_k)+2\|\varphi_k\|^2_\infty\widehat{\mathcal{E}}(f,f)\\
&\quad\le 2n^2\mathcal{E}(\varphi,\varphi)+2\widehat{\mathcal{E}}(f,f)<\infty.
\end{aligned}$$

Therefore, $\{f\varphi_k\}_{k\ge1}$ is $\mathcal{E}$-bounded. Since $\lim_{k\to\infty}f\varphi_k=f\varphi$ a.e. on E, we have $f\varphi$ is in $\mathcal{F}_e$ and therefore in $\mathcal{F}$ since $f\varphi\in L^2(E;m)$. This implies $u_n=f\in\mathcal{F}_{\rm loc}$ and $\mathcal{E}^{\rm ref}(u,u)=\lim_{n\to\infty}\widehat{\mathcal{E}}(u_n,u_n)=\widehat{\mathcal{E}}(u,u)$. The theorem is thus proved. □

Remark 6.4.3. Suppose $(\mathcal{E},\mathcal{F})$ is a regular Dirichlet form. Denote by $\overset{\circ}{\mathcal{F}}_{e,{\rm loc}}$ the space obtained by replacing $\mathcal{F}$ with $\mathcal{F}_e$ in the definition (4.3.31) of $\overset{\circ}{\mathcal{F}}_{\rm loc}$. Apparently the former seems to be larger than the latter, but actually they are the same. To see this, take any $u\in\overset{\circ}{\mathcal{F}}_{e,{\rm loc}}$. By definition, there is an increasing sequence of quasi open sets $\{D_n\}$ with $\cup_{n\ge1}D_n=E$ q.e. and a sequence $\{u_n\}\subset\mathcal{F}_e$ so that $u_n=u$ m-a.e. on D_n for every $n\ge1$. Let $\{G_n\}$ be relatively compact open sets increasing to E. For each $n\ge1$, choose a function $g_n\in\mathcal{F}\cap C_c(E)$ such that $g_n=1$ on G_n. Replacing the quasi open set D_n by a quasi open set $D_n\cap G_n\cap\{|u|>n\}$ and u_n by $((-n)\vee u_n)\wedge n$ if necessary, we may assume that $u_n\in\mathcal{F}_e$ satisfies $-n\le u_n\le n$. Then $v_n=u_n\cdot g_n\in\mathcal{F}$ and $v_n=u_n=u$ on D_n. Hence $u\in\overset{\circ}{\mathcal{F}}_{\rm loc}$. □

Since the notion of $\overset{\circ}{\mathcal{F}}_{\rm loc}$ is transferable under quasi-homeomorphism, we give

DEFINITION 6.4.4. Suppose that $(\mathcal{E},\mathcal{F})$ is a quasi-regular Dirichlet form on $L^2(E;m)$. Its *reflected Dirichlet space* $(\mathcal{E}^{\rm ref},\mathcal{F}^{\rm ref})$ is defined to be

$$\mathcal{F}^{\rm ref}=\left\{u\in\overset{\circ}{\mathcal{F}}_{\rm loc}:\ \widehat{\mathcal{E}}(u,u)<\infty\right\}$$

and

$$\mathcal{E}^{\rm ref}(u,v)=\widehat{\mathcal{E}}(u,v)\qquad\text{for } u,v\in\mathcal{F}^{\rm ref}.$$

Note that by Theorems 2.3.4 and 4.3.3, and by quasi-homeomorphism and Remark 6.4.3, $\mathcal{F}^{\rm ref}\supset\mathcal{F}_e$ and $\mathcal{E}^{\rm ref}(u,u)=\mathcal{E}(u,u)$ for $u\in\mathcal{F}_e$. Clearly, in view of Theorem 4.3.10 and the expression (6.4.1) of $\widehat{\mathcal{E}}$, $(\mathcal{E}^{\rm ref},\mathcal{F}^{\rm ref})$ enjoys the normal contraction property.

THEOREM 6.4.5. *Suppose that $(\mathcal{E},\mathcal{F})$ is a quasi-regular Dirichlet form on $L^2(E;m)$. Let $(\mathcal{E}^{\text{ref}},\mathcal{F}^{\text{ref}})$ be its reflected Dirichlet space and define $\mathcal{F}^{\text{ref}}_a := \mathcal{F}^{\text{ref}} \cap L^2(E;m)$. Then $(\mathcal{E}^{\text{ref}},\mathcal{F}^{\text{ref}}_a)$ is a Dirichlet form on $L^2(E;m)$. If $(\mathcal{E},\mathcal{F})$ is recurrent, then $(\mathcal{E}^{\text{ref}},\mathcal{F}^{\text{ref}}) = (\mathcal{E},\mathcal{F}_e)$.*

Proof. It is clear from Definition 6.4.4 that $(\mathcal{E}^{\text{ref}}_1,\mathcal{F}^{\text{ref}}_a)$ is the reflected (and hence the active reflected) Dirichlet space of the transient Dirichlet form $(\mathcal{E}_1,\mathcal{F})$ on $L^2(E;m)$. So we may assume that $(\mathcal{E},\mathcal{F})$ is transient. By Theorem 1.4.3, $(\mathcal{E},\mathcal{F})$ is quasi-homeomorphic to a regular Dirichlet space $(\widetilde{\mathcal{E}},\widetilde{\mathcal{F}})$ on $L^2(\widetilde{E};\widetilde{m})$ through a quasi-homeomorphic map j. Then $(\widetilde{\mathcal{E}}^{\text{ref}},\widetilde{\mathcal{F}}^{\text{ref}})$ is the image space of $(\mathcal{E}^{\text{ref}},\mathcal{F}^{\text{ref}})$ under the same quasi-homeomorphic map j. So without loss of generality, we may and do assume that $(\mathcal{E},\mathcal{F})$ is a regular transient Dirichlet form on $L^2(E;m)$. Then by Theorems 6.2.14 and 6.4.2, $(\mathcal{E}^{\text{ref}},\mathcal{F}^{\text{ref}}_a)$ is a Dirichlet form on $L^2(E;m)$. When $(\mathcal{E},\mathcal{F})$ is recurrent, by Theorem 6.3.2 we have $\mathcal{F}^{\text{ref}} = \mathcal{F}_e$. In this case, $(\mathcal{E}^{\text{ref}},\mathcal{F}^{\text{ref}}_a) = (\mathcal{E},\mathcal{F})$, which is a Dirichlet form on $L^2(E;m)$. □

When $(\mathcal{E},\mathcal{F})$ is a general regular Dirichlet form, we stress that we adopt Definition 6.4.4 as the definition of the reflected Dirichlet space $(\mathcal{F}^{\text{ref}},\mathcal{E}^{\text{ref}})$ of $(\mathcal{E},\mathcal{F})$ regardless of the transience or recurrence of the latter. If the latter is either transient or recurrent, this definition coincides with the previous one by virtue of Theorem 6.4.2.

The next proposition asserts the invariance of the reflected Dirichlet space of a regular Dirichlet form under a full support time change.

PROPOSITION 6.4.6. *Suppose that $(\mathcal{E},\mathcal{F})$ is a regular Dirichlet form on $L^2(E;m)$. For any $\mu \in \overset{\circ}{S}_E$, let $(\check{\mathcal{E}},\check{\mathcal{F}})$ be the Dirichlet form on $L^2(E;\mu)$ of the time-changed process $\check{X}$ with respect to the PCAF of X with Revuz measure μ. Denote by $(\check{\mathcal{E}}^{\text{ref}},\check{\mathcal{F}}^{\text{ref}})$ the reflected Dirichlet space of $(\check{\mathcal{E}},\check{\mathcal{F}})$. Then $(\check{\mathcal{E}}^{\text{ref}},\check{\mathcal{F}}^{\text{ref}}) = (\mathcal{E}^{\text{ref}},\mathcal{F}^{\text{ref}})$.*

Proof. By Corollary 5.2.10, $(\check{\mathcal{E}},\check{\mathcal{F}})$ is a regular Dirichlet form on $L^2(E;\mu)$. In view of Corollary 5.2.12 the extended Dirichlet space of $(\check{\mathcal{E}},\check{\mathcal{F}})$ coincides with the original extended Dirichlet space of $(\mathcal{E},\mathcal{F})$. Furthermore, the notion of nests and in particular the notion of quasi open sets are invariant under this time change by virtue of Theorem 5.2.11. Thus we get the desired conclusion on account of Definition 6.4.4 and Remark 6.4.3. □

Applying Theorem 6.2.14 to the regular Dirichlet form $(\check{\mathcal{E}},\check{\mathcal{F}})$ on $L^2(E;\mu)$, we are led from Proposition 6.4.6 to the first statement of the following:

COROLLARY 6.4.7. *Let $(\mathcal{E},\mathcal{F})$ be a regular Dirichlet form on $L^2(E;m)$. Then for any $\mu \in \overset{\circ}{S}_E$*

$$(\mathcal{E}^{\text{ref}},\ \mathcal{F}^{\text{ref}} \cap L^2(E;\mu)) \tag{6.4.2}$$

is a Dirichlet form on $L^2(E;\mu)$.

Suppose further that $\mu(E) < \infty$. Then (6.4.2) is recurrent if and only if $\kappa = 0$.

In fact, if $\mu \in \mathring{S}_E$ is finite, then $1 \in L^2(E;\mu)$. Moreover, we see from Definition 6.4.4 that $1 \in \mathcal{F}^{\rm ref} \cap L^2(E;\mu)$ and $\mathcal{E}^{\rm ref}(1,1) = 0$ if and only if $\kappa = 0$. This establishes the last statement of Corollary 6.4.7 by virtue of Theorem 2.1.8.

Remark 6.4.8. For $\mu \in \mathring{S}_E$, the active reflected Dirichlet form (6.4.2) on $L^2(E;\mu)$ may not be regular. Among various $\mu \in \mathring{S}_E$, the corresponding Dirichlet forms (6.4.2) may not be related to each other by means of time changes. Even when $\kappa = 0$, they are not necessarily recurrent unless $\mu(E) < \infty$, as we shall see in various examples in the next section. □

When $(\mathcal{E}, \mathcal{F})$ is a transient quasi-regular Dirichlet form on $L^2(E;m)$, just as in the regular Dirichlet form setting in Section 6.2, we can identify its reflected Dirichlet form $\mathcal{F}^{\rm ref}$ as the orthogonal sum of $\mathcal{F}_e$ and **HN**, the space of harmonic functions on E having finite energy.

In the remainder of this section, $(\mathcal{E}, \mathcal{F})$ is a transient quasi-regular Dirichlet form on $L^2(E;m)$, where E is a Radon space and X is its properly associated m-symmetric right process on E whose lifetime is denoted as ζ. Let $\{\mathcal{F}_t, t \geq 0\}$ be the minimal admissible augmented filtration generated by X. See Section A.3.3(**3**) for definitions of predictable and totally inaccessible stopping times.

LEMMA 6.4.9. *Let $A := \{\zeta < \infty$ and $X_{\zeta-}$ exists and takes value in $E\}$. Define*

$$\zeta_i(\omega) := \begin{cases} \zeta(\omega) & \text{if } \omega \in A \\ \infty & \text{if } \omega \notin A \end{cases} \quad \text{and} \quad \zeta_p(\omega) := \begin{cases} \zeta(\omega) & \text{if } \omega \notin A \\ \infty & \text{if } \omega \in A. \end{cases}$$

Then for q.e. $x \in E$, $\mathbf{P}_x$-a.s., ζ_i is a totally inaccessible stopping time with respect to $\{\mathcal{F}_t, t \geq 0\}$ and ζ_p is a predictable stopping time with respect to $\{\mathcal{F}_t, t \geq 0\}$. The stopping times ζ_i and ζ_p are called, respectively, the totally inaccessible part and the predictable part of the lifetime ζ.

Proof. By Theorem 1.4.3, $(\mathcal{E}, \mathcal{F})$ is related to a regular Dirichlet form $(\widehat{\mathcal{E}}, \widehat{\mathcal{F}})$ on a locally compact separable metric space $\widehat{E}$ by a quasi-homeomorphism j. Let $\widehat{X} = (\widehat{X}_t, \widehat{\mathbf{P}}_x, \widehat{\zeta})$ be a Hunt process on $\widehat{E}$ associated with $(\widehat{\mathcal{E}}, \widehat{\mathcal{F}})$. On account of Theorem 3.1.13, $X|_{E\setminus N}$ can be identified with the special standard process $j^{-1}(\widehat{X}|_{\widehat{E}\setminus\widehat{N}})$ where $\widehat{N}$ is a certain properly exceptional set for $\widehat{X}$ and $N = j^{-1}(\widehat{N})$. In particular, $X_t = j^{-1}(\widehat{X}_t)$ for $t < \widehat{\zeta}$ and $\zeta = \widehat{\zeta}$. Since $\widehat{X}$ is a Hunt process in $\widehat{E}$, $\widehat{X}_{\widehat{\zeta}-} \in \widehat{E} \cup \{\widehat{\partial}\}$, where $\widehat{\partial}$ is added to $\widehat{E}$ as a one-point compactification. Define

$\widehat{A} := \{\widehat{\zeta} < \infty \text{ and } X_{\widehat{\zeta}-} \in \widehat{E}\}$. Clearly

$$\widehat{\zeta}_i := \widehat{\zeta} \cdot \mathbf{1}_{\widehat{A}} + \infty \cdot \mathbf{1}_{\widehat{A}^c} \quad \text{and} \quad \widehat{\zeta}_p := \widehat{\zeta} \cdot \mathbf{1}_{\widehat{A}^c} + \infty \cdot \mathbf{1}_{\widehat{A}}$$

are the totally inaccessible and predictable parts of $\widehat{\zeta}$, respectively. By the quasi-homeomorphism j, it follows that

$$\zeta_i = \zeta \cdot \mathbf{1}_A + \infty \cdot \mathbf{1}_{A^c} \quad \text{and} \quad \zeta_p = \zeta \cdot \mathbf{1}_{A^c} + \infty \cdot \mathbf{1}_A$$

are the totally inaccessible and predictable parts of ζ for $X|_{E\setminus N}$, respectively. □

DEFINITION 6.4.10. (i) We call a random variable $\varphi = \varphi(\omega)$ on Ω a *terminal random variable* of X if

(a) φ is $\mathcal{F}_\infty$-measurable and $\mathbf{P}_x$-integrable for q.e. $x \in E$,
(b) $\varphi(\theta_t\omega) = \varphi(\omega)$ for every $t < \zeta(\omega)$ $\mathbf{P}_x$-a.s. for q.e. $x \in E$, and
(c) $\{\varphi = 0\} \supset \{\zeta_i < \infty\} \cup \{\zeta = 0\}$ $\mathbf{P}_x$-a.s. for q.e. $x \in E$.

(ii) A function h on E is called *harmonic* if for any quasi open subset D with compact closure in E,

$$\mathbf{E}_x\left[\left|h(X_{\tau_D})\right|\right] < \infty \quad \text{and} \quad h(x) = \mathbf{E}_x\left[h(X_{\tau_D})\right] \qquad \text{for q.e. } x \in E,$$

where $\tau_D := \inf\{t > 0 : X_t \notin D\}$ denotes the first exit time from D by X.

Note that the above definition is invariant under quasi-homeomorphism between Dirichlet forms. It is the exact analogy of Definitions 6.1.3 and 6.1.4 in the regular Dirichlet form context in the sense that modulo an exceptional set, φ is a terminal random variable for X if and only if it is a terminal random variable for $\widehat{X}$, and h is a harmonic function for X if and only if $h \circ j$ is a harmonic function for $\widehat{X}$. Here j and $\widehat{X}$ are as in the proof of Lemma 6.4.9.

LEMMA 6.4.11. (i) *Let φ be a terminal random variable. Then*

$$h(x) := \mathbf{E}_x[\varphi], \qquad x \in E,$$

is harmonic in E. Moreover, for any $\mathcal{E}$-nest $\{F_k, k \geq 1\}$ consisting of compact sets, $h(X_{\tau_{F_k}})$ converges to φ $\mathbf{P}_x$-a.s. and in $L^1(\mathbf{P}_x)$ for q.e. $x \in E$.
(ii) *For any $f \in \mathcal{F}_e$ and for any $\mathcal{E}$-nest $\{F_k, k \geq 1\}$ as in* (i), $\lim_{k\to\infty} f(X_{\tau_{F_k}}) = 0$ $\mathbf{P}_x$-a.s. *for q.e. $x \in E$.*

Proof. (i) can be proved in the same way as that for Lemma 6.1.5 by taking Theorem 3.1.13 into account. (ii) follows from Corollary 3.5.3 and Theorem 3.1.13. □

THEOREM 6.4.12. *Suppose that $(\mathcal{E}, \mathcal{F})$ is a transient quasi-regular Dirichlet form on $L^2(E; m)$ and $X = (X_t, \mathbf{P}_x)$ is a properly associated right process on E, where E is a Radon space. Let*

$$\mathbf{N} := \{\varphi : \text{ terminal random variable of } X \text{ with } \mathbf{E}_x[\varphi^2] < \infty \text{ q.e. on } E \text{ and } L(g, 1) < \infty\},$$

where g is defined by (6.1.10) *for φ and L is the energy functional defined by* (5.4.4). *Define*

$$\mathbf{HN} := \{h : h(x) = \mathbf{E}_x[\varphi] \text{ for } \varphi \in \mathbf{N}\}.$$

Then $\mathcal{F}^{\text{ref}} = \mathcal{F}_e + \mathbf{HN}$, and for $u = f + h \in \mathcal{F}^{\text{ref}}$, where $f \in \mathcal{F}_e$ and $h = \mathbf{E}_\cdot[\varphi]$ with $\varphi \in \mathbf{N}$, we have

$$\mathcal{E}^{\text{ref}}(u, u) = \mathcal{E}(f, f) + \mathbf{e}(M^h) + \frac{1}{2}\int_E h(x)^2 \kappa(dx), \tag{6.4.3}$$

where $\mathbf{e}(M^h)$ is the energy of the MAF M^h given by (6.1.7) *and κ is the killing measure appearing in Proposition 6.4.1.*

Furthermore, the statements in Lemma 6.2.3 hold true for the space $(\mathbf{HN}, \mathcal{E}^{\text{ref}})$.

Proof. Lemma 6.4.11 ensures that any function $u \in \mathcal{F}_e + \mathbf{HN}$ is expressed uniquely as the sum of functions in these two spaces. Those statements of this theorem have been established for a transient regular Dirichlet form $(\widehat{\mathcal{E}}, \widehat{\mathcal{F}})$ and an associated Hunt process $\widehat{X}$ in Theorem 6.2.13 and Theorem 6.4.2. They are clearly transferable to the present ones for $(\mathcal{E}, \mathcal{F})$ and X by the quasi-homeomorphism j, as appeared in the proof of Lemma 6.4.9. □

6.5. EXAMPLES

We collect various examples of reflected Dirichlet spaces of regular Dirichlet forms.

(1°) Symmetric Lévy processes

Consider the symmetric translation-invariant transition probability (2.2.10) on $\mathbb{R}^n$ and its Dirichlet form $(\mathcal{E}, \mathcal{F})$ on $L^2(\mathbb{R}^n)$. $(\mathcal{E}, \mathcal{F})$ is then regular and the associated Hunt process X on $\mathbb{R}^n$ is called a *symmetric Lévy process*.

Assume that $S = 0$ in the corresponding Lévy-Khinchin formula (2.2.9). Then $(\mathcal{E}, \mathcal{F})$ admits an explicit expression (2.2.16) in terms of the Lévy measure J. Let $(\mathcal{F}^{\text{ref}}, \mathcal{E}^{\text{ref}})$ be the reflected Dirichlet space of $(\mathcal{E}, \mathcal{F})$. We shall show that

$$(\mathcal{F}^{\text{ref}}, \mathcal{E}^{\text{ref}}) = (\mathcal{G}, \mathcal{A}). \tag{6.5.1}$$

where

$$\begin{cases} \mathcal{G} = \{u \in \mathcal{B}(\mathbb{R}^n) : |u| < \infty \text{ a.e., } \mathcal{A}(u,u) < \infty\}, \\ \mathcal{A}(u,v) = \frac{1}{2}\int_{\mathbb{R}^n \times \mathbb{R}^n} (u(x+y) - u(x))(v(x+y) - v(x))J(dy)dx. \end{cases} \tag{6.5.2}$$

In view of (2.2.16) and Theorem 4.3.3, the jumping measure $J(dx, dy)$ of X is given by $J(dx, dy) = J(dy - x)dx$, the killing measure κ of X vanishes, and the strongly local part $\mathcal{E}^{(c)}$ of $\mathcal{E}$ also vanishes. Accordingly, Definition 6.4.4 reads as follows:

$$\begin{cases} \mathcal{F}^{\text{ref}} = \left\{ u \in \overset{\circ}{\mathcal{F}}_{\text{loc}} : \mathcal{A}(u,u) < \infty \right\}, \\ \mathcal{E}^{\text{ref}}(u,u) = \mathcal{A}(u,u) \qquad \text{for } u \in \mathcal{F}^{\text{ref}}. \end{cases}$$

So we see that $\mathcal{F}^{\text{ref}} \subset \mathcal{G}$. Conversely, take any $u \in \mathcal{G}$. By considering u^+ and u^- separately, without loss of generality, we may and do assume that $u \geq 0$. For any relatively compact open set $U \subset \mathbb{R}^n$, there exists $\varphi \in \mathcal{F} \cap C_c(\mathbb{R}^n)$ such that $\varphi = 1$ on U because $(\mathcal{E}, \mathcal{F})$ is regular. Then $\tau_k u \cdot \varphi \in L^2(\mathbb{R}^n)$ and $\|\tau_k u \cdot \varphi\|_{\mathcal{A}} \leq k\|\varphi\|_{\mathcal{A}} + \|u\|_{\mathcal{A}}\|\varphi\|_\infty < \infty$. We thus conclude by (2.2.16) that $\tau_k u \cdot \varphi \in \mathcal{F}$. This implies that $\tau_k u \in \mathcal{F}_{\text{loc}}$ and so it admits a quasi continuous version $\widetilde{u}_k$. Note that $\widetilde{u}_k \leq \widetilde{u}_{k+1}$ q.e. so that $\widetilde{u}_k$ increases q.e. to a function $\widetilde{u}$ on E. For every $c > 0$, there is an integer $k > c$ and accordingly the set $\{\widetilde{u} < c\}$, which equals $\{\widetilde{u}_k < c\}$ q.e., is quasi open. Consequently, $\widetilde{u}$ is a quasi continuous version of u. Moreover, $u \in \overset{\circ}{\mathcal{F}}_{\text{loc}}$ as $\{\widetilde{u} < k\} = \{\widetilde{u}_k < k\}$ for every $k \geq 1$ and $\cup_{k \geq 1}\{\widetilde{u} < k\} = E$ q.e. This proves that $u \in \mathcal{F}^{\text{ref}}$ and so (6.5.1) is established.

From (2.2.16) and (6.5.1), we have

$$(\mathcal{E}, \mathcal{F}) = (\mathcal{A}, \mathcal{G} \cap L^2(\mathbb{R}^n)). \tag{6.5.3}$$

Let $(\mathcal{F}_e, \mathcal{E})$ be the extended Dirichlet space of $(\mathcal{E}, \mathcal{F})$. Note that $1 \in \mathcal{G}$. It follows that the identity

$$(\mathcal{F}_e, \mathcal{E}) = (\mathcal{G}, \mathcal{A}) \tag{6.5.4}$$

holds if and only if $(\mathcal{E}, \mathcal{F})$ is recurrent. Indeed, the "if" part follows from Theorem 6.3.2, while (6.5.4) implies that $1 \in \mathcal{F}_e$ with $\mathcal{E}(1,1) = 0$, yielding the recurrence of $(\mathcal{E}, \mathcal{F})$.

(2°) Symmetric step processes

We now consider the symmetric regular step process of Section 2.2.1. Let E be a locally compact separable metric space, m_0 be a positive Radon measure on it with $\text{supp}[m_0] = E$, and $Q(x, dy)$ be an m_0-symmetric Markovian kernel with $Q(x, \{x\}) = 0$ for every $x \in E$. Let Z be the regular step process on E with road map Q and speed function 1. The process Z is then m_0-symmetric and its Dirichlet form $(\mathcal{E}^Z, \mathcal{F}^Z)$ on $L^2(E; m_0)$ is given by (2.2.5) with $\mathcal{F}^Z = L^2(E; m_0)$, which is obviously regular.

Just as in the preceding example (**1°**), we can verify the following: The reflected Dirichlet space of $(\mathcal{E}^Z, \mathcal{F}^Z)$ coincides with $(\mathcal{G}, \mathcal{A})$ where

$$\mathcal{G} = \{u \in \mathcal{B}(E) \colon |u| < \infty \; m_0\text{-a.e.}, \quad \mathcal{A}(u,u) < \infty\},$$

$$\mathcal{A}(u,v) = \int_{E\times E} ((u(x)-u(y))(v(x)-v(y))Q(x,dy)m_0(dx) \\ + \int_E u(x)v(y)(1-Q(x,E))m_0(dx). \tag{6.5.5}$$

If $(\mathcal{E}^Z, \mathcal{F}^Z)$ is recurrent, then the extended Dirichlet space $\mathcal{F}^Z_e$ is identical with $\mathcal{G}$ and Q is a probability kernel. Conversely, if $\mathcal{F}^Z_e = \mathcal{G}$ and Q is a probability kernel, then $(\mathcal{E}^Z, \mathcal{F}^Z)$ is recurrent.

Let λ be a positive Borel function on E which is locally bounded away from 0 and X be the regular step process with road map Q and speed function λ. The process X is then symmetric with respect to $m(dx) = \frac{1}{\lambda(x)} m_0(dx)$ and its Dirichlet form $(\mathcal{E}, \mathcal{F})$ on $L^2(E;m)$ admits the expression

$$\begin{cases} \mathcal{F} = \mathcal{F}^Z_e \cap L^2(E;m), \\ \mathcal{E}(u,v) = \mathcal{A}(u,v), \quad u,v \in \mathcal{F}, \end{cases} \tag{6.5.6}$$

because X can be obtained from Z by a time change with respect to the PCAF $A_t = \int_0^t \frac{1}{\lambda(Z_s)} ds$ and (5.2.17) applies.

$(\mathcal{E}, \mathcal{F})$ is regular in view of Corollary 5.2.10. On account of Proposition 6.4.6, the reflected Dirichlet space of $(\mathcal{E}, \mathcal{F})$ is given by (6.5.5) again.

(3°) One-dimensional diffusions

Let $I = (r_1, r_2)$ be an open interval and (s, m) be a pair of a canonical scale and canonical measure on I. Based on s, the form $\mathcal{E}^{(s)}$ and the space $\mathcal{F}^{(s)}$ are defined by (2.2.29) and (2.2.28), respectively. By Theorem 2.2.11(i),

$$(\mathcal{E}, \mathcal{F}) = (\mathcal{E}^{(s)}, \mathcal{F}^{(s)} \cap L^2(I;m)) \tag{6.5.7}$$

is a regular strongly local irreducible Dirichlet form on $L^2(I^*;m)$, where I^* is obtained from I by adding those boundary points r_i that are regular with respect to (s,m). The measure m is extended to I^* by setting $m(I^* \setminus I) = 0$. In Example 3.5.7, the diffusion process X on I^* associated with $(\mathcal{E}, \mathcal{F})$ is called the reflecting diffusion, while the part process X^0 of X on I is called the absorbing diffusion. The Dirichlet form $(\mathcal{E}^0, \mathcal{F}^0)$ of X^0 on $L^2(E;m)$ is regular and given by

$$\mathcal{F}^0 = \{u \in \mathcal{F} : u = 0 \text{ on } I^* \setminus I\}, \quad \mathcal{E}^0 = \mathcal{E} \text{ on } \mathcal{F}^0 \times \mathcal{F}^0.$$

In view of Section 2.2.3, $\overset{\circ}{\mathcal{F}}{}^{0}_{\rm loc} = \mathcal{F}^0_{\rm loc}$ and $\overset{\circ}{\mathcal{F}}_{\rm loc} = \mathcal{F}_{\rm loc}$ so that we immediately get from Definition 6.4.4

$$(\mathcal{E}^{0,{\rm ref}}, (\mathcal{F}^0)^{\rm ref}) = (\mathcal{E}^{\rm ref}, \mathcal{F}^{\rm ref}) = (\mathcal{E}^{(s)}, \mathcal{F}^{(s)}). \tag{6.5.8}$$

As in Example (**1**°), we can further see that the identity

$$(\mathcal{E}, \mathcal{F}_e) = (\mathcal{E}^{(s)}, \mathcal{F}^{(s)}) \tag{6.5.9}$$

holds if and only if $(\mathcal{E}, \mathcal{F})$ is recurrent. But this fact was verified in Theorem 2.2.11(ii) already.

Corollary 6.4.7 says that the Dirichlet form $(\mathcal{E}^{(s)}, \mathcal{F}^{(s)} \cap L^2(I;m))$ on $L^2(I;m)$ is recurrent if $m(I) < \infty$. Actually it is recurrent if and only if m is finite in a neighborhood of the boundary r_i whenever r_i is approachable, as is shown in Theorem 2.2.11.

(4°) Absorbing Brownian motion

Let $n \geq 1$ and $D \subset \mathbb{R}^n$ a connected open set. Recall the function spaces ${\rm BL}(D)$, $H^1(D)$, and $H^1_0(D)$ as well as the Dirichlet integral $\mathbf{D}$ from Section 2.2.4. In this example, we assume that $H^1_0(D) \neq H^1(D)$. Consider $(\mathcal{E}, \mathcal{F}) = (\frac{1}{2}\mathbf{D}, H^1_0(D))$. As is seen in Example 3.5.9, $(\mathcal{E}, \mathcal{F})$ is a regular irreducible transient Dirichlet form on $L^2(D)$ and the associated process X^D is the absorbing Brownian motion on D, namely, the part process on D of the n-dimensional Brownian motion. It then holds that

$$\begin{cases} \mathcal{F}^{\rm ref} = {\rm BL}(D) (= \{u \in L^2_{\rm loc}(D) : \mathbf{D}(u,u) < \infty\}), \\ \mathcal{E}^{\rm ref}(u,v) = \frac{1}{2}\mathbf{D}(u,v), \quad u, v \in \mathcal{F}^{\rm ref}. \end{cases} \tag{6.5.10}$$

Indeed, by Theorem 6.2.13, we have

$$\mathcal{F}^{\rm ref} = \left\{ u : \text{ finite a.e., } \tau_k u \in H^1_{0,{\rm loc}}(D), \ \sup_k \mathbf{D}(\tau_k u, \tau_k u) < \infty \right\}. \tag{6.5.11}$$

Clearly $\mathcal{F}^{\rm ref} \cap L^\infty(D) \subset {\rm BL}(D)$. For any $u \in \mathcal{F}^{\rm ref}$, the Cesàro mean sequence $\{f_n, n \geq 1\}$ of a suitable subsequence of $\{\tau_k u, k \geq 1\} (\subset {\rm BL}(D))$ is $\mathbf{D}$-Cauchy by Theorem A.4.1 and converges pointwise to u. On account of (BL.1) of Section 2.2.4, $\{f_n\} \subset {\rm BL}(D)$ admits constants $\{c_n\}$ and $w \in {\rm BL}(D)$ such that $f_n + c_n$ converges to w a.e. on D as $n \to \infty$. Hence $c = \lim_{n\to\infty} c_n$ exists and $u = w - c \in {\rm BL}(D)$, proving $\mathcal{F}^{\rm ref} \subset {\rm BL}(D)$.

To get the converse inclusion, it suffices to show that any bounded $u \in {\rm BL}(D)$ is in $H^1_{0,{\rm loc}}(D)$. For any compact set $K \subset D$, choose a relatively compact open set U with $K \subset U \subset \overline{U} \subset D$ and $g \in C_c^\infty(D)$ taking value 1 on K and 0 on $D \setminus U$. Using a mollifier, we can construct a sequence $u_k \in C_c^\infty(D)$ such that $\{u_k\}$ is uniformly bounded on D, convergent to u a.e. on D, and, for each $1 \leq i \leq n$, $\{\frac{\partial u_k}{\partial x_i}\}$ is L^2-convergent to $\frac{\partial u}{\partial x_i}$ on U. Then **D**-norm and $L^2(D)$-norm of $\{gu_k\} \subset C_c^\infty(D)$ are uniformly bounded and, by invoking Banach-Saks Theorem (Theorem A.4.1), we conclude that $gu \in H_0^1(D)$ as was to be proved.

Since $\frac{1}{2}\mathbf{D}(u,u) = \lim_{k\to\infty} \frac{1}{2}\mathbf{D}(\tau_k u, \tau_k u)$ for $u \in \mathrm{BL}(D)$, we get (6.5.10).

The active reflected Dirichlet space $(\mathcal{F}_a^{\mathrm{ref}} = \mathcal{F}^{\mathrm{ref}} \cap L^2(D), \mathcal{E}^{\mathrm{ref}})$ is the same as the space $(H^1(D), \frac{1}{2}\mathbf{D})$.

(5°) Reflecting Brownian motion and its time changes

Let $n \geq 1$ and $D \subset \mathbb{R}^n$ a connected open set having continuous boundary. Consider $(\mathcal{E}, \mathcal{F}) = (\frac{1}{2}\mathbf{D}, H^1(D))$. As is seen in Example 3.5.9, $(\mathcal{E}, \mathcal{F})$ is a regular irreducible Dirichlet form on $L^2(\overline{D})$ and the associated diffusion process X^r on $\overline{D}$ is the so-called reflecting Brownian motion. X^r is always conservative but it can be either transient or recurrent depending on the geometric nature of D.

However, using Theorem 6.2.13 in the transient case and the definition (6.3.1), (6.3.2) in the recurrent case, we get the description of the reflected Dirichlet space $\mathcal{F}^{\mathrm{ref}}$ of $(\mathcal{E}, \mathcal{F}) = (\frac{1}{2}\mathbf{D}, H^1(D))$ just by replacing $H^1_{0,\mathrm{loc}}(D)$ with $H^1_{\mathrm{loc}}(D)$ on the right hand side of (6.5.11). As a result, we can immediately identify $\mathcal{F}^{\mathrm{ref}}$ with $\mathrm{BL}(D)$. Thus the reflecting Brownian motion X^r and the absorbing Brownian motion X^D have a common reflected Dirichlet space $(\mathrm{BL}(D), \frac{1}{2}\mathbf{D})$ and a common active reflected Dirichlet space $(H^1(D), \frac{1}{2}\mathbf{D})$ on $L^2(D)$.

When X^r is recurrent, in other words, when $(\frac{1}{2}\mathbf{D}, H^1(D))$ is recurrent, we have from Theorem 6.3.2

$$\left(H_e^1(D), \frac{1}{2}\mathbf{D}\right) = \left(\mathrm{BL}(D), \frac{1}{2}\mathbf{D}\right), \tag{6.5.12}$$

where $H_e^1(D)$ denotes the extended Sobolev space. This has been shown in special recurrent cases in Theorem 2.2.13 and Corollary 2.2.15 but without assuming any regularity of ∂D.

If X^r is transient (see the last part of Example 3.5.9 for such cases), then $H_e^1(D)$ is a proper subspace of $\mathrm{BL}(D)$ because the former does not contain non-zero constant functions due to the transience while the latter always does. This distinction becomes more meaningful if we make a time change of X^r. Consider a positive Radon measure m on $\overline{D}$ charging no polar set possessing full quasi support with respect to $(\frac{1}{2}\mathbf{D}, H^1(D))$. Then $(\mathcal{E}, \mathcal{F}) = (\frac{1}{2}\mathbf{D}, H_e^1(D) \cap L^2(D; m))$ is a regular transient Dirichlet form on $L^2(\overline{D}; m)$ associated with the diffusion process $X^{r,m}$ on $\overline{D}$ obtained from X^r by the time change with respect to its PCAF with Revuz measure m in view of (5.2.17) and Theorem 5.2.5.

By Proposition 6.4.6, this Dirichlet form admits $(\mathrm{BL}(D), \frac{1}{2}\mathbf{D})$ again as its reflected Dirichlet space, and accordingly $(\frac{1}{2}\mathbf{D}, \mathrm{BL}(D) \cap L^2(D; m))$ as its active reflected Dirichlet space. If $m(D) < \infty$, then $\mathcal{F}$ is a proper subspace of this active reflected Dirichlet space. Moreover, the lifetime of $X^{r,m}$ is finite a.s. for q.e. starting point because $1 \in L^1(D;m)$ and Proposition 2.1.3 applies.

(6°) Censored stable process

Let $n \geq 1$, $0 < \alpha < 2$, and $D \subset \mathbb{R}^n$ be a (not necessarily connected) open set. Denote by m or dx the Lebesgue measure on $\mathbb{R}^n$. Fix $c > 0$ and define

$$\begin{cases} \mathbf{B}(u,v) := c\int_{D\times D} \frac{(u(x)-u(y))(v(x)-v(y))}{|x-y|^{n+\alpha}}\,dxdy, \\ \mathcal{W}(D) = \{u \in \mathcal{B}(D) : |u| < \infty \text{ a.e.}, \ \mathbf{B}(u,u) < \infty\}. \end{cases} \tag{6.5.13}$$

Using Fatou's lemma, we readily see that

$$(\mathcal{E},\mathcal{F}) = (\mathbf{B}, \mathcal{W}(D) \cap L^2(D;m)) \tag{6.5.14}$$

is a Dirichlet form on $L^2(D;m)$. When $D = \mathbb{R}^n$ and $c = \frac{1}{2}\mathcal{A}(n,-\alpha)$, this is just the Dirichlet form of the rotationally symmetric α-stable process Z on $\mathbb{R}^n$ on account of (2.2.16) and (2.2.19).

Let $\mathcal{F}^0$ be the closure of $C_c^\infty(D)$ in the Dirichlet space (6.5.14) and let $\mathcal{E}^0 = \mathbf{B}|_{\mathcal{F}^0\times\mathcal{F}^0}$. Clearly $(\mathcal{E}^0,\mathcal{F}^0)$ is a regular Dirichlet form on $L^2(D;m)$. The associated symmetric Hunt process X^0 on D is called a *censored α-stable process* on D. In view of (5.2.31) and a statement following it, X^0 is nothing but the resurrected Hunt process of the part process Z^D on D of the rotationally symmetric α-stable process Z on $\mathbb{R}^n$. Detailed properties of X^0 are studied in [14].

By (6.5.13), (4.3.11), and Proposition 4.3.2, we see that the Lévy system of X^0 is given by $(N(x,dy),dt)$ where

$$N(x,dy) = 2c\,|x-y|^{-(n+\alpha)}dy,$$

from which we can deduce the irreducibility of X^0.

In general, the Dirichlet form (6.5.14) on $L^2(D;m)$ cannot be regarded as a regular Dirichlet form on $L^2(\overline{D};m)$. However, when D is an open *n-set* in $\mathbb{R}^n$, that is, there exists a constant $C_1 > 0$ such that

$$m(B(x,r)) \geq C_1\,r^n \qquad \text{for all } x \in D \text{ and } 0 < r \leq 1,$$

then the space (6.5.14) can be identified with a certain Sobolev space $W^{\alpha/2,2}(D)$ of fractional order so that it becomes a regular Dirichlet form on $L^2(\overline{D};m)$ possessing $C_c^\infty(\mathbb{R}^n)|_{\overline{D}}$ as its core (cf. [14]). Its associated process X is called a *reflecting α-stable process* on $\overline{D}$. By virtue of Theorem 3.3.9, the Dirichlet form $(\mathcal{E}^0,\mathcal{F}^0)$ defined above is the part form of (6.5.14) on D, and accordingly the censored stable process X^0 coincides with the part process of X on D. Note that every uniformly Lipschitz domain in $\mathbb{R}^n$ is an n-open set. But an n-set can have very rough boundary. For example, any n-set with a closed set of zero Lebesgue measure removed is still an n-set.

Let $((\mathcal{F}^0)^{\rm ref}, \mathcal{E}^{0,{\rm ref}})$ be the reflected Dirichlet space of the Dirichlet form $(\mathcal{E}^0, \mathcal{F}^0)$ defined above for an arbitrary open set $D \subset \mathbb{R}^n$. Then it holds that

$$\left((\mathcal{F}^0)^{\rm ref}, \mathcal{E}^{0,{\rm ref}}\right) = (\mathcal{W}(D), \mathbf{B}). \tag{6.5.15}$$

Recall that by Definition 6.4.4

$$(\mathcal{F}^0)^{\rm ref} = \left\{u \in \overset{\circ}{\mathcal{F}}{}^0_{\rm loc}: \ \mathbf{B}(u,u) < \infty\right\}$$

and $\mathcal{E}^{0,{\rm ref}} = \mathbf{B}$ on $(\mathcal{F}^0)^{\rm ref}$. The inclusion "$\subset$" in (6.5.15) follows as in Example (**1**°). Conversely, for any non-negative $u \in \mathcal{W}(D)$, integer $k \geq 1$ and any relatively compact open set $U \subset D$ with smooth boundary ∂D, we can find $v \in \mathcal{F}^0$ such that $v = \tau_k u$ on U by making use of the fact that $\tau_k u|_U \in W^{\alpha/2,2}(U)$. (See [14, Theorem 2.2] for details.) This implies that $\tau_k u \in \mathcal{F}^0_{\rm loc}$. By the same argument as that in the paragraph preceding (6.5.3), we conclude that u admits a quasi continuous version and $u \in \overset{\circ}{\mathcal{F}}{}^0_{\rm loc}$. Consequently, $u \in (\mathcal{F}^0)^{\rm ref}$ and (6.5.15) follows.

When D is an n-set, we can readily show that

$$(\mathcal{F}^{\rm ref}, \mathcal{E}^{\rm ref}) = (\mathcal{W}(D), \mathbf{B}) \tag{6.5.16}$$

for the reflected Dirichlet form of $(\mathcal{E}, \mathcal{F})$ given by (6.5.14) because $\mathcal{F}^{\rm ref}$ can be described as above but with $\overset{\circ}{\mathcal{F}}_{\rm loc}$ in place of $\overset{\circ}{\mathcal{F}}{}^0_{\rm loc}$.

As in Example (**1**°), the identity

$$(\mathcal{F}_e, \mathcal{E}) = (\mathcal{W}(D), \mathbf{B}) \tag{6.5.17}$$

holds if and only if $(\mathcal{E}, \mathcal{F})$ is recurrent.

6.6. SILVERSTEIN EXTENSIONS

Throughout this section, $(\mathcal{E}, \mathcal{F})$ is a quasi-regular symmetric Dirichlet form on $L^2(E; m)$.

Definition 6.6.1. A Dirichlet form $(\widetilde{\mathcal{E}}, \widetilde{\mathcal{F}})$ on $L^2(E; m)$ (not necessarily quasi-regular) is said to be a *Silverstein extension* of $(\mathcal{E}, \mathcal{F})$ if $\mathcal{F}_b$ is an ideal in $\widetilde{\mathcal{F}}_b$ (that is, $\mathcal{F}_b \subset \widetilde{\mathcal{F}}_b$ and $fg \in \mathcal{F}_b$ for every $f \in \mathcal{F}_b$ and $g \in \widetilde{\mathcal{F}}_b$) and $\mathcal{E} = \widetilde{\mathcal{E}}$ on $\mathcal{F}_b \times \mathcal{F}_b$.

Remark 6.6.2. We note that $\mathcal{F}_b \subset \widetilde{\mathcal{F}}_b$ and $\mathcal{E} = \widetilde{\mathcal{E}}$ on $\mathcal{F}_b$ imply that $\mathcal{F} \subset \widetilde{\mathcal{F}} \subset \overset{\circ}{\mathcal{F}}_{\rm loc}$ and $\mathcal{E} = \widetilde{\mathcal{E}}$ on $\mathcal{F}$. To verify the inclusion $\widetilde{\mathcal{F}} \subset \overset{\circ}{\mathcal{F}}_{\rm loc}$, we consider a quasi-homeomorphism j from $(\mathcal{E},\mathcal{F})$ to a regular Dirichlet form $(\widehat{\mathcal{E}},\widehat{\mathcal{F}})$ on $L^2(\widehat{E},\widehat{m})$. Let $\{\widehat{D}_n, n \geq 1\}$ be an increasing sequence of relatively compact open subsets of $\widehat{E}$ that increases to $\widehat{E}$. For each integer $n \geq 1$, choose $\widehat{g}_n \in \widehat{\mathcal{F}}_b$ with $\widehat{g}_n = 1$ on D_n. Then $g_n = j^{-1}\widehat{g}_n$ is an element of $\mathcal{F}_b$ taking values 1 on $D_n = j^{-1}\widehat{D}_n$. Note that $\{D_n\}$ is a sequence of quasi open sets increasing to E q.e. For any $f \in \widetilde{\mathcal{F}}_b$, since $g_n f \in \mathcal{F}_b$ and $g_n f = f$ on D_n for every $n \geq 1$, we have $f \in \overset{\circ}{\mathcal{F}}_{\rm loc}$.

For a general $f \in \widetilde{\mathcal{F}}$, we can use truncations $\tau_k f \in \overset{\circ}{\mathcal{F}}_{\rm loc}$, $k \geq 1$, to obtain $f \in \overset{\circ}{\mathcal{F}}_{\rm loc}$ by the same argument as in Section 6.5 (**1°**). □

THEOREM 6.6.3. *Suppose that $(\mathcal{E},\mathcal{F})$ is a quasi-regular Dirichlet form on $L^2(E;m)$. Then its active reflected Dirichlet space $(\mathcal{E}^{\rm ref},\mathcal{F}^{\rm ref}_a)$ is a Silverstein extension of $(\mathcal{E},\mathcal{F})$.*

Proof. Recall that it follows immediately from Definition 6.4.4 that $(\mathcal{E}^{\rm ref}_1,\mathcal{F}^{\rm ref}_a)$ is the reflected (and hence active reflected) Dirichlet space of the transient Dirichlet form $(\mathcal{E}_1,\mathcal{F})$. So in view of Theorem 1.4.3, without loss of generality we may and do assume that $(\mathcal{E},\mathcal{F})$ is a regular transient Dirichlet form on a locally compact separable metric space E. By virtue of Theorem 6.2.13, it holds that

$$\mathcal{F}_b \subset b\mathcal{F}^{\rm ref}_a \quad \text{and} \quad \mathcal{E} = \mathcal{E}^{\rm ref} \quad \text{on} \quad \mathcal{F}_b \times \mathcal{F}_b. \tag{6.6.1}$$

Let $f \in \mathcal{F}_b$ and $g \in b\mathcal{F}^{\rm ref}_a$. By Theorem 6.2.13, $g \in \mathcal{F}_{\rm loc}$. Choose a sequence $\{f_k, k \geq 1\} \subset \mathcal{F} \cap C_c(E)$ that is $\mathcal{E}_1$-convergent to f as well as q.e. convergent to f on E and with $\|f_k\|_\infty \leq \|f\|_\infty$. Then $f_k g \in \mathcal{F}_b$ and obtain from Exercise 1.1.10 being applied to the active reflected Dirichlet form $(\mathcal{E}^{\rm ref},\mathcal{F}^{\rm ref}_a)$ and (6.6.1) that

$$\|f_k g\|_{\mathcal{E}_1} = \|f_k g\|_{\mathcal{E}^{\rm ref}_1} \leq \|f_k\|_\infty \|g\|_{\mathcal{E}^{\rm ref}} + \|g\|_\infty \|f_k\|_{\mathcal{E}_1} + \|f_k\|_\infty \|g\|_2.$$

Therefore, $\sup_{k\geq 1} \|f_k g\|_{\mathcal{E}_1} < \infty$. Since $f_k g$ converges to fg on E, it follows that $fg \in \mathcal{F}_b$. This proves that $\mathcal{F}_b$ is an ideal in $b\mathcal{F}^{\rm ref}_a$. Thus $(\mathcal{E}^{\rm ref},\mathcal{F}^{\rm ref}_a)$ is a Silverstein extension of $(\mathcal{E},\mathcal{F})$. □

In fact, we also have the following.

PROPOSITION 6.6.4. *Suppose that $(\mathcal{E},\mathcal{F})$ is a quasi-regular Dirichlet form on $L^2(E;m)$. Then $b\mathcal{F}_e$ is an ideal of $b\mathcal{F}^{\rm ref}$. In particular, if $1 \in \mathcal{F}_e$, then $(\mathcal{E},\mathcal{F}_e) = (\mathcal{E}^{\rm ref},\mathcal{F}^{\rm ref})$.*

Proof. By making use of a quasi-homeomorphism as in the proof of Theorem 6.4.5, we may well assume that $(\mathcal{E},\mathcal{F})$ is a regular Dirichlet form. According to Corollary 5.2.12 and Proposition 6.4.6, the extended Dirichlet form and the reflected Dirichlet form are unchanged under a full support time change. Hence replacing m by $f(x)m(dx)$ for some strictly positive bounded function g with $\int_E g(x)m(dx) < \infty$ if necessary, we may and do assume that $m(E) < \infty$. In this case, $b\mathcal{F}_e = b\mathcal{F}$ and $b\mathcal{F}^{\rm ref} = b\mathcal{F}^{\rm ref}_a$. So the conclusion of this proposition follows directly from Theorem 6.6.3. □

The next theorem gives the probabilistic meaning of a Silverstein extension.

Theorem 6.6.5. *Suppose $(\mathcal{E}, \mathcal{F})$ is a quasi-regular Dirichlet form on $L^2(E;m)$ and $(\widetilde{\mathcal{E}}, \widetilde{\mathcal{F}})$ is another symmetric Dirichlet form on $L^2(E;m)$ (not necessarily quasi-regular). Then the following are equivalent.*
(i) *$(\widetilde{\mathcal{E}}, \widetilde{\mathcal{F}})$ is a Silverstein extension of $(\mathcal{E}, \mathcal{F})$.*
(ii) *There is a locally compact separable metric space $\widehat{E}$ and a measurable map $j: E \to \widehat{E}$ so that, by letting $\widehat{m} = m \circ j^{-1}$, j^* is a unitary map from $L^2(\widehat{E};\widehat{m})$ onto $L^2(E;m)$ and the image Dirichlet form $(\widehat{\mathcal{E}}, \widehat{\mathcal{F}}) := j(\widetilde{\mathcal{E}}, \widetilde{\mathcal{F}})$ is a regular Dirichlet form on $L^2(\widehat{E}, \widehat{m})$. Moreover, there is an $\widehat{\mathcal{E}}$-quasi-open Borel subset $\widehat{E}_0$ of $\widehat{E}$ so that $\widehat{m}(\widehat{E} \setminus \widehat{E}_0) = 0$ and j is the quasi-homeomorphism that maps the quasi-regular Dirichlet form $(\mathcal{E}, \mathcal{F})$ on $L^2(E;m)$ onto the quasi-regular Dirichlet form $(\widehat{\mathcal{E}}, \widehat{\mathcal{F}}_{\widehat{E}_0})$ on $L^2(\widehat{E}_0;\widehat{m})$. Here $(\widehat{\mathcal{E}}, \widehat{\mathcal{F}}_{\widehat{E}_0})$ is the part Dirichlet form of $(\widehat{\mathcal{E}}, \widehat{\mathcal{F}})$ on the $\widehat{\mathcal{E}}$-quasi-open subset $\widehat{E}_0$.*

Proof. (ii) $\Rightarrow$ (i): Since $\widehat{\mathcal{F}}_{\widehat{E}_0} = \{u \in \widehat{\mathcal{F}} : u = 0 \ \widehat{\mathcal{E}}\text{-q.e. on } \widehat{E} \setminus \widehat{E}_0\}$, $b\widehat{\mathcal{F}}_{\widehat{E}_0}$ is an ideal of $b\widehat{\mathcal{F}}$. Consequently, $b\mathcal{F}$ is an ideal of $b\widetilde{\mathcal{F}}$ and $\widetilde{\mathcal{E}} = \mathcal{E}$ on $b\mathcal{F} \times b\mathcal{F}$; that is, $(\widetilde{\mathcal{E}}, \widetilde{\mathcal{F}})$ is a Silverstein extension of $(\mathcal{E}, \mathcal{F})$.

(i) $\Rightarrow$ (ii): Since $(\mathcal{E}, \mathcal{F})$ is quasi-regular on $L^2(E;m)$ and $\widetilde{\mathcal{F}} \subset \mathcal{F}_{\rm loc}$ by Remark 6.6.2, every function in $\widetilde{\mathcal{F}}$ has an $\mathcal{E}$-quasi-continuous version on E. As $L^2(E;m)$ is separable, there exists countable set $B_0 = \{f_n, n \geq 1\}$ of bounded functions in $\widetilde{\mathcal{F}}$ that is dense in $(\widetilde{\mathcal{F}}, \widetilde{\mathcal{E}}_1)$ such that

(1) B_0 is an algebra over the rational numbers.
(2) There is a function $h \in \mathcal{F}$ and an $\mathcal{E}$-nest $\{F_k, k \geq 1\}$ consisting of compact sets such that $1/k \leq h \leq k$ on F_k, $B_0 \subset C(\{F_k\})$ and B_0 separates points of $\cup_{k\geq 1} F_k$.

We let functions in B_0 take value zero on $E \setminus \cup_{k\geq 1} F_k$. Define $B := \overline{B_0}^{\|\cdot\|_\infty}$, which is a commutative Banach algebra. We will use the Gelfand transform to construct a regularizing space $\widehat{E}$ for Dirichlet form $(\widetilde{\mathcal{E}}, \widetilde{\mathcal{F}})$ and the quasi-homeomorphic map j which is very similar to the one given in the proof of Theorem 1.4.3. Notice, however, that we now use an $\mathcal{E}$-nest $\{F_k\}$ instead of an $\widetilde{\mathcal{E}}$-nest in taking quasi continuous versions of functions in an $\widetilde{\mathcal{E}}_1$-dense set $B_0 (\subset \widetilde{\mathcal{F}})$ in the above. For readers' convenience, we provide the details of a proof here.

Step 1. Construct a locally compact separable metric space $\widehat{E}$.

Let $\widehat{E}$ be a collection of non-trivial real-valued functionals γ on B which satisfy for $f, g \in B$ and for rational numbers a and b,

$$|\gamma(f)| \leq \|f\|_\infty, \quad \gamma(fg) = \gamma(f)\gamma(g), \quad \gamma(af + bg) = a\gamma(f) + b\gamma(g).$$

We equip $\widehat{E}$ with the weakest topology so that the function $\Phi_f : \gamma \mapsto \gamma(f)$ is continuous for every $f \in B$. $\widehat{E}$ is then a separable locally compact metrizable space and $\{\Phi_f, f \in B\} = C_\infty(\widehat{E})$. $\widehat{E}$ is compact if and only if $1 \in B$.

Let j be the unique map from $\cup_{k\geq 1} F_k$ into $\widehat{E}$ such that

$$(jx)(f) := f(x) \qquad \text{for } f \in B \text{ and } x \in \cup_{k\geq 1} F_k.$$

By (2) above, j is a continuous one-to-one map on each F_k. Hence $\widehat{F}_k := j(F_k)$ is compact in $\widehat{E}$ and j is a topological homeomorphism from F_k onto $\widehat{F}_k$ for every $k \geq 1$. Note that $j : \cup_{k\geq 1} F_k \to \widehat{E}$ is Borel measurable and $m(E \setminus \cup_{k\geq 1} F_k) = 0$. Define $\widehat{m} := m \circ j^{-1}$. Clearly $\widehat{m}(\widehat{E} \setminus \cup_{k\geq 1} \widehat{F}_k) = 0$. It follows from the m-integrability of functions in B_0 that $\widehat{m}$ is a Radon measure and further $\text{supp}[\widehat{m}] = \widehat{E}$. Since B_0 is dense in $L^2(E; m)$, j^* is unitary from $L^2(\widehat{E}; \widehat{m})$ onto $L^2(E; m)$.

Step 2. The image Dirichlet form $j(\widetilde{\mathcal{E}}, \widetilde{\mathcal{F}})$ is regular on $L^2(\widehat{E}; \widehat{m})$.

Let $(\widehat{\mathcal{E}}, \widehat{\mathcal{F}}) := j(\widetilde{\mathcal{E}}, \widetilde{\mathcal{F}})$. Then $(\widehat{\mathcal{E}}, \widehat{\mathcal{F}})$ is clearly a Dirichlet form on $L^2(\widehat{E}, \widehat{m})$. Since $\widehat{\mathcal{F}} \cap C_\infty(\widehat{E}) \supset \Phi(B_0)$ and the latter is uniformly dense in $C_\infty(\widehat{E})$ and $\widehat{\mathcal{E}}_1$-dense in $\widehat{\mathcal{F}}$, $(\widehat{\mathcal{E}}, \widehat{\mathcal{F}})$ is a regular Dirichlet form on $L^2(\widehat{E}; \widehat{m})$.

Step 3. Define $\widehat{E}_1 := \cup_{k\geq 1} \widehat{F}_k$. There is some $\mathcal{E}$-polar set $N \subset E$ so that $\widehat{E}_0 := \widehat{E}_1 \setminus j(N)$ is an $\widehat{\mathcal{E}}$-quasi-open subset of $\widehat{E}$ and that j is a quasi-homeomorphism from $(\mathcal{E}, \mathcal{F})$ on $L^2(E; m)$ onto $(\widehat{\mathcal{E}}, \widehat{\mathcal{F}}_{\widehat{E}_0})$ on $L^2(\widehat{E}_0; \widehat{m})$.

Note that for every $k \geq 1$, $h_k := k(h \wedge (1/k)) \in \mathcal{F}_b$ and $h_k = 1$ on F_k for every $k \geq 1$. Since $\mathcal{F}_b$ is an ideal in $\widetilde{\mathcal{F}}_b$ with $\mathcal{E} = \widetilde{\mathcal{E}}$ on $\mathcal{F}_b$, it follows that $b\mathcal{F}_{F_k} = b\widetilde{\mathcal{F}}_{F_k}$ and so

$$j^* \widehat{\mathcal{F}}_{\widehat{F}_k} = \widetilde{\mathcal{F}}_{F_k} = \mathcal{F}_{F_k} \quad \text{with} \quad j^* \widehat{\mathcal{E}} = \mathcal{E} \text{ on } \mathcal{F}_{F_k}. \tag{6.6.2}$$

Let $\widehat{X} = (\widehat{X}_t, \widehat{\mathbf{P}}_x)$ be the symmetric Hunt process on $\widehat{E}$ associated with the regular Dirichlet form $(\widehat{\mathcal{E}}, \widehat{\mathcal{F}})$ on $L^2(\widehat{E}; \widehat{m})$ and let

$$\widehat{R}_\alpha^{\widehat{E}_1} f(x) = \widehat{\mathbf{E}}_x \left[\int_0^{\sigma_{\widehat{E} \setminus \widehat{E}_1}} e^{-\alpha t} f(\widehat{X}_t) dt \right], \quad x \in \widehat{E}, \ f \in \mathcal{B}_+(\widehat{E}).$$

In view of Corollary 3.2.3, $\widehat{R}_\alpha^{\widehat{E}_1}$ is related to the space $(\widehat{\mathcal{E}}, \widehat{\mathcal{F}}_{E_1})$ by

$$\widehat{R}_\alpha^{\widehat{E}_1} f \in \widehat{\mathcal{F}}_{\widehat{E}_1}, \quad \widehat{\mathcal{E}}_\alpha(\widehat{R}_\alpha^{\widehat{E}_1} f, v) = \int_{\widehat{E}_1} f(x) v(x) \widehat{m}(dx), \tag{6.6.3}$$

for any $f \in \mathcal{B}(\widehat{E}) \cap L^2(\widehat{E};\widehat{m})$ and $v \in \widehat{\mathcal{F}}_{\widehat{E}_1}$, where

$$\widehat{\mathcal{F}}_{\widehat{E}_1} = \left\{ f \in \widehat{\mathcal{F}} : f = 0 \quad \widehat{\mathcal{E}}\text{-q.e. on } \widehat{E} \setminus \widehat{E}_1 \right\}.$$

By (6.6.3), $\{\widehat{R}_\alpha^{\widehat{E}_1} f; f \in L^2(\widehat{E};\widehat{m})\}$ is dense in $(\widehat{\mathcal{F}}_{\widehat{E}_1}, \widehat{\mathcal{E}}_1)$. Similarly, we define

$$\widehat{R}_\alpha^{\widehat{F}_k} f(x) = \widehat{\mathbf{E}}_x \left[\int_0^{\sigma_{\widehat{E}\setminus \widehat{F}_k}} e^{-\alpha t} f(X_t) dt \right], \quad x \in \widehat{E},\ f \in \mathcal{B}_+(\widehat{E}).$$

As $\cup_{k\geq 1} \widehat{F}_k = \widehat{E}_1$, $\lim_{k\to\infty} \widehat{R}_\alpha^{\widehat{F}_k} f(x) = \widehat{R}_\alpha^{\widehat{E}_1} f(x)$ $\widehat{\mathcal{E}}$-q.e. on $\widehat{E}$ for every $f \in \mathcal{B}_+(\widehat{E})$. It follows that $\cup_k \widehat{\mathcal{F}}_{\widehat{F}_k}$ is $\widehat{\mathcal{E}}_1$-dense in $\widehat{\mathcal{F}}_{\widehat{E}_1}$. Since $\{F_k\}$ is an $\mathcal{E}$-nest, we conclude that $(\widehat{\mathcal{E}}, \widehat{\mathcal{F}}_{E_1})$ is the image Dirichlet form of $(\mathcal{E}, \mathcal{F})$ under the quasi-homeomorphism j from E to $\widehat{E}_1$. For emphasis, we denote this Dirichlet form $(\widehat{\mathcal{E}}, \widehat{\mathcal{F}}_{\widehat{E}_1})$ by $\widehat{\mathcal{E}}^{\widehat{E}_1}$.

On the other hand, since $(\mathcal{E}, \mathcal{F})$ is quasi-regular, it admits a properly associated special standard process X on E by virtue of Theorem 1.5.2. Denote by $\{\widehat{R}_\alpha^0; \alpha > 0\}$ the resolvent kernel of the image process jX. It follows from (6.6.2) that $\widehat{R}_\alpha^0 f = \widehat{R}_\alpha^{E_1} f$, $\widehat{m}$-a.e. for every $f \in L^2(\widehat{E}_1;\widehat{m})$. For every $f \in B_0$, $\Phi f \in b\mathcal{B}(\widehat{E}) \cap L^2(\widehat{E};\widehat{m})$. $\widehat{R}_\alpha^0(\Phi f)$ is $\widehat{\mathcal{E}}^{\widehat{E}_1}$-quasi-continuous, and so is $\widetilde{R}_\alpha^{\widehat{E}_1}(\Phi f)|_{\widehat{E}_1}$ because $\widehat{R}_\alpha^{\widehat{E}_1}(\Phi f)$ is $\widehat{\mathcal{E}}$-quasi-continuous and the restriction to $\widehat{E}_1$ of any $\widehat{\mathcal{E}}$-nest is an $\widehat{\mathcal{E}}^{\widehat{E}_1}$-nest. Since they coincide $\widehat{m}$-a.e. on E_1 and B_0 is countable, there is an X-properly exceptional set $N \subset E$ such that

$$\widehat{R}_\alpha^{\widehat{E}_1}(\Phi f)(x) = \widehat{R}_\alpha^0(\Phi f)(x)$$

for every $x \in \widehat{E}_1 \setminus j(N)$ and every $f \in B_0$.

As the linear span of $\{\Phi f; f \in B_0\}$ is uniformly dense in $C_\infty(\widehat{E})$, the above display holds for every $g \in C_\infty(\widehat{E})$ in place of Φf. We thus have for every $x \in \widehat{E}_1 \setminus j(N)$, $\{\widehat{X}_t, t < \sigma_{\widehat{E}\setminus\widehat{E}_1}\}$ under $\widehat{\mathbf{P}}_x$ has the same distribution as that of the image process $\{jX_t, t < \zeta\}$ starting from x. This in particular implies that $\widehat{\mathbf{P}}_x(\tau_{\widehat{E}_1\setminus j(N)} > 0) = 1$ for every $x \in \widehat{E}_1 \setminus j(N)$, namely, the set $\widehat{E}_0 := \widehat{E}_1 \setminus j(N)$ is $\widehat{X}$-finely open and hence $\widehat{\mathcal{E}}$-quasi-open by Theorem 3.3.3(i). Now the part process $\widehat{X}^{\widehat{E}_0}$ of $\widehat{X}$ on the $\widehat{\mathcal{E}}$-quasi-open $\widehat{E}_0$ is associated with the Dirichlet form $\widehat{\mathcal{E}}^{\widehat{E}_1}$; that is, $\widehat{\mathcal{F}}_{\widehat{E}_0} = \widehat{\mathcal{F}}_{\widehat{E}_1}$. Since N is $\mathcal{E}$-polar, there is an $\mathcal{E}$-nest $\{K_n, n \geq 1\}$ so that $N \subset \cap_{n\geq 1}(E \setminus K_n)$. By Theorem 1.2.13(iii), $\{F_n \cap K_n, n \geq 1\}$ is an $\mathcal{E}$-nest consisting of compact sets and clearly j is a topological homeomorphism from $F_n \cap K_n$ to $j(F_n \cap K_n)$ for every $n \geq 1$. It follows that j is a quasi-homeomorphism between the quasi-regular Dirichlet form $(\mathcal{E}, \mathcal{F})$ on $L^2(E;m)$ and the part Dirichlet form $(\widehat{\mathcal{E}}, \widehat{\mathcal{F}}_{\widehat{E}_0})$ on $L^2(\widehat{E}_0;\widehat{m})$. This completes the proof of the theorem. □

The following is an immediate consequence of Theorems 6.6.3 and 6.6.5.

COROLLARY 6.6.6. *Suppose that $(\mathcal{E},\mathcal{F})$ is a quasi-regular Dirichlet form on $L^2(E;m)$ such that its associated process is conservative. Then*

$$(\mathcal{E}^{\rm ref},\mathcal{F}^{\rm ref}_a)=(\mathcal{E},\mathcal{F}).$$

Remark 6.6.7. (i) Let X be an m-symmetric standard process on E properly associated with the quasi-regular Dirichlet form $(\mathcal{E},\mathcal{F})$ on $L^2(E;m)$. Identifying E with E_0 and X with $\widehat{X}^{E_0}$ by the quasi-homeomorphism j, the above theorem says that $(\widetilde{\mathcal{E}},\widetilde{\mathcal{F}})$ is a Silverstein extension of $(\mathcal{E},\mathcal{F})$ if and only if there is an m-symmetric Hunt process $\widetilde{X}$ associated with Dirichlet form $(\widetilde{\mathcal{E}},\widetilde{\mathcal{F}})$ that extends X to some state space $\widetilde{E}$ which contains E as an $\widetilde{\mathcal{E}}$-quasi-open subset of $\widetilde{E}$ up to an $\mathcal{E}$-polar set, where the measure m is extended to $\widetilde{E}$ by setting $m(\widetilde{E}\setminus E)=0$. The last condition amounts to saying that the process $\widetilde{X}$ spends zero Lebesgue amount of time on the "boundary" $\widetilde{E}\setminus E$.
(ii) When $(\mathcal{E},\mathcal{F})$ is a regular Dirichlet form on $L^2(E;m)$, the set $\widehat{E}_0$ in Theorem 6.6.5(ii) can be taken to be an open subset of $\widehat{E}$ (see [138] and [73, Theorem A.4.4]) so that $(\widehat{\mathcal{E}},\widehat{\mathcal{F}}_{\widehat{E}_0})$ is a regular Dirichlet form on $L^2(\widehat{E}_0;\widehat{m})$ by Theorem 3.3.9. □

We define a partial ordering on Dirichlet forms on $L^2(E;m)$.

DEFINITION 6.6.8. For two Dirichlet forms $(\mathcal{E},\mathcal{F})$ and $(\widetilde{\mathcal{E}},\widetilde{\mathcal{F}})$ on $L^2(E;m)$, we write $(\mathcal{E},\mathcal{F})\preceq(\widetilde{\mathcal{E}},\widetilde{\mathcal{F}})$ if

$$\mathcal{F}\subset\widetilde{\mathcal{F}}\quad\text{and}\quad\mathcal{E}(u,u)\geq\widetilde{\mathcal{E}}(u,u)\quad\text{for every } u\in\mathcal{F}.$$

We next show that $(\mathcal{E}^{\rm ref},\mathcal{F}^{\rm ref}_a)$ is the maximum among the Silverstein extensions of $(\mathcal{E},\mathcal{F})$ on $L^2(E;m)$ with respect to the above partial ordering $\preceq$.

THEOREM 6.6.9. *Suppose that $(\mathcal{E},\mathcal{F})$ is a quasi-regular Dirichlet form on $L^2(E;m)$ and that $(\widetilde{\mathcal{E}},\widetilde{\mathcal{F}})$ is a Dirichlet form on $L^2(E;m)$ that is a Silverstein extension of $(\mathcal{E},\mathcal{F})$. Then $(\widetilde{\mathcal{E}},\widetilde{\mathcal{F}})\preceq(\mathcal{E}^{\rm ref},\mathcal{F}^{\rm ref}_a)$.*

Proof. By Theorem 6.6.5 and Remark 6.6.7, there is a locally compact separable metric space $\widetilde{E}$ on which $(\widetilde{\mathcal{E}},\widetilde{\mathcal{F}})$ is a regular Dirichlet form, and modulo a quasi-homeomorphism and an $\mathcal{E}$-polar set, we can and do assume that E is an $\widetilde{\mathcal{E}}$-quasi-open subset of $\widetilde{E}$ and the measure m is extended to space $\widetilde{E}$ by letting $m(\widetilde{E}\setminus E)=0$. Let $\widetilde{X}$ be the symmetric Hunt process on $\widetilde{E}$ associated with the regular Dirichlet form $(\widetilde{\mathcal{E}},\widetilde{\mathcal{F}})$ on $L^2(\widetilde{E};m)=L^2(E;m)$. Without loss of generality, we may assume that E is an $\widetilde{X}$-finely open subset of $\widetilde{E}$. Then the part process $\widetilde{X}^E$ of $\widetilde{X}$ on E is a symmetric standard process properly associated with the quasi-regular Dirichlet form $(\mathcal{E},\mathcal{F})$ on $L^2(E;m)$.

Let $\widetilde{\mu}^c_{\langle f\rangle}$, $\widetilde{J}$ and $\widetilde{\kappa}$ denote the energy measure of the continuous part of the MAF $\widetilde{M}^{[f]}$, the jumping measure and killing measure of $\widetilde{X}$. The corresponding counterparts of $\widetilde{X}^E$ are denoted as $\mu^c_{\langle f\rangle}$, J, and κ. In view of Theorem 3.3.8, $(\mathcal{E},\mathcal{F})$ can be identified with a subspace of $(\widetilde{\mathcal{E}},\widetilde{\mathcal{F}})$ as

$$\mathcal{F}=\{f\in\widetilde{\mathcal{F}}:\widetilde{f}=0\ \widetilde{\mathcal{E}}\text{-q.e. on }\widetilde{E}\setminus E\},\quad \mathcal{E}=\widetilde{\mathcal{E}}\text{ on }\mathcal{F}\times\mathcal{F}.$$

Therefore, we can readily deduce the Beurling-Deny decomposition (Theorem 4.3.3) for $(\mathcal{E}.\mathcal{F})$ from that for $(\widetilde{\mathcal{E}},\widetilde{\mathcal{F}})$ yielding that

$$J(dx,dy)=\widetilde{J}(dx,dy)\big|_{E\times E}\ \text{ and }\ \kappa(dx)=\widetilde{\kappa}(dx)\big|_E+\widetilde{J}(dx,\widetilde{E}\setminus E)\big|_E. \qquad (6.6.4)$$

Further, $\mu^c_{\langle f\rangle}=\widetilde{\mu}^c_{\langle f\rangle}$ for $f\in\mathcal{F}$. $\mu^c_{\langle f\rangle}$ can be extended to $f\in\overset{\circ}{\mathcal{F}}_{\rm loc}$. But since $\widetilde{\mathcal{F}}\subset\overset{\circ}{\mathcal{F}}_{\rm loc}$ in view of Remark 6.6.2, we have by Theorem 4.3.10(i)

$$\mu^c_{\langle f\rangle}=\widetilde{\mu}^c_{\langle f\rangle}\big|_E \qquad \text{for every } f\in\widetilde{\mathcal{F}}. \qquad (6.6.5)$$

Let $\widehat{\mathcal{E}}$ be defined by (6.4.1) with respect to $(\mathcal{E},\mathcal{F})$ in terms of $\mu^c_{\langle f\rangle}$, J, and κ. Then on account of (6.6.4)–(6.6.5), we have for every $f\in\widetilde{\mathcal{F}}$, $\widehat{\mathcal{E}}(f,f)\le\widetilde{\mathcal{E}}(f,f)<\infty$. This proves that $\widetilde{\mathcal{F}}\subset\mathcal{F}^{\rm ref}$ and $\widetilde{\mathcal{E}}(f,f)\ge\mathcal{E}^{\rm ref}(f,f)$ for $f\in\widetilde{\mathcal{F}}$; that is, $(\widetilde{\mathcal{E}},\widetilde{\mathcal{F}})\preceq(\mathcal{E}^{\rm ref},\mathcal{F}^{\rm ref}_a)$. □

It follows from the last statement in Theorem 6.4.5 that when a quasi-regular Dirichlet $(\mathcal{E},\mathcal{F})$ is recurrent, $\mathcal{F}^{\rm ref}$ is the extended Dirichlet space of $(\mathcal{E}^{\rm ref},\mathcal{F}^{\rm ref}_a)$. The following theorem deals with the transient case.

THEOREM 6.6.10. *Suppose that $(\mathcal{E},\mathcal{F})$ is a transient quasi-regular Dirichlet form on $L^2(E;m)$. Then*
(i) *$\left(\mathcal{F}^{\rm ref}_a\right)_e\subset\mathcal{F}^{\rm ref}$ and the restriction of $\mathcal{E}^{\rm ref}$ on $(\mathcal{F}^{\rm ref}_a)_e$ gives the extended Dirichlet form.*
(ii) *When $m(E)<\infty$, $(\mathcal{F}^{\rm ref}_a)_e=\mathcal{F}^{\rm ref}$.*

Proof. (i) Just as in the proof of Theorem 6.4.5, using a quasi-homeomorphism if necessary, we may and do assume that $(\mathcal{E},\mathcal{F})$ is a transient regular Dirichlet form on $L^2(E;m)$. Suppose that $u\in(\mathcal{F}^{\rm ref}_a)_e$. Then there is an $\mathcal{E}^{\rm ref}$-Cauchy sequence $\{u_k,k\ge1\}\subset\mathcal{F}^{\rm ref}_a$ that converges to u m-a.e. on E. For each $k\ge1$, there are unique $f_k\in\mathcal{F}_e$ and $h_k\in\mathbf{HN}$ so that $f_k+h_k=u_k$. It follows from (6.2.4) that $\{f_k,k\ge1\}$ is an $\mathcal{E}$-Cauchy sequence in $\mathcal{F}_e$ and $\{h_k,k\ge1\}$ is an $\mathcal{E}^{\rm ref}$-Cauchy sequence in $\mathbf{HN}$. Since $(\mathcal{F}_e,\mathcal{E})$ is a Hilbert space, f_k is $\mathcal{E}$-convergent to some $f\in\mathcal{F}_e$. Moreover, by Theorem 2.1.5, there is a subsequence $\{f_{n_k},k\ge1\}$ that converges to f m-a.e. on E. Now by Lemma 6.2.3, h_{n_k} is $\mathcal{E}^{\rm ref}$-convergent to $u-f\in\mathbf{HN}$. Consequently, $u=f+(u-f)\in\mathcal{F}^{\rm ref}$ and u_{n_k} and hence u_k is $\mathcal{E}^{\rm ref}$-convergent to u. This proves that $(\mathcal{F}^{\rm ref}_a)_e\subset\mathcal{F}^{\rm ref}$.

(ii) When $m(E)<\infty$, then for every $u\in\mathcal{F}^{\rm ref}$ and integer $k\ge1$, $u_k:=((-k)\vee u)\wedge k\in\mathcal{F}^{\rm ref}_a$ and $\mathcal{E}^{\rm ref}(u_k,u_k)\le\mathcal{E}^{\rm ref}(u,u)$. As $\lim_{k\to\infty}u_k=u$, we have by the Banach-Saks Theorem (Theorem A.4.1) that $u \in (\mathcal{F}_a^{\text{ref}})_e$. So in this case we have established $\mathcal{F}^{\text{ref}} \subset \left(\mathcal{F}_a^{\text{ref}}\right)_e$. □

Remark 6.6.11. (i) Let $(\mathcal{E}, \mathcal{F})$ be a quasi-regular Dirichlet form on $L^2(E; m)$ and $(\mathcal{F}^{\text{ref}}, \mathcal{E}^{\text{ref}})$ be its reflected Dirichlet space. By Theorem 6.6.3, Theorem 6.6.5, and Remark 6.6.7, the active reflected Dirichlet form $(\mathcal{E}^{\text{ref}}, \mathcal{F}_a^{\text{ref}})$ can be represented as a regular Dirichlet form on $L^2(\widetilde{E}, \widetilde{m})$, where $\widetilde{E}$ is some locally compact separable metric space containing E as an $\mathcal{E}^{\text{ref}}$-quasi-open set up to an $\mathcal{E}$-polar set and $\widetilde{m}$ is the extension of m to $\widetilde{E}$ by $\widetilde{m}(\widetilde{E} \setminus E) = 0$. Therefore, using Definition 6.4.4 again, we can further define the reflected Dirichlet space of $(\mathcal{E}^{\text{ref}}, \mathcal{F}_a^{\text{ref}})$ being regarded as a regular Dirichlet form on $L^2(\widetilde{E}, \widetilde{m})$.

Actually this reflected Dirichlet space coincides with $(\mathcal{F}^{\text{ref}}, \mathcal{E}^{\text{ref}})$. In fact, the expression $\widehat{\mathcal{E}}(u, u)$ in Definition 6.4.4 can be viewed as the Beurling-Deny formula for the regular Dirichlet form $(\mathcal{E}^{\text{ref}}, \mathcal{F}_a^{\text{ref}})$ on $L^2(\widetilde{E}, \widetilde{m})$, and accordingly we get the stated coincidence by using the same definition again. In particular, when $(\mathcal{E}^{\text{ref}}, \mathcal{F}_a^{\text{ref}})$ is recurrent, we have by Theorem 6.4.5 that $(\mathcal{F}_a^{\text{ref}})_e = \mathcal{F}^{\text{ref}}$.
(ii) We keep the setting of (i). When $(\mathcal{E}^{\text{ref}}, \mathcal{F}_a^{\text{ref}})$ is transient, we have by Theorem 6.4.12 that $\mathcal{F}^{\text{ref}}$ is the linear span of $(\mathcal{F}_a^{\text{ref}})_e$ and $\mathbf{HN}^{\text{ref}}$, where $\mathbf{HN}^{\text{ref}}$ denotes the space of harmonic functions of $(\mathcal{E}^{\text{ref}}, \mathcal{F}_a^{\text{ref}})$ having finite $\mathcal{E}^{\text{ref}}$-energy. Example ($\mathbf{5}^\circ$) of Section 6.5 illustrates that the linear space $\mathbf{HN}^{\text{ref}}$ can be non-trivial. In fact, Remark 3.7 of [24] suggests that for any integer $k \geq 1$, there exists a regular Dirichlet form $(\mathcal{E}, \mathcal{F})$ so that $\mathbf{HN}^{\text{ref}}$ has dimension k. □

Example 6.6.12. Let $D \subset \mathcal{B}^n$ be a bounded Lipschitz domain and X be the absorbing Brownian motion on D. Its associated Dirichlet form on $L^2(D)$ is $(\mathcal{E}, \mathcal{F})$, where $\mathcal{F} = H_0^1(D)$ and $\mathcal{E}(u, v) = \frac{1}{2}\int_D \nabla u(x) \cdot \nabla v(x)\, dx$. The reflected Dirichlet form $(\mathcal{E}^{\text{ref}}, \mathcal{F}^{\text{ref}})$ of $(\mathcal{E}, \mathcal{F})$ is given by (6.5.10). Its active reflected Dirichlet form $(\mathcal{E}^{\text{ref}}, \mathcal{F}_a^{\text{ref}}) = (\mathcal{E}, H^1(D))$ is a regular Dirichlet form on $L^2(\overline{D}) = L^2(\overline{D}; \mathbf{1}_D dx)$ and its associated process is the normally reflecting Brownian motion X^r on $\overline{D}$.
(i) Let μ be a positive smooth measure of X^r whose quasi support is contained in ∂D (e.g., the surface measure σ on ∂D is a smooth measure of finite energy integral of $\overline{D}$). Let $\widetilde{X}$ be the reflecting Brownian motion killed by means of a PCAF with Revuz measure μ. Its associated Dirichlet form $(\widetilde{\mathcal{E}}, \widetilde{\mathcal{F}})$ is given by $\widetilde{\mathcal{F}} = H^1(D) \cap L^2(\partial D; \mu)$ and

$$\widetilde{\mathcal{E}}(f,f) = \frac{1}{2}\int_D |\nabla f(x)|^2\, dx + \int_{\partial D} f(x)^2 \mu(dx) \qquad \text{for } f \in \widetilde{\mathcal{F}}.$$

Clearly $(\widetilde{\mathcal{E}}, \widetilde{\mathcal{F}})$ is a Silverstein extension of $(\mathcal{E}, \mathcal{F})$ and $(\widetilde{\mathcal{E}}, \widetilde{\mathcal{F}}) \preceq (\mathcal{E}^{\text{ref}}, \mathcal{F}_a^{\text{ref}})$.

In Section 7.4, a lateral condition will be given to characterize the infinitesimal generator of the L^2-semigroup of such an extension as $\widetilde{X}$ of this example in a more general context.

(ii) Let F be a subset of ∂D that is not $\mathcal{E}^{\rm ref}$-polar. Let $\widehat{X}$ be the part process of the reflecting Brownian motion killed upon hitting F. Its associated Dirichlet form $(\widehat{\mathcal{E}}, \widehat{\mathcal{F}})$ is given by

$$\widehat{\mathcal{F}} := \left\{u \in H^1(D) : \ u = 0 \ \mathcal{E}^{\rm ref}\text{-q.e. on } F\right\}$$

and $\widehat{\mathcal{E}} = \mathcal{E}^{\rm ref}$ on $\widehat{\mathcal{F}}$. Clearly $(\widehat{\mathcal{E}}, \widehat{\mathcal{F}})$ is a Silverstein extension of $(\mathcal{E}, \mathcal{F})$ and $(\widehat{\mathcal{E}}, \widehat{\mathcal{F}}) \preceq (\mathcal{E}^{\rm ref}, \mathcal{F}^{\rm ref}_a)$. □

6.7. EQUIVALENT NOTIONS OF HARMONICITY

It is well-known that a harmonic function u in a domain $D \subset \mathbb{R}^d$ can be defined or characterized by $\Delta u = 0$ in D in the distributional sense, that is, $u \in H^1_{\rm loc}(D) := \{v \in L^2_{\rm loc}(D) : \ \nabla v \in L^2_{\rm loc}(D)\}$ so that

$$\int_{\mathbb{R}^d} \nabla u(x) \cdot \nabla v(x) dx = 0 \qquad \text{for every } v \in C_c^\infty(D).$$

It is equivalent to the following averaging property by running a Brownian motion X: for every relatively compact subset D_1 of D,

$$u(X_{\tau_{D_1}}) \in L^1(\mathbf{P}_x) \quad \text{and} \quad u(x) = \mathbf{E}_x\left[u(X_{\tau_{D_1}})\right] \text{ for every } x \in D_1.$$

Here $\tau_{D_1} := \inf\{t \geq 0 : X_t \notin D_1\}$. It is a natural and fundamental question whether the above two notions of harmonicity remain equivalent in a more general context or not, such as for discontinuous symmetric processes including symmetric Lévy processes and for symmetric diffusions on fractals.

The probabilistic notion of harmonicity was briefly touched upon in Definition 6.1.3. It is a special case of the more general Definition 6.7.5 below. (See Theorems 6.7.9 and 6.7.13 for the connection.) In this section, we address the equivalence of the analytic and probabilistic notions of harmonicity in the context of symmetric Hunt processes on local compact separable metric spaces. In short, the answer is yes. See Theorem 6.7.13 below. The material in this section will not be needed for the rest of the book.

Let $X = (\Omega, \mathcal{F}_\infty, \mathcal{F}_t, X_t, \zeta, \mathbf{P}_x, x \in E)$ be an m-symmetric Hunt Markov process on a locally compact separable metric space E such that its associated Dirichlet form $(\mathcal{E}, \mathcal{F})$ is regular on $L^2(E; m)$. Here m is a positive Radon measure on E with full topological support. As before, denote by $E_\partial := E \cup \{\partial\}$ the one-point compactification of E. Let Ω be the totality of right continuous, left-limited sample paths from $[0, \infty[$ to E_∂ that hold the value ∂ once attaining it. For any $\omega \in \Omega$, we set $X_t(\omega) := \omega(t)$. Let $\zeta(\omega) := \inf\{t \geq 0 : X_t(\omega) = \partial\}$ be the lifetime of X. As usual, $\mathcal{F}_\infty$ and $\mathcal{F}_t$ are the minimal augmented σ-algebras obtained from $\mathcal{F}^0_\infty := \sigma\{X_s : 0 \leq s < \infty\}$ and $\mathcal{F}^0_t := \sigma\{X_s : \ 0 \leq s \leq t\}$ under $\{\mathbf{P}_\nu : \nu \in \mathcal{P}(E)\}$. For a Borel subset B of E, $\tau_B := \inf\{t > 0 : X_t \notin B\}$ (the *exit time* of B) is an $(\mathcal{F}_t)$-stopping time. Recall that we use the convention that $X_\infty := \partial$ and every function u defined on E is extended to E_∂ by setting $u(\partial) = 0$. Moreover, any function in $\mathcal{F}_e$ is taken to be quasi continuous. Note that the Hunt process X is stochastically continuous: $\mathbf{P}_x(X_t = X_{t-}) = 1$ for every $t > 0$ and $x \in E$.

We emphasize here that to ensure a wide scope of applicability, unless otherwise stated, we do *not* assume that the process X (or equivalently, its associated Dirichlet form $(\mathcal{E}, \mathcal{F})$ is irreducible.

Let us first prepare two general theorems.

For $t > 0$, let r_t denote the time-reversal operator defined on the path space Ω of X as follows: For $\omega \in \{t < \zeta\}$,

$$r_t(\omega)(s) = \begin{cases} \omega((t-s)-), & \text{if } 0 \le s < t, \\ \omega(0), & \text{if } s \ge t. \end{cases}$$

Exercise 6.7.1. Show that the restriction of the measure $\mathbf{P}_m$ to $\mathcal{F}_t$ is invariant under r_t on $\Omega \cap \{\zeta > t\}$; that is,

$$\mathbf{E}_m [\xi;\, t < \zeta] = \mathbf{E}_m [\xi \circ r_t;\, t < \zeta]$$

for every non-negative random variable $\xi \in \mathcal{F}_t$.

THEOREM 6.7.2. (Lyons-Zheng's forward and backward martingale decomposition) *For every $u \in \mathcal{F}_e$ and $t > 0$, $\mathbf{P}_m$-a.e. on $\{t < \zeta\}$,*

$$u(X_t) - u(X_0) = \frac{1}{2} M_t^{[u]} - \frac{1}{2} M_t^{[u]} \circ r_t, \tag{6.7.1}$$

where $M^{[u]}$ is the MAF of finite energy in the Fukushima's decomposition (Theorem 4.2.6) of $u(X_t) - u(X_0)$.

Proof. Denote by R_1 the 1-resolvent kernel of X. If $u = R_1 f$ for some nearly Borel function $f \in L^2(E; m)$, then we have

$$u(X_t) - u(X_0) = M_t^{[u]} + \int_0^t \mathcal{L}u(X_s)ds,$$

where $\mathcal{L}u = u - f$. Since X is stochastically continuous, applying time reversal operator r_t to above display, we have $\mathbf{P}_m$-a.e. on $\{t < \zeta\}$,

$$u(X_0) - u(X_t) = M_t^{[u]} \circ r_t + \int_0^t \mathcal{L}u(X_s)ds$$

Thus we have

$$u(X_t) - u(X_0) = \frac{1}{2} M_t^{[u]} - \frac{1}{2} M_t^{[u]} \circ r_t \quad \mathbf{P}_m\text{-a.e. on } \{t < \zeta\} \tag{6.7.2}$$

For general $u \in \mathcal{F}_e$, just as in the proof of Theorem 4.2.6, there is a sequence of $\{u_n = R_1 f_n,\ n \geq 1\}$ with $f_n \in L^2(E; m)$ so that $\lim_{n\to\infty} \mathcal{E}(u_n - u, u_n - u) = 0$ and $\lim_{n\to\infty} u_n = u$ m-a.e. on E. Note that $\lim_{n\to\infty} \mathbf{e}(M^{[u_n]} - M^{[u]}) \leq \lim_{n\to\infty} \mathcal{E}(u_n - u, u_n - u) = 0$. Thus by Theorems 3.5.4 and 4.2.5, there is a subsequence $\{n_k, k \geq 1\}$ so that $u_{n_k}(X) - u_{n_k}(X_0)$ and $M^{[u_{n_k}]}$ converges uniformly on each compact time interval to $u(X) - u(X_0)$ and $M^{[u]}$, respectively, $\mathbf{P}_x$-a.s. for q.e. $x \in E$. Since (6.7.2) holds for every u_{n_k} in place of u, we arrive at the conclusion of the theorem by passing $k \to \infty$. □

The next lemma is a consequence of Theorem 6.7.2.

LEMMA 6.7.3. *If $u \in \mathcal{F}_e$ satisfies $\mathcal{E}(u, u) = 0$, then*

$$\mathbf{P}_x\left(u(X_t) = u(X_0) \text{ for every } t \geq 0\right) = 1 \qquad \text{for q.e. } x \in E.$$

In other words, for q.e. $x \in E$, $E_x := \{y \in E : u(y) = u(x)\}$ is an invariant set with respect to the process X in the sense that $\mathbf{P}_x(X[0,\infty) \subset E_x) = 1$.

Proof. Recall that in (4.3.15) $\mu_{\langle u\rangle}$ is the energy measure of u, which is the Revuz measure of $\langle M^{[u]}\rangle$, and κ is the killing measure of $(\mathcal{E}, \mathcal{F})$. As $\mu_{\langle u\rangle}(E) \leq 2\mathcal{E}(u, u) = 0$, we have $M^{[u]} = 0$ and thus, by Theorem 6.7.2, $u(X_t) = u(X_0)$ $\mathbf{P}_m$-a.e. on $\{t < \zeta\}$ for every $t > 0$. This implies by Fukushima's decomposition that $N^{[u]} = 0$ on $[0, \zeta)$ and hence on $[0, \infty)$ $\mathbf{P}_m$-a.e. Consequently, $u(X_t) - u(X_0) = M^{[u]} + N^{[u]} = 0$ for every $t \geq 0$ $\mathbf{P}_x$-a.s. for q.e. $x \in E$, which proves the lemma. □

The next theorem accompanies a new notation and extends Theorem 2.1.12 as well as (2.3.18). Note that the process X is not assumed to be transient.

THEOREM 6.7.4. *Suppose that ν is a smooth measure on E, whose corresponding PCAF of X is denoted by A^ν. Define $U\nu(x) := \mathbf{E}_x[A^\nu_\zeta]$. Then $U\nu \in \mathcal{F}_e$ if and only if $\int_E U\nu(x)\nu(dx) < \infty$. In this case,*

$$\mathcal{E}(U\nu, u) = \int_E u(x)\nu(dx) \qquad \text{for every } u \in \mathcal{F}_e. \tag{6.7.3}$$

Proof. First assume that $m(E) < \infty$. It is easy to check directly that $\{x \in E : \mathbf{E}_x[A^\nu_\zeta] > j\}$ is finely open for every integer $j \geq 1$. So $K_j := \{U\nu \leq j\}$ is finely closed. Define $\nu_j := \mathbf{1}_{K_j}\nu$. Clearly for $x \in K_j$, $U\nu_j(x) \leq U\nu(x) \leq j$, while for $x \in K_j^c$,

$$U\nu_j(x) = \mathbf{E}_x\left[\int_0^\zeta \mathbf{1}_{K_j}(X_s)dA^\nu_s\right] = \mathbf{E}_x\left[U\nu_j(X_{\sigma_{K_j}})\right] \leq j.$$

So $f_j := U\nu_j \leq j$ on E and hence f_j is in $L^2(E; m)$.

Assume that $U\nu \in \mathcal{F}_e$. Then $f_j \le j \wedge U\nu \in \mathcal{F}$. Since f_j is excessive, we have by Lemma 1.2.3 that $f_j \in \mathcal{F}$ and

$$\mathcal{E}_\alpha(f_j, f_j) \le \mathcal{E}_\alpha(j \wedge U\nu, j \wedge U\nu) \qquad \text{for every } \alpha > 0.$$

Taking $\alpha \to 0$ yields that for every $j \ge 1$,

$$\mathcal{E}(f_j, f_j) \le \mathcal{E}(j \wedge U\nu, j \wedge U\nu) \le \mathcal{E}(U\nu, U\nu) < \infty.$$

On the other hand, by Theorem 4.1.1,

$$\mathcal{E}(f_j, f_j) = \lim_{t\to 0} \frac{1}{t}(f_j - P_t f_j, f_j)_m = \lim_{t\to 0} \frac{1}{t} \mathbf{E}_{f_j \cdot m}\left[A_t^{\nu_j}\right]$$

$$= \int_E f_j(x)\nu_j(dx) = \int_{K_j} f_j(x)\nu(dx).$$

Since $U\nu < \infty$ q.e. on E, we have $\nu(E \setminus \cup_{j=1}^\infty K_j) = 0$ and f_j increases to $U\nu$ as $j \to \infty$. Thus by the monotone convergence theorem,

$$\int_E U\nu(x)\nu(dx) = \lim_{j\to\infty} \int_{K_j} f_j(x)\nu(dx) \le \mathcal{E}(U\nu, U\nu) < \infty.$$

Now assume that $\int_E U\nu(x)\nu(dx) < \infty$. In this case, as $U\nu < \infty$ ν-a.e. on E, we have $\nu(E \setminus \cup_{j=1}^\infty K_j) = 0$. Since by Theorem 4.1.1

$$\lim_{t\to 0} \frac{1}{t}(f_j - P_t f_j, f_j)_m = \lim_{t\to 0} \frac{1}{t} \mathbf{E}_{f_j \cdot m}\left[A_t^{\nu_j}\right] = \int_E f_j(x)\nu_j(dx)$$

$$\le \int_E U\nu(x)\nu(dx) < \infty, \tag{6.7.4}$$

we have $f_j \in \mathcal{F}$ with $\mathcal{E}(f_j, f_j) \le \int_E U\nu(x)\nu(dx)$. The same calculation shows that for $i > j$, $f_i - f_j = \mathbf{E}_x\left[A_\zeta^{\mathbf{1}_{K_i \setminus K_j}\cdot \nu}\right]$ and

$$\mathcal{E}(f_i - f_j, f_i - f_j) = \int_{K_i \setminus K_j} (f_i - f_j)(x)\nu(dx) \le \int_{K_i \setminus K_j} U\nu(x)\nu(dx),$$

which tends to zero as $i, j \to \infty$; that is, $\{f_j, j \ge 1\}$ is an $\mathcal{E}$-Cauchy sequence in $\mathcal{F}$. As $\lim_{j\to\infty} f_j = U\nu$ on E, we conclude that $U\nu \in \mathcal{F}_e$.

We have just proved that $U\nu \in \mathcal{F}_e$ if and only if $\int_E U\nu(x)\nu(dx) < \infty$ under the assumption that $m(E) < \infty$. In either cases, we deduce from (6.7.4) that

$$\mathcal{E}(U\nu, U\nu) = \lim_{j\to\infty} \mathcal{E}(f_j, f_j) = \int_E U\nu(x)\nu(dx). \tag{6.7.5}$$

Moreover, for $u \in b\mathcal{F}_+$, by Theorem 4.1.1 and monotone convergence theorem, we have

$$\begin{aligned}\mathcal{E}(U\nu, u) &= \lim_{j\to\infty} \mathcal{E}(f_j, u) = \lim_{j\to\infty}\lim_{t\to 0} \frac{1}{t}(f_j - P_t f_j, u) \\ &= \lim_{j\to\infty} \int_E u(x)\mathbf{1}_{K_j}(x)\nu(dx) = \int_E u(x)\nu(dx).\end{aligned}$$

Since the linear span of $b\mathcal{F}_+$ is $\mathcal{E}$-dense in $\mathcal{F}_e$, we have established (6.7.3).

For a general σ-finite measure m, take a strictly positive m-integrable Borel measurable function g on E and define $\mu = g \cdot m$. Then μ is a finite measure on E. Let Y be the time change of X via measure μ; that is, $Y_t = X_{\tau_t}$, where $\tau_t = \inf\{s > 0 : \int_0^s g(X_s)ds > t\}$. The time-changed process Y is μ-symmetric. Let $(\mathcal{E}^Y, \mathcal{F}^Y)$ be the Dirichlet form of Y on $L^2(E; \mu)$. Then by Corollary 5.2.12, $\mathcal{F}^Y_e = \mathcal{F}_e$ and $\mathcal{E}^Y = \mathcal{E}$ on $\mathcal{F}_e$. The measure ν is also a smooth measure with respect to process Y in view of Theorem 5.2.11. It is easy to verify that the PCAF $A^{Y,\nu}$ of Y corresponding to ν is related to corresponding PCAF A^ν of X by

$$A^{Y,\nu}_t = A^\nu_{\tau_t} \qquad \text{for } t \geq 0.$$

In particular, we have $U^Y\nu(x) = U\nu$ on E. As we just proved that the theorem holds for Y, we conclude that the theorem also holds for X. □

Definition 6.7.5. Let D be an open subset of E. We say a function u is *harmonic* in D (with respect to the process X) if for every relatively compact open subset D_1 of D, $t \mapsto u(X_{t\wedge\tau_{D_1}})$ is a uniformly integrable $\mathbf{P}_x$-martingale for q.e. $x \in D_1$.

Let $(N(x, dy), H)$ be a Lévy system of X and let

$$J(dx, dy) = N(x, dy)\mu_H(dx) \qquad \text{and} \qquad \kappa(dx) = N(x, \{\partial\})\mu_H(dx),$$

where μ_H is the Revuz measure of the positive continuous additive functional H of X. Since $(\mathcal{E}, \mathcal{F})$ is a regular Dirichlet form on $L^2(E; m)$, for any relatively compact open sets D_1, D_2 with $\overline{D}_1 \subset D_2$, there is $\phi \in \mathcal{F} \cap C_c(E)$ so that $\phi = 1$ on D_1 and $\phi = 0$ on D_2^c. We then have from Theorem 4.3.3

$$J(D_1, D_2^c) = \int_{D_1\times D_2^c} (\phi(x) - \phi(y))^2 J(dx, dy) \leq 2\mathcal{E}(\phi, \phi) < \infty. \tag{6.7.6}$$

For open set $D \subset E$, let $X^D = (X^D_t, \zeta^D, \mathbf{P}_x)$ denote the part process of X on the open set D. The Dirichlet form of X^D is $(\mathcal{E}, \mathcal{F}^D)$, where $\mathcal{F}^D := \{u \in \mathcal{F} : u = 0 \text{ q.e. on } E \setminus D\}$. (For notational convenience, in this section we use notation $\mathcal{F}^D$ instead of $\mathcal{F}_D$.) By Theorem 3.3.9, $(\mathcal{E}, \mathcal{F}^D)$ is a regular Dirichlet form in $L^2(D; m)$. A function f is said to be locally in $\mathcal{F}^D$, denoted as $f \in \mathcal{F}^D_{\rm loc}$, if for every relatively compact subset D_1 of D, there is a function $g \in \mathcal{F}^D$ such that $f = g$ m-a.e. on D_1. By Theorem 4.3.10(i), $\mu^c_{\langle u,v\rangle}$ is well defined on D for every $u, v \in \mathcal{F}^D_{\rm loc}$. Further, for a smooth measure μ on D with respect to $(\mathcal{E}, \mathcal{F}^D)$, $U^D\mu$ will denote the function $\mathbf{E}_\cdot[A^\mu_{\zeta^D}]$ for the PCAF A^μ of X^D with Revuz measure μ in accordance with Theorem 6.7.4.

For an open set $D \subset E$, consider the following two conditions for a measurable function u on E. For any relatively compact open sets D_1, D_2 with $\overline{D}_1 \subset D_2 \subset \overline{D}_2 \subset D$,

$$\int_{D_1\times(E\setminus D_2)} |u(y)| J(dx, dy) < \infty \tag{6.7.7}$$

and

$$f_u \in \mathcal{F}^{D_1}_e. \tag{6.7.8}$$

Here the function f_u is defined by

$$f_u(x) = \mathbf{1}_{D_1}(x)\mathbf{E}_x\left[\big((1-\phi_{D_2})|u|\big)(X_{\tau_{D_1}})\right], \quad x \in E, \tag{6.7.9}$$

using a function ϕ_{D_2} satisfying

$$\phi_{D_2} \in C_c(D) \cap \mathcal{F} \text{ with } 0 \le \phi_{D_2} \le 1 \text{ and } \phi_{D_2} = 1 \text{ on } D_2. \tag{6.7.10}$$

Note that both conditions (6.7.7) and (6.7.8) are automatically satisfied when X is a diffusion since in this case the jumping measure J vanishes and $X_{\tau_{D_1}} \in \partial D_1$ on $\{\tau_{D_1} < \zeta\}$. In view of (6.7.6), every bounded function u satisfies the condition (6.7.7). In fact by the following lemma, every bounded function u also satisfies the condition (6.7.8).

LEMMA 6.7.6. *Suppose that u is a measurable function on E and D_1 and D_2 are two relatively compact open sets D_1, D_2 such that $\overline{D}_1 \subset D_2 \subset \overline{D}_2 \subset D$. Let ϕ_{D_2} be a function satisfying* (6.7.10) *and f_u be defined by* (6.7.9). *Define*

$$\mu_u(dx) := \mathbf{1}_{D_1}(x)\int_{E\setminus D_2} (1-\phi_{D_2}(y))|u(y)| N(x, dy)\mu_H(dx), \tag{6.7.11}$$

which is a smooth measure of X^{D_1}. Then $f_u = U^{D_1}\mu_u$. Furthermore, $f_u \in \mathcal{F}^{D_1}_e$ if and only if

$$\int_{D_1} f_u(x)\mu_u(dx) < \infty. \tag{6.7.12}$$

Suppose u satisfies condition (6.7.7) *and the condition that*

$$\sup_{x\in D_1} \mathbf{E}_x\left[\big(\mathbf{1}_{D_2^c}|u|\big)(X_{\tau_{D_1}})\right] < \infty. \tag{6.7.13}$$

Then (6.7.12) *holds for u and so does* (6.7.8).

Proof. Let ϕ_{D_2} be a function satisfying (6.7.10), and f_u be defined by (6.7.9). Note that $1-\phi_{D_2}=0$ on D_2. Taking $T=\tau_{D_1}$, $g(s)=1$, and $f(x,y)=|(1-\phi_{D_2}(y))|u|(y)-(1-\phi_{D_2}(x))|u|(x)|$ in the Lévy system formula (A.3.33), we get

$$f_u(x)=\mathbf{E}_x\left[\int_0^{\tau_{D_1}}\left(\int_{E\setminus D_2}(1-\phi_{D_2}(y))|u|(y)N(X_s,dy)\right)dH_s\right]$$

for $x\in E$. Note that by Proposition 4.1.10, the Revuz measure μ_u on D_1 of the PCAF

$$t\mapsto\int_0^{t\wedge\tau_{D_1}}\left(\int_{E\setminus D_2}(1-\phi_{D_2}(y))|u|(y)N(X_s,dy)\right)dH_s$$

of X^{D_1} is given by (6.7.11) and so $f_u=U^{D_1}\mu_u$. Applying Theorem 6.7.4 to X^{D_1}, we conclude that $f_u\in\mathcal{F}_e^{D_1}$ if and only if $\int_{D_1}f_u(x)\mu_u(dx)<\infty$.

Now assume that conditions (6.7.7) and (6.7.13) hold. Note that under condition (6.7.13), f_u is a bounded function. On the other hand, under condition (6.7.7),

$$\begin{aligned}\mu_u(D_1)&=\int_{D_1}\left(\int_{E\setminus D_2}(1-\phi_{D_2}(y))|u(y)|N(x,dy)\right)\mu_H(dx)\\&\le\int_{D_1}\left(\int_{E\setminus D_2}|u(y)|N(x,dy)\right)\mu_H(dx)<\infty.\end{aligned}$$

So we have $\int_{D_1}U^{D_1}\mu_u(x)\mu_u(dx)\le\|f_u\|_\infty\,\mu_u(D_1)<\infty$; that is, (6.7.12) holds for u. □

In many concrete cases such as in Examples 6.7.14–6.7.16 below, one can show that condition (6.7.7) implies condition (6.7.13).

LEMMA 6.7.7. *Let D be an open subset of E. Every $u\in\mathcal{F}_e$ that is locally bounded on D satisfies conditions* (6.7.7) *and* (6.7.8).

Proof. Let $u\in\mathcal{F}_e$ be locally bounded on D. For any relatively compact open sets D_1, D_2 with $\overline{D}_1\subset D_2\subset\overline{D}_2\subset D$, take $\phi\in\mathcal{F}\cap C_c(D)$ such that $\phi=1$ on D_1 and $\phi=0$ on D_2^c. Then $u\phi\in\mathcal{F}_e$ and

$$\begin{aligned}&\int_{D_1\times(E\setminus D_2)}u(y)^2J(dx,dy)\\&=\int_{D_1\times(E\setminus D_2)}(((1-\phi)u)(x)-((1-\phi)u)(y))^2\,J(dx,dy)\\&\le 2\mathcal{E}(u-u\phi,u-u\phi)<\infty.\end{aligned}$$

This together with (6.7.6) implies (6.7.7).

Let f_u be defined by (6.7.9). Note that $|u| \in \mathcal{F}_e$ is locally bounded on D and so $(1-\phi_{D_2})|u| = |u| - \phi_{D_2}|u| \in \mathcal{F}_e$. Thus it follows from Theorem 3.4.8 that

$$f_u(x) = \mathbf{E}_x\left[\left((1-\phi_{D_2})|u|\right)(X_{\tau_{D_1}})\right] - (1-\phi_{D_2}(x))|u|(x)$$

for q.e. $x \in E$ and $f_u \in \mathcal{F}_e^{D_1}$. □

LEMMA 6.7.8. *Let D be a relatively compact open set of E. Suppose u is a function in $\mathcal{F}^D_{\rm loc}$ that is locally bounded on D and satisfies condition* (6.7.7). *Then for every $v \in C_c(D) \cap \mathcal{F}$, the expression*

$$\frac{1}{2}\mu^c_{\langle u,v\rangle}(D) + \frac{1}{2}\int_{E\times E}(u(x)-u(y))(v(x)-v(y))J(dx,dy)$$
$$+\int_D u(x)v(x)\kappa(dx)$$

is well-defined and finite, and it will still be denoted as $\mathcal{E}(u,v)$.

Proof. Clearly the first and the third terms are well defined and finite. To see that the second term is also well defined, let D_1 be a relatively compact open subset of D such that $\text{supp}[v] \subset D_1$. By assumption, there is some $f \in \mathcal{F}$ so that $f = u$ m-a.e. and hence q.e. on D_1. Then $u\phi \in \mathcal{F}^D$ for $\phi \in C_c(D) \cap \mathcal{F}$ with $\phi = 1$ on D_1. Using property (6.7.6),

$$\begin{aligned}
&\int_{E\times E}|(u(x)-u(y))(v(x)-v(y))|J(dx,dy)\\
&= \int_{D_1\times D_1}|(u(x)-u(y))(v(x)-v(y))|J(dx,dy)\\
&\quad +2\int_{D_1\times(E\setminus D_1)}|u(x)v(x)|J(dx,dy) + 2\int_{D_1}|v(x)|\int_{E\setminus D_1}|u(y)|J(dx,dy)\\
&\le \int_{D_1\times D_1}|(u(x)\phi(x)-u(y)\phi(y))(v(x)-v(y))|J(dx,dy)\\
&\quad +2\|uv\|_\infty J(\text{supp}[v], D_1^c) + 2\|v\|_\infty\int_{\text{supp}[v]\times D_1^c}|u(y)|J(dx,dy)\\
&< \infty.
\end{aligned}$$

This proves the lemma. □

Recall Definition 1.3.17 of a special standard core $\mathcal{C}(\subset \mathcal{F}\cap C_c(E))$ of $(\mathcal{E},\mathcal{F})$. $\mathcal{F}\cap C_c(E)$ is a particular example of a special standard core. For a special standard core $\mathcal{C}$ of $(\mathcal{E},\mathcal{F})$ and an open subset $D \subset E$, we let

$$\mathcal{C}_D = \{u \in \mathcal{C} : \text{supp}[u] \subset D\}.$$

THEOREM 6.7.9. *Let D be an open subset of E. Suppose that $u \in \mathcal{F}^D_{\rm loc}$ is locally bounded on D satisfying conditions* (6.7.7)–(6.7.8) *and that*

$$\mathcal{E}(u, v) = 0 \qquad \textit{for every } v \in \mathcal{C}_D \tag{6.7.14}$$

for some special standard core $\mathcal{C}$ of $(\mathcal{E}, \mathcal{F})$. Then u is harmonic in D. If D_1 is a relatively compact open subset of D such that $\mathbf{P}_x(\tau_{D_1} < \infty) > 0$ for q.e. $x \in D_1$, then $u(x) = \mathbf{E}_x[u(X_{\tau_{D_1}})]$ for q.e. $x \in D_1$.

Proof. Take a relatively compact open set D_2 with $\overline{D}_1 \subset D_2 \subset \overline{D}_2 \subset D$ and a function $\phi := \phi_{D_2}$ as in (6.7.10). Then $\phi u \in \mathcal{F}^D$. So by Theorems 3.4.8 and 3.4.9, $h_1(x) := \mathbf{E}_x[(\phi u)(X_{\tau_{D_1}})] \in \mathcal{F}_e$ and $\phi u - h_1 \in \mathcal{F}_e^{D_1}$. Moreover,

$$\mathcal{E}(h_1, v) = 0 \qquad \text{for every } v \in \mathcal{F}_e^{D_1}. \tag{6.7.15}$$

Let $h_2(x) := \mathbf{E}_x\left[((1-\phi)u)(X_{\tau_{D_1}})\right]$, which is well defined by condition (6.7.8).

Define f_{u^+} (resp. f_{u^-}) by (6.7.9) with $u^+ = u \vee 0$ (resp. $u^- = (-u) \vee 0$) in place of u. Then $\mathbf{1}_{D_1} \cdot h_2 = f_{u^+} - f_{u^-}$. Using a Lévy system formula (A.3.33) as in the proof of Lemma 6.7.6, we have $f_{u^+} = U^{D_1}\mu^+$ and $f_{u^-} = U^{D_1}\mu^-$, where μ^+ (resp. μ^-) is defined by (6.7.11) with u^+ (resp. u^-) in place of u. Thus we obtain

$$\mathbf{1}_{D_1} \cdot h_2 = U^{D_1}\mu^+ - U^{D_1}\mu^-.$$

We claim that $\mathbf{1}_{D_1} h_2 \in \mathcal{F}_e^{D_1}$ and that for every $v \in \mathcal{F}_e^{D_1}$,

$$\mathcal{E}(\mathbf{1}_D h_2,\ v) = \int_E v(x)\mathbf{1}_{D_1}(x)\left(\int_{D_2^c} \big((1-\phi)u\big)(z)N(x, dz)\right)\mu_H(dx). \tag{6.7.16}$$

Clearly $U^{D_1}\mu^+ \leq U^{D_1}\mu$. For $j \geq 1$, let $F_j := \{x \in D_1 : U^{D_1}\mu^+(x) \leq j\}$, which is a finely closed subset of D_1. Define $\nu_j := \mathbf{1}_{F_j}\mu^+$. Then for $x \in F_j$, $U^{D_1}\nu_j(x) \leq U^{D_1}\mu^+(x) \leq j$, while for $x \in D_1 \setminus F_j$,

$$U^{D_1}\nu_j(x) = \mathbf{E}_x\left[U^{D_1}\nu_j(X_{\sigma_{F_j}})\right] \leq j.$$

In other words, we have $U^{D_1}\nu_j \leq j \wedge U^{D_1}\mu^+ \leq j \wedge f$. As both $U^{D_1}\nu_j$ and $j \wedge f$ are excessive functions of X^{D_1} and $m(D_1) < \infty$, we have by Theorem 1.1.5 and Lemma 1.2.3 that $\{U^{D_1}\nu_j,\ j \wedge U^{D_1}\mu\} \subset \mathcal{F}^{D_1}$ with

$$\mathcal{E}(U^{D_1}\nu_j,\ U^{D_1}\nu_j) \leq \mathcal{E}(j \wedge f, j \wedge f) \leq \mathcal{E}(f, f) < \infty.$$

Moreover, for each $j \geq 1$, we have by Theorem 4.1.1 that

$$\begin{aligned}
&\mathcal{E}(U^{D_1}\nu_j,\, U^{D_1}\nu_j) \\
&= \lim_{t\to 0}\frac{1}{t}\int_E U^{D_1}(\nu_j(x) - P_t^{D_1}U^{D_1}\nu_j(x))U^{D_1}\nu_j(x)m(dx) \\
&= \lim_{t\to 0}\frac{1}{t}\int_E \mathbf{E}_x\left[A^{\nu_j}_{t\wedge\tau_{D_1}}\right]U^{D_1}\nu_j(x)m(dx) \\
&= \int_{D_1} U^{D_1}\nu_j(x)\,\mathbf{1}_{F_j}(x)\mu_1(dx),
\end{aligned}$$

which increases to $\int_{D_1} U^{D_1}\mu^+(x)\mu^+(dx)$. Consequently,

$$\int_{D_1} U^{D_1}\mu^+(x)\mu^+(dx) \leq \mathcal{E}(f,f) < \infty.$$

So we have by Theorem 6.7.4 applied to X^{D_1} that $U^{D_1}\mu^+ \in \mathcal{F}_e^{D_1}$ with $\mathcal{E}(U^{D_1}\mu^+, v) = \int_{D_1} v(x)\mu^+(dx)$ for every $v \in \mathcal{F}_e^{D_1}$. Similarly, we have $U^{D_1}\mu^- \in \mathcal{F}_e^{D_1}$ with $\mathcal{E}(U^{D_1}\mu^-, v) = \int_{D_1} v(x)\mu_2(dx)$ for every $v \in \mathcal{F}_e^{D_1}$. It follows that $\mathbf{1}_{D_1}h_2 = U^{D_1}\mu_1 - U^{D_1}\mu_2 \in \mathcal{F}_e^{D_1}$ and claim (6.7.16) is established.

As $h_2 = \mathbf{1}_{D_1}h_2 + (1-\phi)u$ and $(1-\phi)u$ satisfies condition (6.7.7), we have by Lemma 6.7.8 and (6.7.16) that for every $v \in C_c(D_1)\cap\mathcal{F}$,

$$\begin{aligned}
\mathcal{E}(h_2, v) &= \mathcal{E}(\mathbf{1}_{D_1}h_2, v) + \mathcal{E}((1-\phi)u, v) \\
&= \int_{E\times E} v(x)(1-\phi(y))u(y)N(x,dy)\mu_H(dx) \\
&\quad - \int_{E\times E} v(x)(1-\phi(y))u(y)N(x,dy)\mu_H(dx) \\
&= 0.
\end{aligned} \tag{6.7.17}$$

This combining with (6.7.15) and condition (6.7.14) proves that

$$\mathcal{E}(u - h_1 - h_2, v) = 0 \qquad \text{for every } v \in \mathcal{C}_{D_1}. \tag{6.7.18}$$

Since $u - (h_1 + h_2) = (\phi u - h_1) - \mathbf{1}_{D_1}h_2 \in \mathcal{F}_e^{D_1}$ and $\mathcal{C}_{D_1}$ is $\mathcal{E}$-dense in $\mathcal{F}_e^{D_1}$ by Theorem 3.3.9(iii), the above display holds for every $v \in \mathcal{F}_e^{D_1}$. In particular, we have

$$\mathcal{E}(u - h_1 - h_2,\, u - h_1 - h_2) = 0. \tag{6.7.19}$$

By Lemma 6.7.3, $u(X_t) - h_1(X_t) - h_2(X_t)$ is a bounded $\mathbf{P}_x$-martingale for q.e. $x \in E$. As

$$h_1(x) + h_2(x) = \mathbf{E}_x\left[u(X_{\tau_{D_1}})\right] \qquad \text{for } x \in D_1,$$

the above implies that $t \mapsto u(X_{t\wedge\tau_{D_1}})$ is a uniformly integrable $\mathbf{P}_x$-martingale for q.e. $x \in D_1$. If $\mathbf{P}_x(\tau_{D_1} < \infty) > 0$ for q.e. $x \in D_1$, applying Lemma 6.7.3 to the Dirichlet form $(\mathcal{E}, \mathcal{F}^{D_1})$, we have $u - h_1 - h_2 = 0$ q.e. on D_1 because $u(X_t^D) - h_1(X_t^D) - h_2(X_t^D) = 0$ for $t > \tau_{D_1}$. So $u(x) = \mathbf{E}_x[u(X_{\tau_{D_1}})]$ for q.e. $x \in D_1$. This completes the proof of the theorem. □

Remark 6.7.10. (i) The principal difficulty in the above proof is establishing (6.7.18) and that $u - (h_1 + h_2) \in \mathcal{F}_e^{D_1}$ for general $u \in \mathcal{F}_{\text{loc}}^D$ satisfying conditions (6.7.7) and (6.7.8). If u is assumed a priori to be in $\mathcal{F}_e$, these facts and therefore the theorem itself are then much easier to establish. Note that when $u \in \mathcal{F}_e$, it follows immediately from Theorem 3.4.8 that $h_1 + h_2 = \mathbf{E}_x[u(X_{\tau_{D_1}})] \in \mathcal{F}_e$ enjoys property (6.7.18) and $u - (h_1 + h_2) \in \mathcal{F}_e^{D_1}$. Therefore, (6.7.19) holds and consequently u is harmonic in D.
(ii) If we assume that the process X (or equivalently $(\mathcal{E}, \mathcal{F})$) is m-irreducible and that D_1^c is not m-polar, then by Theorem 3.5.6, $\mathbf{P}_x(\tau_{D_1} < \infty) > 0$ for q.e. $x \in D$. □

THEOREM 6.7.11. *Suppose D is an open set of E with $m(D) < \infty$ and u is a function on E satisfying the condition* (6.7.7) *such that $u \in L^\infty(D; m)$ and that $\{u(X_{t\wedge\tau_D}), t \geq 0\}$ is a uniformly integrable $\mathbf{P}_x$-martingale for q.e. $x \in E$. Then*

$$u \in \mathcal{F}_{\text{loc}}^D \quad \textit{and} \quad \mathcal{E}(u, v) = 0 \ \textit{for every } v \in C_c(D) \cap \mathcal{F}. \tag{6.7.20}$$

Proof. As for q.e. $x \in E$, $\{u(X_{t\wedge\tau_D}), t \geq 0\}$ is a uniformly integrable $\mathbf{P}_x$-martingale, $u(X_{t\wedge\tau_D})$ converges in $L^1(\mathbf{P}_x)$ as well as $\mathbf{P}_x$-a.s. to some random variable ξ. By considering ξ^+, ξ^- and $u_+(x) := \mathbf{E}_x[\xi^+]$, $u_-(x) := \mathbf{E}_x[\xi^-]$ separately, we may assume without loss of generality that $u \geq 0$. Note that $\xi \mathbf{1}_{\{\tau_D < \infty\}} = u(X_{\tau_D})$. Define $u_1(x) := \mathbf{E}_x[u(X_{\tau_D})]$ and $u_2(x) := \mathbf{E}_x[\xi \mathbf{1}_{\{\tau_D = \infty\}}] = u(x) - u_1(x)$.

Let $\{P_t^D, t \geq 0\}$ denote the transition function of the part process X^D of X on D. Then for q.e. $x \in D$ and every $t > 0$, by the Markov property of X^D,

$$P_t^D u_2(x) = \mathbf{E}_x\left[u_2(X_t); t < \tau_D\right] = \mathbf{E}_x\left[\xi \mathbf{1}_{\{\tau_D = \infty\}} \circ \theta_t; t < \tau_D\right] = u_2(x).$$

Since $u_2 \in L^2(D; m)$, by (1.1.4)–(1.1.5),

$$u_2 \in \mathcal{F}^D \quad \text{with} \quad \mathcal{E}(u_2, u_2) = 0. \tag{6.7.21}$$

On the other hand,

$$P_t^D u(x) = \mathbf{E}_x\left[u(X_t); t < \tau_D\right] = \mathbf{E}_x\left[u(X_{\tau_D}) : t < \tau_D\right] \leq u(x).$$

Let $\{D_n, n \geq 1\}$ be an increasing sequence of relatively compact open subsets of D with $\cup_{n\geq 1} D_n = D$ and define

$$\sigma_n := \inf\left\{t \geq 0 \colon X_t^D \in D_n\right\}.$$

Let $e_n(x) = \mathbf{E}_x[e^{-\sigma_n}]$, $x \in D$, be the 1-equilibrium potential of D_n with respect to the part process X^D. Clearly $e_n \in \mathcal{F}^D$ is 1-excessive with respect to the process X^D, $e_n(x) = 1$ q.e. on D_n. Let $a := \|\mathbf{1}_D u\|_\infty$. Then for every $t > 0$,

$$e^{-t} P_t^D((ae_n) \wedge u)(x) \leq ((ae_n) \wedge u)(x) \qquad \text{for q.e. } x \in D.$$

By Lemma 1.2.3, we have $(ae_n) \wedge u \in \mathcal{F}^D$ for every $n \geq 1$. Since $(ae_n) \wedge u = u$ m-a.e. on D_n, we have $u \in \mathcal{F}^D_{\text{loc}}$.

Let D_1 be a relatively compact open subset of D. Let $\phi \in C_c(D) \cap \mathcal{F}$ so that $0 \leq \phi \leq 1$ and $\phi = 1$ in an open neighborhood D_2 of $\overline{D_1}$. Define for $x \in E$

$$h_1(x) := \mathbf{E}_x\left[(\phi u)(X_{\tau_{D_1}})\right] \quad \text{and} \quad h_2(x) := \mathbf{E}_x\left[((1-\phi)u)(X_{\tau_{D_1}})\right].$$

Then $u_1 = h_1 + h_2$ on E. Since $\phi u \in \mathcal{F}$, we know as in (6.7.15) that $h_1 \in \mathcal{F}_e$ and

$$\mathcal{E}(h_1, v) = 0 \qquad \text{for every } v \in \mathcal{F}_e^{D_1}.$$

By the same argument as that for (6.7.17), we have

$$\mathcal{E}(h_2, v) = 0 \qquad \text{for every } v \in \mathcal{F}_e^{D_1}.$$

These together with (6.7.21) in particular imply that

$$\mathcal{E}(u, v) = \mathcal{E}(h_1 + h_2 + u_2, v) = 0 \qquad \text{for every } v \in C_c(D_1) \cap \mathcal{F}.$$

Since D_1 is an arbitrary relatively compact subset of D, we have

$$\mathcal{E}(u, v) = 0 \qquad \text{for every } v \in C_c(D) \cap \mathcal{F}.$$

This completes the proof. □

Remark 6.7.12. The principal difficulty for the proof of the above theorem is establishing that a function u harmonic in D is in $\mathcal{F}^D_{\text{loc}}$ with $\mathcal{E}(u, v) = 0$ for every $v \in \mathcal{F} \cap C_c(D)$. If a priori u is assumed to be in $\mathcal{F}_e$, then Theorem 6.7.11 is easy to establish. In this case, it follows from Theorem 3.4.8 that $u_1 = h_1 + h_2 = \mathbf{E}_x[u(X_{\tau_{D_1}})] \in \mathcal{F}_e$ and that u_1 is $\mathcal{E}$-orthogonal to $\mathcal{F} \cap C_c(D)$. This together with (6.7.21) immediately implies that u enjoys (6.7.20). □

Combining Theorems 6.7.9 and 6.7.11, we have the following.

THEOREM 6.7.13. *Let D be an open subset of E. Suppose that u is a function on E that is locally bounded on D and satisfies conditions* (6.7.7)–(6.7.8).

(i) *u is harmonic in D if and only if $u \in \mathcal{F}^D_{\rm loc}$ and the condition* (6.7.14) *holds for some special standard core $\mathcal{C}$ of $(\mathcal{E}, \mathcal{F})$.*
(ii) *Assume that for every relatively compact open subset D_1 of D, $\mathbf{P}_x(\tau_{D_1} < \infty) > 0$ for q.e. $x \in D_1$. (By Remark 6.7.10, this condition is satisfied if $(\mathcal{E}, \mathcal{F})$ is irreducible.) Then u is harmonic in D if and only if for every relatively compact subset D_1 of D, $u(X_{\tau_{D_1}}) \in L^1(\mathbf{P}_x)$ and $u(x) = \mathbf{E}_x[u(X_{\tau_{D_1}})]$ for q.e. $x \in D_1$.*

Example 6.7.14. (Stable-like process on $\mathbb{R}^n$) Consider the following Dirichlet form $(\mathcal{E}, \mathcal{F})$, where

$$\mathcal{F} := W^{\alpha/2,2}(\mathbb{R}^n)$$

$$:= \left\{ u \in L^2(\mathbb{R}^n) : \int_{\mathbb{R}^n \times \mathbb{R}^n} \frac{(u(x) - u(y))^2}{|x - y|^{n+\alpha}} \, dxdy < \infty \right\}$$

and for $u, v \in \mathcal{F}$,

$$\mathcal{E}(u, v) = \frac{1}{2} \int_{\mathbb{R}^n \times \mathbb{R}^n} (u(x) - u(y))(v(x) - v(y)) \frac{c(x, y)}{|x - y|^{n+\alpha}} \, dxdy.$$

Here $n \geq 1$, $\alpha \in (0, 2)$, and $c(x, y)$ is a symmetric function in (x, y) that is bounded between two positive constants. In view of Section 2.2.2, $(\mathcal{E}, \mathcal{F})$ is a regular Dirichlet form on $L^2(\mathbb{R}^n)$. Its associated symmetric Hunt process X is called a *symmetric α-stable-like process* on $\mathbb{R}^n$, which is studied in [29]. $C_c^\infty(\mathbb{R}^n)$ is a special standard core of $(\mathcal{E}, \mathcal{F})$. The process X has a strictly positive jointly continuous transition density function $p(t, x, y)$ and hence is irreducible. Moreover, there is constant $c > 0$ such that

$$p(t, x, y) \leq c\, t^{-n/\alpha} \qquad \text{for } t > 0 \text{ and } x, y \in \mathbb{R}^n \tag{6.7.22}$$

and consequently by [33, Theorem 1],

$$\sup_{x \in D} \mathbf{E}_x[\tau_D] < \infty \tag{6.7.23}$$

for any open set D having finite Lebesgue measure. When $c(x, y)$ is constant, the process X is nothing but the rotationally symmetric α-stable process on $\mathbb{R}^n$. In this example, the jumping measure

$$J(dx, dy) = \frac{c(x, y)}{|x - y|^{n+\alpha}} \, dxdy.$$

Hence for any non-empty open set $D \subset \mathbb{R}^n$, condition (6.7.7) is satisfied if and only if $(1 \wedge |x|^{-n-\alpha}) u(x) \in L^1(\mathbb{R}^n)$. Moreover, for such a function u and relatively compact open sets D_1, D_2 with $\overline{D}_1 \subset D_2 \subset \overline{D}_2 \subset D$, we can use

Lévy system formula (A.3.33) to obtain

$$\begin{aligned}
&\sup_{x\in D_1} \mathbf{E}_x\left[(\mathbf{1}_{D_2^c}|u|)(X_{\tau_{D_1}})\right] \\
&\quad = \sup_{x\in D_1} \mathbf{E}_x\left[\int_0^{\tau_{D_1}} \left(\int_{D_2^c} \frac{c(X_s,y)\,|u(y)|}{|X_s-y|^{n+\alpha}}dy\right) ds\right] \\
&\quad \le \left(c\int_{\mathbb{R}^n}(1\wedge |y|^{-n-\alpha})|u(y)|dy\right)\sup_{x\in D_1}\mathbf{E}_x[\tau_{D_1}] < \infty. \qquad (6.7.24)
\end{aligned}$$

In other words, for this example, condition (6.7.13) and hence (6.7.8) is a consequence of (6.7.7). So Theorem 6.7.13 says that for an open set D and a function u on $\mathbb{R}^n$ that is locally bounded on D with $(1\wedge |x|^{-n-\alpha})u(x) \in L^1(\mathbb{R}^n)$, the following are equivalent:

(i) u is harmonic in D.
(ii) For every relatively compact subset D_1 of D, $u(X_{\tau_{D_1}}) \in L^1(\mathbf{P}_x)$ and $u(x) = \mathbf{E}_x[u(X_{\tau_{D_1}})]$ for q.e. $x \in D_1$.
(iii) $u \in \mathcal{F}^D_{\rm loc} = W^{\alpha/2,2}_{\rm loc}(D)$ and for every $v \in C_c^\infty(D)$,

$$\int_{\mathbb{R}^n\times\mathbb{R}^n} (u(x)-u(y))(v(x)-v(y))\frac{c(x,y)}{|x-y|^{n+\alpha}}\,dxdy = 0.$$

It was shown in [29] that every bounded harmonic function in D admits a (locally) Hölder continuous version. □

Example 6.7.15. (Diffusion process on a locally compact separable metric space) Let $(\mathcal{E},\mathcal{F})$ be a local regular Dirichlet form on $L^2(E;m)$, where E is a locally compact separable metric space, and X is its associated Hunt process. In this case, X has continuous sample paths and so the jumping measure J is null in view of Theorem 4.3.4. Hence conditions (6.7.7)–(6.7.8) are automatically satisfied. Let D be an open subset of E and u be a function on E that is locally bounded in D. Then by Theorem 6.7.13, u is harmonic in D if and only if $u \in \mathcal{F}^D_{\rm loc}$ and (6.7.14) holds for some special standard core $\mathcal{C}$ of $(\mathcal{E},\mathcal{F})$.

Now consider the following special case: $E = \mathbb{R}^n$ with $n \ge 1$, $m(dx)$ is the Lebesgue measure dx on $\mathbb{R}^n$, $\mathcal{F} = H^1(\mathbb{R}^n)$, and

$$\mathcal{E}(u,v) = \frac{1}{2}\sum_{i,j=1}^n \int_{\mathbb{R}^n} a_{ij}(x)\frac{\partial u(x)}{\partial x_i}\frac{\partial v(x)}{\partial x_j}\,dx \qquad \text{for } u,v \in W^{1,2}(\mathbb{R}^n),$$

where $(a_{ij}(x))_{1\le i,j\le n}$ is an $n\times n$-matrix valued measurable function on $\mathbb{R}^n$ that is uniformly elliptic and bounded. Then $(\mathcal{E},\mathcal{F})$ is a regular local Dirichlet form on $L^2(\mathbb{R}^n)$ possessing $C_c^\infty(\mathbb{R}^n)$ as its special standard core and its associated Hunt process X is a conservative diffusion on $\mathbb{R}^n$ having jointly continuous transition density function. Let D be an open set in $\mathbb{R}^n$. Then by Theorem 6.7.13, the following are equivalent for a locally bounded function u on D:

(i) u is harmonic in D.
(ii) For every relatively compact open subset D_1 of D, $u(X_{\tau_{D_1}}) \in L^1(\mathbf{P}_x)$ and $u(x) = \mathbf{E}_x[u(X_{\tau_{D_1}})]$ for every $x \in D_1$.
(iii) $u \in H^1_{\rm loc}(D)$ and for every $v \in C_c^\infty(D)$,

$$\sum_{i,j=1}^n \int_{\mathbb{R}^n} a_{ij}(x) \frac{\partial u(x)}{\partial x_i} \frac{\partial v(x)}{\partial x_j}\, dx = 0.$$

In fact, in this case, it can be shown (see, e.g., [83]) that every (locally bounded) function satisfying condition (iii) has a continuous version. Therefore, in condition (ii), the statement "for q.e. $x \in D_1$" can be strengthened into "for every $x \in D_1$" because $\mathbf{E}_x[u(X_{\tau_{D_1}})]$ is a difference of two bounded excessive functions in $x \in D_1$ relative to the part process of X on D_1 which has an absolutely continuous transition function. □

Example 6.7.16. (Diffusions with jumps on $\mathbb{R}^n$) Consider the following Dirichlet form $(\mathcal{E}, \mathcal{F})$, where $\mathcal{F} = H^1(\mathbb{R}^n) \cap W^{\alpha/2,2}(\mathbb{R}^n)$ and for $u, v \in H^1(\mathbb{R}^n) \cap W^{\alpha/2,2}(\mathbb{R}^n)$,

$$\mathcal{E}(u, v) = \frac{1}{2} \sum_{i,j=1}^n \int_{\mathbb{R}^n} a_{ij}(x) \frac{\partial u(x)}{\partial x_i} \frac{\partial v(x)}{\partial x_j}\, dx$$

$$+ \frac{1}{2} \int_{\mathbb{R}^n \times \mathbb{R}^n} (u(x) - u(y))(v(x) - v(y)) \frac{c(x, y)}{|x - y|^{n+\alpha}}\, dxdy.$$

Here $n \geq 1$, $(a_{ij}(x))_{1 \leq i,j \leq n}$ is an $n \times n$-matrix valued measurable function on $\mathbb{R}^n$ that is uniformly elliptic and bounded, $\alpha \in (0, 2)$, and $c(x, y)$ is a symmetric function in (x, y) that is bounded between two positive constants. It is easy to check that $(\mathcal{E}, \mathcal{F})$ is a regular Dirichlet form on $L^2(\mathbb{R}^n)$ possessing $C_c^\infty(\mathbb{R}^n)$ as a special standard core. Its associated symmetric Hunt process X has both the diffusion and jumping components. Such a process has recently been studied in [30]. It is shown there that the process X has strictly positive jointly continuous transition density function $p(t, x, y)$ and hence is irreducible. Moreover, a sharp two-sided estimate is obtained in [30] for $p(t, x, y)$. In particular, there is a constant $c > 0$ such that

$$p(t, x, y) \leq c \left(t^{-n/\alpha} \wedge t^{-n/2} \right) \qquad \text{for } t > 0 \text{ and } x, y \in \mathbb{R}^n.$$

Note that when $(a_{ij})_{1\leq i,j\leq n}$ is the identity matrix and $c(x,y)$ is constant, the process X is nothing but the symmetric Lévy process that is the independent sum of a Brownian motion and a rotationally symmetric α-stable process on $\mathbb{R}^n$. In this example, the jumping measure

$$J(dx,dy)=\frac{c(x,y)}{|x-y|^{n+\alpha}}\,dxdy.$$

Hence for any non-empty open set $D\subset\mathbb{R}^n$, condition (6.7.7) is satisfied if and only if $(1\wedge|x|^{-n-\alpha})u(x)\in L^1(\mathbb{R}^n)$. By the same reasoning as that for (6.7.24), we see that for this example, condition (6.7.13) and hence (6.7.8) is implied by condition (6.7.7). So Theorem 6.7.13 says that for an open set D and a function u on $\mathbb{R}^n$ that is locally bounded on D with $(1\wedge|x|^{-n-\alpha})u(x)\in L^1(\mathbb{R}^n)$, the following are equivalent:

(i) u is harmonic in D with respect to X.
(ii) For every relatively compact subset D_1 of D, $u(X_{\tau_{D_1}})\in L^1(\mathbf{P}_x)$ and $u(x)=\mathbf{E}_x[u(X_{\tau_{D_1}})]$ for q.e. $x\in D_1$.
(iii) $u\in H^1_{\rm loc}(D)\cap W^{\alpha/2,2}_{\rm loc}(D)$ and for every $v\in C_c^\infty(D)$,

$$\sum_{i,j=1}^n\int_{\mathbb{R}^n}a_{ij}(x)\frac{\partial u(x)}{\partial x_i}\frac{\partial v(x)}{\partial x_j}\,dx$$
$$+\int_{\mathbb{R}^n\times\mathbb{R}^n}(u(x)-u(y))(v(x)-v(y))\frac{c(x,y)}{|x-y|^{n+\alpha}}\,dxdy=0.$$

It was shown in [30] that every bounded harmonic function in D admits a (locally) Hölder continuous version. □

Remark 6.7.17. It is possible to extend the results of this section to a general m-symmetric right process X on a Lusin space, where m is a positive σ-finite measure with full topological support on E. In this case, the Dirichlet $(\mathcal{E},\mathcal{F})$ of X is a quasi-regular Dirichlet form on $L^2(E;m)$. By Theorem 1.4.3, $(\mathcal{E},\mathcal{F})$ is quasi-homeomorphic to a regular Dirichlet form on a locally compact separable metric space. So the results of this section can be extended to the quasi-regular Dirichlet form setting, by using this quasi-homeomorphism. However, since the notion of open set is not invariant under quasi-homeomorphism, some modifications are needed. We need to replace open set D in Definition 6.7.5 by quasi open set D. Similar modifications are needed for conditions (6.7.7) and (6.7.8) as well. We say a function u is harmonic in a quasi open set $D\subset E$ if for every quasi open subset $D_1\subset D$ so that $\overline{D_1}\cap F_k\subset D$ for every $k\geq 1$, where $\{F_k,k\geq 1\}$ is an $\mathcal{E}$-nest consisting of compact sets, $t\mapsto u(X_{t\wedge\tau_{D_1\cap F_k}})$ is a uniformly integrable $\mathbf{P}_x$-martingale for q.e. $x\in D_1\cap F_k$, and for every $k\geq 1$.

The local Dirichlet space $\mathcal{F}^D_{\rm loc}$ needs to be replaced by $\overset{\circ}{\mathcal{F}}{}^D_{\rm loc}$, which is defined by (4.3.31) but with $(\mathcal{E}, \mathcal{F}^D)$ in place of $(\mathcal{E}, \mathcal{F})$. Condition (6.7.20) should be replaced by

$$u \in \overset{\circ}{\mathcal{F}}{}^D_{\rm loc} \quad \text{and} \quad \mathcal{E}(u, v) = 0 \text{ for every } v \in \mathcal{F} \text{ with } \mathcal{E}\text{-supp}[v] \subset D. \qquad (6.7.25)$$

Here $\mathcal{E}$-supp$[u]$ is the smallest quasi closed set outside which $u = 0$ m-a.e. We leave the details to interested readers. □

Chapter Seven

BOUNDARY THEORY FOR SYMMETRIC MARKOV PROCESSES

Let X^0 be an m_0-symmetric right process on a state space E_0. The boundary theory of symmetric Markov processes concerns all possible symmetric extensions of X^0 to some state space $\widetilde{E}$ containing E_0 as an intrinsic open subset so that they admit no-sojourn (that is, spend zero Lebesgue amount of time) at $\widetilde{E} \setminus E_0$. The no-sojorn condition ensures that the extension process is m_0-symmetric after the measure m_0 is extended to $\widetilde{E}$ by setting $m_0(\widetilde{E} \setminus E_0) = 0$. The initial "minimal" process X^0 can be taken as the part process on E_0 of a symmetric Hunt process X on E killed upon leaving a quasi open subset E_0 and this is the viewpoint we take in this chapter. Of course, one can just start with X^0 with no reference to X.

As we see from (5.6.7), the Douglas integral described by the Feller measures determined by X^0 occupies a principal part of the trace Dirichlet form of X on $\widetilde{E} \setminus E_0$. On the other hand, in view of Theorem 6.6.5, we may well expect that the reflected Dirichlet space $(\mathcal{E}^{0,\mathrm{ref}}, (\mathcal{F}^0)^{\mathrm{ref}})$ of X^0 will play an important role in the boundary theory. One of the goals of this chapter is to demonstrate how these two tools can be used in the study of possible symmetric extensions of X^0.

On account of the quasi-homeomorphism theorem (Theorem 1.4.3), the three theorems in Section 1.5 and Theorem 3.1.13, without loss of generality, we may and do assume throughout this chapter that E is a locally compact separable metric space, m is a positive Radon measure on E with $\mathrm{supp}[m] = E$, $(\mathcal{E}, \mathcal{F})$ is a regular irreducible symmetric Dirichlet form in $L^2(E; m)$, and $X = (\Omega, \mathcal{M}, X_t, \zeta, \mathbf{P}_x)$ is an m-symmetric Hunt process associated with $(\mathcal{E}, \mathcal{F})$. Let F be a quasi closed subset of E that is not $\mathcal{E}$-polar and X^0 be the part process of X on $E_0 := E \setminus F$. The process X^0 is then symmetric with respect to the measure $m_0 := m|_{E_0}$. Let $(\mathcal{E}^0, \mathcal{F}^0)$ be the Dirichlet form of X^0 on $L^2(E_0; m_0)$. We know that $\mathcal{F}^0 = \{u \in \mathcal{F} : u = 0 \ \mathcal{E}\text{-q.e. on } F\}$ and its extended Dirichlet space $\mathcal{F}^0_e$ equals $\{u \in \mathcal{F}_e : u = 0 \ \mathcal{E}\text{-q.e. on } F\}$. We shall consider the reflected Dirichlet space $((\mathcal{F}^0)^{\mathrm{ref}}, \mathcal{E}^{0,\mathrm{ref}})$ and the active reflected Dirichlet space $(\mathcal{F}^0)^{\mathrm{ref}}_a = (\mathcal{F}^0)^{\mathrm{ref}} \cap L^2(E_0; m_0)$ of the quasi-regular Dirichlet form $(\mathcal{E}^0, \mathcal{F}^0)$ as defined in Section 6.4.

In Section 7.1, we investigate the relationship between the space $(\mathcal{F}_e, \mathcal{E})$ (respectively, $(\mathcal{F}, \mathcal{E}_\alpha)$) and the space $((\mathcal{F}^0)^{\rm ref}, \mathcal{E}^{0,{\rm ref}})$ (respectively, $((\mathcal{F}^0)^{\rm ref}_a, \mathcal{E}^{0,{\rm ref}}_\alpha)$). We will see that the former is always dominated by the latter.

In Section 7.2 we focus our attention on the restricted spaces $\mathcal{F}_e\big|_F, \mathcal{F}\big|_F$ and their descriptions in terms of the Feller measures U, V, and U_α introduced in Section 5.4 and the Douglas integrals defined by them. We give conditions to ensure that those spaces coincide with function spaces on F with finite Douglas integrals. We also use 1-order Feller measure U_1 to introduce an intrinsic measure μ_0 on F such that $\mathcal{F}\big|_F = \mathcal{F}_e\big|_F \cap L^2(F; \mu_0)$. These results are then applied to identifying the trace Dirichlet space of the reflecting extension X of a symmetric standard process X^0 with the function space on F of finite Douglas integrals. The process X will be called a reflecting extension of X^0 when the condition that $(\mathcal{E}, \mathcal{F}) = (\mathcal{E}^{0,{\rm ref}}, (\mathcal{F}^0)^{\rm ref}_a)$ is fulfilled among others. It will then be shown that the active reflected Dirichlet space of any quasi-regular Dirichlet form admits such a representation on its regularizing boundary $\widehat{F}$.

The infinitesimal generator $\mathcal{A}$ of the L^2-semigroup of X is a linear operator $\mathcal{L}$, to be introduced in Section 7.3, satisfying a lateral (boundary) condition. The linear operator $\mathcal{L}$ is defined through the active reflected Dirichlet form $(\mathcal{E}^{0,{\rm ref}}, (\mathcal{F}^0)^{\rm ref}_a)$ of $(\mathcal{E}^0, \mathcal{F}^0)$. The aforementioned lateral (boundary) condition uses the notion of flux functional $\mathcal{N}$ that is also defined in terms of the reflected Dirichlet space $(\mathcal{E}^{0,{\rm ref}}, (\mathcal{F}^0)^{\rm ref})$. The trace Dirichlet form on $L^2(F; \mu_0)$ will play a role in the description of the lateral condition.

In Section 7.4, we study the case where the set F consists of countably many points that are located in an invariant way under a quasi-homeomorphism. We shall show that not only the lateral condition but also the Dirichlet form and the resolvent of X can be described in quite tractable ways. In particular, when X admits no killing on F or jump from F to F, the trace Dirichlet form is the subspace of the space of finite Douglas integrals spanned by finitely supported functions, and accordingly the resolvent of X is uniquely determined by X^0. The Dirichlet space $\mathcal{F}$ of X can be also characterized as a subspace of the active reflected Dirichlet space of X^0 spanned by $\mathcal{F}^0$ and α-order hitting distributions to F.

When F consists of only a single point a, X may be called a one-point extension of X^0. In Section 7.5, the uniqueness of a one-point extension of a symmetric standard process will be established under a certain condition without requiring the regularity of the associated Dirichlet forms. A construction of a one-point extension of a symmetric standard process X^0 has been carried out in [75] and [28] by making use of a Poisson point process taking values in the space of excursions of X^0 around the point a. Consider a symmetric Hunt process X on a state space E, a closed set $K \subset E$, and the part process X^0 of X on $E_0 = E \setminus K$. The above-mentioned probabilistic construction is robust enough to enable us to produce a one-point extension of X^0 by collapsing the set K into a single point a^*, as will be shown in Section 7.5. While a one-point extension is always irreducible, the symmetric process X^0 we start with may not be irreducible, so X^0 may have more than one symmetrizing measure. Accordingly, X^0 may admit a variety of skew extensions as will be shown in Section 7.5. In Section 7.6, we collect various concrete examples of X^0 and identify their one-point extensions.

In Section 7.7, we shall generalize the uniqueness theorem and existence theorem of the one-point extension established in Section 7.5 to the countably many points extension. The uniqueness is deduced from a theorem in Section 7.4 via quasi-homeomorphism. The construction will be carried out by repeating the one-point extensions of Section 7.5. Section 7.8 will present several concrete examples including a two-point reflecting extension in higher dimensions and darning of countably many holes.

Recall that the following convention is in force: any numerical function f defined on a subset of E is extended to ∂ by setting $f(\partial) = 0$.

7.1. REFLECTED DIRICHLET SPACE FOR PART PROCESSES

From Section 7.1 to Section 7.3 except for the last part of Section 7.2, we shall adopt the same setting as in Section 5.5. Thus E is a locally compact separable metric space, m is a positive Radon measure on E with $\mathrm{supp}[m] = E$, $(\mathcal{E}, \mathcal{F})$ is a regular irreducible symmetric Dirichlet form in $L^2(E; m)$, and $X = (\Omega, \mathcal{M}, X_t, \zeta, \mathbf{P}_x)$ is an m-symmetric Hunt process associated with $(\mathcal{E}, \mathcal{F})$. The resolvent of X will be denoted by $\{G_\alpha; \alpha > 0\}$. Let F be a quasi closed subset of E that is not $\mathcal{E}$-polar. We denote by (N, H) a Lévy system of the Hunt process X on E. The jumping measure and the killing measure for the Dirichlet form $(\mathcal{E}, \mathcal{F})$ of X are given by

$$J(dx, dy) := N(x, dy)\mu_H(dx) \quad \text{and} \quad \kappa(dx) := N(x, \{\partial\})\mu_H(dx).$$

As before, each element of the extended Dirichlet space $\mathcal{F}_e$ will be represented by its quasi continuous version and we will assume without loss of generality that F is nearly Borel and finely closed. Then $E_0 = E \setminus F$ is finely open. Recall that $\tau_0 := \tau_{E_0}$, defined by (5.5.2), is the first exit time from $E_0 = E \setminus F$ by X. The part process X^0 of X on E_0 can then be realized as

$$X^0 = \{\Omega, \mathcal{M}^0, X_t^0, \zeta^0, \mathbf{P}_x\}_{x \in E_0},$$

where

$$\zeta^0 = \tau_0 \quad \text{and} \quad X_t^0 = \begin{cases} X_t & \text{for } 0 \le t < \zeta^0, \\ \partial & \text{for } t \ge \zeta^0, \end{cases}$$

and $\mathcal{M}^0$ is the σ-field generated by X_t^0 with a usual augmentation by null sets. For emphasis, the law of X^0 under $\mathbf{P}_x$ and its expectation will be denoted as $\mathbf{P}_x^0$

and $\mathbf{E}_x^0$, respectively. The process X^0 is a standard process on E_0. $\{G_\alpha^0; \alpha > 0\}$ and G_0^0 will denote the resolvent and the 0-order resolvent of X^0, respectively. By Theorems 3.5.6 and 3.3.8 the Dirichlet form $(\mathcal{E}^0, \mathcal{F}^0)$ on $L^2(E_0; m_0)$ defined by (5.5.4) is transient and quasi-regular with which X^0 is properly associated. In view of Theorem 3.3.8, X^0 can be considered as a special Borel standard process by restricting it to outside of an m-inessential set if necessary. In general $(\mathcal{E}^0, \mathcal{F}^0)$ is not regular on $L^2(E_0; m_0)$.

Let $\mathcal{S}(E_0)$ (resp. $\mathcal{S}^{\rm pur}(E_0)$) be the space of q.e. excessive functions (resp. q.e. purely excessive functions) of X^0 on E_0 and $L^0(f, g)$ be the energy functional of X^0 for $f \in \mathcal{S}^{\rm pur}(E_0), g \in \mathcal{S}(E_0)$ defined by (5.4.4). As $(\mathcal{E}^0, \mathcal{F}^0)$ is a quasi-regular Dirichlet form on $L^2(E_0; m_0)$, one can define its reflected Dirichlet space $(\mathcal{E}^{0,\rm ref}, (\mathcal{F}^0)^{\rm ref})$ by Definition 6.4.4 but with X^0 in place of X. Let us define the notion of the terminal random variable by Definition 6.4.10 and put

$$\mathbf{N} = \Big\{\Phi : \ \Phi \text{ is a terminal random variable of } X^0 \text{ with } \mathbf{E}_x^0[\Phi^2] < \infty \\ \text{for q.e. } x \in E_0 \text{ and } L^0\left(\mathbf{E}_\cdot^0[\Phi^2] - (\mathbf{E}_\cdot^0[\Phi])^2,\ 1\right) < \infty\Big\}.$$

By Theorem 6.4.12, it holds that $(\mathcal{F}^0)^{\rm ref} = \mathcal{F}_e^0 + \mathbf{H}^0\mathbf{N}$, where

$$\mathbf{H}^0\mathbf{N} := \left\{h : \ h(x) = \mathbf{E}_x^0[\Phi] \text{ for q.e. } x \in E_0 \text{ with} \Phi \in \mathbf{N}\right\}.$$

Moreover, for $f = f_0 + h \in (\mathcal{F}^0)^{\rm ref}$, where $f_0 \in \mathcal{F}_e^0$ and $h = \mathbf{E}_\cdot^0[\Phi]$ with $\Phi \in \mathbf{N}$,

$$\mathcal{E}^{0,\rm ref}(f,f) = \mathcal{E}^0(f_0,f_0) + \frac{1}{2}L^0(\mathbf{E}_\cdot^0[\Phi^2] - h(\cdot)^2, 1) + \frac{1}{2}\int_{E_0} h(x)^2\kappa_0(dx), \tag{7.1.1}$$

where κ_0 is the killing measure for X^0. $f = f_0 + h$ as above is a unique $\mathcal{E}^{0,\rm ref}$-orthogonal decomposition of $f \in (\mathcal{F}^0)^{\rm ref}$.

Exercise 7.1.1. Prove the identity

$$\kappa_0(B) := \kappa(B) + J(B, F), \quad B \in \mathcal{B}(E_0). \tag{7.1.2}$$

The *active reflected Dirichlet space* $(\mathcal{F}^0)_a^{\rm ref}$ of $(\mathcal{E}^0, \mathcal{F}^0)$ is defined by

$$(\mathcal{F}^0)_a^{\rm ref} := (\mathcal{F}^0)^{\rm ref} \cap L^2(E_0; m_0).$$

By Theorem 6.4.5, $(\mathcal{E}^{0,\rm ref}, (\mathcal{F}^0)_a^{\rm ref})$ is a Dirichlet form on $L^2(E_0; m_0)$. In this section, we first establish a unique $\mathcal{E}_\alpha^{0,\rm ref}$-orthogonal decomposition of $f \in (\mathcal{F}^0)_a^{\rm ref}$. To this end, we put for $\alpha > 0$ and a terminal random variable Φ of X^0

$$\mathbf{H}^0\Phi(x) := \mathbf{E}_x^0[\Phi] \quad \text{and} \quad \mathbf{H}_\alpha^0\Phi(x) := \mathbf{E}_x[e^{-\alpha\zeta^0}\Phi], \qquad x \in E_0,$$

whenever the expectations make sense. We also define for terminal random variables Φ, Ψ of X^0

$$U_\alpha(\Phi, \Psi) := \alpha(\mathbf{H}^0_\alpha \Phi, \mathbf{H}^0 \Psi).$$

Here and in what follows, we denote by (u, v) the integral $\int_{E_0} u(x) \cdot v(x) m_0(dx)$.

Lemma 7.1.2. *If Φ is a non-negative terminal random variable of X^0 with $\mathbf{H}^0\Phi(x) < \infty$, then*

$$\mathbf{H}^0\Phi(x) - \mathbf{H}^0_\alpha \Phi(x) = \alpha G^0_0 \mathbf{H}^0_\alpha \Phi(x),$$

$$\mathbf{H}^0_\beta \Phi(x) - \mathbf{H}^0_\alpha \Phi(x) = (\alpha - \beta) G^0_\alpha \mathbf{H}^0_\beta \Phi(x) \qquad \text{for } \alpha,\ \beta > 0.$$

If Φ is a non-negative terminal random variable of X^0, then $U_\alpha(\Phi, \Phi)$ is increasing in $\alpha > 0$ and $\frac{1}{\alpha} U_\alpha(\Phi, \Phi)$ is decreasing in $\alpha > 0$.

Proof. We have

$$\begin{aligned}
\mathbf{H}^0\Phi(x) - \mathbf{H}^0_\alpha \Phi(x) &= \alpha \mathbf{E}^0_x \left[\int_0^{\zeta^0} e^{-\alpha(\zeta^0 - s)} ds \cdot \Phi \right] \\
&= \alpha \mathbf{E}^0_x \left[\int_0^\infty e^{-\alpha \zeta^0(\theta_s)} \Phi(\theta_s \omega) \mathbf{1}_{\{s < \zeta^0\}} ds \right] \\
&= \alpha \mathbf{E}^0_x \left[\int_0^\infty \mathbf{E}^0_{X^0_s} \left[e^{-\alpha \zeta^0} \Phi \right] \mathbf{1}_{\{s < \zeta^0\}} ds \right] \\
&= \alpha G^0_0 \mathbf{H}^0_\alpha \Phi.
\end{aligned}$$

The second identity can be obtained similarly. Suppose $U_\beta(\Phi, \Phi) < \infty$ for $\beta > 0$ and a non-negative terminal random variable Φ. Then it follows from the second identity and the symmetry of G^0_α that, for $\beta > \alpha$,

$$U_\beta(\Phi, \Phi) - U_\alpha(\Phi, \Phi) = (\beta - \alpha) \left((\mathbf{H}^0_\beta \Phi, \mathbf{H}^0 \Phi) - (\mathbf{H}^0_\beta \Phi, \alpha G^0_\alpha \mathbf{H}^0 \Phi) \right),$$

which is non-negative because $\alpha G^0_\alpha \mathbf{H}^0 \Phi \leq \mathbf{H}^0 \Phi$. The decreasing property of $\frac{1}{\alpha} U_\alpha(\Phi, \Phi)$ is obvious. □

We introduce a subspace of $\mathbf{N}$ by

$$\mathbf{N}_1 = \{\Phi \in \mathbf{N} : \ U_1(|\Phi|, |\Phi|) < \infty\}. \tag{7.1.3}$$

Proposition 7.1.3. *A terminal random variable Φ of X^0 is in $\mathbf{N}_1$ if and only if $\mathbf{H}^0_\alpha |\Phi| \in (\mathcal{F}^0)^{\text{ref}}_a$ for some and hence for all $\alpha > 0$. In this case,*

$$\Phi = \mathbf{1}_{\{\zeta^0 < \infty\}} \Phi, \qquad \mathbf{P}^0_x\text{-a.e. for q.e. } x \in E_0 \tag{7.1.4}$$

and

$$\mathbf{H}^0_\alpha \Phi = \mathbf{H}^0 \Phi - \alpha G^0_0 \mathbf{H}^0_\alpha \Phi \tag{7.1.5}$$

represents the unique decomposition of $\mathbf{H}^0_\alpha \Phi$ *as a sum of elements of* $\mathbf{H}^0\mathbf{N}$ *and* $\mathcal{F}^0_e$. *Furthermore,*

$$\mathcal{E}^{0,\mathrm{ref}}_\alpha(\mathbf{H}^0_\alpha \Phi, \mathbf{H}^0_\alpha \Phi) = \mathcal{E}^{0,\mathrm{ref}}(\mathbf{H}^0 \Phi, \mathbf{H}^0 \Phi) + U_\alpha(\Phi, \Phi), \tag{7.1.6}$$

where $\mathcal{E}^{0,\mathrm{ref}}_\alpha(u, v) := \mathcal{E}^{0,\mathrm{ref}}(u, v) + \alpha(u, v)$ *for* $u, v \in (\mathcal{F}^0)^{\mathrm{ref}}_a$.

Proof. It suffices to prove this proposition for a non-negative terminal random variable Φ of X^0. By Lemma 7.1.2,

$$\mathbf{H}^0 \Phi = \alpha G^0_0 \mathbf{H}^0_\alpha \Phi + \mathbf{H}^0_\alpha \Phi.$$

Suppose $\Phi \in \mathbf{N}_1$. The above identity implies not only (7.1.5) but also

$$\alpha(\mathbf{H}^0_\alpha \Phi, \mathbf{H}^0_\alpha \Phi) + \alpha^2(\mathbf{H}^0_\alpha \Phi, G^0_0 \mathbf{H}^0_\alpha \Phi) = U_\alpha(\Phi, \Phi). \tag{7.1.7}$$

Since $U_\alpha(\Phi, \Phi) < \infty$, this means first that $\mathbf{H}^0_\alpha \Phi \in L^2(E_0; m_0)$ and second that $\mathbf{H}^0_\alpha \Phi$ is of finite energy integral with respect to G^0_0 and accordingly $G^0_0 \mathbf{H}^0_\alpha \Phi \in \mathcal{F}^0_e$. Hence $\mathbf{H}^0_\alpha \Phi \in (\mathcal{F}^0)^{\mathrm{ref}}_a$ by (7.1.5). Furthermore, (7.1.5) and (7.1.7) yield (7.1.6).

Conversely, suppose $\mathbf{H}^0_\alpha \Phi \in (\mathcal{F}^0)^{\mathrm{ref}}_a$. Then, in exactly the same manner as the proof of Theorem 6.2.15 for $u := \mathbf{H}^0_\alpha \Phi$, but by taking an $\mathcal{E}^0$-nest in place of $\{D_k\}$ there and by taking Theorem 3.1.13 into account, we can conclude that Φ has the property (7.1.4) and furthermore $f_0 = \mathbf{H}^0 \Phi - \mathbf{H}^0_\alpha \Phi \in \mathcal{F}^0_e$. By Lemma 7.1.2, $f_0 = -G^0_0 \mathbf{H}^0_\alpha \Phi \in \mathcal{F}^0_e$. We note that if $G^0_0 u \in \mathcal{F}^0_e$ for some non-negative measurable function u on E_0, then

$$\mathcal{E}^0(G^0_0 u, G^0_0 u) = (u, G^0_0 u) < \infty \quad \text{and} \quad \mathcal{E}^0(G^0_0 u, w) = (u, w) \tag{7.1.8}$$

for every $w \in \mathcal{F}^0_e$. The above can be proved by approximate u by functions $u_n = \mathbf{1}_{\{g \geq 1/n\}} (u \wedge n)$, where g is a reference function for the transient Dirichlet form $(\mathcal{E}^0, \mathcal{F}^0)$ (see Theorem 2.1.5(i)). Therefore, we have

$$(\mathbf{H}^0_\alpha \Phi, G^0_0 \mathbf{H}^0_\alpha \Phi) < \infty.$$

This combined with (7.1.7) shows that $U_\alpha(\Phi, \Phi) < \infty$, namely, $\Phi \in \mathbf{N}_1$. □

THEOREM 7.1.4. *Define the subspace* $\mathbf{N}_1$ *of* $\mathbf{N}$ *by* (7.1.3). *Then for every* $\alpha > 0$,

$$(\mathcal{F}^0)^{\mathrm{ref}}_a = \mathbf{H}^0_\alpha \mathbf{N}_1 + \mathcal{F}^0 = \{\mathbf{H}^0_\alpha \Phi + f_0 : \Phi \in \mathbf{N}_1, f_0 \in \mathcal{F}^0\}. \tag{7.1.9}$$

The above decomposition is an $\mathcal{E}^{0,\mathrm{ref}}_\alpha$*-orthogonal decomposition. Moreover, for* $f = \mathbf{H}^0_\alpha \Phi + f_0 \in (\mathcal{F}^0)^{\mathrm{ref}}_a$ *with* $\Phi \in \mathbf{N}_1$ *and* $f_0 \in \mathcal{F}^0$,

$$\mathcal{E}^{0,\mathrm{ref}}_\alpha(f, f) = \mathcal{E}^{0,\mathrm{ref}}(\mathbf{H}^0 \Phi, \mathbf{H}^0 \Phi) + U_\alpha(\Phi, \Phi) + \mathcal{E}^0_\alpha(f_0, f_0). \tag{7.1.10}$$

Proof. Again in the same way as the proof of Theorem 6.2.15, any non-negative $u \in (\mathcal{F}^0)^{\text{ref}}_a$ can be seen to be expressed as $u = \mathbf{H}^0_\alpha \Phi + f_0$ with $\mathbf{H}^0_\alpha \Phi \in (\mathcal{F}^0)^{\text{ref}}_a$, $\Phi \in \mathbf{N}$, $\Phi \geq 0$ and $f_0 \in \mathcal{F}^0$, which combined with the preceding proposition proves that $(\mathcal{F}^0)^{\text{ref}}_a \subset \mathcal{F}^0 + \mathbf{H}_\alpha \mathbf{N}_1$. The converse inclusion is obvious.

The $\mathcal{E}^{\text{ref}}_\alpha$-orthogonality of the decomposition (7.1.9) can be also derived from (7.1.5) and (7.1.8):

$$\mathcal{E}^{0,\text{ref}}_\alpha(\mathbf{H}^0_\alpha \Phi, f_0) = -\alpha\mathcal{E}^0(G^0_0 \mathbf{H}^0_\alpha \Phi, f_0) + \alpha(\mathbf{H}^0_\alpha \Phi, f_0) = 0$$

for $\Phi \in \mathbf{N}_1$ and $f_0 \in \mathcal{F}^0$. Expression (7.1.10) then follows from (7.1.6). □

We now explore the relationship between $(\mathcal{F}^0)^{\text{ref}}$ and $\mathcal{F}_e$. Throughout the remainder of this section, we shall assume that X admits no jumps from E_0 to F; that is,

$$J(E_0 \times F) = 0. \tag{7.1.11}$$

For a Borel measurable function φ on F and $x \in E$, we put

$$\mathbf{H}\varphi(x) := \mathbf{E}_x\left[\varphi(X_{\sigma_F});\ \sigma_F < \infty\right]$$

and

$$\mathbf{H}_\alpha\varphi(x) := \mathbf{E}_x\left[e^{-\alpha\sigma_F}\varphi(X_{\sigma_F}); \sigma_F < \infty\right],$$

whenever they make sense.

LEMMA 7.1.5. *Assume that condition* (7.1.11) *holds. For* $\varphi \in \mathcal{B}(F)$, *we let*

$$\Phi = \mathbf{1}_{\{\tau_0 < \infty, X_{\tau_0-} \in F\}}\varphi(X_{\tau_0-}). \tag{7.1.12}$$

Then Φ *is a terminal random variable for* X^0 *and it holds* $\mathbf{P}_x$*-a.s. for q.e.* $x \in E_0$ *that*

$$\Phi = \mathbf{1}_{\{\sigma_F < \infty\}}\,\varphi(X_{\sigma_F}) \tag{7.1.13}$$

In particular, $\mathbf{H}\varphi\big|_{E_0} = \mathbf{H}^0\Phi$ *(resp.* $\mathbf{H}_\alpha\varphi\big|_{E_0} = \mathbf{H}^0_\alpha\Phi$*) if* $\mathbf{H}|\varphi| < \infty$ *(resp. if* $\mathbf{H}_\alpha|\varphi| < \infty$*) q.e. on* E_0.

Proof. Clearly Φ satisfies the three conditions in Definition 6.4.10 of a terminal random variable of X^0.

For a set $B \subset E$, define $\widehat{\sigma}_B = \inf\{t > 0 : X_{t-} \in B\}$. By Lemma 5.7.2, it then holds that $\mathbf{P}_x(\sigma_F \leq \widehat{\sigma}_F) = 1$, $x \in E_0$. Due to the condition (7.1.11),

$$\mathbf{P}_x(\sigma_F < \widehat{\sigma}_F) \leq \mathbf{P}_x(\sigma_F < \infty,\ X_{\sigma_F-} \notin F) = 0, \quad x \in E_0,$$

so that $\mathbf{P}_x(\sigma_F = \widehat{\sigma}_F) = 1, x \in E_0$. Therefore,

$$\begin{aligned}\Phi &= \Phi \mathbf{1}_{\{\sigma_F<\infty\}} + \Phi \mathbf{1}_{\{\sigma_F=\infty\}} \\ &= \mathbf{1}_{\{\sigma_F<\infty, X_{\sigma_F-}\in F\}}\varphi(X_{\sigma_F-}) + \mathbf{1}_{\{\zeta<\infty, X_{\zeta-}\in F\}}\varphi(X_{\zeta-})\mathbf{1}_{\{\widehat{\sigma}_F=\infty\}} \\ &= \mathbf{1}_{\{\sigma_F<\infty, X_{\sigma_F}\in F\}}\varphi(X_{\sigma_F}) = \mathbf{1}_{\{\sigma_F<\infty\}}\varphi(X_{\sigma_F}).\end{aligned}$$

□

Let us introduce the space $\mathcal{N}$ of functions on F by

$$\mathcal{N} = \big\{\varphi \in \mathcal{B}(F) : \mathbf{H}(\varphi^2)(x) < \infty \text{ for q.e. } x \in E_0 \text{ and } L^0(\mathbf{H}(\varphi^2) - (\mathbf{H}\varphi)^2, \mathbf{1}) < \infty\big\}. \tag{7.1.14}$$

Two functions φ and ψ in $\mathcal{N}$ will be regarded the same if $\varphi(X_{\sigma_F}) = \psi(X_{\sigma_F})$ $\mathbf{P}_x$-a.s. for q.e. $x \in E_0$. On account of the above lemma and Theorem 6.4.12, we have $\{\mathbf{H}\varphi\big|_{E_0} : \varphi \in \mathcal{N}\} \subset \mathbf{H}^0\mathbf{N}$.

THEOREM 7.1.6. *Assume that condition* (7.1.11) *holds.*
(i) *The following inclusions hold:*

$$\mathcal{F}_e\big|_F \subset \mathcal{N}, \quad \big\{\mathbf{H}\varphi\big|_{E_0} : \varphi \in \mathcal{N}\big\} \subset \mathbf{H}^0\mathbf{N}, \text{ and } \quad \mathcal{F}_e\big|_{E_0} \subset (\mathcal{F}^0)^{\text{ref}}. \tag{7.1.15}$$

(ii) *For $u \in \mathcal{F}_e$, it holds that*

$$\mathcal{E}(u,u) = \mathcal{E}^{0,\text{ref}}(u|_{E_0}, u|_{E_0}) + \frac{1}{2}\mu_{\langle u\rangle}(F) + \frac{1}{2}\int_F u(x)^2\kappa(dx), \tag{7.1.16}$$

$$\mu_{\langle u\rangle}(F) = \mu^c_{\langle u\rangle}(F) + \int_{F\times F}(u(\xi)-u(\eta))^2 J(d\xi, d\eta) + \int_F u(\xi)^2\kappa(d\xi), \tag{7.1.17}$$

and

$$\mathcal{E}^{0,\text{ref}}(u\big|_{E_0}, u\big|_{E_0}) = \mathcal{E}^0(u_0, u_0) + \mathbf{C}(u\big|_F, u\big|_F) \tag{7.1.18}$$

where $u_0 := u - \mathbf{H}u$, $\mathbf{C}$ is the Douglas integral defined by (5.6.6).
(iii) *For $u \in \mathcal{F}_e$, $\mathcal{E}(u, u) = \mathcal{E}^{0,\text{ref}}(u|_{E_0}, u|_{E_0})$ if and only if*

$$\mu_{\langle \mathbf{H}u\rangle}(F) = 0. \tag{7.1.19}$$

Proof. In order to prove the first inclusion of (i), take any $u \in b\mathcal{F}_e$. Define $h := \mathbf{H}u$ and $w := \mathbf{H}u^2 - h^2$. By virtue of Lemma 5.5.2, w is the 0-order potential for the Dirichlet form $(\mathcal{E}^0, \mathcal{F}^0)$ of the measure $\mu_{\langle h\rangle}\big|_{E_0}$. Consequently, by Theorem 5.4.3,

$$L^0(w, 1) = \mu_{\langle h \rangle}(E_0) < \infty$$

and so $u\big|_F \subset \mathcal{N}$. Due to the assumption (7.1.11) and Exercise 7.1.1, the killing measure for $(\mathcal{E}^0, \mathcal{F}^0)$ of the part process X^0 equals $\kappa\big|_{E_0}$. Hence we further obtain

$$\mathcal{E}^{0,\mathrm{ref}}(h|_{E_0}, h|_{E_0}) = \frac{1}{2}\mu_{\langle h \rangle}(E_0) + \frac{1}{2}\int_{E_0}(h(x))^2\kappa(dx) \leq \mathcal{E}(h, h) \qquad (7.1.20)$$

on account of (7.1.1) and (4.3.15). Now take any $u \in \mathcal{F}_e$ and let $h = \mathbf{H}u$. For $n \geq 1$, put $u_n := ((-n) \vee u) \wedge n$ and $h_n := \mathbf{H}u_n$. Since u_n is $\mathcal{E}$-convergent and pointwise convergent to u and so is h_n to h as $n \to \infty$, we see from (7.1.20) for h_n and the last statement of Theorem 6.4.12 that $h\big|_{E_0} \in \mathbf{H}^0\mathbf{N}$ and $u\big|_F \subset \mathcal{N}$, and further (7.1.20) remains valid for such h.

The second inclusion of (i) is shown already. The third one is now clear from the first and the second together with the decomposition of $u \in \mathcal{F}_e$:

$$u = u_0 + \mathbf{H}u \qquad \text{with} \quad u_0 = u - \mathbf{H}u \in \mathcal{F}_e^0.$$

Moreover, (4.3.15) and (7.1.20) yield (7.1.16).

On the other hand, by the assumption (7.1.11) and the symmetry of J,

$$J(F \times E_0 \setminus d) = 0, \qquad (7.1.21)$$

so that we get the identity (7.1.17) from (4.3.7) and (4.3.12). Furthermore, by virtue of (5.2.4), we have for $u \in \mathcal{F}_e$,

$$\mathcal{E}(u, u) = \mathcal{E}^0(u_0, u_0) + \mathcal{E}(\mathbf{H}u, \mathbf{H}u) = \mathcal{E}^0(u_0, u_0) + \check{\mathcal{E}}(u\big|_F, \big|_F),$$

and (7.1.18) follows from (7.1.16), (7.1.17), and (5.6.7).

Note that for $u \in \mathcal{F}_e$,

$$\frac{1}{2}\int_F u(x)^2\kappa(dx) = \frac{1}{2}\int_F \mathbf{H}u(x)^2\kappa(dx) \leq \mu_{\langle \mathbf{H}u \rangle}(F)$$

and so (iii) also follows from (7.1.16). □

Remark 7.1.7. (i) Condition (7.1.11) is needed for

$$\left\{\mathbf{H}u\big|_{E_0} : u \in \mathcal{F}_e\right\} \subset \mathbf{H}^0\mathbf{N} \quad \text{and} \quad \mathcal{F}_e\big|_{E_0} \subset (\mathcal{F}^0)^{\mathrm{ref}}$$

to hold. For example, let X be a rotationally symmetric α-stable process in $\mathbb{R}^n$ for $0 < \alpha < 2$ and D is a bounded Lipschitz domain in $\mathbb{R}^n$. Define $\tau_D := \inf\{t > 0 : X_t \notin D\}$. It is well-known that

$$\mathbf{P}_x(X_{\tau_D -} \in D \text{ and } \tau_D < \infty) = 1 \qquad \text{for every } x \in D,$$

and so condition (7.1.11) fails with $E = \mathbb{R}^n$ and $E_0 := D$. Further, for the part process X^0 of X on D, any terminal random variable vanishes.

The Dirichlet form $(\mathcal{E}, \mathcal{F})$ of X on $L^2(\mathbb{R}^n; dx)$ is given by

$$\begin{cases} \mathcal{E}(u,u) = \frac{1}{2}\mathcal{A}(n,-\alpha)\int_{\mathbb{R}^n\times\mathbb{R}^n} \frac{(u(x)-u(y))^2}{|x-y|^{n+\alpha}}dxdy, \\ \mathcal{F} = \{u \in L^2(\mathbb{R}^n; dx) : \mathcal{E}(u,u) < \infty\}, \end{cases}$$

where $\mathcal{A}(n,-\alpha)$ is given by (2.2.19). Let $\mathcal{F}_e$ be its extended Dirichlet space. Then the extended Dirichlet space $\mathcal{F}^0_e$ of the Dirichlet form $(\mathcal{E}^0, \mathcal{F}^0)$ on $L^2(D)$ of X^0 is given by

$$\mathcal{F}^0_e = \{u \in \mathcal{F}_e : u = 0 \text{ q.e. on } \mathbb{R}^n \setminus D\}$$

in view of Theorem 3.4.9. Since $\mathbf{H}^0\mathbf{N} = \{0\}$, we have $(\mathcal{F}^0)^{\mathrm{ref}} = \mathcal{F}^0_e\big|_D$, which is a proper subset of $\mathcal{F}_e\big|_D$.

(ii) Under assumption (7.1.11), condition (7.1.19) holding for every $u \in \mathcal{F}_e$ is equivalent to

$$\mu_{\langle u\rangle}(F) = 0 \quad \text{for every } u \in \mathcal{F}_e. \tag{7.1.22}$$

Clearly (7.1.22) implies (7.1.19). Assume conversely that (7.1.19) holds. For every $u \in \mathcal{F}_e$, let $u_0 = u - \mathbf{H}u$, which is in $\mathcal{F}^0_e$. It can be shown just as in the proof of Theorem 5.6.2 that $\mu^c_{\langle u_0\rangle}(F) = 0$. This together with (7.1.17) implies that $\mu_{\langle u_0\rangle}(F) = 0$. Hence

$$0 \le \mu_{\langle u\rangle}(F) = \mu_{\langle u_0+\mathbf{H}u\rangle}(F) \le \left(\sqrt{\mu_{\langle u_0\rangle}(F)} + \sqrt{\mu_{\langle \mathbf{H}u\rangle}(F)}\right)^2 = 0.$$

□

Combining Theorem 7.1.6 with Proposition 7.1.3, we can formulate a relationship between the space $(\mathcal{F}, \mathcal{E}_\alpha)$ and the active reflected Dirichlet space $((\mathcal{F}^0)^{\mathrm{ref}}_a, \mathcal{E}^{0,\mathrm{ref}}_\alpha)$. Since $\mathcal{F} = \mathcal{F}_e \cap L^2(E; m)$, we first deduce from Theorem 7.1.6 that

$$\mathcal{F}\big|_{E_0} \subset (\mathcal{F}^0)^{\mathrm{ref}}_a. \tag{7.1.23}$$

We introduce the subspace $\mathcal{N}_1$ of $\mathcal{N}$ by

$$\mathcal{N}_1 = \{\varphi \in \mathcal{N} : U_1(|\varphi|, |\varphi|) < \infty\}, \tag{7.1.24}$$

where

$$U_\alpha(\varphi, \psi) = \alpha(\mathbf{H}_\alpha\varphi, \mathbf{H}\psi), \quad \varphi, \psi \in \mathcal{B}(F). \tag{7.1.25}$$

Obviously $\{\mathbf{H}\varphi|_{E_0} : \varphi \in \mathcal{N}_1\} \subset \mathbf{H}^0\mathbf{N}_1$. In view of Proposition 7.1.3, we can see that $\varphi \in \mathcal{B}_+(F)$ is in $\mathcal{N}_1$ if and only if $\mathbf{H}_\alpha\varphi \in (\mathcal{F}^0)^{\mathrm{ref}}_a$ for $\alpha > 0$.

For $u \in \mathcal{F}$, we have the $\mathcal{E}_\alpha$-orthogonal decomposition

$$u = \mathbf{H}_\alpha u + f_0 \qquad \text{with } f_0 \in \mathcal{F}^0. \tag{7.1.26}$$

Since $\mathbf{H}_\alpha u\big|_{E_0} \in (\mathcal{F}^0)^{\rm ref}_a$ by (7.1.23), we conclude that $\mathcal{F}\big|_F \subset \mathcal{N}_1$.

THEOREM 7.1.8. *Assume that condition* (7.1.11) *holds. Fix* $\alpha > 0$. *We have the inclusions*

$$\mathcal{F}\big|_F \subset \mathcal{N}_1, \quad \left\{(\mathbf{H}_\alpha\varphi)\big|_{E_0} : \varphi \in \mathcal{N}_1\right\} \subset \mathbf{H}^0_\alpha \mathbf{N}_1, \quad \textit{and} \quad \mathcal{F}\big|_{E_0} \subset (\mathcal{F}^0)^{\rm ref}_a. \tag{7.1.27}$$

For $u \in \mathcal{F}$, *it holds that*

$$\begin{aligned}\mathcal{E}_\alpha(u,u) = {}& \mathcal{E}^{0,\rm ref}_\alpha(u|_{E_0}, u|_{E_0}) + \frac{1}{2}\mu^c_{\langle \mathbf{H}u\rangle}(F) + \frac{1}{2}\int_{F\times F}(u(\xi)-u(\eta))^2 J(d\xi, d\eta)\\ &+ \int_F u(\xi)^2\kappa(d\xi) + \alpha\int_F u(\xi)^2 m(d\xi),\end{aligned} \tag{7.1.28}$$

$$\mathcal{E}_\alpha(\mathbf{H}_\alpha u, \mathbf{H}_\alpha u) = \check{\mathcal{E}}(u|_F, u|_F) + U_\alpha(u|_F, u|_F) + \alpha\int_F u(\xi)^2 m(d\xi), \tag{7.1.29}$$

and, for $u_0 := u - \mathbf{H}_\alpha u$,

$$\mathcal{E}^{0,\rm ref}_\alpha(u|_{E_0}, u|_{E_0}) = \mathcal{E}^0_\alpha(u_0, v_0) + \mathbf{C}(u|_F, u|_F) + U_\alpha(u|_F, u|_F). \tag{7.1.30}$$

Proof. Identity (7.1.28) follows from (7.1.16) and (7.1.17). Then (7.1.29) is a consequence of (7.1.6), (7.1.28), and (5.2.4). Formula (7.1.30) follows from (7.1.18) and (7.1.25). □

7.2. DOUGLAS INTEGRALS AND REFLECTING EXTENSIONS

In the first half of this section, we continue to study under the same setting as in Section 5.5 and Section 7.1. Thus X is an m-symmetric Hunt process whose Dirichlet form $(\mathcal{E}, \mathcal{F})$ on $L^2(E; m)$ is regular irreducible and $F \subset E$ is a non-$\mathcal{E}$-polar quasi closed set. But we make an additional assumption (7.1.11) that X admits no jump from $E_0 = E \setminus F$ to F. As before, each element of the extended Dirichlet space $\mathcal{F}_e$ of X will be represented by its $\mathcal{E}$-quasi-continuous version.

We shall focus our attention on the restricted spaces $\mathcal{F}_e\big|_F$ and $\mathcal{F}\big|_F$. We are concerned with describing them in terms of the quantities related to the Feller measure U on $F \times F$, the supplementary Feller measure V on F defined by (5.5.7), and the α-order Feller measure U_α defined by (5.5.13). The obtained results will be applied in the latter half of this section to a characterization of a reflecting extension of a symmetric standard process.

We let

$$\mathcal{G}_0 = \{\varphi \in \mathcal{B}(F)\colon \varphi \text{ is finite q.e. on } F,\ \mathbf{C}(\varphi,\varphi) < \infty\}, \tag{7.2.1}$$

where $\mathbf{C}(\varphi,\varphi)$ is the Douglas integral of φ introduced at the end of Section 5.6:

$$\mathbf{C}(\varphi,\varphi) = \frac{1}{2}\int_{F\times F}(\varphi(x)-\varphi(y))^2U(dx,dy) + \int_F \varphi(x)^2V(dx).$$

We observed there that this integral makes sense for every $\varphi \in \mathcal{B}(F)$, which is finite q.e., and gives the same value for q.e. equivalent functions.

Our first aim in this section is to give a sufficient condition for a general Dirichlet form $(\mathcal{E},\mathcal{F})$ so that the space $\mathcal{F}_e\big|_F$ can be identified with a space of functions of finite Douglas integrals. We introduce the space $\mathcal{G}$ by

$$\mathcal{G} = \{\varphi \in \mathcal{G}_0 :\ \mathbf{H}\varphi^2(x) < \infty \ m\text{-a.e. on } E_0\}. \tag{7.2.2}$$

Two functions ϕ and ψ in $\mathcal{G}$ will be regarded the same if $\phi(X_{\sigma_F}) = \psi(X_{\sigma_F})$ $\mathbf{P}_x$-a.s for q.e. $x \in E_0$. Let $\mathcal{N}$ be the function space defined by (7.1.14) in the preceding section.

THEOREM 7.2.1. (i) *It holds that*

$$\mathcal{F}_e\big|_F \subset \mathcal{G} \subset \mathcal{N}. \tag{7.2.3}$$

(ii) $\mathcal{E}(\mathbf{H}\varphi,\mathbf{H}\varphi) \geq \mathbf{C}(\varphi,\varphi)$ *for every* $\varphi \in \mathcal{F}_e\big|_F$. *The equality holds for every* $\varphi \in \mathcal{F}\big|_F$ *if and only if*

$$\mu_{\langle u\rangle}(F) = 0 \quad \textit{for every} \quad u \in \mathcal{F}_e. \tag{7.2.4}$$

(iii) *Assume that*

$$\mathcal{F}_e\big|_{E_0} = (\mathcal{F}^0)^{\mathrm{ref}}. \tag{7.2.5}$$

Then

$$\mathcal{F}_e\big|_F = \mathcal{G}, \tag{7.2.6}$$

$$\mathbf{H}^0\mathbf{N} = \{\mathbf{H}\varphi\big|_{E_0}\colon \varphi \in \mathcal{G}\}, \tag{7.2.7}$$

and

$$\mathcal{E}^{0,\mathrm{ref}}(\mathbf{H}\varphi\big|_{E_0}, \mathbf{H}\varphi\big|_{E_0}) = \mathbf{C}(\varphi,\varphi) \ \textit{for every}\ \varphi \in \mathcal{G}. \tag{7.2.8}$$

Proof. (i) If $\varphi \in \mathcal{F}_e\big|_F$, then φ is of finite Douglas integral in view of (5.6.7) and further $\mathbf{H}\varphi^2(x) < \infty$ for q.e. $x \in E_0$ by virtue Theorem 7.1.6(i). Hence $\mathcal{F}_e\big|_F \subset \mathcal{G}_0$.

To prove the second inclusion in (7.2.3), take any $\varphi \in \mathcal{G}$. Note first that $\mathbf{H}\varphi^2(x)$ is finite q.e. on E_0 on account of the X^0-excessiveness of $\mathbf{H}\varphi^2$,

Theorem A.2.13, and Theorem 3.1.3. In particular, $\mathbf{H}\varphi^2 - (\mathbf{H}\varphi)^2 \in \mathcal{S}^{\text{pur}}(E_0)$ by Lemma 7.1.5 and Theorem 6.1.8. As in the proof for (ii)–(iii) of Theorem 5.4.3, we can find an increasing sequence of non-negative universally measurable functions $\{g_\ell\}$ such that

$$L^0(f, 1) = \lim_{\ell\to\infty} (f, g_\ell) \quad \text{for any } f \in \mathcal{S}^{\text{pur}}(E_0).$$

Next put

$$\varphi_n = (-n) \vee (\varphi \wedge n) \qquad \text{and} \qquad f_n = \mathbf{H}\varphi_n^2 - (\mathbf{H}\varphi_n)^2 \quad \text{for } n \geq 1.$$

Using Theorem 5.4.3, (5.5.7), and Lemma 5.5.3, we obtain, for each $n \geq 1$, $\ell \geq 1$,

$$\begin{aligned}(f_n, g_\ell) &\leq L^0(f_n, 1) = L^0(f_n, \mathbf{H}1) + L^0(f_n, q) \\ &\leq L^0(\mathbf{H}1, f_n) + L^0(\mathbf{H}\varphi_n^2, q) = \lim_{\alpha\to\infty} \alpha(\mathbf{H}_\alpha 1, f_n) + \int_F \varphi_n^2 dV \\ &\leq \int_{F\times F} (\varphi_n(\xi) - \varphi_n(\eta))^2 U(d\xi, d\eta) + 2\int_F \varphi_n^2 dV \leq 2\mathbf{C}(\varphi, \varphi).\end{aligned}$$

We first let $n \to \infty$ using Fatou's lemma and then let $\ell \to \infty$ to get

$$\frac{1}{2} L^0(\mathbf{H}(\varphi^2) - (\mathbf{H}\varphi)^2,\ 1) \leq \mathbf{C}(\varphi, \varphi) < \infty, \tag{7.2.9}$$

which means that the function φ belongs to the space $\mathcal{N}$.
(ii) This is obvious from Theorem 7.1.6 and Remark 7.1.7.
(iii) The present assumption (7.2.5) is equivalent to

$$\{\mathbf{H}\varphi\big|_{E_0} : \varphi \in \mathcal{F}_e\big|_F\} = \mathbf{H}^0\mathbf{N},$$

which combined with (7.2.3) and the second inclusion of Theorem 7.1.6(i) leads us to (7.2.6) and (7.2.7). Property (7.2.8) follows from (7.1.18). □

As a consequence of Theorem 7.2.1(ii), we have

Corollary 7.2.2. *Assume that $(\mathcal{E}, \mathcal{F})$ is transient, condition* (7.2.4) *is fulfilled, and the supplementary Feller measure V vanishes. Then $\mathcal{F}_e\big|_F$ is a proper subspace of $\mathcal{G}$.*

Proof. We know from Section 5.2 that $(\check{\mathcal{F}}_e, \check{\mathcal{E}})$ with $\check{\mathcal{F}}_e = \mathcal{F}_e\big|_F$ and $\check{\mathcal{E}}(\varphi, \varphi) = \mathcal{E}(\mathbf{H}\varphi, \mathbf{H}\varphi)$ is the extended Dirichlet space of the time-changed process of X by a PCAF having F as its support. By virtue of Theorem 5.2.5, $(\check{\mathcal{E}}, \check{\mathcal{F}}_e)$ is then transient. If $1 \in \check{\mathcal{F}}_e$, then $\check{\mathcal{E}}(1, 1) = 0$ by Theorem 7.2.1(ii) contradicting Theorem 2.1.9. So $1 \notin \check{\mathcal{F}}_e$, while 1 is an element of $\mathcal{G}$. □

We turn to the study of the space $\mathcal{F}\big|_F$ in terms of the Feller measures. Recall the α-order Feller measure defined by the integral $U_\alpha(\varphi, \psi)$, $\varphi, \psi \in \mathcal{B}_+(F)$

of (7.1.25). We can define a measure μ_0 on F by

$$\int_F \psi(\eta)\mu_0(d\eta) = U_1(1,\psi), \quad \psi \in \mathcal{B}_+(F). \tag{7.2.10}$$

μ_0 is extended to E by setting $\mu_0(E \setminus F) = 0$ and called an *intrinsic measure* on F owing to its role, which will be revealed in Theorem 7.2.4.

We next introduce the space $\mathcal{G}_1$ by

$$\mathcal{G}_1 = \mathcal{G}_0 \cap L^2(F;\mu_0). \tag{7.2.11}$$

Since X is irreducible and F is not $\mathcal{E}$-polar, $\mathbf{H}_1 1(x) = \mathbf{E}_x[e^{-\alpha\sigma_F}] > 0$ for m-a.e. $x \in E$ by Theorem 3.5.6. Hence $\varphi \in L^2(F;\mu_0)$ satisfies $\mathbf{H}\varphi^2(x) < \infty$ for m-a.e. $x \in E_0$ so that

$$\mathcal{G}_1 = \mathcal{G} \cap L^2(F;\mu_0), \tag{7.2.12}$$

where $\mathcal{G}$ is the space defined by (7.2.2).

Let $\overset{\circ}{S}_F$ denote the space defined by (5.2.15), which is the collection of Radon measures on E charging no $\mathcal{E}$-polar set with quasi support being equal to F.

LEMMA 7.2.3. (i) *It holds that*

$$\mathcal{G}_1 = \{\varphi \in \mathcal{G}_0 : \ U_1(|\varphi|,|\varphi|) < \infty\}. \tag{7.2.13}$$

For each $\alpha > 0$, *we let*

$$\mathbf{C}^{[\alpha]}(\varphi,\varphi) = \mathbf{C}(\varphi,\varphi) + U_\alpha(\varphi,\varphi). \tag{7.2.14}$$

Then

$$\mathbf{C}^{[\alpha]}(\varphi,\varphi) \geq U_\alpha(1,\varphi^2) \geq U_\alpha(|\varphi|,|\varphi|), \quad \varphi \in \mathcal{G}_1. \tag{7.2.15}$$

In particular, $\mathbf{C}^{[\alpha]}(\varphi,\varphi)$ *defines a metric on* $\mathcal{G}_1$ *equivalent to* $\mathbf{C}(\varphi,\varphi) + \|\varphi\|^2_{L^2(F;\mu_0)}$.

(ii) $\mu_0 \in \overset{\circ}{S}_F$.

Proof. (i) Using (5.5.14), we obtain for $\varphi \in \mathcal{B}_+(F)$

$$\mathbf{C}^{[\alpha]}(\varphi,\varphi) \geq \frac{1}{2}\int_{F\times F}(\varphi(\xi)-\varphi(\eta))^2 U_\alpha(d\xi,d\eta) + U_\alpha(\varphi,\varphi) = U_\alpha(1,\varphi^2).$$

Other assertions are obvious.

(ii) In view of Theorem 7.1.8, it holds for $u \in \mathcal{F}$ that

$$\begin{aligned}\mathcal{E}_1(\mathbf{H}_1 u, \mathbf{H}_1 u) = & \mathbf{C}^{[1]}(u\big|_F, u\big|_F) + \frac{1}{2}\mu_{\langle \mathbf{H}u \rangle}(F) \\ & + \frac{1}{2}\int_{F\times F}(u(\xi) - u(\eta))^2 J(d\xi, d\eta) \\ & + \int_F u(\xi)^2 \kappa(d\xi) + \int_F u(\xi)^2 m(d\xi). \end{aligned} \tag{7.2.16}$$

In particular, we have from (7.2.15)

$$\int_F u(\xi)^2 \mu_0(d\xi) \le \mathcal{E}_1(\mathbf{H}_1 u, \mathbf{H}_1 u) < \infty, \quad u \in \mathcal{F}. \tag{7.2.17}$$

For any compact set $K \subset E$, taking a function $u \in \mathcal{F} \cap C_c(E)$ greater than 1 on K, we get $\mu_0(K) = \mu_0(F \cap K) < \infty$, which proves that μ_0 is a Radon measure on E. Since $\mathbf{H}_1 1(x) > 0$ m-a.e. on E as we have observed already, $\int_F \psi d\mu_0 = \mathbf{E}_{\mathbf{H}_1 1 \cdot m}\left[\psi(X_{\sigma_F}); \sigma_F < \infty\right]$ belongs to the type of measure exhibited in the proof of Lemma 5.2.9. Therefore, $\mu_0 \in \overset{\circ}{S}_F$. □

Theorem 7.2.4. *Assume that*

$$m(F) = 0. \tag{7.2.18}$$

It then holds that

$$\mathcal{F}\big|_F = \mathcal{F}_e\big|_F \cap L^2(F; \mu_0). \tag{7.2.19}$$

More specifically, let $\check{X}$ be the time-changed process of X by means of the PCAF with Revuz measure μ_0 and $(\check{\mathcal{E}}, \check{\mathcal{F}})$ be the Dirichlet form of $\check{X}$ on $L^2(F; \mu_0)$. Then $\check{\mathcal{F}} = \mathcal{F}\big|_F$ and $\check{\mathcal{E}}_1(\varphi, \varphi)$ on $\check{\mathcal{F}}$ is equivalent to $\mathcal{E}_1(\mathbf{H}_1\varphi, \mathbf{H}_1\varphi)$ $(= \check{\mathcal{E}}(\varphi, \varphi) + U_1(\varphi, \varphi))$ on $\mathcal{F}\big|_F$.

Proof. We know by (5.2.5) that $\check{\mathcal{F}} = \mathcal{F}_e\big|_F \cap L^2(F; \mu_0)$. By (5.6.7), we have, for $\varphi \in \check{\mathcal{F}}$,

$$\begin{aligned}\check{\mathcal{E}}_1(\varphi, \varphi) = & \mathbf{C}(\varphi, \varphi) + \|\varphi\|^2_{L^2(F;\mu_0)} + \frac{1}{2}\mathcal{M}\varphi_{\langle \mathbf{H}\varphi \rangle}(F) \\ & + \frac{1}{2}\int_{F\times F}(\varphi(\xi) - \varphi(\eta))^2 J(d\xi, d\eta) + \int_F \varphi(\xi)^2 \kappa(d\xi). \end{aligned} \tag{7.2.20}$$

Let $\mathcal{C}$ be a special standard core of $(\mathcal{E}, \mathcal{F})$. By virtue of Theorem 5.2.8, $\mathcal{C}|_F$ is $\check{\mathcal{E}}_1$-dense in $\check{\mathcal{F}}$. Since $\mathcal{C}$ is $\mathcal{E}_1$-dense in $\mathcal{F}$, $\mathcal{C}|_F$ is dense in $\mathcal{F}_F$ with metric $\sqrt{\mathcal{E}_1(\mathbf{H}_1\varphi, \mathbf{H}_1\varphi)}$, $\varphi \in \mathcal{F}_F$. On account of (7.2.15), (7.2.16), (7.2.20), and the assumption (7.2.18), we have for any $\varphi \in \mathcal{C}_F$

$$\mathcal{E}_1(\mathbf{H}_1\varphi, \mathbf{H}_1\varphi) \le \check{\mathcal{E}}_1(\varphi, \varphi) \le 2\mathcal{E}_1(\mathbf{H}_1\varphi, \mathbf{H}_1\varphi).$$

This leads us to the desired identity (7.2.19). Therefore, we get $\check{\mathcal{F}} = \mathcal{F}_F$ and the metric equivalence. □

We can now formulate a counterpart of Theorem 7.2.1 for $\mathcal{F}$ in place of $\mathcal{F}_e$.

Theorem 7.2.5. (i) *It holds that*

$$\mathcal{F}\big|_F \subset \mathcal{G}_1 \subset \mathcal{N}_1 \tag{7.2.21}$$

and

$$\mathcal{E}_\alpha(\mathbf{H}_\alpha\varphi, \mathbf{H}_\alpha\varphi) \ge \mathbf{C}(\varphi, \varphi) + U_\alpha(\varphi, \varphi), \qquad \varphi \in \mathcal{F}\big|_F. \tag{7.2.22}$$

(ii) *Suppose*

$$\mathcal{F}\big|_{E_0} = (\mathcal{F}^0)_a^{\text{ref}}, \tag{7.2.23}$$

then

$$\mathcal{F}\big|_F = \mathcal{G}_1 = \mathcal{N}_1, \tag{7.2.24}$$

$$\mathbf{H}_\alpha^0\mathbf{N}_1 = \{\mathbf{H}_\alpha\varphi\big|_{E_0} : \varphi \in \mathcal{G}_1\}, \quad \alpha > 0, \tag{7.2.25}$$

and

$$\mathcal{E}_\alpha^{0,\text{ref}}(\mathbf{H}_\alpha\varphi\big|_{E_0}, \mathbf{H}_\alpha\varphi\big|_{E_0}) = \mathbf{C}(\varphi, \varphi) + U_\alpha(\varphi, \varphi), \quad \varphi \in \mathcal{G}_1,\ \alpha > 0, \tag{7.2.26}$$

$$\mathcal{E}^{0,\text{ref}}(\mathbf{H}\varphi\big|_{E_0}, \mathbf{H}\varphi\big|_{E_0}) = \mathbf{C}(\varphi, \varphi), \quad \varphi \in \mathcal{G}_1. \tag{7.2.27}$$

Proof. (i) From (7.2.17), we get the inclusion $\mathcal{F}\big|_F \subset \mathcal{F}_e\big|_F \cap L^2(F; \mu_0)$. By Theorem 7.2.1(i), we have $\mathcal{F}_e\big|_F \subset \mathcal{G}$, from which we can deduce the first inclusion in (7.2.21) by taking intersections with $L^2(F; \mu_0)$ and noting (7.2.12).

Next take any $\varphi \in \mathcal{G}_1$. Then $\varphi \in \mathcal{G} \subset \mathcal{N}$ by (7.2.12) and Theorem 7.2.1(i). Since $U_1(|\varphi|, |\varphi|) < \infty$ by (7.2.13), we get $\varphi \in \mathcal{N}_1$, proving the second inclusion in (7.2.21). The inequality (7.2.22) follows from (7.1.30).

(ii) Combining the assumption (7.2.23) with the inclusion in (i) and the second inclusion in (7.1.27), we are led to the identities (7.2.24) and (7.2.25). Relations (7.2.26) and (7.2.27) then follow from (7.1.30) and (7.1.18), respectively. □

We remark that the condition (7.2.5) implies the condition (7.2.23), but the converse implication is not necessarily true as we saw for the reflecting Brownian motion in Section 6.5(**5°**).

So far in this chapter, we have employed the setting that we are given a symmetric Hunt process X on E whose Dirichlet form is regular irreducible and we have studied the properties of X in relation to the part process X^0 of X on a quasi open set E_0 of E. We can regard X as an extension of X^0 from E_0 to E.

DEFINITION 7.2.6. A symmetric Hunt process X is said to be a *reflecting extension* of a symmetric standard process X^0 if the following holds:

(RE.1) E is a locally compact separable metric space, m is an everywhere dense positive Radon measure on E, and X is an m-symmetric Hunt process on E whose Dirichlet form $(\mathcal{E}, \mathcal{F})$ on $L^2(E; m)$ is regular.

(RE.2) X^0 is the part process of X on a non-$\mathcal{E}$-polar, $\mathcal{E}$-quasi-open subset E_0 of E whose Dirichlet form $(\mathcal{E}^0, \mathcal{F}^0)$ on $L^2(E_0; m\big|_{E_0})$ is irreducible. Further, $\mathcal{F}^0$ is a proper subset of $\mathcal{F}$.

(RE.3) $m(F) = 0$ where $F = E \setminus E_0$.

(RE.4) Denote by $\big(\mathcal{E}^{0,\mathrm{ref}}, (\mathcal{F}^0)^{\mathrm{ref}}_a\big)$ the active reflected Dirichlet form of $(\mathcal{E}^0, \mathcal{F}^0)$. Then $(\mathcal{E}, \mathcal{F}) = \big(\mathcal{E}^{0,\mathrm{ref}}, (\mathcal{F}^0)^{\mathrm{ref}}_a\big)$.

LEMMA 7.2.7. *Suppose a symmetric Hunt process X is a reflecting extension of a symmetric standard process X^0. Let μ_0 be the intrinsic measure on F defined by* (7.2.10). *Then*
(i) *X admits no jump from E_0 to F in the sense of* (7.1.11).
(ii) *The Dirichlet form $(\mathcal{E}, \mathcal{F})$ of X on $L^2(E; m)$ is irreducible.*
(iii) *The set F is non-$\mathcal{E}$-polar.*

Proof. (i) Fix $\alpha > 0$ and take any $f \in \mathcal{F}$. Since $\mathbf{H}_\alpha f$ is in the $\mathcal{E}_\alpha$-orthogonal complement of $\mathcal{F}^0$, we see from **(RE.4)** and Theorem 7.1.4 that $\mathbf{H}_\alpha f = \mathbf{H}^0_\alpha \Phi$ for some $\Phi \in \mathbf{N}_1$. Let $\{A_k\}$ be an $\mathcal{E}^0$-nest consisting of compact subsets of E_0. Then $\tau_{A_k} \to \sigma_F \wedge \zeta$, $k \to \infty$, $\mathbf{P}_x$-a.s. for q.e. $x \in E_0$ in view of Theorem 3.1.13. By taking Proposition 7.1.3, its proof, Theorem 3.1.7, and the quasi-left-continuity of the Hunt process X into account, we can get

$$\begin{aligned}\Phi = \Phi \cdot \mathbf{1}_{\{\sigma_F \wedge \zeta < \infty\}} &= \lim_{k\to\infty} \mathbf{H}_\alpha f(X_{\tau_{A_k}}) \cdot \mathbf{1}_{\{\lim_{k\to\infty} \tau_{A_k} < \infty\}} \\ &= f(X_{\sigma_F}) \cdot \mathbf{1}_{\{\sigma_F < \infty\}}, \qquad \mathbf{P}_x\text{-a.s. for q.e. } x \in E_0.\end{aligned}$$

As $\Phi \cdot \mathbf{1}_{\{X_{\sigma_F \wedge \zeta -} \in E_0, \sigma_F \wedge \zeta < \infty\}} = 0$ by Definition 6.4.10, we arrive at

$$f(X_{\sigma_F}) \cdot \mathbf{1}_{\{X_{\sigma_F -} \in E_0,\ \sigma_F < \infty\}} = 0, \qquad \mathbf{P}_x\text{-a.s. for q.e. } x \in E_0,$$

proving (7.1.11).

(ii) Denote by $\{P_t; t \geq 0\}$ and $\{P_t^0; t \geq 0\}$ the transition functions of X and X^0, respectively. If $A \in \mathcal{B}(E)$ satisfies $\int_A P_t(x, E \setminus A)m(dx) = 0$, then $\int_{A\cap E_0} P_t^0(x, E_0 \setminus A)m(dx) = 0$, which implies either $m(A) = m(A \cap E_0) = 0$ or $m(E \setminus A) = m(E_0 \setminus A) = 0$ by **(RE.2), (RE.3)**. Hence $(\mathcal{E}, \mathcal{F})$ is irreducible.
(iii) If F is $\mathcal{E}$-polar, there exists an $\mathcal{E}$-nest $\{A_k\}$ such that $F \subset \cap_k A_k^c$. Then $\cup_k \mathcal{F}_{A_k} \subset \mathcal{F}^0$ is $\mathcal{E}_1$-dense in $\mathcal{F}$, which means $\mathcal{F}^0 = \mathcal{F}$ a contradiction to **(RE.2)**.
□

COROLLARY 7.2.8. *Suppose a symmetric Hunt process X is a reflecting extension of a symmetric standard process X^0.*
(i) *For each $\alpha > 0$, $\mathcal{F}$ admits an $\mathcal{E}_\alpha$-orthogonal decomposition as*

$$\mathcal{F} = \mathcal{F}^0 \oplus \{\mathbf{H}_\alpha \varphi : \varphi \in \mathcal{G}_1\}, \tag{7.2.28}$$

and for $u = u_0 + \mathbf{H}_\alpha \varphi$ with $u_0 \in \mathcal{F}^0$ and $\varphi \in \mathcal{G}_1$,

$$\mathcal{E}_\alpha(u, u) = \mathcal{E}_\alpha^0(u_0, u_0) + \mathbf{C}(\varphi, \varphi) + U_\alpha(\varphi, \varphi). \tag{7.2.29}$$

Here $\mathbf{C}$ and $\mathcal{G}_1$ are defined by (7.2.1) and (7.2.11), respectively, in terms of the Feller measures U, V, and U_α determined by the hitting distributions $\mathbf{H}$ and $\mathbf{H}_\alpha$ of X from E_0 to F.
(ii) *$(\mathbf{C}, \mathcal{G}_1)$ is a regular Dirichlet form on $L^2(F^*; \mu_0)$ associated with the time-changed process of X by means of a PCAF with Revuz measure μ_0, where F^* is the topological support of μ_0 on E.*

Proof. (i) On account of **(RE.1)**, **(RE.2)**, and Lemma 7.2.7, $(\mathcal{E}, \mathcal{F})$, F, and X fulfill all requirements imposed at the beginning of this section. Therefore, we get (7.2.28) and (7.2.29) from **(RE.4)** and Theorem 7.2.5 with $\mathcal{E}_\alpha^{0,\mathrm{ref}} = \mathcal{E}_\alpha$.
(ii) This follows from (7.2.19), (7.2.28), **(RE.4)**, (7.2.27), and Theorem 5.2.13.
□

Finally, we shall start with a quasi-regular Dirichlet form $(\mathcal{E}^0, \mathcal{F}^0)$ and, by invoking those theorems in Section 6.6, construct a reflecting extension of a symmetric standard process associated with another quasi-regular Dirichlet form which is quasi-homeomorphic to $(\mathcal{E}^0, \mathcal{F}^0)$.

Let E_0 be a Hausdorff topological space whose Borel σ-field $\mathcal{B}(E_0)$ is generated by continuous functions on E_0, m_0 be an everywhere dense σ-finite measure on E_0, and $(\mathcal{E}^0, \mathcal{F}^0)$ be a quasi-regular Dirichlet form on $L^2(E_0; m_0)$. Denote by $(\mathcal{E}^{0,\mathrm{ref}}, (\mathcal{F}^0)_a^{\mathrm{ref}})$ the active reflected Dirichlet form of $(\mathcal{E}^0, \mathcal{F}^0)$. We assume that

$$(\mathcal{E}^0, \mathcal{F}^0) \text{ is irreducible} \tag{7.2.30}$$

and

$$\mathcal{F}^0 \text{ is a proper subspace of } (\mathcal{F}^0)_a^{\mathrm{ref}}. \tag{7.2.31}$$

By virtue of Theorem 6.6.3, $(\mathcal{E}^{0,\mathrm{ref}}, (\mathcal{F}^0)^{\mathrm{ref}}_a)$ is a Silverstein extension of $(\mathcal{E}^0, \mathcal{F}^0)$ so that Theorem 6.6.5 applies to the pair $(\mathcal{E}^0, \mathcal{F}^0)$ and $(\tilde{\mathcal{E}}, \tilde{\mathcal{F}}) = (\mathcal{E}^{0,\mathrm{ref}}, (\mathcal{F}^0)^{\mathrm{ref}}_e)$ of Dirichlet forms on $L^2(E_0; m_0)$.

As a result, there exist a locally compact separable metric space $\widehat{E}$ and a measurable map $j : E_0 \to \widehat{E}$ satisfying the following:

(1) Let $\widehat{m} = m_0 \circ j^{-1}$ be the image measure of m. The map j^* sends $L^2(\widehat{E}; \widehat{m})$ onto $L^2(E_0; m_0)$. The image form $(\widehat{\mathcal{E}}, \widehat{\mathcal{F}}) = j(\mathcal{E}^{0,\mathrm{ref}}, (\mathcal{F}^0)^{\mathrm{ref}}_a)$ is a regular Dirichlet form on $L^2(\widehat{E}; \widehat{m})$.

(2) There exists an $\widehat{\mathcal{E}}$-quasi-open set $\widehat{E}_0 \subset \widehat{E}$ such that $\widehat{m}(\widehat{E} \setminus \widehat{E}_0) = 0$ and the quasi-regular Dirichlet form $(\mathcal{E}^0, \mathcal{F}^0)$ on $L^2(E_0; m_0)$ is quasi-homeomorphic to the quasi-regular Dirichlet form $(\widehat{\mathcal{E}}^0, \widehat{\mathcal{F}}^0)$ on $L^2(\widehat{E}_0, \widehat{m})$ under the map j. Here $(\widehat{\mathcal{E}}^0, \widehat{\mathcal{F}}^0)$ denotes the part of the regular Dirichlet form $(\widehat{\mathcal{E}}, \widehat{\mathcal{F}})$ on the quasi open set $\widehat{E}_0$.

We call $(\widehat{E}, \widehat{m}, \widehat{\mathcal{E}}, \widehat{\mathcal{F}}, \widehat{E}_0)$ satisfying **(1)** and **(2)** above a *regularization* of the active reflected Dirichlet form $(\mathcal{E}^{0,\mathrm{ref}}, (\mathcal{F}^0)^{\mathrm{ref}}_a)$.

THEOREM 7.2.9. *Let $(\widehat{E}, \widehat{m}, \widehat{\mathcal{E}}, \widehat{\mathcal{F}}, \widehat{E}_0)$ be a regularization of the active reflected Dirichlet form $(\mathcal{E}^{0,\mathrm{ref}}, (\mathcal{F}^0)^{\mathrm{ref}}_a)$ of the given quasi-regular Dirichlet form $(\mathcal{E}^0, \mathcal{F}^0)$ on $L^2(E_0; m_0)$ satisfying* (7.2.30) *and* (7.2.31). *Let $\widehat{X}$ be an $\widehat{m}$-symmetric Hunt process on $\widehat{E}$ associated with a regular Dirichlet form $(\widehat{\mathcal{E}}, \widehat{\mathcal{F}})$ on $L^2(\widehat{E}; \widehat{m})$ and $\widehat{X}^0$ be its part process on $\widehat{E}_0$.*

Then $\widehat{X}$ is a reflecting extension of $\widehat{X}^0$. For each $\alpha > 0$, $\widehat{\mathcal{F}}$ admits an $\widehat{\mathcal{E}}_\alpha$-orthogonal decomposition as

$$\widehat{\mathcal{F}} = \widehat{\mathcal{F}}^0 \oplus \{\widehat{\mathbf{H}}_\alpha \varphi : \varphi \in \widehat{\mathcal{G}}_1\} \tag{7.2.32}$$

and

$$\widehat{\mathcal{E}}_\alpha(u, u) = \widehat{\mathcal{E}}^0_\alpha(u_0, u_0) + \widehat{\mathbf{C}}(\varphi, \varphi) + \widehat{U}_\alpha(\varphi, \varphi) \tag{7.2.33}$$

for $u = u_0 + \widehat{\mathbf{H}}_\alpha \varphi$ with $u_0 \in \widehat{\mathcal{F}}^0$ and $\varphi \in \widehat{\mathcal{G}}_1$. Here $\widehat{\mathbf{C}}$ and $\widehat{\mathcal{G}}_1$ are defined by (7.2.1) *and* (7.2.11), *respectively, in terms of the Feller measure $\widehat{U}$, $\widehat{V}$, and $\widehat{U}_\alpha$ determined by the hitting distributions $\widehat{\mathbf{H}}$ and $\widehat{\mathbf{H}}_\alpha$ of $\widehat{X}$ from $\widehat{E}_0$ to $\widehat{E} \setminus \widehat{E}_0$.*

Proof. In view of Corollary 7.2.8, it suffices to show that $\widehat{X}$ and $\widehat{X}^0$ enjoy the properties **(RE.1)–(RE.4)**. **(RE.1)** and **(RE.3)** are fulfilled. The irreducibility of $(\widehat{\mathcal{E}}^0, \widehat{\mathcal{F}}^0)$ follows from (7.2.30) because of the quasi-homeomorphism.

It remains to prove the property **(RE.4)** that

$$(\widehat{\mathcal{E}}, \widehat{\mathcal{F}}) = (\widehat{\mathcal{E}}^{0,\mathrm{ref}}, (\widehat{\mathcal{F}}^0)^{\mathrm{ref}}_a), \tag{7.2.34}$$

where the right hand side denotes the active reflected Dirichlet form of $(\widehat{\mathcal{E}}^0, \widehat{\mathcal{F}}^0)$. Indeed, since (7.2.31) implies that $\widehat{\mathcal{F}}^0$ is a proper subspace of $\widehat{\mathcal{F}}$, it follows from (7.2.34) that $\widehat{\mathcal{F}}^0$ is a proper subspace of $(\widehat{\mathcal{F}}^0)^{\mathrm{ref}}_a$ yielding the property **(RE.2)**.

To show (7.2.34), observe that $\widehat{X}^0$ is a standard process on $\widehat{E}_0$ properly associated with the quasi-regular Dirichlet form $(\widehat{\mathcal{E}}^0, \widehat{\mathcal{F}}^0)$ by virtue of Theorem 3.3.8. Its pull back $X^0 = j^{-1}\widehat{X}^0$ by the quasi-homeomorphism j is a standard process on E_0 properly associated with the quasi-regular Dirichlet form $(\mathcal{E}^0, \mathcal{F}^0)$. It is readily seen that $\widehat{X}^0$ shares in common with X^0 the space of terminal random variables as well as its subspace $\mathbf{N}_1$ defined by (7.1.3).

We denote those quantities associated with $\widehat{X}^0$ with superscript $\widehat{\ }$. Then, for $\Phi \in \mathbf{N}_1$, $h(\widehat{x}) = \widehat{E}^0_{\widehat{x}}[\Phi]$, and $g(\widehat{x}) = \widehat{E}^0_{\widehat{x}}[\Phi^2] - h(\widehat{x})^2, \widehat{x} \in \widehat{E}_0$, we have

$$
\begin{aligned}
&\frac{1}{2}L^0(j^*g, 1) + \frac{1}{2}\int_E (j^*h)^2(x)\kappa_0(dx) + U_\alpha(\Phi, \Phi) \\
&= \frac{1}{2}\widehat{L}^0(g, 1) + \frac{1}{2}\int_{\widehat{E}_0} h(\widehat{x})^2\widehat{\kappa}_0(d\widehat{x}) + \widehat{U}_\alpha(\Phi, \Phi),
\end{aligned}
$$

which together with Theorem 7.1.4 and (7.1.1) implies

$$
j(\mathcal{E}^{0,\text{ref}}, (\mathcal{F}^0)^{\text{ref}}_a) = (\widehat{\mathcal{E}}^{0,\text{ref}}, (\widehat{\mathcal{F}}^0)^{\text{ref}}_a).
$$

As the left hand side equals $(\widehat{\mathcal{E}}, \widehat{\mathcal{F}})$, we arrive at (7.2.34). □

Example 7.2.10. Let D be a domain in $\mathbb{R}^n$ with $n \geq 1$. We consider the function spaces $\text{BL}(D), H^1(D)$, $H^1_e(D)$, and $H^1_0(D)$ as well as the Dirichlet integral $\mathbf{D}$ defined in Section 2.2.4. We assume that

(I) $(\frac{1}{2}\mathbf{D}, H^1(D))$ is a regular Dirichlet form on $L^2(\overline{D}; \mathbf{1}_D(x)dx)(= L^2(D))$.

(II) $H^1_0(D)$ is a proper subspace of $H^1(D)$.

Condition **(I)** amounts to assuming that $C_c(\overline{D}) = C_c(\mathbb{R}^n)\big|_{\overline{D}}$ is dense in $H^1(D)$ with metric induced by $\frac{1}{2}\mathbf{D}(u, u) + \|u\|^2_{L^2(D)}$. **(I)** is fulfilled, for instance, when D has a continuous boundary (see Section 2.2.4).

We let $E = \overline{D}$, $F = \partial D$, $E_0 = D$, $m(dx) = \mathbf{1}_D(x)dx$, and $(\mathcal{E}, \mathcal{F}) = (\frac{1}{2}\mathbf{D}, H^1(D))$. $(\mathcal{E}, \mathcal{F})$ is then an irreducible strongly local regular Dirichlet form on $L^2(E; m)$ and the associated diffusion process X^r is, by definition, the reflecting Brownian motion on $\overline{D}$. The process X^r is conservative. It satisfies (7.1.11). See Example 3.5.9 for some other properties of X^r. The part process X^D of X^r on D is the absorbing Brownian motion on D associated with the Dirichlet form $(\mathcal{E}^0, \mathcal{F}^0) = (\frac{1}{2}\mathbf{D}, H^1_0(D))$ on $L^2(D)$. As in the proof of Lemma 7.2.7, condition **(II)** implies that ∂D is non-$\mathcal{E}$-polar. Each function in the extended Sobolev space $H^1_e(D)$ will be represented by its $\mathcal{E}$-quasi-continuous version on $\overline{D}$.

In view of Example (**4**°) of Section 6.5, we have $(\mathcal{F}^0)^{\text{ref}} = \text{BL}(D)$ so that

$$
(\mathcal{F}^0)^{\text{ref}}_a = \text{BL}(D) \cap L^2(D) = H^1(D) = \mathcal{F}\big|_D,
$$

namely, condition (7.2.23) is fulfilled. By virtue of Theorem 7.2.5, we have

$$H^1(D)\big|_{\partial D} = \left\{ \varphi \in L^2(\partial D; \mu_0) : \frac{1}{2} \int_{\partial D \times \partial D} (\varphi(\xi) - \varphi(\eta))^2 U(d\xi, d\eta) \right.$$
$$\left. + \int_{\partial D} \varphi(\xi)^2 V(d\xi) < \infty \right\}. \tag{7.2.35}$$

X^r is a reflecting extension of X^0 in the sense of Definition 7.2.6.

If we assume an additional condition that

(III) X^r is recurrent,

then $H^1_e(D) = \mathrm{BL}(D)$ by Example (**5**°) of Section 6.5 so that

$$(\mathcal{F}^0)^{\mathrm{ref}} = \mathrm{BL}(D) = \mathcal{F}_e\big|_D,$$

namely, condition (7.2.5) is fulfilled. According to Theorem 7.2.1, we have

$$H^1_e(D)\big|_{\partial D} = \left\{ \varphi : \mathbf{H}\varphi^2(x) < \infty \text{ a.e. on } D \text{ and} \right. \tag{7.2.36}$$
$$\left. \int_{\partial D \times \partial D} (\varphi(\xi) - \varphi(\eta))^2 U(d\xi, d\eta) < \infty \right\},$$

provided that X^r is recurrent. In this case, since X is irreducible recurrent and ∂D is not $\mathcal{E}$-polar, the function $q(x) = \mathbf{P}_x(\sigma_{\partial D} = \infty)$ vanishes for q.e. $x \in D$ by Theorem 3.5.6 and therefore the supplementary Feller measure V vanishes.

Let us consider a special case that D is the upper half-space:

$$D = \{(x', x_n) : x' \in \mathbb{R}^{n-1},\ x_n > 0\}.$$

We saw in Example (**4**°) of Section 5.3 that X^r is obtained by reflecting the n-dimensional Brownian motion $X_t = (X'_t, X^{(n)}_t)$: $X^r_t = (X'_t, |X^{(n)}_t|)$. By Lemma 7.2.3, the intrinsic measure μ_0 on $\partial D = \{x = (x', 0) : x' \in \mathbb{R}^{n-1}\}$ is σ-finite. It is shift invariant in view of its definition (7.2.10). Hence μ_0 equals the Lebesgue measure σ on ∂D up to a multiplicative positive constant. On account of (7.2.19), the space $H^1(D)\big|_{\partial D}$ coincides with the Dirichlet space on $L^2(\partial D; \sigma)$ of the time-changed process of X^r by means of the PCAF with Revuz measure σ. The latter was identified in Example (**4**°) of Section 5.3, with the Dirichlet space (5.3.17) of the symmetric Cauchy process on $\mathbb{R}^{n-1}$. Thus we conclude that

$$H^1(D)\big|_{\partial D} =$$
$$\left\{ \varphi \in L^2(\partial D; \sigma) : \int_{\partial D \times \partial D} (\varphi(\xi) - \varphi(\eta))^2 |\xi - \eta|^{-n} d\xi d\eta < \infty \right\}. \tag{7.2.37}$$

Since $\mathbf{H}\varphi^2(x)$ appearing on the right hand side of (7.2.36) is an everywhere defined positive harmonic function on D, it is finite at some specific point $x_0 \in D$ if and only if it is so at every point $x \in D$. So the right hand side of (7.2.36) is reduced to

$$\left\{\varphi : \mathbf{H}\varphi^2(x_0) < \infty, \int_{\partial D\times\partial D} (\varphi(\xi) - \varphi(\eta))^2|\xi - \eta|^{-n} d\xi d\eta < \infty\right\}. \quad (7.2.38)$$

When $n = 2$, X^r is recurrent so that the space $H^1_e(D)\big|_{\partial D}$ is identical with the above space. When $n \geq 3$, X^r is transient and $H^1_e(D)\big|_{\partial D}$ is a proper subspace of (7.2.38) because the former is a subspace of the latter containing no non-zero constants.

Suppose now that D is bounded and has a Lipschitz continuous boundary. Then all conditions **(I), (II), (III)** are fulfilled so that both (7.2.35) and (7.2.36) hold true (with $V = 0$). Since the reflecting Brownian motion X then has a Hölder continuous transition density function with respect to $\mathbf{1}_D(x)dx$ by [5], X can be refined to start from every point in $\overline{D}$.

Fix some $x_0 \in D$. For $x \in D$, let $K(x, \xi)$ be the density of the harmonic measure $\mathbf{P}_x(X_{\sigma_{\partial D}} \in d\xi)$ with respect to the base harmonic measure $\nu(d\xi) := \mathbf{P}_{x_0}(X_{\sigma_{\partial D}} \in d\xi)$. The function $K(x, \xi)$ is called the classical Poisson kernel for $\frac{1}{2}\Delta$ in D, which is continuous on $D \times \partial D$ and is harmonic in $x \in D$ for each fixed $\xi \in \partial D$. So for any bounded function φ on ∂D,

$$\mathbf{H}\varphi(x) = \mathbf{E}_x[\varphi(X_{\sigma_{\partial D}})] = \int_{\partial D} K(x, \xi)\varphi(\xi)\nu(d\xi).$$

By the Harnack inequality for positive harmonic functions in D, the condition that $\mathbf{H}\varphi^2(x) < \infty$ for some (and hence for all) $x \in D$ is equivalent to $\varphi \in L^2(\partial D; \nu)$. So the condition that $\mathbf{H}\varphi^2(x) < \infty$ a.e. on D on the right hand side of (7.2.36) can be replaced by the condition $\varphi \in L^2(\partial D; \nu)$.

Denoting $K(\cdot, \xi)$ by K^ξ, we define for $\xi, \eta \in \partial D$,

$$K^\xi_\alpha := K^\xi - \alpha R^0_\alpha K^\xi, \qquad U_\alpha(\xi, \eta) := \alpha(K^\xi_\alpha, K^\eta)$$

and

$$U(\xi, \eta) :=\uparrow \lim_{\alpha\uparrow\infty} U_\alpha(\xi, \eta).$$

As we have from (5.4.6) and (5.5.8) the identity

$$L^0(\mathbf{H}\varphi, \mathbf{H}\psi) = \int_{\partial D\times\partial D} U(\xi, \eta)\varphi(\xi)\psi(\eta)\nu(d\xi)\nu(d\eta),$$

$U(\xi, \eta)\nu(d\xi)\nu(d\eta)$ gives the Feller measure $U(d\xi, d\eta)$.

The intrinsic measure $\mu_0(\in \overset{\circ}{S}_{\partial D})$ appearing in (7.2.35) has a density function

$$U_1 1(\xi) = (K_1^{\xi}, 1) = (K^{\xi}, \mathbf{H}_1 1)$$

with respect to ν. Since $U_1 1(\xi)$ is lower semicontinuous and strictly positive on ∂D, we have the inclusion $L^2(\partial D; \mu_0) \subset L^2(\partial D; \nu)$.

We notice that the boundary ∂D satisfies condition (5.2.16). This is because, by a result of Dahlberg [35], the harmonic measure from each point of D is mutually absolutely continuous with respect to the surface measure σ on ∂D. Clearly ∂D is the topological support of σ. On the other hand, the surface measure σ is mutually absolutely continuous with respect to the measure μ constructed in the proof of Lemma 5.2.9(i) with $g \equiv 1$ and therefore $\sigma \in \overset{\circ}{S}_{\partial D}$. In particular, ∂D is a quasi support of σ, which proves that the condition (5.2.16) holds. Consequently, any measure $\mu \in \overset{\circ}{S}_{\partial D}$ admits ∂D as its topological support and the associated time-changed Dirichlet form $(\check{\mathcal{E}}, H_e^1(D)\big|_{\partial D} \cap L^2(\partial D; \mu)$ on $L^2(\partial D; \mu)$ is regular by Corollary 5.2.10.

Dahlberg [35] proved that there is an $\varepsilon > 0$ such that $f(\xi) := \frac{\nu(d\xi)}{\sigma(d\xi)}$ is locally in $L^{2+\varepsilon}(\partial D, \sigma)$ and showed that this result cannot be improved in general. However, when D is a bounded $C^{1,1}$ domain in $\mathbb{R}^n$, by using the two-sided Green function estimates in D it can be shown that f is bounded between two positive constants and furthermore the function $U_1 1(\xi)$ defined above is upper bounded on ∂D. Hence when D is a bounded $C^{1,1}$ domain in $\mathbb{R}^n$, it holds that

$$\begin{aligned} H^1(D)\big|_{\partial D} &= H_e^1(D)\big|_{\partial D} \\ &= \left\{ \varphi \in L^2(\partial D; \sigma) : \int_{\partial D \times \partial D} (\varphi(\xi) - \varphi(\eta))^2 U(d\xi, d\eta) < \infty \right\}. \end{aligned}$$

We saw in (**1°**) of Section 5.8 that the Feller kernel with respect to $\sigma \times \sigma$ admits a more explicit expression (5.8.2) when ∂D is smoother. This together with the above identification of the trace space recovers the corresponding statements in Example (**3°**) of Section 5.3 in the case of the planar unit disk. □

Example 7.2.11. Let $n \geq 1$, $0 < \alpha < 2$, $D \subset \mathbb{R}^n$ be an open set and m be the Lebesgue measure on $\mathbb{R}^n$. In Example (**6°**) of Section 6.5, we considered the following function space $(\mathbf{B}, \mathcal{W})$:

$$\begin{cases} \mathbf{B}(u, v) := c \displaystyle\int_{D \times D} \frac{(u(x) - u(y))(v(x) - v(y))}{|x - y|^{n+\alpha}} \, dxdy, \\ \mathcal{W}(D) = \{u : |u| < \infty \text{ a.e. and } \mathbf{B}(u, u) < \infty\}. \end{cases}$$

$(\mathcal{E}, \mathcal{F}) = (\mathbf{B}, \mathcal{W}(D) \cap L^2(D; m))$ is then a Dirichlet form on $L^2(D; m)$. The closure of $C_c^\infty(D)$ in this Dirichlet space is denoted by $\mathcal{F}^0$ and we let

$\mathcal{E}^0 = \mathcal{E}\big|_{\mathcal{F}^0 \times \mathcal{F}^0}$. $(\mathcal{E}^0, \mathcal{F}^0)$ is then a regular irreducible Dirichlet form on $L^2(E; m)$ and the associated Hunt process X^0 on D is, by definition, the censored α-stable process.

Let us assume that

(I) $(\mathcal{E}, \mathcal{F})$ is a regular Dirichlet form on $L^2(\overline{D}; m)$ and $m(\partial D) = 0$.

(II) $\mathcal{F}^0$ is a proper subspace of $\mathcal{F}$.

Under the condition **(I)**, the m-symmetric Hunt process X on $\overline{D}$ associated with $(\mathcal{E}, \mathcal{F})$ is, by definition, the reflecting α-stable process. It has no killing: $\kappa = 0$. The censored α-stable process is the part process of X on D. Condition **(I)** is satisfied if D is an n-set as is defined in Example (**6**°) of Section 6.5. If further $1 < \alpha < 2$ and the boundary ∂D has locally finite $(n-1)$-dimensional Hausdorff measure, then condition **(II)** is also fulfilled (cf. [14]).

In view of (6.5.15), we have

$$((\mathcal{F}^0)^{\mathrm{ref}}_a, \mathcal{E}^{0,\mathrm{ref}}) = (\mathcal{W}(D) \cap L^2(D; m), \mathbf{B}) = (\mathcal{F}, \mathcal{E}).$$

In particular, condition (7.2.23) is fulfilled and we see by Theorem 7.2.5 that

$$\mathcal{F}\big|_{\partial D} = \left\{ \varphi \in L^2(\partial D; \mu_0) : \frac{1}{2} \int_{\partial D \times \partial D} (\varphi(\xi) - \varphi(\eta))^2 U(d\xi, d\eta) \right.$$
$$\left. + \int_{\partial D} \varphi(\xi)^2 V(d\xi) < \infty \right\}. \tag{7.2.39}$$

Furthermore, X is a reflecting extension of the censored stable process X^0 in the sense of Definition 7.2.6.

If we assume an additional condition that **(III)** X is recurrent,

then the identity (6.5.17) in Example (**6**°) of Section 6.5 is fulfilled, which combined with (6.5.15) leads us to the property (7.2.5) for $(\mathcal{E}, \mathcal{F})$. Therefore, by virtue of Theorem 7.2.1, we have the identification

$$\mathcal{F}_e\big|_{\partial D} = \left\{ \varphi : \mathbf{H}\varphi^2 < \infty \text{ a.e. on } D \text{ and} \right.$$
$$\left. \int_{\partial D \times \partial D} (\varphi(\xi) - \varphi(\eta))^2 U(d\xi, d\eta) < \infty \right\}, \tag{7.2.40}$$

provided that X is recurrent. **(III)** is satisfied if $m(D) < \infty$ for instance. Note that since X is irreducible recurrent and ∂D is non-$\mathcal{E}$-polar, $q(x) = \mathbf{P}_x(\sigma_{\partial D} = \infty) = 0$ and so the supplementary Feller measure V vanishes.

It is shown in [29] that X has Hölder continuous transition density functions with respect to the Lebesgue measure dx on $\overline{D}$ and therefore X can be refined to start from every point in $\overline{D}$. Let us assume additionally that D is a bounded $C^{1,1}$-domain. By [108, Theorem 3.14], the surface measure σ

on ∂D is mutually absolutely continuous with respect to the X^0-harmonic measure $\mathbf{P}_x(X_{\sigma_{\partial D}} \in d\xi)$ for every $x \in D$. Moreover, for each $x \in D$, the Radon-Nikodym derivative of $\mathbf{P}_x(X_{\sigma_{\partial D}} \in d\xi)$ with respect to $\sigma(d\xi)$ is bounded between two positive constants. Consequently, just as in Example 7.2.10, $\sigma \in \overset{\circ}{S}_{\partial D}$. In particular, ∂D is a quasi support of σ. Since clearly ∂D is the topological support of the surface measure σ, condition (5.2.16) holds. Therefore, the trace Dirichlet form is regular on $L^2(\partial D; \mu)$ for any choice of $\mu \in \overset{\circ}{S}_{\partial D}$. Furthermore, the condition that $\mathbf{H}\varphi^2(x) < \infty$ for some (and hence for all) $x \in D$ is equivalent to $\varphi \in L^2(\partial D; \sigma)$. So in this case, we can replace the condition that $\mathbf{H}\varphi^2(x) < \infty$ a.e. on D in (7.2.40) by the condition $\varphi \in L^2(\partial D; \sigma)$. By an argument similar to that of Example 7.2.10, it can be shown further that there are density functions $U(\xi, \eta)$, $U_1 1(\xi)$ such that $U(d\xi, d\eta) = U(\xi, \eta)\sigma(d\xi)\sigma(d\eta)$, $\mu_0(d\xi) = U_1 1(\xi)\sigma(d\xi)$. □

7.3. LATERAL CONDITION FOR L^2-GENERATORS

In this section, we employ the same setting as in the first half of Section 7.2. Thus X is an m-symmetric Hunt process whose Dirichlet form $(\mathcal{E}, \mathcal{F})$ on $L^2(E; m)$ is regular and irreducible, F is a non-$\mathcal{E}$-polar quasi closed subset of E, and X is assumed to be of no jump from $E_0 = E \setminus F$ to F in the sense of (7.1.11). We aim at characterizing the infinitesimal generator $\mathcal{A}$ of the L^2-semigroup of X directly in terms of quantities based on the part process X^0 of X on E_0 and some others.

Every function in the extended Dirichlet space $\mathcal{F}_e$ is taken to be $\mathcal{E}$-quasi-continuous. Accordingly, every function in $\mathcal{F}^0$ is $\mathcal{E}^0$-quasi-continuous by Theorem 3.3.8(iv) and q.e. finely continuous with respect to X^0 by Theorem 3.1.7.

In view of Theorem 7.1.4, every function $f \in (\mathcal{F}^0)^{\rm ref}_a$ admits $\Phi \in \mathbf{N}_1$ and $f_0 \in \mathcal{F}^0$ with $f = \mathbf{H}^0_\alpha \Phi + f_0$ m_0-a.e. From (7.1.5) and the proof of Theorem 6.1.6, $\mathbf{H}^0_\alpha \Phi$ is a difference of X^0-q.e. excessive functions and hence X^0-q.e. finely continuous. In what follows, any function in $(\mathcal{F}^0)^{\rm ref}_a$ will be represented by an X^0-q.e. finely continuous version. Such a version always exists. Notice that if two X^0-q.e. finely continuous functions coincide m_0-a.e. on E_0, then they are equal q.e. on E_0 by Theorem A.2.13.

An X^0-q.e. finely continuous function f on E_0 is said to have an *X^0-fine limit function* on F if there exists a Borel measurable function ψ on F such that

$$\mathbf{P}^0_x\left(\lim_{t\uparrow\zeta^0} f(X^0_t)\right) = \psi(X^0_{\zeta^0-}) \mid \zeta^0 < \infty \text{ and } X^0_{\zeta^0-} \in F) = 1$$

for q.e. $x \in E_0$. In this case, we write ψ as γf and call γf an X^0-fine limit function of f on F.

We denote by $\mathcal{QC}$ the family of all $\mathcal{E}$-quasi-continuous functions on E. For $f \in \mathcal{QC}, f\big|_{E_0}$ is X^0-q.e. finely continuous.

LEMMA 7.3.1. (i) *For $f \in \mathcal{QC}, f\big|_{E_0}$ admits an X^0-fine limit function $f|_F$ on F.* (ii) *If $f \in \mathcal{F}_e$, then $f|_{E_0}$ admits an X^0-fine limit function $f|_F$ on F. If $f \in \mathcal{F}_e^0$, then $f\big|_{E_0}$ admits zero X^0-fine limit function on F.*

Proof. (i) By Theorem 3.1.7,

$$\mathbf{P}_x\Big(\lim_{t' \uparrow t} f(X_{t'}) = f(X_{t-}) \text{ for every } t \in [0, \zeta) \Big) = 1 \quad \text{for q.e. } x \in E.$$

Because of the inclusion $\{\tau_0 < \infty, \ \sigma_F = \infty\} \subset \{\tau_0 < \infty, \ X_{\tau_0-} \in E_0 \cup \{\partial\}\}$, it holds that

$$\{\tau_0 < \infty, \ X_{\tau_0-} \in F\} \subset \{\tau_0 = \sigma_F < \zeta\}. \tag{7.3.1}$$

Hence $\lim_{t' \uparrow \tau_0} f(X_{t'}) = f(X_{\tau_0-})$ $\mathbf{P}_x$-a.s. on $\{\tau_0 < \infty, \ X_{\tau_0-} \in F\}$ for q.e. $x \in E_0$. Replacing $\mathbf{P}_x$, τ_0, X_t by $\mathbf{P}_x^0$, ζ^0, X_t^0, respectively, we arrive at (i).

(ii) The first assertion is immediate from (i). The second follows from the fact that any function in $\mathcal{F}_e^0$ can be regarded as a quasi continuous function in $\mathcal{F}_e$ vanishing q.e. on F. □

Recall the function space $\mathcal{N}_1$ defined by (7.1.24). By (7.1.27), we have for each $\alpha > 0$,

$$\mathcal{F}\big|_F \subset \mathcal{N}_1 \quad \text{and} \quad \mathbf{H}_\alpha \mathcal{N}_1\big|_{E_0} \subset \mathbf{H}_\alpha^0 \mathbf{N}_1 \subset (\mathcal{F}^0)_a^{\text{ref}}. \tag{7.3.2}$$

LEMMA 7.3.2. (i) *For $\Phi \in \mathbf{N}$, let $h(x) = \mathbf{E}_x^0[\Phi]$, $x \in E$. Then*

$$\lim_{t \uparrow \zeta^0} h(X_t^0) = \Phi \quad \mathbf{P}_x^0\text{-a.s. on } \Big\{\zeta^0 < \infty, \ X_{\zeta^0-}^0 \in F\Big\} \tag{7.3.3}$$

for q.e. $x \in E_0$.
(ii) *For $\psi \in \mathcal{N}_1$, the function $\mathbf{H}_\alpha \psi$ is X^0-q.e. finely continuous and has the X^0-fine limit function ψ on F for $\alpha > 0$.*

Proof. (i) By virtue of Lemma 6.1.7, $\{M_t^h\}_{t \geq 0}$ defined by (6.1.7) for X^0 is a $\mathbf{P}_x^0$-square integrable martingale and hence $h(X_t)$ admits a left limit at ζ^0 $\mathbf{P}_x^0$-a.s. on $\{\zeta^0 < \infty\}$ for q.e. $x \in E_0$. Hence (7.3.3) follows from Lemma 6.4.11 and (7.3.1).
(ii) By (7.1.5) and Lemma 7.1.5,

$$\mathbf{H}_\alpha \psi - \mathbf{H} \psi = -\alpha G_{0+}^0 \mathbf{H}_\alpha \psi \in \mathcal{F}^0.$$

Hence we get from (7.3.3), Lemma 7.1.5, and Lemma 7.3.1(ii) that $\mathbf{H}_\alpha \psi$ is X^0-q.e. finely continuous and

$$\lim_{t \uparrow \zeta^0} \mathbf{H}_\alpha \psi(X_t^0) = \psi(X_{\zeta^0-}^0), \qquad \mathbf{P}_x^0\text{-a.s. on } \{\zeta^0 < \infty, \ X_{\zeta^0-}^0 \in F\}$$

for q.e. $x \in E_0$, namely, $\mathbf{H}_\alpha \psi$ admits ψ as an X^0-fine limit function on F. □

We now consider the following condition:

$$\text{If } f \in (\mathcal{F}^0)_a^{\text{ref}} \text{ admits an } X^0\text{-fine limit function } 0 \text{ on } F, \text{ then } f \in \mathcal{F}^0. \tag{7.3.4}$$

Remark 7.3.3. (i) Let X^0 be the absorbing Brownian motion on the interval $E_0 = (0, 1)$. Then $\mathcal{F}^0 = H_0^1(0, 1)$ and $(\mathcal{F}^0)_a^{\text{ref}} = H^1(0, 1)$. Condition (7.3.4) is satisfied if $F = \{0, 1\}$ (as in the case that X is the reflecting Brownian motion on $E = [0, 1]$) but it is not satisfied when $F = \{0\}$ (as in the case that X is the Brownian motion on $E = [0, 1)$ reflected at 0 and absorbed at 1). If E is the one-point compactification of $E_0 = (0, 1)$, then (7.3.4) is fulfilled by the one-point set $F = E \setminus E_0$.
(ii) Assume that

$$\mathbf{P}_x^0 \left(X_{\zeta^0-} \in F \mid \zeta^0 < \infty \right) = 1 \qquad \text{for q.e. } x \in E_0, \tag{7.3.5}$$

then (7.3.4) is fulfilled. To verify this, suppose $f \in (\mathcal{F}^0)_a^{\text{ref}}$ and $\gamma f = 0$. By Theorem 7.1.4 and (7.1.5), f can be decomposed as $f(x) = \mathbf{E}_x^0[\Phi] + f_0(x)$ for q.e. $x \in E_0$ with $\Phi \in \mathbf{N}_1$ and $f_0 \in \mathcal{F}^0$. By Lemma 7.3.1 (ii), $\gamma f_0 = 0$ and so $\Phi \cdot \mathbf{1}_{\{X^0_{\zeta^0-} \in F,\, \zeta^0 < \infty\}} = 0$ $\mathbf{P}_x$-a.s. for q.e. $x \in E_0$ by (7.3.3). Equations (7.1.4) and (7.3.5) then yield that $\Phi = 0$ and consequently $f = f_0$ q.e.
(iii) Condition (7.3.4) is also fulfilled if

$$\mathcal{F} = (\mathcal{F}^0)_a^{\text{ref}}, \tag{7.3.6}$$

because (7.3.4) is clearly satisfied by $f \in \mathcal{F}$ on account of Lemma 7.3.1 and (7.1.26). □

Let us introduce a linear operator $\mathcal{L}$ on $L^2(E_0; m_0)$ specified:

$$f \in \mathcal{D}(\mathcal{L}) \quad \text{with} \quad \mathcal{L}f = g \ (\in L^2(E_0; m_0))$$

if and only if

$$f \in (\mathcal{F}^0)_a^{\text{ref}} \quad \text{with} \quad \mathcal{E}^{0,\text{ref}}(f, v) = -(g, v) \quad \text{for every } v \in \mathcal{F}^0. \tag{7.3.7}$$

Any function $f \in \mathcal{D}(\mathcal{L})(\subset (\mathcal{F}^0)_a^{\text{ref}})$ will be represented by its X^0-q.e. finely continuous version.

LEMMA 7.3.4. *Assume that condition (7.3.4) holds. Suppose $u \in \mathcal{D}(\mathcal{L})$ having an X-fine limit function $\gamma u \in \mathcal{N}_1$ on F and*

$$\mathcal{L}u = \alpha u \qquad \textit{for some } \alpha > 0.$$

Then $u = \mathbf{H}_\alpha(\gamma u)$ on E_0.

Proof. By (7.3.7), $u \in (\mathcal{F}^0)^{\rm ref}_a$ and $\mathcal{E}^{0,{\rm ref}}_\alpha(u,w) = 0$ for any $w \in \mathcal{F}^0$. Define $\psi := \gamma u$ and $u_0 := u - \mathbf{H}_\alpha\psi$. Then $u_0 \in (\mathcal{F}^0)^{\rm ref}_a$ by (7.3.2) and $\gamma(u_0) = \psi - \psi = 0$ by Lemma 7.3.2. Hence by assumption (7.3.4), $u_0 \in \mathcal{F}^0$.

Since by (7.3.2) $\mathbf{H}_\alpha\psi \in \mathbf{H}^0_\alpha\mathbf{N}_1$, we have by Theorem 7.1.4 that

$$\mathcal{E}^{0,{\rm ref}}_\alpha(\mathbf{H}_\alpha\psi, w) = 0 \qquad \text{for every } w \in \mathcal{F}^0.$$

It follows then that

$$\mathcal{E}^{\rm ref}_\alpha(u_0, w) = \mathcal{E}^{0,{\rm ref}}_\alpha(u - \mathbf{H}_\alpha\psi, w) = 0 \qquad \text{for every } w \in \mathcal{F}^0.$$

Taking $w = u_0$ we get $u_0 = 0$ and therefore $u = \mathbf{H}_\alpha\psi = \mathbf{H}_\alpha(\gamma u)$. □

For $f \in \mathcal{D}(\mathcal{L})$ and $\psi \in \mathcal{N}_1$, define

$$\mathcal{N}(f)(\psi) := \mathcal{E}^{0,{\rm ref}}(f, \mathbf{H}_\alpha\psi) + (\mathcal{L}f, \mathbf{H}_\alpha\psi), \quad \alpha > 0. \tag{7.3.8}$$

Note that for α and $\beta > 0$, $\mathbf{H}_\alpha\psi - \mathbf{H}_\beta\psi \in \mathcal{F}^0$ by Lemma 7.1.2 and Lemma 7.1.5. Hence $\mathcal{N}(f)(\psi)$ defined by (7.3.8) is independent of the choice of $\alpha > 0$ in view of (7.3.7). We call $\mathcal{N}(f)$ the *flux functional* of f being regarded as a linear functional on the space $\mathcal{N}_1$.

In the remainder of this section, we assume that

$$m(F) = 0. \tag{7.3.9}$$

Denote by $\mathcal{A}$ the L^2-infinitesimal generator of X. That is, $\mathcal{A}$ is the self-adjoint operator on $L^2(E;m)$ $(= L^2(E_0;m_0))$ such that

$$f \in \mathcal{D}(\mathcal{A}) \text{ with } \mathcal{A}f = g \text{ if and only if } f \in \mathcal{F} \text{ with}$$
$$\mathcal{E}(f,v) = -(g,v) \text{ for every } v \in \mathcal{F}. \tag{7.3.10}$$

Recall the operator $\mathcal{L}$ is defined by (7.3.7). We see from Theorem 7.1.6 that $\mathcal{L}$ is an extension of $\mathcal{A}$ in the sense that

$$\mathcal{D}(\mathcal{A}) \subset \mathcal{D}(\mathcal{L}) \quad \text{and} \quad \mathcal{A}f = \mathcal{L}f \ \text{ for } f \in \mathcal{D}(\mathcal{A}). \tag{7.3.11}$$

We are in a position to formulate a *lateral* (*boundary*) *condition* that gives a characterization of a function in $\mathcal{D}(\mathcal{L})$ to be in $\mathcal{D}(\mathcal{A})$. To this end, we consider the intrinsic measure $\mu_0 (\in \overset{\circ}{S}_F)$ on F defined by (7.2.10): $\int_F \psi\, d\mu_0 = U_1(1,\psi)$, $\psi \in \mathcal{B}_+(F)$. We note that if $B \in \mathcal{B}(F)$ is μ_0-negligible, then

$$\mathbf{P}^0_x\left(X^0_{\zeta^0-} \in B \mid \zeta^0 < \infty,\ X^0_{\zeta^0-} \in F\right) = 0 \quad \text{for q.e. } x \in E_0, \tag{7.3.12}$$

because $U_1(1, \mathbf{1}_B) = 0$ implies $\mathbf{P}_x(\sigma_B < \infty, X_{\sigma_B} \in B) = \mathbf{H1}_B(x) = 0$ for m-a.e. $x \in E_0$ and hence for q.e. $x \in E_0$, due to the X^0-fine continuity of $\mathbf{H1}_B$. We then get (7.3.12) from Lemma 7.1.5.

Let $(\check{\mathcal{E}}, \check{\mathcal{F}})$ be the Dirichlet form on $L^2(F; \mu_0)$ associated with the time-changed process $\check{X}$ of X by means of its PCAF with Revuz measure μ_0. By virtue of Theorem 7.2.4, we have

$$\check{\mathcal{F}} = \mathcal{F}\big|_F, \tag{7.3.13}$$

which means that $\check{\mathcal{F}}$ consists of those μ_0-equivalence classes of μ_0-measurable functions on F admitting the restrictions to F of functions in $\mathcal{F}$ as their representatives.

Suppose a function $f \in (\mathcal{F}^0)^{\mathrm{ref}}_a$ admits a μ_0-measurable function $\varphi \in \check{\mathcal{F}}$ as its X^0-fine limit function on F. Then, on account of (7.3.12), f also has as its X^0-fine limit function on F any function which is μ_0-equivalent to φ, so that φ can be taken from $\mathcal{F}\big|_F$.

THEOREM 7.3.5. (i) *Suppose $f \in \mathcal{D}(\mathcal{A})$. Then $f \in \mathcal{D}(\mathcal{L})$ and f satisfies the lateral conditions that*

$$f \text{ admits an } X^0\text{-fine limit function } \gamma f \in \check{\mathcal{F}}, \tag{7.3.14}$$

and for every $\psi \in \check{\mathcal{F}}$,

$$\begin{aligned} &\mathcal{N}(f)(\psi) + \frac{1}{2}\mu^c_{\langle \mathbf{H}(\gamma f), \mathbf{H}\psi\rangle}(F) \\ &\quad + \frac{1}{2}\int_{F\times F} ((\gamma f)(\xi) - (\gamma f)(\eta))(\psi(\xi) - \psi(\eta))J(d\xi, d\eta) \\ &\quad + \int_F (\gamma f)(\xi)\psi(\xi)\kappa(d\xi) = 0. \end{aligned} \tag{7.3.15}$$

(ii) *Assume that the condition* (7.3.4) *holds. If $f \in \mathcal{D}(\mathcal{L})$ satisfies the lateral conditions* (7.3.14) *and* (7.3.15) *holding for every $\psi \in \check{\mathcal{D}}$, then $f \in \mathcal{D}(\mathcal{A})$. Here $\check{\mathcal{D}}$ is any fixed $\check{\mathcal{E}}_1$-dense subspace of $\check{\mathcal{F}}$.*

Proof. (i) Suppose $f \in \mathcal{D}(\mathcal{A})$. Then for $\alpha > 0$, $f = G_\alpha g$ with $g = (\alpha - \mathcal{A})f \in L^2(E; m_0)$. Here G_α denotes the α-resolvent of X. Since $f \in \mathcal{F}$, f admits the X^0-fine limit function $\gamma f = f|_F \in \check{\mathcal{F}}$ by Lemma 7.3.1(ii) and (7.3.13). For $\psi \in \check{\mathcal{F}} (= \mathcal{F}\big|_F)$, $\mathbf{H}_\alpha\psi \in \mathcal{F}$. Since $\mathcal{L}$ is an extension of $\mathcal{A}$, we have from (7.3.10)

$$\mathcal{E}(f, \mathbf{H}_\alpha\psi) + (\mathcal{L}f, \mathbf{H}_\alpha\psi) = 0,$$

whose left hand side coincides with the left hand side of (7.3.15) in view of (7.3.8), (7.1.16), and (7.1.17).

(ii) Suppose that $f \in \mathcal{D}(\mathcal{L})$ satisfies (7.3.14) and (7.3.15) for every $\psi \in \check{\mathcal{D}}$. Let $g = (\alpha - \mathcal{L})f$ and $w = f - G_\alpha g\big|_{E_0}$. Then by the preceding remark, (7.3.2), and Lemma 7.3.1(ii), w satisfies $\gamma w \in \mathcal{F}\big|_F \subset \mathcal{N}_1$ and moreover $(\alpha - \mathcal{L})w = 0$ on account of (7.3.11). Consequently, $w = \mathbf{H}_\alpha(\gamma w) \in \mathcal{F}$ by virtue of Lemma 7.3.4.

As $w \in \mathcal{D}(\mathcal{L})$ and $(\alpha - \mathcal{L})w = 0$,

$$\mathcal{N}(w)(\psi) = \mathcal{E}^{0,\mathrm{ref}}(w, \mathbf{H}_\alpha \psi) + (\mathcal{L}w, \mathbf{H}_\alpha \psi) = \mathcal{E}_\alpha^{0,\mathrm{ref}}(w, \mathbf{H}_\alpha \psi)$$

for every $\psi \in \mathcal{N}_1$. On the other hand, we see by (i) that $G_\alpha g \in \mathcal{D}(\mathcal{A})$ satisfies equation (7.3.15) and so does w. It follows then for every $\varphi \in \check{\mathcal{D}}$,

$$\begin{aligned}
&\mathcal{E}_\alpha^{0,\mathrm{ref}}(\mathbf{H}_\alpha(\gamma w), \mathbf{H}_\alpha \varphi) + \frac{1}{2}\mu^c_{\langle \mathbf{H}(\gamma w), \mathbf{H}\varphi \rangle}(F) \\
&+ \frac{1}{2}\int_{F\times F} (\gamma w(\xi) - \gamma w(\eta))(\varphi(\xi) - \varphi(\eta)) J(d\xi, d\eta) \\
&+ \int_F \gamma w(\xi)\varphi(\xi)\kappa(d\xi) = 0.
\end{aligned}$$

Since $w \in \mathcal{F}$, we see by (7.1.28) that the above identity is equivalent to $\mathcal{E}_\alpha(\mathbf{H}_\alpha(\gamma w), \mathbf{H}_\alpha \varphi) = 0$ for every $\varphi \in \check{\mathcal{D}}$, which extends to every $\varphi \in \check{\mathcal{F}}$ because $\check{\mathcal{D}}$ is $\check{\mathcal{E}}_1$-dense in $\check{\mathcal{F}}$ and Theorem 7.2.4 applies. Taking $\varphi = \gamma w$, we obtain $w = \mathbf{H}_\alpha(\gamma w) = 0$ and consequently, $f = G_\alpha g \in \mathcal{D}(\mathcal{A})$. □

Remark 7.3.6. (i) Let $\mathcal{H}$ be any $\mathcal{E}_1$-dense subspace of $\mathcal{F}$. Then a set $\check{\mathcal{D}}$ satisfying $\mathcal{H}|_F \subset \check{\mathcal{D}} \subset \check{\mathcal{F}}$ is $\check{\mathcal{E}}_1$-dense in $\check{\mathcal{F}}$ in view of Theorem 7.2.4.
(ii) The function space $\check{\mathcal{F}}$ is involved in the lateral condition (7.3.14). In order to make Theorem 7.3.5 meaningful as a characterization of the domain $\mathcal{D}(\mathcal{A})$ of the L^2-generator of X, we need to have some explicit description of $\check{\mathcal{F}}$. □

In regard to the second remark above, we recall some properties of the Dirichlet form $(\check{\mathcal{E}}, \check{\mathcal{F}})$ on $L^2(F; \mu_0)$ studied in Sections 5.2 and 5.6. By virtue of Theorem 5.2.7, $(\check{\mathcal{E}}, \check{\mathcal{F}})$ is a quasi-regular Dirichlet form on $L^2(F; \mu_0)$ and, in view of (5.6.7), it admits the Beurling-Deny decomposition

$$\begin{aligned}
\check{\mathcal{E}}(\varphi, \varphi) =& \frac{1}{2}\mu^c_{\langle \mathbf{H}\varphi \rangle}(F) + \frac{1}{2}\int_{F\times F} (\varphi(\xi) - \varphi(\eta))^2 (U + J)(d\xi, d\eta) \\
&+ \int_F \varphi(\xi)^2 (V + \kappa)(d\xi), \qquad \varphi \in \check{\mathcal{F}}.
\end{aligned} \tag{7.3.16}$$

Denote by $(\check{\mathcal{F}}_a^{\mathrm{ref}}, \check{\mathcal{E}}^{\mathrm{ref}})$ the active reflected Dirichlet space of $(\check{\mathcal{E}}, \check{\mathcal{F}})$. Explicit descriptions of reflected Dirichlet spaces can be given for various concrete examples as in Section 6.5 and also in the case where the underlying space is discretely countable as will be considered in the next section.

Lemma 7.3.7. (i) *If we let $\check{\mathcal{E}}_1^{\mathrm{ref}}(\varphi, \varphi) = \check{\mathcal{E}}^{\mathrm{ref}}(\varphi, \varphi) + U_1(1, \varphi^2)$, then, for any $\check{\mathcal{E}}_1$-dense subspace $\check{\mathcal{D}}$ of $\check{\mathcal{F}}$,*

$$\check{\mathcal{F}} = \textit{ the } \check{\mathcal{E}}_1^{\mathrm{ref}}\textit{-closure of } \check{\mathcal{D}} \textit{ in } \check{\mathcal{F}}_a^{\mathrm{ref}}. \tag{7.3.17}$$

(ii) *Assume that X is recurrent. Then* $\check{\mathcal{F}} = \check{\mathcal{F}}_a^{\rm ref}$ *and* $V = 0$, $\kappa = 0$.

Proof. (i) is obvious from Theorem 7.2.4.
(ii) If X is recurrent, then the time-changed process $\check{X}$ is also recurrent by Theorem 5.2.5 and hence $(\check{\mathcal{F}}, \check{\mathcal{E}}) = (\check{\mathcal{F}}_a^{\rm ref}, \check{\mathcal{E}}^{\rm ref})$ on account of Theorem 6.3.2, and further $V + \kappa = 0$ because otherwise $\check{\mathcal{E}}(1, 1)$ cannot be zero. □

For later use, we deduce here a direct characterization of the resolvent $\{G_\alpha; \alpha > 0\}$ of X.

THEOREM 7.3.8. *For* $\alpha > 0$ *and* $g \in L^2(E; m)$, *let* $f = G_\alpha g$. *Then* $\varphi = f\big|_F$ *is the unique element of* $\check{\mathcal{F}}$ *satisfying the following equation:*

$$\check{\mathcal{E}}(\varphi, \psi) + U_\alpha(\varphi, \psi) = (g, \mathbf{H}_\alpha \psi) \quad \textit{for any } \psi \in \check{\mathcal{D}}, \tag{7.3.18}$$

where $\check{\mathcal{D}}$ *is any fixed* $\check{\mathcal{E}}_1$*-dense subspace of* $\check{\mathcal{F}}$.

Proof. Since $(\alpha - \mathcal{L})f = g$, we have from (7.1.30)

$$\mathcal{N}(f)(\psi) = \mathcal{E}_\alpha^{0,\rm ref}(f, \mathbf{H}_\alpha \psi) - (g, \mathbf{H}_\alpha \psi) = \mathbf{C}(\varphi, \psi) + U_\alpha(\varphi, \psi) - (g, \mathbf{H}_\alpha \psi),$$

for any $\psi \in \check{\mathcal{D}}$, which combined with the lateral condition (7.3.15) for f and the identity (7.3.16) gives (7.3.18). If $\varphi \in \check{\mathcal{F}}$ satisfies $\check{\mathcal{E}}(\varphi, \psi) + U_\alpha(\varphi, \psi) = 0$ for any $\psi \in \check{\mathcal{D}}$, then $\varphi = 0$ on account of Theorem 7.2.4 and so the uniqueness statement follows. □

Theorem 7.3.8 says that for $\alpha > 0$ and $g \in L^2(E; m)$, the resolvent $G_\alpha g$ can be recovered as follows: find the solution φ of (7.3.18) and then let

$$G_\alpha g = G_\alpha^0 g + \mathbf{H}_\alpha \varphi, \tag{7.3.19}$$

where $\{G_\alpha^0; \alpha > 0\}$ is the resolvent of X^0.

In the rest of this section, we exhibit the lateral condition appearing in Theorem 7.3.5 in special cases that X is a reflecting extension and its perturbation by a measure supported by F.

Consider a reflecting extension X on E of a symmetric standard process X^0 on E_0 in the sense of Definition 7.2.6. The crucial condition **(RE.4)** is

$$(\mathcal{E}, \mathcal{F}) = (\mathcal{E}^{0,\rm ref}, (\mathcal{F}^0)_a^{\rm ref}). \tag{7.3.20}$$

Lemma 7.2.7 ensures that all the conditions on X and $F = E \setminus E_0$ imposed in the first half of Section 7.2 as well as in Section 7.3 are satisfied. We can readily derive the following simple characterization of the domain of the generator $\mathcal{A}$ of the L^2-semigroup of X directly from the defining formulas (7.3.7) for $\mathcal{L}$, (7.3.8) for $\mathcal{N}$, and (7.3.10) for $\mathcal{A}$:

$$f \in \mathcal{D}(\mathcal{A}) \quad \Longleftrightarrow \quad f \in \mathcal{D}(\mathcal{L}), \quad \mathcal{N}(f) = 0. \tag{7.3.21}$$

Exercise 7.3.9. Derive (7.3.21) from (7.3.20).

The right hand side of (7.3.21) does not involve the condition that $\gamma f \in \check{\mathcal{F}}(= \mathcal{F}|_F)$ as appeared in Theorem 7.3.5 because the condition $f \in \mathcal{F}$ is already implied by $f \in \mathcal{D}(\mathcal{L})$ under (7.3.20).

We remark that

$$\mathcal{F}|_F = \mathcal{G}_1 \tag{7.3.22}$$

by virtue of Theorem 7.2.5, and that $(\mathbf{C}, \mathcal{G}_1)$ is a regular Dirichlet form on $L^2(F^*; \mu_0)$ which is associated with the time-changed process of X by means of the PCAF with Revuz measure μ_0 in view of Corollary 7.2.8. On account of Theorem 5.2.8, a subset of F is $\mathcal{E}$-polar if and only if it is $\mathbf{C}$-polar. Let us denote by $\tilde{\mathcal{G}}_1$ the family of all $\mathbf{C}$-quasi-continuous functions in $\mathcal{G}_1$.

We next take any positive Radon measure κ charging no $\mathcal{E}$-polar set such that

$$\kappa(E_0) = 0. \tag{7.3.23}$$

The quasi support of κ is then contained in F. Let $(\mathcal{E}^\kappa, \mathcal{F}^\kappa)$ be the perturbed Dirichlet form of $(\mathcal{E}, \mathcal{F})$ defined by

$$\begin{cases} \mathcal{F}^\kappa = \mathcal{F} \cap L^2(E; \kappa), \\ \mathcal{E}^\kappa(u, v) = \mathcal{E}(u, v) + (u, v)_\kappa, \quad u, v \in \mathcal{F}^\kappa. \end{cases} \tag{7.3.24}$$

As we saw in Section 5.1, $(\mathcal{E}^\kappa, \mathcal{F}^\kappa)$ is a regular Dirichlet form on $L^2(E; m)$ sharing the quasi notions in common with $(\mathcal{E}, \mathcal{F})$. The associated Hunt process X^κ on E is obtained from X by killing by means of the PCAF with Revuz measure κ. In particular, X^κ still admits no jump from E_0 to F.

The parts of $(\mathcal{E}^\kappa, \mathcal{F}^\kappa)$ and X^κ on E_0 are equal to $(\mathcal{E}^0, \mathcal{F}^0)$ and X^0, respectively, because of the property (7.3.23). Owing to the irreducibility condition imposed in **(RE.2)** on $(\mathcal{E}^0, \mathcal{F}^0)$, we can see as in the proof of Lemma 7.2.7 that $(\mathcal{E}^\kappa, \mathcal{F}^\kappa)$ is irreducible.

From (7.3.22) and (7.3.23) and the remarks made above, we get the identity of the spaces

$$\mathcal{F}^\kappa|_F = \tilde{\mathcal{G}}_1 \cap L^2(F; \kappa). \tag{7.3.25}$$

Furthermore, we have from (7.3.20) and (7.3.23) the equality

$$\mathcal{E}^\kappa(u, u) = \mathcal{E}^{0,\mathrm{ref}}(u, u) + \int_F u^2 d\kappa, \quad u \in \mathcal{F}^\kappa, \tag{7.3.26}$$

which particularly means that $\mu^c_{\langle \mathbf{H}u \rangle}(F) = 0$ for $u \in \mathcal{F}^\kappa$, the restriction to $F \times F$ of the jumping measure of X^κ vanishes and the restriction to F of the killing measure of X^κ equals κ in view of Theorem 7.1.6.

The condition in **(RE.2)** that $\mathcal{F}^0$ is a proper subspace of $\mathcal{F}$ and (7.3.22) imply that $\mathcal{G}_1 \neq \{0\}$. Then $\mathcal{F}^\kappa\big|_F \neq \{0\}$ because $(\mathbf{C}, \mathcal{G}_1)$ is regular. Hence $\mathcal{F}^0$ is again a proper subspace of $\mathcal{F}^\kappa$ and we get the non-$\mathcal{E}^\kappa$-polarity of F as in the proof of Lemma 7.2.7. Thus all conditions imposed in the preceding section are fulfilled by X^κ and F. Moreover, the condition (7.3.4) is met in view of (7.3.20) and Remark 7.3.3(iii).

Therefore, we are led from Theorem 7.3.5, (7.3.25), (7.3.26), and Theorems 5.1.6, 5.2.8 to the following theorem.

THEOREM 7.3.10. *Let $\mathcal{A}^\kappa$ be the infinitesimal generator of the L^2-semigroup of X^κ. $f \in \mathcal{D}(\mathcal{A}^\kappa)$ if and only if $f \in \mathcal{D}(\mathcal{L})$,*

$$f \text{ admits an } X^0\text{-fine limit function } \gamma f \in \tilde{\mathcal{G}}_1 \cap L^2(F;\kappa), \tag{7.3.27}$$

and

$$\mathcal{N}(f)(\psi) + \int_F \gamma f(\xi)\psi(\xi)\kappa(d\xi) = 0, \quad \text{for any } \psi \in \mathcal{C}\big|_F, \tag{7.3.28}$$

where $\mathcal{C}$ is any fixed special standard core of $(\mathcal{E}, \mathcal{F})$.

Example 7.3.11. We maintain the setting in Example 7.2.10. Thus D is a domain in $\mathbb{R}^n$ with $n \geq 1$ satisfying **(I)** and **(II)**. We consider the case that $E = \overline{D}$, $F = \partial D$, $E_0 = D$, $m(dx) = \mathbf{1}_D(x)dx$, and $(\mathcal{E}, \mathcal{F}) = (\frac{1}{2}\mathbf{D}, H^1(D))$. The part $(\mathcal{E}^0, \mathcal{F}^0)$ of $(\mathcal{E}, \mathcal{F})$ on D equals $(\frac{1}{2}\mathbf{D}, H_0^1(D))$. As we saw in Example 7.2.10,

$$(\mathcal{E}^{0,\mathrm{ref}}, (\mathcal{F}^0)_a^{\mathrm{ref}}) = \left(\frac{1}{2}\mathbf{D}, H^1(D)\right)$$

and the condition (7.3.20) is met. Hence the reflecting Brownian motion X^r on $\overline{D}$ that is associated with $(\mathcal{E}, \mathcal{F})$ is a reflecting extension of its part X^0 on D, the absorbing Brownian motion.

The linear operator $\mathcal{L}$ on $L^2(D) = L^2(D;m)$ defined by (7.3.7) is now described as

$$\mathcal{L} = \frac{1}{2}\Delta, \quad \mathcal{D}(\mathcal{L}) = \{f \in H^1(D) : \Delta f \in L^2(D)\}, \tag{7.3.29}$$

where Δ denotes the distribution derivative $\sum_{i=1}^n \frac{\partial^2}{\partial^2 x_i}$ in the Schwartz distribution sense. In view of Theorem 7.2.5, $\mathcal{N}_1 = \mathcal{G}_1$. So the flux functional $\mathcal{N}(f)$, $f \in \mathcal{D}(\mathcal{L})$, defined by (7.3.8), is a linear functional on $\mathcal{G}_1$ specified by

$$\mathcal{N}(f)(\psi) = \frac{1}{2}\mathbf{D}(f, \mathbf{H}_\alpha\psi) + \frac{1}{2}(\Delta f, \mathbf{H}_\alpha\psi), \quad \psi \in \mathcal{G}_1, \tag{7.3.30}$$

which is independent of $\alpha > 0$. When $\partial D, f$, and $\mathbf{H}_\alpha \psi$ are smooth enough, the right hand side of the above is equal to

$$\frac{1}{2}\int_{\partial D} \frac{\partial f(\xi)}{\partial \mathbf{n}_\xi} \cdot \psi(\xi)\sigma(d\xi)$$

by the Gauss-Green formula, where $\mathbf{n}_\xi$ is the unit inward normal vector at $\xi \in \partial D$ and $\sigma(d\xi)$ is the surface measure on ∂D. Therefore, $\mathcal{N}(f)$ is a general substitute of the inward normal derivative $\frac{1}{2}\frac{\partial f}{\partial \mathbf{n}}$.

Let $\mathcal{A}$ be the infinitesimal generator of the semigroup on $L^2(D)$ of X^r. According to (7.3.21),

$$f \in \mathcal{D}(\mathcal{A}) \iff f \in \mathcal{D}(\mathcal{L}),\ \mathcal{N}(f)(\psi) = 0 \text{ for all } \psi \in C_c^\infty(\overline{D}), \tag{7.3.31}$$

$$\mathcal{A}f = \frac{1}{2}\Delta f, \qquad f \in \mathcal{D}(\mathcal{A}). \tag{7.3.32}$$

Let κ be a positive Radon measure on $\overline{D}$ charging no $\mathcal{E}$-polar set satisfying $\kappa(D) = 0$ and let X^κ be the Hunt process on $\overline{D}$ being killed by means of the PCAF with Revuz measure κ. The part process of X^κ on D is still the absorbing Brownian motion on D. The domain $\mathcal{D}(\mathcal{A}^\kappa)$ of the generator of $\mathcal{A}^\kappa$ of the L^2-semigroup of X^κ is characterized as Theorem 7.3.10 with $F = \partial D$, $\mathcal{C}\big|_F = C_c^\infty(\overline{D})\big|_{\partial D}$, and $\mathcal{L}$ being given by (7.3.29). It holds that $\mathcal{A}^\kappa f = \frac{1}{2}\Delta f$, $f \in \mathcal{D}(\mathcal{A}^\kappa)$.

As another example of the extension of the absorbing Brownian motion X^0 on D, consider a non-$\mathcal{E}$-polar closed subset K of ∂D and the part process $X^{r,K}$ of the reflecting Brownian motion X^r on $\overline{D} \setminus K$. Its Dirichlet form $(\mathcal{E}^K, \mathcal{F}^K)$ on $L^2(\overline{D} \setminus K; m) = L^2(D)$ is given by

$$\mathcal{F}^K = \left\{u \in H^1(D) : u = 0\ \mathcal{E}\text{-q.e. on } K\right\} \quad \text{and} \quad \mathcal{E}^K = \frac{1}{2}\mathbf{D}.$$

Its part on D is still given by $(\mathcal{E}^0, \mathcal{F}^0) = (\frac{1}{2}\mathbf{D}, H_0^1(D))$ so that

$$(\mathcal{E}^{0,\text{ref}}, (\mathcal{F}^0)_a^{\text{ref}}) = \left(\frac{1}{2}\mathbf{D}, H^1(D)\right).$$

In this case

$$E = \overline{D} \setminus K, \quad E_0 = D, \quad F = \partial D \setminus K,$$

and the condition (7.3.4) is violated. The L^2-generator $\mathcal{A}^K$ of $X^{r,K}$ cannot be characterized in a way of Theorem 7.3.5 using only the X^0 fine limit function on $\partial D \setminus K$. □

7.4. COUNTABLE BOUNDARY

In this section, we study the case where the set F is countable. The trace Dirichlet form $(\check{\mathcal{E}}, \check{\mathcal{F}})$ on $L^2(F;\mu_0)$ then admits a simple explicit expression so that not only the lateral condition but also the Dirichlet form and resolvent of X can be described in quite tractable ways.

Let E be a locally compact separable metric space, m be a positive Radon measure on X with full support, and $X = (X_t, \zeta, \mathbf{P}_x)$ be an m-symmetric Hunt process on X whose Dirichlet form $(\mathcal{E}, \mathcal{F})$ on $L^2(E;m)$ is regular and irreducible. Recall that any function in $\mathcal{F}_e$ is taken to be quasi continuous. Two subsets A and B of E are called *quasi separated* if there exist quasi open sets U, V such that $A \subset U$, $B \subset V$ and $U \cap V = \emptyset$. We assume that $F = \{a_1, a_2, \ldots, a_i, \ldots\}$ is a finite or countably infinite subset of E satisfying the following:

(F.1) F is quasi closed, $m(F) = 0$, and $\{a_i\}$ is not $\mathcal{E}$-polar for every $i \geq 1$.
(F.2) For each $i \geq 1$, the one-point set $\{a_i\}$ and the set $F \setminus \{a_i\}$ are quasi separated.
(F.3) There exists an $\mathcal{E}$-nest $\{K_n, n \geq 1\}$ such that $F \cap K_n$ is a finite set for each $n \geq 1$.

The above three conditions are invariant under a quasi-homeomorphism. If a countable closed set F contains no accumulation point, then condition **(F.2)** is fulfilled. In this case, any sequence $\{K_n, n \geq 1\}$ of compact sets increasing to E satisfies **(F.3)**.

Let $E_0 := E \setminus F$. We assume that the Hunt process X on E satisfies the condition (7.1.11). We let X^0 be the part process of X on E_0 and U, V be the Feller measures on F relative to X^0.

For $i, j \geq 1$, let

$$U^{ij} := U(\{a_i\}, \{a_j\}) \quad \text{for } i \neq j, \qquad V^i := V(\{a_i\})$$

and $U^{ij}_\alpha := U_\alpha(\{a_i\}, \{a_j\})$.

Define for $x \in E$ and $i \geq 1$,

$$\varphi^{(i)}(x) := \mathbf{P}_x(\sigma_F < \infty,\ X_{\sigma_F} = a_i)$$

and

$$u^{(i)}_\alpha(x) := \mathbf{E}_x\left[e^{-\alpha\sigma_F};\ X_{\sigma_F} = a_i\right],$$

which does not vanish m_0-a.e. on E_0 because $\{a_i\}$ is also non-polar relative to the part of $\mathcal{E}$ on the set $E_0 \cup \{a_i\}$ in view of Theorem 3.3.8(iii).

Since $\varphi^{(i)} = \mathbf{H}\mathbf{1}_{\{a_i\}}$ and $u^{(i)}_\alpha = \mathbf{H}_\alpha \mathbf{1}_{\{a_i\}}$, we see that for $i, j \geq 1$,

$$U^{ij} = L^0(\varphi^{(i)}, \varphi^{(j)}),\ i \neq j,\ V^i = L^0(\varphi^{(i)}, 1 - \mathbf{H}1),\ U^{ij}_\alpha = \alpha(u^{(i)}_\alpha, \varphi^{(j)}).$$

Moreover, as X admits no jumps from E_0 to F, we have from Lemma 7.1.5

$$\varphi^{(i)}(x) = \mathbf{P}_x^0\Big(\zeta^0 < \infty \text{ and } X^0_{\zeta^0-} = a_i\Big)$$

and

$$u_\alpha^{(i)}(x) = \mathbf{E}_x^0\Big[e^{-\alpha\zeta^0}; X^0_{\zeta^0-} = a_i\Big].$$

The intrinsic measure μ_0 on F defined by (7.2.10) is a measure assigning each point $a_i \in F$ a positive mass

$$\mu_0(\{a_i\}) = U_1 1(i) = \sum_{j\geq 1} U_1^{ij}, \tag{7.4.1}$$

which is extended to E by setting $\mu_0(E_0) = 0$. Clearly F is a quasi support of μ_0. Let $(\check{\mathcal{E}}, \check{\mathcal{F}})$ be the Dirichlet form on $L^2(F;\mu_0)$ associated with the time-changed process $\check{X}$ of X by means of PCAF with Revuz measure μ_0. $(\check{\mathcal{E}}, \check{\mathcal{F}})$ is quasi-regular by Theorem 5.2.7 admitting the expression (7.3.16). We denote by $(\check{\mathcal{F}}^{\rm ref}, \check{\mathcal{E}}^{\rm ref})$ the reflected Dirichlet space of $(\check{\mathcal{E}}, \check{\mathcal{F}})$ and by $\check{\mathcal{F}}_a^{\rm ref}$ its active reflected Dirichlet space.

For a real-valued function ψ on F, we put

$$\mathbf{C}(\psi,\psi) = \frac{1}{2}\sum_{i,j\geq 1: i\neq j}(\psi(a_i)-\psi(a_j))^2 U^{ij} + \sum_{i\geq 1}\psi(a_i)^2 V^i, \tag{7.4.2}$$

$$\mathbf{B}(\psi,\psi) = \mathbf{C}(\psi,\psi) + \frac{1}{2}\sum_{i,j\geq 1: i\neq j}(\psi(a_i)-\psi(a_j))^2 J_{ij} + \sum_{i\geq 1}\psi(a_i)^2\kappa_i, \tag{7.4.3}$$

where $J_{ij} := J(\{a_i\},\{a_j\})$ and $\kappa_i := \kappa(\{a_i\})$. Recall that the form $\mathbf{C}$ has already appeared as the Douglas integral in Section 7.2. Let $\mathcal{B}_0(F)$ be the space of functions on F vanishing except on finite many points.

PROPOSITION 7.4.1. (i) *The strongly local term $\mu^c_{\langle \mathbf{H}\psi\rangle}(F)$ in the expression (7.3.16) of $\check{\mathcal{E}}(\psi,\psi)$ vanishes for any $\psi \in \check{\mathcal{F}}$ and*

$$\check{\mathcal{E}}(\psi,\psi) = \mathbf{B}(\psi,\psi), \qquad \psi \in \check{\mathcal{F}}. \tag{7.4.4}$$

(ii) *$\mathcal{B}_0(F)$ is an $\check{\mathcal{E}}_1$-dense subspace of $\check{\mathcal{F}}$.*

Proof. (i) The energy measure $\mu^c_{\langle u\rangle}$ for $u \in b\mathcal{F}$ does not charge any single point owing to the energy image density property formulated in Theorem 4.3.8 and in particular $\mu^c_{\langle \mathbf{H}\psi\rangle}(a_i) = 0$ for any $i \geq 1$ and $\psi \in \check{\mathcal{F}}$.

(ii) Fix $i \geq 0$ and consider the function

$$v(x) = \mathbf{E}_x \left[\int_0^{\tau_{U_i}} e^{-\alpha s} f(X_s) ds \right], \quad x \in E,$$

for an m-integrable strictly positive bounded continuous function f on E, where U_i is a quasi neighborhood of a_i appearing in condition **(F.2)**. Then $v \in \mathcal{F}$, $v(a_i) > 0$, and $v(a_j) = 0$ for any $j \neq i$. Since the one-point set $\{a_j\}$ is non-$\mathcal{E}$-polar, a_j is regular for itself for every $j \geq 1$ in view of Theorem 3.1.10. Therefore, $\mathbf{1}_{\{a_i\}} = \frac{1}{v(a_i)} \mathbf{H}v\big|_F \in \check{\mathcal{F}}$. Consequently, $\mathcal{B}_0(F) \subset \check{\mathcal{F}}$.

Let $\{K_n\}$ be an $\mathcal{E}$-nest appearing in condition **(F.3)** and $\mathcal{F}_{K_n} = \{u \in \mathcal{F} : u = 0$ m-a.e. on $E \setminus K_n\}$. We can modify each function $u \in \mathcal{F}_{K_n}$ to vanish identically on $E \setminus K_n$ so that $\mathcal{F}_{K_n}\big|_F \subset \mathcal{B}_0(F)$ in view of the condition **(F.3)**. Since $\cup_n \mathcal{F}_{K_n}$ is $\mathcal{E}_1$-dense in $\mathcal{F}$, $\mathcal{B}_0(F)$ is $\check{\mathcal{E}}_1$-dense in $\check{\mathcal{F}}$ in view of Remark 7.3.6. □

PROPOSITION 7.4.2. (i) *It holds that*

$$\check{\mathcal{F}}^{\text{ref}} = \{\psi : \mathbf{B}(\psi, \psi) < \infty\}, \quad \check{\mathcal{E}}^{\text{ref}}(\psi, \psi) = \mathbf{B}(\psi, \psi), \ \psi \in \check{\mathcal{F}}^{\text{ref}}, \tag{7.4.5}$$

and $\check{\mathcal{F}}_a^{\text{ref}} = \{\psi \in L^2(F; \mu_0) : \mathbf{B}(\psi, \psi) < \infty\}$.
(ii) *For* $\psi \in \check{\mathcal{F}}_a^{\text{ref}}$, *let* $\mathbf{B}_1(\psi, \psi) := \mathbf{B}(\psi, \psi) + \sum_{i \geq 1} \psi(a_i)^2 U_1 1(i)$. *Then*

$$\check{\mathcal{F}} = \textit{the } \mathbf{B}_1\textit{-closure of } \mathcal{B}_0(F) \textit{ in } \check{\mathcal{F}}_a^{\text{ref}}. \tag{7.4.6}$$

(iii) *For each* $i \geq 1$,

$$\sum_{j \geq 1, j \neq i} (U^{ij} + J_{ij}) < \infty. \tag{7.4.7}$$

Proof. (i) On account of Proposition 7.4.1, condition **(F.3)**, and Theorem 5.2.6, the local Dirichlet space defined by (4.3.31) for the space $\check{\mathcal{F}}$ consists of all real-valued functions on F. Hence (7.4.5) follows from Definition 6.4.4.
(ii) This follows from (i), Lemma 7.3.7(ii), and Proposition 7.4.1(ii).
(iii) We get this from (7.4.2) with $\psi = \mathbf{1}_{\{a_i\}} \in \check{\mathcal{F}}$. □

We can define for $f \in \mathcal{D}(\mathcal{L})$, the flux of f at a_i by

$$\mathcal{N}(f)(a_i) := \mathcal{N}(f)(\mathbf{1}_{\{a_i\}}) = \mathcal{E}^{0,\text{ref}}(f, u_\alpha^{(i)}) + (\mathcal{L}f, u_\alpha^{(i)}). \tag{7.4.8}$$

Theorem 7.3.5 and Lemma 7.3.7 now read as follows:

THEOREM 7.4.3. (i) *Assume that condition* (7.3.4) *is fulfilled.* $f \in \mathcal{D}(\mathcal{A})$ *if and only if* $f \in \mathcal{D}(\mathcal{L})$ *and* f *satisfies the lateral conditions that*

$$f \textit{ admits an } X^0\textit{-fine limit function } \gamma f \in \check{\mathcal{F}}, \tag{7.4.9}$$

for the space $\check{\mathcal{F}}$ *specified by* (7.4.6), *and, for every* $i \geq 1$,

$$\mathcal{N}(f)(a_i) + \frac{1}{2} \sum_{j \geq 1, j \neq i} ((\gamma f)(a_i) - (\gamma f)(a_j))J_{ij} + (\gamma f)(a_i)\kappa_i = 0. \tag{7.4.10}$$

(ii) *If* X *is recurrent, then* $V = \kappa = 0$ *and*

$$\check{\mathcal{F}} = \left\{ \psi : \frac{1}{2} \sum_{i \neq j} (\psi(a_i) - \psi(a_j))^2 (U^{ij} + J_{ij}) + \sum_{i \geq 1} \psi(a_i)^2 U_1 1(i) < \infty \right\}. \tag{7.4.11}$$

Taking Proposition 7.4.1(ii) into account, we can rewrite Theorem 7.3.8 and (7.3.19) as follows:

THEOREM 7.4.4. *For* $\alpha > 0$ *and* $g \in L^2(E; m)$, *let* $f = G_\alpha g$. $\psi = f|_F$ *is then the unique element of* $\check{\mathcal{F}}$ *such that*

$$\mathbf{B}(\psi, \mathbf{1}_{a_i}) + \sum_{j \geq 1} U_\alpha^{ij} \psi(a_j) = (u_\alpha^{(i)}, g), \quad i \geq 1. \tag{7.4.12}$$

Moreover, $G_\alpha g$ *admits the representation*

$$G_\alpha g(x) = G_\alpha^0 g(x) + \sum_{i \geq 1} u_\alpha^{(i)}(x) \psi(a_i), \quad x \in E. \tag{7.4.13}$$

In the remainder of this section, we assume that the symmetric Hunt process X on E admits no killing on F or jumps from F to F in the following sense:

$$\kappa_i = 0 \quad \text{and} \quad J_{ij} = 0 \quad \text{for every } i, j \geq 1. \tag{7.4.14}$$

This condition is equivalent to $\mathbf{B} = \mathbf{C}$. Recall the space $\mathcal{G}_1$ of functions on F with finite Douglas integrals introduced by (7.2.11):

$$\mathcal{G}_1 = \{\psi \in L^2(F; \mu_0) : \mathbf{C}(\psi, \varphi) < \infty\}.$$

THEOREM 7.4.5. *Assume that* X *satisfies condition* (7.4.14). *Then the following hold true:*

(i) $\check{\mathcal{F}}_a^{\text{ref}} = \mathcal{G}_1$ *and* $\check{\mathcal{E}}^{\text{ref}}(\psi, \psi) = \mathbf{C}(\psi, \psi)$ *for* $\psi \in \check{\mathcal{F}}_a^{\text{ref}}$.

(ii) *Let* $\mathbf{C}_1(\psi, \psi) = \mathbf{C}(\psi, \psi) + \sum_{i \geq 1} \psi(a_i)^2 \mu_0(\{a_i\})$. *Then*

$$\begin{cases} \check{\mathcal{F}} = \text{ the } \mathbf{C}_1\text{-closure of } \mathcal{B}_0(F) \text{ in } \mathcal{G}_1, \\ \check{\mathcal{E}}(\psi, \psi) = \mathbf{C}(\varphi, \psi) \qquad \text{for } \psi \in \check{\mathcal{F}}. \end{cases} \tag{7.4.15}$$

(iii) $\sum_{j \geq 1, j \neq i} U^{ij} < \infty$ *for every* $i \geq 1$.

(iv) *For* $\alpha > 0$ *and* $g \in L^2(E;m)$, *let* $f = G_\alpha g$. *Then* $\psi = f\big|_F$ *is the unique element of* $\check{\mathcal{F}}$ *such that*

$$\mathbf{C}(\psi, \mathbf{1}_{a_i}) + \sum_{j\geq 1} U_\alpha^{ij}\psi(a_j) = (u_\alpha^{(i)}, g), \quad i \geq 1. \tag{7.4.16}$$

Moreover, $G_\alpha g$ *admits the representation* (7.4.13).
(v) $\mathcal{F} \subset (\mathcal{F}^0)^{\mathrm{ref}}_a$, $\mathcal{E}_\alpha(u,v) = \mathcal{E}_\alpha^{0,\mathrm{ref}}(u,v)$ *for* $u, v \in \mathcal{F}$, *and*

$$\mathcal{F} = \mathcal{F}^0 \oplus \mathcal{H}_\alpha, \tag{7.4.17}$$

where $\mathcal{H}_\alpha$ *is the* $\mathcal{E}_\alpha^{0,\mathrm{ref}}$*-closure of the linear span of* $\{u_\alpha^{(i)}; i \geq 1\}$ *and* (7.4.17) *is an* $\mathcal{E}_\alpha^{0,\mathrm{ref}}$*-orthogonal decomposition.*
(vi) $\mathcal{F}_e \subset (\mathcal{F}^0)^{\mathrm{ref}}$, $\mathcal{E}(u,v) = \mathcal{E}^{0,\mathrm{ref}}(u,v)$ *for* $u, v \in \mathcal{F}_e$. *Furthermore,* $\mathcal{F}_e^0$ *and the linear span of* $\{\varphi^{(i)} : i \geq 1\}$ *are subsets of* $\mathcal{F}_e$ *which are* $\mathcal{E}^{0,\mathrm{ref}}$*-orthogonal to each other.*
(vii) *When* F *consists of finite number of points* $\{a_1, a_2, \ldots, a_N\}$, $\mathcal{F}_e$ *is the linear subspace of* $(\mathcal{F}^0)^{\mathrm{ref}}$ *spanned by* $\mathcal{F}_e^0$ *and* $\{\varphi^{(i)},\ 1 \leq i \leq N\}$, *while* $\mathcal{F}$ *is the linear subspace of* $(\mathcal{F}^0)^{\mathrm{ref}}_a$ *spanned by* $\mathcal{F}^0$ *and* $\{u_\alpha^{(i)},\ 1 \leq i \leq N\}$.
(viii) *Assume that condition* (7.3.4) *holds. Then* $f \in \mathcal{D}(\mathcal{A})$ *if and only if* $f \in \mathcal{D}(\mathcal{L})$, (7.4.9) *is satisfied for the space* $\check{\mathcal{F}}$ *specified by* (7.4.15), *and*

$$\mathcal{N}(f)(a_i) = 0 \qquad \textit{for every } i \geq 1. \tag{7.4.18}$$

Proof. (v) follows from (ii) and Theorem 7.1.8 by noting Lemma 7.2.3. (vi) uses Theorem 7.1.6. Other assertions are restatements of the preceding propositions and theorems under the assumption (7.4.14). □

Keeping the assumption (7.4.14), we consider the special case where the set F consists of finite number of points: $F = \{a_1, \ldots, a_N\}$. Let $\check{\mathcal{A}}$ be the $N \times N$ matrix with entry $\check{\mathcal{A}}_{ij}$ being given by

$$\begin{aligned} \check{\mathcal{A}}_{ij} &= U^{ij}, \quad 1 \leq i, j \leq N, \quad i \neq j, \\ \check{\mathcal{A}}_{ii} &= -\sum_{k\geq 1,\, k\neq i} U^{ik} - V^i, \quad 1 \leq i \leq N. \end{aligned} \tag{7.4.19}$$

U_α denotes the $N \times N$ matrix with entry U_α^{ij}. We then let

$$R_\alpha(x,y) = -\mathbf{u}_\alpha(x) \cdot (\check{\mathcal{A}} - U_\alpha)^{-1}\ {}^t\mathbf{u}_\alpha(y), \quad x, y \in E, \tag{7.4.20}$$

where $\mathbf{u}_\alpha(x) = (u_\alpha^{(1)}(x), \ldots, u_\alpha^{(N)}(x))$ and ${}^t\mathbf{u}_\alpha(y)$ is the transpose of $\mathbf{u}_\alpha(y)$.

The next decomposition formula of the resolvent follows from (7.4.16) and (7.4.13):

$$G_\alpha g(x) = G^0_\alpha g(x) + \int_{E_0} R_\alpha(x,y)g(y)m(dy), \quad x \in E. \tag{7.4.21}$$

We finally consider the case where the set F consists of only one point.

Theorem 7.4.6. *Suppose F consists of one point a and X admits no killing at a.*

(i) *The resolvent $\{G_\alpha;\alpha > 0\}$ of X admits the expression for $g \in L^2(E;m)$:*

$$G_\alpha g(a) = \frac{(u_\alpha, g)}{\alpha(u_\alpha,\varphi) + L^0(\varphi, 1-\varphi)} \tag{7.4.22}$$

$$G_\alpha g(x) = G^0_\alpha g(x) + u_\alpha(x)G_\alpha g(a), \quad x \in E,$$

where

$$u_\alpha(x) = \mathbf{E}_x\left[e^{-\alpha\sigma_a};\sigma_a < \infty\right], \quad \varphi(x) = \mathbf{P}_x(\sigma_a < \infty), \quad x \in E.$$

(ii) *$\mathcal{F}_e$ is the linear subspace of $(\mathcal{F}^0)^{\mathrm{ref}}$ spanned by $\mathcal{F}^0_e$ and φ.*
$\mathcal{F}$ is the linear subspace of $(\mathcal{F}^0)^{\mathrm{ref}}_a$ spanned by $\mathcal{F}^0$ and u_α.
For $f = f_0 + c\varphi$, $f_0 \in \mathcal{F}^0_e$, $c \in \mathbb{R}$,

$$\mathcal{E}(f,f) = \mathcal{E}^{0,\mathrm{ref}}(f,f) = \mathcal{E}^0(f_0,f_0) + c^2L^0(\varphi, 1-\varphi). \tag{7.4.23}$$

(iii) *Assume that condition (7.3.4) is fulfilled. Let $\mathcal{A}$ be the L^2-generator of X. $f \in \mathcal{D}(\mathcal{A})$ if and only if $f \in \mathcal{D}(\mathcal{L})$, f admits an X^0-fine limit value $\gamma f(a)$ at a and*

$$\mathcal{N}(f)(a) = 0. \tag{7.4.24}$$

Here $\mathcal{N}(f)(a)$ is the flux of $f \in \mathcal{D}(\mathcal{L})$ at a defined by

$$\mathcal{N}(f)(a) = \mathcal{E}^{0,\mathrm{ref}}(f, u_\alpha) + (\mathcal{L}f, u_\alpha). \tag{7.4.25}$$

Proof. The formulas (7.4.20) and (7.4.21) for $N = 1$ reduce to (7.4.22). (ii) follows from Theorem 7.1.6 and Theorem 7.1.8 by noting that $V(\{a\}) = L^0(\varphi, 1-\varphi)$. (iii) is a special case of Theorem 7.4.5. □

7.5. ONE-POINT EXTENSIONS

In this section, every sample path of a right process X on a state space E will be assumed to possess left limit X_{t-} in E for all $t \in (0, \zeta)$, where ζ is the lifetime of X. A right process $X = (X_t, \mathbf{P}_x, \zeta)$ is said to be of *no killing inside* if

$$\mathbf{P}_x(\zeta < \infty,\ X_{\zeta-} \in E) = 0 \quad \text{for any } x \in E,$$

where $\{\zeta < \infty,\ X_{\zeta-} \in E\}$ means the event that the left limit of X_t at $t = \zeta < \infty$ exists and belongs to E.

Let E be a locally compact separable metric space and m be a positive Radon measure on E with $\text{supp}[m] = E$. We fix a non-isolated point $a \in E$ with $m(\{a\}) = 0$. Put $E_0 = E \setminus \{a\}$. $E_\partial = E \cup \{\partial\}$ denotes the one-point compactification of E. $(E_0)_\partial = E_0 \cup \{\partial\}$ is regarded as a topological subspace of E_∂.

Let $X^0 = (X_t^0, \mathbf{P}_x^0, \zeta^0)$ be an m-symmetric Borel standard process on E_0 satisfying the following condition:

(A.1) X^0 admits no killing inside and

$$\mathbf{P}_x^0\left(\zeta^0 < \infty,\ X_{\zeta^0-}^0 = a\right) > 0 \quad \text{for every } x \in E_0. \tag{7.5.1}$$

Then for $x \in E_0$ and $\alpha > 0$, we let

$$\varphi(x) = \mathbf{P}_x^0\left(\zeta^0 < \infty,\ X_{\zeta^0-}^0 = a\right), \quad u_\alpha(x) = \mathbf{E}_x^0\left[e^{-\alpha\zeta^0}; X_{\zeta^0-}^0 = a\right]. \tag{7.5.2}$$

DEFINITION 7.5.1. A right process $X = (X_t,\ \mathbf{P}_x,\ \zeta)$ on E is called a *one-point extension* of X^0 if X is m-symmetric and of no killings on $\{a\}$, and the part process of X on E_0 is X^0.

Exercise 7.5.2. Suppose a one-point extension $X = (X_t, \mathbf{P}_x)$ of X^0 is a Borel standard process. Show that X satisfies the following properties. Denote by σ_a the hitting time of the set $\{a\}$ by X.

(i) For any $x \in E_0$ and $\alpha > 0$,

$$\varphi(x) = \mathbf{P}_x(\sigma_a < \infty), \quad u_\alpha(x) = \mathbf{E}_x\left[e^{-\alpha\sigma_a}; \sigma_a < \infty\right]. \tag{7.5.3}$$

(ii) X admits no jump from E_0 *to* $\{a\}$:

$$\mathbf{P}_x\left(X_{t-} \in E_0,\ X_t = a \text{ for some } t > 0\right) = 0,\ x \in E. \tag{7.5.4}$$

LEMMA 7.5.3. *Suppose a one-point extension X of X^0 is a Hunt process whose Dirichlet form $(\mathcal{E}, \mathcal{F})$ is regular. Then the point a is non-$\mathcal{E}$-polar and regular for itself with respect to X in the sense that $\mathbf{P}_a(\sigma_a = 0) = 1$. Further, X is irreducible.*

Proof. By (7.5.1) and (7.5.3), $\{a\}$ is not m-polar with respect to X and hence non-$\mathcal{E}$-polar. On account of Theorem 3.1.10, $\{a\}$ is then regular for itself.

Denote by $\{G_\alpha;\alpha>0\}$ and $\{G^0_\alpha;\alpha>0\}$ the resolvents of X and X^0, respectively. Take any non-negative $g\in C_c(E)$ and let $w=G_\alpha g$. Then $w\in\mathcal{F}$ admits two types of decomposition

$$w=G^0_\alpha g+cu_\alpha=f_0+c\varphi \quad \text{with } c=w(a),\ f_0=w-c\varphi,$$

as a sum of $\mathcal{E}_\alpha$-orthogonal elements in $\mathcal{F}$ and a sum of $\mathcal{E}$-orthogonal elements in $\mathcal{F}_e$. On account of Exercise 2.1.13 and Lemma 2.1.15, we can find a uniformly bounded sequence v_n of functions in $\mathcal{F}$ that is $\mathcal{E}$-convergent as well as m-a.e. convergent to φ. Letting $n\to\infty$ in the equation $(g,v_n)=\mathcal{E}_\alpha(w,v_n)=\mathcal{E}(f_0+c\varphi,v_n)+\alpha(G^0_\alpha g+cu_\alpha,v_n)$, we get $c\,[\mathcal{E}(\varphi,\varphi)+\alpha(u_\alpha,\varphi)]=(g,\varphi)-\alpha(G^0_\alpha g,\varphi)=(u_\alpha,g)$, and accordingly

$$0<\mathcal{E}(\varphi,\varphi)+\alpha(u_\alpha,\varphi)<\infty,\quad G_\alpha g(a)=\frac{(u_\alpha,g)}{\mathcal{E}(\varphi,\varphi)+\alpha(u_\alpha,\varphi)},$$

which extends to $g\in\mathcal{B}(E)$. Therefore, for any $A,B\in\mathcal{B}(E)$ with positive m-measures, we have $(\mathbf{1}_A,G_\alpha\mathbf{1}_B)\geq(\mathbf{1}_A,u_\alpha)G_\alpha\mathbf{1}_B(a)>0$, the irreducibility of X. □

THEOREM 7.5.4. *Let X^0 be an m-symmetric Borel standard process on E_0 satisfying condition* **(A.1)**. *A one-point extension of X^0 is then unique in law. More specifically, let $X=(X_t,\mathbf{P}_x,\zeta)$ be a one-point extension of X^0. Then X enjoys the following properties. Denote by $\{G^0_\alpha;\alpha>0\}$ and $(\mathcal{E}^0,\mathcal{F}^0)$ (resp. $\{G_\alpha;\alpha>0\}$ and $(\mathcal{E},\mathcal{F})$) the resolvent and the Dirichlet form of X^0 (resp. X) on $L^2(E_0;m)(=L^2(E;m))$.*

(i) *The point $\{a\}$ is non-m-polar and regular for itself with respect to X. The process X is irreducible.*
(ii) *For any bounded $g\in L^2(E;m)$,*

$$G_\alpha g(a)=\frac{(u_\alpha,g)}{\alpha(u_\alpha,\varphi)+L^0(\varphi,1-\varphi)} \tag{7.5.5}$$

$$G_\alpha g(x)=G^0_\alpha g(x)+u_\alpha(x)G_\alpha g(a),\quad x\in E, \tag{7.5.6}$$

where L^0 is the energy functional for X^0.
(iii) *$\mathcal{F}_e$ is the linear subspace of $(\mathcal{F}^0)^{\rm ref}$ spanned by $\mathcal{F}^0_e$ and φ. $\mathcal{F}$ is the linear subspace of $(\mathcal{F}^0)^{\rm ref}_a$ spanned by $\mathcal{F}^0$ and u_α.*

$$\mathcal{E}(f,f)=\mathcal{E}^{0,{\rm ref}}(f,f),\quad f\in\mathcal{F}_e. \tag{7.5.7}$$

Further, for $f=f_0+c\varphi$, $f_0\in\mathcal{F}^0_e$, $c\in\mathbb{R}$,

$$\mathcal{E}^{0,{\rm ref}}(f,f)=\mathcal{E}^0(f_0,f_0)+c^2L^0(\varphi,1-\varphi). \tag{7.5.8}$$

(iv) *Assume that condition* (7.3.4) *is fulfilled. Let* $\mathcal{A}$ *be the* L^2*-generator of* X. $f \in \mathcal{D}(\mathcal{A})$ *if and only if* $f \in \mathcal{D}(\mathcal{L})$, f *admits an* X^0*-fine limit value at* a, *and*

$$\mathcal{N}(f)(a) = 0, \tag{7.5.9}$$

where $\mathcal{N}(f)(a)$ *is the flux of* f *at* a *defined by* (7.4.25). *It then holds that*

$$\mathcal{A}f = \mathcal{L}f, \quad f \in \mathcal{D}(\mathcal{A}). \tag{7.5.10}$$

We remark that since X^0 is an m-symmetric standard process, $(\mathcal{E}^0, \mathcal{F}^0)$ is quasi-regular by Theorem 1.5.3 and so its reflected Dirichlet space $((\mathcal{F}^0)^{\mathrm{ref}}, \mathcal{E}^{0,\mathrm{ref}})$ is well defined.

Proof. The uniqueness follows from (ii). We prove (i)~(iv) by a transfer method. Since X is an m-symmetric right process on a locally compact separable metric space E, $(\mathcal{E}, \mathcal{F})$ is quasi-regular with which X is properly associated by Theorem 1.5.3. According to Theorem 1.4.3, $(\mathcal{E}, \mathcal{F})$ is quasi-homeomorphic to a regular Dirichlet form $(\widehat{\mathcal{E}}, \widehat{\mathcal{F}})$ on $L^2(\widehat{E}; \widehat{m})$ for some locally compact separable metric space $\widehat{E}$ by a quasi-homeomorphism j. In view of Theorem 3.1.13, X is equivalent in law to a pull back $\tilde{X}$ by j of a Hunt process $\widehat{X} = (\widehat{X}_t, \widehat{\mathbf{P}}_x, \widehat{\zeta})$ on $\widehat{E}$ associated with $(\widehat{\mathcal{E}}, \widehat{\mathcal{F}})$ outside some Borel properly exceptional set.

To be more precise, j is a quasi-homeomorphism between an $\mathcal{E}$-nest $\{F_k\}$ and an $\widehat{\mathcal{E}}$-nest $\{\widehat{F}_k\}$ both consisting of compact sets so that j is a one-to-one map from $E_1 = \cup_k F_k$ onto $\widehat{E}_1 = \cup_k \widehat{F}_k$. There exists then an $\widehat{X}$-properly exceptional Borel set $\widehat{N}$ containing $\widehat{E} \setminus \widehat{E}_1$ such that $\widehat{\mathbf{P}}_{\widehat{x}}(\lim_{k\to\infty} \tau_{\widehat{F}_k} = \widehat{\zeta}) = 1$ for all $\widehat{x} \in \widehat{E} \setminus \widehat{N}$, and if we define $N \subset E$ by $E \setminus N = j^{-1}(\widehat{E}_1 \cap \widehat{N})$, then $X_{E\setminus N}$ is equivalent to the Borel special standard process $\tilde{X} = j^{-1}(\widehat{X}|_{\widehat{E}\setminus\widehat{N}})$. In particular, N is properly exceptional for X in the sense of Theorem 3.1.13.

Since X extends X^0 satisfying **(A.1)**, the set $\{a\}$ must be located in $E \setminus N$. We can then apply Exercise 7.5.2 to the standard process $X\big|_{E\setminus N}$ extending $X^0\big|_{E^0\setminus N}$ to conclude that a is not m-polar for X and $X\big|_{E\setminus N}$ admits no jump from $E_0 \setminus N$ to a.

Therefore, $\widehat{a} := ja \in \widehat{E} \setminus \widehat{N}$ is non-$\widehat{m}$-polar for the Hunt process $\widehat{X}' := \widehat{X}|_{\widehat{E}'}$ on $\widehat{E}' := \widehat{E} \setminus \widehat{N}$. Obviously $\widehat{m}(\{\widehat{a}\}) = 0$. Let $\widehat{X}^0$ be the part process of $\widehat{X}'$ on $\widehat{E} \setminus \widehat{N} \setminus \{\widehat{a}\}$, which is an $\widehat{m}$-symmetric standard process satisfying **(A.1)**. It is now clear that $\widehat{X}'$ is a one-point extension of $\widehat{X}_0$ admitting no jump from $\widehat{E}' \setminus \{\widehat{a}\}$ to $\{\widehat{a}\}$. Since $(\widehat{\mathcal{E}}, \widehat{\mathcal{F}})$ is regular, $\{\widehat{a}\}$ is regular for itself with respect to $\widehat{X}'$ and $\widehat{X}'$ is irreducible in view of Lemma 7.5.3. In particular, $\widehat{X}'$ and $\{\widehat{a}\}$ satisfy all the conditions imposed in Section 7.4 so that Theorem 7.4.6 applies.

Consequently, (i), (iii), (iv) as well as (7.5.5) hold for $\widehat{X}'$, $\{\widehat{a}\}$, and $\widehat{X}^0$. It is not hard to verify that those properties are honestly inherited by X, a, and

X^0 through the quasi-homeomorphic map j^{-1}. For instance, if we denote by $\{\widehat{G}_\alpha; \alpha > 0\}$ and $\{\widehat{G}^0_\alpha; \alpha > 0\}$ the resolvents of $\widehat{X}'$ and $\widehat{X}^0$, respectively, and by $\widehat{u}_\alpha$ and $\widehat{\varphi}$ the α-order and 0-order hitting probabilities of $\widehat{X}'$ for $\widehat{a}$, then

$$\widehat{G}_\alpha \widehat{g}(\widehat{a}) = \frac{(\widehat{u}_\alpha, \widehat{g})_{\widehat{m}}}{\alpha(\widehat{u}_\alpha, \widehat{\varphi})_{\widehat{m}} + \widehat{L}^0(\widehat{\varphi}, 1 - \widehat{\varphi})},$$

where $\widehat{L}^0$ is the energy functional relative to $\widehat{X}^0$. Since $j^* \widehat{G}_\alpha \widehat{g} = G_\alpha j^* \widehat{g}$, $j^* \widehat{G}^0_\alpha \widehat{g} = G^0_\alpha j^* \widehat{g}$, $j^* \widehat{u}_\alpha = u_\alpha$, $j^* \widehat{\varphi} = \varphi$, and $\int_E j^* \widehat{f} dm = \int_{\widehat{E}} \widehat{f} d\widehat{m}$, we can remove the sign $\widehat{\ }$ from both sides of the above identity to get the desired equality (7.5.5). As $\widehat{a}$ is regular for itself with respect to $\widehat{X}'$, so is $a = j^{-1}\widehat{a}$ with respect to $X\big|_{E \setminus N}$, and accordingly with respect to X. Hence the identity (7.5.6) is valid for every $x \in E$ because X is an extension of X^0.

Similarly, we can conclude just as in the proof of Theorem 7.2.9 that the reflected Dirichlet form $(\mathcal{E}^{0,\mathrm{ref}}, (\mathcal{F}^0)^{\mathrm{ref}})$ of X^0 is the image by j^{-1} of the reflected Dirichlet form of $\widehat{X}^0$. □

A one-point extension X of an m-symmetric Hunt process X^0 can be constructed under some of the following additional conditions for X^0:

(A.2) $\int_E u_\alpha(x) m(dx) < \infty$ for $\alpha > 0$, and for every $x \in E_0$,

$$\mathbf{P}^0_x\Big(\zeta^0 < \infty,\ X^0_{\zeta^0-} \in \{a, \partial\}\Big) = \mathbf{P}^0_x(\zeta^0 < \infty). \tag{7.5.11}$$

(A.2)′ $\mathbf{P}^0_x(X^0_{\zeta^0-} \in \{a, \partial\}) = 1$ for every $x \in E_0$ (regardless the length of the lifetime $\zeta^0 \in (0, \infty]$).

(A.3) The exists a neighborhood U of a such that $\inf_{x \in V} G^0_1 \varphi(x) > 0$ for any compact set $V \subset U \setminus \{a\}$.

(A.4) Either $E \setminus U$ is compact for any neighborhood U of a, or for any open neighborhood U_1 of a in E, there exists an open neighborhood U_2 of a in E with $\overline{U}_2 \subset U_1$ such that $J_0(U_2 \setminus \{a\}, E_0 \setminus U_1) < \infty$. Here J_0 denotes the jumping measure of X^0.

We do not need condition **(A.4)** when X^0 is a diffusion.

Lemma 7.5.5. *Assume that X^0 is an m-symmetric Hunt process on E_0 satisfying the condition* **(A.1)**.
(i) *If X^0 is a diffusion, then it satisfies the condition* (7.5.11).
(ii) *If X^0 is a diffusion and the Dirichlet form of X^0 is regular, then condition* **(A.2)′** *is fulfilled for q.e. $x \in E_0$.*
(iii) *If X^0 admits a one-point extension X on E, then X^0 satisfies the integrability condition in* **(A.2)**.
(iv) *If either $G^0_1 f$ or $G^0_{0+} f$ is lower semicontinuous on E_0 for any $f \in \mathcal{B}_+(E_0)$, then X^0 satisfies condition* **(A.3)**.

Proof. (i) Since X^0 admits no killing inside E_0, the quasi-left-continuity of X^0 implies that $X^0_{\zeta^0-} = \partial^0$ $\mathbf{P}_x$-a.s. for every $x \in E_0$, where ∂^0 denotes the point at infinity of E_0 for the one-point compactification of E_0. By the continuity of the sample paths of X^0, this means that for any relatively compact neighborhood U of a and any compact set $K \supset U$, the entire portion $\{X^0_t : t \in [T, \zeta^0)\}$ of the path is included either in $U \setminus \{a\}$ or $E \setminus K$ for some $T \in (0, \zeta^0)$ $\mathbf{P}_x$-a.s. for $x \in E_0$.

(ii) By **(A.1)** and Proposition 2.1.10, X^0 is transient. Theorem 3.5.2 then implies that $\lim_{t \to \zeta^0} X^0_t = \partial^0$ regardless the length of ζ^0 $\mathbf{P}_x$-a.s. for q.e. $x \in E_0$. By the same reasoning as that in (i), one can conclude that **(A.2)**$'$ holds for q.e. $x \in E_0$.

(iii) Denote by $\{G_\alpha; \alpha > 0\}$ the resolvent of X and choose a non-negative $f \in C_c(E)$ with $f(a) > 0$. Then $G_\alpha f(a) > 0$ and $u_\alpha(x) G_\alpha f(a) \le G_\alpha f(x)$ for $x \in E_0$, which yields the m-integrability of u_α.

(iv) It suffices to note the identity $G^0_1 \varphi = G^0_{0+} u_1$. □

THEOREM 7.5.6. *Let X^0 be an m-symmetric Hunt process on E_0 satisfying conditions* **(A.1)** *and* **(A.3)** *as well as* **(A.4)** *in a non-diffusion case. Assume also that either* **(A.2)** *or* **(A.2)**$'$ *is satisfied by X^0. Then there exists a one-point extension $X = (X_t, \mathbf{P}_x, \zeta)$ of X^0 from E_0 to E such that X admits no jumps to or from a and*

$$\mathbf{P}_x(X_{\zeta-} = \partial,\ \zeta < \infty) = \mathbf{P}_x(\zeta < \infty), \quad x \in E. \tag{7.5.12}$$

When X^0 is a diffusion, so is its one-point extension X.

Sketch of proof. A construction of a one-point extension X of X^0 can be carried out probabilistically by using a Poisson point process of excursions of X^0 around the point a, as will now be explained briefly.

We first assume that X^0 satisfies the conditions **(A.1)**, **(A.2)**, and **(A.3)** as well as **(A.4)** in non-diffusion cases. Let $\{P^0_t; t \ge 0\}$ be the transition function of X^0. Recall that a system $\{\nu_t; t > 0\}$ of σ-finite measures on E_0 is said to be X^0-*entrance law* if $\nu_s P^0_t = \nu_{s+t}$ for every $s, t > 0$. By **(A.1)**, X^0 is transient and, by virtue of Lemma 5.7.1, there exists a unique X^0-entrance law $\{\nu_t; t > 0\}$ such that

$$\int_0^\infty \nu_t dt = \varphi \cdot m. \tag{7.5.13}$$

Due to the integrability of u_α in the assumption **(A.2)** and (5.7.4), each ν_t is a finite measure. Denote by W the space of cadlàg paths w defined on a time interval $(0, \zeta(w))$ taking values in E_0 with $w_{0+} = a$ and $w_{\zeta-} \in \{a, \partial\}$ whenever

$\zeta(w) < \infty$. We can define a σ-finite measure $\mathbf{n}$ on W so that

$$\int_W f_1(w_{t_1})f_2(w_{t_2})\cdots f_n(w_{t_n})\mathbf{n}(dw)$$

$$= \nu_{t_1}f_1P^0_{t_2-t_1}f_2\cdots P^0_{t_{n-1}-t_{n-2}}f_{n-1}P^0_{t_n-t_{n-1}}f_n \qquad (7.5.14)$$

for $0 < t_1 < t_2 < \cdots < t_n$ and $f_1, f_2, \ldots, f_n \in \mathcal{B}_+(E_0)$. Put $W^+ = \{w \in W : \zeta(w) < \infty,\ w_{\zeta-} = a\}$, $W^- = W \setminus W^+$ and denote by $\mathbf{n}^+, \mathbf{n}^-$ the restrictions of $\mathbf{n}$ to W^+, W^-, respectively. Owing to the assumptions **(A.3)**, **(A.4)**, it can be shown that

$$\mathbf{n}(\sigma_{E_0\setminus U} < \zeta) < \infty \quad \text{for any neighborhood } U \text{ of } a, \qquad (7.5.15)$$

from which it follows that $\mathbf{n}(W^-) < \infty$.

Let $\mathbf{p} = \{\mathbf{p}_t, t \geq 0\}$ be the Poisson point process taking values in W with characteristic measure $\mathbf{n}$ on an appropriate probability space (Ω, P). Clearly, $\mathbf{p}$ is a sum of independent Poisson point processes $\mathbf{p}^+$ and $\mathbf{p}^-$ with characteristic measures $\mathbf{n}^+$ and $\mathbf{n}^-$, respectively. We then create a path X_a starting at a by piecing together the returning excursions $\mathbf{p}^+$ until the first occurrence time T of a non-returning excursion $\mathbf{p}^-$ and then adjoining $\mathbf{p}^-_T$ to finish the construction of X_a. We note that

$$P(T > t) = e^{-\mathbf{n}(W^-)t} \text{ and the distribution of } \mathbf{p}^-_T \text{ is } \mathbf{n}(W^-)^{-1}\,\mathbf{n}^-. \qquad (7.5.16)$$

The resulting path X_a is not only cadlàg but also continuous at those moments t when $X_a(t) = a$ on account of the property (7.5.15) of $\mathbf{n}$.

The one-point extension X can be eventually constructed by joining X^0 to X_a. We refer the readers to [75] (when X^0 is a diffusion) and [28] (when X^0 is a general symmetric Markov process) as well as [21, Remark 3.2(ii)] for more details of the proof sketched above.

The assumption **(A.2)** can be replaced by **(A.2)**$'$ using a method of a time change (cf. [28]). Indeed, consider a continuous function $\gamma(x)$ on E_0 such that $0 < \gamma \leq 1$, $\int_E \gamma dm < \infty$ with $\gamma = 1$ on a neighborhood of U of a and let $Y^0 = (X^0_{\tau_t}, \mathbf{P}^0_x)$ be the time-changed process of X^0 where τ_t is the inverse of the PCAF $A_t = \int_0^t \gamma(X^0_s)ds$ of X^0. This time-changed process Y^0 is a $\gamma \cdot m$-symmetric Hunt process on E_0 and it is easy to see that properties **(A.1)** and **(A.3)** remain valid for Y^0.

Furthermore, the jumping measure J_0 is invariant under a time change by a strictly increasing PCAF A_t and so **(A.4)** also remains valid for Y^0. To see this, let (N, H) be a Lévy system of X^0 and μ_H be the Revuz measure of H with respect to m. Substitute $T = \tau_t$, $h(s) = 1$ in (A.3.33) for X^0 and make a change of variable $s \mapsto \tau_s$ to obtain (A.3.31) with Y^0_t, H_{τ_s} in place of X^0_t, H_s. This means that Y^0 admits $(N, H_{\tau_\cdot})$ as its Lévy system. Since the Revuz measure of the PCAF $H_{\tau_\cdot}$ of Y^0 with respect to $\gamma \cdot m$ coincides with μ_H (cf. [73, Lemma 6.2.9]), the jumping measure for Y^0 equals $N(x, dy)\mu_H(dx) = J_0$.

Y^0 trivially satisfies the integrability in **(A.2)**, whose second condition also follows from **(A.2)**$'$. Therefore, a one-point extension Y of Y^0 exists by the method described above. We then do a time change of Y by means of its PCAF $\int_0^t \gamma(Y_s)^{-1}ds$, which is evidently a one-point extension of X^0. □

Remark 7.5.7. Let X^0 be an m-symmetric Hunt process on E_0 satisfying **(A.1)**. Suppose we are given a one-point extension $X = (X_t, \mathbf{P}_x)$ of X^0 such that X is a Hunt process whose Dirichlet form is regular and X admits no jumps from or to a.

By applying Theorem 5.7.5 to X and $F = \{a\}$, we readily see that the entrance law $\{Q_t^*\}$ induced by the exit system for $(X, \{a\})$ is necessarily characterized by equation (7.5.13).

In fact, one can say more ([28]). Since $\{a\}$ is not m-polar with respect to X, there exists a PCAF ℓ_t of X with Revuz measure $\delta_{\{a\}}$. Let τ_t be its right continuous inverse and $(\mathbf{p}, \mathbf{P}_a)$ be a point process taking values in W defined by

$$\mathcal{D}_{\mathbf{p}(\omega)} = \{s \in (0,\infty) : \tau_{s-}\omega < \tau_s\omega\}, \quad \mathbf{p}_s(\omega) = i_{\tau_{s-}}\omega \text{ for } s \in \mathcal{D}_{\mathbf{p}(\omega)}.$$

Here the sample space $\Omega = \{\omega\}$ is taken to be of function space type and $i_t = k_{\sigma_a} \circ \theta_t$ for the *killing operator* k_t specified by $X_s(k_t\omega)$ being equal to $X_s(\omega)$ if $s < t$ and to Δ if $s \geq t$. Define $\mathbf{n}$ by (7.5.13) and (7.5.14). Then $(\mathbf{p}, \mathbf{P}_a)$ is a Poisson point process with characteristic measure $\mathbf{n}$ absorbed at the random time T as is described above. Furthermore, $\mathbf{n} = k_{\sigma_a}\mathbf{P}_a^*$ where $(\mathbf{P}_a^*, \ell)$ is an exit system for $(X, \{a\})$.

The above-mentioned proof of Theorem 7.5.6 asserts that the converse procedure starting from X^0 and constructing X is possible. □

We call the X^0-entrance law $\{\nu_t; t > 0\}$ characterized by (7.5.13) the *entrance law for* the one-point extension X. According to (5.7.5), it admits the following explicit expression: for $t > 0$, $B \in \mathcal{B}_+(E_0)$,

$$\int_0^t \nu_s(B)ds = \int_B \mathbf{P}_x^0\left(\zeta^0 \leq t,\ X_{\zeta^0 -} = a\right) m(dx). \tag{7.5.17}$$

Kiyosi Itô [94] introduced the notion of the Poisson point process of excursions around one point a in the state space of a standard Markov process X just in a way of Remark 7.5.7. He was motivated by giving systematic constructions of Markovian extensions of the absorbing diffusion process X^0 on the half-line $(0, \infty)$ subjected to Feller's general boundary conditions [96]. Itô had constructed the most general jump-in process from the exit boundary 0 by using the Poisson point process in his unpublished lecture notes [93] that preceded [94]. In that case, a is just the point 0. However, recent work ([75], [21], [28]) reveals that Itô's program works equally well in constructing a one-point extension by conceiving a certain set K as a single point a^*.

To be precise, let E be a locally compact separable metric space and m be an everywhere dense positive Radon measure E. Consider a closed subset K of E and put $E_0 = E \setminus K$. We assume that either K is compact or E_0 is relatively compact in E. Let us extend the topological space E_0 to $E^* = E_0 \cup \{a^*\}$ by adding an extra point a^* to E_0 whose topology is prescribed as follows: a subset U of E^* containing the point a^* is an open neighborhood of a^* if there is an open set $U_1 \subset E$ containing K such that $U_1 \cap E_0 = U \setminus \{a^*\}$. In other words, E^* is obtained from E_0 by identifying K into one point a^*. Notice that in the special case that $\overline{E}_0$ is compact in E, $E^* = E \cup \{a^*\}$ is nothing but the one-point compactification of E_0. The restriction of the measure m to E_0 will be denoted by m_0, which is then extended to E^* by setting $m_0(\{a^*\}) = 0$. (f, g) will denote the integral of $f \cdot g$ on E_0 against the measure m_0.

Let $X = (X_t, \mathbf{P}_x, \zeta)$ be an m-symmetric Hunt process on E whose Dirichlet form $(\mathcal{E}, \mathcal{F})$ on $L^2(E; m)$ is regular. Let $X^0 = (X_t^0, \mathbf{P}_x^0, \zeta^0)$ be the part process of X on E_0. X^0 is an m_0-symmetric Hunt process on E_0 (cf. Exercise 3.3.7). The resolvent of X^0 is denoted by $\{G_\alpha^0; \alpha > 0\}$. The energy functional for X^0 is denoted by L^0 again. We aim at creating a q.e. one-point extension of X^0 from E_0 to E^*.

We shall assume once and for all that X satisfies the following conditions:

(B.1) X is irreducible.

(B.2) $m_0(U \cap E_0)$ is finite for some neighborhood U of K.

(B.3) X admits no killings inside or jumps from E_0 to K.

Define the functions $\{\varphi^K,\ u_\alpha^K,\ \alpha > 0\}$ on E_0 by

$$\varphi^K(x) = \mathbf{P}_x^0\Big(\zeta^0 < \infty,\ X_{\zeta^0 -}^0 \in K\Big),\ u_\alpha^K(x) = \mathbf{E}_x^0\Big[e^{-\alpha\zeta^0};\ X_{\zeta^0 -}^0 \in K\Big]. \quad (7.5.18)$$

Just as in Exercise 7.5.2, we can then verify that for $x \in E_0$

$$\varphi^K(x) = \mathbf{P}_x(\sigma_K < \infty) \quad \text{and} \quad u_\alpha^K(x) = \mathbf{E}_x\left[e^{-\alpha\sigma_K}\right] \quad (7.5.19)$$

Recall the conditions **(A.1)–(A.4)** imposed on X^0 in Theorem 7.5.6. Let us denote by **(A°.1)**, **(A°.3)**, and **(A°.4)** the conditions on the present X^0 obtained by replacing a with K in (7.5.1), **(A.3)**, and **(A.4)**, respectively.

DEFINITION 7.5.8. A Borel right process $X^* = (X_t^*,\ \mathbf{P}_x^*,\ \zeta^*)$ on E^* is called a q.e. *one-point extension* of X^0 if X^* is m_0-symmetric and has no killings inside, and the part process of X^* on E_0 coincides with X^0 q.e., namely, outside some m-polar set for X^0.

THEOREM 7.5.9. *Assume that conditions* **(B.1)–(B.3)** *hold for X and that* X^0 *satisfies* **(A°.1)**, **(A°.3)** *as well as* **(A°.4)** *in the non-diffusion case. Then the following hold.*

(i) *There exists a q.e. one-point extension* X^* *of* X^0 *from* E_0 *to* E^*. *Such an extension process is unique in law.*
(ii) *Denote by* $\{G^*_\alpha; \alpha > 0\}$ *the resolvent of* X^*. *Then* (7.5.5) *holds with* G_α, φ *and* u_α *being replaced by* G^*_α, φ^K *and* u^K_α, *respectively. With this replacement,* (7.5.6) *holds for q.e.* $x \in E_0$.
(iii) *Denote by* $(\mathcal{E}^*, \mathcal{F}^*)$ *the Dirichlet form of* X^* *on* $L^2(E^*; m_0)$. *Then the statement* (iii) *of Theorem 7.5.4 holds with* $\mathcal{F}^*$, $\mathcal{E}^*$, φ^K *and* u^K_α *in place of* $\mathcal{F}$, $\mathcal{E}$, φ, *and* u_α, *respectively.*
(iv) *Denote by* $\mathcal{A}^*$ *the generator of the* L^2*-semigroup of* X^*. *Define the flux* $\mathcal{N}(f)(a^*)$ *at* a^* *by* (7.4.25) *with* u^K_α *in place of* u_α. *Then the statement* (iv) *of Theorem 7.5.4 holds true with* $X^*, \mathcal{A}^*$ *and* a^* *in place of* X, $\mathcal{A}$ *and* a, *respectively.*

If in addition X *satisfies the condition* **(AC)** *in Definition A.2.16, then "q.e." in statements* (i) *and* (ii) *above can be dropped.*

Proof. (i) It follows from **(B.3)** that X^0 admits no killing inside E_0. This together with **(A°.1)** implies that X^0 satisfies **(A.1)** with a^* in place of a. The properties **(A.3)** and **(A.4)** for X^0 with a^* in place of a follow from **(A°.3)** and **(A°.4)**, respectively. By virtue of Theorem 7.5.6, it now suffices to prove that X^0 satisfies the condition **(A.2)**$'$ with a^* in place of a holding for q.e. $x \in E_0$.

We note that $\mathbf{P}_x(X_{\sigma_K-} \in K,\ \sigma_K < \infty) = \mathbf{P}_x(\sigma_K < \infty)$ for $x \in E$, as X admits no jumps from E_0 to K by **(B.3)**. Moreover, the function φ^K defined by (7.5.18) coincides with the hitting probability of K by X so that K is non-m-polar with respect to X by the assumption **(A°.1)**. By the irreducibility assumption **(B.1)**, X is either recurrent or transient. In the recurrent case, $\mathbf{P}_x(\sigma_K < \infty) = 1$ for q.e. $x \in E$, by Theorem 3.5.6(ii). Consequently, $\mathbf{P}_x(X_{\sigma_K-} \in K,\ \sigma_K < \infty) = 1$ and $\mathbf{P}^0_x(X_{\zeta^0-} \in K, \zeta^0 < \infty) = 1$ for q.e. $x \in E$. When X is transient, $X_{\sigma_K-} \in K$, $\mathbf{P}_x$-a.s. on $\{\sigma_K < \infty\}$, while $X_{\zeta-} = \partial$, $\mathbf{P}_x$-a.s. for q.e. $x \in E$, in view of Theorem 3.5.2 and the condition **(B.3)**. Therefore, $\mathbf{P}^0_x(X_{\zeta^0-} \in K \cup \{\partial\}) = 1$ for q.e. $x \in E_0$, as was to be proved.

Notice that if in addition X satisfies the absolute continuity condition **(AC)**, then so does X^0 and we can then replace "for q.e. $x \in E_0$" by "for every $x \in E_0$" in the above conclusion.

That the law of X is unique follows directly from Theorem 7.5.4.

(ii), (iii), and (iv) are consequences of Theorem 7.5.4. □

We shall call such a procedure of obtaining X^* from X (or from X^0) *darning a hole* K or *collapsing a hole* K. The entrance law $\{\nu^K_t\}$ of X^0 taking part in darning a hole K is characterized by

$$\int_0^\infty \nu^K_t dt = \varphi^K \cdot m_0. \tag{7.5.20}$$

In view of (5.7.5), it admits the following explicit expression: for $t > 0$, $B \in \mathcal{B}_+(E_0)$,

$$\int_0^t \nu_s^K(B)ds = \int_B \mathbf{P}_x(\sigma_K \le t)\, m_0(dx), \tag{7.5.21}$$

in terms of the original process $X = (X_t, \mathbf{P}_x)$.

Assume that the originally given Hunt process X on E as above satisfies an additional condition that it admits no jumps from K to $E_0 = E \setminus K$. We assume that the part process X^0 on E_0 satisfies **(A°.1)**. Then the closed set K is not m-polar with respect to X and the exit system $(\mathbf{P}_x^*, L)$ for (X, K) is well defined by Definition 5.7.4. Note that the jump component J in the exit system vanishes due to the above-mentioned assumption on X.

Let $Q_t^*(x, \cdot)$, $t > 0$, $x \in K$, be the associated entrance law on E_0 defined by (5.7.13). By putting $f = \mathbf{1}_K$ in equation (5.7.14) of Theorem 5.7.5, we are immediately led to the following theorem from (7.5.20):

THEOREM 7.5.10. *The entrance law $\{\nu_t^K(\cdot)\}$ for the darning of the hole K is related to the entrance law $\{Q_t^*(x, \cdot)\}$ induced by the exit system $(\mathbf{P}^*, L)$ for (X, K) as*

$$\nu_t^K(B) = \int_K Q_t^*(x, B)\mu_L(dx), \quad t > 0,\ B \in \mathcal{B}(E_0), \tag{7.5.22}$$

where μ_L is the Revuz measure of the PCAF L.

We notice that the relation (7.5.22) holds for any process $\widetilde{X}$ on a state space $\widetilde{E}$ as above whose part process on $E_0 \subset \widetilde{E}$ equals X^0. For instance, consider a compact set $K \subset \mathbb{R}^n$ with continuous boundary and the absorbing Brownian motion X^0 on $D = \mathbb{R}^n \setminus K$. Then we can take as $\widetilde{X}$ either the Brownian motion on $\mathbb{R}^n$ or the RBM on $\overline{D}$.

A q.e. one-point extension X^* of an m_0-symmetric standard process X^0 is irreducible in view of Theorem 7.5.4. However, X^0 may not be irreducible so that if we change m_0 by multiplying an arbitrary positive constant on each irreducible component, X^0 is still symmetric with respect to the changed measure and may admit a different one-point extension.

Let us consider the assumption that

$$E_0 = E_{01} \cup \cdots \cup E_{0k} \text{ for some disjoint open sets } E_{0i},\ 1 \le i \le k, \text{ and each } E_{0i} \text{ is } X^0\text{-invariant.} \tag{7.5.23}$$

Assumption (7.5.23) means that

$$\mathbf{P}_x(\Omega_{E_{0i}}) = 1 \qquad \text{for every } x \in E_{0i} \text{ and every } i = 1, \ldots, k,$$

where

$$\Omega_{E_{0i}} := \left\{\omega \in \Omega : \{X^0_t(\omega),\ X^0_{t-}(\omega)\} \subset E_{0i} \text{ for every } t \in [0, \zeta^0)\right\}.$$

Condition (7.5.23) is equivalent to saying that X^0 does not travel over two different sets E_{0i} and E_{0j}, $i \neq j$, $1 \le i, j \le k$. If X^0 is a diffusion, then the second condition in (7.5.23) is automatically satisfied. The restrictions of functions and measures on E_0 to E_{0i} will be designated by the superscript ^{i} for $1 \le i \le k$.

Choosing any k-vector $\mathbf{p}$ with positive entries:

$$\mathbf{p} = (p_1, \dots, p_k) \quad \text{with } p_1, \dots, p_k > 0,$$

we define a new measure $\widetilde{m}_0$ on E_0 by

$$\widetilde{m}^i_0 = p_i \cdot m^i_0, \qquad 1 \le i \le k. \tag{7.5.24}$$

The measure $\widetilde{m}_0$ will be also designated by $m^{\mathbf{p}}_0$ to indicate its dependence on $\mathbf{p}$.

Clearly X^0 is $\widetilde{m}_0$-symmetric and we extend $\widetilde{m}_0$ to $E^* = E \cup \{a^*\}$ by setting $\widetilde{m}_0(\{a^*\}) = 0$.

PROPOSITION 7.5.11. *Assume that the same conditions as in Theorem 7.5.9 as well as the condition* (7.5.23) *are satisfied. Let $\widetilde{m}_0$ be the measure defined by* (7.5.24).

(i) *There exists an $\widetilde{m}_0$-symmetric Borel right process $\widetilde{X}^*$ on E^* that is a q.e. one-point extension of the process X^0 being regarded as an $\widetilde{m}_0$-symmetric Hunt process on E_0.*

(ii) *The entrance laws $\{\widetilde{\nu}_t, t > 0\}$ for $\widetilde{X}^*$ and $\{\nu_t, t > 0\}$ for X^* are related by*

$$\widetilde{\nu}^i_t = p_i \cdot \nu^i_t \qquad \text{for } 1 \le i \le k.$$

(iii) *The q.e. one-point extensions $\widetilde{X}^*$ and $\widetilde{X}'^*$ corresponding to two different k-vectors $\mathbf{p}$ and $\mathbf{p}'$ are equivalent in law if and only if*

$$\mathbf{p} = \lambda\, \mathbf{p}' \qquad \text{for some } \lambda > 0.$$

Proof. (i) Follows from Theorem 7.5.9 with $\widetilde{m}_0$ in place of m_0 there. (ii) is immediate from the characterization (7.5.13) of the entrance law. By substituting $m^{\mathbf{p}}_0$ and $m^{\mathbf{p}'}_0$ in (7.5.5), the corresponding one-point extensions can be seen to have the same resolvents if and only if the condition in (iii) is valid, □

Under the assumption (7.5.23), the Dirichlet form $(\mathcal{E}^0, \mathcal{F}^0)$ of X^0 on $L^2(E_0, m_0)$ and its reflected Dirichlet space $(\mathcal{E}^{0,\text{ref}}, (\mathcal{F}^0)^{\text{ref}})$ can be described as follows. For each $1 \le i \le k$, define the restriction $(\mathcal{E}^{0i}, \mathcal{F}^{0i})$ of $(\mathcal{E}^0, \mathcal{F}^0)$ to E_{0i} by $\mathcal{F}^{0i} = \mathcal{F}^0\big|_{E_{0i}}$ and

$$\mathcal{E}^{0i}(u|_{E_{0i}},\ v|_{E_{0i}}) = \mathcal{E}^0(u\mathbf{1}_{E_{0i}},\ v\mathbf{1}_{E_{0i}}) \qquad \text{for } u, v \in \mathcal{F}^0.$$

This is a transient Dirichlet form on $L^2(E_{0i}; m_0^i)$, whose reflected Dirichlet space will be denoted by $((\mathcal{F}^{0i})^{\rm ref}, \mathcal{E}^{{\rm ref},i})$. It holds then that

$$\mathcal{F}^0 = \{u : \ u|_{E_{0i}} \in \mathcal{F}^{0i} \quad \text{for } 1 \le i \le k\},$$

$$\mathcal{E}^0(u,v) = \sum_{i=1}^k \mathcal{E}^{0i}(u|_{E_{0i}}, v|_{E_{0i}}) \qquad \text{for } u, v \in \mathcal{F}^0,$$

$$(\mathcal{F}^0)^{\rm ref} = \{u : \ u|_{E_{0i}} \in (\mathcal{F}^{0i})^{\rm ref} \quad \text{for } 1 \le i \le k\}, \quad \text{and}$$

$$\mathcal{E}^{0,{\rm ref}}(u,v) = \sum_{i=1}^k \mathcal{E}^{{\rm ref},i}(u|_{E_{0i}}, u|_{E_{0i}}) \quad \text{for } u, v \in (\mathcal{F}^0)^{\rm ref}. \tag{7.5.25}$$

Now, for measure $\widetilde{m}_0$ defined by (7.5.24), we denote by $(\widetilde{\mathcal{E}}^0, \widetilde{\mathcal{F}}^0)$ the Dirichlet form of X^0 on $L^2(E_0, \widetilde{m}_0)$ and by $((\widetilde{\mathcal{F}}^0)^{\rm ref}, \widetilde{\mathcal{E}}^{0,{\rm ref}})$ its reflected Dirichlet space. We then readily see that

$$\widetilde{\mathcal{F}}^0 = \mathcal{F}^0 \quad \text{and} \quad \widetilde{\mathcal{E}}^0(u,v) = \sum_{i=1}^k p_i \, \mathcal{E}^{0i}(u^i, v^i) \quad \text{for } u, v \in \mathcal{F}^0, \tag{7.5.26}$$

$$(\widetilde{\mathcal{F}}^0)^{\rm ref} = (\mathcal{F}^0)^{\rm ref} \quad \text{and} \quad \widetilde{\mathcal{E}}^{0,{\rm ref}}(u,v) = \sum_{i=1}^k p_i \, \mathcal{E}^{{\rm ref},i}(u^i, v^i) \tag{7.5.27}$$

for $u, v \in (\mathcal{F}^0)^{\rm ref}$.

Applying Theorem 7.5.9(iii) to $\widetilde{m}_0$ and $\widetilde{X}^*$ and taking (7.5.26) and (7.5.27) into account, we arrive at the next theorem.

THEOREM 7.5.12. *Under the same assumptions as in Proposition 7.5.11, let $\widetilde{m}_0$ be the measure defined by (7.5.24) and $\widetilde{X}^*$ be the $\widetilde{m}_0$-symmetric Borel right process on E^* as appeared in Proposition 7.5.11(i).*

(i) *Let $(\widetilde{\mathcal{E}}^*, \widetilde{\mathcal{F}}^*)$ be the Dirichlet form of $\widetilde{X}^*$ on $L^2(E^*; \widetilde{m}_0)$. Then*
$\widetilde{\mathcal{F}}^*_e$ is a linear subspace of $(\mathcal{F}^0)^{\rm ref}$ spanned by $\mathcal{F}^0_e$ and φ,
$\widetilde{\mathcal{F}}^*$ is a linear subspace of $(\mathcal{F}^0)^{\rm ref}_a$ spanned by $\mathcal{F}^0$ and u_α, and

$$\widetilde{\mathcal{E}}^*(u,v) = \sum_{i=1}^k p_i \, \mathcal{E}^{{\rm ref},i}(u|_{E_{0i}}, u|_{E_{0i}}) \quad \textit{for } u, v \in \widetilde{\mathcal{F}}^*. \tag{7.5.28}$$

(ii) *Assume further that condition (7.3.4) is satisfied. Let $\widetilde{\mathcal{A}}^*$ be the $L^2(E^*; \widetilde{m}_0)$-infinitesimal generator of $\widetilde{X}^*$. Then, $f \in \mathcal{D}(\widetilde{\mathcal{A}}^*)$ if and only if*

$$f|_{E_{0i}} \in \mathcal{D}(\mathcal{L}^i) \ \textit{for } 1 \le i \le k, \quad f \textit{ admits an } X^0\textit{-fine limit at } a^*$$

and

$$\sum_{i=1}^{k} p_i \, \mathcal{N}^i(f|_{E_{0i}})(a^*) = 0,$$

where, for $1 \le i \le k$, $\mathcal{L}^i$ *is the linear operator defined by* (7.3.7) *with* $L^2(E_0, m_0)$, $\mathcal{F}^0$, $(\mathcal{E}^{0,\mathrm{ref}}, (\mathcal{F}^0)^{\mathrm{ref}})$ *being replaced by* $L^2(E_{0i}, m_0^i)$, $\mathcal{F}^{0i}$ *and* $(\mathcal{E}^{\mathrm{ref},i}, (\mathcal{F}^{0i})^{\mathrm{ref}})$, *respectively, and* $\mathcal{N}^i$ *denotes a flux at* a^* *defined by* (7.4.25) *with* $\mathcal{L}$, $\mathcal{E}^{0,\mathrm{ref}}$ *and* u_α *being replaced by* $\mathcal{L}^i$, $\mathcal{E}^{\mathrm{ref},i}$ *and* $u_\alpha^{K,i}$, *respectively.*

As compared to X^*, we may call $\widetilde{X}^*$ a *skew extension* of X^0.

7.6. EXAMPLES OF ONE-POINT EXTENSIONS

In this section, we shall consider various concrete examples of X^0 and exhibit their one-point extensions.

(1°) One-dimensional absorbing Brownian motion

Let $\mathbb{R}$ be the real line, I be its open subset and m be the Lebesgue measure on it. The restriction of m to I is denoted by m_0. We introduce function spaces by

$$\mathrm{BL}(I) := \left\{ u : \text{ absolutely continuous on } I \text{ with } \int_I (u')^2 dx < \infty \right\},$$

$$H^1_{0e}(I) := \{u \in \mathrm{BL}(I) : \ u = 0 \ \text{ at the finite boundary points of } I\},$$

$$H^1(I) := \mathrm{BL}(I) \cap L^2(I; m_0) \quad \text{and} \quad H^1_0(I) := H^1_{0e}(I) \cap L^2(I; m_0),$$

$$\mathbf{D}^I(u, v) := \int_I u'(x)v'(x)m_0(dx) \quad \text{and} \quad (u, v) := \int_I u(x)v(x)m_0(dx).$$

Let $X^0(I)$ be the absorbing Brownian motion on I and $(\mathcal{E}^0, \mathcal{F}^0)$ be its Dirichlet form on $L^2(I; m_0)$. It holds then that

$$(\mathcal{E}^0, \mathcal{F}^0) = (\tfrac{1}{2}\mathbf{D}^I, H^1_0(I)), \quad \mathcal{F}^0_e = H^1_{0e}(I), \quad \text{and} \quad (\mathcal{F}^0)^{\mathrm{ref}} = \mathrm{BL}(I). \tag{7.6.1}$$

The linear operator $\mathcal{L}_I$ on $L^2(I; m_0)$ introduced by (7.3.7) reads as follows:

$$\mathcal{D}(\mathcal{L}_I) = \{f \in H^1(I) : f' \text{ has an absolutely continuous version with } f'' \in L^2(I; m_0)\}, \tag{7.6.2}$$

$$\mathcal{L}_I f = \frac{1}{2} f'' \qquad \text{for } f \in \mathcal{D}(\mathcal{L}).$$

We let

$$\mathcal{N}_I(f) = \frac{1}{2}\mathbf{D}^I(f, u_\alpha) + (\mathcal{L}_I f, u_\alpha)$$

for $f \in \mathcal{D}(\mathcal{L}_I)$ and $u_\alpha(x) = \mathbf{E}_x^0[e^{-\alpha\zeta^0}; \zeta^0 < \infty]$, $x \in I$. An integration by parts gives

$$\mathcal{N}_I(f) = \begin{cases} f'(b-) - f'(a+) & \text{when } I = (a,b); \\ f'(a-) & \text{when } I = (-\infty, a); \\ -f'(a+) & \text{when } I = (a, \infty). \end{cases} \tag{7.6.3}$$

In fact, the existence of the limit of f' at finite end points of I is clear from (7.6.1). When $I = (a, \infty)$,

$$\mathcal{N}_I(f) = \lim_{\xi\to\infty} f'(\xi)u_\alpha(\xi) - f'(a+)$$

and, if the first term of the right hand side does not vanish, then $f \notin \mathrm{BL}(I)$.

We shall examine the one-point extension of $X^0(I)$ in several cases of I.

(i) Reflecting and circular Brownian motions

Let

$$I = (0, \infty), \quad E_0 = I, \quad E = [0, \infty) = I \cup \{0\}.$$

Note that the point 0 is approachable in finite time by $X^0(I)$ with probability 1; that is, $\varphi(x) := \mathbf{P}_x(\sigma_0 < \infty) = 1$ for every $x \in E_0$. So $X^0(I)$ satisfies **(A.1)** and **(A.3)**. It also satisfies **(A.2)**$'$ on account of Lemma 7.5.5(ii) and the fact that q.e. is a synonym of "everywhere" for $X^0(I)$ (see Example 3.5.7). So by Theorem 7.5.6, $X^0(I)$ has a one-point extension X to E.

Denote by $(\mathcal{E}, \mathcal{F})$ the Dirichlet form of X on $L^2(E; m_0)(= L^2(E_0; m_0))$. We conclude from Theorem 7.5.4(iii) and (7.6.1) that

$$\mathcal{F} = H^1(I) \qquad \text{and} \qquad \mathcal{E} = \frac{1}{2}\mathbf{D}^I,$$

namely, X is the reflected Brownian motion of $E = [0, \infty)$. This follows also from the uniqueness statement of Theorem 7.5.4 because the reflecting Brownian motion on $[0, \infty)$ is obviously a one-point extension of $X^0(I)$.

Since $\mathcal{F}^0 = \{u \in (\mathcal{F}^0)_a^{\mathrm{ref}} : u(0+) = 0\}$, the condition (7.3.4) is satisfied for $F = \{0\}$. By Theorem 7.5.4(iv) and (7.6.3), the L^2-generator $\mathcal{A}$ of X can be described as

$$\mathcal{D}(\mathcal{A}) = \{f \in \mathcal{D}(\mathcal{L}_I) : f'(0+) = 0\} \quad \text{and} \quad \mathcal{A}f(x) = \frac{1}{2}f''(x),\ x \in I,$$

where $\mathcal{D}(\mathcal{L}_I)$ is defined by (7.6.2). We can also get the Skorohod equation for $X = (X_t, \mathbf{P}_x)$:

$$X_t = X_0 + B_t + \ell_t, \quad t > 0, \quad \mathbf{P}_x\text{-a.s.}, \quad x \in [0, \infty),$$

where B_t is a Brownian motion with $B_0 = 0$ and ℓ_t is the positive continuous additive functional of X with Revuz measure $\delta_{\{0\}}$.

Next let $I = (0, 1)$, $E_0 = I$ and $E = I \cup \{a\}$ be the one-point compactification of I. As in the preceding case, we can apply Theorem 7.5.6 to $X^0(I)$ to get its one-point extension X to E. Let $(\mathcal{E}, \mathcal{F})$ and $\mathcal{A}$ be the Dirichlet form and the L^2-generator of X. Since

$$\mathcal{F}^0 = \{u \in (\mathcal{F}^0)_a^{\text{ref}} : u(0+) = u(1-) = 0\},$$

condition (7.3.4) is satisfied for $F = \{0, 1\}$. In the same way as above, we can conclude that

$$\mathcal{F} = H_0^1(I) \cup \{\text{constant functions}\} \quad \text{and} \quad \mathcal{E} = \frac{1}{2}\mathbf{D}^I,$$

$$\mathcal{D}(\mathcal{A}) = \{f \in \mathcal{D}(\mathcal{L}_I) : f(0+) = f(1-), \quad f'(0+) = f'(1-)\}, \quad \text{and}$$

$$\mathcal{A}f(x) = \frac{1}{2}f''(x) \qquad \text{for } f \in \mathcal{D}(\mathcal{A}) \text{ and } x \in I.$$

Consequently, X is the Brownian motion on the circle E, which can also be obtained by wrapping the Brownian motion on $\mathbb{R}$ to $[0, 1)$ (more precisely, by modulo 1).

(ii) Skew Brownian motion

Let

$$I = (-\infty, 0) \cup (0, \infty), \quad E = \mathbb{R}, \quad E_0 = I, \quad K = \{0\}, \quad E^* = \mathbb{R}.$$

We now apply Theorem 7.5.9 and Proposition 7.5.11 to Brownian motion X on $\mathbb{R}$ and absorbing Brownian motion $X^0(I)$ to get one-point extensions of the latter to $E^* = \mathbb{R}$ by darning the hole $K = \{0\}$. Obviously X satisfies the conditions **(B.1)**, **(B.2)**, and **(B.3)** and $X^0(I)$ satisfies the conditions **(A°.1)** and **(A°.3)**, so Theorem 7.5.9 and Proposition 7.5.11 are applicable. Note that $\mathbb{R}^+ = (0, \infty)$ and $\mathbb{R}^- = (-\infty, 0)$ are two invariant sets for X^0. The restrictions of functions and measures on $\mathbb{R}$ to $\mathbb{R}^+$, $\mathbb{R}^-$ will be denoted by putting superscript $+$ and $-$, respectively. We can then rewrite expression (7.6.1) as

$$\begin{aligned} \mathcal{F}_e^0 &= \left\{u : u^\pm \in H_{0e}^1(\mathbb{R}^\pm)\right\}, \\ \mathcal{E}^0(u, v) &= \frac{1}{2}\mathbf{D}^{\mathbb{R}^+}(u^+, v^+) + \frac{1}{2}\mathbf{D}^{\mathbb{R}^-}(u^-, v^-) \end{aligned} \tag{7.6.4}$$

for $u, v \in \mathcal{F}^0_e$, and we get

$$(\mathcal{F}^0)^{\rm ref} = \left\{u : u^\pm \in {\rm BL}(\mathbb{R}^\pm)\right\},$$
$$\mathcal{E}^{0,{\rm ref}}(u, v) = \frac{1}{2}\mathbf{D}^{\mathbb{R}^+}(u^+, v^+) + \frac{1}{2}\mathbf{D}^{\mathbb{R}^-}(u^-, v^-) \qquad (7.6.5)$$

for $u, v \in (\mathcal{F}^0)^{\rm ref}$. Note that the functions in $(\mathcal{F}^0)^{\rm ref}$ may not be continuous at 0.

For any $p_+ > 0$ and $p_- > 0$, let $\widetilde{m}_0$ be the measure on $\mathbb{R}$ defined by $\widetilde{m}_0(dx) = p_+dx$ on $(0, \infty)$, $\widetilde{m}_0(dx) = p_-dx$ on $(-\infty, 0)$ and $\widetilde{m}_0(\{0\}) = 0$. Then $X^0(I)$ can be regarded as an $\widetilde{m}_0$-symmetric diffusion on $\mathbb{R}_0$ whose Dirichlet form $(\widetilde{\mathcal{E}}^0, \widetilde{\mathcal{F}}^0)$ on $L^2(\mathbb{R}_0, \widetilde{m}_0)$ is described as

$$\widetilde{\mathcal{F}}^0_e = \mathcal{F}^0_e, \quad \widetilde{\mathcal{E}}^0(u, v) = \frac{p_+}{2}\mathbf{D}^{\mathbb{R}^+}(u^+, v^+) + \frac{p_-}{2}\mathbf{D}^{\mathbb{R}^-}(u^-, v^-), \ u, v \in \widetilde{\mathcal{F}}^0_e,$$

and accordingly,

$$(\widetilde{\mathcal{F}}^0)^{\rm ref} = (\mathcal{F}^0)^{\rm ref},$$
$$\widetilde{\mathcal{E}}^{0,{\rm ref}}(u, v) = \frac{p_+}{2}\mathbf{D}^{\mathbb{R}^+}(u^+, v^+) + \frac{p_-}{2}\mathbf{D}^{\mathbb{R}^-}(u^-, v^-) \quad \text{for } u, v \in (\widetilde{\mathcal{F}}^0)^{\rm ref}.$$

Since $\widetilde{\mathcal{F}}^0 = \{u \in (\widetilde{\mathcal{F}}^0)^{\rm ref}_a : u^-(0-) = u^+(0+) = 0\}$, condition (7.3.4) is satisfied for $K = \{0\}$.

Let m_0 be the Lebesgue measure and X^* be the m_0-symmetric extension of $X^0(I)$ to $E^* = \mathbb{R}$ by Theorem 7.5.9, namely, X^* is constructed based on the entrance law $\{\mu_t, t > 0\}$ for $X^0(I)$ specified by $\int_0^\infty \mu_t\, dt = m_0$. By (7.5.21), it is given by

$$\mu_t^\pm(B) = \int_B \mathbf{P}_x(\sigma_0 \in dt)dx, \quad B \in \mathcal{B}(\mathbb{R}^\pm), \qquad (7.6.6)$$

in terms of the Brownian motion $X = (X_t, \mathbf{P}_x)$ on $\mathbb{R}$.

By Theorem 7.5.9, the Dirichlet form $(\mathcal{E}^*, \mathcal{F}^*)$ of X^* on $L^2(\mathbb{R}; m_0)(= L^2(E_0; m_0))$ is given by

$$\mathcal{F}^*_e = \{f = f_0 + c : f_0 \in \mathcal{F}^0,\ c \in \mathbb{R}\} \quad \text{and} \quad \mathcal{E}^*(u, v) = \mathcal{E}^{\rm ref}(u, v) \qquad (7.6.7)$$

for $u, v \in \mathcal{F}^*_e$. In particular, we see from (7.6.7) that

$$\text{every } u \in \mathcal{F}^* \text{ is continuous at } 0. \qquad (7.6.8)$$

We can conclude from (7.6.4), (7.6.5), (7.6.7), and (7.6.8) that

$$(\mathcal{F}^*_e, \mathcal{E}^*) = ({\rm BL}(\mathbb{R}), \tfrac{1}{2}\mathbf{D}).$$

Hence X^* is nothing but the Brownian motion on $\mathbb{R}$.

On the other hand, let $\widetilde{X}^*$ be the $\widetilde{m}$-symmetric extension of X^0 in Proposition 7.5.11: namely, $\widetilde{X}^*$ is constructed based on the entrance law $\widetilde{\mu}_t$ for X^0 given by

$$\mu_t^{\pm}(B) = p_{\pm} \int_B \mathbf{P}_x(\sigma_0 \in dt)dx, \quad B \in \mathcal{B}(\mathbb{R}^{\pm}).$$

By Theorem 7.5.12, the Dirichlet form $(\widetilde{\mathcal{E}}^*, \widetilde{\mathcal{F}}^*)$ of $\widetilde{X}^*$ on $L^2(\mathbb{R}; \widetilde{m}_0)$ is given by

$$\widetilde{\mathcal{F}}^* = \mathcal{F}^* = H^1(\mathbb{R}),$$

$$\widetilde{\mathcal{E}}^*(u, v) = \widetilde{\mathcal{E}}^{\text{ref}}(u, v) \tag{7.6.9}$$

$$= \frac{1}{2} p_+ \mathbf{D}^{\mathbb{R}^+}(u^+, v^+) + \frac{1}{2} p_- \mathbf{D}^{\mathbb{R}^-}(u^-, v^-)$$

for $u, v \in H^1(\mathbb{R})$.

Let $\mathcal{A}^*$ be the infinitesimal generator of $\widetilde{X}^*$ on $L^2(\mathbb{R}; \widetilde{m}_0)$. We then see from Theorem 7.5.12(ii) and (7.6.3) that $f \in \mathcal{D}(\widetilde{\mathcal{A}}^*)$ if and only if

$$f^{\pm} \in \mathcal{D}(\mathcal{L}_{\mathbb{R}^{\pm}}) \quad \text{with} \quad f(0-) = f(0+) \text{ and } p_- f'(0-) = p_+ f'(0+) \tag{7.6.10}$$

and $\mathcal{A}^* f(x) = \frac{1}{2} f''(x)$, $x \in E_0$. Here $\mathcal{D}(\mathcal{L}_{\mathbb{R}^{\pm}})$ is defined by (7.6.2).

In accordance with Harrison and Shepp [84], we call a real-valued process Y *a skew Brownian motion* on $\mathbb{R}$ with parameter $\beta \in (-1, 1)$ if

$$Y_t = Y_0 + B_t + \beta L_t, \qquad t \geq 0, \tag{7.6.11}$$

where B is Brownian motion on $\mathbb{R}$ and L is the symmetric local time of Y at 0, that is,

$$L_t = \lim_{\varepsilon \to 0} \frac{1}{2\varepsilon} \int_0^t \mathbf{1}_{\{|Y_s| \leq \varepsilon\}} ds.$$

Actually the process $\widetilde{X}^*$ can be shown to be a skew Brownian motion with parameter $\frac{p_+ - p_-}{p_+ + p_-}$ in this sense. Furthermore, the local time L in equation (7.6.11) can be verified to be the positive continuous additive functional of Y having Revuz measure $(p_+ + p_-)\delta_0$. We refer the readers to [21, §5] for the proof.

(iii) One-point skew extensions of $\mathbf{X}^0$ obtained by identifying multi-points

Choose any $k-1$ points $-\infty < a_1 < a_2 < \cdots < a_{k-1} < \infty$ and let

$$I = \mathbb{R} \setminus \{a_1, a_2, \ldots, a_{k-1}\} = \bigcup_{i=1}^{k} I_i,$$

where

$$I_1 := (-\infty, a_1), \quad I_j := (a_j, a_{j+1}),\ 1 \le j \le k-1, \quad I_k := (a_{k-1}, \infty).$$

We consider the case that

$$E = \mathbb{R}, \quad E_0 = I, \quad K = \bigcup_{i=1}^{k-1} \{a_i\}, \quad E^* = I \cup \{a^*\},$$

where E^* is obtained from I by identifying the compact set K as one point a^* in the way described in Section 7.5. Measure m_0 is the restriction of the Lebesgue measure to I. The restrictions of functions and measures on I to the interval I_i will be designated by using the superscript ^{i}.

As in the preceding case, the Brownian motion X on $\mathbb{R}$ and the absorbing Brownian motion $X^0(I)$ satisfy the conditions imposed in Theorem 7.5.9. Let $\mathbf{p} = (p_1, p_2, \ldots, p_k)$ be a k-vector with positive entries. The absorbing Brownian motion $X^0(I)$ is then symmetric with respect to the measure

$$\widetilde{m}_0 = \sum_{i=1}^{k} p_i\, m_0^i,$$

so that we can construct its $\widetilde{m}_0$-symmetric extension $\widetilde{X}^*$ to E^* according to Proposition 7.5.11. The Dirichlet form $(\widetilde{\mathcal{E}}^*, \widetilde{\mathcal{F}}^*)$ on $L^2(E^*; \widetilde{m}_0)$ of $\widetilde{X}^*$ then admits the description

$$\widetilde{\mathcal{F}}^* = \{f \in H^1(\mathbb{R}) : f(a_1) = f(a_2) = \cdots = f(a_{k-1})\}, \quad \text{and}$$

$$\widetilde{\mathcal{E}}^*(f, g) = \sum_{i=1}^{k} \frac{1}{2}\, p_i\, \mathbf{D}^{I_i}(f^i, g^i) \qquad \text{for } f, g \in \widetilde{\mathcal{F}}^*. \tag{7.6.12}$$

Let $\mathcal{A}^*$ be the L^2-infinitesimal generator of $\widetilde{X}^*$ on $L^2(\mathbb{R}; \widetilde{m}_0)$. As in previous cases, condition (7.3.4) is satisfied. Accordingly, $f \in \mathcal{D}(\widetilde{\mathcal{A}}^*)$ if and only if

$$f|_{I_i} \in \mathcal{D}(\mathcal{L}_{I_i}) \quad \text{for } 1 \le i \le k \quad \text{with}$$
$$f(a_1\pm)=f(a_2\pm)=\cdots=f(a_{k-1}\pm) \quad \text{and} \tag{7.6.13}$$
$$p_1 f'(a_1+)+\sum_{j=2}^{k-1} p_j\left(f'(a_j-)-f'(a_{j-1}+)\right) - p_k f'(a_{k-1}-)=0.$$

Both (7.6.12) and (7.6.13) can be shown in the same way as in the previous cases by using Theorem 7.5.12 and (7.6.3).

(2°) Absorbing diffusion on a half-line

We consider an open half-line $I=(0,\infty)$, a pair (s,m) of a canonical scale and a canonical measure on I, and associated diffusions studied in Example 3.5.7. Let $\mathcal{E}^{(s)}(u,v)=\int_I \frac{du}{ds}\frac{dv}{ds}ds$ and $\mathcal{F}^{(s)}$ be the space of those functions u on I which are absolutely continuous with respect to s and $\mathcal{E}^{(s)}(u,u)<\infty$. We assume that the left boundary 0 of I is regular but the right boundary ∞ is non-regular.

Then

$$(\mathcal{E},\mathcal{F})=(\mathcal{E}^{(s)},\mathcal{F}^{(s)}\cap L^2(I;m)) \tag{7.6.14}$$

is a strongly local regular Dirichlet form on $L^2([0,\infty);m)$ and the associated diffusion $X=(X_t,\mathbf{P}_x)$ on $[0,\infty)$ is by definition the reflecting diffusion, which satisfies $\mathbf{P}_x(\sigma_0<\infty)>0$ for any $x\in I$. The part process $X^0=(X^0_t,\mathbf{P}^0_x,\zeta^0)$ of X on I is the absorbing diffusion satisfying

$$\mathbf{P}^0_x\left(X_{\zeta^0-}=0,\ \zeta^0<\infty\right)>0 \qquad \text{for any } x\in I. \tag{7.6.15}$$

The Dirichlet form $(\mathcal{E}^0,\mathcal{F}^0)$ of X^0 is given by

$$\mathcal{F}^0=\{u\in\mathcal{F}: u(0+)=0\}, \quad \mathcal{E}^0=\mathcal{E}^{(s)}. \tag{7.6.16}$$

X is a one-point extension of X^0. In particular, X^0 satisfies the integrability condition in **(A.2)** on account of Lemma 7.5.5(iii).

In view of (6.5.8),

$$\left(\mathcal{E}^{0,\mathrm{ref}},(\mathcal{F}^0)^{\mathrm{ref}}\right)=\left(\mathcal{E}^{(s)},\mathcal{F}^{(s)}\right), \quad (\mathcal{F}^0)^{\mathrm{ref}}_a=\mathcal{F}^{(s)}\cap L^2(I;m). \tag{7.6.17}$$

Characterization (7.3.7) now reads

$$\mathcal{D}(\mathcal{L})=\left\{f\in\mathcal{F}^{(s)}\cap L^2(I;m): d\frac{df}{ds} \text{ is absolutely continuous with respect to } m,\ \mathcal{L}f\in L^2(I;m)\right\}, \tag{7.6.18}$$

$$\mathcal{L}f = \frac{d}{dm}\frac{d}{ds}f \qquad \text{for } f \in \mathcal{D}(\mathcal{L}). \tag{7.6.19}$$

Furthermore, from (7.6.14), (7.6.16), and (7.6.17), we get

$$\mathcal{F}^0 = \{(\mathcal{F}^0)^{\text{ref}}_a : f(0+) = 0\},$$

which means that the condition (7.3.4) is satisfied for $F = \{0\}$. Let us compute the flux $\mathcal{N}(f)(0)$ for $f \in \mathcal{D}(\mathcal{L})$ at 0 defined by (7.4.25).

LEMMA 7.6.1. *We have*

$$\mathcal{N}(f)(0) = -\frac{df}{ds}(0+) \qquad \text{for } f \in \mathcal{D}(\mathcal{L}).$$

Proof. For $f \in \mathcal{D}(\mathcal{L})$, the integration by parts gives

$$\int_0^x \frac{df}{ds}\frac{du_\alpha}{ds}ds + \int_0^x d\frac{df}{ds}u_\alpha = \frac{df}{ds}(x)u_\alpha(x) - \frac{df}{ds}(0+).$$

Since the left hand side converges to $\mathcal{N}(f)(0)$ as $x \to \infty$, the finite limit

$$c = \lim_{x\to\infty} u_\alpha(x) \cdot \frac{df}{ds}(x)$$

exists and hence it suffices to prove $c = 0$.

Since ∞ is assumed to be non-regular, either m or s diverges near ∞. Suppose m diverges near ∞. Then $\lim_{x\to\infty} u_\alpha(x) = 0$ because u_α is non-increasing in x and m-integrable by the preceding observation. If c were not 0, then $\frac{df}{ds}$ diverges near ∞, violating the property that $f \in \mathcal{F}^{(s)}$. Next suppose s diverges near ∞. Then the same property of f implies $\lim_{x\to\infty} \frac{df}{ds}(x) = 0$ and we get $c = 0$. □

Let $\mathcal{A}$ be the infinitesimal generator of the L^2-semigroup of the reflecting diffusion X. By virtue of Theorem 7.5.4, we conclude that $f \in \mathcal{D}(\mathcal{A})$ if and only if $f \in \mathcal{D}(\mathcal{L})$, the limit $f(0+)$ exists, and

$$\frac{df}{ds}(0+) = 0. \tag{7.6.20}$$

In this case we have $\mathcal{A}f = \frac{d}{dm}\frac{d}{ds}f$, $f \in \mathcal{D}(\mathcal{A})$.

In view of (7.6.14) and (7.6.17), X is a reflecting extension of X^0 in the sense of Definition 7.2.6, so the above characterization of $\mathcal{A}$ also follows from (7.3.21).

(3°) Diffusions on half-lines merging at one point

We consider a finite number of disjoint rays ℓ_i, $i = 1, \ldots, k$, on $\mathbb{R}^2$ merging at a point $a \in \mathbb{R}^2$. Each ray ℓ_i is homeomorphic to the open half-line $(0, \infty)$

and the point $a \in \mathbb{R}^2$ is the boundary of each ray at 0-side. We put

$$E_0 = \cup_{i=1}^k \ell_i, \qquad E = E_0 \cup \{a\}.$$

E is endowed with the induced topology as a subset of $\mathbb{R}^2$.

Let m be a positive Radon measure on E such that $\mathrm{supp}[m] = E$ and $m(\{a\}) = 0$. The restriction of m to ℓ_i is denoted by m^i. For any function g on E_0, its restriction to ℓ_i will be denoted by g^i. We consider a diffusion process $X^0 = \{X_t^0, \zeta^0, \mathbf{P}_x^0\}$ on E_0 such that its restriction $X^{0,i}$ to each open half-line $\ell_i \sim (0, \infty)$ is the absorbing diffusion governed by the speed measure m^i and a canonical scale, say s^i, which is assumed to satisfy

$$s^i(0+) > -\infty, \qquad 1 \le i \le k.$$

Since $m^i((0,1)) < \infty$, $1 \le i \le k$, 0 is a regular boundary for each $X^{0,i}$, $1 \le i \le k$. We shall also assume that ∞ is non-regular for each $X^{0,i}$, $1 \le i \le k$.

In view of the observation made in the preceding example, each $X^{0,i}$ satisfies (7.6.15) and the integrability condition in **(A.2)**. Since $R_1^{0,i}(L^2(\ell_i; m^i)) \subset \mathcal{F}^{(s_i)}$, $R_1^{0,i} f^i$ is lower semicontinuous for any non-negative Borel function f on E_0. Therefore, by taking Lemma 7.5.5(i) into account, we conclude that the diffusion X^0 meets all conditions **(A.1)**, **(A.2)**, and **(A.3)** in Theorem 7.5.6 yielding a one-point diffusion extension X of X^0 from E_0 to E.

For each $i \le i \le k$, denote by $(\mathcal{E}^{0,i}, \mathcal{F}^{0,i})$ the Dirichlet form of $X^{0,i}$ on $L^2(\ell_i; m^i)$ and by $((\mathcal{F}^{0,i})^{\mathrm{ref}}, \mathcal{E}^{\mathrm{ref},i})$ the reflected Dirichlet space. We use $(\mathcal{E}^{(s^i)}, \mathcal{F}^{(s^i)})$ to denote the function space considered in **(i)** but with ℓ_i and s^i in place of $I = [0, \infty)$ and s there. It follows from **(i)** that

$$\left(\mathcal{E}^{\mathrm{ref},i}, (\mathcal{F}^{0,i})^{\mathrm{ref}}\right) = \left(\mathcal{E}^{(s^i)}, \mathcal{F}^{(s^i)}\right), \tag{7.6.21}$$

$$\mathcal{F}^{0,i} = \left\{v \in \mathcal{F}^{(s^i)} \cap L^2((0,\infty); m^i) : \ v(0+) = 0\right\}, \tag{7.6.22}$$

and

$$\mathcal{E}^{0,i}(v_1, v_2) = \mathcal{E}^{(s^i)}(v_1, v_2) \qquad \text{for } v_1, v_2 \in \mathcal{F}^{0,i}.$$

By (7.6.21) and (7.6.22), we see that the condition (7.3.4) is fulfilled again.

Now let $(\mathcal{E}, \mathcal{F})$ be the Dirichlet form on $L^2(E : m)$ of the one-extension X of X^0 to E as given by Theorem 7.5.6. Since X^0 has the property (7.5.23) with $E_{0i} = \ell_i$, $1 \le i \le k$, we have from (7.5.25) and Theorem 7.5.4(iii)

$$\mathcal{F} = \{f = f_0 + c\,u_\alpha : f_0^i \in \mathcal{F}^{0,i}, \ 1 \le i \le k, \ c \in \mathbb{R}\} (\subset (\mathcal{F}^0)_a^{\mathrm{ref}})$$

$$\mathcal{E}(u, v) = \sum_{i=1}^k \mathcal{E}^{s^i}(u^i, v^i) \qquad \text{for } u, v \in \mathcal{F},$$

where $\mathcal{F}^{0,i}$ are specified by (7.6.22).

The entrance law $\{\mu_t, t > 0\}$ from a for X is the sum of its restriction μ_t^i to ℓ_i, which is describable as

$$\mu_t^i(f)dt = \mathbf{P}^{0,i}_{f\cdot m^i}\left(X^{0,i}_{\zeta^{0,i}-} = 0 \text{ and } \zeta^{0,i} \in dt\right). \tag{7.6.23}$$

We next choose any k-vector $\mathbf{p} = (p_1, \ldots, p_k)$ with positive entries and define a new measure $\widetilde{m}$ on E_0 by

$$\widetilde{m}^i = p_i m^i, \qquad 1 \le i \le k,$$

which is extended to E by setting $\widetilde{m}(\{a\}) = 0$. Since X^0 is also $\widetilde{m}$-symmetric, we can construct by Proposition 7.5.11 a unique $\widetilde{m}$-symmetric diffusion $\widetilde{X}$ on E with no sojorn or killing at a that extends X^0. By virtue of Theorem 7.5.12(i), the Dirichlet form $(\widetilde{\mathcal{E}}, \widetilde{\mathcal{F}})$ on $L^2(E; \widetilde{m})$ of $\widetilde{X}$ can be described as follows:

$$\widetilde{\mathcal{F}} = \mathcal{F} = \left\{f = f_0 + c\, u_\alpha : f_0^i \in \mathcal{F}^{0,i} \text{ for } 1 \le i \le k, \text{ and } c \in \mathbb{R}\right\},$$

$$\widetilde{\mathcal{F}}_e = \mathcal{F}_e = \left\{f = f_0 + c\, \varphi : f_0^i \in \mathcal{F}_e^{0,i} \text{ for } 1 \le i \le k, \text{ and } c \in \mathbb{R}\right\},$$

$$\widetilde{\mathcal{E}}(u, v) = \sum_{i=1}^k p_i\, \mathcal{E}^{s^i}(u^i, v^i) \qquad \text{for } u, v \in \widetilde{\mathcal{F}}_e.$$

Let $\widetilde{\mathcal{A}}$ be the $L^2(E; \widetilde{m}_0)$-infinitesimal generator of $\widetilde{X}$. Combining Theorem 7.5.12(ii) with (7.6.18), (7.6.19), and Lemma 7.6.1, we can see that $f \in \mathcal{D}(\widetilde{\mathcal{A}})$ if and only if the following conditions are satisfied:

$$f^i \in \mathcal{F}^{(s^i)} \cap L^2(\ell_i; m^i)$$

$$\frac{df^i}{ds^i} \text{ is absolutely continuous with respect to } m^i,$$

$$\mathcal{L}^i f^i = \frac{d}{dm^i}\frac{d}{ds^i} f^i \in L^2(\ell_i; m^i) \qquad \text{for } 1 \le i \le k,$$

$$-\infty < f^1(0+) = \cdots = f^k(0+) < \infty, \qquad \sum_{i=1}^k p_i \frac{df^i}{ds^i}(0+) = 0.$$

We have in this case

$$\widetilde{\mathcal{A}} f(x) = \mathcal{L}^i f^i(x) \quad \text{if } x \in \ell_i \qquad \text{for } 1 \le i \le k.$$

The entrance law $\{\widetilde{\mu}_t, t > 0\}$ from a for $\widetilde{X}$ is given by

$$\widetilde{\mu}_t^i = p_i \mu_t^i, \qquad 1 \le i \le k,$$

where $\{\mu_t^i, t > 0\}$ is given by (7.6.23).

Clearly Example (**1°**) (**ii**) of this section may be considered as a special case of the present one with $k = 2$ and $E = \mathbb{R}$.

When $m^i(dx) = dx$ and $s^i(x) = x$ for all $i = 1, \ldots, k$, it is easy to see that $\widetilde{X}$ is a Walsh Brownian motion [7] with k number of random markers.

We remark that in this example, one can also replace diffusions on some half-lines by discontinuous Markov processes that can approach 0 such as censored α-stable processes on $(0, \infty)$ with $\alpha > 1$. The latter will be described in (**5°**) below.

(4°) Brownian motion on $\mathbb{R}^n$

Let $X = (X_t, \mathbf{P}_x)$ be the Brownian motion on the Euclidean space $\mathbb{R}^n$ for $n \geq 2$, K be a non-polar compact subset of $\mathbb{R}^n$, and $X^0 = (X_t^0, \mathbf{P}_x^0, \zeta^0)$ be the absorbing Brownian motion on $E_0 = \mathbb{R}^n \setminus K$.

Conditions **(B.1)**, **(B.2)**, and **(B.3)** are trivially satisfied by X. The irreducibility of X and the non-polarity of K imply that $\mathbf{P}_x(\sigma < \infty) > 0$ for a.e. $x \in \mathbb{R}^n$ and consequently for every x because the transition function of X is absolutely continuous. Hence X^0 satisfies **(A°.1)**. Since the resolvent G_α^0 of X^0 is related to the resolvent G_α of X by

$$G_\alpha^0 f(x) = G_\alpha f(x) - \mathbf{E}_x \left[e^{-\alpha\sigma_K} G_\alpha f(X_{\sigma_K}) \right]$$

and the second term of the right hand side is α-harmonic in $x \in E_0$, X^0 is strong Feller in the sense that $G_\alpha^0(b\mathcal{B}(E_0)) \subset bC(E_0)$ and so the condition **(A°.3)** is fulfilled by X^0.

By Theorem 7.5.9, a unique m_0-symmetric diffusion X^* extending X^0 to $E^* = E_0 \cup \{a^*\}$ can be constructed by darning the hole K. Here E^* is the one-point extension of E_0 by regarding the set K as one point a^* and m_0 is the Lebesgue measure on E_0 extended to E by setting $m_0(\{a^*\}) = 0$.

By (7.5.21), the entrance law for X^0 taking part in the darning is given by

$$\int_0^t \nu_s^K(B)ds = \int_B \mathbf{P}_x(\sigma_K \leq t)dx, \quad t > 0,\ B \in \mathcal{B}(E_0), \tag{7.6.24}$$

in terms of the Brownian motion $X = (X_t, \mathbf{P}_x)$. When K is a ball centered at the origin, ν_t^K is therefore a spherically symmetric measure on E_0.

For the open set $D \subset \mathbb{R}^n$, we consider the function spaces $\mathrm{BL}(D)$, $H^1(D)$, $H_0^1(D)$ and $H_{0,e}^1(D)$ as well as the Dirichlet integral $\mathbf{D}$ introduced in Section 2.2.4. The Dirichlet form $(\mathcal{E}^0, \mathcal{F}^0)$ of X^0 on $L^2(E_0)$ equals $(\frac{1}{2}\mathbf{D}, H_0^1(E_0))$. In view of Example (**4°**) of Section 6.5, the reflected Dirichlet space of $(\mathcal{E}^0, \mathcal{F}^0)$ is equal to $(\mathrm{BL}(E_0), \frac{1}{2}\mathbf{D})$.

The linear operator $\mathcal{L}$ on $L^2(E_0)$ specified by (7.3.7) and the flux $\mathcal{N}(f)(a^*)$ specified by Theorem 7.5.9(iv) are

$$\mathcal{L} = \tfrac{1}{2}\Delta \qquad \text{with} \qquad \mathcal{D}(\mathcal{L}) = \left\{ f \in H^1(E_0) : \Delta f \in L^2(E_0) \right\},$$

$$\mathcal{N}(f)(a^*) = \tfrac{1}{2}\mathbf{D}(f, u_\alpha^K) + \frac{1}{2}(\Delta f, u_\alpha^K) \qquad \text{for } f \in \mathcal{D}(\mathcal{L}).$$

By Theorem 7.5.9, the Dirichlet form $(\mathcal{E}^*, \mathcal{F}^*)$ of X^* on $L^2(E^*; m_0)$ and its extended Dirichlet space $\mathcal{F}_e^*$ can be expressed as follows:

$$\begin{aligned} \mathcal{F}^* &= \{f = f_0 + c u_\alpha^K : f_0 \in H_0^1(E_0),\ c \in \mathbb{R}\}, \\ \mathcal{F}_e^* &= \{f = f_0 + c \varphi^K : f_0 \in H_{0,e}^1(E_0),\ c \in \mathbb{R}\}, \end{aligned} \tag{7.6.25}$$

$$\mathcal{E}^*(u, v) = \frac{1}{2}\mathbf{D}(u, v) \qquad \text{for } u, v \in \mathcal{F}_e^*. \tag{7.6.26}$$

Note that $\mathcal{F}^* \subset H^1(E_0)$ and $\mathcal{F}_e^* \subset \mathrm{BL}(E_0)$. We know that u_α (resp. φ) is $\mathcal{E}_\alpha^*$-orthogonal (resp. $\mathcal{E}^*$-orthogonal) to the space $H_0^1(E_0)$ (resp. $H_{0,e}^1(E_0)$).

The process X^* is an irreducible diffusion by Theorem 7.5.4(i) and Theorem 7.5.6. When $n = 2$, X^* is recurrent by Theorem 2.1.8 because $\varphi^K = 1$ on E_0. When $n \geq 3$, X^* is transient but still conservative. Indeed, $1 - \varphi^K(x) = \mathbf{P}_x^0(\zeta^0 = \infty)$ is P_t^0-invariant and so for every $t, s > 0$, $\langle \nu_{t+s}, 1 - \varphi \rangle = \langle \nu_t, P_s^0(1 - \varphi) \rangle = \langle \nu_t, 1 - \varphi \rangle$; that is, $\langle \nu_t, 1 - \varphi \rangle$ is a constant function in $t > 0$. Thus by (ii) and (iii) of Lemma 5.7.1, for each fixed $\alpha > 0$,

$$L^0(\varphi^K, 1 - \varphi^K) = \lim_{t \downarrow 0} \langle \nu_t^K, 1 - \varphi^K \rangle = \alpha \langle \bar{\nu}_\alpha^K, 1 - \varphi^K \rangle = \alpha (u_\alpha^K, 1 - \varphi^K).$$

Now it follows from Theorem 7.5.9(ii) that $\alpha G_\alpha^* \mathbf{1}(a^*) = 1$ and, consequently, $\alpha G_\alpha^* \mathbf{1} = \mathbf{1}$ on E^*. This shows that X^* is conservative.

Remark 7.6.2. The above proof shows that under the condition of Theorem 7.5.9, if X is conservative, then so is X^*. □

Clearly X^0 satisfies

$$\mathbf{P}_x^0 \left(X_{\zeta^0 -}^0 \in K \middle| \zeta^0 < \infty \right) = 1, \quad x \in E_0,$$

and condition (7.3.4) is fulfilled on account of Remark 7.3.3. By Theorem 7.5.9(iv), the generator $\mathcal{A}^*$ of X^* on $L^2(E^*; m_0)$ can be characterized as

$$f \in \mathcal{D}(\mathcal{A}^*) \Longleftrightarrow f \in \mathcal{D}(\mathcal{L}),\ f \text{ admits } X^0\text{-fine limit at } a^* \text{ and } \mathcal{N}(f)(a^*) = 0$$

and

$$\mathcal{A}^* f = \frac{1}{2}\Delta \qquad \text{for } f \in \mathcal{D}(\mathcal{A}^*).$$

When ∂K and f are smooth enough and f was of compact support, then the Gauss-Green formula yields

$$\mathcal{N}(f)(a^*) = -\frac{1}{2}\int_{\partial K} \frac{\partial f}{\partial \mathbf{n}}(\xi)\sigma(d\xi),$$

where $\mathbf{n}$ denotes the outward normal for ∂K and σ is the surface measure on ∂K. In this sense, $\mathcal{N}(f)(a^*)$ may be interpreted as the flux of the vector field $\frac{1}{2}\nabla f$ at a^* or into K.

Note that $(\mathcal{E}^*, \mathcal{F}^*)$ is a quasi-regular Dirichlet form on $L^2(E^*; m_0)$ but may not be a regular Dirichlet form unless every point of ∂K is a regular boundary point of E_0 with respect to the Dirichlet problem for $(\alpha - \frac{1}{2}\Delta)$ on E_0. Therefore, we cannot construct X^* by using the theory of the regular Dirichlet form in general.

Like planar reflecting Brownian motion (cf. Example **(2°)** of Section 5.3), X^* enjoys the following conformal invariance when $n = 2$. Note that the compact sets K and $\widehat{K}$ in the next theorem can be disconnected.

THEOREM 7.6.3. *Suppose $n = 2$, K and $\widehat{K}$ are two non-polar compact subsets of $\mathbb{R}^2$. Let X^* be the diffusion process on $E^* := (\mathbb{R}^2 \setminus K) \cup \{a^*\}$ obtained from X by darning the hole K into a single point a^*. Suppose that ϕ is a conformal map from $\mathbb{R}^2 \setminus K$ onto $\mathbb{R}^2 \setminus \widehat{K}$ that maps ∞ to ∞. Identify the compact set $\widehat{K}$ with a single point $\widehat{a}^*$ and equip $\widehat{E}^* := (\mathbb{R}^2 \setminus \widehat{K}) \cup \{\widehat{a}^*\}$ the topology induced from $\mathbb{R}^2$ by collapsing $\widehat{K}$ into $\widehat{a}^*$. Define $\phi(a^*) = \widehat{a}^*$. Then ϕ is a topological homeomorphism from E^* onto $\widehat{E}^*$, and $\phi(X^*)$ is, up to a time change, the diffusion process obtained from a Brownian motion on $\mathbb{R}^2$ by darning the hole $\widehat{K}$.*

Proof. Denote $\mathbb{R}^2 \setminus K$ and $\mathbb{R}^2 \setminus \widehat{K}$ by E_0 and $\widehat{E}_0$, respectively. Since ϕ maps ∞ to ∞, it maps the E_0-portion of any neighborhood of K into the $\widehat{E}_0$-portion of a neighborhood of $\widehat{K}$, and vice versa. Hence ϕ is a topological homeomorphism from $E^* = E_0 \cup \{a^*\}$ onto $\widehat{E}^* = \widehat{E}_0 \cup \{\widehat{a}^*\}$. The Lebesgue measure on $\mathbb{R}^2$ will be denoted by λ. As is noted in the above, $X^* = (X_t^*, \mathbf{P}_z^*)_{z \in E^*}$ is an irreducible recurrent diffusion on E^*. Further, its transition function is absolutely continuous with respect to $\lambda\big|_D$ because so is its resolvent kernel by Theorem 7.5.9(ii) and consequently Corollary 3.1.14 applies.

We let

$$m(dz) = |\phi'(z)|^2 \mathbf{1}_{E_0}(z)dz, \quad A_t = \int_0^t |\phi'(X_s^*)|^2 \mathbf{1}_{E_0}(X_s^*)ds. \tag{7.6.27}$$

Just as in Example **(2°)** of Section 5.3, we see that for every $z \in E^*$, $\mathbf{P}_z^*$-a.s. A_t is strictly increasing to ∞, as $t \to \infty$, and so is its inverse τ_t. The time-changed process $\check{X}^* = (X_{\tau_t}^*, \mathbf{P}_z^*)$ is an m-symmetric conservative diffusion on E^*. In fact, X^* is recurrent in view of Theorem 5.2.5.

Define a process $\widehat{Y}^* = (\widehat{Y}^*_t, \widehat{\mathbf{P}}^*_w)_{w\in\widehat{E}^*}$ by

$$\widehat{Y}^*_t = \phi(X^*_{\tau_t}), \quad \widehat{\mathbf{P}}^*_w = \mathbf{P}^*_{\phi^{-1}(w)}, \ w \in \widehat{E}^*. \tag{7.6.28}$$

$\widehat{Y}^*$ is then a conservative diffusion on $\widehat{E}^*$ and, as in the above cited example, it is symmetric with respect to the zero extension of λ on $\widehat{E}_0$ to $\widehat{E}^*_0$.

We claim that $\widehat{Y}^*$ has the same law as the diffusion process obtained from Brownian motion on $\mathbb{R}^2$ by darning the hole $\widehat{K}$. In view of Theorem 7.5.4 on the uniqueness one-point extension process, it suffices to show that the part process $\widehat{Y}^{*,0}$ of $\widehat{Y}^*$ in $\widehat{E}_0$ is an absorbing Brownian motion in $\widehat{E}_0$. Notice that $\widehat{Y}^{*,0}$ is the image by ϕ of the part process on E_0 of the time-changed process $(X^*_{\tau_t}, \mathbf{P}^*_z)$. The latter equals the part process of X^* on E_0, namely, the absorbing Brownian motion X^0 on E_0, being time-changed by the inverse of its PCAF $A^0_t = \int_0^{t\wedge\zeta_0} |\phi'(X^0_s)|^2 ds$. The desired property of $\widehat{Y}^{*,0}$ is now a direct consequence of Theorem 5.3.1. □

Remark 7.6.4. (i) When $n = 2$ and K is a disk in $\mathbb{R}^2$, let $\{\nu^K_t; t > 0\}$ be the entrance law for the absorbing Brownian motion X^0 on $E_0 := \mathbb{R}^2 \setminus K$ that takes part in darning the hole K for the Brownian motion X on $\mathbb{R}^2$. As is noted right after Theorem 7.5.10, we can then apply this theorem to the RBM $\widetilde{X}$ on $\overline{E}_0$ so that

$$\nu^K_t(B) = \int_{\partial K} Q^*_t(x, B)\mu_L(dx), \quad t > 0, \ B \in \mathcal{B}(E_0)$$

Here $\{Q^*_t(x, \cdot), t > 0, x \in \partial K\}$ is the entrance law induced by an exit system $(\mathbf{P}^*, L)$ for the RBM $\widetilde{X}$ on the boundary set ∂K and μ_L is the Revuz measure of a PCAF (boundary local time) L of $\widetilde{X}$ on ∂K. While exit system $(\mathbf{P}^*, L)$ of $\widetilde{X}$ on ∂K (and hence the pair $(\{Q^*_t(x, \cdot), t > 0, x \in \partial K\}, L)$) is not unique, $\{Q^*_t(x, \cdot)\mu_L(dx), t > 0\}$ is unique by (5.7.11). Since ∂K is a circle, we can and do take μ_L as the uniform measure on ∂K. The corresponding $\{Q^*_t(x, \cdot), t > 0, x \in \partial K\}$ is then the commonly used entrance law of reflecting Brownian motion $\widetilde{X}$ on ∂K in literature (cf. [90, 115]).

Hence the process X^* can also be obtained in the following heuristic way. Run a reflecting Brownian motion Y on $\mathbb{R}^2 \setminus K$. When Y hits the boundary ∂K, rotate the next excursion of Y away from ∂K by a random angle uniform over $[0, 2\pi)$, and then continue this process. Collapsing the hole K into a single point $\{a\}$ then results in a continuous process on E^* that has the same distribution as X^*. It is now easy to see that the process X^* obtained from X by darning the hole K as given in Theorem 7.5.9 can be identified with the *excursion-reflected Brownian motion* coined in Lawler [115]. The latter arose in the study of SLE in multiply-connected planar domains. Theorem 7.6.3 extends such an identification to any compact set K for which $\mathbb{R}^2 \setminus K$ is conformally equivalent to the complement of a closed ball in $\mathbb{R}^2$.

(ii) Same consideration can be made with the absorbing Brownian motion on a domain $D \subset \mathbb{R}^n$ in place of Brownian motion on $\mathbb{R}^n$. In this case, K is a compact subset of the domain D. When $n = 2$, Theorem 7.6.3 (conformal invariance of X^*) remains true for those conformal maps ϕ that map the $(D \setminus K)$-portion of any neighborhood of K into the $(D \setminus \widehat{K})$-portion of a neighborhood of $\widehat{K}$. See Example (**1**°) of Section 7.8 for darning multiple numbers of holes. □

We next consider the case where the closed set K is the complement of a bounded open set $E_0 \subset \mathbb{R}^n$. In this case, $E^* = E_0 \cup \{a^*\}$ is just the one-point compactification of E_0. The symmetric diffusion X^* extending the absorbing Brownian motion $X^0 = (X_t^0, \zeta^0, \mathbf{P}_x^0)$ on E_0 to E^* has the Dirichlet form $(\mathcal{E}^*, \mathcal{F}^*)$ on $L^2(E^*; m_0)$ expressible as

$$\mathcal{F}^* = H_0^1(E_0) + \{\text{constant functions on } E^*\},$$

$$\mathcal{E}^*(w_1, w_2) = \frac{1}{2}\mathbf{D}(f_1, f_2) \qquad \text{for } w_i = f_i + c_i \text{ with } f_i \in H_0^1(E_0)$$

$$\text{and } c_i \in \mathbb{R},\ i = 1, 2,$$

which is a regular, strongly local and irreducible recurrent Dirichlet form as has been studied in [75, §3]. Hence we can construct the symmetric diffusion X^* on E^* by a direct use of the Dirichlet form theory in this case. The L^2-generator of X^* can be characterized exactly in the same way as the preceding case.

Finally, let $n \geq 3$ and $m(dx) = m(x)dx$ be a measure on $\mathbb{R}^n$ with density m being strictly positive, bounded continuous, and integrable on $\mathbb{R}^n$. Let $Y = (Y_t, \mathbf{P}_x, \zeta^Y)$ be the time change of the Brownian motion $X = (X_t, \mathbf{P}_x)$ on $\mathbb{R}^n$ by means of its PCAF $A_t = \int_0^t m(X_s)ds$. As we saw in Example (**1**°) of Section 5.3, Y is m-symmetric and its Dirichlet form $(\mathcal{E}^Y, \mathcal{F}^Y)$ on $L^2(\mathbb{R}^n; m)$ is given by

$$\left(\mathcal{E}^Y,\ \mathcal{F}^Y\right) = \left(\frac{1}{2}\mathbf{D},\ \widetilde{H}_e^1(\mathbb{R}^n) \cap L^2(\mathbb{R}^n; m)\right). \tag{7.6.29}$$

The process Y is transient and its 0-order resolvent $R^Y f$ for $f \in b\mathcal{B}(\mathbb{R}^n)$ has the expression $R^Y f = 2I_2 * (fm)$ with the Newtonian convolution kernel $2I_2$ defined by (2.2.26). We notice that $I_2 * g \in C_\infty(\mathbb{R}^n)$ for any $g \in bL^1(\mathbb{R}^n)$. Since X_t converges to the point ∂ at infinity of $\mathbb{R}^n$ as $t \to \infty$ $\mathbf{P}_x$-a.s. for any $x \in \mathbb{R}^n$ and $R^Y 1(x) < \infty$ for any $x \in \mathbb{R}^n$, we have

$$\varphi(x) = \mathbf{P}_x\left(\zeta^Y < \infty, \quad Y_{\zeta^Y -} = \partial\right) = 1 \quad \text{for every } x \in \mathbb{R}^n. \tag{7.6.30}$$

We can now apply Theorem 7.5.6 to

$$E_0 = \mathbb{R}^n,\ X^0 = Y,\ a = \partial,\ E = \mathbb{R}^n \cup \{\partial\}$$

in getting a one-point extension X of Y from $\mathbb{R}^n$ to $\mathbb{R}^n \cup \{\partial\}$ because conditions **(A.1)**, **(A.2)** are clearly satisfied and **(A.3)** follows from the lower semicontinuity of $R^Y f$ for $f \in \mathcal{B}_+(\mathbb{R}^n)$ and Lemma 7.5.5(iv). By setting $m(\{\partial\}) = 0$, X is an m-symmetric recurrent diffusion on $\mathbb{R}^n \cup \{\partial\}$ and its Dirichlet form $(\mathcal{E}, \mathcal{F})$ on $L^2(\mathbb{R}^n; m)$ is given by

$$(\mathcal{E},\ \mathcal{F}) = \left(\frac{1}{2}\mathbf{D},\ \mathrm{BL}(\mathbb{R}^n) \cap L^2(\mathbb{R}^n; m)\right) \tag{7.6.31}$$

in view of (7.6.30), Theorem 7.5.4(iii), and Theorem 2.2.12.

On account of Example (**5°**) of Section 6.5, the active reflected Dirichlet form of (7.6.29) equals (7.6.31). We can draw from this the following two conclusions.

First, X is a reflecting extension of Y in the sense of Definition 7.2.6. Therefore, X is not only a one-point extension but also the maximal Silverstein extension of Y. Based on this fact, X has been proved to be the unique genuine symmetric extension of Y in [24].

Second, the associated operator $\mathcal{L}$ specified by (7.3.7) is given by

$$\begin{cases} \mathcal{D}(\mathcal{L}) = \{f \in \mathrm{BL}(\mathbb{R}^n) \cap L^2(\mathbb{R}^n; m) : \frac{1}{m}\Delta f \in L^2(\mathbb{R}^n; m)\}, \\ \mathcal{L}f(x) = \dfrac{1}{2m(x)}\ \Delta f(x), \quad x \in \mathbb{R}^n, \quad f \in \mathcal{D}(\mathcal{L}). \end{cases} \tag{7.6.32}$$

The function $u_\alpha(x) = \mathbf{E}_x[e^{-\alpha\zeta^Y}; Y_{\zeta^Y -}]$, $x \in \mathbb{R}^n$, satisfies $u_\alpha = 1 - \alpha R^Y u_\alpha$. Hence $\lim_{x\to\partial} u_\alpha(x) = 1$. Accordingly, the flux $\mathcal{N}(f)(\partial)$, $f \in \mathcal{D}(\mathcal{L})$, specified by

$$\mathcal{N}(f)(\partial) = \frac{1}{2}\mathbf{D}(f, u_\alpha) + (\mathcal{L}f, u_\alpha)_{L^2(\mathbb{R}^n;m)}$$

reads via the Gauss-Green formula as

$$\mathcal{N}(f)(\partial) = -\frac{1}{2}\lim_{r\to\infty} r^{n-1}\int_{\Sigma_1} f_r(r\xi)\sigma(d\xi),$$

provided that $f \in \mathcal{D}(\mathcal{L}) \cap C^1(\mathbb{R}^n)$. Here $\Sigma_1 = \{x \in \mathbb{R}^n : |x| = 1\}$ and σ is the surface measure on Σ_1. The infinitesimal generator of the L^2-semigroup of X can be characterized as Theorem 7.5.4(iv) in terms of $\mathcal{L}$ and $\mathcal{N}(f)(\partial)$ as above.

The entrance law $\{\nu_t; t > 0\}$ taking part in the above construction of X is, in view of (7.5.17), given by

$$\int_0^t \nu_s(B)ds = \int_B \mathbf{P}_x(\zeta^Y \le t)m(x)dx, \quad t > 0,\ B \in \mathcal{B}(\mathbb{R}^d).$$

If m is spherically symmetric, then so is ν_t. An alternative construction of X by means a Dirichlet form is possible as in the case of the one-point extension of the absorbing Brownian motion on a bounded open set.

(5°) Censored stable process

We consider a case where X^0 is of pure jump type and admits no killings inside E_0. A typical example of such a process is a censored stable process on a Euclidean open set, as previously seen in Example (**6°**) of Section 6.5 and Example 7.2.11.

Let D be an open *n-set* in $\mathbb{R}^n$ as is defined in Example (**6°**) of Section 6.5. D can be disconnected. Fix $0 < \alpha < 2$ and $c > 0$, and consider the function space $(\mathcal{W}(D), \mathbf{B})$ defined by (6.5.13). We let $W^{\alpha/2,2}(D) = \mathcal{W}(D) \cap L^2(D; m)$, where m denotes the Lebesgue measure. Then the bilinear form defined by

$$(\mathcal{E}, \mathcal{F}) = (\mathbf{B}, W^{\alpha/2,2}(D))$$

is a regular irreducible Dirichlet form on $L^2(\overline{D}; m)$ and the associated Hunt process X on $\overline{D}$ is a *reflecting α-stable process* by definition. It is shown in [29] that X has Hölder continuous transition density functions with respect to the Lebesgue measure m on $\overline{D}$ and therefore X can be refined to start from every point in $\overline{D}$. X admits no killing inside $\overline{D}$. Moreover, X admits no jump from D to ∂D or from ∂D to ∂D.

The part process $X^0 = (X_t^0, \mathbf{P}_x^0, \zeta^0)$ of X on D is identical with the *censored α-stable process* on D. Its Dirichlet form on $L^2(D; m)$ is given by $(\mathcal{E}, W_0^{\alpha/2,2}(D))$, where $W_0^{\alpha/2,2}(D)$ is the closure of $C_c^\infty(D)$ in $\mathcal{F}$ with respect to $\mathcal{E}_1 := \mathcal{E} + (\cdot,\cdot)_{L^2(D;m)}$. Note that the censored stable process X^0 has no killings inside D. The extended Dirichlet space of $(\mathcal{E}, W_0^{\alpha/2,2}(D))$ is denoted by $W_{0,e}^{\alpha/2,2}(D)$.

Let us assume that $D \subset \mathbb{R}^n$ is a proper open n-set, ∂D is a compact set of a positive d-dimensional Hausdorff measure with $\alpha > n - d$ when $n \geq 2$ and $\alpha > 1$ when $n = 1$. Let $D^* = D \cup \{a\}$ be the topological space obtained from $\overline{D} = D \cup \partial D$ by regarding ∂D as one point $\{a\}$ in the way prescribed in Section 7.5. We consider the extension of the censored stable process X^0 to D^*. When D is bounded, D^* is just the one-point compactification of D.

We now apply Theorem 7.5.9 to the case that $E = \overline{D}$, $K = \partial D$. By the above-mentioned properties of the reflecting stable process $X = (X_t, \mathbf{P}_x)$ on $\overline{D}$, it clearly satisfies conditions **(B.1)**, **(B.2)**, **(B.3)**. Note that $\varphi(x) := \mathbf{P}_x(\sigma_{\partial D} < \infty) = 1$ for $x \in D$, when D is bounded, and $0 < \varphi < 1$ on D when D is unbounded with compact boundary in view of [14]. Hence the condition **(A°.1)** for X^0 is also satisfied.

Any bounded measurable function f on D is extended to $\overline{D}$ by defining $f(x) = 0$ on ∂D. By [29], $G_\alpha f(x) := \mathbf{E}_x\left[\int_0^\infty e^{-\beta t} f(X_t)dt\right]$ is a continuous function on $\overline{D}$. Let $\tau_D := \inf\{t > 0 : X_t \notin D\}$. Then we have for $G_\beta^0 f(x) := \mathbf{E}_x\left[\int_0^{\tau_D} e^{-\beta t} f(X_t)dt\right]$,

$$G_\beta^0 f(x) = G_\beta f(x) - \mathbf{E}_x\left[e^{-\beta\tau_D} G_\beta f(X_{\tau_D})\right] \qquad \text{for } x \in D.$$

Since $x \mapsto \mathbf{E}_x\left[e^{-\beta\tau_D} G_\beta f(X_{\tau_D})\right]$ is a β-harmonic function of X^0 and so it is continuous on D (see [14, (3.8)]), we conclude that $G^0_\beta f$ is continuous on D. Hence the condition **(A°.3)** is always satisfied for censored α-stable process X^0 in any open n-set D. A Lévy system of X^0 is given by $(N(x,dy), dt)$ with

$$N(x,dy) = c\ |x-y|^{-(n+\alpha)}dy$$

and the condition **(A°.4)** is clearly satisfied.

By Theorem 7.5.9, we can thus construct a unique symmetric extension X^* on D^* of X^0 by darning the hole ∂D. The entrance law $\{\nu_t; t>0\}$ taking part in the darning is given by

$$\int_0^t \nu_s(B)ds = \int_B \mathbf{P}_x(\sigma_{\partial D} \le t)m(dx), \quad t>0,\ B \in \mathcal{B}(D),$$

in terms of the reflecting stable process X on $\overline{D}$.

Let $u_1 := \mathbf{E}_x\left[e^{-\tau_D}\right]$. By virtue of (6.5.15), the reflected Dirichlet space of $(\mathcal{E}^0, \mathcal{F}^0)$ equals $(\mathcal{W}(D), \mathbf{B})$ and it follows from Theorem 7.5.9 that the Dirichlet form $(\widetilde{\mathcal{E}}, \widetilde{\mathcal{F}})$ and its extended Dirichlet form $(\widetilde{\mathcal{E}}, \widetilde{\mathcal{F}}_e)$ is given by

$$\widetilde{\mathcal{F}} = \left\{f = f_0 + cu_1 : f_0 \in W_0^{\alpha/2,2}(D) \text{ and } c \in \mathbb{R}\right\},$$

$$\widetilde{\mathcal{F}}_e = \left\{f = f_0 + c : f_0 \in W_{0,e}^{\alpha/2,2}(D) \text{ and } c \in \mathbb{R}\right\},$$

$$\widetilde{\mathcal{E}}(f,g) = \mathbf{B}(f,g), \quad f,g \in \widetilde{\mathcal{F}}_e.$$

7.7. MANY-POINT EXTENSIONS

We now generalize the uniqueness theorem and existence theorem of the one-point extension established in Section 7.5 to a countably many points extension. As in Section 7.5, every sample path of a right process X on a state space E will be assumed to possess the left limit X_{t-} in E for all $t \in (0,\zeta)$, where ζ is the lifetime of X.

Let E be a locally compact separable metric space and m be a positive Radon measure on E with supp$[m] = E$. We fix a closed set $F = \{a_1, a_2, \dots, a_i, \dots\}$ consisting of finite or countably many non-isolated points of E such that F possesses no accumulating point and $m(F) = 0$.

Put $E_0 = E \setminus F$. As usual, let $E_\partial = E \cup \{\partial\}$ denote the one-point compactification of E. $(E_0)_\partial = E_0 \cup \{\partial\}$ is regarded as a topological subspace of E_∂.

Let $X^0 = (X^0_t, \mathbf{P}^0_x, \zeta^0)$ be an m-symmetric Borel standard process on E_0 satisfying the following condition:

(M.1) X^0 admits no killing inside and

$$\mathbf{P}_x^0\big(\zeta^0 < \infty,\ X_{\zeta^0-}^0 = a_i\big) > 0 \quad \text{for every } x \in E_0 \text{ and } i \geq 1. \tag{7.7.1}$$

We then define, for $x \in E_0$, $i \geq 1$, and $\alpha > 0$,

$$\varphi^{(i)}(x) = \mathbf{P}_x^0\big(\zeta^0 < \infty,\ X_{\zeta^0-}^0 = a_i\big),\ \ u_\alpha^{(i)}(x) = \mathbf{E}_x^0\big[e^{-\alpha\zeta^0}; X_{\zeta^0-}^0 = a_i\big]. \tag{7.7.2}$$

DEFINITION 7.7.1. A right process $X = (X_t,\ \mathbf{P}_x,\ \zeta)$ on E is called an *extension* of X^0 if X is m-symmetric, having no killings on F, admitting no jumps from F to F, and with X^0 the part process of X on E_0.

Suppose an extension $X = (X_t, \mathbf{P}_x)$ of X^0 is a Borel standard process. Then it can be verified exactly as in Exercise 7.5.2 that the following holds. For any $x \in E_0$, $i \geq 1$, and $\alpha > 0$,

$$\varphi^{(i)}(x) = \mathbf{P}_x\left(\sigma_F < \infty,\ X_{\sigma_F} = a_i\right),\ u_\alpha^{(i)}(x) = \mathbf{E}_x\left[e^{-\alpha\sigma_F};\ X_{\sigma_F} = a_i\right], \tag{7.7.3}$$

and X admits no jump from E_0 to F:

$$\mathbf{P}_x\left(X_{t-} \in E_0,\ X_t \in F \text{ for some } t > 0\right) = 0 \quad \text{for } x \in E. \tag{7.7.4}$$

LEMMA 7.7.2. *Suppose an extension X of X^0 is a Hunt process whose Dirichlet form $(\mathcal{E}, \mathcal{F})$ is regular. Then each point a_i is non-$\mathcal{E}$-polar and regular for itself with respect to X in the sense that $\mathbf{P}_{a_i}(\sigma_{a_i} = 0) = 1$. Moreover, X is irreducible.*

Proof. Fix $i \geq 1$ and let $X^{(i)}$ be the part process of X on the set $E_0 \cup \{a_i\}$. Then $X^{(i)}$ is a one-point extension of X^0. By Theorem 3.3.9, $X^{(i)}$ is a Hunt process with the associated Dirichlet form being regular. Therefore, $X^{(i)}$ is irreducible by Lemma 7.5.3. Since this holds for every $i \geq 1$, it follows that X is irreducible. The proof of other assertions is the same as the one for Lemma 7.5.3. □

THEOREM 7.7.3. *Let X^0 be an m-symmetric Borel standard process on E_0 satisfying condition* **(M.1)**. *An extension of X^0 is then unique in law. More specifically, let $X = (X_t, \mathbf{P}_x, \zeta)$ be an extension of X^0. Denote by $\{G_\alpha^0; \alpha > 0\}$ and $(\mathcal{E}^0, \mathcal{F}^0)$ (resp. $\{G_\alpha; \alpha > 0\}$ and $(\mathcal{E}, \mathcal{F})$) the resolvent and the Dirichlet form of X^0 (resp. X) on $L^2(E_0; m) = L^2(E; m)$. Let $\mathcal{B}_0(F)$ be the space of functions on F vanishing except on finite many points. The energy functional relative to X^0 is denoted by L^0.*

(i) *Each point $\{a_i\}$ is non-m-polar and regular for itself with respect to X. The process X is irreducible.*

(ii) *Define for* $i,j \geq 1$,

$$U^{ij} = L^0(\varphi^{(i)}, \varphi^{(j)}) \text{ for } i \neq j, \qquad U_\alpha^{ij} = \alpha(u_\alpha^{(i)}, \varphi^{(j)}),$$

$$V^i = L^0\left(\varphi^{(i)}, 1 - \sum_{k\geq 1}\varphi^{(k)}\right), \quad \mu_0(\{a_i\}) = \sum_{k\geq 1} U_1^{ik}, \tag{7.7.5}$$

and define a function space on F *by*

$$\begin{cases} \mathbf{C}(\psi,\psi) = \frac{1}{2}\sum_{i,j\geq 1, i\neq j}(\psi(a_i)-\psi(a_j))^2 U^{ij} + \sum_{i\geq 1}\psi(a_i)^2 V^i, \\ \mathcal{G}_1 = \{\psi \in L^2(F;\mu_0) : \mathbf{C}(\psi,\psi) < \infty\}. \end{cases} \tag{7.7.6}$$

It holds that

$$\mathcal{B}_0(F) \subset \mathcal{F}\big|_F, \quad \mathcal{F}\big|_F = \text{ the } \mathbf{C}_1\text{-closure of } \mathcal{B}_0(F) \text{ in } \mathcal{G}_1, \tag{7.7.7}$$

where $\mathbf{C}_1(\psi,\psi) = \mathbf{C}(\psi,\psi) + \|\psi\|^2_{L^2(F;\mu_0)}$.
(iii) *For any* $\alpha > 0$ *and* $g \in L^2(E;m)$, *let* $\psi = G_\alpha g\big|_F$. *Then* ψ *is an element of* $\mathcal{F}\big|_F$ *such that*

$$\mathbf{C}(\psi, \mathbf{1}_{a_i}) + \sum_{j\geq 1} U_\alpha^{ij}\psi(a_j) = (u_\alpha^{(i)}, g), \quad i \geq 1. \tag{7.7.8}$$

Moreover, $G_\alpha g$ *admits the representation*

$$G_\alpha g(x) = G_\alpha^0 g(x) + \sum_{i\geq 1} u_\alpha^{(i)}(x)\psi(a_i), \quad x \in E. \tag{7.7.9}$$

(iv) $\mathcal{F} \subset (\mathcal{F}^0)_a^{\mathrm{ref}}$, $\mathcal{E}_\alpha(u,v) = \mathcal{E}_\alpha^{0,\mathrm{ref}}(u,v)$ *for* $u,v \in \mathcal{F}$, *and*

$$\mathcal{F} = \mathcal{F}^0 \oplus \mathcal{H}_\alpha,$$

where $\mathcal{H}_\alpha$ *is* $\mathcal{E}_\alpha^{0,\mathrm{ref}}$*-closure of the linear span of* $\{u_\alpha^{(i)}; i \geq 1\}$ *and the above is an* $\mathcal{E}_\alpha^{0,\mathrm{ref}}$*-orthogonal decomposition.*
(v) $\mathcal{F}_e \subset (\mathcal{F}^0)^{\mathrm{ref}}$, $\mathcal{E}(u,v) = \mathcal{E}^{0,\mathrm{ref}}(u,v)$ *for* $u,v \in \mathcal{F}_e$. $\mathcal{F}_e^0$ *and the linear span of* $\{\varphi^{(i)} : i \geq 1\}$ *are subsets of* $\mathcal{F}_e$, *which are* $\mathcal{E}^{0,\mathrm{ref}}$*-orthogonal to each other.*
(vi) *When* F *consists of a finite number of points* $\{a_1, a_2, \ldots, a_N\}$, $\mathcal{F}_e$ *is the linear subspace of* $(\mathcal{F}^0)^{\mathrm{ref}}$ *spanned by* $\mathcal{F}_e^0$ *and* $\varphi^{(i)}$, $1 \leq i \leq N$. $\mathcal{F}$ *is the linear subspace of* $(\mathcal{F}^0)_a^{\mathrm{ref}}$ *spanned by* $\mathcal{F}^0$ *and* $u_\alpha^{(i)}$, $1 \leq i \leq N$.
(vii) *Assume that the condition* (7.3.4) *is fulfilled.* $f \in \mathcal{D}(\mathcal{A})$ *if and only if* $f \in \mathcal{D}(\mathcal{L})$, (7.4.9) *is satisfied for the space* $\check{\mathcal{F}}$ *specified by* (7.4.15), *and*

$$\mathcal{N}(f)(a_i) = 0 \qquad \text{for every } i \geq 1. \tag{7.7.10}$$

Proof. The uniqueness follows from (ii) and (iii). To see this, suppose two functions on F satisfy equation (7.7.8) and denote by ψ their difference. Then $\mathbf{C}(\psi,\eta)+U_\alpha(\psi,\eta)=0$ for any $\eta\in\mathcal{B}_0(F)$, which implies $\mathbf{C}_1(\psi,\psi)=0$ and $\psi=0$ by virtue of (7.7.7) and Lemma 7.2.3. Therefore, the solution of (7.7.8) is unique and so is the resolvent $\{G_\alpha;\alpha>0\}$ of X in view of (7.7.9).

The assertions (i), (ii), and (iii) can be deduced from Theorem 7.4.5 with the help of a quasi-homeomorphism j that has been used in the proof of Theorem 7.5.4. In fact, since X is an m-symmetric right process on the locally compact separable metric space E, we can start with the setting of the first two paragraphs in the proof of Theorem 7.5.4 using the same notations appearing there. Since X extends X^0 satisfying **(M.1)**, each set $\{a_i\}$ must be located outside the properly exceptional set N for X. Applying (7.7.3) and (7.7.4) to the Borel standard process $X\big|_{E\setminus N}$ extending $X^0\big|_{E_0\setminus N}$, we conclude that each a_i is not m-polar for X and $X\big|_{E\setminus N}$ admits no jump from $E_0\setminus N$ to F.

Therefore, each $\widehat{a}_i:=ja_i\in\widehat{E}\setminus\widehat{N}$ is non-$\widehat{m}$-polar for the Hunt process $\widehat{X}'=\widehat{X}\big|_{\widehat{E}'}$ on $\widehat{E}'=\widehat{E}\setminus\widehat{N}$. Let $\widehat{F}=\{\widehat{a}_1,\widehat{a}_2,\dots\}$. Evidently $\widehat{m}(\widehat{F})=0$. Moreover, $\widehat{F}$ satisfies the conditions **(F.1)**, **(F.2)**, and **(F.3)** of Section 7.4 with respect to the Dirichlet form $(\widehat{\mathcal{E}},\widehat{\mathcal{F}})$, because these properties are invariant under the quasi-homeomorphism j and the set F trivially satisfies these conditions with respect to $(\mathcal{E},\mathcal{F})$. Let $\widehat{X}^0$ be the part process of $\widehat{X}'$ on $\widehat{E}'\setminus\widehat{F}$, which is an $\widehat{m}$-symmetric standard process satisfying **(M.1)**. Clearly $\widehat{X}'$ is an extension of $\widehat{X}^0$ admitting no jumps from $\widehat{E}'\setminus\widehat{F}$. Since $(\widehat{\mathcal{E}},\widehat{\mathcal{F}})$ is regular, each $\widehat{a}_i$ is regular for itself with respect to $\widehat{X}$ and $\widehat{X}'$ is irreducible in view of Lemma 7.7.2. Now $\widehat{X}'$, $(\widehat{\mathcal{E}},\widehat{\mathcal{F}})$, $\widehat{F}$ and $\widehat{X}^0$ satisfy all the conditions imposed in Section 7.4, and accordingly Theorem 7.4.5 holds true for them.

All quantities defined by (7.7.5) and (7.7.6) for $\widehat{X}'$ and $\widehat{X}^0$ will be designated by the superscript $\widehat{\ }$. Then by Theorem 7.4.5(ii),

$$\mathcal{B}_0(\widehat{F})\subset\widehat{\mathcal{F}}\big|_{\widehat{F}},\quad \widehat{\mathcal{F}}\big|_{\widehat{F}}=\text{the }\widehat{\mathbf{C}}_1\text{-closure of }\mathcal{B}_0(\widehat{F})\text{ in }\widehat{\mathcal{G}}_1. \tag{7.7.11}$$

The restriction to F of the quasi-homeomorphism j is denoted by $\check{j}$. By noting that $j^*\widehat{G}^0_\alpha\widehat{g}=G^0_\alpha j^*\widehat{g}$, $j^*\widehat{u}^{(i)}_\alpha=u^{(i)}_\alpha$, $j^*\widehat{\varphi}^{(i)}=\varphi^{(i)}$, and the relation $\int_E j^*\widehat{f}dm=\int_{\widehat{E}}\widehat{f}d\widehat{m}$, we have $U^{ij}=\widehat{U}^{ij}$, $U^{ij}_\alpha=\widehat{U}^{ij}_\alpha$, $V^i=\widehat{V}^i$ so that $\mathbf{C}_1(\psi,\psi)=\widehat{\mathbf{C}}_1(\widehat{\psi},\widehat{\psi})$ for $\psi=\check{j}^*\widehat{\psi}$ and $\mathcal{G}_1=\check{j}^*\widehat{\mathcal{G}}_1$. Since $\mathcal{F}\big|_F=\check{j}^*(\widehat{\mathcal{F}}\big|_{\widehat{F}})$, (7.7.7) follows from (7.7.11).

We see also from Theorem 7.4.5(i) that for $\widehat{g}\in L^2(\widehat{E};\widehat{m})$, the function $\widehat{\psi}=\widehat{G}_\alpha\widehat{g}\big|_{\widehat{F}}$ satisfies

$$\widehat{\mathbf{C}}(\widehat{\psi},\mathbf{1}_{\widehat{a}_i})+\sum_{j\ge 1}\widehat{U}^{ij}_\alpha\widehat{\psi}(\widehat{a}_j)=(\widehat{u}^{(i)}_\alpha,\widehat{g})_{\widehat{m}},\quad i\ge 1. \tag{7.7.12}$$

For $g\in L^2(E;m)$ with $g=j^*\widehat{g}$, $G_\alpha g=j^*\widehat{G}_\alpha\widehat{g}$ so that $\psi=\check{j}^*\widehat{\psi}$ for $\psi=G_\alpha g\big|_F$ and $\widehat{\psi}=\widehat{G}_\alpha\widehat{g}\big|_{\widehat{F}}$. By noting that $(u^{(i)}_\alpha,g)=(\widehat{u}^{(i)}_\alpha,\widehat{g})_{\widehat{m}}$, we are led from (7.7.12) to (7.7.8).

As each $\widehat{a}_i$ is regular for itself with respect to $\widehat{X}'$, so is $a_i = j^{-1}\widehat{a}_i$ with respect to $X\big|_{E\setminus N}$, and accordingly with respect to X. Hence the identity (7.7.9) is valid because X is an extension of X^0.

(iv), (v), (vi), and (vii) also follow from Theorem 7.4.5(v), (vi), (vii), and (viii), respectively, by using the quasi-homeomorphism map j. □

Many-point extension X of an m-symmetric Hunt process X^0 can be constructed under some of the following additional conditions for X^0:

(M.2) $\int_E u_\alpha^{(i)}(x)m(dx) < \infty$ for every $i \geq 1$ and

$$\mathbf{P}_x^0(\zeta^0 < \infty, X^0_{\zeta^0-} \in F \cup \{\partial\}) = \mathbf{P}_x^0(\zeta^0 < \infty), \quad x \in E_0. \tag{7.7.13}$$

(M.2)$'$ $\mathbf{P}_x^0(X^0_{\zeta^0-} \in F \cup \{\partial\}) = 1$ for every $x \in E_0$ (regardless the length of the lifetime $\zeta^0 \in (0, \infty]$).

(M.3) For every $i \geq 1$, there exists a neighborhood of U_i of a_i such that $\inf_{x \in V} G_1^0 \varphi^{(i)}(x) > 0$ for any compact set $V \subset U_i \cap E_0$.

(M.4) For each $i \geq 1$ and every open neighborhood U_1 of a_i in E, there exists an open neighborhood U_2 of a_i in E with $\overline{U_2} \subset U_1$ such that $J_0(E_0 \cap U_2 \setminus \{a_i\}, E_0 \setminus U_1) < \infty$. Here J_0 denotes the jumping measure of X^0.

We do not need condition **(M.4)** when X^0 is a diffusion.

Theorem 7.7.4. *Let X^0 be an m-symmetric Hunt process on E_0 satisfying conditions* **(M.1)** *and* **(M.3)** *as well as* **(M.4)** *in a non-diffusion case. We also assume that either* **(M.2)** *or* **(M.2)$'$** *is satisfied by X^0. Then there exists an extension X of X^0 from E_0 to E. X admits no jump to or from the set F. When X^0 is a diffusion, so is X.*

Proof. For $n \geq 1$, we let $E_n = E_0 \cup \{a_1, a_2, \ldots, a_n\}$ and $F^{(n)} = E \setminus E_n = F \setminus \{a_1, a_2, \ldots, a_n\}$. First we can construct a one-point extension of X^0 from E_0 to E_1. Indeed, under the stated conditions on X^0, the corresponding conditions imposed in Theorem 7.5.6 are satisfied for $\{a_1\}$ and $F^{(1)} \cup \{\partial\}$ in place of $\{a\}$ and ∂, respectively. Therefore, by defining the spaces of excursion paths with this replacement, the proof of Theorem 7.5.6 works to produce a one-point extension $X^1 = (X_t^1, \mathbf{P}_x^1, \zeta^1)$ of X^0 from E_0 to E_1 such that

$$\mathbf{P}_x^1(\zeta^1 < \infty, X^1_{\zeta^1-} \in F^{(1)} \cup \{\partial\}) = \mathbf{P}_x^1(\zeta^1 < \infty), \quad x \in E_1. \tag{7.7.14}$$

Lemma 7.5.5(iii) then guarantees the integrability $\int_E u_\alpha^{(1)}(x)m(dx) < \infty$. In the same way, the integrability of $u_\alpha^{(i)}$ is ensured for every $i \geq 1$.

Next we verify that the process X^1 on E_1 satisfies the conditions **(A.1)**, **(A.2)**, **(A.3)**, and **(A.4)** relative to the point a_2. Define

$$\varphi^{12}(x) = \mathbf{P}_x^1(\zeta^1 < \infty, X^1_{\zeta^1-} = a_2), \quad x \in E_2.$$

Since the part process of X^1 on E_0 equals X^0, we have $\varphi^{12}(x) \geq \varphi^{(2)}(x) > 0$ for $x \in E_0$. We also have

$$\varphi^{12}(a_1) > 0. \tag{7.7.15}$$

To see this, we use the same notations $\nu_t, W, W^+, W^-, \mathbf{n}$ as in the proof of Theorem 7.5.6 for $\{a_1\}$ and $F^{(1)} \cup \{\partial\}$ in place of $\{a\}$ and ∂. By (7.5.14),

$$\mathbf{n}^-\{w \in W^- : t < \zeta,\ w_{\zeta-} = a_2\} = \int_{E_0} \nu_t(dx)\varphi^{(2)}(x),$$

which is strictly positive for some $t > 0$ in view of (7.5.13) (with $\varphi^{(1)}$ in place of φ) and the assumption **(M.1)**. By letting $t \downarrow 0$, we get $\mathbf{n}^-\{w_{\zeta-} = a_2\} > 0$. This together with (7.5.16) leads us to (7.7.15). Condition **(A.1)** is verified.

The integrability condition in **(A.2)** is verifiable because for $x \in E_0$,

$$\begin{aligned}
&\mathbf{E}_x^1\left[e^{-\alpha\zeta^1}; X^1_{\zeta^1-} = a_2\right] \\
&= \mathbf{E}_x^1\left[e^{-\alpha\zeta^1};\ X^1_{\zeta^1-} = a_2, \zeta^1 < \sigma_{a_1}\right] + \mathbf{E}_x^1\left[e^{-\alpha\zeta^1};\ X^1_{\zeta^1-} = a_2, \zeta^1 > \sigma_{a_1}\right] \\
&\leq u_\alpha^{(2)}(x) + \mathbf{E}_x^1\left[e^{-\alpha\sigma_{a_1}}\right] = u_\alpha^{(2)}(x) + u_\alpha^{(1)}(x).
\end{aligned}$$

The second condition (7.5.11) of **(A.2)** is also met in view of (7.7.14). Since the resolvent of X^1 dominates that of X^0 and $\varphi^{12} \geq \varphi^{(2)}$, the condition **(A.3)** for X^1 and a_2 follows from **(M.1)**, **(M.3)**. As X^1 admits no jump from E_0 to a_1, the condition **(A.4)** for X^1 and a_2 follows from **(M.4)**.

We can now apply Theorem 7.5.6 to X^1 and a_2 in constructing a one-point extension X^2 of X^1 from E_1 to E_2. In the same way, we can construct a one-point extension X^3 of X^2 from E_2 to E_3. For instance, the integrability condition in **(A.2)** for $X^2 = (X_t^2, \mathbf{P}_x^2, \zeta^2)$ relative to a_3 can be verified as for $x \in E_0$,

$$\begin{aligned}
\mathbf{E}_x^2\left[e^{-\alpha\zeta^2};\ X^2_{\zeta-} = a_3\right] &= \mathbf{E}_x^2\left[e^{-\alpha\zeta^2};\ X^2_{\zeta-} = a_3, \zeta^2 < \sigma_{a_1} \wedge \sigma_{a_2}\right] \\
&\quad + \mathbf{E}_x^2\left[e^{-\alpha\zeta^2};\ X^2_{\zeta-} = a_3, \zeta^2 > \sigma_{a_1} \wedge \sigma_{a_2}\right] \\
&\leq u_\alpha^{(3)}(x) + \mathbf{E}_x^1\left[e^{-\alpha\sigma_{a_1} \wedge \sigma a_2}\right] \\
&\leq u_\alpha^{(3)}(x) + u_\alpha^{(1)}(x) + u_\alpha^{(2)}(x).
\end{aligned}$$

By continuing this procedure, we get a sequence of symmetric right processes $X^n = (X_t^n, \mathbf{P}_x^n, \zeta^n)$ on E_n such that X^n is a one-point extension of X^{n-1}

from E_{n-1} to E_n, $n \geq 1$. Each X^n admits no jump from or to $\{a_1, a_2, \ldots, a_n\}$. We can define the sequence on a common probability space. Now define $X_t := X_t^n$ if $t < \zeta^n$ and $X_t := \partial$ for $t \geq \zeta := \lim_{n\to\infty} \zeta^n$. Note that for every $k \geq j$ and $x \in E_j$, $\mathbf{P}_x^k = \mathbf{P}_x^j$ on $\mathcal{F}_{\zeta^j}$. Therefore, for every $x \in E$, there is a unique probability measure $\mathbf{P}_x$ on $\mathcal{F}_\infty$ so that $\mathbf{P}_x = \mathbf{P}_x^n$ on $\mathcal{F}_{\zeta^n}$ for every $n \geq j$, where $x \in E_j$. It is easy to see that $\{X_t, \mathbf{P}_x, x \in E\}$ is an extension of X^0 from E_0 to E that admits no jumps from E to F or from F to E. □

To formulate a counterpart of Theorem 7.5.9, let E be a locally compact separable metric space and m be an everywhere dense positive Radon measure E. We consider a closed subset K of E such that either

(K.1) $K = \cup_i K_i$, where $\{K_i\}$ are finite or countably infinite disjoint compact sets which are locally finite in the sense that any compact set intersects only with finite many of K_i's; or

(K.2) $K = K_1 \cup \cdots \cup K_N$, where $\{K_i\}_{1\leq i\leq N}$, are disjoint, $K_1, \ldots, K_{N-1}$ are compact and $E \setminus K_N$ is relatively compact.

We put $E_0 = E \setminus K$, $F^* := \bigcup_i \{a_i^*\}$ and let $E^* := E_0 \cup F^*$ be the topological Hausdorff space obtained by adding to E_0 extra points $\{a_1^*, a_2^*, \ldots\}$, whose topology is prescribed as follows: for each $i \in \Lambda$, a subset U of E^* containing the point a_i^* is a neighborhood of a_i^* if there is an open set $\widetilde{U} \subset E$ containing K_i such that $\widetilde{U} \cap E_0 = U \setminus \{a_i^*\}$. In other words, E^* is obtained from E by identifying each closed set K_i with the point $\{a_i^*\}$ for every $i \in \Lambda$. We denote by m_0 the restriction of the measure m on E to E_0. The measure m_0 is then extended to E^* by setting $m_0(F^*) = 0$. In the remainder of this section, (f, g) will denote the integral of $f \cdot g$ on E_0 against the measure m_0.

Let $X = (X_t, \mathbf{P}_x, \zeta)$ be an m-symmetric Hunt process on E whose Dirichlet form $(\mathcal{E}, \mathcal{F})$ on $L^2(E; m)$ is regular. Let $X^0 = (X_t^0, \mathbf{P}_x^0, \zeta^0)$ be the part process of X on E_0, which is an m_0-symmetric Hunt process on E_0 (cf. Exercise 3.3.7). Denote by τ_0 the lifetime of X^0; that is, $\tau_0 = \inf\{t > 0 : X_t \notin E_0\}$. The resolvent of X^0 is denoted by $\{G_\alpha^0; \alpha > 0\}$. We aim at constructing a q.e. many-point extension of X^0 from E_0 to E^*, which can be intuitively viewed as the process obtained from X by collapsing holes K_i's into a_i^*.

We impose the following conditions on X:

(C.1) X is irreducible.

(C.2) $m_0(U \cap E_0)$ is finite for some neighborhood U of K.

(C.3) X admits no killings inside or jumps from $E \setminus K_i$ to K_i for each $i \in \Lambda$.

We define, for $\alpha > 0$ and $i \in \Lambda$, the functions $\varphi^{(i)}$, $u_\alpha^{(i)}$ on E_0 by

$$\varphi^{(i)}(x) = \mathbf{P}_x^0\big(\zeta^0 < \infty \text{ and } X^0_{\zeta^0-} \in K_i\big),$$

$$u_\alpha^{(i)}(x) = \mathbf{E}_x^0\big[e^{-\alpha\zeta^0};\ X^0_{\zeta^0-} \in K_i\big].$$

Due to the condition **(C.3)**, these two functions can also be expressed as

$$\varphi^{(i)}(x) = \mathbf{P}_x\left(\sigma_K < \infty, X_{\sigma_K} \in K_i\right)$$

and

$$u_\alpha^{(i)}(x) = \mathbf{E}_x\left[e^{-\alpha\sigma_K};\ X_{\sigma_K} \in K_i\right].$$

Let us consider the following conditions on X^0:

(M°.1) For each $i \in \Lambda$, $\varphi^{(i)}(x) > 0$ for every $x \in E_0$.

(M°.3) For each $i \in \Lambda$, there is some neighborhood U_i of K_i such that $\sup_{x\in V} G_1^0\varphi^{(i)}(x) < \infty$ for every compact set $V \subset U_i \cap E_0$.

(M°.4) Either $E \setminus U$ is compact for any neighborhood U of K in E, or, for any open neighborhood U_1 of K in E, there exists an open neighborhood U_2 of K in E with $\overline{U}_2 \subset U_1$ such that $J_0(U_2 \setminus K, E_0 \setminus U_1) < \infty$, where J_0 denotes the jumping measures of X^0.

These three conditions are the counterparts of (7.7.1), **(M.3)**, and **(M.4)**, respectively, with a_i being replaced by K_i.

A right process X^* on $E^* = E_0 \cup F^*$ is called a q.e. *extension* of X^0 if X^* is m_0-symmetric, having no killings on F admitting no jumps from F^* to F^*, and the part process of X^* on E_0 coincides with X^0 q.e., namely, outside some m_0-polar set for X^0.

THEOREM 7.7.5. *Assume that X satisfies the conditions* **(C.1)**, **(C.2)** *and* **(C.3)**. *Assume further that X^0 satisfies conditions* **(M°.1)** *and* **(M°.3)** *as well as condition* **(M°.4)** *when X^0 is not a diffusion. Then there exists an m_0-symmetric Borel right process X^* on E^* which is a q.e. extension of X^0 on E_0. Such extension is unique in law. The process X^* admits no jumps to or from the set F^*. If X^0 is a diffusion, then so is X^*.*

If X satisfies the condition **(AC)** *of Definition A.2.16 additionally, "q.e." can be removed from the above conclusion.*

Proof. We shall give the proof only for the case where K is of the form **(K.1)** because the second case **(K.2)** can be treated in a similar way.

It follows from **(C.3)** that X^0 admits no killing inside E_0, which together with **(M°.1)** means that X^0 satisfies **(M.1)** with a_i^* in place of a_i. The properties

(M.3), **(M.4)** for X^0 with a_i^* in place of a_i follow from **(M°.3)**, **(M°.4)**, respectively.

Furthermore, exactly in the same way as the proof of Theorem 7.5.9, we can obtain the following property of X^0:

(M°.2)′ $\mathbf{P}_x^0(X_{\zeta^0-} \in K \cup \{\partial\}) = 1$ for q.e. $x \in E_0$.

This implies that X^0 also satisfies the condition **(M.2)′** with $F^* = \bigcup_i \{a_i^*\}$ in place of F holding for q.e. $x \in E_0$.

If, in addition, X satisfies the condition **(AC)**, then so does X^0 and we can replace "for q.e." by "for every" in the above assertion.

We can now deduce the desired conclusion from Theorem 7.7.4. As an extension of X^0 from E_0 to E^* in the sense of Definition 7.7.1, X^* is unique in law by virtue of Theorem 7.7.3. □

We call the above procedure of obtaining X^* from X (or from X^0) *darning each hole* K_i into one point a_i^*, $i \geq 1$. X^* enjoys those properties listed in Theorem 7.7.3 but with $\{a_1^*, a_2^*, \cdots\}$ in place of $\{a_1, a_2, \cdots\}$.

7.8. EXAMPLES OF MANY-POINT EXTENSIONS

We present several concrete examples of many-point extensions.

(1°) Darning multiple holes for multidimensional Brownian motion

We consider the Brownian motion $X = (X_t, \mathbf{P}_x)$ on $\mathbb{R}^n$ with $n \geq 2$. Let $K = \bigcup_{i \geq 1} K_i$ where $\{K_i\}$ is a collection of finite or countably infinite number of disjoint non-polar compact subsets of $\mathbb{R}^n$, which is locally finite. Let $X^0 = (X_t^0, \mathbf{P}_x^0, \zeta^0)$ be the absorbing Brownian motion on $E_0 = \mathbb{R}^n \setminus K$.

Conditions **(C.1)**, **(C.2)**, **(C.3)** are trivially satisfied by X. By Theorem 3.3.8, each set K_i is non-polar relative to the part process X^{0i} of X on $\mathbb{R}^n \setminus \bigcup_{j \neq i} K_j$. As $n \geq 2$, X^{0i} is irreducible and it has a transition density with respect to the Lebesgue measure. Therefore, the condition **(M°.1)** is fulfilled by X^0. The condition **(M°.3)** for X^0 can be verified just as in Example **(4°)** of Section 7.6. Thus we can use Theorem 7.7.5 to obtain a diffusion extension X^* of X^0 from E_0 to $E^* = E_0 \cup F^*$ with $F^* = \cup_{i \geq 1}\{a_i^*\}$ by darning each hole K_i into a point a_i^*.

Assume that $n = 2$ and consider the case that the number of holes K_i is finite: $K = \cup_{i=1}^N K_i$ for disjoint non-polar compact subsets $\{K_1, K_2, \ldots, K_N\}$ of $\mathbb{R}^2$. X^* then enjoys the conformal invariance property analogous to those in Theorem 7.6.3 and Remark 7.6.4(ii), as will be formulated below.

THEOREM 7.8.1. *Let $\widehat{K} = \cup_{i=1}^N \widehat{K}_i$, where $\{\widehat{K}_i; 1 \leq i \leq N\}$ is a second set of disjoint non-polar compact subsets of $\mathbb{R}^2$. Suppose that ϕ is a conformal map from $\mathbb{R}^2 \setminus K$ onto $\mathbb{R}^2 \setminus \widehat{K}$ that maps ∞ to ∞ and, for each $i \geq 1$, ϕ maps the $\mathbb{R}^2 \setminus K$-portion of any neighborhood of K_i into the $\mathbb{R}^2 \setminus \widehat{K}$-portion of a neighborhood of $\widehat{K}_i$, and vice versa. Identify the compact set $\widehat{K}_i$ with a single* *point* $\widehat{a}_i^*$ *and equip* $\widehat{E}^* := (\mathbb{R}^2 \setminus \widehat{K}) \cup \{\widehat{a}_1^*, \widehat{a}_2^*, \ldots, \widehat{a}_N^*\}$ *the topology induced from* $\mathbb{R}^2$ *by collapsing each set* $\widehat{K}_i$ *into one point* $\widehat{a}_i^*$. *Define* $\phi(a_i^*) = \widehat{a}_i^*$, $1 \le i \le N$. *Then* ϕ *is a topological homeomorphism from* E^* *onto* $\widehat{E}^*$. *Moreover,* $\phi(X^*)$ *is, up to a time change, the diffusion process obtained from a Brownian motion on* $\mathbb{R}^2$ *by darning each hole* $\widehat{K}_i^*$ *into* $\widehat{a}_i^*$ *with* $1 \le i \le N$.

Proof. Let m_0 be the Lebesgue measure on $E_0 = \mathbb{R}^2 \setminus K$ extended to E^* by setting $m_0(\{a_i^*\}) = 0$, $1 \le i \le N$. The corresponding quantity on $\widehat{E}^*$ is denoted by $\widehat{m}_0$. $X^* = (X_t^*, \mathbf{P}_z^*)$ is an extension of the absorbing Brownian motion in E_0 to E^* and m_0-symmetric. By Theorem 7.7.3(vi), the extended Dirichlet space $(\mathcal{F}_e^*, \mathcal{E}^*)$ of $X^* = (X_t^*, \mathbf{P}_z^*)$ is given by

$$\begin{cases} \mathcal{F}_e^* = \{f + \sum_{i=1}^N c_i \varphi^{(i)} : f \in H^1_{0,e}(E_0),\ c_i \in \mathbb{R}\} \subset \mathrm{BL}(E_0), \\ \mathcal{E}^*(u, v) = \frac{1}{2}\mathbf{D}_{E_0}(u, v), \quad u, v \in \mathcal{F}_e^*, \end{cases}$$

where $\varphi^{(i)}(x) := \mathbf{P}_x(X_{\sigma_K -} \in K_i)$ for $x \in \mathbb{R}^2 \setminus K$. By the recurrence of X, we have $\sum_{i=1}^N \varphi^{(i)}(x) = \mathbf{P}_x(\sigma_K < \infty) = 1$ for every $x \in E_0$. This in particular implies that $1 \in \mathcal{F}_e^*$ with $\mathcal{E}^*(1, 1) = 0$ and therefore, by Theorem 2.1.8, X^* is recurrent. The process X^* is irreducible by Theorem 7.7.3(i). Its resolvent kernel is absolutely continuous with respect to m_0 on account of Theorem 7.7.3(iii) and so is its transition function by virtue of Corollary 3.1.14.

We can now proceed exactly along the same line as in the proof of Theorem 7.6.3. The PCAF A_t of X^* defined by (7.6.27) is strictly increasing to ∞ as $t \to \infty$ $\mathbf{P}_z^*$-a.s. for any $z \in E^*$. Let $\check{X}^*$ be the time change of X^* by means of the inverse of A_t and $\widehat{Y}^*$ be the image of $\check{X}^*$ by ϕ defined as (7.6.28). Since ϕ is a topological homeomorphism from E^* to $\widehat{E}^*$, $\widehat{Y}^*$ is a conservative (in fact, recurrent) diffusion process on $\widehat{E}^*$. It is $\widehat{m}_0$-symmetric. Furthermore, as in the last paragraph in the proof of Theorem 7.6.3, we see by using Theorem 5.3.1 that the part process of $\widehat{Y}^*$ in $\widehat{E}_0$ is an absorbing Brownian motion in $\widehat{E}_0$. Thus $\widehat{Y}^*$ is an extension of the absorbing Brownian motion in $\widehat{E}_0$ to $\widehat{E}^*$ in the sense of Definition 7.7.1 and so we can apply Theorem 7.7.3 on uniqueness to obtain the desired conclusion. □

Remark 7.8.2. Theorem 7.8.1 remains valid in the general case where X is the absorbing Brownian motion on a domain $D \subset \mathbb{R}^n$ and $\{K_i\}$ is a collection of countably infinite number of disjoint non-polar compact subsets of D, which is locally finite. In this general case, the PCAF A_t of X^* defined by (7.6.27) is strictly increasing in $t \in [0, \zeta^*)$ $\mathbf{P}_z^*$-a.s. for any $z \in E^*$, where ζ^* is the lifetime of X^*. The image process $\widehat{Y}^*$ defined by (7.6.28) is an $\widehat{m}_0$-symmetric diffusion process on $\widehat{E}^*$ with lifetime A_{ζ^*} admitting no killing at $\widehat{a}_i$ for every $i \ge 1$. Therefore, $\widehat{Y}^*$ is an extension of the absorbing Brownian motion on $\widehat{E}_0$ in the sense of Definition 7.7.1 and the uniqueness theorem applies. □

(2°) Darning multiple holes for a reflecting stable process

Suppose that D is an open *n-set* in $\mathbb{R}^n$ and m denotes the Lebesgue measure on $\overline{D}$. Let $X = (X_t, \mathbf{P}_x, \zeta)$ be the reflecting α-stable process on $\overline{D}$ considered in Example (**5°**) of Section 7.6. But this time we let $K = \bigcup_{i \in \Lambda} K_i$ be the union of a finite or countably infinite number of disjoint non-trivial compact subsets of ∂D which are locally finite. In view of the properties of X explained in Example (5°) of Section 7.6, we readily see that conditions **(C.1)**–**(C.3)** are satisfied by X for $E = \overline{D}$ and $K = \cup_{i \in \Lambda} K_i$.

Suppose each compact set $K_i \subset \partial D$ has finite and strictly positive d_i-dimensional Hausdorff measure when $n \geq 2$ and is non-empty when $n = 1$. Let

$$\sigma_K := \inf\{t \geq 0 : X_t \in K\} \wedge \zeta.$$

It follows from [14, Theorem 2.5 and Remark 2.2(i)] that

$$\varphi^{(i)}(x) := \mathbf{P}_x(\sigma_K < \infty \text{ and } X_{\sigma_K -} \in K_i) > 0 \tag{7.8.1}$$

for every $x \in \overline{D} \setminus K$ if and only if $\alpha > n - d_i$ when $n \geq 2$ and $\alpha > 1$ when $n = 1$.

Now we assume that $\alpha > n - d_i$ for each $i \geq 1$ when $n \geq 2$ and $\alpha > 1$ when $n = 1$. Let $X^0 = (X_t^0, \mathbf{P}_x^0, \zeta^0)$ of X killed upon hitting K. X^0 then satisfies the condition **(M°.1)**. It is easy to see that X^0 has a symmetric transition density function $p^0(t, x, y)$, which can be represented using the Hölder continuous transition density $p(t, x, y)$ of X as

$$p^0(t, x, y) = p(t, x, y) - \mathbf{E}_x \left[p\left(t - \tau_D, X_{\tau_D}, y\right) ; \tau_D < t \right]$$

for $t > 0$ and $x, y \in \overline{D} \setminus K$. Moreover, the density function $p^0(t, x, y)$ is continuous on $(0, \infty) \times (\overline{D} \setminus K) \times (\overline{D} \setminus K)$. Thus X^0 satisfies also the condition **(M°.3)**. The property **(M°.4)** of X^0 can be readily verified as in Example (**5°**) of Section 7.6.

Hence we can apply Theorem 7.7.5 to get the symmetric extension X^* of X^0 by darning the holes $\{K_i, i \in \Lambda\}$ into single points $\{a_i^*, i \in \Lambda\}$. □

In the above two examples, the uniqueness of the extension X^*, the description of its Dirichlet form, and the characterization of its L^2-generator by the zero flux condition can be derived from Theorem 7.7.3 by replacing points a_i with a_i^*.

We say that a symmetric right process X *admits a reflecting extension* if

$$\mathcal{F} \subsetneqq \mathcal{F}_a^{\text{ref}}, \tag{7.8.2}$$

where $\mathcal{F}$ is the Dirichlet space of X and $\mathcal{F}_a^{\text{ref}}$ is its active reflected Dirichlet space. If X admits a reflecting extension, then X is non-conservative by

Corollary 6.6.6. In the following two examples, we consider such a situation together with all possible symmetric extensions with no killing or jump on the boundary.

(3°) One-dimensional minimal diffusion

Let $I = (r_1, r_2)$ be a one-dimensional interval, X^0 be a minimal diffusion on I with no killing inside I in the sense that it satisfies the properties **(d.1)**, **(d.2)**, **(d.3)**, and **(d.4)** of Example 3.5.7, and (s, m) be a pair of a canonical scale and a canonical measure corresponding to X^0. X^0 is then m-symmetric and, by Theorem 3.5.8, the Dirichlet form $(\mathcal{E}^0, \mathcal{F}^0)$ of X^0 on $L^2(I; m)$ is given by

$$\begin{cases} \mathcal{F}^0 = \{u \in \mathcal{F}^{(s)} \cap L^2(I; m) : u(r_i) = 0 \text{ if } r_i \text{ is regular}\} \\ \mathcal{E}^0 = \mathcal{E}^{(s)}\big|_{\mathcal{F}^0 \times \mathcal{F}^0}. \end{cases}$$

In view of (6.5.8),

$$(\mathcal{F}^0)_a^{\text{ref}} = \mathcal{F}^{(s)} \cap L^2(I; m), \quad \mathcal{E}^{0,\text{ref}} = \mathcal{E}^{(s)},$$

and consequently X^0 admits a reflecting extension if and only if either r_1 or r_2 is regular.

In this case, let I^* be the interval obtained from I by adding the boundary r_i whenever r_i is regular. Then $(\mathcal{E}, \mathcal{F}) = (\mathcal{E}^{(s)}, \mathcal{F}^{(s)} \cap L^2(I; m))$ is a regular strongly local Dirichlet form on $L^2(I^*; \mathbf{1}_I \cdot m)$ and the associated diffusion process X on I^* is by definition the reflecting diffusion (Example 3.5.7). X is a symmetric extension of X^0 and actually a reflecting extension in the sense of Definition 7.2.6.

When both r_1 and r_2 are regular, then the reflecting extension X is a symmetric two-point extension of X^0 from I to $I^* = [r_1, r_2]$. X^0 admits three other kinds of symmetric (one-point) extensions: the extension to $[r_1, r_2)$ reflecting only at r_1, the extension to $(r_1, r_2]$ reflecting only at r_2, and an extension $\dot{X}$ to the one-point compactification $\dot{I} = I \cup \{\partial\}$ of I. Extend m to $\dot{m}$ on $\dot{I}$ by setting $\dot{m}(\{\partial\}) = 0$. Then the Dirichlet form $(\dot{\mathcal{E}}, \dot{\mathcal{F}})$ of $\dot{X}$ on $L^2(\dot{I}; \dot{m})$ is regular and it can be described as

$$\dot{\mathcal{E}} = \mathcal{E}^{(s)} \quad \text{and} \quad \dot{\mathcal{F}} = \{u \in \mathcal{F}^{(s)} : u(r_1) = u(r_2)\}.$$

Without using Dirichlet forms, these three one-point extensions of X^0 can be constructed by means of Theorem 7.5.6, while the two-point extension to $[r_1, r_2]$ can be done by means of Theorem 7.7.4.

(4°) Time-changed transient reflecting Brownian motion

Let D be a domain of $\mathbb{R}^n$ with a continuous boundary and $Z = (Z_t, \mathbf{Q}_x)$ be the reflecting Brownian motion on $\overline{D}$. The Dirichlet form of Z on $L^2(D)$ is $(\frac{1}{2}\mathbf{D}, H^1(D))$. In what follows, we assume the transience of Z so that n must be more than 3 and D must be unbounded. If further D contains an infinite cone, then Z is transient as has been seen in Example 3.5.9.

More specifically, we assume that D is an unbounded Lipschitz domain in $\mathbb{R}^n$ with $n \geq 3$ in the following sense: there are constants $\delta > 0$, $M > 0$ and a locally finite open covering $\{U_j\}_{j \in J}$ of ∂D such that, for each $j \in J$, the set $D \cap U_j$ is expressed as (2.2.55) in some coordinate system by some Lipschitz continuous function F_j with Lipschitz constant bounded by M, and $\partial D \subset \bigcup_{j \in J} \{x \in U_j : \text{dist}(x, \partial U_j) > \delta\}$. By [76], there exists then a reflecting Brownian motion (RBM) $Z = (Z_t, \mathbf{Q}_x)$ on $\overline{D}$ whose resolvent $\{G^Z_\alpha; \alpha > 0\}$ has the strong Feller property in the sense that

$$G^Z_\alpha(bL^1(D)) \subset bC(\overline{D}). \tag{7.8.3}$$

In particular, Z satisfies the absolute continuity condition **(AC)** in Definition A.2.16. Such a process Z had been constructed in [5] for a bounded Lipschitz domain.

The process Z is always conservative but we see from Theorem 3.5.2 that Z escapes to infinity as $t \to \infty$:

$$\mathbf{Q}_x \left(\lim_{t \to \infty} Z_t = \partial \right) = 1 \quad \text{for every } x \in \overline{D}, \tag{7.8.4}$$

where ∂ is the point at infinity of $\overline{D}$. Any function in the space $\text{BL}(D)$ is represented by its quasi continuous version. By virtue of Corollary 3.5.3, any $u \in H^1_e(D)$ then satisfies

$$\mathbf{Q}_x \left(\lim_{t \to \infty} u(Z_t) = 0 \right) = 1 \quad \text{for every } x \in \overline{D}. \tag{7.8.5}$$

The above two statements hold "for q.e. $x \in \overline{D}$", which are now strengthened to "for every $x \in \overline{D}$" owing to the property **(AC)** of Z.

By Section 6.5(**5**°), the reflected Dirichlet space of the Dirichlet form of $(\frac{1}{2}\mathbf{D}, H^1(D))$ is identical with the space $(\frac{1}{2}\mathbf{D}, \text{BL}(D))$ of BL functions. The extended Sobolev space $H^1_e(D)$ is a subspace of $\text{BL}(D)$, and indeed a proper subspace because the former does not contain a non-zero constant function due to the transience assumption while the latter does.

Let $\mathcal{H}^\star(D)$ be the space of functions in $\text{BL}(D)$ which are $\mathbf{D}$-orthogonal to $H^1_e(D)$. Then $\mathcal{H}^\star(D) \subset \mathcal{H}(D)$, where $\mathcal{H}(D)$ denotes the space of harmonic functions on D with finite Dirichlet integral. We will be concerned with the condition that

$$\mathcal{H}^\star(D) \text{ consists of constant functions on } D. \tag{7.8.6}$$

Condition (7.8.6) holds true when $D = \mathbb{R}^n$, $n \geq 3$, in view of Theorem 2.2.12. This property remains valid for any unbounded uniform domain. A domain D is called a *uniform domain* if there exists $C > 1$ such that for every $x, y \in D$, there is a rectifiable curve γ in D connecting x and y with $\text{length}(\gamma) \leq C|x - y|$

and moreover

$$\min\{|x-z|,\ |z-y|\} \le C\,\mathrm{dist}(z, D^c) \qquad \text{for every } z \in \gamma.$$

An infinite cone is a special unbounded uniform domain. We refer the readers to [24, Proposition 3.6] for a proof of the following implication:

$$D \setminus \overline{B_r(0)} \text{ is a unbounded uniform domain}$$
$$\Longrightarrow \qquad \text{(7.8.6) holds.} \tag{7.8.7}$$

Here $B_r(0)$ denotes the open ball with center 0 and radius $r > 0$.

Since the active reflected Dirichlet space of Z coincides with $\mathrm{BL}(D) \cap L^2(D) = H^1(D)$, condition (7.8.2) fails and the reflecting Brownian motion Z does not admit a reflecting extension. But if we make a time change of Z, then the situation may change radically.

Let m be a positive Radon measure on $\overline{D}$ charging no polar set possessing full quasi support with respect to the Dirichlet form $(\frac{1}{2}\mathbf{D}, H^1(D))$. For instance, $m(dx) = f(x)dx$ for a strictly positive $f \in L^1_{\mathrm{loc}}(D)$ has these properties. Let $X = (X_t, \zeta, \mathbf{P}_x)$ be the time-changed process of the reflecting Brownian motion Z on $\overline{D}$ by means of its PCAF A with Revuz measure m. X is then m-symmetric, and the Dirichlet form of X on $L^2(\overline{D}; m)$ and its active reflected Dirichlet space are given by

$$\Big(\frac{1}{2}\mathbf{D}, H^1_e(D) \cap L^2(\overline{D}; m)\Big), \qquad \Big(\frac{1}{2}\mathbf{D}, \mathrm{BL}(D) \cap L^2(\overline{D}; m)\Big), \tag{7.8.8}$$

respectively, as was seen in Section 6.5(**5°**).

We see that X admits a reflecting extension if $m(\overline{D}) < \infty$, because then the two Dirichlet forms in (7.8.8) obviously differ. If $m(\overline{D}) = \infty$, then $1 \notin L^2(D; m)$ and they coincide under (7.8.6). Thus

PROPOSITION 7.8.3. *Assume that a domain satisfies condition* (7.8.6). *Then X admits a reflecting extension if and only if $m(\overline{D}) < \infty$.*

We now take a strictly positive continuous and integrable function f on D and let $dm = fdx$. $A_t = \int_0^t f(Z_s)ds$ is the associated PCAF of Z in the strict sense with full support $\overline{D}$. The time-changed process $X = (X_t, \mathbf{P}_x)$ with $X_t = Z_{\tau_t}$, $\tau = A^{-1}, \zeta = A_\infty$ is a diffusion process on $\overline{D}$ whose 0-order resolvent is given by

$$G^X_{0+}u(x) = G^Z_{0+}(uf)(x), \quad x \in \overline{D}, \quad u \in \mathcal{B}_+(\overline{D}).$$

This in particular implies that X also satisfies the condition **(AC)** and $G^X_{0+}u$ is lower semicontinuous for any $u \in \mathcal{B}_+(\overline{D})$ in view of (7.8.3).

As was observed in Section 6.5(**5°**), the finiteness of $m(\overline{D})$ implies that ζ is finite a.s. and so we get from (7.8.4) and the above observation

$$\mathbf{P}_x(\zeta < \infty,\ X_{\zeta-} = \partial) = 1 \quad \text{for every } x \in \overline{D}. \tag{7.8.9}$$

Therefore, the conditions **(A.1)**, **(A.2)**, **(A.3)** of Section 7.5 with X and ∂ in place of X^0 and a, respectively, are fulfilled so that X admits a unique one-point symmetric extension $X^\star$ to the one-point compactification $\overline{D} \cup \{\partial\}$ of $\overline{D}$ by Theorems 7.5.4 and 7.5.6. By Theorem 7.5.4, its extended Dirichlet space $(\mathcal{F}_e^\star, \mathcal{E}^\star)$ can be described as

$$\mathcal{F}_e^\star = H_e^1(D) \oplus \{c:\ c \in \mathbb{R}\} \subset \mathrm{BL}(D), \qquad \mathcal{E}^\star = \frac{1}{2}\mathbf{D}.$$

In particular, $X^\star$ is recurrent. When condition (7.8.6) is satisfied, this extension is a reflecting extension of X. The special case that $D = \mathbb{R}^n$, $n \geq 3$, was considered in Example (**4°**) of Section 7.6.

If an unbounded domain D is not a uniform domain, the dimension of the space $\mathcal{H}^*(D)$ may exceed 2 as we shall see in the following example. Let

$$D = B_1(0) \cup \left\{ x = (x_1, x_2, \ldots, x_n) : x_d^2 > \sum_{k=1}^{n-1} x_k^2 \right\}, \quad n \geq 3. \tag{7.8.10}$$

D contains the upper cone C_+ and lower cone C_- where

$$C_+ = B_1(0)^c \cap \left\{ x_n > \left(\sum_{k=1}^{n-1} x_k^2 \right)^{1/2} \right\}, \quad C_- = B_1(0)^c \cap \left\{ x_n < -\left(\sum_{k=1}^{n-1} x_k^2 \right)^{1/2} \right\}$$

so that D is transient as it contains an infinite cone C_+ but D cannot be a uniform domain because it has a bottleneck $B_1(0)$. Note that D is a Lipschitz domain.

The point at infinity of $\overline{D}$ at the upper end (lower end) is denoted by ∂_+ (∂_-). Let $Z = (Z_t, \mathbf{Q}_x)$ be the RBM on $\overline{D}$ and

$$\varphi^+(x) = \mathbf{Q}_x\left(\lim_{t\to\infty} Z_t = \partial_+ \right), \quad \varphi^-(x) = \mathbf{Q}_x\left(\lim_{t\to\infty} Z_t = \partial_- \right).$$

We then see from (7.8.4) that

$$\varphi^+(x) + \varphi^-(x) = 1 \quad \text{for every } x \in \overline{D}. \tag{7.8.11}$$

Furthermore,

$$\varphi^+(x) > 0, \quad \varphi^-(x) > 0 \quad \text{for every } x \in \overline{D}, \tag{7.8.12}$$

because $\varphi^+(x)$ is either identically positive or identically zero due to the irreducibility, so the above follows from the symmetry of the domain D.

Choose again a strictly positive integrable function f on $\overline{D}$ and let $dm = fdx$. Consider the time-changed diffusion process $X = (X_t, \mathbf{P}_x, \zeta)$ on $\overline{D}$ obtained from Z by the time change with respect to the PCAF $A_t = \int_0^t f(Z_s)ds$. X is m-symmetric and has the property (7.8.9). In particular, it holds for $x \in \overline{D}$ that

$$\varphi^+(x) = \mathbf{P}_x(\zeta < \infty,\ X_{\zeta-} = \partial_+),\ \varphi^-(x) = \mathbf{P}_x(\zeta < \infty,\ X_{\zeta-} = \partial_-). \quad (7.8.13)$$

We put for $\alpha > 0, x \in \overline{D}$

$$u_\alpha^+(x) = \mathbf{E}_x\left[e^{-\alpha\zeta};\ X_{\zeta-} = \partial_+\right],\ u_\alpha^-(x) = \mathbf{E}_x\left[e^{-\alpha\zeta};\ X_{\zeta-} = \partial_-\right].$$

Proposition 7.8.4. (i) *There exists a unique m-symmetric two-point extension $X^\star$ of X from $\overline{D}$ to $\overline{D} \cup \{\partial_+\} \cup \{\partial_-\}$. $X^\star$ is recurrent. Denote by $(\mathcal{E}^\star, \mathcal{F}^\star)$ and $\mathcal{F}_e^\star$ the Dirichlet form of $X^\star$ on $L^2(\overline{D} \cup \{\partial_+\} \cup \{\partial_-\}; m)(= L^2(D; m))$ and its extended Dirichlet space, respectively. Then*

$$\mathcal{F}^\star = H^1(D) \oplus \{c_+ u_\alpha^+ + c_- u_\alpha^- \,:\, c_+, c_- \in \mathbb{R}\} \subset \mathrm{BL}(D) \cap L^2(D; m),$$

$$\mathcal{F}_e^\star = H_e^1(D) \oplus \{c_+\varphi^+ + c_-\varphi^- \,:\, c_+, c_- \in \mathbb{R}\} \subset \mathrm{BL}(D), \quad (7.8.14)$$

$$\mathcal{E}^\star(u, v) = \frac{1}{2}\mathbf{D}(u, v), \qquad u, v \in \mathcal{F}_e^\star. \quad (7.8.15)$$

(ii) *For any $u \in \mathrm{BL}(D)$, there exist constants c_+, c_- such that*

$$\begin{cases} \mathbf{Q}_x(Z_{\infty-} = \partial_+, \lim_{t\to\infty} u(Z_t) = c_+) = \mathbf{Q}_x(Z_{\infty-} = \partial_+), \\ \mathbf{Q}_x(Z_{\infty-} = \partial_-, \lim_{t\to\infty} u(Z_t) = c_-) = \mathbf{Q}_x(Z_{\infty-} = \partial_-). \end{cases} \quad (7.8.16)$$

If $c_+ = c_- = 0$, then $u \in \mathbf{H}_e^1(D)$.

Proof. (i) We check conditions **(M.1)**, **(M.2)**, **(M.3)** for X on $\overline{D}$ and ∂_+, ∂_- in place of X^0 on E_0 and a_1, a_2, respectively. **(M.1)** is ensured by (7.8.12) and (7.8.13). **(M.2)** is trivially true because $m(\overline{D}) < \infty$. **(M.3)** follows from the lower semicontinuity of $G_{0+}^X u$ for $u \in \mathcal{B}_+(\overline{D})$ proved in the paragraph below Proposition 7.8.3 and an obvious counterpart of Lemma 7.5.5(iv). Therefore, X admits a two-point extension $X^\star$ by virtue of Theorem 7.7.4. By Theorem 7.7.3, we get the uniqueness of $X^\star$ and the properties (7.8.14) and (7.8.15) of its Dirichlet form. Properties (7.8.11) and (7.8.14) imply the recurrence of $X^\star$ because $1 \in \mathcal{F}_e^\star$ and $\mathcal{E}^\star(1, 1) = 0$.
(ii) Define, for $n \geq 1$,

$$D_n = B_1(0) \cup C_+ \cup (C_- \cap B_n(0)), \quad \Gamma_n = C_- \cap \{|x| = n\}.$$

Then D_n is a uniform Lipschitz domain and D_n increases to D as $n \to \infty$.

For any $u \in \mathrm{BL}(D)$, $u|_{D_n} \in \mathrm{BL}(D_n)$ which is a sum of a function in $H_e^1(D_n)$ and some constant c_+ in view of (7.8.7). Let $Z^n = (Z_t^n, \mathbf{Q}_x^n)$ be the reflecting Brownian motion on $\overline{D}_n$, namely, a diffusion associated with the Dirichlet form $(\frac{1}{2}\mathbf{D}, H^1(D_n))$ on $L^2(D_n)$. On account of (7.8.4) and (7.8.5), we have

$$\mathbf{Q}_x^n\left(\lim_{t\to\infty} u(Z_t^n) = c_+,\ Z_{\infty-}^n = \partial_+\right) = 1, \quad x \in \overline{D}_n,$$

and we see that c_+ is independent of n.

Since the part processes of Z and Z^n on $\overline{D}_n \setminus \Gamma_n$ are identical in law, we further obtain

$$\begin{aligned}
&\mathbf{Q}_x\left(Z_{\infty-} = \partial_+,\ \lim_{t\to\infty} u(Z_t) = c_+\right) \\
&= \lim_{n\to\infty} \mathbf{Q}_x\left(\sigma_{\Gamma_n} = \infty,\ Z_{\infty-} = \partial_+,\ \lim_{t\to\infty} u(Z_t) = c_+\right) \\
&= \lim_{n\to\infty} \mathbf{Q}_x^n\left(\sigma_{\Gamma_n} = \infty,\ Z_{\infty-}^n = \partial_+,\ \lim_{t\to\infty} u(Z_t^n) = c_+\right) \\
&= \lim_{n\to\infty} \mathbf{Q}_x^n(\sigma_{\Gamma_n} = \infty) = \lim_{n\to\infty} \mathbf{Q}_x(\sigma_{\Gamma_n} = \infty) = \mathbf{Q}_x(Z_{\infty-} = \partial_+),
\end{aligned}$$

completing the proof of the first identity of (7.8.16). The second one can be shown similarly.

If $c_+ = c_- = 0$ for $u \in \mathrm{BL}(D)$, then the restriction of u to $B_1(0) \cup C_+$ belongs to the space $H_e^1(B_1(0) \cup C_+)$ in view of the above proof. Similarly, the restriction of u to $B_1(0) \cup C_-$ is an element of $H_e^1(B_1(0) \cup C_-)$. Hence $u \in H_e^1(D)$. □

PROPOSITION 7.8.5. *It holds that*

$$\mathcal{H}^\star(D) = \{c_+\varphi^+ + c_-\varphi^- : c_+,\ c_- \in \mathbb{R}\}$$

and the extension $X^\star$ of Proposition 7.8.4 is a reflecting extension of the time-changed RBM X on $\overline{D}$ in the sense of Definition 7.2.6.

Proof. By (7.8.14), $\varphi_+,\ \varphi_- \in \mathrm{BL}(D)$ and so each of them has an associated pair of constants (c_+, c_-) according to (ii) of Proposition 7.8.4. Let τ_n be the exit time of Z from the set $D \cap B_n(0)$, $n \geq 1$. Then $\{\varphi_+(Z_{\tau_n})\}$ is a bounded $\mathbf{Q}_x$-martingale and possesses an a.s. limit Φ with $\varphi_+(x) = \mathbf{E}^{\mathbf{Q}_x}[\Phi]$. Then

$$\Phi \mathbf{1}_{\{Z_{\infty-}=\partial_+\}} = c_+ \mathbf{1}_{\{Z_{\infty-}=\partial_+\}},\ \ \Phi \mathbf{1}_{\{Z_{\infty-}=\partial_-\}} = c_- \mathbf{1}_{\{Z_{\infty-}=\partial_-\}}.$$

But

$$c_+ \mathbf{1}_{\{Z_{\infty-}=\partial_+\}} = \lim_{n\to\infty} \varphi_+(Z_{\tau_n})\mathbf{1}_{\{Z_{\infty-}\circ\theta_{\tau_n}=\partial_+\}}$$

$$= \lim_{n\to\infty} \mathbf{Q}_x \left(Z_{\infty-} \circ \theta_{\tau_n} = \partial_+,\ Z_{\infty-} \circ \theta_{\tau_n} = \partial_+ \middle| \mathcal{F}_{\tau_n}\right)$$

$$= \lim_{n\to\infty} \mathbf{Q}_x \left(Z_{\infty-} \circ \theta_{\tau_n} = \partial_+ \middle| \mathcal{F}_{\tau_n}\right) = \lim_{n\to\infty} \varphi_+(Z_{\tau_n}) = \Phi.$$

By taking the $\mathbf{Q}_x$-expectation, we get $c_+ = 1$. In the same way, we get $c_- = 0$. We have shown that the associated pair with φ_+ is $(1, 0)$. Similarly, we see that the associated pair with φ_- is $(0, 1)$.

Take now any $u \in \mathrm{BL}(D)$ and let (c_+, c_-) be the associated pair with u. Define $u_0 = u - (c_+\varphi_+ + c_-\varphi_-)$. Then, by the above observation, $u_0 \in \mathrm{BL}(D)$ and the associated pair with u_0 equals $(0, 0)$. Therefore, $u_0 \in H^1_e(D)$ by virtue of the second assertion of Proposition 7.8.4.

The second statement follows from the first one and (7.8.14). □

Just as in the preceding example of the one-dimensional diffusion with two regular boundaries, the time-changed reflecting Brownian motion X on $\overline{D}$ admits three other kinds of one-point symmetric extensions: the extension to $\overline{D} \cup \{\partial_+\}$ reflecting only at ∂_+, the extension to $\overline{D} \cup \{\partial_-\}$ reflecting only at ∂_-, and the extension to the one-point compactification $\overline{D} \cup \{\partial\}$. The last one was constructed in the second paragraph after Proposition 7.8.3. Other two can be constructed in similar ways.

(5°) One-dimensional Brownian motion with countable boundary

We investigate the uniqueness and existence of the extension of a one-dimensional absorbing Brownian motion with a countable boundary.

Let $a_0 = 0$ and $\{a_n\}_{n\geq 1}$ be a sequence of positive numbers strictly decreasing to 0. Set

$$F := \{a_n\}_{n\geq 0}, \quad I_0 := (-\infty, 0), \quad I_1 := (a_1, \infty), \quad I_n := (a_n, a_{n-1})$$

for $n \geq 2$ and $E_0 := \mathbb{R} \setminus F = \cup_{n=0}^{\infty} I_n$. The Lebesgue measure on $\mathbb{R}$ is denoted by λ.

Let X^0 be the absorbing Brownian motion on E_0, namely, the Brownian motion being killed upon hitting the set F. Since $a_0 = 0$ is an accumulation point in F, we cannot use Theorem 7.7.3 in characterizing the extension of X^0 to $\mathbb{R}$. But we can combine Theorem 7.1.8 with three other general theorems in getting its unique existence.

In accordance with Example 3.5.7, a Markov process on an open interval is called a *minimal diffusion* with no killing inside if it satisfies the properties **(d.1)**, **(d.2)**, **(d.3)**, and **(d.4)** stated in Example 3.5.7.

PROPOSITION 7.8.6. *Let X be a minimal diffusion on $\mathbb{R}$ with no killing inside such that X is symmetric with respect to λ and the part process of X on E_0 coincides with X^0. Then X is the Brownian motion on $\mathbb{R}$.*

Proof. Let (s,m) be a pair of a canonical scale and a canonical measure associated with X. Then X is symmetric with respect to m so that m must be a constant time of λ by a uniqueness theorem of a symmetrizing measure due to Ying-Zhao [153] and a strong irreducibility **(d.4)** of X. We may take $m=\lambda$. It then suffices to show

$$ds = 2\lambda. \tag{7.8.17}$$

By virtue of Theorem 3.5.8, the Dirichlet form $(\mathcal{E},\mathcal{F})$ of X on $L^2(\mathbb{R};\lambda)$ is identical with (3.5.9) expressed in terms of the current pair (s,λ). Especially $(\mathcal{E},\mathcal{F})$ is a regular strongly local Dirichlet form. Accordingly,

$$J=0, \quad \kappa=0, \tag{7.8.18}$$

in its Beurling-Deny representation. Furthermore, the energy measure $\mu^c_{\langle u\rangle}$ for $u\in b\mathcal{F}$ does not charge any single point owing to the energy image density property formulated in Theorem 4.3.8, and in particular

$$\mu^c_{\langle \mathbf{H}u\rangle}(\{a_i\})=0 \qquad \text{for every } i\geq 0 \text{ and for any } u\in\mathcal{F}. \tag{7.8.19}$$

Denote by $(\mathcal{E}^0,\mathcal{F}^0)$ and $(\mathcal{E}^{0,\mathrm{ref}},(\mathcal{F}^0)^{\mathrm{ref}}_a)$ the Dirichlet form of X^0 on $L^2(E_0;\lambda)$ and its active reflected Dirichlet space, respectively. For a function f on $\mathbb{R}$ and $n\geq 0$, we let $f_n := f|_{I_n}$. Using the notations in Example (**1**°) of Section 7.6, we then have

$$\mathcal{F}^0=\left\{f\in L^2(\mathbb{R};\lambda):\ f_n\in H^1_0(I_n) \text{ for every } n\geq 0\right\},$$
$$(\mathcal{F}^0)^{\mathrm{ref}}_a=\left\{f\in L^2(\mathbb{R};\lambda):\ f_n\in H^1(I_n) \text{ for every } n\geq 0\right\},$$
$$\mathcal{E}^{0,\mathrm{ref}}(f,f)=\sum_{n=0}^{\infty}\frac{1}{2}\,\mathbf{D}^{I_n}(f_n,f_n) \qquad \text{for } f\in(\mathcal{F}^0)^{\mathrm{ref}}_a. \tag{7.8.20}$$

We now apply Theorem 7.1.8 to the current X^0 and X. On account of (7.8.18) and (7.8.19), we get

$$\mathcal{F}^0\subset \mathcal{F}\big|_{E_0}\subset(\mathcal{F}^0)^{\mathrm{ref}}_a, \tag{7.8.21}$$

$$\mathcal{E}(u,u)=\mathcal{E}^{0,\mathrm{ref}}\left(u\big|_{E_0},u\big|_{E_0}\right) \quad \text{for } u\in\mathcal{F}. \tag{7.8.22}$$

Taking the expression (3.5.9) of $(\mathcal{E},\mathcal{F})$ into account, we conclude from (7.8.21) that ds and $\lambda=dx$ is mutually absolutely continuous. If we let $\frac{ds}{dx}=f>0$, then we get from (7.8.22)

$$\int_{I_n} u'(s(x))^2 f(x)dx = \frac{1}{2}\int_{I_n} u'(s(x))^2 f(x)^2 dx$$

for any $u \in C_0^1(s(I_n))$ and $n \geq 0$, which means $f = 2$ and the desired (7.8.17). □

Next let us take positive numbers $\{p_n\}_{n\geq 0}$ such that

$$\alpha \leq p_n \leq \beta, \qquad n = 0, 1, 2, \ldots$$

for some positive constants α, β, and we let

$$\widetilde{m}(dx) := \sum_{n=0}^{\infty} p_n \mathbf{1}_{I_n}(x)dx. \tag{7.8.23}$$

The absorbing Brownian motion X^0 is symmetric with respect to the Lebesgue measure λ but it can also be viewed as an $\widetilde{m}$-symmetric diffusion on E_0, as observed in Section 7.5 already. Let $(\widetilde{\mathcal{E}}^0, \widetilde{\mathcal{F}}^0)$ and $(\widetilde{\mathcal{E}}^{0,\text{ref}}, (\widetilde{\mathcal{F}}^0)_a^{\text{ref}})$ be the Dirichlet form of X^0 on $L^2(E_0; \widetilde{m})$ and its active reflected Dirichlet space, respectively. They are given by

$$\widetilde{\mathcal{F}}^0 = \left\{f \in L^2(\mathbb{R}; \widetilde{m}) : f_n \in H_0^1(I_n) \text{ for every } n \geq 0\right\},$$

$$(\widetilde{\mathcal{F}}^0)_a^{\text{ref}} = \left\{f \in L^2(\mathbb{R}; \widetilde{m}) : f_n \in H^1(I_n) \text{ for every } n \geq 0\right\},$$

$$\widetilde{\mathcal{E}}^{0,\text{ref}}(f,f) = \sum_{n=0}^{\infty} \frac{p_n}{2}\, \mathbf{D}^{I_n}(f_n, f_n) \qquad \text{for } f \in (\widetilde{\mathcal{F}}^0)_a^{\text{ref}}.$$

In exactly the same way as the proof of the preceding proposition, we can prove the following:

Proposition 7.8.7. *Let X be a minimal diffusion on $\mathbb{R}$ with no killing inside such that X is symmetric with respect to $\widetilde{m}$ and the part process of X on E_0 coincides with X^0. Then X is the diffusion associated with the canonical scale*

$$\widetilde{s}(dx) = 2\sum_{n=0}^{\infty} p_n^{-1} \mathbf{1}_{I_n}(x)dx \tag{7.8.24}$$

and the canonical measure $\widetilde{m}$.

By repeating the one-point skew extensions formulated in Theorem 7.5.9, the process X of Proposition 7.8.7 can be constructed from the Brownian motion directly as follows. Let B^- and B^+ be the absorbing Brownian motions on $\mathbb{R}_- = (-\infty, 0)$ and $\mathbb{R}_+ = (0, \infty)$, respectively. Let X^{01} be the subprocess of B^+ on $\mathbb{R}_+ \setminus \{a_1\}$ killed upon hitting a_1. The process X^{01} is symmetric with respect to the measure

$$m_1(dx) = p_2 \mathbf{1}_{(0,a_1)}(x)dx + p_1 \mathbf{1}_{(a_1,\infty)}(x)dx,$$

and we can apply Theorem 7.5.9 to construct a unique m_1-symmetric diffusion X^1 on $\mathbb{R}_+$ extending X^{01} by darning the hole a_1 with entrance law μ_t^1 determined by $\int_0^\infty \mu_t^1 dt = m_1$.

We next consider the subprocess X^{02} of X^1 on $\mathbb{R}_+ \setminus \{a_2\}$ being killed upon hitting the point a_2. X^{02} is symmetric with respect to the measure

$$\begin{aligned} m_2(dx) &= \frac{p_3}{p_2} \mathbf{1}_{(0,a_2)} m_1(dx) + \mathbf{1}_{(a_2,\infty)} m_1(dx) \\ &= p_3 \mathbf{1}_{(0,a_2)} dx + p_2 \mathbf{1}_{(a_2,a_1)} dx + p_1 \mathbf{1}_{(a_1,\infty)} dx, \end{aligned}$$

and we can construct a unique m_2-symmetric diffusion X^2 on $\mathbb{R}_+$ extending X^{02} just as above.

Repeating this procedure and taking the limit as in [22, §3], we get a diffusion X^+ on $\mathbb{R}_+$ satisfying the following: X^+ is symmetric with respect to the measure

$$m_+(dx) = \mathbf{1}_{\mathbb{R}_+}(x)\widetilde{m}(dx) = \sum_{n=1}^{\infty} p_n \mathbf{1}_{I_n}(x)dx$$

and it is actually an m_+-symmetric extension of the subprocess $X^{0,+}$ of X^0 on $E_0 \cap (0, \infty) = \cup_{n=1}^{\infty} I_n$. The process X^+ has a finite lifetime and approaches 0 almost surely.

We finally piece X^+ together with B^- at 0 via Proposition 7.5.11 to get a desired diffusion X on $\mathbb{R}$ which is symmetric with respect to

$$\widetilde{m}(dx) = \mathbf{1}_{\mathbb{R}_-}(x)dx + m_+(dx)$$

and actually an $\widetilde{m}$-symmetric extension of X^0.

Thus we may call a diffusion associated with the pair $(\widetilde{s}, \widetilde{m})$ of (7.8.24) and (7.8.23) a *skew Brownian motion* on $\mathbb{R}$. Even if the boundary set F has no accumulation point, the absorbing Brownian motion does not satisfy the condition (7.7.1) so that Theorem 7.7.3 is not applicable. Nevertheless, a skew Brownian motion can be characterized and constructed as above.

For instance, for any sequence $\mathbf{q} = \{q_n : n \geq 0\}$ of positive numbers, the diffusion process $X^{\mathbf{q}}$ on $\mathbb{R}$ associated with the pair

$$\begin{cases} ds^{\mathbf{q}} = 2q_0^{-1}\mathbf{1}_{(-\infty,0)}(x)dx + 2\sum_{n=1}^{\infty} q_n^{-1}\mathbf{1}_{(n-1,n)}(x)dx, \\ dm^{\mathbf{q}} = q_0\mathbf{1}_{(-\infty,0)}(x)dx + \sum_{n=1}^{\infty} q_n\mathbf{1}_{(n-1,n)}(x)dx \end{cases}$$

is a skew Brownian motion characterized and constructed in a similar manner to the above. When $s^{\mathbf{q}}(0,\infty)$ is finite, $+\infty$ is approachable and $X^{\mathbf{q}}$ becomes transient. But $X^{\mathbf{q}}$ is always conservative because

$$\int_0^n m^{\mathbf{q}}(0,x)s^{\mathbf{q}}(dx) \geq \sum_{k=1}^{n} \int_{k-1}^{k} m^{\mathbf{q}}(k-1,x)s^{\mathbf{q}}(dx) = n \to \infty, \ n \to \infty,$$

namely, $+\infty$ is non-exit. Thus $X^{\mathbf{q}}$ admits no reflecting extension.

Appendix A

ESSENTIALS OF MARKOV PROCESSES

A.1. MARKOV PROCESSES

In this section, we introduce various concepts of Markov processes, such as the Markov process, right process, standard process, and Hunt process.

A.1.1. Preliminaries

In this subsection, we collect some preliminaries for Sections A.1, A.2, and A.3. Let Ω be a set. A class $\mathcal{M}$ of subsets of Ω is called a *σ-field* if $\Omega \in \mathcal{M}$ and it is closed under the operations of taking a countable union and complement. The couple $(\Omega, \mathcal{M})$ is said to be a *measurable space* in this case. Given a measurable space $(\Omega, \mathcal{M})$ and a subset $\widetilde{\Omega}$ of Ω, the *trace* of $\mathcal{M}$ on $\widetilde{\Omega}$ is defined by $\{\Lambda \cap \widetilde{\Omega} : \Lambda \in \mathcal{M}\}$ and denoted by $\mathcal{M}|_{\widetilde{\Omega}}$. $(\widetilde{\Omega}, \mathcal{M}|_{\widetilde{\Omega}})$ is then a measurable space. For a family $\mathcal{C}$ of subsets of Ω, $\sigma(\mathcal{C})$ will denote the smallest σ-field containing $\mathcal{C}$.

A probability measure $\mathbf{P}$ on a measurable space $(\Omega, \mathcal{M})$ is a map $\mathbf{P} : \mathcal{M} \mapsto [0, 1]$ such that $\mathbf{P}(\Omega) = 1$ and $\mathbf{P}$ is countably additive in the sense that

$$\mathbf{P}\left(\bigcup_{n=1}^{\infty} \Lambda_n\right) = \sum_{n=1}^{\infty} \mathbf{P}(\Lambda_n)$$

for $\Lambda_n \in \mathcal{M}$ such that $\Lambda_j \cap \Lambda_j = \emptyset$ for $i \neq j$. In this case, the triple $(\Omega, \mathcal{M}, \mathbf{P})$ is called a *probability space*. Denote by $\mathcal{P}(\Omega)$ the family of all probability measures on a measurable space $(\Omega, \mathcal{M})$. For $\mathbf{P} \in \mathcal{P}(\Omega)$, we put

$$\mathcal{N}^{\mathbf{P}} := \{N \subset \Omega : \exists \Lambda \in \mathcal{M} \text{ such that } \mathbf{P}(\Lambda) = 0 \text{ and } N \subset \Lambda\},$$

$$\mathcal{M}^{\mathbf{P}} := \{\Lambda \Delta N : \Lambda \in \mathcal{M},\ N \in \mathcal{N}^{\mathbf{P}}\}, \quad \mathbf{P}(\Lambda \Delta N) := \mathbf{P}(\Lambda).$$

The σ-field $\mathcal{M}^{\mathbf{P}}$ so defined is called the *completion* of $\mathcal{M}$ with respect to $\mathbf{P}$. $N \in \mathcal{N}^{\mathbf{P}}$ is called a *null set* in $\mathcal{M}^{\mathbf{P}}$. The σ-field $\mathcal{M}^*$ defined by $\mathcal{M}^* = \cap_{\mathbf{P} \in \mathcal{P}(\Omega)} \mathcal{M}^{\mathbf{P}}$ is called the *universal completion* of $\mathcal{M}$.

Let $\mathbb{R}_+ = [0, \infty)$ and $\mathcal{B}$ be the family of Borel subsets of $\mathbb{R}_+$, namely, $\mathcal{B}$ is the σ-field generated by all open subsets of $\mathbb{R}_+$. Given a measurable space $(\Omega, \mathcal{M})$, we denote by $\mathcal{B} \times \mathcal{M}$ the product σ-field of the product space $\mathbb{R}_+ \times \Omega$, namely, the σ-field generated by sets of the type $B \times \Lambda$ with $B \in \mathcal{B}$ and $\Lambda \in \mathcal{M}$.

PROPOSITION A.1.1. *Let* $A \in \mathcal{B} \times \mathcal{M}$ *and* $\pi(A) = \{\omega \in \Omega : \exists t \geq 0, (t, \omega) \in A\}$. *Then* $\pi(A) \in \mathcal{M}^*$.

This proposition follows from the observations that the notion of analytic sets is stable under projections ([37, III:13]) and any $\mathcal{M}$-analytic subset of Ω belongs to $\mathcal{M}^{\mathbf{P}}$ for any $\mathbf{P} \in \mathcal{P}(\Omega)$ ([37, III:33]).

We next state monotone class theorems. For a set Ω, a class $\mathcal{C}$ of its subsets is said to be *a π-system* if it is closed under finite intersections. A class $\mathcal{D}$ of its subsets is called a *Dynkin class* if (i) $\Omega \in \mathcal{D}$, (ii) if $A \subset B$ are both from $\mathcal{D}$, then so is $B \setminus A$, (iii) if $\{A_j, j \geq 1\}$ is a countable disjoint sets from $\mathcal{D}$, then so is $\cup_{j=1}^{\infty} A_j$.

PROPOSITION A.1.2. *Let $\mathcal{C}$ be a π-system on a set Ω.*
(i) *If $\mathcal{D}$ is a Dynkin class containing $\mathcal{C}$, then $\sigma(\mathcal{C}) \subset \mathcal{D}$.*
(ii) *Suppose H is a linear space consisting of bounded real functions on Ω such that $1 \in H$, $\mathbf{1}_A \in H$ for every $A \in \mathcal{C}$, and H is closed under the operation of taking uniformly bounded increasing limits. Then H contains any bounded $\sigma(\mathcal{C})$-measurable function.*

Proof. (i) See [13, Chapter 0, (2.2)]. (ii) follows from (i) because $\mathcal{D} = \{A \subset \Omega : \mathbf{1}_A \in H\}$ is a Dynkin class. □

Let E be a topological space. The *Borel σ-field* $\mathcal{B}(E)$ of E is by definition the σ-field generated by all open subsets of E. Denote by $\mathcal{P}(E)$ the family of all probability measures on the measurable space $(E, \mathcal{B}(E))$. We let $\mathcal{B}^*(E) = \cap_{\mu \in \mathcal{P}(E)} \mathcal{B}^{\mu}(E)$ where $\mathcal{B}^{\mu}(E)$ is the μ-completion of $\mathcal{B}(E)$. Each element of $\mathcal{B}(E)$ (resp. $\mathcal{B}^*(E)$) is called a *Borel* (resp. *universally measurable*) subset of E.

PROPOSITION A.1.3. *Let (F, d) be a compact metric space equipped with a metric d and E be an arbitrary subset of F. We endow E with the relative topology and let $\mathcal{B}(E)$ be the Borel σ-field of E with respect to this topology. We denote by $C_u(E)$ the collection of all d-uniformly continuous real functions on E.*

Suppose H is a linear space consisting of bounded real functions on E such that $1 \in H$, $C_u(E) \subset H$ and H is closed under the operation of taking uniformly bounded increasing limits. Then H contains any bounded $\mathcal{B}(E)$-measurable function.

Proof. For the family $\mathcal{O}$ of all open subsets of F, its trace $\mathcal{O}|_E$ on E is the family of all open subsets of E (with respect to the relative topology on E).

Accordingly, $\mathcal{B}(E)$ is the trace of $\mathcal{B}(F)$ on E:

$$\mathcal{B}(E) = \mathcal{B}(F)\big|_E. \tag{A.1.1}$$

Notice that $C_u(E) = C(F)|_E$ and in particular $C_u(E) \subset bC(E)$.

Let $H(F) = \{f \in b\mathcal{B}(F) : f|_E \in H\}$. Then $1 \in H(F)$, $C(F) \subset H(F)$ and $H(F)$ is closed under uniformly bounded increasing limits. For the distance function $d(x, y)$ on F and $A \subset F$, we put $d(x, A) := \inf_{y \in A} d(x, y)$. If we let, for any $U \in \mathcal{O}, f(x) = d(x, F \setminus U) \wedge 1$, $x \in F$, then $f \in C(F)$, $f^{1/n}(\in C(F)) \uparrow \mathbf{1}_U$, $n \to \infty$, and hence $\mathbf{1}_U \in H(F)$. As $\mathcal{O}$ is a π-system, we have by Proposition A.1.2 that $H(F)$ coincides with the space of all bounded $\mathcal{B}(F)$-measurable functions. □

Let $(E, \mathcal{B}(E))$ be a measurable space and denote by T either $\mathbb{R}_+ = [0, \infty)$ or $[0, \infty]$. A quadruplet $X = (\Omega, \mathcal{M}, \{X_t\}_{t \in T}, \mathbf{P})$ is called a *stochastic process* with *time parameter set* T and *state space* $(E, \mathcal{B}(E))$ if $(\Omega, \mathcal{M}, \mathbf{P})$ is a probability space and, for each $t \in T$, X_t is a measurable map from $(\Omega, \mathcal{M})$ to $(E, \mathcal{B}(E))$. In this book, the last property is designated by $X_t \in \mathcal{M}/\mathcal{B}(E)$. We call Ω a *sample space*, $\omega \in \Omega$ a *sample*, $\Lambda \in \mathcal{M}$ an *event* and $\mathbf{P}(\Lambda)$ its *probability*, respectively, of X. For each $t \in T, X_t = X_t(\omega)$ is a function of $\omega \in \Omega$ with value in E. For each fixed $\omega \in \Omega$, the function $X_\cdot(\omega) : T \to E$ is called the *sample path* of ω.

A family $\{\mathcal{M}_t;\ t \geq 0\}$ of sub-σ-fields of $\mathcal{M}$ with parameter $\mathbb{R}_+$ is said to be a *filtration* if $\mathcal{M}_s \subset \mathcal{M}_t$ for $0 \leq s < t$. In this case, we always extend the parameter of the filtration to $[0, \infty]$ by setting $\mathcal{M}_\infty = \sigma\{\mathcal{M}_t, t \geq 0\}$. For a stochastic process X and a filtration $\{\mathcal{M}_t; t \geq 0\}$, the former (resp. latter) is called *adapted to* (resp. *admissible for*) the latter (resp. former) if $X_t \in \mathcal{M}_t/\mathcal{B}(E)$ for every $t \geq 0$. For a filtration $\{\mathcal{M}_t; t \geq 0\}$, we put $\mathcal{M}_{t+} = \bigcap_{t' > t} \mathcal{M}_{t'}$, $t \geq 0$. A filtration is called *right continuous* if $\mathcal{M}_t = \mathcal{M}_{t+}$ for every $t \geq 0$.

Definition A.1.4. Let $\{\mathcal{M}_t\}_{t \geq 0}$ be a filtration. A function $\sigma : \Omega \mapsto [0, \infty]$ is called an $\{\mathcal{M}_t\}$-*stopping time* if $\{\sigma \leq t\} \in \mathcal{M}_t$ for every $t \geq 0$.

Obviously, a non-random constant time $\sigma(\omega) = r(\geq 0)$ for every $\omega \in \Omega$ is an $\{\mathcal{M}_t\}$-stopping time.

Exercise A.1.5. Prove the following.

(i) A function $\sigma : \Omega \mapsto [0, \infty]$ is an $\{\mathcal{M}_{t+}\}$-stopping time if and only if $\{\sigma < t\} \in \mathcal{M}_t$ for every $t \geq 0$.
(ii) If σ is an $\{\mathcal{M}_t\}$-stopping time and $c \geq 0$, then $\sigma + c$ is also an $\{\mathcal{M}_t\}$-stopping time.
(iii) Let σ_n, $n = 1, 2, \ldots$, be a sequence of $\{\mathcal{M}_t\}$-stopping times. If $\sigma_n(\omega) \uparrow \sigma(\omega)$ as $n \to \infty$ for $w \in \Omega$, then σ is an $\{\mathcal{M}_t\}$-stopping time.

If $\sigma_n(\omega) \downarrow \sigma(\omega)$ as $n \to \infty$ for $w \in \Omega$, then σ is an $\{\mathcal{M}_{t+}\}$-stopping time.
(iv) For an $\{\mathcal{M}_{t+}\}$-stopping time σ and for $n \in \mathbb{N}$, put

$$\sigma_n(\omega) = \begin{cases} \sum_{k=1}^{\infty} k2^{-n}\mathbf{1}_{\{(k-1)2^{-n} \le \sigma(\omega) < k2^{-n}\}} & \text{if } \sigma(\omega) < \infty, \\ \infty & \text{if } \sigma(\omega) = \infty. \end{cases} \tag{A.1.2}$$

σ_n is then an $\{\mathcal{M}_t\}$-stopping time and for $\omega \in \Omega$, $\sigma_n(\omega) \downarrow \sigma(\omega)$ as $n \uparrow \infty$.

In the rest of this subsection, we only consider real-valued stochastic processes X with time parameter $\mathbb{R}_+$ so that $T = \mathbb{R}_+$ and $(E, \mathcal{B}(E)) = (\mathbb{R}, \mathcal{B}(\mathbb{R}))$ where $\mathcal{B}(\mathbb{R})$ is the Borel σ-field of $\mathbb{R}$. More specifically, we fix a probability space $(\Omega, \mathcal{M}, \mathbf{P})$ and a filtration $\{\mathcal{M}_t;\ t \ge 0\}$ which is assumed to be right continuous and complete in the sense that $\mathcal{M}_0$ contains all null sets in $\mathcal{M}_\infty^{\mathbf{P}}$. A real-valued stochastic process $\{X_t\}_{t\in\mathbb{R}_+}$ defined on $(\Omega, \mathcal{M})$ and adapted to $\{\mathcal{M}_t\}$ is simply called a stochastic process. Two stochastic processes $X = \{X_t\}$ and $Y = \{Y_t\}$ are called *indistinguishable* if $\mathbf{P}(X_t \ne Y_t$ for some $t \ge 0) = 0$. A stochastic process $X = \{X_t\}$ is said to be *right* (resp. *left*) *continuous* if its sample path $X_\cdot(\omega)$ is right (resp. left) continuous for each $\omega \in \Omega$.

Let us regard a stochastic process $X = \{X_t\}_{t\ge 0}$ as a real-valued function of two variables $(t, \omega) \in \mathbb{R}_+ \times \Omega$. If it is measurable with respect to the product σ-field $\mathcal{B}(\mathbb{R}_+) \times \mathcal{M}$, then X is called *measurable*. If X is right continuous, then it is measurable because $X_t(\omega)$ is a limit as $n \to \infty$ of $X_t^{(n)}(\omega) = \sum_{k=1}^{\infty} X_{k2^{-n}} \cdot \mathbf{1}_{[(k-1)2^{-n}, k2^{-n})}(t)$, which is obviously $\mathcal{B}(\mathbb{R}_+) \times \mathcal{M}$-measurable as a function of (t, ω). Similarly, any left continuous process is measurable.

Denote by $\mathcal{O}$ (resp. $\mathcal{P}$) the smallest sub-σ-field of $\mathcal{B}(\mathbb{R}_+) \times \mathcal{M}$ making all right (resp. left) continuous processes measurable. We call a stochastic process $X = \{X_t\}_{t\in\mathbb{R}_+}$ *optional* (resp. *predictable*) if it is $\mathcal{O}$ (resp. $\mathcal{P}$) measurable as a real-valued function of (t, ω). The next theorem is a useful application of the optional section theorem due to C. Dellacherie and P. A. Meyer [37, Chapter IV].

Theorem A.1.6. *Assume that a stochastic process $X = \{X_t\}_{t\in\mathbb{R}_+}$ is indistinguishable from an optional process.*
(i) *If the limit $\lim_{n\to\infty} \mathbf{E}[X_{\sigma_n}]$ exists for any decreasing sequence of bounded $\{\mathcal{M}_t\}$-stopping times $\{\sigma_n\}$, then the sample path $X_\cdot(\omega)$ has right limits for* $\mathbf{P}$-a.e. $\omega \in \Omega$.
(ii) *If furthermore $\lim_{n\to\infty} \mathbf{E}[X_{\sigma_n}] = \mathbf{E}[X_\sigma]$ holds for $\sigma = \lim_{n\to\infty} \sigma_n$, then X is indistinguishable from a right continuous process.*

Proof. See Theorem 48 in Chapter VI of [37]. □

A stochastic process $X = \{X_t\}_{t\in\mathbb{R}_+}$ is said to be a *martingale* if $\mathbf{E}[|X_t|] < \infty$ for every $t \geq 0$, and

$$\mathbf{E}\left[X_t \mid \mathcal{M}_s\right] = X_s \quad \text{for every } 0 \leq s \leq t \quad \mathbf{P}\text{-a.s.}$$

$X = \{X_t\}_{t\in\mathbb{R}_+}$ is said to be a *supermartingale* (resp. *submartingale*) if in the above the equality is replaced by $\leq$ (resp. $\geq$).

THEOREM A.1.7. *Let $\{X^n\}$ be an increasing sequence of non-negative right continuous supermartingales. For $t \geq 0$, define $X_t := \lim_{n\to\infty} X^n_t$. Then $X = \{X_t\}_{t\in\mathbb{R}_+}$ is indistinguishable from a right continuous (extended real-valued) process.*

Proof. Fix $k \in \mathbb{N}$. We put $Y^n_t = X^n_t \wedge k$ and $Y_t := \lim_{n\to\infty} Y^n_t = X_t \wedge k$. It suffices to show that the stochastic process $Y = \{Y_t\}_{t\in\mathbb{R}_+}$ is indistinguishable from a right continuous process.

Y is optional as a limit of right continuous stochastic processes. For any $\{\mathcal{M}_t\}$-stopping times σ and τ with $\sigma \leq \tau < \infty$, by the optional sampling theorem $Y^n_\sigma \geq \mathbf{E}[Y^n_\tau | \mathcal{F}_\sigma]$. Letting $n \to \infty$, we get $Y_\sigma \geq \mathbf{E}[Y_\tau \mid \mathcal{F}_\sigma]$, which particularly means that Y is a supermartingale.

Let $\{\sigma_i\}$ be a decreasing sequence of bounded $\{\mathcal{M}_t\}$-stopping times with limit σ. Since $\{Y^n_t\}_{t\in\mathbb{R}_+}$ is bounded and right continuous, we have $\mathbf{E}[Y^n_\sigma] = \lim_{i\to\infty} \mathbf{E}[Y^n_{\sigma_i}]$ for each n and consequently $\mathbf{E}[Y_\sigma] \leq \liminf_{i\to\infty} \mathbf{E}[Y_{\sigma_i}]$. Combining this with the above supermartingale inequality, we get the identity $\mathbf{E}[Y_\sigma] = \lim_{i\to\infty} \mathbf{E}[Y_{\sigma_i}]$ and Theorem A.1.6 applies. □

A.1.2. Markov Processes

Given a measurable space $(E, \mathcal{B}(E))$, we add a cemetery point ∂ to it and define $E_\partial := E \cup \{\partial\}$, $\mathcal{B}(E_\partial) := \mathcal{B}(E) \cup \{B \cup \{\partial\} : B \in \mathcal{B}(E)\}$.

DEFINITION A.1.8. A quadruplet $X = \left(\Omega, \mathcal{M}, \{X_t\}_{t\in[0,\infty]}, \{\mathbf{P}_x\}_{x\in E_\partial}\right)$ is called a *Markov process* on $(E, \mathcal{B}(E))$ if it satisfies the following five conditions:

- **(X.1)** For each $x \in E_\partial$, $(\Omega, \mathcal{M}, \{X_t\}_{t\in[0,\infty]}, \mathbf{P}_x)$ is a stochastic process with state space $(E_\partial, \mathcal{B}(E_\partial))$ and time parameter set $[0, \infty]$, and $X_\infty(\omega) = \partial$ for every $\omega \in \Omega$.
- **(X.2)** For each $t \geq 0$ and $B \in \mathcal{B}(E_\partial)$, $\mathbf{P}_x(X_t \in B)$ is $\mathcal{B}(E_\partial)$-measurable as a function of $x \in E_\partial$.
- **(X.3)** There exists an admissible filtration $\{\mathcal{M}_t\}_{t\in[0,\infty]}$ such that, for any $x \in E_\partial,\ s, t \geq 0,\ B \in \mathcal{B}(E_\partial)$,

$$\mathbf{P}_x\left(X_{s+t} \in B \middle| \mathcal{M}_t\right) = \mathbf{P}_{X_t}(X_s \in B), \quad \mathbf{P}_x\text{-a.s.} \tag{A.1.3}$$

- **(X.4)** $\mathbf{P}_\partial(X_t = \partial) = 1$ for every $t \geq 0$.
- **(X.5)** $\mathbf{P}_x(X_0 = x) = 1$ for every $x \in E_\partial$.

(X.5) is called the *normality* of a Markov process X and it indicates that the probability measure $\mathbf{P}_x$ governs the behaviors of the sample path starting from x at time 0. A point $x \in E_\partial$ is called a *trap* of X if $\mathbf{P}_x(X_t = x) = 1$ for every $t \geq 0$. **(X.4)** means that the cemetery point ∂ is a trap of X and the sample path starting at ∂ never enters into E. **(X.3)** is called the *Markov property* of X with respect to an admissible filtration $\{\mathcal{M}_t\}_{t\geq 0}$. $\mathcal{M}_t$ contains all information about $\{X_s\}$ for s up to time t. The identity (A.1.3) requires that, conditioned on the state of $X_t(\omega)$ at *present* time t, the probability law of X at *future* time $t+s$ is the same as the law of the sample path at time s starting afresh from $X_t(\omega)$, independently of the *past* history $\{X_s; s < t\}$ of X.

For a Markov process X, we define

$$P_t(x, B) = \mathbf{P}_x(X_t \in B), \quad t \geq 0,\ x \in E,\ B \in \mathcal{B}(E). \tag{A.1.4}$$

which is called the *transition function* of X. For each $t \geq 0$, P_t is a Markovian kernel on $(E, \mathcal{B}(E))$ in the sense of Section 1.1 by virtue of **(X.2)**. It is not conservative in general because for $x \in E$,

$$P_t(x, E) = \mathbf{P}_x(X_t \in E_\partial) - \mathbf{P}_x(X_t = \partial) = 1 - \mathbf{P}_x(X_t = \partial) \leq 1. \tag{A.1.5}$$

For a Markov process X as above, we let

$$\mathcal{F}^0_\infty := \sigma\{X_s;\ s < \infty\} \quad \text{and} \quad \mathcal{F}^0_t := \sigma\{X_s; s \leq t\} \ \text{ for } t < \infty. \tag{A.1.6}$$

$\{\mathcal{F}^0_t\}_{t\in[0,\infty]}$ can be said to be the *minimum admissible filtration* generated by X.

THEOREM A.1.9. *Let X be a Markov process.*
(i) *The transition function $\{P_t; t \geq 0\}$ of X enjoys the properties* **(t.1)** *and* **(t.3)** *in Definition 1.1.13.*
(ii) *For any n, $t \geq 0$, $0 \leq s_1 < s_2 < \cdots < s_n$, any functions $f_1, f_2, \ldots, f_n \in b\mathcal{B}(E_\partial)$ and any $x \in E_\partial$, the following holds $\mathbf{P}_x$-a.s.:*

$$\begin{aligned} &\mathbf{E}_x\left[f_1(X_{t+s_1})f_2(X_{t+s_2})\cdots f_n(X_{t+s_n})\middle|\mathcal{M}_t\right] \\ &\quad = \mathbf{E}_{X_t}\left[f_1(X_{s_1})f_2(X_{s_2})\cdots f_n(X_{s_n})\right]. \end{aligned} \tag{A.1.7}$$

(iii) *X has the Markov property with respect to the minimum admissible filtration $\{\mathcal{F}^0_t\}_{t\geq 0}$ generated by X.*
(iv) *For any $\Lambda \in \mathcal{F}^0_\infty$, $\mathbf{P}_x(\Lambda)$ is a $\mathcal{B}(E_\partial)$-measurable function of $x \in E_\partial$.*

Proof. (i) is a consequence of **(X.3)**, **(X.4)**, and **(X.5)**. (ii) can be derived from (A.1.3) by induction. Since $\mathcal{F}^0_t \subset \mathcal{M}_t$, $X_t \in \mathcal{F}^0_t/\mathcal{B}(E_\partial)$, we obtain (iii) by taking the conditional expectation with respect to $\mathcal{F}^0_t$ of both sides of (A.1.3).

To prove (iv), we denote by $\mathcal{G}$ the family of all bounded $\mathcal{F}^0_\infty$-measurable functions F on Ω such that $\mathbf{E}_x[F]$ are $\mathcal{B}(E_\partial)$-measurable in $x \in E_\partial$. By **(X.2)**,

(A.1.7), and induction, we can readily show that for any n, $0 \le s_1 < s_2 < \cdots < s_n$, and for any functions $f_1, f_2, \ldots, f_n \in b\mathcal{B}(E_\partial)$ on E_∂,

$$F = f_1(X_{s_1})f_2(X_{s_2})\cdots f_n(X_{s_n}) \in \mathcal{G}.$$

Since $\mathcal{G}_+$ is closed under the operation of taking a uniformly bounded increasing limit, we conclude that $\mathcal{G}$ coincides with the space of all bounded $\mathcal{F}^0_\infty$-measurable functions on Ω by virtue of Proposition A.1.2. □

Properly (A.1.7) implies the following expression of the finite dimensional distribution of a Markov process X:

COROLLARY A.1.10. *Let X be a Markov process with transition function $\{P_t; t \ge 0\}$ on E. Under the convention that any function f on E is extended to a function on E_∂ by setting $f(\partial) = 0$, we have, for any n, $0 \le s_1 < s_2 < \cdots < s_{n-1} < s_n$, for any functions $f_1, f_2, \ldots, f_{n-1}, f_n \in b\mathcal{B}(E)$ and for any $x \in E$,*

$$\mathbf{E}_x\left[\prod_{j=1}^n f_j(X_{s_j})\right] = \int_E \cdots \int_E P_{s_1}(x, dy_1)f_1(y_1)P_{s_2-s_1}(y_1, dy_2)f_2(y_2)\cdots$$

$$\cdot P_{s_{n-1}-s_{n-2}}(y_{n-2}, dy_{n-1})f_{n-1}(y_{n-1})P_{s_n-s_{n-1}}(y_{n-1}, dy_n)f_n(y_n). \qquad \text{(A.1.8)}$$

The integrals of the right hand side are performed starting with the last variable y_n and then successively with forward variables.

We next introduce a specific completion of the minimum admissible filtration (A.1.6) generated by a Markov process X. By Theorem A.1.9(iv), for every $\mu \in \mathcal{P}(E_\partial)$, the integral

$$\mathbf{P}_\mu(\Lambda) = \int_{E_\partial} \mathbf{P}_x(\Lambda)\mu(dx), \quad \Lambda \in \mathcal{F}^0_\infty,$$

defines a probability measure $\mathbf{P}_\mu$ on $(\Omega, \mathcal{F}^0_\infty)$, which is called the probability law of the Markov process X with *initial distribution* μ because by **(X.5)**

$$\mathbf{P}_\mu(X_0 \in B) = \int_{E_\partial} \mathbf{1}_B(x)\mu(dx) = \mu(B) \quad \text{for } B \in \mathcal{B}(E_\partial).$$

Denote by $\mathcal{F}^\mu_\infty$ the $\mathbf{P}_\mu$-completion of $\mathcal{F}^0_\infty$ and by $\mathcal{N}$ the family of all null sets in $\mathcal{F}^\mu_\infty$. We then let $\mathcal{F}^\mu_t = \sigma(\mathcal{F}^0_t, \mathcal{N})$ for each $t \ge 0$. The filtration $\{\mathcal{F}^\mu_t\}_{t\in[0,\infty]}$ so obtained is called the $\mathbf{P}_\mu$*-augmentation of* the filtration $\{\mathcal{F}^0_t\}_{t\in[0,\infty]}$ *in* $\mathcal{F}^0_\infty$. We further let $\mathcal{F}_t = \bigcap_{\mu\in\mathcal{P}(E_\partial)} \mathcal{F}^\mu_t$, $t \in [0, \infty]$.

$\{\mathcal{F}_{t+}\}$, $\{\mathcal{F}^\mu_t\}$, and $\{\mathcal{F}_t\}$ are all admissible filtrations of X. Especially we may call $\{\mathcal{F}_t\}_{t\in[0,\infty]}$ the *minimum augmented admissible filtration* of X. We note that $\mathcal{F}^\mu_t$ is in general larger than the $\mathbf{P}_\mu$-completion of $\mathcal{F}^0_t$ itself. The next lemma indicates an advantage of augmented filtrations.

Lemma A.1.11. *If a Markov process X has the Markov property with respect to the filtration $\{\mathcal{F}^0_{t+}\}$, then the filtrations $\{\mathcal{F}^\mu_t\}$ and $\{\mathcal{F}_t\}$ are both right continuous.*

Proof. See [73, Lemma A.2.2]. □

The concept of an $\{\mathcal{M}_t\}$-stopping time has been introduced by Definition A.1.4. For an $\{\mathcal{M}_t\}$-stopping time σ, we let

$$\mathcal{M}_\sigma = \{\Lambda \in \mathcal{M}_\infty : \Lambda \cap \{\sigma \le t\} \in \mathcal{M}_t \text{ for every } t \ge 0\}, \tag{A.1.9}$$

which is easily verified to be a sub-σ-field of $\mathcal{M}_\infty$. When σ is a non-random time $r \ge 0$, $\mathcal{M}_\sigma = \mathcal{M}_r$. $\mathcal{M}_\sigma$ can be considered as the information of X up to the stopping time σ. For an $\{\mathcal{M}_{t+}\}$-stopping time σ, we let

$$\mathcal{M}_{\sigma+} = \{\Lambda \in \mathcal{M}_\infty : \Lambda \cap \{\sigma \le t\} \in \mathcal{M}_{t+} \text{ for every } t \ge 0\}.$$

We state a useful lemma due to Dynkin.

Lemma A.1.12. *Let $\mu \in \mathcal{P}(E_\partial)$. For any $\{\mathcal{F}^\mu_t\}$-stopping time σ, there exists an $\{\mathcal{F}^0_{t+}\}$-stopping time σ' such that $\mathbf{P}_\mu(\sigma \ne \sigma') = 0$. Furthermore, for any $\Lambda \in \mathcal{F}^\mu_\sigma$, there exists $\Lambda' \in \mathcal{F}^0_{\sigma'+}$ such that $\mathbf{P}_\mu(\Lambda \Delta \Lambda') = 0$.*

Proof. See [73, Lemma A.2.3]. □

A.1.3. Right Processes, Standard Processes, and Hunt Processes

In the preceding subsection, the state space $(E, \mathcal{B}(E))$ of a Markov process X is only assumed to be a measurable space. From now on, we make a topological assumption on E so that we can discuss the continuity, right continuity, and existence of left limits of the sample path $X_\cdot(\omega) : [0, \infty] \to E_\partial$.

In this subsection, we are primarily concerned with a Markov process X whose state space E is a Lusin topological space equipped with the Borel σ-field $\mathcal{B}(E)$, which already covers a large family of Markov processes on both finite and infinite dimensional state spaces. However, by some basic transformations of X such as time changes and killings, we are inevitably led to a Markov process on a more general Radon space E equipped with the σ-field $\mathcal{B}^*(E)$ of universally measurable subsets, which will be formulated in the second part of this subsection.

In what follows, we will adopt the convention that any numerical function f on E is extended to E_∂ by setting $f(\partial) = 0$.

Let E be a Lusin space, namely, E is homeomorphic to a Borel subset of some compact metric space F. By identifying E with its image under this homeomorphism, we may assume that $E \in \mathcal{B}(F)$ and the topology of E is its relative topology as a subspace of F. By (A.1.1), the Borel σ-field $\mathcal{B}(E)$ of E consists of all elements of $\mathcal{B}(F)$ that are contained in E. Without loss of generality, we may assume that $F \setminus E$ is non-empty so that we may choose a point ∂ from it and consider a topological subspace $E_\partial = E \cup \{\partial\}$ of F. The Borel σ-field $\mathcal{B}(E_\partial)$ of E_∂ then coincides with $\mathcal{B}(E) \cup \{B \cup \{\partial\} : B \in \mathcal{B}(E)\}$.

Let $X = (\Omega, \mathcal{M}, \{X_t\}_{t\in[0,\infty]}, \{\mathbf{P}_x\}_{x\in E_\partial})$ be a Markov process on a Lusin space $(E, \mathcal{B}(E))$ that satisfies conditions **(X.1)** $\sim$ **(X.5)** of Definition A.1.8. In what follows, we shall impose on X the following additional condition $(\mathbf{X.6})_\mathbf{r}$:

$(\mathbf{X.6})_\mathbf{r}$ The pair $(\Omega, \{X_t\})$ satisfies the following conditions:

- **(i)** $X_t(\omega) = \partial$ for every $t \geq \zeta(\omega)$, where $\zeta(\omega) = \inf\{t \geq 0 : X_t(\omega) = \partial\}$. (with the convention that $\inf \emptyset = \infty$).
- **(ii)** For each $t \geq 0$, there exists a map $\theta_t : \Omega \to \Omega$ such that $X_s \circ \theta_t = X_{s+t}$ for every $s \geq 0$. Moreover, $\theta_0 \omega = \omega$, $\theta_\infty(\omega) = [\partial]$ for every $\omega \in \Omega$, where $[\partial]$ denotes a specific element of Ω such that $X_t([\partial]) = \partial$ for every $t \in T$.
- **(iii)** For each $\omega \in \Omega$, the sample path $t \mapsto X_t(\omega) \in E_\partial$ is right continuous on $[0, \infty)$.

The random variable $\zeta(\omega)$ in **(i)** is called the *lifetime* of the sample path of ω. The map $\theta_t : \Omega \to \Omega$ in **(ii)** is called the *shift operator* on Ω. We say that X is *conservative* if

$$\mathbf{P}_x(\zeta < \infty) = 0 \quad \text{for every } x \in E, \tag{A.1.10}$$

which is equivalent to the conservativeness of the transition function of X in view of (A.1.5).

We state some consequences of the condition $(\mathbf{X.6})_\mathbf{r}$ added to a Markov process X. First, using θ_t, we can extend the Markov property (A.1.3) into the following much more general relation. For any $t \geq 0$, $x \in E_\partial$ and for any random variable $F \in b\mathcal{F}^0_\infty$ on Ω, we have $\mathbf{P}_x$-a.s.

$$F \circ \theta_t \in b\mathcal{F}^0_\infty \quad \text{and} \quad \mathbf{E}_x[F \circ \theta_t | \mathcal{M}_t] = \mathbf{E}_{X_t}[F]. \tag{A.1.11}$$

When F is of the type $F(\omega) = \prod_{j=1}^n f_j(X_{s_j}(\omega))$, the above relation is nothing but (A.1.7) derived from (A.1.3) so that we can apply Proposition A.1.2 to the family of all functions $F \in b\mathcal{F}^0_\infty$ satisfying the relation (A.1.11) to identify it with $b\mathcal{F}^0_\infty$.

The transition function $\{P_t; t \geq 0\}$ of a Markov process X defined by (A.1.4) is a transition function on $(E, \mathcal{B}(E))$ in the sense of Definition 1.1.13. Conditions **(t.1)**, **(t.3)** are verified in Theorem A.1.9 already. **(t.4)**$'$ follows from $(\mathbf{X.6})_\mathbf{r}$ **(iii)** and the expression

$$P_t f(x) := \mathbf{E}_x[f(X_t)], \quad t \geq 0, \ x \in E, \ f \in b\mathcal{B}(E), \tag{A.1.12}$$

so that Lemma 1.1.14(i) applies.

We denote by $\mathcal{B}_t$ the Borel σ-field of subsets of the time interval $[0,t]$ for each $t \geq 0$ and by $\mathcal{B}_\infty$ the Borel σ-field of $[0,\infty)$. For an admissible filtration $\{\mathcal{M}_t\}_{t\in[0,\infty]}$ of a Markov process X and for each $t \in [0,\infty]$, we consider the map

$$\Phi_t \ : \ (s,\omega) \in [0,t] \times \Omega \ \longmapsto \ X_s(\omega) \in E_\partial.$$

We say that X is *progressive* with respect to the filtration $\{\mathcal{M}_t\}$ if $\Phi_t \in \mathcal{B}_t \times \mathcal{M}_t/\mathcal{B}(E_\partial)$ for each $t \in [0,\infty]$, where $\mathcal{B}_t \times \mathcal{M}_t$ denotes the product σ-field. If this condition is required only for $t=\infty$, then X is simply said to be *measurable*.

LEMMA A.1.13. (i) *A Markov process X is progressive with respect to any admissible filtration of X.*
(ii) *For an admissible filtration $\{\mathcal{M}_t\}$ of X and an $\{\mathcal{M}_t\}$-stopping time σ, consider the composite map $X_\sigma : \omega \in \Omega \mapsto X_{\sigma(\omega)}(\omega) \in E_\partial$. Then $X_\sigma \in \mathcal{M}_\sigma/\mathcal{B}(E_\partial)$.*

Proof. (i) Let $\{\mathcal{M}_t\}$ be an admissible filtration of X. We put, for $n \in \mathbb{N}$,

$$X_s^{(n)}(\omega) = X_{k2^{-n}}(\omega), \quad (k-1)2^{-n} \leq s < k2^{-n}, \quad k = 1,2,\ldots.$$

For an arbitrarily fixed $t \geq 0$ and for any $\varepsilon > 0$, we define a map $\widetilde{X}_s^{(n,\varepsilon)}(\omega)$ from $[0,t]\times\Omega$ to E_∂ by

$$\widetilde{X}_s^{n,\varepsilon}(\omega) = X_{s\wedge(t-\varepsilon)}^{(n)}(\omega) \ \text{for}\ 0 \leq s < t \quad \text{and} \quad \widetilde{X}_t^{n,\varepsilon}(\omega) = X_t(\omega).$$

Since $\{X_t\}$ is adapted to $\{\mathcal{M}_t\}$, this map can be seen to be $\mathcal{B}_t \times \mathcal{M}_t$-measurable for sufficiently large n. By the right continuity of the sample path **(X.6)$_\mathbf{r}$ (iii)**,

$$\lim_{n\to\infty} \widetilde{X}_s^{n,\varepsilon}(\omega) = \begin{cases} X_{s\wedge(t-\varepsilon)}(\omega), & 0 \leq s < t, \\ X_t(\omega), & s = t. \end{cases}$$

Since $\varepsilon > 0$ is arbitrary, we get the desired measurability of the map Φ_t.
(ii) Take any $t \geq 0$. As a composition of two maps $\omega \in \Omega \mapsto (\sigma(\omega)\wedge t, \omega)$ and $(s,\omega) \in [0,t]\times\Omega \mapsto X_s(\omega) \in E_\partial$, we get $X_{\sigma\wedge t} \in \mathcal{M}_t/E_\partial$. Therefore, $\{X_\sigma \in B\} \cap \{\sigma \leq t\} = \{X_{\sigma\wedge t} \in B\} \cap \{\sigma \leq t\} \in \mathcal{M}_t$ for $B \in \mathcal{B}(E_\partial)$. □

Remark A.1.14. For an admissible filtration $\{\mathcal{M}_t\}$ of X and an $\{\mathcal{M}_t\}$-stopping time σ, define the σ-fields $\mathcal{M}_{\sigma-}$ of events strictly prior to time σ by

$$\mathcal{M}_{\sigma-} = \sigma\{\Lambda \cap \{s < \sigma\} \ : \ \Lambda \in \mathcal{M}_s,\ s \geq 0\}. \tag{A.1.13}$$

If a real-valued $\{\mathcal{M}_t\}$-adapted stochastic process $\{Z_t\}$ is predictable, then

$$Z_\sigma \cdot \mathbf{1}_{\{\sigma<\infty\}} \in \mathcal{M}_{\sigma-}. \tag{A.1.14}$$

This is a useful analogue to Lemma A.1.13 and we refer to [37, IV:67] for its proof.

DEFINITION A.1.15. A Markov process X is called *strong Markov* if there exists a right continuous admissible filtration $\{\mathcal{M}_t\}$ for which the following property holds. For any $\{\mathcal{M}_t\}$-stopping time σ and for $\mu \in \mathcal{P}(E_\partial)$, $s \geq 0$, $B \in \mathcal{B}(E_\partial)$,

$$\mathbf{P}_\mu \left(X_{\sigma+s} \in B \middle| \mathcal{M}_\sigma \right) = \mathbf{P}_{X_\sigma}(X_s \in B) \quad \mathbf{P}_\mu\text{-a.s.} \tag{A.1.15}$$

Notice that the right continuity of an admissible filtration is required in the above definition of a strong Markov property. On account of Lemma A.1.13, (A.1.15) is equivalent to the following identity:

$$\mathbf{P}_\mu \left(\{X_{\sigma+s} \in B\} \cap \Lambda \right) = \mathbf{E}_\mu \left[\mathbf{P}_{X_\sigma}(X_s \in B);\ \Lambda \right] \quad \text{for } \Lambda \in \mathcal{M}_\sigma. \tag{A.1.16}$$

Exercise A.1.16. Using Exercise A.1.5(ii) and Lemma A.1.13, derive the following from the identity (A.1.15). For any n, $t \geq 0$, $0 \leq s_1 < s_2 < \cdots < s_n$, for any functions $f_1, f_2, \ldots, f_n \in b\mathcal{B}(E_\partial)$ on E_∂ and for any $\mu \in \mathcal{P}(E_\partial)$, the next identity holds $\mathbf{P}_\mu$-a.s.:

$$\mathbf{E}_\mu \left[\prod_{j=1}^{n} f_j(X_{\sigma+s_j}) \middle| \mathcal{M}_\sigma \right] = \mathbf{E}_{X_\sigma} \left[\prod_{j=1}^{n} f_j(X_{s_j}) \right]. \tag{A.1.17}$$

Denote by θ_σ the mapping on Ω sending $\omega \in \Omega$ to $\theta_{\sigma(\omega)}(\omega) \in \Omega$, which is obtained by inserting $t = \sigma$ in $\theta_t \omega$. By using this notation, (A.1.15) can be generalized as follows. For any $\mu \in \mathcal{P}(E_\partial)$ and for any $F \in b\mathcal{F}^0_\infty$,

$$F \circ \theta_\sigma \in b\mathcal{M}_\infty \quad \text{and} \quad \mathbf{E}_\mu \left[F \circ \theta_\sigma \middle| \mathcal{M}_\sigma \right] = \mathbf{E}_{X_\sigma}[F] \quad \mathbf{P}_\mu\text{-a.s.} \tag{A.1.18}$$

Indeed, if we let $\mathcal{G}$ be the set of all functions $F \in \mathcal{F}^0_\infty$ satisfying the above relation, then we can readily conclude by Proposition A.1.2 and (A.1.17) that $\mathcal{G} = \mathcal{F}^0_\infty$.

We remark that if $F \in b\mathcal{F}^0_t$, $t \geq 0$, then $F \circ \theta_\sigma \in b\mathcal{M}_{\sigma+t}$.

DEFINITION A.1.17. X is called a *Borel right process* if it is a strong Markov process on a Lusin space $(E, \mathcal{B}(E))$ and satisfies $(\mathbf{X.6})_\mathbf{r}$.

The term "Borel" indicates that the state space of X is $(E, \mathcal{B}(E))$ equipped with the Borel σ-field $\mathcal{B}(E)$ so that the transition function $P_t f$ is Borel measurable for every $f \in b\mathcal{B}(E)$.

The above definition of a Borel right process apparently depends on an arbitrary choice of a right continuous admissible filtration $\{\mathcal{M}_t\}$ for X describing the strong Markov property. The next theorem says that it actually depends only on the minimum admissible filtration $\{\mathcal{F}^0_t\}$ for X.

THEOREM A.1.18. *Let $(E, \mathcal{B}(E))$ be a Lusin space. For a Markov process $X = (\Omega, \mathcal{M}, X_t, \mathbf{P}_x)$ on $(E, \mathcal{B}(E))$ satisfying condition* $(\mathbf{X.6})_\mathbf{r}$, *the following three conditions are mutually equivalent:*
(α) X is a Borel right process.
(β) The minimum augmented admissible filtration $\{\mathcal{F}_t\}$ of X is right continuous and X is strong Markov with respect to it.
(γ) For each $\mu \in \mathcal{P}(E_\partial)$, the filtration $\{\mathcal{F}_t^\mu\}$ is right continuous and the identity (A.1.15) *holds for any $\{\mathcal{F}_t^\mu\}$-stopping time σ and for the probability measure $\mathbf{P}_\mu$.*

Proof. Since the implications $(\gamma) \Rightarrow (\beta) \Rightarrow (\alpha)$ are clear, it suffices to show $(\alpha) \Rightarrow (\gamma)$. Suppose X is strong Markov with respect to a right continuous admissible filtration $\{\mathcal{M}_t\}$. Since $\mathcal{F}_{t+}^0 \subset \mathcal{M}_{t+} \subset \mathcal{M}_t$, X is also strong Markov with respect to the filtration $\{\mathcal{F}_{t+}^0\}$. By Lemma A.1.11, the filtration $\{\mathcal{F}_t^\mu\}$ is right continuous for each $\mu \in \mathcal{P}(E_\partial)$. Take any $\{\mathcal{F}_t^\mu\}$-stopping time σ and any $\Lambda \in \mathcal{F}_\sigma^\mu$. By Lemma A.1.12, there exist an $\{\mathcal{F}_{t+}^0\}$-stopping time σ' and $\Lambda' \in \mathcal{F}_{\sigma'+}^0$ such that $\mathbf{P}_\mu(\sigma = \sigma') = 0$ and $\mathbf{P}_\mu(\Lambda \Delta \Lambda') = 0$. Since for σ' and Λ', the strong Markov relation (A.1.16) of X is valid with respect to $\{\mathcal{F}_{t+}^0\}$, the same relation holds for σ and Λ, yielding the property (γ) for X. □

For a Borel right process X and a Borel set $B \in \mathcal{B}(E_\partial)$, the *hitting time* σ_B and the *entrance time* $\dot{\sigma}_B$ of B are defined as follows:

$$\sigma_B(\omega) = \inf\{t > 0 : X_t(\omega) \in B\}, \tag{A.1.19}$$

$$\dot{\sigma}_B(\omega) = \inf\{t \geq 0 : X_t(\omega) \in B\}. \tag{A.1.20}$$

When B is an open set, we have by the right continuity of the sample path

$$\{\sigma_B(\omega) < t\} = \{\dot{\sigma}_B(\omega) < t\} = \bigcup_{r \in \mathbf{Q} \cap (0,t)} \{X_r \in B\} \in \mathcal{F}_t^0, \quad t > 0,$$

and hence both σ_B and $\dot{\sigma}_B$ are $\{\mathcal{F}_{t+}^0\}$-stopping times. Since $\mathcal{F}_{t+}^0 \subset \mathcal{F}_{t+} = \mathcal{F}_t$ by the preceding theorem, σ_B, $\dot{\sigma}_B$ are $\{\mathcal{F}_t\}$-stopping times. The following theorem asserts that the same is true for a general Borel set B.

THEOREM A.1.19. *Let X be a Borel right process. For any Borel set $B \in \mathcal{B}(E_\partial)$, both the hitting time σ_B and the entrance time $\dot{\sigma}_B$ are $\{\mathcal{F}_t\}$-stopping times.*

Proof. Take any $\mu \in \mathcal{P}(E_\partial)$. It suffices to show that σ_B and $\dot{\sigma}_B$ are both $\{\mathcal{F}_t^\mu\}$-stopping times. For a fixed $t > 0$, let us consider a subset

$$A = \{(s, \omega) \in (0, t) \times \Omega : X_s(\omega) \in B\}$$

of the product space $[0,\infty) \times \Omega$. Since $\{\mathcal{F}_t^0\}$ is an admissible filtration of X, we have by Lemma A.1.13

$$A = \bigcup_{k=[1/t]+1}^{\infty} \left\{\Phi_{t-(1/k)}^{-1}(B) \setminus \Phi_{1/k}^{-1}(B)\right\} \in \mathcal{B}_t \times \mathcal{F}_t^0 \subset \mathcal{B}_\infty \times \mathcal{F}_t^0.$$

Here for $a \in \mathbb{R}$, $[a]$ denotes the largest integer that does not exceed a. Let Λ be the projection of A on $\Omega : \Lambda = \{\omega \in \Omega : \exists s \in [0,t),\ (s,\omega) \in A\}$. On account of Proposition A.1.1, Λ belongs to the $\mathbf{P}_\mu$-completion $(\mathcal{F}_t^0)^{\mathbf{P}_\mu}$ of the σ-field $\mathcal{F}_t^0$. Since $(\mathcal{F}_t^0)^{\mathbf{P}_\mu} \subset \mathcal{F}_t^\mu$, we get $\Lambda \in \mathcal{F}_t^\mu$.

On the other hand, we have clearly $\{\sigma_B(\omega) < t\} = \Lambda$ and hence $\{\sigma_B < t\} \in \mathcal{F}_t^\mu$. Since this is true for every $t > 0$ and $\{\mathcal{F}_t^\mu\}$ is right continuous by Theorem A.1.18, σ_B is an $\{\mathcal{F}_t^\mu\}$-stopping time. The proof for $\dot{\sigma}_B$ is the same. □

According to Theorem A.1.18 and (A.1.18), a Borel right process X enjoys the following property: for any $\{\mathcal{F}_t\}$-stopping time σ, $F \in b\mathcal{F}_\infty^0$, and $\mu \in \mathcal{P}(E_\partial)$,

$$F \circ \theta_\sigma \in b\mathcal{F}_\infty \quad \text{and} \quad \mathbf{E}_\mu\left[F \circ \theta_\sigma \middle| \mathcal{F}_\sigma\right] = \mathbf{E}_{X_\sigma}[F] \quad \mathbf{P}_\mu\text{-a.s.} \tag{A.1.21}$$

If $F \in b\mathcal{F}_t^0$ for some $t \geq 0$, then $F \circ \theta_\sigma \in b\mathcal{F}_{\sigma+t}$.

The relation (A.1.21) can be further extended to $F \in b\mathcal{F}_\infty$.

Exercise A.1.20. Show the following:
(i) If $F \in b\mathcal{F}_\infty$, then $\mathbf{E}_x[F]$ is $\mathcal{B}^*(E_\partial)$-measurable as a function of $x \in E_\partial$.
(ii) If σ is an $\{\mathcal{F}_t\}$-stopping time, then

$$\bigcap_{\mu \in \mathcal{P}(E_\partial)} (\mathcal{F}_\sigma)^{\mathbf{P}_\mu} = \mathcal{F}_\sigma, \qquad X_\sigma \in \mathcal{F}_\sigma / \mathcal{B}^*(E_\partial), \tag{A.1.22}$$

where $(\mathcal{F}_\sigma)^{\mathbf{P}_\mu}$ denotes the $\mathbf{P}_\mu$-completion of $\mathcal{F}_\sigma$.

THEOREM A.1.21. *Let X be a Borel right process and σ be an $\{\mathcal{F}_t\}$-stopping time. For any $F \in b\mathcal{F}_\infty$ and any $\mu \in \mathcal{P}(E_\partial)$,*

$$F \circ \theta_\sigma \in b\mathcal{F}_\infty, \quad \mathbf{E}_\mu\left[F \circ \theta_\sigma \middle| \mathcal{F}_\sigma\right] = \mathbf{E}_{X_\sigma}[F],\ \mathbf{P}_\mu\text{-}a.s. \tag{A.1.23}$$

If $F \in b\mathcal{F}_t$, $t \geq 0$, then $F \circ \theta_\sigma \in b\mathcal{F}_{\sigma+t}$.

Proof. For $F \in b\mathcal{F}_\infty$, we have by Exercise A.1.20,

$$\mathbf{E}_{X_\sigma}[F] \in \mathcal{F}_\sigma. \tag{A.1.24}$$

Take any $\mu \in \mathcal{P}(E_\partial)$ and define $\nu \in \mathcal{P}(E_\partial)$ by $\nu(B) = \mathbf{P}_\mu(X_\sigma \in B)$ for $B \in \mathcal{B}(E_\partial)$. Then, particularly for $F' \in b\mathcal{F}_\infty^0$, we take the $\mathbf{P}_\mu$-expectation on both sides of (A.1.21) to obtain

$$\mathbf{E}_\mu[F' \circ \theta_\sigma] = \mathbf{E}_\mu\left[\mathbf{E}_{X_\sigma}(F')\right] = \mathbf{E}_\nu[F'].$$

Since $F \in b\mathcal{F}_\infty \subset b\mathcal{F}^\nu_\infty$, this identity means that $F \circ \theta_\sigma \in b(\mathcal{F}_\infty)^{\mathbf{P}_\mu} \subset b\mathcal{F}^\mu_\infty$. We can also derive from (A.1.21) the equality

$$\mathbf{E}_\mu[F \circ \theta_\sigma; \Lambda] = \mathbf{E}_\mu\left[\mathbf{E}_{X_\sigma}(F); \Lambda\right] \quad \text{for } \Lambda \in \mathcal{F}_\sigma,$$

which, together with (A.1.24), implies the latter half of (A.1.23). The proof of the last assertion is similar. □

Since any $\{\mathcal{F}_t\}$-stopping time σ is obviously $\mathcal{F}_\sigma$-measurable, we multiply both sides of (A.1.23) by $\mathbf{1}_{\{\sigma<\infty\}}$ to get its useful variant

$$\mathbf{E}_\mu\left[F \circ \theta_\sigma \cdot \mathbf{1}_{\{\sigma<\infty\}} \middle| \mathcal{F}_\sigma\right] = \mathbf{E}_{X_\sigma}[F] \cdot \mathbf{1}_{\{\sigma<\infty\}}, \ \mathbf{P}_\mu\text{-a.s.} \tag{A.1.25}$$

Next we present frequently used formulas on resolvents. Let $\{P_t; t \geq 0\}$ be the transition function of a Borel right process X and $\{R_\alpha; \alpha > 0\}$, R be its resolvent kernel and its 0-order resolvent kernel defined by (1.1.27) and (2.1.3), respectively. Recall the convention that any numerical function f on E is extended to E_∂ by setting $f(\partial) = 0$ is in force unless stated otherwise. We note that the expression $P_t f(x) = \mathbf{E}_x[f(X_t)]$ of (A.1.12) is valid for every $x \in E_\partial$ under this convention.

THEOREM A.1.22. (i) *For $\alpha > 0$ and $f \in b\mathcal{B}^*(E)$, we have $R_\alpha f \in b\mathcal{B}^*(E)$, $\widetilde{F} = \int_0^\infty e^{-\alpha t} f(X_t)dt \in \mathcal{F}_\infty$ and*

$$\langle \mu, R_\alpha f \rangle = \mathbf{E}_\mu\left[\int_0^\infty e^{-\alpha t} f(X_t)dt\right] \quad \text{for } \mu \in \mathcal{P}(E). \tag{A.1.26}$$

Further, for any $\{\mathcal{F}_t\}$-stopping time σ,

$$\mathbf{E}_\mu\left[e^{-\alpha\sigma} R_\alpha f(X_\sigma)\right] = \mathbf{E}_\mu\left[\int_\sigma^\infty e^{-\alpha t} f(X_t)dt\right] \quad \text{for } \mu \in \mathcal{P}(E). \tag{A.1.27}$$

(ii) *For any $f \in \mathcal{B}^*_+(E)$, we have $Rf \in \mathcal{B}^*_+(E)$, $\int_0^\infty f(X_t)dt \in \mathcal{F}_\infty$ and*

$$\langle \mu, Rf \rangle = \mathbf{E}_\mu\left[\int_0^\infty f(X_t)dt\right] \quad \text{for } \mu \in \mathcal{P}(E). \tag{A.1.28}$$

Further, for any $\{\mathcal{F}_t\}$-stopping time σ,

$$\mathbf{E}_\mu\left[Rf(X_\sigma); \ \sigma < \infty\right] = \mathbf{E}_\mu\left[\int_\sigma^\infty f(X_t)dt\right] \quad \text{for } \mu \in \mathcal{P}(E). \tag{A.1.29}$$

Proof. (i) By the expression (A.1.12) of $P_t f(x)$ and the Fubini theorem, we have, for any $f \in b\mathcal{B}(E)$, $x \in E$, $\alpha > 0$,

$$R_\alpha f(x) = \int_0^\infty e^{-\alpha t}\mathbf{E}_x(f(X_t))dt = \mathbf{E}_x\left[\int_0^\infty e^{-\alpha t} f(X_t)dt\right].$$

Here we are using a consequence of Lemma A.1.13 that $f(X_t(\omega))$ is a $\mathcal{B}([0,\infty)) \times \mathcal{F}^0_\infty$-measurable function of (t,ω).

Let $f \in \mathcal{B}^*(E)$. For any $\mu \in \mathcal{P}(E)$, put $\nu(B) = \langle \mu, R_\alpha \mathbf{1}_B \rangle$, $B \in \mathcal{B}(E)$. There exist $f_1, f_2 \in b\mathcal{B}(E)$ such that $f_1 \le f \le f_2$, $\langle \nu, f_2 - f_1 \rangle = 0$. Then both f_1, f_2 satisfy equation (A.1.26) and $\langle \mu, R_\alpha f_1 \rangle = \langle \mu, R_\alpha f_2 \rangle$. Hence $R_\alpha f \in b\mathcal{B}^\mu(E)$, $\widetilde{F} \in \mathcal{F}^{\mathbf{P}_\mu}_\infty \subset \mathcal{F}^\mu_\infty$, and f also satisfies (A.1.26). Since this holds for every $\mu \in \mathcal{P}(E)$, $R_\alpha f \in b\mathcal{B}^*(E)$ and $\widetilde{F} \in \mathcal{F}_\infty$.

We next apply the strong Markov property (A.1.23) to $\widetilde{F}$ and any $\{\mathcal{F}_t\}$-stopping time σ. Since $\int_\sigma^\infty e^{-\alpha t} f(X_t)dt = e^{-\alpha\sigma}\widetilde{F} \circ \theta_\sigma$, the right hand side of (A.1.27) equals

$$\mathbf{E}_\mu\left[e^{-\alpha\sigma}\widetilde{F} \circ \theta_\sigma\right] = \mathbf{E}_\mu\left\{\mathbf{E}_\mu\left[e^{-\alpha\sigma}\widetilde{F} \circ \theta_\sigma \middle| \mathcal{F}_\sigma\right]\right\} = \mathbf{E}_\mu\left[e^{-\alpha\sigma}\mathbf{E}_{X_\sigma}[\widetilde{F}]\right],$$

which coincides with the left hand side of (A.1.27).

(ii) For $f \in \mathcal{B}^*_+(E)$, we first apply (i) to $f \wedge n$, then let $\alpha \downarrow 0$ and $n \to \infty$ successively to get (ii). □

Now let us denote by $(\mathbf{X.6})_\mathbf{s}$, $(\mathbf{X.6})_\mathbf{h}$ the conditions obtained from $(\mathbf{X.6})_\mathbf{r}$ by replacing its third requirement **(iii)** with stronger ones **(iii)**′ and **(iii)**″, respectively:

- **(iii)**′ For each $\omega \in \Omega$, sample path $t \mapsto X_t(\omega)$ is right continuous on $[0,\infty)$ and has left limits on $(0,\zeta(\omega))$ in E.
- **(iii)**″ For each $\omega \in \Omega$, its sample path $t \mapsto X_t(\omega)$ is right continuous on $[0,\infty)$ and has left limits on $(0,\infty)$ in E_∂.

A Markov process X is called *quasi-left-continuous on* $(0,\zeta)$ (resp. $(0,\infty)$) if it possesses the following property (A.1.30) (resp. (A.1.31)): there exists a right continuous admissible filtration $\{\mathcal{M}_t\}$ for X such as, for any increasing $\{\mathcal{M}_t\}$-stopping times $\{\sigma_n\}$ with $\sigma = \lim_{n\to\infty}\sigma_n$,

$$\mathbf{P}_\mu\left(\lim_{n\to\infty} X_{\sigma_n} = X_\sigma,\ \sigma < \zeta\right) = \mathbf{P}_\mu(\sigma < \zeta) \quad \text{for } \mu \in \mathcal{P}(E_\partial), \qquad \text{(A.1.30)}$$

$$\mathbf{P}_\mu\left(\lim_{n\to\infty} X_{\sigma_n} = X_\sigma,\ \sigma < \infty\right) = \mathbf{P}_\mu(\sigma < \infty) \quad \text{for } \mu \in \mathcal{P}(E_\partial). \qquad \text{(A.1.31)}$$

We introduce two subfamilies of Borel right processes.

DEFINITION A.1.23. (i) X is called a *Borel standard process* if X is a Markov process on a Lusin space $(E, \mathcal{B}(E))$ and satisfies $(\mathbf{X.6})_\mathbf{s}$ as well as the strong Markov property and the quasi-left-continuity on $(0,\zeta)$ with respect to an admissible filtration $\{\mathcal{M}_t\}$.

(ii) X is called a *Hunt process* if X is a Markov process on a Lusin space $(E, \mathcal{B}(E))$ and satisfies $(\mathbf{X.6})_\mathbf{h}$ as well as the strong Markov property and the quasi-left-continuity on $(0, \infty)$ with respect to an admissible filtration $\{\mathcal{M}_t\}$.

The next theorem can be proved in exactly the same way as the proof of Theorem A.1.18.

THEOREM A.1.24. *Let $(E, \mathcal{B}(E))$ be a Lusin space. For a Markov process $X = (\Omega, \mathcal{M}, X_t, \mathbf{P}_x)$ on $(E, \mathcal{B}(E))$ satisfying condition $(\mathbf{X.6})_\mathbf{s}$ (resp. $(\mathbf{X.6})_\mathbf{h}$) the following three conditions are mutually equivalent:*
(α) X is a standard (resp. Hunt) process on E.
(β) The minimum augmented admissible filtration $\{\mathcal{F}_t\}$ of X is right continuous and X is strong Markov and quasi-left-continuous on $(0, \zeta)$ (resp. $(0, \infty)$) with respect to it.
(γ) For each $\mu \in \mathcal{P}(E_\partial)$, the filtration $\{\mathcal{F}_t^\mu\}$ is right continuous and the identity (A.1.15) as well as (A.1.30) (resp. (A.1.31)) hold for $\{\mathcal{F}_t^\mu\}$-stopping times and for the probability measure $\mathbf{P}_\mu$.

Remark A.1.25 (*A generalization of state space E*). In defining a Borel right process, a Borel standard process, and a Hunt process in Definitions A.1.17 and A.1.23, respectively, the state space $(E, \mathcal{B}(E))$ is assumed to be a Lusin space so that E is regarded as a Borel subset of a compact metric space F and the Borel σ-field $\mathcal{B}(E)$ is the trace of $\mathcal{B}(F)$ on E. By dropping the Borel measurability assumption on E in these definitions, we can take an *arbitrary* subset E of a compact metric space F, endow E with the relative topology, and take as $\mathcal{B}(E)$ the Borel σ-field with respect to this topology. In what follows, we shall admit this generalization of the state space $(E, \mathcal{B}(E))$ for a Borel right process, a Borel standard process, and a Hunt process, respectively, with no specific mentions. With this generalization, nothing is changed because $\mathcal{B}(E)$ is still the trace of $\mathcal{B}(F)$ on E and the basic monotone class theorem Proposition A.1.3 continues to work. □

Exercise A.1.26. Let X be a Hunt process on a Lusin space $(E, \mathcal{B}(E))$.
(i) Show that $\{X_{t-}(\omega)\}_{t>0}$ is progressive with respect to any admissible filtration $\{\mathcal{M}_t\}$ for X.
(ii) For $B \in \mathcal{B}(E)$, show that $\widehat{\sigma}_B$ defined below is an $\{\mathcal{F}_t\}$-stopping time.

$$\widehat{\sigma}_B(\omega) = \inf\{t > 0 : X_{t-}(\omega) \in B\}, \quad \inf \emptyset = \infty. \tag{A.1.32}$$

As an application of Theorem A.1.18 (resp. Theorem A.1.24), we can formulate a restriction of a Borel right process (resp. Hunt process) to its invariant set, which is one of the most trivial and yet important transformations of a Markov process.

Let $X = (\Omega, \mathcal{M}, \{X_t\}_{t\in[0,\infty]}, \{\mathbf{P}_x\}_{x\in E_\partial})$ be a Borel right process (resp. Hunt process) on a Lusin space $(E, \mathcal{B}(E))$. A Borel set $A \subset E$ is said to be *X-invariant* if

$$\mathbf{P}_x(\sigma_{E\setminus A} < \infty) = 0 \quad \text{for every } x \in A, \tag{A.1.33}$$

$$\Big(\text{resp. } \mathbf{P}_x(\sigma_{E\setminus A} \wedge \widehat{\sigma}_{E\setminus A} < \infty) = 0 \quad \text{for every } x \in A\Big). \tag{A.1.34}$$

Since $\sigma_{E\setminus A}$ (resp. $\sigma_{E\setminus A} \wedge \widehat{\sigma}_{E\setminus A}$) is an $\{\mathcal{F}_t\}$-stopping time by Theorem A.1.19 (resp. Theorem A.1.19 and Exercise A.1.26), we have $\widetilde{\Omega} \in \mathcal{F}_\infty$, where

$$\widetilde{\Omega} = \{\omega \in \Omega : \sigma_{E\setminus A}(\omega) = \infty\}$$

$$(\text{resp. } \widetilde{\Omega} = \{\omega \in \Omega : \sigma_{E\setminus A}(\omega) = \infty \text{ and } \widehat{\sigma}_{E\setminus A} = \infty\}).$$

Let $A \subset E(\subset F)$ be a Borel X-invariant set endowed with the relative topology. Put $A_\partial = A \cup \{\partial\}$ and regard it as a topological subspace of E_∂. Then $\mathcal{B}(A_\partial) = \mathcal{B}(A) \cup \{B \cup \{\partial\} : B \in \mathcal{B}(A)\}$. We then call

$$X|_A = (\widetilde{\Omega}, \mathcal{M}|_{\widetilde{\Omega}}, \{X_t\}_{t\in[0,\infty]}, \{\mathbf{P}_x(\cdot \cap \widetilde{\Omega})\}_{x\in A_\partial}) \tag{A.1.35}$$

the *restriction of* X to its invariant set $(A, \mathcal{B}(A))$.

LEMMA A.1.27. *For a Borel right process (resp. Hunt process) on $(E, \mathcal{B}(E))$, $X|_A$ is a Borel right process (resp. Hunt process) on $(A, \mathcal{B}(A))$.*

Proof. We give a proof only for a Borel right process X because the same proof as below works for a Hunt process X by using Theorem A.1.24 instead of Theorem A.1.18. First $X|_A$ is a Markov process on A, namely, all the five conditions of Definition A.1.8 are fulfilled with $(E_\partial, \mathcal{B}(E_\partial))$ being replaced by $(A_\partial, \mathcal{B}(A_\partial))$. In fact, the measurability condition **(X.2)** for $X|_A$ follows from that for X and

$$\mathbf{P}_x(\{X_t \in B\} \cap \widetilde{\Omega}) = \mathbf{P}_x(X_t \in B) \quad \text{for } x \in A_\partial,\ B \in \mathcal{B}(A_\partial).$$

Four other conditions for $X|_A$ are clear. The property $\mathbf{(X.6)_r}$ for the pair $(\widetilde{\Omega}, X_t)$ follows readily from that for the pair (Ω, X_t).

We now examine the property (γ) of Theorem A.1.18 for $X|_A$. Take any $\mu \in \mathcal{P}(A_\partial)$ and regard it as an element of $\mathcal{P}(E_\partial)$ by taking $\mu(E \setminus A) = 0$. The σ-fields generated by $X|_A$ will be designated by putting $\widetilde{\ }$. It then follows from $\widetilde{\mathcal{F}}^0_t = \mathcal{F}^0_t \cap \widetilde{\Omega}$, $\widetilde{\Omega} \in \mathcal{F}^\mu_\infty$, and $\mathbf{P}_\mu(\widetilde{\Omega}) = 1$ that $\widetilde{\mathcal{F}}^\mu_t = \mathcal{F}^\mu_t \cap \widetilde{\Omega} \subset \mathcal{F}^\mu_t$. Consequently, the property (γ) for X immediately implies that for $X|_A$, yielding that $X|_A$ is a Borel right process by virtue of Theorem A.1.18. □

Suppose that $X = (\Omega, \mathcal{M}, \{X_t\}_{t\in[0,\infty]}, \{\mathbf{P}_x\}_{x\in E_\partial})$ is a Borel right process on a Lusin space $(E, \mathcal{B}(E))$. It will be useful to introduce for X a σ-field $\mathcal{B}^n(E_\partial)$ slightly larger than $\mathcal{B}(E_\partial)$ as follows:

DEFINITION A.1.28. A set $B \subset E_\partial$ is said to be *nearly Borel measurable* if, for any $\mu \in \mathcal{P}(E_\partial)$, there exist sets $B_1, B_2 \in \mathcal{B}(E_\partial)$ such that $B_1 \subset B \subset B_2$, $\mathbf{P}_\mu(X_t \in B_2 \setminus B_1 \text{ for some } t \geq 0) = 0$. The totality of nearly Borel measurable subsets of E_∂ is denoted by $\mathcal{B}^n(E_\partial)$.

Exercise A.1.29. (i) Show that $\mathcal{B}^n(E_\partial)$ is a σ-field contained in $\mathcal{B}^*(E_\partial)$.
(ii) Show that for every $\mu \in \mathcal{P}(E)$ and for a non-negative nearly Borel measurable function f, there are non-negative Borel measurable functions g and h such that $g \leq f \leq h$ and $\mathbf{P}_\mu(g(X_s) < h(X_s) \text{ for some } s \geq 0) = 0$.

The hitting time σ_B and the entrance time $\dot{\sigma}_B$ of $B \in \mathcal{B}(E_\partial)$ are $\{\mathcal{F}_t\}$-stopping times by Theorem A.1.19. This property remains valid for $B \in \mathcal{B}^n(E_\partial)$, because, for any $\mu \in \mathcal{P}(E_\partial)$, we take the set $B_1 \in \mathcal{B}(E_\partial)$ in the above definition to see that for each $t \geq 0$, $\{\sigma_B \leq t\}$ differs from $\{\sigma_{B_1} \leq t\} \in \mathcal{F}_t$ only by a $\mathbf{P}_\mu$-null set and consequently $\{\sigma_B \leq t\} \in \mathcal{F}_t$.

Remark A.1.30 (Restriction to a nearly Borel invariant set). In the above, we assume that an X-invariant set A for a Borel right process (resp. Hunt process) X is a Borel subset of E. But more generally we can take as A a nearly Borel measurable subset of E defined by Definition A.1.28 above. A nearly Borel measurable set is universally measurable but not necessarily Borel measurable. The X-invariance of a nearly Borel measurable set $A \subset E$ is defined by (A.1.33) (resp. (A.1.34)). Note that $\sigma_{E\setminus A}$ (resp. $\sigma_{E\setminus A} \wedge \widehat{\sigma}_{E\setminus A}$) is still an $\{\mathcal{F}_t\}$-stopping time. Hence, for a nearly Borel measurable X-invariant set $A \subset E$, the restriction $X|_A$ of X to A is well defined by (A.1.35). On the other hand, since E is regarded as a subset of a compact metric space F, we can endow A with the relative topology and define $\mathcal{B}(A)$ as the Borel σ-field with respect to this topology. Then, for any $B \in \mathcal{B}(A)$, there is $B_1 \in \mathcal{B}(E)$ so that $B = B_1 \cap A$ and $\mathbf{P}_x(X_t \in B) = \mathbf{P}_x(X_t \in B_1)$, $x \in A$, yielding the measurability **(X.2)**. Thus $X|_A$ is a Borel right process (resp. Hunt process) on $(A, \mathcal{B}(A))$ in the extended sense of Remark A.1.25. □

Exercise A.1.31. Let X be a Borel right process on a Lusin space $(E; \mathcal{B}(E))$ and A be a nearly Borel measurable X-invariant subset of E. Show that for any nearly Borel measurable function f on E, $f|_A$ is nearly Borel measurable with respect to the restricted Borel right process $X|_A$.

LEMMA A.1.32. *Let X be a Hunt process on a Lusin space $(E, \mathcal{B}(E))$ and A be a nearly Borel measurable subset of E. If* (A.1.33) *holds for A, then so does* (A.1.34). *In other words, A is invariant with respect to the Hunt process X whenever it is so with respect to X regarded as a Borel right process.*

Proof. Assume first that $A \in \mathcal{B}(E)$ satisfies (A.1.33). Due to the quasi-left-continuity on $[0, \infty)$ of the Hunt process X, it is known (cf. [73, Theorem A.2.3]) that $\mathbf{P}_x(\sigma_{E\setminus A} \leq \widehat{\sigma}_{E\setminus A}) = 1$ for any $x \in E$, from which the property (A.1.34) for A follows.

If A is a nearly Borel measurable subset of E satisfying (A.1.33), there exists, for any $x \in A$, a Borel set A_1 such that $x \in A_1 \subset A$ and $\mathbf{P}_x(\sigma_{E\setminus A_1} < \infty) = 0$ in view of Definition A.1.28. This combined with the above consideration yields (A.1.34) for A. □

A topological space E is called a *Radon space* if it is homeomorphic to a universally measurable subset of a compact metric space F. By identifying E with its image under this homeomorphism, we can then regard E as a subspace of F with relative topology and as an element of $\mathcal{B}^*(F)$. Any Lusin space is a Radon space. We equip a Radon space E with the universally measurable σ-field $\mathcal{B}^*(E)$.

Exercise A.1.33. Show that for a Radon space E, $\mathcal{B}^*(E)$ is the trace of $\mathcal{B}^*(F)$ on E.

Even if we start with a Borel right process X on a Lusin space $(E, \mathcal{B}(E))$, we are led to Markov processes on Radon spaces by non-trivial transformations of X such as time changes, Feynman-Kac transforms, and killing upon leaving a nearly Borel finely open set, since the transition functions of the transformed Markov processes can hardly make the space of bounded $\mathcal{B}(E)$-measurable functions invariant.

Let $(\Omega, \mathcal{M}, X_t, \mathbf{P}_x)$ be a Markov process on a Radon space $(E, \mathcal{B}^*(E))$. We may then consider two σ-fields of Ω: $\mathcal{F}^{0*}_\infty = \sigma\{X_s^{-1}(B) : B \in \mathcal{B}^*(E),\ s < \infty\}$ and $\mathcal{F}^0_\infty = \sigma\{X_s^{-1}(B) : B \in \mathcal{B}(E),\ s < \infty\}$. Similarly, we have two σ-fields $\mathcal{F}^{0*}_t$ and $\mathcal{F}^0_t$ for each $t \geq 0$. Starting with $\mathcal{F}^{0*}_\infty$, $\mathcal{F}^{0*}_t$, $t \geq 0$, and following the procedure described in Section A.1.1, we can construct the minimum augmented admissible filtration $\{\mathcal{F}_t\}_{t\in[0,\infty]}$ for X.

Exercise A.1.34. Show that the filtration $\{\mathcal{F}_t\}_{t\in[0,\infty]}$ constructed as above from $\mathcal{F}^{0*}_\infty$ and $\{\mathcal{F}^{0*}_t\}$ is the same as the minimum augmented admissible filtration constructed in Section A.1.1 from the filtration $\mathcal{F}^0_\infty$ and $\{\mathcal{F}^0_t,\ t \geq 0\}$.

The transition function $\{P_t; t \geq 0\}$ of X is now a transition function on $(E, \mathcal{B}^*(E))$ (instead of $(E, \mathcal{B}(E))$) in the sense of Definition 1.1.13.

DEFINITION A.1.35. For a Markov process on a Radon space $(E, \mathcal{B}^*(E))$ with a transition function $\{P_t; t \geq 0\}$ and for $\alpha \geq 0$, a $[0, \infty]$-valued function u on E is said to be *α-excessive* if u is $\mathcal{B}^*(E)$-measurable and

$$e^{-\alpha t} P_t u(x) \uparrow u(x) \ \text{ as } \ t \downarrow 0 \quad \text{for every } x \in E.$$

A 0-excessive function will be simply called *excessive*.

DEFINITION A.1.36. $X = (X_t, \mathbf{P}_x)$ is called a *right process* if it is a Markov process on a Radon space $(E, \mathcal{B}^*(E))$ satisfying condition $(\mathbf{X.6})_\mathbf{r}$ as well as the property that for every α-excessive function g for X with $\alpha \geq 0$ and for every $\mu \in \mathcal{P}(E)$,

$$\mathbf{P}_\mu \left(\lim_{t' \downarrow t} g(X_{t'}) = g(X_t) \right) = 1. \tag{A.1.36}$$

The next theorem gives useful criteria for a Markov process on a Radon space to be a right process. It also shows that the notion of a right process on a Radon space is a genuine extension of the notion of a Borel right process on a Lusin space.

THEOREM A.1.37. (i) *Every Borel right process on a Lusin space is a right process; if $X = (X_t, \mathbf{P}_x)$ is a Borel right process on a Lusin space $(E, \mathcal{B}(E))$, then X possesses the property* (A.1.36).
(ii) *The following conditions are mutually equivalent for a Markov process $X = (X_t, \mathbf{P}_x)$ on a Radon space $(E, \mathcal{B}^*(E))$ satisfying condition $(\mathbf{X.6})_\mathbf{r}$. The resolvent kernel for X is denoted by $\{R_\alpha; \alpha > 0\}$.*

- **(a)** *X is a right process.*
- **(b)** *X is strong Markov in the sense of Definition A.1.15, and $\{g(X_t)\}_{t \geq 0}$ is $\mathbf{P}_\mu$-indistinguishable from an $\{\mathcal{F}_t\}$-adapted optional process for any $g \in R_\alpha(C_{u+}(E))$, $\alpha > 0$ and $\mu \in \mathcal{P}(E)$.*
- **(c)** *X has the property* (A.1.36) *for any $g \in R_\alpha(C_{u+}(E))$, $\alpha > 0$ and every $\mu \in \mathcal{P}(E)$.*

Proof. (i) This will be proved in Theorem A.2.2 of Section A.2.1.
(ii) Since $g \in R_\alpha(bC_+(E))$ is α-excessive, the implication **(a)** $\Rightarrow$ **(c)** is trivial.

(c) $\Rightarrow$ **(b)**: Condition **(c)** obviously implies the second condition of **(b)**. We shall derive from condition **(c)** the strong Markov property of X with respect to the right continuous filtration $\{\mathcal{F}^0_{t+}\}$, or more specifically (A.1.16) holding for $B \in \mathcal{B}(E)$ and $\Lambda \in \mathcal{F}^0_{\sigma+}$.

Let σ be an $\{\mathcal{F}^0_{t+}\}$-stopping time and σ_n, $n \geq 1$, be its discrete approximations from the above defined by (A.1.2). For $\Lambda \in \mathcal{F}^0_{\sigma+}$ and for $f \in bC_+(E)$, $\alpha > 0$, we have

$$\int_0^\infty e^{-\alpha t}\mathbf{E}_x\left[f(X_{\sigma+t}); \Lambda\right] dt = \lim_{n\to\infty} \mathbf{E}_x\left[\int_0^\infty e^{-\alpha t} f(X_{\sigma_n+t})dt; \Lambda\right]$$

$$= \lim_{n\to\infty} \sum_{k=1}^\infty \mathbf{E}_x\left[\int_0^\infty e^{-\alpha t} f(X_{t+k2^{-n}})dt; \sigma_n = k2^{-n},\ \Lambda\right].$$

Notice that $\{\sigma_n = k2^{-n}\} \cap \Lambda = \{(k-1)2^{-n} \leq \sigma < k2^{-n}\} \cap \Lambda \in \mathcal{F}^0_{k2^{-n}}$. Using the Markov property of X relative to the filtration $\{\mathcal{F}^0_t\}$ and the assumption (**c**), we see that the above expression equals

$$\lim_{n\to\infty} \sum_{k=1}^\infty \mathbf{E}_x\left[\mathbf{E}_{X_{k2^{-n}}}\left[\int_0^\infty e^{-\alpha t} f(X_t)dt\right]; \sigma_n = k2^{-n},\ \Lambda\right]$$

$$= \lim_{n\to\infty} \mathbf{E}_x\left[R_\alpha f(X_{\sigma_n}); \Lambda\right] = \int_0^\infty e^{-\alpha t}\mathbf{E}_x\left\{\mathbf{E}_{X_\sigma}\left[f(X_t)\right]; \Lambda\right\} dt.$$

Since both $\mathbf{E}_x\left[f(X_{\sigma+t}); \Lambda\right]$ and $\mathbf{E}_x\left\{\mathbf{E}_{X_\sigma}\left[f(X_t)\right]; \Lambda\right\}$ are right continuous in t, they must coincide owing to the uniqueness of the Laplace transformation and we arrive at the desired identity (A.1.16) for the filtration $\{\mathcal{F}^0_{t+}\}$ by using Proposition A.1.3.

(**b**) $\Rightarrow$ (**a**): Assume (**b**). Then by virtue of Theorem A.1.18 and Exercise A.1.34, the minimum augmented admissible filtration $\{\mathcal{F}_t\}$ for X is right continuous and X has the strong Markov property with respect to $\{\mathcal{F}_t\}$, which can be further extended to (A.1.23).

For fixed $\alpha > 0$, let H be the collection of $f \in b\mathcal{B}(E)$ such that the process $\{R_\alpha f(X_t)\}_{t\geq 0}$ is $\mathbf{P}_\mu$-indistinguishable from an $\{\mathcal{F}_t\}$-adapted optional process for each $\mu \in \mathcal{P}(E)$. Then $C_u(E) \subset H$ and H is a linear space closed under uniformly bounded increasing limits. Accordingly, $H = b\mathcal{B}(E)$ by Proposition A.1.3.

For $\alpha \geq 0$, we can then make use of the strong Markov property of X with respect to $\{\mathcal{F}_t\}$ to show that (A.1.36) holds for any α-excessive function g of X in exactly the same way as in the proof of Theorem A.2.2 below. □

Remark A.1.38. We shall prove in Theorem A.2.2 that any α-excessive function of a Borel right process X on a Lusin space satisfies not only the property (A.1.36) but is also nearly Borel measurable. But for a general right process X on a Radon space, the latter property may not hold. □

In Definition A.1.23, we defined a Borel standard process and a Hunt process as specific Borel right processes on a Lusin space with additional properties on the left limits of sample paths. A general standard process will now be defined as a specific right process X on a Radon space.

DEFINITION A.1.39. A right process X on a Radon space $(E, \mathcal{B}^*(E))$ is called a *standard process* if X satisfies additionally **(X.6)$_s$** and the quasi-left-continuity on $(0, \zeta)$.

We end this subsection by quoting two fundamental theorems from Blumenthal-Getoor [13].

A filtration $\{\mathcal{M}_t\}_{t\in[0,\infty]}$ is called *quasi-left-continuous* if, for any increasing sequence $\{\sigma_n\}$ of $\{\mathcal{M}_t\}$-stopping times with limit σ,

$$\bigvee_{n=1}^{\infty} \mathcal{M}_{\sigma_n} = \mathcal{M}_{\sigma}. \tag{A.1.37}$$

Here $\vee_{n=1}^{\infty}\mathcal{M}_{\sigma_n}$ denotes the σ-field generated by $\{\mathcal{M}_{\sigma_n},\ n \geq 1\}$.

THEOREM A.1.40. *The minimum augmented admissible filtration $\{\mathcal{F}_t\}_{t\in[0,\infty]}$ of a Hunt process is quasi-left-continuous.*

Proof. See [13, IV, (4.2)]. □

This theorem indicates an advantage of a Hunt process over a Borel standard process especially in dealing with its martingale additive functionals. If X is a Hunt process, then for any sequence $\{\sigma_n\}$ of $\{\mathcal{F}_t\}$-stopping times increasing to σ,

$$X_\sigma \in \bigvee_{n=1}^{\infty} \mathcal{F}_{\sigma_n}/\mathcal{B}(E_\partial). \tag{A.1.38}$$

This is because $\{\sigma < \infty\} = \cup_k \cap_n \{\sigma_n \leq k\} \in \bigvee_{n=1}^{\infty} \mathcal{F}_{\sigma_n}$ and $X_\infty(\omega) = \partial$ for every $\omega \in \Omega$, and accordingly, the quasi-left-continuity (A.1.31) of X on $[0, \infty)$ implies (A.1.38). Actually the property (A.1.38) is shown in [13, IV, (4.2)] to be equivalent to the quasi-left-continuity of $\{\mathcal{F}_t\}$. However, a standard process may fail to satisfy (A.1.38) because $\{\sigma < \zeta\} = \{X_\sigma \in E\}$ may not be in $\bigvee_{n=1}^{\infty} \mathcal{F}_{\sigma_n}$.

DEFINITION A.1.41. A standard process X is called *special* if its minimum augmented admissible filtration $\{\mathcal{F}_t\}_{[0,\infty]}$ is quasi-left-continuous.

We have verified that the transition function $\{P_t; t \geq 0\}$ of a Borel right process X defined by (A.1.4) is a transition function on a Lusin space $(E, \mathcal{B}(E))$ as an analytic quantity satisfying conditions **(t.1)**–**(t.4)** of Definition 1.1.13. Conversely, given a transition function $\{P_t; t \geq 0\}$ in this analytic sense, it is important to construct an associated nice Markov process X. A well-know theorem of this sort is provided by a Feller transition function on a locally compact separable metric space E.

Let E be a locally compact separable metric space and $E_\partial = E \cup \{\partial\}$ be the one-point compactification of E. We put

$$C_\infty(E) = \{f \in C(E) : \lim_{x \to \partial} f(x) = 0\}.$$

If a family of Markovian kernels $\{P_t; t \geq 0\}$ on $(E, \mathcal{B}(E))$ satisfies **(t.1)**, **(t.3)** and

$$P_t(C_\infty(E)) \subset C_\infty(E),\ t \geq 0,\ \lim_{t \downarrow 0} P_t f(x) = f(x),\ x \in E, f \in C_\infty(E), \quad \text{(A.1.39)}$$

then it is called a *Feller transition function* or *Feller semigroup* regarded as a semigroup of linear operators on $C_\infty(E)$.

Exercise A.1.42. Verify that a Feller transition function is a transition function on $(E, \mathcal{B}(E))$ in the sense of Definition 1.1.13.

THEOREM A.1.43. *For any Feller transition function $\{P_t; t \geq 0\}$ on a locally compact separable metric space E, there exists a Hunt process on E having $\{P_t; t \geq 0\}$ as its transition function. Such a process will be called a Feller process.*

Proof. See (9.4) in Chapter 1 of [13]. □

We remark that, once a Markov process X on E with property $(\mathbf{X.6})_\mathbf{r}$ is constructed having Feller transition function, it is clear then that X satisfies the property (A.1.36) for $u \in R_\alpha(C_{u+}(E))$ and consequently, by Theorem A.1.37, the strong Markov property.

A.2. BASIC PROPERTIES OF BOREL RIGHT PROCESSES

A.2.1. Excessive Functions

Let $X = (\Omega, \mathcal{M}, \{X_t\}_{t \in [0,\infty]}, \{\mathbf{P}_x\}_{x \in E_\partial})$ be a Borel right process on a Lusin space E with transition function $\{P_t, t \geq 0\}$.

Recall that the notions of an α-excessive function and an excessive function were defined in Definition A.1.35 for a general right process.

Suppose a non-negative universally measurable function u on E satisfies the inequality

$$e^{-\alpha t} P_t u(x) \leq u(x) \quad \text{for every } t > 0 \text{ and } x \in E. \quad \text{(A.2.1)}$$

An application of the operator $e^{-\alpha s} P_s$ to both sides of (A.2.1) yields that

$$e^{-\alpha(s+t)} P_{s+t} f(x) \leq e^{-\alpha s} P_s u(x);$$

namely, $e^{-\alpha t}P_t u(x)$ increases as t decreases. Denote by $\widetilde{u}(x)$ its limit as $t \downarrow 0$. Then $\widetilde{u}$ can be easily seen to be an α-excessive function satisfying $\widetilde{u}(x) \leq u(x)$ for every $x \in E$. The function $\widetilde{u}$ will be called the *α-excessive regularization of u*.

A non-negative constant function on E is excessive because of

$$P_t 1(x) = P_t(x, E) \leq 1 \quad \text{for every } t \geq 0 \text{ and } x \in E,$$

and properties **(t.3)–(t.4)** of $\{P_t; t \geq 0\}$. For any non-negative universally measurable function f on E, the resolvent kernel $R_\alpha f$ is α-excessive for every $\alpha \geq 0$ in view of Theorem A.1.22.

LEMMA A.2.1. *Let $\alpha > 0$.*
(i) *For any α-excessive function u, there exists a sequence $\{f_n\}$ of non-negative universally measurable functions such that for every $x \in E$, $R_\alpha f_n(x) \uparrow u(x)$ as $n \uparrow \infty$.*
(ii) *The limit of an increasing sequence of α-excessive functions is again α-excessive.*

Proof. (i) We put $u_n = u \wedge n$. By the resolvent equation (1.1.25),

$$nR_{n+\alpha}u_n = R_\alpha \left(n(u_n - nR_{n+\alpha}u_n)\right).$$

Further,

$$nR_{n+\alpha}u_n(x) = \int_0^\infty e^{-s} g(s, n, x) ds,$$

where $g(s, n, x) = e^{-\alpha \frac{s}{n}} P_{\frac{s}{n}} u_n(x)$. Since u_n satisfies (A.2.1), $g(s, n, x)$ increases to some $a(x) \leq \infty$ as $n \to \infty$. Clearly $e^{-\alpha t}P_t u(x) \leq a(x) \leq u(x)$ for every $t > 0$, and we get $a(x) = u(x)$ by letting $t \downarrow 0$. Hence $nR_{n+\alpha}u_n(x) \uparrow u(x)$ as $n \uparrow \infty$, and $f_n(x) := n(u_n(x) - nR_{n+\alpha}u_n(x))$ for $x \in E$ is the desired non-negative function.
(ii) Let $\{u_n\}$ be a sequence of α-excessive functions increasing to a function u. Then u is universally measurable and satisfies the inequality (A.2.1). Since $e^{-\alpha t}P_t u_n(x)$ increases as t decreases or as n increases, we obtain by exchanging the order of limits that for $x \in E$,

$$\lim_{t \downarrow 0} e^{-\alpha t}P_t u(x) = \lim_{t \downarrow 0} \lim_{n \to \infty} e^{-\alpha t}P_t u_n(x) = \lim_{n \to \infty} \lim_{t \downarrow 0} e^{-\alpha t}P_t u_n(x)$$

$$= \lim_{n \to \infty} u_n(x) = u(x).$$

□

Recall the convention that any numerical function u on E is extended to E_∂ by setting $u(\partial) = 0$.

THEOREM A.2.2. *Let $\alpha \geq 0$ and u be an α-excessive function.*
(i) *u is right continuous along the sample path of X in the sense of* (A.1.36).
(ii) *u is nearly Borel measurable.*

Proof. Since any (0−) excessive function is α-excessive for any $\alpha > 0$, we may assume that $\alpha > 0$.

We start with the case that $u = R_\alpha f$ with $f \in b\mathcal{B}_+(E)$. Since u is Borel measurable and X_t is right continuous on E adapted to the filtration $\{\mathcal{F}_t\}$, $u(X_t)$ is optional in the sense of Section A.1.1. Indeed, for any $v \in C_u(E)$, $\{v(X_t)\}_{t\geq 0}$ is a right continuous real valued process adapted to $\{\mathcal{F}_t\}$ and hence an optional process. Therefore, by Proposition A.1.3, $u(X_t)$ is also optional.

Let σ_n be any sequence of uniformly bounded and decreasing $\{\mathcal{F}_t\}$-stopping times with limit σ. Since $\{\mathcal{F}_t\}$ is right continuous, σ is again an $\{\mathcal{F}_t\}$-stopping time by Exercise A.1.5. By virtue of Theorem A.1.22, we have the identity (A.1.27) with σ being replaced by σ_n. Letting $n \to \infty$, we get

$$\lim_{n\to\infty} \mathbf{E}_\mu \left[e^{-\alpha\sigma_n} u(X_{\sigma_n})\right] = \mathbf{E}_\mu \left[e^{-\alpha\sigma} u(X_\sigma)\right] \quad \text{for every } \mu \in \mathcal{P}(E).$$

Therefore, u satisfies (A.1.36) by Theorem A.1.6.

Next we let $u = R_\alpha f$ with $f \in b\mathcal{B}^*_+(E)$. For any $\mu \in \mathcal{P}(E)$, we define a finite measure ν by $\nu(B) = \langle \mu, R_\alpha \mathbf{1}_B\rangle$, $B \in \mathcal{B}(E)$. There exist $f_1, f_2 \in b\mathcal{B}(E)$ such that $f_1 \leq f \leq f_2$ and $\langle \nu, f_2 - f_1\rangle = 0$. Let $u_i := R_\alpha f_i$, $i = 1, 2$, which are Borel measurable functions with $u_1 \leq u \leq u_2$. The Markov property (A.1.3) yields that for any $t \geq 0$,

$$\mathbf{E}_\mu \left[u_2(X_t) - u_1(X_t)\right] = \langle \mu, P_t R_\alpha (f_2 - f_1)\rangle \leq e^{\alpha t}\langle \nu, f_2 - f_1\rangle = 0.$$

Since $u_i(X_t)$, $i = 1, 2$, are right continuous, we conclude that

$$\mathbf{P}_\mu(u_1(X_t) = u_2(X_t) \text{ for every } t \geq 0) = 1.$$

Consequently, u satisfies (A.1.36) and u is also nearly Borel measurable.

We further let $Y_t = e^{-\alpha t}u(X_t)$. Then $\{Y_t\}_{t\geq 0}$ is an $(\{\mathcal{F}_t\}, \mathbf{P}_\mu)$-supermartingale for any $\mu \in \mathcal{P}(E)$, because, by the Markov property of X, especially the formula (A.1.23) with $\sigma = s$, we have for $t > s \geq 0$

$$\begin{aligned}\mathbf{E}_\mu [Y_t | \mathcal{F}_s] &= e^{-\alpha t}\mathbf{E}_{X_s}[u(X_{t-s})] \\ &= e^{-\alpha s} e^{-\alpha(t-s)} P_{t-s} u(X_s) \leq e^{-\alpha s} u(X_s) = Y_s.\end{aligned}$$

When u is an arbitrary α-excessive function, we choose a sequence $\{f_n\}$ satisfying the condition of Lemma A.2.1(i). Since $u_n = R_\alpha f_n$ satisfies (A.1.36), so does its limit function u by virtue of Theorem A.1.7. Since u_n is nearly Borel measurable, so is u. □

Exercise A.2.3. Show that if u and v are α-excessive, then so is $u \wedge v$.

Let $B \subset E$ be nearly Borel measurable. As we noted in Section A.1, the hitting time σ_B of B is an $\{\mathcal{F}_t\}$-stopping time. We define the *hitting distribution* and *α-order hitting distribution* of B by

$$\mathbf{H}_B(x,A) := \mathbf{P}_x\left(X_{\sigma_B} \in A,\ \sigma_B < \infty\right),$$
$$\mathbf{H}_B^\alpha(x,A) := \mathbf{E}_x\left[e^{-\alpha\sigma_B};\ X_{\sigma_B} \in A\right], \tag{A.2.2}$$

respectively, where $x \in E$ and $A \in \mathcal{B}^*(E)$. Both of them are kernels on $(E, \mathcal{B}^*(E))$ (cf. Exercise A.1.20). We further define

$$p_B(x) = \mathbf{P}_x(\sigma_B < \infty), \quad p_B^\alpha(x) = \mathbf{E}_x\left[e^{-\alpha\sigma_B}\right], \quad x \in E, \tag{A.2.3}$$

the *hitting probability* and *α-order hitting probability* of B, respectively. Clearly $p_B = \mathbf{H}_B 1$ and $p_B^\alpha = \mathbf{H}_B^\alpha 1$.

LEMMA A.2.4. (i) *If u is α-excessive for $\alpha > 0$ and σ is an $\{\mathcal{F}_t\}$-stopping time, then*

$$\mathbf{E}_x\left[e^{-\alpha\sigma}u(X_\sigma)\right] \le u(x), \quad x \in E.$$

(ii) *Let $B \subset E$ be nearly Borel measurable. If u is α-excessive for $\alpha > 0$, then so is $\mathbf{H}_B^\alpha u$. If u is excessive, then so is $\mathbf{H}_B u$. In particular, p_B^α is α-excessive and p_B is excessive.*

Proof. (i) follows immediately from Lemma A.2.1 and the identity (A.1.27). When $B \subset E$ is nearly Borel measurable, we see by the identity (A.1.27) and the Markov property that for $u = R_\alpha f$ with $f \in \mathcal{B}_+^*(E)$ and $\alpha > 0$,

$$\begin{aligned} e^{-\alpha t}P_t\left(\mathbf{H}_B^\alpha u\right)(x) &= e^{-\alpha t}\mathbf{E}_x\left[\int_{\sigma_B\circ\theta_t}^\infty e^{-\alpha s}f(X_s\circ\theta_t)ds\right] \\ &= \mathbf{E}_x\left[\int_{t+\sigma_B\circ\theta_t}^\infty e^{-\alpha s}f(X_s)ds\right], \quad x \in E. \end{aligned}$$

Since $t + \sigma_B \circ \theta_t \downarrow \sigma_B$ as $t \downarrow 0$, $\mathbf{H}_B^\alpha u$ is α-excessive. The same conclusion holds for a general α-excessive function in view of Lemma A.2.1.

If u is excessive, then u is α-excessive for any $\alpha > 0$ and so is $\mathbf{H}_B^\alpha u$. If we let $\alpha \downarrow 0$, then $\mathbf{H}_B^\alpha u \uparrow \mathbf{H}_B u$ and hence $P_t(\mathbf{H}_B u) \le \mathbf{H}_B u$ for every $t \ge 0$. Moreover, interchanging the order of taking monotone limits, we have

$$\lim_{t\downarrow 0} P_t(\mathbf{H}_B u)(x) = \lim_{\alpha\downarrow 0}\lim_{t\downarrow 0} e^{-\alpha t}P_t(\mathbf{H}_B^\alpha u)(x) = \mathbf{H}_B u(x), \ x \in E.$$

□

A.2.2. Fine Topology, Excessive Measures, and Exceptional Sets

Let X be a Borel right process on a Lusin space $(E, \mathcal{B}(E))$ with transition function $\{P_t; t > 0\}$.

LEMMA A.2.5 (Blumenthal's 0-1 law). *For any* $\Lambda \in \mathcal{F}_0$ *and any* $x \in E$, *either* $\mathbf{P}_x(\Lambda) = 0$ *or* $\mathbf{P}_x(\Lambda) = 1$.

Proof. By **(X.6)$_\mathbf{r}$(ii)**, $\theta_0^{-1}\Lambda = \Lambda$ for every $\Lambda \subset \Omega$. When $\Lambda \in \mathcal{F}_0^0$, we have, from the Markov property (A.1.11) of X and **(X.5)**,

$$\mathbf{P}_x(\Lambda) = \mathbf{P}_x(\Lambda \cap \theta_0^{-1}\Lambda) = \mathbf{E}_x\left[\mathbf{P}_{X_0}(\Lambda); \Lambda\right] = (\mathbf{P}_x(\Lambda))^2.$$

Thus $\mathbf{P}_x(\Lambda) = 0$ or $\mathbf{P}_x(\Lambda) = 1$. For $\Lambda \in \mathcal{F}_0$ and $x \in E$, we then choose $\Lambda' \in \mathcal{F}_0^0$ with $\mathbf{P}_x(\Lambda \Delta \Lambda') = 0$ to get $\mathbf{P}_x(\Lambda) = \mathbf{P}_x(\Lambda') \in \{0, 1\}$. □

For any nearly Borel measurable set $B \subset E$, σ_B is an $\{\mathcal{F}_t\}$-stopping time and in particular $\{\sigma_B = 0\} \in \mathcal{F}_0$. Hence, for each $x \in E$, the probability $\mathbf{P}_x(\sigma_B = 0)$ is either 1 or 0 by the above lemma. A point $x \in E$ is called *regular* (resp. *irregular*) for B if this value equals 1 (resp. 0). We denote by B^r the totality of regular points of B. Since $B^r = \{x \in E : p_B^\alpha(x) = 1\}$ for $\alpha > 0$ and p_B^α is nearly Borel measurable in x in view of Theorem A.2.2 and Lemma A.2.4, we see that $B^r \in \mathcal{B}^n(E)$.

We now introduce several notions of smallness of subsets of E relative to the Borel right process X.

DEFINITION A.2.6. (i) $B \in \mathcal{B}^*(E)$ is called *of potential zero* if $R(x, B) = 0$ for every $x \in E$.
(ii) $B \subset E$ is called *polar* if there is a set $D \in \mathcal{B}^n(E)$ such that $B \subset D$ and $\mathbf{P}_x(\sigma_D < \infty) = 0$ for every $x \in E$.
(iii) $B \subset E$ is called *thin* if there is a set $D \in \mathcal{B}^n(E)$ such that $B \subset D$ and D^r is empty, namely, if $\mathbf{P}_x(\sigma_D = 0) = 0$ for every $x \in E$.
(iv) A set $B \subset E$ contained in a countable union of thin sets is called *semipolar*.

Obviously any polar set is thin and any thin set is semipolar.

DEFINITION A.2.7. (i) A set $B \subset E$ is said to be *thin at* a point $x \in E$ if there is a set D with $B \subset D$, $D \in \mathcal{B}^n(E)$ such that x is irregular for D, namely, $\mathbf{P}_x(\sigma_D = 0) = 0$.
(ii) $B \subset E$ is said to be *finely open* if $B^c = E \setminus B$ is thin at each point $x \in B$, namely, there is for each $x \in B$ a set $D(x)$ with $B^c \subset D(x)$, $D(x) \in \mathcal{B}^n(E)$ such that $\mathbf{P}_x(\sigma_{D(x)} > 0) = 1$.

By the right continuity of the sample path, any open set $D \subset E$ in the original topology of E is finely open because $\mathbf{P}_x(\sigma_{D^c} > 0) = 1$ for every $x \in D$.

Exercise A.2.8. Denote by $\mathcal{O}$ the totality of finely open subsets of E. Show that $\mathcal{O}$ satisfies the following axiom of a family of open subsets to determine a topology of E:

(i) $E, \emptyset \in \mathcal{O}$.

(ii) $O_1,\ O_2 \in \mathcal{O} \Rightarrow O_1 \cap O_2 \in \mathcal{O}$.

(iii) $\{O_\lambda; \lambda \in \Lambda\} \subset \mathcal{O} \Rightarrow \cup_{\lambda \in \Lambda} O_\lambda \in \mathcal{O}$.

The topology of E determined by $\mathcal{O}$ is called the *fine topology* with respect to X. A function on E is called *finely continuous* if it is continuous with respect to the fine topology. The fine topology is finer that the original one and any continuous function on E with respect to the original topology is finely continuous.

In what follows, we say that an event $\Lambda \in \mathcal{M}$ occurs *almost surely* or a.s. in abbreviation if $\mathbf{P}_x(\Lambda) = 1$ for every $x \in E$.

THEOREM A.2.9. *A nearly Borel measurable function u on E is finely continuous if and only if the real-valued composite process $t \in [0, \infty) \mapsto u(X_t)$ is right continuous a.s. Here we set $u(\partial) = 0$ as usual.*

Proof. See [13, II.(4.8)] or [73, Theorem A.2.7]. □

In particular, any α-excessive function on E is finely continuous ($\alpha \geq 0$) in view of Theorem A.2.2.

LEMMA A.2.10. *Let $B \in \mathcal{B}^n(E)$. For any $\mu \in \mathcal{P}(E)$,*

$$\mathbf{P}_\mu \left(X_{\sigma_B} \in B \cup B^r,\ \sigma_B < \infty \right) = \mathbf{P}_\mu(\sigma_B < \infty). \tag{A.2.4}$$

In particular, for any $x \in E$, the α-order hitting distribution $\mathbf{H}_B^\alpha(x, \cdot)$ and the hitting distribution $\mathbf{H}_B(x, \cdot)$ of B are concentrated on the set $B \cup B^r$.

Proof. See [13, I, (11.4)] or [73, Lemma A.2.7]. □

If $B \in \mathcal{B}^n(E)$ is finely closed, then any point $x \in B^c$ cannot be regular for B and so

$$\mathbf{P}_\mu \left(X_{\sigma_B} \in B,\ \sigma_B < \infty \right) = \mathbf{P}_\mu(\sigma_B < \infty), \quad \mu \in \mathcal{P}(E). \tag{A.2.5}$$

For a kernel κ on $(E, \mathcal{B}(E))$ and a measure μ on E, we denote by $\mu\kappa$ the measure defined by $\int_E \mu(dx)\kappa(x, B),\ B \in \mathcal{B}(E)$.

Let $\{P_t; t \geq 0\}$ be the transition function of a Borel right process X. A measure m on E is called *excessive* with respect to $\{P_t\}$ if m is σ-finite and satisfies $mP_t \leq m$ for every $t > 0$.

LEMMA A.2.11. *If m is an excessive measure, then for any $B \in \mathcal{B}(E)$,*

$$mP_t(B) \uparrow m(B) \quad \text{as } t \downarrow 0. \tag{A.2.6}$$

Proof. For $B \in \mathcal{B}(E)$, we denote the increasing limit $\lim_{t\downarrow 0} mP_t(B)$ by $nu(B)$. Then $\nu(B) \leq m(B)$ and, for mutually disjoint sets $\{B_n,\ n \geq 1\} \subset \mathcal{B}(E)$, we interchange increasing limits to obtain $\nu\left(\bigcup_{n=1}^{\infty} B_n\right) = \sum_{n=1}^{\infty} \nu(B_n)$, thus ν is a σ-finite measure. We next show that $\nu = m$.

Fix an $\alpha > 0$. If $A \in \mathcal{B}(E)$ satisfies $\nu(A) < \infty$, then, for any $\varepsilon > 0$, it follows from $\alpha\nu R_\alpha(A) \leq \nu(A)$ that

$$\nu\{x \in E : R_\alpha(x, A) > \varepsilon\} \leq (\alpha\varepsilon)^{-1}\nu(A) < \infty.$$

Take a sequence $\{A_n\}$ of Borel sets increasing to E with $\nu(A_n) < \infty$ for every $n \geq 1$. Let $B_n = \left\{x \in E : R_\alpha(x, A_n) > \frac{1}{n}\right\}$. Then $\{B_n\}$ is a sequence of Borel finely open sets increasing to E with $\nu(B_n) < \infty$ for every $n \geq 1$.

For any non-negative continuous function f on E and integer $n \geq 1$, we have by Fatou's lemma

$$\liminf_{t\downarrow 0} P_t(f\mathbf{1}_{B_n}) \geq f\mathbf{1}_{B_n}.$$

By Fatou's lemma again,

$$\langle \nu, f\mathbf{1}_{B_n}\rangle = \lim_{t\downarrow 0}\langle mP_t, f\mathbf{1}_{B_n}\rangle = \lim_{t\downarrow 0}\langle m, P_t(f\mathbf{1}_{B_n})\rangle \geq \langle m, f\mathbf{1}_{B_n}\rangle.$$

Thus $\nu = m$ on B_n and consequently $\nu = m$ on $\cup_{n=1}^{\infty} B_n = E$. □

In the remainder of this section, we shall fix an excessive measure m for a Borel right process X.

DEFINITION A.2.12. (i) A set $N \subset E$ is called *m-polar* if there is a nearly Borel measurable set $\widetilde{N} \supset N$ such that

$$\mathbf{P}_m(\sigma_{\widetilde{N}} < \infty) = 0. \tag{A.2.7}$$

(ii) "*Quasi everywhere*" or "q.e." in abbreviation means "except for an m-polar set."
(iii) A subset $N \subset E$ is called an *m-inessential set* for the Borel right process X if N is an m-negligible nearly Borel measurable set such that $E \setminus N$ is X-invariant.

Clearly any m-inessential set is m-polar.

THEOREM A.2.13. (i) *If f is nearly Borel measurable on E, then there exist Borel measurable functions g, h such that $g \leq f \leq h$ and $g = h$ q.e. Any m-polar set is contained in a Borel m-polar set.*

(ii) *If $B \subset E$ is m-polar, then $m(B) = 0$. If $B \in \mathcal{B}^m(E)$ satisfies $mR_\alpha(B) = 0$ for some $\alpha > 0$, then $m(B) = 0$.*
(iii) *If $B \subset E$ is nearly Borel, finely open, and m-negligible, then B is m-polar.*
(iv) *If functions f, g on E are nearly Borel measurable, finely continuous, and satisfy $f \geq g$ $[m]$, then $f \geq g$ q.e.*
(v) *Let $\alpha \geq 0$. If f is α-excessive and $f < \infty$ m-a.e., then $f < \infty$ q.e.*

Proof. (i) By Exercise A.1.29(ii), there are Borel measurable functions g, h such that $g \leq f \leq h$ and $\mathbf{P}_m(g(X_s) < h(X_s)\ \exists s \geq 0) = 0$, which means $g = h$ q.e. If in particular B is a nearly Borel measurable m-polar set, then there are Borel sets B_1, B_2 such that $B_1 \subset B \subset B_2$ and $B_2 \setminus B_1$ is m-polar, and hence B is contained in the Borel m-polar set $B_2 = B \cup (B_2 \setminus B)$.
(ii) If B is a Borel m-polar set, then $mP_t(B) = \mathbf{P}_m(X_t \in B) = 0$ for every $t > 0$. By Lemma A.2.11, we have $m(B) = 0$. Next note that a set $B \in \mathcal{B}^m(E)$ belongs to the completions of $\mathcal{B}(E)$ with respect to the measures mP_t and mR_α for any $t > 0$ and $\alpha > 0$. From

$$\int_0^\infty e^{-\alpha t} mP_t(B)dt = mR_\alpha(B) = 0,$$

we have $mP_t(B) = 0$ for almost every $t > 0$ and so $m(B) = 0$ by Lemma A.2.11 again.
(iii) By the Fubini theorem and the excessiveness of m,

$$\mathbf{E}_m\left[\int_0^\infty \mathbf{1}_B(X_s)ds\right] = \int_0^\infty mP_t(B)dt = 0.$$

Since B is finely open, it follows from Lemma A.2.10 and the strong Markov property of X that

$$\mathbf{P}_m(\sigma_B < \infty) \leq \mathbf{P}_m\left(\bigcup_{r_1 < r_2,\ r_1, r_2 \in \mathbf{Q}_+} \{X_t \in B,\ \forall t \in (r_1, r_2)\}\right) = 0.$$

Hence B is m-polar.
(iv) Since $\{f < g\}$ is a nearly Borel, finely open m-negligible set, it is m-polar.
(v) Let $N = \{x \in E : f(x) = \infty\}$. N is nearly Borel and it follows from Lemma A.2.4 that

$$\mathbf{E}_x\left[e^{-\alpha\sigma_N} f(X_{\sigma_N})\right] \leq f(x) < \infty, \quad x \in E \setminus N.$$

Since N is finely closed, $X_{\sigma_N} \in N$ a.s. by Lemma A.2.10 and we see from the above that $\mathbf{P}_x(\sigma_N < \infty) = 0$, $x \in E \setminus N$, the m-polarity of N. □

LEMMA A.2.14. *Let $\alpha \geq 0$. If f is α-excessive, then there are Borel measurable α-excessive functions g, h such that $g \leq f \leq h$ and $g = h$ q.e.*

Proof. First we consider the case where $f = R_\alpha q$ with $q \in b\mathcal{B}^*_+(E)$ and $\alpha > 0$. Choose $q_1, q_2 \in b\mathcal{B}_+(E)$ such that

$$q_1 \leq q \leq q_2 \quad \text{and} \quad mR_\alpha(q_1 < q_2) = 0.$$

Then $R_\alpha q_1 \leq R_\alpha q \leq R_\alpha q_2$ and $R_\alpha q_1 = R_\alpha q_2$ $[m]$. Since by Theorem A.2.13, $R_\alpha q_i$, $i = 1, 2$, are Borel measurable α-excessive functions, we have $R_\alpha q_1 = R_\alpha q_2$ q.e. on E.

Next, for a general α-excessive function f, by Lemma A.2.1 there is a sequence $\{f_n\} \subset b\mathcal{B}^*_+(E)$ such that $R_\alpha f_n \uparrow f$ as $n \uparrow \infty$. Then for each $n \geq 1$, we can find Borel measurable α-excessive functions g_n and h_n such that $g_n \leq R_\alpha f_n \leq h_n$ and $g_n = h_n$ q.e. Put $\bar{g} := \liminf_{n\to\infty} g_n$ and $\bar{h} := \liminf_{n\to\infty} h_n$. Clearly $\bar{g} \leq f \leq \bar{h}$, $\bar{g} = \bar{h}$ q.e. and for every $t > 0$, $e^{-\alpha t}P_t\bar{g} \leq \bar{g}$ and $e^{-\alpha t}P_t\bar{h} \leq \bar{h}$.

Let g and h be the α-excessive regularization of $\bar{g}$ and $\bar{h}$, respectively, which are Borel measurable α-excessive functions satisfying $g \leq \bar{g} \leq f \leq h \leq \bar{h}$. But for each $x \in E$, $e^{-\alpha t}P_t\bar{g}(x)$ is a decreasing function of t and hence possesses at most a countable number of discontinuous points. It is easy to see that $P_t g(x) = P_t\bar{g}(x)$ for every continuity point t of $P_t\bar{g}(x)$. Hence $R_\alpha g = R_\alpha \bar{g}$ and consequently $m(g < \bar{g}) = 0$ by Theorem A.2.13(ii). Furthermore, $m(\bar{g} < h) = 0$ by Theorem A.2.13(ii). Thus $m(g < h) = 0$, which implies $g = h$ q.e. by Theorem A.2.13(iv).

Finally, when f is a (0-)excessive function, f is $\frac{1}{n}$-excessive for each $n \geq 1$, and by the preceding result, there exist Borel measurable $\frac{1}{n}$-excessive functions g_n, h_n with

$$g_n \leq f \leq h_n \quad \text{and} \quad g_n = h_n \text{ q.e.}$$

Put $\bar{g} := \liminf_{n\to\infty} g_n$ and $\bar{h} := \liminf_{n\to\infty} h_n$. Then

$$\bar{g} \leq f \leq \bar{h}, \quad \bar{g} = \bar{h} \text{ q.e.}, \quad P_t\bar{g} \leq \bar{g}, \quad P_t\bar{h} \leq \bar{h} \text{ for } t > 0.$$

In the same way as above, the excessive regularization g and h of $\bar{g}$ and $\bar{h}$ are the desired functions. □

THEOREM A.2.15. *Any m-polar set is contained in an m-inessential Borel set.*

Proof. Let B be an m-polar set. By Theorem A.2.13(i), we may assume that $B \in \mathcal{B}(E)$. If we put $\varphi(x) = \mathbf{P}_x(\sigma_B < \infty)$, $x \in E$, then $\varphi = 0$ $[m]$. Since φ is excessive, by Lemma A.2.14 there exists a Borel measurable function g with $\varphi \leq g = 0$ $[m]$. Then

$$\widehat{B} := B \cup C \quad \text{where} \quad C = \{x \in E : g(x) > 0\}$$

is a desired set. In fact, $\widehat{B}$ is a Borel set with $m(\widehat{B}) = 0$ by Theorem A.2.13(ii). It remains to show that $\widehat{B}^c = B^c \cap \{g = 0\}$ is X-invariant. Take any $x \in \widehat{B}^c$. For $x \in E$, since $g(x) = 0$, we have $\varphi(x) = \mathbf{P}_x(\sigma_B < \infty) = 0$. Recall our convention $g(\partial) = 0$. Since $\mathbf{E}_x(g(X_t)) = P_t g(x) \leq g(x) = 0$, the right continuity of $g(X_t)$ yields $\mathbf{P}_x(g(X_t) = 0$ for every $t \geq 0) = 1$, which implies $\mathbf{P}_x(\sigma_C < \infty) = 0$ and consequently $\mathbf{P}_x(\sigma_{\widehat{B}} < \infty) = 0$. □

Definition A.2.16. The following **(AC)** and **(AC)′** are called the *absolute continuity condition* for the transition function $\{P_t; t \geq 0\}$ and the resolvent kernel $\{R_\alpha; \alpha > 0\}$, respectively.

(AC) for any $t > 0$, $x \in E$, the measure $P_t(x, \cdot)$ is absolutely continuous with respect to m.

(AC)′ For some fixed $\alpha > 0$ and for any $x \in E$, the measure $R_\alpha(x, \cdot)$ is absolutely continuous with respect to m.

Theorem A.2.17. (i) **(AC)** *implies* **(AC)′**.
(ii) *Condition* **(AC)′** *holds if and only if every m-polar set is polar.*
(iii) *Assume condition* **(AC)′**. *If f and g are α-excessive for some $\alpha \geq 0$ and $f \geq g$ $[m]$, then $f \geq g$.*

Proof. (i) is trivial.
(ii) For any nearly Borel measurable m-polar set B and $x \in E$, let $\varphi(x) := \mathbf{P}_x(\sigma_B < \infty)$. Then $\varphi = 0$ $[m]$ and so, under the condition **(AC)′**, we have for any $x \in E$

$$\varphi(x) = \lim_{\beta \to \infty} \beta \int_E R_\beta(x, dy)\varphi(y) = 0,$$

namely, B is polar.

Conversely, assume that every m-polar set is polar. If $B \in \mathcal{B}(E)$ satisfies $m(B) = 0$, then by the excessiveness of m we have, for $\alpha > 0$, $mR_\alpha(B) = 0$ and hence $R_\alpha(\cdot, B) = 0$ $[m]$. On account of Theorem A.2.13, the set $N = \{x \in E : R_\alpha(x, B) > 0\}$ is m-polar and hence polar. So for any $x \in E$, $R_\alpha(x, B) = \lim_{t \downarrow 0} \mathbf{E}_x[R_\alpha(X_t, B); X_t \notin N] = 0$. So condition **(AC)′** holds.
(iii) Under the stated conditions, we have for any $x \in E$

$$\begin{aligned} f(x) &= \lim_{\beta \to \infty} \beta \int_E R_{\alpha+\beta}(x, dy) f(y) \\ &\geq \lim_{\beta \to \infty} \beta \int_E R_{\alpha+\beta}(x, dy) g(y) = g(x). \end{aligned}$$

□

Lemma A.2.18. (i) *For any $B \in \mathcal{B}^n(E)$, $B \setminus B^r$ is semipolar.*
(ii) *If $B \subset E$ is semipolar, then $X_t \in B$ occurs for at most countable many $t \geq 0$ a.s.*

Proof. See [13, II, (3.3)–(3.4)]. □

COROLLARY A.2.19. *Let $\alpha \geq 0$ and B be a semipolar set. If f, g are α-excessive functions on E and $f(x) = g(x)$ for every $x \in E \setminus B$, then they are equal identically.*

Proof. Since B is of potential zero by Lemma A.2.18, $\beta R_{\beta+\alpha} f$ and $\beta R_{\beta+\alpha} g$ are equal identically for any $\beta > 0$. It then suffices to let $\beta \to \infty$. □

A.3. ADDITIVE FUNCTIONALS OF RIGHT PROCESSES

A.3.1. Revuz Measures of Positive Continuous Additive Functionals

We continue to consider a Borel right process $X = (\Omega, \mathcal{M}, X_t, \zeta, \mathbf{P}_x)$ on a Lusin space $(E, \mathcal{B}(E))$ with transition function $\{P_t; t \geq 0\}$. Let us fix a σ-finite excessive measure m of $\{P_t; t \geq 0\}$. Any numerical function f on E is extended to E_∂ by setting $f(\partial) = 0$.

DEFINITION A.3.1. A numerical function $A_t(\omega)$ of two variables $t \geq 0$, $\omega \in \Omega$ is called an *additive functional* of X if there exist $\Lambda \in \mathcal{F}_\infty$ and an m-inessential set $N \subset E$ with

$$\mathbf{P}_x(\Lambda) = 1 \text{ for } x \in E \setminus N \quad \text{and} \quad \theta_t \Lambda \subset \Lambda \quad \text{for } t > 0, \tag{A.3.1}$$

and the following conditions are satisfied:

(A.1) For each $t \geq 0$, $A_t|_\Lambda$ is $\mathcal{F}_t|_\Lambda$-measurable.
(A.2) For any $\omega \in \Lambda$, $A_\cdot(\omega)$ is right continuous on $[0, \infty)$ has the left limits on $(0, \zeta(\omega))$, $A_0(\omega) = 0$, $|A_t(\omega)| < \infty$ for $t < \zeta(\omega)$, and $A_t(\omega) = A_{\zeta(\omega)}(\omega)$ for $t \geq \zeta(\omega)$. Further, the additivity

$$A_{t+s}(\omega) = A_t(\omega) + A_s(\theta_t \omega) \quad \text{for every } t, s \geq 0, \tag{A.3.2}$$

is satisfied.

We call Λ and N in the above definition a *defining set* and an *exceptional set* of $A_t(\omega)$, respectively. If in particular the exceptional set N can be taken to be empty, in other words, if a defining set Λ satisfies $\mathbf{P}_x(\Lambda) = 1$ for every $x \in E$, then the additive functional $A_t(\omega)$ is said to be an *additive functional in the strict sense*.

Two additive functionals A, B are called *m-equivalent* if

$$\mathbf{P}_m(A_t \neq B_t) = 0 \qquad \text{for every } t > 0, \tag{A.3.3}$$

and we write $A \sim B$ in this case.

Lemma A.3.2. *If additive functionals A and B are m-equivalent, then there are a common defining set Λ and a common Borel exceptional set N such that*

$$A_t(\omega) = B_t(\omega) \quad \textit{for every } t \geq 0 \textit{ and } \omega \in \Lambda. \tag{A.3.4}$$

Proof. Denote the defining sets and exceptional sets of additive functionals A and B by Λ_A, N_A, Λ_B, and N_B, respectively. Put $N_0 := N_A \cup N_B$, $\Lambda_0 := \Lambda_A \cap \Lambda_B$, $\Lambda_1 := \{A_t = B_t \text{ for every } t > 0\}$, and $\Lambda := \Lambda_0 \cap \Lambda_1$. Then N_0 is m-inessential and it is easy to see that $\theta_t(\Lambda) \subset \Lambda$ for every $t > 0$. Moreover, $\mathbf{P}_x(\Lambda_0^c) = 0$ for $x \in E \setminus N_0$. Hence

$$\mathbf{P}_x(\Lambda^c) \leq \mathbf{P}_x(\Lambda_0^c) + \mathbf{P}_x(\Lambda_0 \setminus \Lambda_1) = \mathbf{P}_x(\Lambda_0 \setminus \Lambda_1).$$

On account of the right continuity and the m-equivalence of A and B, $g(x) = \mathbf{P}_x(\Lambda_0 \setminus \Lambda_1)$ vanishes m-a.e. on E. Further, we can easily see that $g|_{E \setminus N_0}$ is an excessive function of the restricted right process $X_{E \setminus N_0}$. Applying Theorem A.2.9 and Theorem A.2.13 to the right process $X_{E \setminus N_0}$, we get $g = 0$ q.e. on $E \setminus N_0$, and accordingly, $g(x) = 0$ for every $x \in E \setminus N_1$ for some m-polar set $N_1 \supset N_0$. By virtue of Theorem A.2.15, there exists a Borel m-inessential set N containing N_1. Λ and N have the desired properties of this lemma. □

An additive functional A is called *finite càdlàg* if for every ω in its defining set Λ, $t \mapsto A_t(\omega)$ is a real-valued right continuous function of $t \in [0, \infty)$ possessing the left limits on $(0, \infty)$. In Section A.3.3, we shall deal with finite càdlàg additive functionals in the strict sense.

An additive functional A is called a *positive continuous additive functional* (*PCAF* in abbreviation) if for every ω in its defining set Λ, $t \mapsto A_t(\omega)$ is a $[0, \infty]$-valued continuous function of $t \geq 0$. In the rest of this section, we shall be exclusively concerned with PCAFs. Denote by $\mathbf{A}_c^+$ the totality of PCAFs of X. What we are interested in is the family of equivalence classes $\mathbf{A}_c^+ / \sim$ of $\mathbf{A}_c^+$.

A typical example of a PCAF is given by

$$A_t(\omega) = \int_0^t f(X_s(\omega))ds, \quad t \geq 0, \ \omega \in \Omega, \tag{A.3.5}$$

where $f \in b\mathcal{B}_+(E)$. In this case, $A_t(\omega)$ is a PCAF in the strict sense with Ω as its defining set and its additivity (A.3.2) follows from $\int_t^{t+s} f(X_u)du = \int_0^s f(X_u(\theta_t\omega))du$. If we let f equal 1 identically on E, then we have the simplest PCAF, $A_t = t \wedge \zeta$.

For $A \in \mathbf{A}_c^+$ and $f \in \mathcal{B}_+^*(E)$, we define $f \cdot A$ by

$$(f \cdot A)_t(\omega) = \int_0^t f(X_s(\omega))dA_s(\omega), \quad t \geq 0, \ \omega \in \Lambda.$$

Exercise A.3.3. Show that if $A \in \mathbf{A}_c^+$ and $f \in b\mathcal{B}_+(E)$, then $f \cdot A \in \mathbf{A}_c^+$.

LEMMA A.3.4. (i) *For $A \in \mathbf{A}_c^+$, let $\varphi(t) = \mathbf{E}_m[A_t]$, $t \geq 0$. If $\varphi(t)$ is finite for some $t > 0$, then it is finite for all $t > 0$. In this case, $\varphi(t)$ is a continuous concave function of $t \in [0, \infty)$.*
(ii) *For $A \in \mathbf{A}_c^+$, $f \in \mathcal{B}_+(E)$,*

$$\frac{1}{t}\mathbf{E}_m\left[(f \cdot A)_t\right] \text{ is increasing as } t \text{ decreases to } 0. \tag{A.3.6}$$

Proof. (i) If we put $c_t(x) = \mathbf{E}_x[A_t]$, then by (A.3.2)

$$\varphi(t+s) = \varphi(t) + \langle mP_t, c_s \rangle, \qquad t, s \geq 0. \tag{A.3.7}$$

Since m is excessive, $\varphi(t+s) \leq \varphi(t) + \varphi(s)$, $t, s \geq 0$. Combining this with the fact that $\varphi(t)$ is non-decreasing in t, we get the first conclusion of (i). Next suppose $\varphi(t) < \infty$ for every $t \geq 0$. Then by the dominated convergence theorem, $\varphi(t)$ is continuous in $t \geq 0$. Due to the excessiveness of m, the measure mP_t decreases in t and hence, for $0 < t < t'$ and $s > 0$, we have from (A.3.7) $\varphi(t'+s) - \varphi(t') \leq \varphi(t+s) - \varphi(t)$. By taking $t' = t+s$,

$$\frac{1}{2}(\varphi(t+2s) + \varphi(t)) \leq \varphi(t+s) \quad \text{for } s, t > 0.$$

Thus, φ is midpoint concave and, since it is continuous, φ is concave.
(ii) Let $A \in \mathbf{A}_c^+$. If $f \in b\mathcal{B}_+(E)$, then $f \cdot A \in \mathbf{A}_c^+$ by Exercise A.3.3. We put $\varphi(t) = \mathbf{E}_m[(f \cdot A)_t]$. If $\varphi(t) < \infty$ for some $t > 0$, then φ is concave on $[0, \infty)$ by (i) and $\varphi(0) = 0$. Hence, for $0 < s < t$, we have $\varphi(s) \geq \frac{s}{t}\varphi(t)$, namely, (A.3.6) holds. The monotonicity is trivially true when $\varphi(t) = \infty$ for every $t > 0$. For a general $f \in \mathcal{B}_+(E)$, we put $f_n(x) = f(x) \wedge n$, $x \in E$. Then $\mathbf{E}_m[(f_n \cdot A)_t] \uparrow \mathbf{E}_m[(f \cdot A)_t]$ as $n \uparrow \infty$, and consequently (A.3.6) for f_n implies the same for f. □

THEOREM A.3.5. (i) *For $A \in \mathbf{A}_c^+$, there exists a unique measure μ_A on $(E, \mathcal{B}(E))$ satisfying*

$$\int_E f(x)\mu_A(dx) = \lim_{t \downarrow 0} \frac{1}{t}\mathbf{E}_m\left[\int_0^t f(X_s)dA_s\right], \quad \forall f \in \mathcal{B}_+(E). \tag{A.3.8}$$

(ii) *If A, $B \in \mathbf{A}_c^+$, $A \sim B$, then $\mu_A = \mu_B$. For $A \in \mathbf{A}_c^+$, μ_A charges no semipolar set or m-polar set.*
(iii) *For $A \in \mathbf{A}_c^+$, $f \in b\mathcal{B}_+(E)$, the measure corresponding to $f \cdot A \in \mathbf{A}_c^+$ in the sense of* (i) *equals $f \cdot \mu_A$.*
(iv) *It holds for $A \in \mathbf{A}_c^+$, $f \in \mathcal{B}_+(E)$ that*

$$\int_E f(x)\mu_A(dx) = \lim_{\alpha \to \infty} \alpha \mathbf{E}_m\left[\int_0^\infty e^{-\alpha t} f(X_t)dA_t\right]. \tag{A.3.9}$$

Proof. (i) By virtue of Lemma A.3.4, the limit on the right hand of (A.3.8) exists. For $B \in \mathcal{B}(E)$, we define $\mu_A(B)$ to be this value with $f = \mathbf{1}_B$. By interchanging the order of taking increasing limits, we get the complete additivity of μ_A as well as the identity (A.3.8).
(ii) is clear from Lemma A.3.2, property of a semipolar set in Lemma A.2.18 and Definition A.2.12 of an m-polar set.
(iii) follows from (i).
(iv) We may assume that $f \in b\mathcal{B}_+(E)$. We put $\varphi(t) = \mathbf{E}_m[(f \cdot A)_t]$. Using integration by parts and taking expectation, we get

$$\mathbf{E}_m\left[\int_0^t e^{-\alpha s} f(X_s)dA_s\right] = \mathbf{E}_m\left[\int_0^t e^{-\alpha s} d(f \cdot A)_s\right]$$
$$= e^{-\alpha t}\varphi(t) + \alpha \int_0^t e^{-\alpha s}\varphi(s)ds.$$

On account of Lemma A.3.4, if $\varphi(t)$ is finite for some $t > 0$, then so it is for every $t > 0$ and $\varphi(t) \le \varphi(1) \cdot t$ for $t \ge 1$. Therefore, we have in this case

$$\alpha \mathbf{E}_m\left[\int_0^\infty e^{-\alpha s} f(X_s)dA_s\right] = \alpha^2 \int_0^\infty e^{-\alpha s}\varphi(s)ds < \infty. \tag{A.3.10}$$

As $\alpha \to \infty$, the right hand side increases to $\int_E f d\mu_A$ by (A.3.8). When $\varphi(t) = \infty$ for every $t > 0$, (A.3.9) is trivial. □

The measure μ_A determined by (A.3.8) is called the *Revuz measure* of $A \in \mathbf{A}_c^+$. This measure was introduced by D. Revuz [128] in 1970. The Revuz measure of the simplest PCAF $A_t = t \wedge \zeta$ equals m by Lemma A.2.11. Hence, for $f \in b\mathcal{B}_+(E)$, the Revuz measure of the PCAF given by (A.3.5) is equal to $f \cdot m$. However, the Revuz measure of $A \in \mathbf{A}_c^+$ is not generally absolutely continuous with respect to the reference excessive measure m.

A.3.2. Time Change and Killing of a Borel Right Process

Let $X = (\Omega, \mathcal{M}, X_t, \zeta, \mathbf{P}_x)$ be a Borel right process on a Lusin space $(E, \mathcal{B}(E))$. In this subsection, we consider a positive continuous additive functional in the strict sense $A_t(\omega)$ of X and formulate the time change X_{τ_t} of X_t with respect to the right continuous inverse $\tau_t(\omega)$ of A which lives on the support F of A defined below. We do not consider a reference excessive measure m for X or an exceptional set N for A. In the case that we admit an exceptional set N for A, we can apply the consideration of the present subsection to the restricted right process $X|_{E\setminus N}$ for which A can be regarded as a PCAF in the strict sense.

At the end of this subsection, we shall also give a definition of the canonical subprocess of X with respect to the multiplicative functional e^{-A_t} for a PCAF A in the strict sense.

Thus we assume that $A_t(\omega)$ satisfies conditions of Definition A.3.1 with $N = \emptyset$ and that $A_t(\omega)$ is a $[0, \infty]$-valued continuous function in $t \geq 0$ for every ω in the defining set $\Lambda \subset \Omega$. Further, by replacing $(\Omega, \mathcal{M})$ with $(\Lambda, \mathcal{M}|_\Lambda)$ and by restricting $(X_t, \zeta, \mathbf{P}_x)$ to this measurable space, we may and do assume that the defining set of $A_t(\omega)$ is Ω itself.

Define

$$R(\omega) = \inf\{t > 0 : A_t(\omega) > 0\}, \quad F = \{x \in E : \mathbf{P}_x(R = 0) = 1\}. \tag{A.3.11}$$

Observe that $\{R = 0\} \in \mathcal{F}_{0+} = \mathcal{F}_0$ so, by Blumenthal's 0-1 law (Lemma A.2.5), $\mathbf{P}_x(R = 0)$ is either 0 or 1 for $x \in E$. The set F is called the *support* of the PCAF A. If we introduce a function φ_A by $\varphi_A(x) = \mathbf{E}_x\left[e^{-R}\right], x \in E$, then $F = \{x \in E : \varphi_A(x) = 1\}$. Since $t + R(\theta_t\omega) \downarrow R(\omega),\ t \downarrow 0,\ \varphi_A$ is 1-excessive and consequently F is a nearly Borel finely closed set.

Denote by σ_F the hitting time of F and by p_F^1 the 1-order hitting probability of F: $p_F^1(x) = \mathbf{E}_x[e^{-\sigma_F}],\ x \in E$. For each $\omega \in \Lambda$, we introduce the time sets

$$I(\omega) = \{t : A_{t+\varepsilon}(\omega) - A_t(\omega) > 0, \text{ for every } \varepsilon > 0\}$$

and

$$Z(\omega) = \{t : X_t(\omega) \in F\}.$$

$I(\omega)$ is the set of *right increasing points* of $A_\cdot(\omega)$.

PROPOSITION A.3.6. *For every $x \in E$,*

$$\mathbf{P}_x(R = \sigma_F) = 1, \tag{A.3.12}$$

$$\mathbf{P}_x(I \subset Z) = 1, \tag{A.3.13}$$

$$\mathbf{P}_x(A_t = (\mathbf{1}_F \cdot A)_t \quad \text{for every } t > 0) = 1. \tag{A.3.14}$$

Further, each point of F is regular for F.

Proof. Due to the right continuity of $\varphi_A(X_t)$, it holds for any $x \in E$ that $\mathbf{P}_x(X_{\sigma_F} \in F) = 1$ and hence

$$\begin{aligned}\mathbf{P}_x(\sigma_F < R) &= \mathbf{P}_x(\sigma_F < R,\ R \circ \theta_{\sigma_F} > 0)\\ &= \mathbf{E}_x\left[\mathbf{P}_{X_{\sigma_F}}(R > 0);\ \sigma_F < R\right] = 0.\end{aligned}$$

We next show

$$\mathbf{P}_x(R < \sigma_F) = 0 \quad \text{for } x \in F^r \cup (E \setminus F).$$

This is trivially true for $x \in F^r$. If $x \in E \setminus F$, then for any $t > 0$

$$\begin{aligned}\mathbf{P}_x(R < \sigma_F) &= \mathbf{P}_x(A_{R+t} > 0,\ R < \sigma_F) = \mathbf{E}_x\left[\mathbf{P}_{X_R}(A_t > 0); R < \sigma_F\right]\\ &\le \mathbf{E}_x\left[\mathbf{P}_{X_R}(R < t); X_R \in E \setminus F\right],\end{aligned}$$

which tends to zero as $t \downarrow 0$.

Hence we have $\mathbf{P}_x(\sigma_F \ge R) = 1$ for all $x \in E$ and $p_F^1(x) = \varphi_A(x)$ for $x \notin F \setminus F^r$. Since both φ_A and p_F^1 are 1-excessive and the set $F \setminus F^r$ is semipolar, we see from Lemma A.2.18 and Corollary A.2.19 that the last identity holds for all $x \in E$ and consequently (A.3.12) is valid. In particular, $p_F^1(x) = \varphi_A(x) = 1$ for every $x \in F$, and so each $x \in F$ is regular for F.

Next, the continuity of $t \mapsto A_t$ and (A.3.12) imply that for every $x \in E$,

$$\begin{aligned}\mathbf{P}_x(I \not\subset Z) &\le \mathbf{P}_x\left(\bigcup_{t \in I}\{\sigma_F \circ \theta_t > 0\}\right)\\ &\le \mathbf{P}_x\left(\bigcup_{r \in \mathbb{Q}_+}\{\exists\, q \in \mathbb{Q}_+ \text{ with } q > r,\ A_q - A_r > 0,\ r + \sigma_F \circ \theta_r > q\}\right)\\ &\le \sum_{r \in \mathbb{Q}_+} \mathbf{E}_x\left[\mathbf{P}_{X_r}(R < \sigma_F)\right] = 0,\end{aligned}$$

yielding (A.3.13).

Since the set $I(\omega)$ of right increasing times of $A_t(\omega)$ differs from the set of its increasing times by at most a countable set and $A_t(\omega)$ is continuous, we can get from (A.3.13) that, $\mathbf{P}_x$-a.s. for any $x \in E$,

$$\begin{aligned}A_t(\omega) &= \int_{[0,t] \cap I(\omega)} dA_s(\omega)\\ &= \int_{[0,t] \cap I(\omega)} \mathbf{1}_F(X_s(\omega))dA_s(\omega) = (\mathbf{1}_F \cdot A)_t(\omega).\end{aligned}$$

This completes the proof of Proposition A.3.6. □

Here we quote a change of variable formula from [13, V.(2.2)]. Let $a(t)$ be a right continuous, non-decreasing function from $[0, \infty]$ to $[0, \infty]$ with $a(0) = 0,\ a(\infty) = \lim_{t \to \infty} a(t)$. We define

$$\tau(t) = \inf\{s : a(s) > t\},\ t \ge 0, \quad \inf \emptyset = \infty. \tag{A.3.15}$$

The function $\tau : [0, \infty) \to [0, \infty]$ will be called the *inverse* of a. It is right continuous and non-decreasing.

LEMMA A.3.7. (i) *For any non-negative Borel measurable function f on $[0,\infty]$ with $f(\infty)=0$,*

$$\int_0^\infty f(t)da(t) = \int_0^\infty f(\tau(t))dt. \tag{A.3.16}$$

(ii) *Suppose a is continuous, then*

$$\tau(t) = \begin{cases} \max\{s : a(s) = t\} & \text{for } t \in [0, a(\infty)), \\ \infty & \text{for } t \geq a(\infty). \end{cases}$$

In particular, τ is strictly increasing on $[0, a(\infty))$ and $a(\tau(t)) = t$ for $t \in [0, a(\infty))$.

Now, for the PCAF $A_t(\omega)$ of the Borel right process $X = (\Omega, \mathcal{M}, X_t, \theta_t, \zeta, \mathbf{P}_x)$, its inverse is defined for each $\omega \in \Omega$ by

$$\tau_t(\omega) = \begin{cases} \inf\{s : A_s(\omega) > t\} & \text{for } t < A_{\zeta(\omega)-}(\omega), \\ \infty & \text{for } t \geq A_{\zeta(\omega)-}(\omega). \end{cases} \tag{A.3.17}$$

We then let

$$\check{X}_t(\omega) = X_{\tau_t(\omega)}(\omega), \quad \check{\zeta}(\omega) = A_{\zeta(\omega)-}(\omega), \quad t \geq 0, \ \omega \in \Omega. \tag{A.3.18}$$

Recall that $X_\infty(\omega)$ is defined to be ∂ so $\check{X}_t(\omega) = \partial$ for $t \geq \check{\zeta}(\omega)$.

The support F of A defined by (A.3.11) is nearly Borel measurable and hence $F \in \mathcal{B}^*(E)$. We denote the set $F \cup \{\partial\}$ by F_∂, which is regarded as a topological subspace of E_∂.

PROPOSITION A.3.8. (i) *For each $s \geq 0$, τ_s is an $\{\mathcal{F}_t\}$-stopping time.*
(ii) $\tau_{s+t}(\omega) = \tau_s(\omega) + \tau_t(\theta_{\tau_s(\omega)}(\omega)), \quad s \geq 0, \ t \geq 0, \ \omega \in \Omega.$
(iii) *Define*

$$\check{\mathcal{F}}_t = \mathcal{F}_{\tau_t}, \qquad t \geq 0. \tag{A.3.19}$$

$\{\check{\mathcal{F}}_t\}$ is then a right continuous filtration and $\check{X}_t \in \check{\mathcal{F}}_t/\mathcal{B}^*(E_\partial)$ for every $t \geq 0$.
(iv) $\mathbf{P}_x\left(\check{X}_t \in F_\partial \ \text{ for every } t \geq 0\right) = 1$ *for every $x \in E$.*
(v) *If η is an $\{\check{\mathcal{F}}_t\}$-stopping time, then τ_η is an $\{\mathcal{F}_t\}$-stopping time.*
(vi) *If $B \subset E$ is nearly Borel measurable with respect to X, then its hitting time $\check{\sigma}_B = \inf\{t > 0 : \check{X}_t \in B\}$ by the process $\{\check{X}_t\}$ is an $\{\check{\mathcal{F}}_t\}$-stopping time.*

Proof. (i) For any $u > 0$, $\{\tau_s < u\} = \cup_{n=1}^{\infty}\{A_{u-\frac{1}{n}} > s\} \in \mathcal{F}_u$.
(ii) $\tau_s + \tau_t \circ \theta_{\tau_s} = \tau_s + \inf\{u : A_u \circ \theta_{\tau_s} > t\}$. If $\tau_s < \infty$, then $A_u \circ \theta_{\tau_s} = A_{u+\tau_s} - A_{\tau_s} = A_{u+\tau_s} - s$ in view of Lemma A.3.7(ii). Hence $\tau_s + \tau_t \circ \theta_{\tau_s} = \inf\{u + \tau_s : u \geq 0,\ A_{u+\tau_s} > s + t\} = \tau_{s+t}$.
(iii) Since $\{\tau_t,\ t \geq 0\}$ is a non-decreasing sequence of $\{\mathcal{F}_t\}$-stopping times, $\{\check{\mathcal{F}}_t\}$ are clearly non-decreasing. If $\Lambda \in \cap_n \check{\mathcal{F}}_{t+\frac{1}{n}}$, by the right continuity of τ_t, we have for any $u > 0$

$$\Lambda \cap \{\tau_t < u\} = \bigcup_n \{\Lambda \cap \{\tau_{t+\frac{1}{n}} < u\} \in \mathcal{F}_u,$$

which means that $\Lambda \in \check{\mathcal{F}}_t$, yielding the right continuity of $\{\check{\mathcal{F}}_t\}$. The second assertion follows from Exercise A.1.20.
(iv) On account of Lemma A.3.7(ii), we see for every $\omega \in \Omega$ that the time set $\{t < \infty : \tau_u(\omega) = t \text{ for some } u\}$ coincides with the time set $I(\omega)$ of right increase of $A_t(\omega)$. Therefore, the assertion follows from (A.3.13).
(v) Since $\{\eta < t\} \in \check{\mathcal{F}}_t$ for any $t > 0$, we have $\{\eta < t\} \cap \{\tau_t < s\} \in \mathcal{F}_s$ for any $s > 0$ and

$$\{\omega : \tau_{\eta(\omega)}(\omega) < s\} = \bigcup_{r \in \mathbb{Q}_+} \{\omega : \eta(\omega) < r\} \cap \{\omega : \tau_r(\omega) < s\} \in \mathcal{F}_s.$$

(vi) Since $\{\check{X}_t\}$ is a right continuous $\{\check{\mathcal{F}}_t\}$-adapted process and $\check{\mathcal{F}}_t$ is equal to $\bigcap_{\mu \in \mathcal{P}(E_\partial)} (\check{\mathcal{F}}_t)^{\mathbf{P}_\mu}$ according to Exercise A.1.20(ii), the proof of Theorem A.1.19 shows that $\check{\sigma}_B$ is an $\{\check{\mathcal{F}}_t\}$-stopping time provided that $B \in \mathcal{B}(E)$. For $B \in \mathcal{B}^n(E)$ and $\mu \in \mathcal{P}(E)$, the sets $B_1, B_2 \in \mathcal{B}(E)$ in Definition A.1.28 satisfy $\mathbf{P}_\mu(\check{X}_t \in B_2 \setminus B_1 \text{ for some } t \geq 0) = 0$. This means that for each $t \geq 0$, $\Lambda_t = \{\check{\sigma}_B \leq t\}$ differs from $\{\check{\sigma}_{B_1} \leq t\} \in \check{\mathcal{F}}_t$ only by $\mathbf{P}_\mu$-null set and $\Lambda_t \in (\check{\mathcal{F}}_t)^{\mathbf{P}_\mu}$. Hence $\Lambda_t \in \check{\mathcal{F}}_t$. □

We finally let

$$\Omega_0 = \{\omega \in \Omega : \check{X}_t(\omega) \in F_\partial \text{ for every } t \geq 0\}. \tag{A.3.20}$$

Since the set $E \setminus F$ is nearly Borel measurable with respect to X and $\Omega_0 = \{\omega \in \Omega : \check{\sigma}_{E\setminus F}(\omega) = \infty\}$, we see that $\Omega_0 \in \mathcal{F}_\infty$ by Proposition A.3.8(vi) and $\mathbf{P}_x(\Omega_0) = 1$ for any $x \in E$ by Proposition A.3.8(iv).

The restrictions to Ω_0 of those functions on Ω will be denoted by the same notations. The traces on Ω_0 of the σ-fields $\mathcal{M}$, $\mathcal{F}_t$, $t \in [0, \infty]$, will also be denoted by the same notations. We then redefine $\check{\mathcal{F}}_t$, $t \geq 0$, by (A.3.19).

THEOREM A.3.9. $\check{X} = \left(\Omega_0, \mathcal{M}, \check{X}_t, \theta_{\tau_t}, \zeta, \{\mathbf{P}_x\}_{x \in F_\partial}\right)$ *is a strong Markov process on* $(F_\partial, \mathcal{B}^*(F_\partial))$ *satisfying condition* **(X.6)$_\mathbf{r}$**.

Proof. Since $\tau_0 = R$, it holds for $x \in F_\partial$ that $\mathbf{P}_x(\check{X}_0 = x) = \mathbf{P}_x(X_0 = x) = 1$. For $B \in \mathcal{B}^*(F_\partial)$, $\mathbf{P}_x(\check{X}_t \in B)$ is $\mathcal{B}^*(F_\partial)$-measurable in $x \in F_\partial$ by virtue of Proposition A.3.8(iii) and Exercise A.1.20(i). Thus $\check{X}$ satisfies conditions **(X.1)**, **(X.2)**, **(X.4)**, and **(X.5)** in Definition A.1.8.

Obviously $\check{X}_t, \check{\zeta}, \theta_{\tau_t}$ satisfy the condition **(X.6)r** of Section A.1.3. $\{\check{\mathfrak{F}}_t\}$ is a right continuous admissible filtration for $\check{X}_t$ by Proposition A.3.8(iii). Let η be an $\{\check{\mathfrak{F}}_t\}$-stopping time. By Proposition A.3.8(ii), we have $\tau_{\eta+t} = \tau_\eta + \tau_t \circ \theta_{\tau_\eta}$ and $\check{X}_{\eta+t} = X_{\tau_t} \circ \theta_{\tau_\eta}$. On account of Proposition A.3.8(v) and (A.1.23), it then holds for any $f \in b\mathcal{B}^*(F_\partial)$ and $\mu \in \mathcal{P}(F_\partial)$ that

$$\mathbf{E}_\mu\left[f(\check{X}_{\eta+t})|\check{\mathfrak{F}}_\eta\right] = \mathbf{E}_\mu\left[f(X_{\tau_t} \circ \theta_{\tau_\eta})|\mathfrak{F}_{\tau_\eta}\right]$$

$$= \mathbf{E}_{X_{\tau_\eta}}\left[f(X_{\tau_t})\right] = \mathbf{E}_{\check{X}_\eta}[f(\check{X}_t)], \quad \mathbf{P}_\mu\text{-a.s.},$$

proving the strong Markov property of $\check{X}$. □

$\check{X}$ is called the *time-changed process* of the Borel right process X by its PCAF A. The transition function $\{\check{P}_t;\ t \geq 0\}$ and the resolvent kernel $\{\check{R}_p;\ p \geq 0\}$ of $\check{X}$ are given for $f \in b\mathcal{B}^*(F)$, $x \in F$, by

$$\check{P}_t f(x) = \mathbf{E}_x\left[f(\check{X}_t)\right], \quad \check{R}_p f(x) = \mathbf{E}_x\left[\int_0^\infty e^{-pt} f(\check{X}_t)dt\right]. \tag{A.3.21}$$

Here any numerical function f on F is extended to F_∂ by setting $f(\partial) = 0$ as usual. They are a transition function and a resolvent kernel on $(F, \mathcal{B}^*(F))$ in the sense of Definition 1.1.13, respectively.

For $\alpha \geq 0$, $p \geq 0$, we define kernels $U^\alpha_{p,A}$, R^A_α on $(E, \mathcal{B}^*(E))$ by

$$U^\alpha_{p,A} f(x) = \mathbf{E}_x\left[\int_0^\infty e^{-\alpha t - pA_t} f(X_t)dA_t\right] \tag{A.3.22}$$

and

$$R^A_\alpha f(x) = \mathbf{E}_x\left[\int_0^\infty e^{-\alpha t - A_t} f(X_t)dt\right] \tag{A.3.23}$$

for $x \in E$ and $f \in \mathcal{B}^*_+(E)$.

LEMMA A.3.10. *The following identities hold for $p > 0$ and $\alpha > 0$.*

$$\check{R}_p f(x) = U^0_{p,A} f(x), \quad x \in F,\ f \in b\mathcal{B}^*(F). \tag{A.3.24}$$

For $x \in E$ and $f \in b\mathcal{B}^*(E)$,

$$U^0_{p,A} f(x) - U^\alpha_{p,A} f(x) - \alpha R^{pA}_\alpha (U^0_{p,A} f)(x) = 0, \tag{A.3.25}$$

$$U^\alpha_A f(x) - U^\alpha_{p,A} f(x) - p U^\alpha_A U^\alpha_{p,A} f(x) = 0, \tag{A.3.26}$$

$$R^A_\alpha f(x) - R_\alpha f(x) + U^\alpha_A R^A_\alpha f(x) = 0, \tag{A.3.27}$$

where $U^\alpha_A := U^\alpha_{0,A}$.

Proof. Identity (A.3.24) follows from Lemma A.3.7. Equation (A.3.25) follows from $1 - e^{-\alpha t} = \alpha \int_0^t e^{-\alpha s} ds$. Equation (A.3.26) can be derived from the expression $U^{\alpha}_{p,A} f(x) = \mathbf{E}_x \left[\int_0^{\infty} e^{-\alpha\tau_t - pt} f(\check{X}_t) dt\right]$ and $1 - e^{-pt} = pe^{-pt} \int_0^t e^{ps} ds$. Equation (A.3.27) appears in Exercise 4.1.2. □

THEOREM A.3.11. *The time-changed process $\check{X}$ is a right process on $(F, \mathcal{B}^*(F))$.*

Proof. In Theorem A.3.9, we proved that $\check{X} = (\check{X}_t, \{\mathbf{P}_x\}_{x\in F})$ is a strong Markov process on $(F, \mathcal{B}^*(F))$ with the property $(\mathbf{X.6})_\mathbf{r}$. In view of Theorem A.1.37, it suffices to show that the process $\check{R}_p f(\check{X}_t)$ is $\mathbf{P}_\mu$-indistinguishable from an optional process for any $f \in C_{u+}(F)$ and any $\mu \in \mathcal{P}(F) \subset \mathcal{P}(E)$.

For any $\alpha > 0$ and $f \in b\mathcal{B}_+(E)$, let $g = R^A_\alpha f$. By (A.3.27), $U^\alpha_A g$ is then finite on E. Combining (A.3.25), (A.3.26), and (A.3.27), we see that $U^0_{p,A} g$ can be expressed as $U^\alpha_A g + R_\alpha g_1 - U^\alpha_A g_2 - U^\alpha_{pA} g_3$ for $g_1, g_2, g_3 \in b\mathcal{B}_+(E)$, a linear combination of four finite α-excessive functions, and consequently $U^0_{p,A} g(X_t)$ is right continuous in $t \geq 0$ $\mathbf{P}_\mu$-a.s. by virtue of Theorem A.2.2. Therefore, so is the process $\check{R}_p g(\check{X}_t) = \check{R}_p g(X_{\tau_t})$ on account of (A.3.24).

Now take any $f \in C_{u+}(E)$. We have seen that the process $\check{R}_p g_\alpha(\check{X}_t)$ is $\mathbf{P}_\mu$-indistinguishable from an optional process for $g_\alpha = \alpha R^A_\alpha f$. If we let $\alpha \to \infty$, the process converges to $\check{R}_p f(\check{X}_t)$, which must enjoy the desired property accordingly. □

Finally, we take from [13, III.3] a definition of the canonical subprocess of a Borel right process $X = (\Omega, \mathcal{M}, \mathcal{M}_t, \zeta, X_t, \theta_t, \mathbf{P}_x)$ on a Lusin space $(E, \mathcal{B}(E))$ with respect to the multiplicative functional e^{-A_t} for a positive continuous additive functional A of X in the strict sense. Without loss of generality, we assume that the defining set of A is Ω itself. We also make a convention that $A_\infty(\omega) = 0$ for every $\omega \in \Omega$.

Let $\widehat{\Omega} = \Omega \times [0, \infty]$ and write $\widehat{\omega} = (\omega, \lambda)$ for the generic point in $\widehat{\Omega}$. Let $\mathcal{R}$ be the Borel σ-field of $[0, \infty]$ and set $\widehat{\mathcal{M}} = \mathcal{M} \times \mathcal{R}$. Define

$$\widehat{X}_t(\widehat{\omega}) = \begin{cases} X_t(\omega) & \text{if } t < \lambda \\ \partial & \text{if } t \geq \lambda \end{cases} \quad \text{and} \quad \widehat{\zeta}(\widehat{\omega}) = \zeta(\omega) \wedge \lambda. \tag{A.3.28}$$

Note that $\widehat{\zeta} = \inf\{t > 0 : \widehat{X}_t = \partial\}$. Further, define $\widehat{\theta}_t \widehat{\omega} = (\theta_t \omega, (\lambda - t) \vee 0)$ so that $\widehat{X}_t \circ \widehat{\theta}_h = \widehat{X}_{t+h}$ for $t, \ h \geq 0$. Let $\widehat{\Omega}_t = \Omega \times (t, \infty]$ and $\widehat{\mathcal{M}}_t$ be the space that consists of all sets $\widehat{\Lambda} \in \widehat{\mathcal{M}}$ for which there exists a $\Lambda \in \mathcal{M}_t$ such that $\widehat{\Lambda} \cap \widehat{\Omega}_t = \Lambda \times (t, \infty]$.

For each $\omega \in \Omega$, there is a unique probability measure α_ω on $([0, \infty], \mathcal{R})$ such that $\alpha_\omega([t, \infty]) = e^{-A_t(\omega)}$. For $\widehat{\Lambda} \in \widehat{\mathcal{M}}$, denote by $\widehat{\Lambda}^\omega = \{\lambda : (\omega, \lambda) \in \widehat{\Lambda}\}$

its ω-section and define

$$\widehat{\mathbf{P}}_x(\widehat{\Lambda}) = \mathbf{E}_x\left[\alpha_\omega(\widehat{\Lambda}^\omega)\right]. \quad x \in E. \tag{A.3.29}$$

In particular, for $B \in \mathcal{B}^*(E)$, $\{\widehat{\omega} = (\omega, \lambda) : \widehat{X}_t(\widehat{\omega}) \in B\} = \{\omega : X_t(\omega) \in B\} \times (t, \infty]$ and we get from (A.3.29)

$$\widehat{P}_t(x, B) = \widehat{\mathbf{P}}_x(\widehat{X}_t \in B) = \mathbf{E}_x\left[e^{-A_t}; X_t \in B\right].$$

We refer the reader to [13, III. (3.3)] for a proof of the next lemma.

LEMMA A.3.12. $\widehat{X} = (\widehat{\Omega}, \widehat{\mathcal{M}}, \widehat{\mathcal{M}}_t, \widehat{X}_t, \widehat{\zeta}, \widehat{\theta}_t, \widehat{\mathbf{P}}_x)$ *is a Markov process on* $(E, \mathcal{B}^*(E))$ *with transition function determined by*

$$\widehat{P}_t f(x) = \mathbf{E}_x\left[f(X_t)\, e^{-A_t}\right], \quad t \geq 0,\ x \in E,\ f \in b\mathcal{B}(E). \tag{A.3.30}$$

THEOREM A.3.13. (i) $\widehat{X} = (\widehat{\Omega}, \widehat{\mathcal{M}}, \widehat{\mathcal{M}}_t, \widehat{X}_t, \widehat{\zeta}, \widehat{\theta}_t, \widehat{\mathbf{P}}_x)$ *is a right process on* $(E, \mathcal{B}^*(E))$.
(ii) *If X is a Hunt process on $(E, \mathcal{B}(E))$, then $\widehat{X}$ is a Hunt process on E with the $\mathcal{B}(E)$-measurability of the transition function being weakened to $\mathcal{B}^*(E)$-measurability.*

Proof. (i) By Lemma A.3.12, $\widehat{X}$ is a Markov process on $(E, \mathcal{B}^*(E))$ whose resolvent kernel $\widehat{R}_\alpha f(x) = \int_0^\infty e^{-\alpha t} \widehat{P}_t f(x) dt$, $x \in E$, coincides with the function $R^A_\alpha f(x)$, $x \in E$, defined by (A.3.23). Obviously $\widehat{X}$ satisfies the property $(\mathbf{X.6})_\mathbf{r}$.

On account of Theorem A.1.37, it only remains to show that $u(\widehat{X}_t)$ is right continuous in $t \geq 0$ $\widehat{\mathbf{P}}_\mu$-a.s. for $u = R^A_\alpha f$, $f \in b\mathcal{B}_+(E)$. But in view of (A.3.27), u is then a difference of two bounded excessive functions with respect to X and hence the process $u(X_t(\omega))$ is right continuous in $t \geq 0$ $\mathbf{P}_\mu$-a.s. by virtue of Theorem A.2.2. Choose $\Omega_1 \in \mathcal{M}$ such that $\mathbf{P}_\mu(\Omega_1) = 1$ and this statement is true for all $\omega \in \Omega_1$.

Since $u(\widehat{X}_t(\widehat{\omega}))$ is equal to $u(X_t(\omega))$ when $t < \widehat{\zeta}(\widehat{\omega})$ and to 0 otherwise, it is right continuous in $t \geq 0$ for any $\widehat{\omega} \in \Omega_1 \times [0, \infty]$ which is of $\widehat{\mathbf{P}}_\mu$-measure 1.

(ii) It can be shown that the quasi-left-continuity of X is inherited by $\widehat{X}$ (cf. [13, III, (3.13)]). □

$\widehat{X}$ is called the *canonical subprocess* of X with respect to the multiplicative functional e^{-A_t}. It is also called the process obtained from X by *killing* with respect to the positive continuous additive functional A.

A.3.3. Review of Martingale Additive Functionals

In this subsection, we list some basic facts on martingale additive functionals in the strict sense and related additive functionals in the strict sense that are utilized in the main text. They can be deduced from the corresponding facts in the general theory of square integrable martingales and this deduction is summarized in [73, A.3]. An excellent reference that contains more details is [34].

Let $X = (\Omega, \mathcal{M}, \{X_t\}, \theta_t, \zeta, \{\mathbf{P}_x\}_{x\in E_\partial})$ be a Hunt process on a Lusin space $(E, \mathcal{B}(E))$ and $\{\mathcal{F}_t\}$ be its minimum augmented admissible filtration. $\{\mathcal{F}_t\}$ is quasi-left-continuous by Theorem A.1.40. A statement concerning $\omega \in \Omega$ is said to hold a.s. if it is true $\mathbf{P}_x$-a.s. for every $x \in E$. By a *stochastic process*, we mean in this subsection a real-valued $\{\mathcal{F}_t\}$-adapted process whose sample paths are right continuous on $[0, \infty)$ and have left limits on $(0, \infty)$ a.s. Two stochastic processes are called *equivalent* if they are $\mathbf{P}_x$-indistinguishable for every $x \in E$. In this case, we regard them as identical.

We call a stochastic process $\{A_t\}$ an *additive functional* (*AF* in abbreviation) if almost surely $A_{s+t} = A_s + A_t \circ \theta_s$ for every $s, t \geq 0$, and $A_t = A_\zeta$ for every $t \geq \zeta$. Thus an additive functional in the present sense is a finite càdlàg additive functional in the strict sense in the terminology of Section A.3.1.

Let us introduce several classes of additive functionals:

$$\mathcal{V}^+ := \{\text{AF } Z : \ Z_t \geq 0 \text{ for every } t \geq 0 \text{ a.s.}\},$$

$$\mathcal{V} := \mathcal{V}^+ - \mathcal{V}^+ = \{A - B : A, B \in \mathcal{V}^+\},$$

$$\mathcal{A} = \left\{A \in \mathcal{V} : \mathbf{E}_x\Big[\int_0^t |dA|_s\Big] < \infty \text{ for every } t > 0 \text{ and } x \in E\right\},$$

$$\mathcal{PA} = \{A \in \mathcal{A} : A \text{ is predictable}\},$$

$$\mathcal{CA} = \{A \in \mathcal{A} : A \text{ is continuous}\},$$

$$\mathbb{M} = \left\{\text{AF } M : \ \mathbf{E}_x[M_t^2] < \infty, \ \mathbf{E}_x[M_t] = 0 \text{ for all } t \geq 0 \text{ and } \ x \in E\right\},$$

$$\mathbb{M}^c = \{M \in \mathbb{M} : M \text{ is continuous}\}.$$

See Section A.2.1 for the definition of a stochastic process being predictable. Obviously $\mathcal{CA} \subset \mathcal{PA} \subset \mathcal{A}$. $M \in \mathbb{M}$ is called a *martingale additive functional* (*MAF* in abbreviation) because it is easily seen to be a martingale with respect to $(\{\mathcal{F}_t\}, \mathbf{P}_x)$ for every $x \in E$. The space $\mathbb{M}$ is equipped with the seminorm $\eta_{t,x} = \mathbf{E}_x[M_t^2]$, $t > 0$, $x \in E$.

(1) For any $M \in \mathbb{M}$, there exists a unique $\langle M \rangle \in \mathcal{CA}^+$ such that

$$\mathbf{E}_x[\langle M \rangle_t] = \mathbf{E}_x[M_t^2], \quad \forall t > 0, \ \forall x \in E,$$

or, equivalently, $M_t^2 - \langle M \rangle_t$ is a $\mathbf{P}_x$-martingale for every $x \in E$. $\langle M \rangle$ is called the *predictable quadratic variation* of M. For $M, N \in \mathbb{M}$, we define their covariation by

$$\langle M, N \rangle := \frac{1}{2}(\langle M + N \rangle - \langle M \rangle - \langle N \rangle) \in \mathcal{CA}.$$

(2) For any $A \in \mathcal{A}$, there exists a unique $A^p \in \mathcal{PA}$ such that $A - A^p$ is a $\mathbf{P}_x$-martingale for every $x \in E$. A^p is called the *dual predictable projection* of A.
(3) We let

$$\mathbb{M}^d = \{M \in \mathbb{M} : \langle M, N \rangle = 0 \text{ for every } N \in \mathbb{M}^c\}.$$

Any $M \in \mathbb{M}$ admits a unique decomposition

$$M = M^c + M^d, \quad M^c \in \mathbb{M}^c, \quad M^d \in \mathbb{M}^d.$$

M^c is called the *continuous part* of M, while M^d is called the *purely discontinuous part* of M. It is known (see [85]) that M^d admits the expression

$$M_t^d = \lim_{\varepsilon \to 0} \left(B_t^\varepsilon - (B^\varepsilon)_t^p\right), \qquad t \geq 0,$$

where $B_t^\varepsilon := \sum_{0<s\leq t}(M_s - M_{s-})\mathbf{1}_{\{|M_s - M_{s-}| > \varepsilon\}}$, and the convergence is in probability as well as in L^2 with respect to $\mathbf{P}_x$ for every $x \in E$.

By a stopping time, we mean an $\{\mathcal{F}_t\}$-stopping time. A stopping time T is said to be *predictable* if there exists an increasing sequence of stopping times $\{T_n,\ n \geq 1\}$ such that $T_n < T$ and $\lim_{n\to\infty} T_n = T$ a.s. on $\{T > 0\}$. A stopping time T is called *totally inaccessible* if, for any predictable stopping time S, $\mathbf{P}_x(T = S < \infty) = 0$ for every $x \in E$.

If T is a stopping time such that $X_{T-} \neq X_T$ a.s. on $\{T < \infty\}$, then T is totally inaccessible owing to the quasi-left-continuity of X. Moreover, we can find a sequence of totally inaccessible stopping times $\{T_n\}$ such that

$$\{(t,\omega) \in (0,\infty) \times \Omega : \Delta X_t \neq 0\} = \bigcup_{n=1}^{\infty} \{(T_n(\omega), \omega) : \omega \in \Omega\},$$

the right hand side being a disjoint union.

For $A \in \mathcal{A}$, $A^p \in \mathcal{CA}$ if and only if $\mathbf{E}_x[A_T] = \mathbf{E}_x[A_{T-}]$ for every $x \in E$ and every predictable stopping time T.

In particular, if $A \in \mathcal{A}$ jumps only when X jumps, namely, the set $\{(t,\omega) : \Delta A_t(\omega) \neq 0\}$ is contained in the left hand side of the above identity, then $A^p \in \mathcal{CA}$.

(4) For $M \in \mathbb{M}$, we define $[M] \in \mathcal{A}^+$ by

$$[M]_t = \langle M^c \rangle_t + \sum_{s \le t} (\Delta M_s)^2, \quad t \ge 0,$$

which is called the *quadratic variation* of M. Here $\Delta M_s := M_s - M_{s-}$. For $M, N \in \mathbb{M}$, we let $[M, N] = \frac{1}{2}([M+N] - [M] - [N])$. It then holds that

$$[M]^p = \langle M \rangle, \quad M \in \mathbb{M}.$$

Furthermore, for $M, N \in \mathbb{M}$, the approximation

$$[M, N]_t = \lim_{n \to \infty} \sum_{j=1}^{2^n} \left(M_{t_j^n} - M_{t_{j-1}^n} \right) \left(N_{t_j^n} - N_{t_{j-1}^n} \right), \quad t_j^n = jt/2^n,$$

takes place in $L^1(\Omega, \mathbf{P}_x)$ for every $x \in E$. Moreover, the above convergence is uniform in probability with respect to $\mathbf{P}_x$ on every compact time interval for every $x \in E$ (cf. [85]).
(5) Let $M \in \mathbb{M}$. For any stopping time T, define

$$M_t^T = \Delta M_T \cdot \mathbf{1}_{\{T \le t\}} - (\Delta M_T \cdot \mathbf{1}_{\{T \le t\}})^p, \quad t \ge 0.$$

Then $M^T \in \mathbb{M}^d$.

Let $\{T_n\}$ be the sequence of totally inaccessible stopping times appearing in **(3)**. Then $\langle M^{T_n}, M^{T_\ell} \rangle = 0$, $n \neq \ell$, and $\sum_{n=1}^{\infty} M^{T_n}$ converges to the purely discontinuous part M^d of M in the topology of $\mathbb{M}$.

A.3.4. Lévy system of a Hunt Process

We keep the same setting as in the preceding subsection.

Let $X = (\Omega, \{\mathcal{F}_t\}, \{X_t\}, \{\theta_t\}, \zeta, \{\mathbf{P}_x\}_{x \in E_\partial})$ be a Hunt process on a Lusin space $(E, \mathcal{B}(E))$, where $\{\mathcal{F}_t\}_{t \ge 0}$ is the minimum augmented admissible filtration. In this subsection, we introduce the notion of a Lévy system for X that describes the jump behaviors of the sample path X_t completely.

The pair (N, H) of a kernel $N(x, dy)$ on $(E_\partial, \mathcal{B}(E_\partial))$ and a positive continuous additive functional H in the strict sense of X in the terminology of Section A.3.1 is called a *Lévy system* of X if the following equation holds for any $f \in \mathcal{B}_+(E_\partial \times E_\partial)$ with $f(x, x) = 0$, $x \in E_\partial$, and for every $t > 0$ and $x \in E$:

$$\mathbf{E}_x \left[\sum_{0 < s \le t} f(X_{s-}, X_s) \right] = \mathbf{E}_x \left[\int_0^t \left(\int_{E_\partial} f(X_s, y) N(X_s, dy) \right) dH_s \right]. \quad \text{(A.3.31)}$$

When H is a specific PCAF arising in relation to a boundary problem of a Markov process, M. Motoo [123] had called such a pair (N, H) a Lévy system, but it was S. Watanabe [149] who first formulated a Lévy system in the above fashion and proved its existence for a general Hunt process X. A kind of the absolute continuity condition called (L) for X was imposed in [149], and was later removed by A. Benveniste and J. Jacod [9] by using a method of Ray compactification.

There are many choices of a Lévy system (N, H). For example, if (N, H) is a Lévy system of X, then so is $(\widehat{N}, \widetilde{H})$ for any strictly positive bounded Borel function f on E, where

$$\widetilde{N}(x, dy) := f(x)^{-1} N(x, dy) \quad \text{and} \quad \widetilde{H}_t := \int_0^t f(X_s) dH_s.$$

Nevertheless, according to (A.3.31), the measure $N(x, dy)\mu_H(dx)$ on $E \times E_\partial$ is uniquely determined by X and its excessive measure m, where μ_H is the Revuz measure of the PCAF of H relative to m. In this book, we take and fix one Lévy system (N, H) of X and we take its existence for granted.

The defining formula (A.3.31) of a Lévy system is known to be equivalent to the following more general one (cf. [137, p. 346]): for any non-negative predictable process $\{Y_s\}$, any $f \in \mathcal{B}_+(E_\partial \times E_\partial)$ with $f(x, x) = 0, \ x \in E_\partial$, and for every $x \in E_\partial$,

$$\mathbf{E}_x\left[\sum_{s>0} Y_s f(X_{s-}, X_s)\right] = \mathbf{E}_x\left[\int_0^\infty Y_s \left(\int_{E_\partial} f(X_s, y) N(X_s, dy)\right) dH_s\right]. \tag{A.3.32}$$

In particular, for any stopping time T and for any non-negative Borel function $g(s)$ on $(0, \infty)$, we can substitute $Y_s = \mathbf{1}_{(0,T]}(s) \cdot g(s), \ s \geq 0$, in the above formula to obtain

$$\begin{aligned} &\mathbf{E}_x\left[\sum_{0<s\leq T} g(s) f(X_{s-}, X_s)\right] \\ &= \mathbf{E}_x\left[\int_0^T g(s) \left(\int_{E_\partial} f(X_s, y) N(X_s, dy)\right) dH_s\right]. \end{aligned} \tag{A.3.33}$$

This identity is frequently utilized in the main text of the present book.

In fact, the formula (A.3.32) is proved in [137] for a general right process X but by interpreting X_{t-} as the left limit of X on E relative to a Ray topology induced on E. When X is a Hunt process, however, this limit coincides with the left limit of X_t relative to the original topology (cf. [137, (47.10)]). More generally, (A.3.32) is known to be true relative to the original topology (with a certain interpretation of $X_{\zeta-}$) for any special standard process X (see [27, §2.4]). If X is a Hunt process, its transition function sends Borel measurable functions into Borel measurable functions, and consequently the kernel N appearing in the formula (A.3.32) can be taken to be a kernel on $(E_\partial, \mathcal{B}(E_\partial))$ (cf. [137, (17.12) and (73.5)]).

We mention an important viewpoint on the identity (A.3.31). If the left hand side of (A.3.31) is finite for every $t > 0$ and $x \in E$, then $A_t^f = \sum_{s \le t} f(X_{s-}, X_s)$, $t \ge 0$, can be viewed as an element of the class $\mathcal{A}^+$ of additive functionals for X introduced in the preceding subsection. Moreover, equation (A.3.31) implies that the positive continuous additive functional $(A^f)^p$ in the strict sense defined by

$$(A^f)_t^p = \int_0^t \left(\int_{E_\partial} f(X_s, y) N(X_s, dy) \right) dH_s, \ t > 0. \tag{A.3.34}$$

is the dual predictable projection of A^f in the sense of **(3)** of Section A.3.3.

A.3.5. Itô's Formula

In this section, we present Itô's formula for semimartingales. We will restrict our presentation to a subclass of semimartingales that is used in this book. A good reference on this subject is [85].

Let $(\Omega, \mathcal{F}, \mathbf{P})$ be a probability space with filtration $\{\mathcal{F}_t\}_{t \ge 0}$ satisfying the usual condition. In the following, all the processes are assumed to be adapted to the filtration $\{\mathcal{F}_t\}_{t \ge 0}$. Martingale property and dual predictable projections are related to this filtration as well.

Let M be a square integrable martingale. Its predictable quadratic variation process $\langle M \rangle$ is the unique predictable process A with $A_0 = 0$ so that $M^2 - A$ is a martingale. M can be uniquely decomposed as $M = M^c + M^d$, where M^c is a continuous square integrable martingale and

$$M_t^d = \lim_{\varepsilon \to 0} \left(B_t^\varepsilon - (B^\varepsilon)_t^p \right), \qquad t \ge 0,$$

where $B_t^\varepsilon := \sum_{0 < s \le t} (M_s - M_{s-}) \mathbf{1}_{\{|M_s - M_{s-}| > \varepsilon\}}$, and the convergence is in probability as well as in L^2 with respect to $\mathbf{P}$. M^c and M^d are called the continuous part and purely discontinuous part of M, respectively. For two semimartingales, $X = M + A$ and $Y = N + C$, where M and N are square integrable martingales and A and C are processes of finite variations with $A_0 = C_0 = 0$, we define

$$[X, Y]_t = \langle M^c, N^c \rangle_t + \sum_{0 < s \le t} (X_s - X_{s-})(Y_s - Y_{s-}), \quad t \ge 0, \tag{A.3.35}$$

with $\langle M^c, N^c \rangle := \frac{1}{4} \left(\langle M^c + N^c \rangle - \langle M^c - N^c \rangle \right)$. For simplicity, we will define $X^c := M^c$ and $Y^c := N^c$, as the semimartingale decomposition of X and Y into such $M + A$ and $N + C$ is unique. For a square integrable martingale M, both $[M]$ and $\langle M \rangle$ are non-decreasing processes and $\langle M \rangle$ is the dual predictable projection of $[M]$ ([85, Theorem 6.26]). Thus we have $\mathbf{E}[M_t^2] = \mathbf{E}[M]_t = \mathbf{E}[\langle M \rangle_t]$ for every $t \ge 0$. If the square integrable martingale M has $M_t = M_T$ for every $t \ge T$, where T is an $\{\mathcal{F}_t\}_{t \ge 0}$-stopping time, then $\langle M \rangle_t = \langle M \rangle_T$ and $[M]_t = [M]_T$ for every $t \ge T$ (see [85, Theorem 6.31]).

The following Doob's maximal inequality holds for every square integrable martingale M:

$$\mathbf{E}\Big[\Big(\sup_{s\leq t}|M_s|\Big)^2\Big] \leq 4\,\mathbf{E}[M_t^2], \quad t\geq 0. \tag{A.3.36}$$

Moreover, for every stopping time T and $t>0$,

$$[M]_{t\wedge T} = \lim_{n\to\infty}\sum_{k=1}^{n}\Big(M_{\frac{kt}{n}\wedge T} - M_{\frac{(k-1)t}{n}\wedge T}\Big)^2, \tag{A.3.37}$$

where the convergence is in probability with respect to $\mathbf{P}$ (see the Remark after Theorem 9.33 in [85]). For a square integrable martingale M and a stopping time T,

$$K_t := \Delta M_T \cdot \mathbf{1}_{\{t\geq T\}} - (\Delta M_T \cdot \mathbf{1}_{\{t\geq T\}})^p$$

is a purely discontinuous square integrable martingale. So by (A.3.35), we have in particular

$$[M-K,\, K]_t = 0 \quad \text{for } t\geq 0. \tag{A.3.38}$$

A stochastic integral $K_t := \int_0^t H_s dM_s$ can be defined for a predictable integrand H and for a square integrable martingale M. If $\mathbf{E}[\int_0^t H_s^2 d\langle M\rangle_s] < \infty$, then $K = \int_0^\cdot H_s dM_s$ is also a square integrable martingale. For a predictable process H, a semimartingale $X = M + A$, where M is a square integrable martingale and A is a process of finite variation, the stochastic integral $\int_0^t H_s dX_s$ is defined to be $\int_0^t H_s dM_s + \int_0^t H_s dA_s$. Here $\int_0^t H_s dA_s$ is the Lebesgue-Stieltjes integral.

For $f \in C^2(\mathbb{R}^2)$ and two semimartingales $X^{(i)} = M + A$ and $Y^{(2)} = N + C$ as above, the following Itô's formula holds (see [85, Theorem 9.35]): $\mathbf{P}$-a.s. for $t\geq 0$,

$$\begin{aligned} f(X_t^{(1)}, X_t^{(2)}) = {} & f(X_0^{(1)}, X_0^{(2)}) + \sum_{k=1}^{2}\int_0^t \frac{\partial}{\partial x_k} f(X_{s-}^{(1)}, X_{s-}^{(2)}) dX_s^{(k)} \\ & + \frac{1}{2}\sum_{k,j=1}^{2}\int_0^t \frac{\partial^2}{\partial x_k \partial x_j} f(X_{s-}^{(1)}, X_{s-}^{(2)}) d\langle X^{(k),c}, X^{(j),c}\rangle_s \\ & + \sum_{0<s\leq t}\Big(f(X_s^{(1)}, X_s^{(2)}) - f(X_{s-}^{(1)}, X_{s-}^{(2)}) \\ & - \sum_{k=1}^{2} \frac{\partial}{\partial x_k} f(X_{s-}^{(1)}, X_{s-}^{(2)})(X_s^{(k)} - X_{s-}^{(k)})\Big). \end{aligned} \tag{A.3.39}$$

In particular, we have the following product rule:

$$X_tY_t = X_0Y_0 + \int_0^t Y_{s-}dX_s + \int_0^t X_{s-}dY_s + [X, Y]_t, \quad t \geq 0. \tag{A.3.40}$$

A.4. REVIEW OF SYMMETRIC FORMS

Let H be a real Hilbert space with inner product $(\cdot, \cdot)$. The domain of a linear operator S on H is denoted by $\mathcal{D}(S)$. A linear operator S on H is said to be *symmetric* if $\mathcal{D}(S) = H$, $(Sf, g) = (f, Sg)$, $f, g \in H$. A symmetric form $(\mathcal{E}, \mathcal{D}(\mathcal{E}))$ on H and its closedness are defined at the beginning of Section 1.1. The notions of a strongly continuous contraction semigroup $\{T_t; t > 0\}$ of symmetric linear operators on H and a strongly continuous contraction resolvent $\{G_\alpha; \alpha > 0\}$ on H are also introduced there. The latter is obtained from the former by taking the Laplace transform (1.1.1), which is called the *resolvent of* a semigroup $\{T_t : t > 0\}$.

A self-adjoint linear operator $\mathcal{A}$ on H is called *non-negative definite* if $(\mathcal{A}f, f) \geq 0$ for every $f \in \mathcal{D}(\mathcal{A})$. The generator of a strongly continuous contraction resolvent $\{G_\alpha; \alpha > 0\}$ is defined by

$$\mathcal{A}f = \alpha f - G_\alpha^{-1}f, \quad \mathcal{D}(\mathcal{A}) = G_\alpha(H),$$

and $-\mathcal{A}$ is easily seen to be a non-negative definite self-adjoint operator independent of $\alpha > 0$. The generator $\mathcal{A}$ of a strongly continuous contraction semigroup $\{T_t; t > 0\}$ is defined by

$$\mathcal{A}f = \lim_{t \downarrow 0} \frac{T_t f - f}{t}, \quad \mathcal{D}(\mathcal{A}) = \{f \in H : \mathcal{A}f \text{ exists as a strong limit}\},$$

which can be verified to be identical with the generator of the resolvent of $\{T_t; t > 0\}$.

A symmetric linear operator S on H satisfying $S^2 = S$ is called a *projection operator*. A family $\{E_\lambda;\ -\infty < \lambda < \infty\}$ of projection operators on H is said to be *spectral family* if it satisfies the following conditions: $E_\lambda E_\mu = E_\lambda$, $\lambda \leq \mu$, $\lim_{\lambda' \downarrow \lambda} E_{\lambda'} f = E_\lambda f$, $\lim_{\lambda \to -\infty} E_\lambda f = 0$, $\lim_{\lambda \to \infty} E_\lambda f = f$, $f \in H$. It holds then that $0 \leq (E_\lambda f, f) \uparrow (f, f)$, $\lambda \uparrow \infty$, $f \in H$. Further, $(E_\lambda f, g)$ is of bounded variation in λ for $f, g \in H$.

Given a spectral family $\{E_\lambda;\ \lambda \in \mathbb{R}\}$ and a continuous real-valued function φ on $\mathbb{R}$, there exists a unique self-adjoint operator $\mathcal{A} = \int_{-\infty}^{\infty} \varphi(\lambda) dE_\lambda$ on H characterized by

$$\begin{cases} (\mathcal{A}f, g) &= \displaystyle\int_{-\infty}^{\infty} \varphi(\lambda) d(E_\lambda f, g), \quad \forall g \in H \\ \mathcal{D}(\mathcal{A}) &= \{f \in H : \displaystyle\int_{-\infty}^{\infty} \varphi(\lambda)^2 d(E_\lambda f, f) < \infty\}. \end{cases} \tag{A.4.1}$$

Any non-negative definite self-adjoint operator $-\mathcal{A}$ on H admits a unique spectral family $\{E_\lambda; \lambda \in \mathbb{R}\}$ with $E_\lambda = 0$ for every $\lambda < 0$, such that $-\mathcal{A} = \int_0^\infty \lambda dE_\lambda$ (cf. [154]). In this case, for a non-negative continuous real function φ on $[0, \infty)$, the self-adjoint operator determined by $\int_0^\infty \varphi(\lambda) dE_\lambda$ will be denoted by $\varphi(-\mathcal{A})$, which is also non-negative definite.

We now state mutual correspondences among the four objects below:

(a) The totality of closed symmetric forms $\mathcal{E}$ on H.

(b) The totality of self-adjoint operators on H with $-\mathcal{A}$ being non-negative definite.

(c) The totality of strongly continuous contraction semigroups $\{T_t; t > 0\}$ of symmetric operators on H.

(d) The totality of strongly continuous contraction resolvents $\{G_\alpha; \alpha > 0\}$ of symmetric operators on H.

A more detailed proof can be found in [73, §1.3].

(i) **(b)**, **(c)** and **(d)** are mutually in one-to-one correspondence.

The correspondences **(b)** $\Rightarrow$ **(c)**, **(b)** $\Rightarrow$ **(d)** are defined by $T_t = \exp(t\mathcal{A})$, $G_\alpha = (\alpha - \mathcal{A})^{-1}$, respectively.

The correspondences **(c)** $\Rightarrow$ **(b)**, **(d)** $\Rightarrow$ **(b)** are defined by the operations of taking generators, respectively.

The correspondences **(c)** $\Rightarrow$ **(d)**, **(d)** $\Rightarrow$ **(c)** are defined by (1.1.1), (1.1.2), respectively.

(ii) **(a)** and **(b)** are in one-to-one correspondence. The correspondence **(b)** $\Rightarrow$ **(a)** is defined by

$$\mathcal{D}(\mathcal{E}) = \mathcal{D}(\sqrt{-\mathcal{A}}), \quad \mathcal{E}(f,g) = (\sqrt{-\mathcal{A}}f, \sqrt{-\mathcal{A}}g), \quad f,g \in \mathcal{D}(\mathcal{E}). \tag{A.4.2}$$

The correspondence **(a)** $\Rightarrow$ **(b)** is characterized by a direct relation

$$\mathcal{D}(\mathcal{A}) \subset \mathcal{D}(\mathcal{E}), \quad \mathcal{E}(f,g) = -(\mathcal{A}f, g), \quad \forall f \in \mathcal{D}(\mathcal{A}), \forall g \in \mathcal{D}(\mathcal{E}). \tag{A.4.3}$$

Expression (A.4.2) can be rewritten using (A.4.1) as

$$\begin{cases} \mathcal{D}(\mathcal{E}) = \left\{ f \in H : \int_0^\infty \lambda d(E_\lambda f, f) < \infty \right\}, \\ EE(f,g) = \int_0^\infty \lambda d(E_\lambda f, g),\ f,g \in \mathcal{D}(\mathcal{E}). \end{cases} \tag{A.4.4}$$

(iii) The one-to-one correspondence of **(a), (d)** is characterized by a direct relation (1.1.6).

(iv) The direct correspondence **(c)** $\Rightarrow$ **(a)** is described by (1.1.4), (1.1.5), where $\mathcal{E}^{(t)}$ is an approximating form defined by (1.1.3) for $\{T_t; t > 0\}$. For $f \in H$, $\mathcal{E}^{(t)}(f,f)$ increases as t decreases.

The direct correspondence (**d**) $\Rightarrow$ (**a**) is described by

$$\mathcal{D}(\mathcal{E}) = \{f \in H : \lim_{\beta\to\infty} \mathcal{E}^{(\beta)}(f,f) < \infty\}, \ \ \mathcal{E}(f,g) = \lim_{\beta\to\infty} \mathcal{E}^{(\beta)}(f,g),$$

where $\mathcal{E}^{(\beta)}(f,g) = \beta(f - \beta G_\beta f, g)$ for $f,g \in H$. For $f \in H$, $\mathcal{E}^{(\beta)}(f,f)$ increases as β increases.
(**v**) Let $\{T_t; t>0\}$ and $\{G_\alpha; \alpha > 0\}$ be related to $\mathcal{E}$ as above. Then $T_t(H) \subset \mathcal{D}(\mathcal{E})$, $\|T_t g\|_{\mathcal{E}}^2 \le \frac{1}{2t}\|g\|^2$ for every $g \in H$, and furthermore $\lim_{t\downarrow 0} \|T_t f - f\|_{\mathcal{E}_1} = 0$, $\lim_{\alpha\to\infty} \|\alpha G_\alpha f - f\|_{\mathcal{E}_1} = 0$ for every $f \in \mathcal{D}(\mathcal{E})$.

Using the spectral family expressing $\mathcal{E}$ as (A.4.4), the first assertion of (**iv**) follows from $\mathcal{E}^{(t)}(f,g) = \frac{1}{t}\int_0^\infty (1-e^{-t\lambda})d(E_\lambda f, g)$, $f,g \in H$. The proof of the second is similar. The first assertion of (**v**) is obtained by integrating the inequality $e^{-2t\lambda}\lambda \le \frac{1}{2t}$ by $d(E_\lambda g, g)$. The second follows from the expression $\|\alpha G_\alpha f - f\|_{\mathcal{E}_1}^2 = \int_0^\infty \left(\frac{\lambda}{\alpha+\lambda}\right)^2 (\lambda+1) d(E_\lambda f, f)$. The next theorem is frequently used in the main text.

Theorem A.4.1 (Banach-Saks). (i) *Let $\mathcal{H}$ be a real Hilbert space with inner product $(\cdot,\cdot)$ and norm $\|\cdot\|$. If, for $f_n \in \mathcal{H}$, $n \ge 1$, $\sup_n \|f_n\| = M$ is finite, then the Cesàro mean sequence of a suitable subsequence of $\{f_n\}$ converges strongly to an element of $\mathcal{H}$.*
(ii) *Let $\mathcal{H}$ be a real linear space and $\mathcal{C}(f,g)$, $f,g \in \mathcal{H}$, be a non-negative definite symmetric bilinear form on $\mathcal{H}$. If $f_n \in \mathcal{H}$, $n \ge 1$, satisfies $\sup_n \mathcal{C}(f_n, f_n) < \infty$, then the Cesàro mean sequence of a suitable subsequence of $\{f_n\}$ is $\mathcal{C}$-Cauchy.*

Proof. (i) Under the present assumption, a suitable subsequence of $\{f_n\}$ converges weakly to some element $f \in \mathcal{H}$. Denoting the difference of the subsequence and f by $\{f_n\}$ again, $\{f_n\}$ converges to zero and hence we can choose its subsequence n_k as follows. Let $n_1 = 1$. If $n_1, \dots, n_N$ are chosen, we select $n_{N+1} > n_N$ such that $|(f_{n_1}, f_{n_{N+1}})| \le \frac{1}{N}, \dots, |(f_{n_N}, f_{n_{N+1}})| < \frac{1}{N}$. Then the Cesàro mean $g_N = \frac{1}{N}\sum_{k=1}^N f_{n_k}$ of $\{f_{n_k}\}$ satisfies

$$\|g_N\|^2 = \frac{1}{N^2}\sum_{k=1}^N \|f_{n_k}\|^2 + \frac{2}{N^2}\sum_{1\le i<k\le N} (f_{n_i}, f_{n_k})$$

$$\le \frac{M^2}{N} + \frac{2}{N^2}\sum_{k=1}^N \frac{k-1}{k} \le \frac{M^2+2}{N} \to 0, \ N \to \infty.$$

(ii) For $\mathcal{F}, g \in \mathcal{H}$, let $f \sim g$ if $\mathcal{C}(f-g, f-g) = 0$. Then $\mathcal{C}$ is a pre-Hilbertian inner product on the quotient space $\mathcal{H}/\sim$ of $\mathcal{H}$ with respect to this equivalence relation. It suffices to apply (i) to the Hilbert space $(\mathcal{H}^*, \mathcal{C})$ obtained by completing $(\mathcal{H}/\sim, \mathcal{C})$. □

Appendix B

SOLUTIONS TO EXERCISES

◇ **1.1.10**: Take approximating sequences $\{f_n\}$, $\{g_n\} \subset \mathcal{F}$ for f, g. On account of the proof of Theorem 1.1.5(ii), we may assume that $|f_n| \le \|f\|_\infty$, $|g_n| \le \|g\|_\infty$. Since

$$|f_n(x)g_n(x) - f_n(y)g_n(y)| \le \|g\|_\infty |f_n(x) - f_n(y)| + \|f\|_\infty |g_n(x) - g_n(y)|,$$

and $|f_n(x)g_n(x)| \le \|g\|_\infty |f_n(x)| + \|f\|_\infty |g_n(x)|$, we get from (1.1.13) that $\mathcal{A}_{T_t}(f_n g_n)^{1/2} \le \|g\|_\infty \mathcal{A}_{T_t}(f_n)^{1/2} + \|f\|_\infty \mathcal{A}_{T_t}(g_n)^{1/2}$. Dividing by t and letting $t \downarrow 0$, we have $\|f_n g_n\|_{\mathcal{E}} \le \|g\|_\infty \|f_n\|_{\mathcal{E}} + \|f\|_\infty \|g_n\|_{\mathcal{E}}$. Since $\|f_n g_n\|_{\mathcal{E}}$ is uniformly bounded and $f_n g_n$ converges to fg as $n \to \infty$, we can obtain the desired conclusion using Theorem A.4.1 and Corollary 1.1.4.

◇ **1.3.13**: Choose $\{g_n\} \subset \mathcal{F} \cap C_c(E)$ such that $\|g_n - f\|_\infty < \frac{1}{n}$ for $n \ge 1$. Let $f_n = \varphi_n(g_n)$ for φ_n defined by (1.3.1). Then $\text{supp}[f_n] \subset \text{supp}[f]$ and $\|f_n - f\|_\infty \le \frac{2}{n}$ for every $n \ge 1$.

◇ **1.3.16**: See [73, Lemma 2.1.6 and Theorem 2.1.4].

◇ **1.4.2**: (i) Define $(\Phi f)(\widehat{x}) = f(j^{-1}\widehat{x})$, $\widehat{x} \in \widehat{E}$. Then $\Phi\mathcal{F} = \widehat{\mathcal{F}}$ and $\widehat{\mathcal{E}}_1(\Phi f, \Phi f) = \mathcal{E}_1(f, f)$, $f \in \mathcal{F}$. Clearly an increasing sequence $\{E_k\}$ of closed subsets of E is an $\mathcal{E}$-nest if and only if so is $\{E_k \cap F_k\}$, which is in turn equivalent to the condition that $\{j(E_k \cap F_k)\}$ is an $\widehat{\mathcal{E}}$-nest because of $\Phi\mathcal{F}_{E_k \cap F_k} = \widehat{\mathcal{F}}_{j(E_k \cap \mathbf{F}_k)}$ and the above isometry by Φ.

(ii) and (iii) are consequences of (i).

◇ **2.1.13**: This can be proved in exactly the same way as the proof of Lemma 1.1.11 by making use of Theorem 2.1.9 and Theorem 2.1.12.

◇ **2.2.5**: $\psi(x) = \int_{\mathbb{R}^n} (1 - \cos\langle x, y\rangle) J(dy) = \frac{2}{n} \sum_{k=1}^n \sin^2(x_k/2)$. Since $(2r^2)/(\pi^2 n) \le \psi(x) \le r^2/(2n)$ for $|x| < \pi$,

$$\int_{\{|x|<\pi\}} \frac{dx}{\psi(x)} \sim \int_0^\pi \frac{1}{r^{3-n}}\, dr \begin{cases} = \infty, & n = 1, 2, \\ < \infty, & n \ge 3. \end{cases}$$

◇ **2.3.3**: For $f \in L^2_+(\mathbb{R}^n)$, let $f_\ell = 1_{B_\ell(0)} \cdot (f \wedge \ell)$. As $\ell \to \infty$, $I_{\alpha/2} * f_\ell \in C_\infty(\mathbb{R}^n) \cap \mathcal{F}_e$ is $\mathcal{E}$-convergent and pointwisely convergent to $I_{\alpha/2} * f$. Hence by Theorem 2.3.2, $I_{\alpha/2} * f$ is quasi continuous.

◇ **2.3.8**: By (2.3.6), we have for any $g \in \mathcal{F} \cap C_c(E)$, $\mathcal{E}_\alpha(U_\alpha\mu - U_\beta\mu, v) + (\alpha - \beta)(U_\beta\mu, v) = 0$. Formula (2.3.8) then follows from $(U_\beta\mu, v) = \mathcal{E}_\alpha(G_\alpha U_\beta\mu, v)$.

◇ **3.1.16**: Substitute $A_t = \gamma \int_0^t \mathbf{1}_C(X_s)ds$ in (4.1.4) and (4.1.5) to obtain $U_A^\alpha f = \gamma R_\alpha(\mathbf{1}_C \cdot f)$ and $R_\alpha^A f = R_\alpha^{\gamma C} f$ so that (4.1.7) reads

$$R_\alpha^{\gamma C} f - R_\alpha f + \gamma R_\alpha(\mathbf{1}_C \cdot R_\alpha^{\gamma C} f) = 0.$$

An operation of $(\beta - \alpha)R_\beta$ then yields

$$\begin{aligned} 0 &= (\beta - \alpha)R_\beta R_\alpha^{\gamma C} f - R_\alpha f + R_\beta f + \gamma R_\alpha(\mathbf{1}_C \cdot R_\alpha^{\gamma C} f) - \gamma R_\beta(\mathbf{1}_C \cdot R_\alpha^{\gamma C} f) \\ &= (\beta - \alpha)R_\beta R_\alpha^{\gamma C} f - R_\alpha^{\gamma C} f + R_\beta f - \gamma R_\beta(\mathbf{1}_C \cdot R_\alpha^{\gamma C} f). \end{aligned}$$

Note that the proof of Lemma 3.1.15 implies that $\mathcal{E}_\alpha(R_\alpha^{F'} f, R_\alpha^F f) = (f, R_\alpha^F f)$ if F and F' are nearly Borel finely closed and $F \subset F'$. For any open set $D \subset E$, choose $\{F_n\}$ of closed sets increasing to D and denote $R_\alpha^{F_n} f$ by $R_\alpha^n f$. Then, for $m < n$, $\|R_\alpha^n f - R_\alpha^m f\|_{\mathcal{E}_\alpha}^2 = (f, R_\alpha^n f - R_\alpha^m f)$, $f \in C_c(E)$, so that $\{R_\alpha^n f\}$ is $\mathcal{E}_\alpha$-Cauchy. It then suffices to note that $\lim_{n\to\infty} R_\alpha^n f(x) = R_\alpha^D f(x)$ by the quasi-left-continuity of X.

◇ **3.3.2**: (i) Let $\widetilde{E}$ be q.e. equivalent to a quasi open set E. Choose nests $\{F_k^{(1)}\}$, $\{F_k^{(2)}\}$ such that $\widetilde{E} \cap F_k^{(1)} = E \cap F_k^{(1)}$, $k \geq 1$, and $E \cap F_k^{(2)}$ is relatively open for every k. Let $F_k = F_k^{(1)} \cap F_k^{(2)}$, $k \geq 1$. Then $\{F_k\}$ is a nest and $\widetilde{E} \cap F_k = E \cap F_k$ is relatively open for every k.
(ii) The "only if" part is obvious. Suppose f has the stated property. Then, in the same way as the proof of Lemma 1.3.1, a Cap_1-nest $\{F_k\}$ can be constructed so that, for each $k \geq 1$, the sets $\{x \in F_k : f(x) \geq r\}$, $\{x \in F_k : f(x) \leq r\}$ are closed sets for every $r \in \mathbb{Q}$. $f\big|_{F_k}$ is then continuous for each $k \geq 1$.
(iii) Let $\{F_k\}$ be a nest such that $f \in C(\{F_k\})$ and $D \cap F_k$ is relatively open for each $k \geq 1$. Then $\widehat{F}_k = \mathrm{supp}[\mathbf{1}_{F_k} \cdot m]$ defines an m-regular nest $\{\widehat{F}_k\}$ by Lemma 1.3.6 and so $f(x) \geq 0$ for any $x \in D \cap (\cup_k \widehat{F}_k)$.

◇ **3.3.7**: (i) X^D is a Markov process on D relative to the filtration $\{\mathcal{F}_t\}$ because, for $x \in D$ and $A \in \mathcal{B}(D)$,

$$\begin{aligned} \mathbf{P}_x(X_{t+s}^D \in A \big| \mathcal{F}_t) &= \mathbf{P}_x(X_s \circ \theta_t \in A,\ t < \tau_D,\ s < \tau_D \circ \theta_t \big| \mathcal{F}_t) \\ &= \mathbf{1}_{\{t<\tau_D\}} \mathbf{P}_{X_t}(X_s \in A,\ s < \tau_D) = \mathbf{P}_{X_t^D}(X_s^D \in A). \end{aligned}$$

For $f \in b\mathcal{B}(D)$, the resolvent kernel $R_\alpha^D f$ defined by (3.3.2) equals $R_\alpha f - \mathbf{H}_\alpha R_\alpha f$ on D, which is a difference of bounded α-excessive functions with respect to X. Consequently, $R_\alpha^D f(X_t^D)$ is right continuous in $t \geq 0$ $\mathbf{P}_x$-a.s. for $x \in D$, and so X^D is a right process on D by virtue of Theorem A.1.37. The quasi-left-continuity of X^D on $(0, \zeta^D)$ follows from the quasi-left-continuity on $(0, \infty)$ of X.

(ii) When D is an open subset of E, we define X^D by taking ∂ as the point at infinity of D. For the quasi-left-continuity of X^D on $(0, \infty)$, it suffices to show that if $\{\sigma_n\}$ is a sequence of $\{\mathcal{F}_t\}$-stopping times increasing to $\tau_D < \infty$ with $\sigma_n < \tau_D$ for every $n \geq 1$, then $\lim_{n\to\infty} X^D_{\sigma_n} = \partial$. But in this case $Y := \lim_{n\to\infty} X_{\sigma_n} \in \overline{D} \cup \Delta$ and $Y = X_{\tau_D} \in D^c \cup \Delta$ by the quasi-left-continuity of X, where Δ denotes the point at infinity of E. This means that $\lim_{n\to\infty} X^D_{\sigma_n} = \partial$.
Let $\{D_n\}$ be a sequence of relatively compact open sets increasing to D with $\overline{D}_n \subset D_{n+1}$, $n \geq 1$. The quasi-left-continuity of X implies that $\tau_D = \lim_{n\to\infty} \tau_{\overline{D}_n}$ $\mathbf{P}_x$-a.s. for $x \in D$. Since $\tau_{\overline{D}_n}$ are $\{\mathcal{F}^0_{t+}\}$-stopping times, $\{P^D_t\}$ defined by (3.3.2) makes the space $b\mathcal{B}(D)$ invariant. Thus X^D is a Hunt process on D.

◇ **4.1.2**: Observe that

$$\begin{aligned}
&\frac{1}{\alpha-\beta}(U^\beta_A f(x) - U^\alpha_A f(x)) \\
&= \mathbf{E}_x\left[\int_0^\infty e^{-\alpha s}\int_0^\infty e^{-\beta t} f(X_t\circ\theta_s)dA_t\circ\theta_s ds\right] \\
&= \mathbf{E}_x\left[\int_0^\infty e^{-\alpha s} U^\beta_A f(X_s)ds\right] = R_\alpha U^\beta_A f(x),
\end{aligned}$$

while

$$\begin{aligned}
&R_\alpha f(x) - R^A_\alpha f(x) \\
&= \mathbf{E}_x\left[\int_0^\infty e^{-\alpha t}e^{-A_t}(e^{A_t}-1)f(X_t)dt\right] \\
&= \mathbf{E}_x\left[\int_0^\infty e^{-\alpha s}\mathbf{E}_{X_s}\left[\int_0^\infty e^{-\alpha t}e^{-A_t}f(X_t)dt\right]dA_s\right] = U^\alpha_A R^A_\alpha f(x).
\end{aligned}$$

◇ **4.1.7**: Rewrite the left hand side of (4.1.21) as $I + II$, where I (resp. II) is obtained by replacing the interval $[0, \infty)$ of integration with respect to s in the left hand side with $[0, t)$ (resp. $[t, \infty)$). Then

$$\begin{aligned}
II &= \mathbf{E}_x\left[\int_0^\infty e^{-t}e^{-t}\int_0^\infty e^{-s}dB_s\circ\theta_t dA_t\right] \\
&= \mathbf{E}_x\left[\int_0^\infty e^{-2t}\mathbf{E}_{X_t}\left[\int_0^\infty e^{-s}dB_s\right]dA_t\right] = U^2_A(U^1_B 1)(x).
\end{aligned}$$

As for I, we interchange the order of integrations in s, t to obtain $I = U^2_B(U^1_A 1)(x)$.

◇ **4.1.9**: (i) We denote τ_D as τ. If $s+t<\tau$, then $X_t^D(\theta_s^0\omega)=X_t^D(\theta_s\omega)=X_t(\theta_s\omega)=X_{s+t}(\omega)=X_{s+t}^D(\omega)$. If $s<\tau$, $s+t\geq\tau$, then $t\geq\tau-s=\tau(\theta_s\omega)=\tau(\theta_s^0\omega)$ and $X_t^D(\theta_s^0\omega)=\partial=X_{s+t}^D(\omega)$. If $\tau\leq s$, then $\theta_s^0\omega=\omega_\partial$ and $X_t^D(\theta_s^0\omega)=X_t^D(\omega_\partial)=\partial=X_{s+t}^D(\omega)$.
(ii) $\mathcal{G}_t^0$ is generated by $\{X_s^D\in B\}=\{X_s\in B\}\cap\{s<\tau\}$ for $B\in\mathcal{B}(D)$, $s\leq t$. Let $\Lambda:=\{X_s^D\in B\}\cap\{\tau\wedge s\leq u\}$. Then $\Lambda=\{X_s\in B\}\cap\{s<\tau\}\in\mathcal{F}_s^\mu\subset\mathcal{F}_u^\mu$ if $s\leq u$, while $\Lambda=\emptyset$ if $s>u$. Therefore, $\{X_s^D\in B\}\in\mathcal{F}_{\tau\wedge s}^\mu$ and $\mathcal{G}_t^\mu\subset\mathcal{F}_{t\wedge\tau}^\mu$.
On the other hand, by Lemma A.1.12, $\mathcal{F}_{(\tau\wedge t)-}^\mu\subset\sigma\{\{X_u\in B\}\cap\{u<\tau\}: u<t\}\vee\mathcal{N}=\sigma\{\{X_u^D\in B\}:u<t\}\vee\mathcal{N}=\mathcal{G}_{t-}^u$, so that $\mathcal{F}_{(\tau\wedge t)-}^\mu\subset\mathcal{G}_{t-}^\mu\subset\mathcal{G}_t^\mu$.
The Markov property (4.1.24) holds for $F=\mathbf{1}_{\{X_s^D\in A\}}$, $A\in\mathcal{B}(D_\partial)$, by Exercise 3.3.7, which extends to a general $F\in\mathcal{G}_\infty^0$ by induction and the monotone class theorem as in the paragraph below (A.1.11). The extension from $\mathcal{G}_\infty^0$ to $\mathcal{G}_\infty$ can be carried out just as the proof of Theorem A.1.21.
(iii) When $t+s<\tau$, $B_s+B_t\circ\theta_s^0=A_s+A_t\circ\theta_s=A_{s+t}=B_{s+t}$. When $s<\tau(\omega)$, $t+s>\tau(\omega)$, $t>\tau(\omega)-s=\tau(\theta_s^0\omega)$, $B_s=A_s$ and $B_t(\theta_s^0\omega)=A_{t\wedge\tau(\theta_s^0\omega)}(\theta_s^0\omega)=A_{\tau(\theta_s^0\omega)}(\theta_s^0\omega)=A_{\tau(\theta_s\omega)}(\theta_s\omega)$ so that $B_s+B_t(\theta_s^0\omega)=A_s+A_{\tau(\theta_s\omega)}(\theta_s\omega)=A_\tau$, $B_{s+t}=A_{(s+t)\wedge\tau}=A_\tau$. When $\tau(\omega)\leq s$, $B_s=A_{s\wedge\tau}=A_\tau$, $B_{s+t}=A_{(s+t)\wedge\tau}=A_\tau$, while $\theta_s^0\omega=\omega_\partial$ and $B_t(\theta_s^0\omega)=A_{t\wedge\tau(\omega_\partial)}(\omega_\partial)=A_0(\omega_\partial)=0$. When A is continuous, it is predictable so that $B_t\in\mathcal{F}_{(\tau_0\wedge t)-}\subset\mathcal{G}_t$ by (A.1.14) and (ii).

◇ **4.3.12**: Since $\mathcal{E}^{(c)}(u,v)=\frac{1}{2}\mu_{\langle u,v\rangle}^c(E)$ for $u,v\in\mathcal{F}_e$, (4.3.34) follows from Lemma 4.3.6 immediately.

◇ **5.1.2**: Note that

$$\begin{aligned}P_{s+t}f(x)&=\mathbf{E}_x\left[e^{-A_s}e^{-A_t\circ\theta_s}f(X_t\circ\theta_s)\right]\\&=\mathbf{E}_x\left[e^{-A_s}\mathbf{E}_{X_s}\left[e^{-A_t}f(X_t)\right]\right]=P_sP_tf(x).\end{aligned}$$

(t.3) and **(t.4)**$'$ are obvious.

◇ **5.4.1**: From (5.4.5), we see that $e(t)$ is non-decreasing as $t\uparrow$ and further the subadditivity $e(t+s)\leq e(t)+e(s)$ holds. Therefore, if $e(t)$ is finite for some $t>0$, then so it is for any $t>0$. Suppose $e(t)<\infty$ for any $t>0$. Since $0\leq f-P_tf$ decreases to zero as $t\downarrow 0$, we get by the dominated convergence theorem that $e(t)\downarrow 0$, $t\downarrow 0$. For $t<t'$, we have from (5.4.5) $e(t')=e(t)+(f-P_{t'-t}f,P_tg)$. The second term of the right hand side is dominated by $(f-P_{t'-t}f,g)=e(t'-t)$, which decreases to zero as $t'\downarrow t$ or $t\uparrow t'$, so that $e(t)$ is continuous in $t>0$.
We obtain also from (5.4.5) the inequality $e(t'+s)-e(t')\leq e(t+s)-e(t)$ for $t<t'$. Substitution $t'=t+s$ then yields $\frac{1}{2}\{e(t)+e(t+2s)\}\leq e(t+s)$, namely, e is midpoint concave. Since it is continuous on $[0,\infty)$ with $e(0)=0$, we have $e(s)\geq\frac{s}{t}e(t)$ for $s<t$, that is to say, $\frac{e(t)}{t}\uparrow$ as $t\downarrow 0$.

◇ **5.5.1**: We denote τ_0 by τ.

$$\begin{aligned}
&\frac{1}{\alpha-\beta}(\mathbf{H}^{\beta}f(x)-\mathbf{H}^{\alpha}f(x))\\
&=\mathbf{E}_x\left[\int_0^{\tau} e^{-(\alpha-\beta)t}e^{-\beta\tau}f(X_\tau)dt\right]\\
&=\mathbf{E}_x\left[\int_0^{\infty} e^{-\alpha t}\mathbf{1}_{\{t<\tau\}}e^{-\beta\tau\circ\theta_t}f(X_\tau\circ\theta_t)dt\right]\\
&=\mathbf{E}_x\left[\int_0^{\infty} e^{-\alpha t}\mathbf{1}_{\{t<\tau\}}\mathbf{E}_{X_t^0}[e^{-\beta\tau}f(X_\tau)]dt\right]=R_\alpha^0\mathbf{H}^{\beta}f(x).
\end{aligned}$$

◇ **5.5.7**: It suffices to prove this for $v\in b\mathcal{F}_e$. In the proof of Theorem 5.5.6, the only property that is used for $v=\mathbf{H}u$ with $u\in b\mathcal{F}_e$ beyond $v\in\mathcal{F}_e$ is the fact that $M_t^{[v]}=v(X_t)-v(X_0)$ for $t\le\tau_0$. For $v=R_\alpha f$ for some bounded $f\in L^2(E;m)$, $M_t^{[v]}=v(X_t)-v(X_0)-\int_0^t(\alpha u-f)(X_s)ds$. Since $\int_0^t|(\alpha u-f)(X_s)|ds\le ct$, the computation involving I in the proof of Theorem 5.5.6 goes through. It then follows that (5.5.21) holds for such v. Since such v is $\mathcal{E}$-dense in $\mathcal{F}_e$, it follows from Lemma 5.5.4 that (5.5.21) holds for general $v\in b\mathcal{F}_e$.

◇ **6.7.1**: Using the m-symmetry of the transition semigroup $\{P_t,t\ge 0\}$, we see that

$$\mathbf{E}_m[\xi;t<\zeta]=\mathbf{E}_m[\xi\circ r_t;t<\zeta] \qquad \text{(B.0.1)}$$

holds for random variable $\xi=\prod_{j=1}^n f_j(X_{t_j})$, where $f_j\in\mathcal{B}_+(E)$ and $0\le t_1<t_2<\cdots<t_n\le t$. Consequently, (B.0.1) holds for any non-negative $\xi\in\mathcal{F}_t^0$ and hence for every non-negative $\xi\in\mathcal{F}_t$.

◇ **7.1.1**: Take any $h,f\in\mathcal{B}_+(E_0)$. By Lemma 4.2.2 and a quasi-homeomorphism, we have $\langle G_\alpha^0 h,f\cdot\kappa_0\rangle=\mathbf{E}_{h\cdot m_0}[e^{-\alpha\zeta^0}f(X^0_{\zeta^0-})]$. By (A.3.33) for $g(s)=e^{-\alpha s}$, $T=\tau_0$, the right hand side equals

$$\begin{aligned}
\mathbf{E}_{h\cdot m_0}[e^{-\alpha\tau_0}f(X_{\tau_0-})]&=\mathbf{E}_{h\cdot m_0}\left[\sum_{s\le\tau_0}e^{-\alpha s}f(X_{s-})\mathbf{1}_{F\cup\{\partial\}}(X_s)\right]\\
&=\mathbf{E}_{h\cdot m_0}\left[\int_0^{\tau_0}e^{-\alpha s}f(X_s)N(X_s,F\cup\{\partial\})dH_s\right],
\end{aligned}$$

which equals $\langle G_\alpha^0 h,f\kappa+fJ(\cdot,F)\rangle$ on account of (4.1.26).

◇ **7.3.9** "⇒": By Theorem 7.2.4, it holds for any $\psi\in\mathcal{N}_1(=\mathcal{F}\big|_F)$ that $\mathbf{H}_\alpha\psi\in\mathcal{F}$ and

$$\mathcal{N}(f)(\psi)=\mathcal{E}^{0,\mathrm{ref}}(f,\mathbf{H}_\alpha\psi)+(\mathcal{L}f,\mathbf{H}_\alpha\psi)=\mathcal{E}(f,\mathbf{H}_\alpha\psi)+(\mathcal{A}f,\mathbf{H}_\alpha\psi)=0.$$

"⇐": Let $(\alpha-\mathcal{L})f=g$. Then $\mathcal{E}_\alpha(f,v)=(g,v)$, $\forall v\in\mathcal{F}^0$. From $\mathcal{N}(f)=0$, we get $\mathcal{E}_\alpha(f,\mathbf{H}_\alpha\psi)=(g,\mathbf{H}_\alpha\psi)$ for any $\psi\in\mathcal{N}_1$. Therefore, $\mathcal{E}_\alpha(f,w)=(g,w)$ for any $w\in\mathcal{F}$ so that $f=G_\alpha g\in\mathcal{D}(\mathcal{A})$.

◇ **7.5.2**: (i) Follow the proof of Lemma 7.1.5 by noting that X^0 admits no killing inside.
(ii) Take any open subset O of E_0 with positive distance from a and define stopping times $\{\sigma_a^n : n \geq 0\}$, $\{\eta^n : n \geq 0\}$ recursively by
$\eta^0 = 0$, $\sigma_a^0 = \sigma_a$, $\eta^n = \sigma_a^{n-1} + \sigma_O \circ \theta_{\sigma_a^{n-1}}$, $\sigma_a^n = \eta^n + \sigma_a \circ \theta_{\eta^n}$. Then

$$\{t \in (0, \zeta(\omega)) : X_{t-}(\omega) \in O,\ X_t(\omega) = a\}$$

$$\subset \bigcup_{n=1}^{\infty} \left\{\sigma_a^n(\omega) : X_{\sigma_a^n(\omega)-}(\omega) \in O,\ X_{\sigma_a^n(\omega)}(\omega) = a\right\}$$

and

$$\begin{aligned}\mathbf{P}_x\left(X_{\sigma_a^n-} \in O,\ X_{\sigma_a^n} = a\right) &= \mathbf{P}_x\left(X_{\sigma_a-} \circ \theta_{\eta_n} \in O,\ X_{\sigma_a} \circ \theta_{\eta_n} = a\right)\\ &= \mathbf{E}_x\left[\mathbf{P}_{X_{\eta n}}\left(X_{\sigma_a-} \in O,\ X_{\sigma_a} = a\right)\right] = 0.\end{aligned}$$

◇ **A.1.5**: (i) Sufficiency: $\{\sigma \leq t\} = \cap_n\{\sigma < t + \frac{1}{n}\} \in \cap_n \mathcal{M}_{t+\frac{1}{n}} = \mathcal{M}_{t+}$. Necessity: $\{\sigma < t\} = \cup_n\{\sigma \leq t - \frac{1}{n}\} \in \mathcal{M}_{(t-\frac{1}{n})+} \subset \mathcal{M}_t$.
(ii) If $s < c$, then $\{\sigma + c \leq s\} = \emptyset$. If $c \leq s$, then $\{\sigma + c \leq s\} = \{\sigma \leq s - c\} \in \mathcal{M}_{s-c} \subset \mathcal{M}_t$.
(iii) If $\sigma_n \uparrow \sigma$, then $\{\sigma < t\} = \cap_n\{\sigma_n \leq t\} \in \mathcal{M}_t$. If $\sigma_n \downarrow \sigma$, then $\{\sigma < t\} = \cup_n\{\sigma_n < t\} \in \mathcal{M}_t$.
(iv) When $(k-1)2^{-n} \leq t < k2^{-n}$, $\{\sigma_n \leq t\} = \{\sigma_n \leq (k-1)2^{-n}\} = \{\sigma < (k-1)2^{-n}\} \in \mathcal{M}_{(k-1)2^{-n}} \subset \mathcal{M}_t$.

◇ **A.1.16**: When $n = 1$, (A.1.17) for $s_1 = s$, $f_1 = \mathbf{1}_B$ is nothing but (A.1.15) so that (A.1.17) holds for any $f_1 \in b\mathcal{B}(E_\partial)$. Assume that (A.1.17) is valid for $n-1$. Then

$$\begin{aligned}&\mathbf{E}_\mu\left[\prod_{j=1}^n f_j(X_{\sigma+s_j})\big|\mathcal{M}_\sigma\right]\\ &= \mathbf{E}_\mu\left[f_1(X_{\sigma+s_1})\mathbf{E}_\mu\left[\prod_{j=2}^n f_j(X_{\sigma+s_j})\Big|\mathcal{M}_{\sigma+s_1}\right]\Big|\mathcal{M}_\sigma\right]\\ &= \mathbf{E}_\mu\left[f_1(X_{\sigma+s_1})\mathbf{E}_{X_{\sigma+s_1}}\left[\prod_{j=2}^n f_j(X_{s_j-s_1})\right]\Big|\mathcal{M}_\sigma\right]\\ &= \mathbf{E}_\mu\left[f_1(X_{\sigma+s_1})g(X_{\sigma+s_1})\big|\mathcal{M}_\sigma\right] = \mathbf{E}_{X_\sigma}\left[(f_1 \cdot g)(X_{s_1})\right],\end{aligned}$$

where $g(x) = \mathbf{E}_x[\prod_{j=2}^n f_j(X_{s_j-s_1})]$. The last expectation is equal to the right hand side of (A.1.17) on account of (A.1.7).

◇ **A.1.20**: (i) Let $F \in b\mathcal{F}_\infty$. For each $x \in E_\partial$, $F \in b\mathcal{F}_\infty^{\mathbf{P}_x}$ so that $\mathbf{E}_x[F]$ is well defined. For any $\mu \in \mathcal{P}(E_\partial)$, there exist $F_1,\ F_2 \in b\mathcal{F}_\infty^0$ with $F_1 \le F \le F_2$, $\int_{E_\partial} (\mathbf{E}_x[F_2] - \mathbf{E}_x[F_1])\,\mu(dx) = \mathbf{E}_\mu[F_2 - F_1] = 0$. Therefore, $\mathbf{E}_\cdot[F] \in \mathcal{B}^\mu(E_\partial)$.
(ii) For $\mu \in \mathcal{P}(E_\partial)$, $(\mathcal{F}_\sigma)^{\mathbf{P}_\mu} \subset (\mathcal{F}_\sigma^\mu)^{\mathbf{P}_\mu} = \mathcal{F}_\sigma^\mu$ by definition. Let $\nu(A) = \mathbf{P}_\mu(X_\sigma \in A)$, $A \in \mathcal{B}(E_\partial)$. For any $B \in \mathcal{B}^*(E_\partial)$, there exist $B_1,\ B_2 \in \mathcal{B}(E_\partial)$ such that $B_1 \subset B \subset B_2$, $\nu(B_2 \setminus B_1) = 0$. Since $\{X_\sigma \in B_i\} \in \mathcal{F}_\sigma$ by Lemma A.1.13(iii), $\{X_\sigma \in B\} \in (\mathcal{F}_\sigma)^{\mathbf{P}_\mu} \subset \mathcal{F}_\sigma^\mu$. $\mu \in \mathcal{P}(E_\partial)$ being arbitrary, $\{X_\sigma \in B\} \in \bigcap_{\mu\in\mathcal{P}(E_\partial)} \mathcal{F}_\sigma^\mu = \mathcal{F}_\sigma$.

◇ **A.1.26**: (i) Define $X_s^{(n)}(\omega) = X_{(k-1)2^{-n}}(\omega)$, $(k-1)2^{-n} < s \le k2^{-n}$. Then $\{X_s^{(n)}(\omega); (s,\omega) \in (0,t) \times \Omega\}$ is $\mathcal{B}_t \times \mathcal{M}_t$-measurable and $X_{s-}(\omega) = \lim_{n\to\infty} X_s^{(n)}(\omega)$.
(ii) Define $A = \{(s,\omega) \in (0,t)\times\Omega : X_{s-}(\omega) \in B\}$. Then $A \in \mathcal{B}_t, \times \mathcal{F}_t^0$ by (i) and $\Lambda = \{\widehat{\sigma}_B(\omega) < t\}$ is the projection of A on Ω, which is in $\mathcal{F}_t$ by virtue of Proposition A.1.1.

◇ **A.1.29**: (i) For $B \in \mathcal{B}^n(E_\partial)$ and for any $\mu \in \mathcal{P}(E_\partial)$, let $B_1,\ B_2 \in \mathcal{B}(E_\partial)$ be as in Definition A.1.28. Since $\mu(B_2 \setminus B_1) = 0$, $B \in \mathcal{B}^\mu(E_\partial)$ and consequently $B \in \mathcal{B}^*(E_\partial)$. It follows from

$$(\cup_k B_2^k) \setminus (\cup_k B_1^k) \subset \cup_k (B_2^k \setminus B_1^k), \quad B_1^c \setminus B_2^c = B_2 \setminus B_1,$$

that $\mathcal{B}^n(E_\partial)$ is closed under the operations of taking a countable union and complement.
(ii) This is clear for $f = \sum_{i=1}^n a_i \mathbf{1}_{B_i}$, $B_i \in \mathcal{B}^n(E_\partial)$, $a_i \in \mathbb{R}$, $B_i \cap B_j = \emptyset$, $i \ne j$.

◇ **A.1.31**: For any $\mu \in \mathcal{P}(A_\partial)$, there exists $f_1,\ f_2 \in \mathcal{B}(E)$ and $B_1,\ B_2 \in \mathcal{B}(E)$ with $f_1 \le f \le f_2$, $\mathbf{P}_\mu(f_1(X_s) < f_2(X_s),\ \exists s \ge 0) = 0$, $B_1 \subset A \subset B_2$, $\mathbf{P}_\mu(X_s \in B_2 \setminus B_1, \exists \ge 0) = 0$ so that $\mathbf{P}_\mu(X_s \in A \setminus B_1, \exists s \ge 0) = 0$. Let $g_1 = f_1 \cdot \mathbf{1}_{B_1}\big|_A \in \mathcal{B}(A)$ and $g_2 = f_2\big|_A \in \mathcal{B}(A)$. Then $g_1 \le f\big|_A \le g_2$ and $\mathbf{P}_\mu(g_1(X_s) < g_2(X_s),\ \exists s \ge 0) = \mathbf{P}_\mu(g_1(X_s) < g_2(X_s),\ X_s \in B_1,\ \exists s \ge 0) \le \mathbf{P}_\mu(f_1(X_s) < f_2(X_s),\ \exists s \ge 0) = 0$.

◇ **A.1.33**: See [80, (8.4)].

◇ **A.1.34**: See [80, p. 52].

◇ **A.1.42**: For a Feller transition function $\{P_t; t \ge 0\}$, $P_t f(x)$ is right continuous in $t \ge 0$ for any $f \in C_\infty(E)$ and $x \in E$, so that the property **(t.2)** is verified in the same way as the proof of Lemma 1.1.14(i). Take any $f \in C_b(E)$ and, for a fixed $x \in E$, choose $g \in C_c(E)$ with $g(x) = 1$ and $0 \le g \le 1$ on E. Set $f_1 := fg$, which is in $C_c(E)$), and $f_2 := f(1-g)$. Then

$$\lim_{t\to 0} |P_t f_2(x)| \le \lim_{t\to 0} \|f\|_\infty P_t(1-g)(x) \le \lim_{t\to 0} \|f\|_\infty (1 - P_t g(x)) = 0.$$

Therefore, $\lim_{t\to 0} P_t f(x) = \lim_{t\to 0}(P_t f_1(x) + P_t f_2(x)) = f(x)$, verifying **(t.4)**.

- ◇ **A.2.3**: For $w = u \wedge v$, $P_t w \leq w$ and $\lim_{t\downarrow 0} P_t w(x) \leq w(x)$ for any $x \in E$. From Theorem A.2.2 and Fatou's lemma, it follows that $w(x) = \mathbf{E}_x[\lim_{t\downarrow 0} w(X_t)] \leq \lim_{t\downarrow 0} P_t w(x)$.

- ◇ **A.2.8**: For O_1, $O_2 \in \mathcal{O}$, $x \in O_1 \cap O_2$, there exist $D_i(x) \in \mathcal{B}^n(E)$ with $O_i^c \subset D_i(x)$ so that $\mathbf{P}_x\left(\sigma_{D_i(x)} > 0\right) = 1$ for $i = 1, 2$. Then $(O_1 \cap O_2)^c \subset D_1(x) \cup D_2(x)$ and $\mathbf{P}_x\left(\sigma_{D_1(x)\cup D_2(x)} = \sigma_{D_1(x)} \wedge \sigma_{D_2(x)} > 0\right) = 1$. Consequently, $O_1 \cap O_2 \in \mathcal{O}$. Other properties are obvious.

- ◇ **A.3.3**: Let Λ and N be a defining set and an exceptional set, respectively, for $A \in \mathbf{A}_c^+$. For any $f \in b\mathcal{B}_+(E)$, the process $Y(t,\omega) = f(X_t(\omega))$ is progressive with respect to $\{\mathcal{F}_t, t \geq 0\}$ in view of Lemma A.1.13, and so is the restriction of Y to $[0,\infty) \times \Lambda$ relative to $\{\mathcal{F}_t\big|_\Lambda, t \geq 0\}$. $\int_0^t f(X_s(\omega))dA_s(\omega)$, $\omega \in \Lambda$, is therefore well defined and is $\mathcal{F}_t\big|_\Lambda$-measurable for each $t \geq 0$. It is then easily seen to be a PCAF with defining set Λ and exceptional set N.

Notes

NOTES ON CHAPTERS 1 TO 4

Many materials in Sections 1.1 and 2.1, Chapter 3, and Chapter 4 are based on a recently published Japanese book by Fukushima-Takeda [74]. The notion of h-capacity and quasi-regular Dirichlet form were first introduced in Albeverio and Ma [2] for symmetric Dirichlet forms. It was later extended to non-symmetric Dirichlet forms by Albeverio, Ma, and Röckner in [4]. The presentation in Sections 1.2 and 1.3 are based on the lecture notes of a graduate course the first author taught in the Autumn semester of 1995 at Cornell University. Capacitary characterization of an $\mathcal{E}$-nest was also presented in the book of Ma-Röckner [119] under a more general non-symmetric setting. Section 1.4 is based on Chen-Ma-Röckner [31]. The analytic potential theory presented in Section 2.3 goes back to J. Deny [39] and its content is not much different from that of Fukushima-Oshima-Takeda [73, §2.1, §2.2]. The first half of Section 3.5 on probabilistic features of transience is mostly taken from Chen-Fukushima [21, §2].

In Section 1.1 the extended Dirichlet space is formulated with no topological assumption on the underlying space following an idea of B. Schumland [134], while the Dirichlet space itself has been defined without topological assumption in the books by Bouleau-Hirsch [15] and by Ma-Röckner [119].

Analogous results to those in Section 1.5 hold for the non-symmetric sectorial Dirichlet form as well. In particular, the book [119] presents a direct construction of a Borel special standard process associated with a quasi-regular non-symmetric Dirichlet form as well as the necessity of the quasi-regularity. Further extensions to more general non-symmetric situation have been carried out by W. Stannat [141].

Theorem 2.1.11 on the irreducibility characterization is due to M. Takeda.

All examples in Section 2.2 are finite dimensional. They are collected here mainly to identify explicitly the associated extended Dirichlet spaces and reflected Dirichlet spaces (later in Section 6.6).

The theory of Dirichlet forms brings the analytic and probabilistic potential theories together in a natural and coherent way. Its use and application is wide ranging. It is especially better suited for the study involving non-smooth data or irregular or infinite dimensional state spaces. The Dirichlet form theory has played an important role in the investigation of various subjects in probability theory, analysis, and statistical physics, in finite as well as infinite dimensional spaces. See [5, 14, 17, 20, 29, 30, 36, 43, 59, 60, 100, 114, 118, 129, 130, 143, 144] and references therein for a sample. The concept of Dirichlet forms has also been a powerful tool in the analytic and probabilistic study of fractals. See the book by J. Kigami [105] and a recent paper by Barlow-Bass-Kumagai-Teplyaev [6].

When $(\mathcal{C}, \mathcal{F})$ is a positivity preserving coercive form, the equivalence of local properties for $(\mathcal{C}, \mathcal{F})$ in analogy to that stated in Theorem 2.4.2 was established by B. Schmuland [133]. The formulation of Theorem 2.4.2 in this book is new; it is applicable to the strongly local part $\mathcal{E}^c$ of a Dirichlet form $(\mathcal{E}, \mathcal{F})$. Theorem 2.4.3 on strongly local property is new. See Proposition I.5.1.3 of Bouleau-Hirsch [15] for other equivalent characterizations of strongly local property of a Dirichlet form.

Many assertions made in Chapter 3 overlap with those in the book [73, Chapter 4] but some of them are proved quite differently. Especially the current proof of Theorem 3.3.9 on the regularity of the part of the Dirichlet form on an open set is new and probabilistic, while the proof in [73] is given by reinterpreting the celebrated spectral synthesis theorem in the analytic potential theory due to J. Deny [39, Chapter 4].

The methods of the proof employed in Chapter 4 are different from those in the book [73, §5.1, §5.2, §5.3] in various respects. Moreover, the probabilistic characterizations of the local and strongly local properties in Theorem 4.3.4 are obtained as direct consequences of the probabilistic descriptions of the Beurling-Deny formula contrarily to [73, §4.5]. Theorem 4.3.8 is an extension of Theorems I.5.2.3 and I.7.1.1 in Bouleau-Hirsch [15], where the same result is established for the local Dirichlet forms.

The stochastic calculus in terms of Dirichlet forms has applications to infinite dimensional quasi-regular situations. For instance, the concept of BV functions over the abstract Wiener space has been formulated and studied in this way (cf. M. Fukushima [67, 68], Fukushima-Hino [72], M. Hino [87]). See L. Zambotti [155] for a similar approach to a SPDE with reflection.

NOTES ON CHAPTER 5

The contents of Sections 5.1 and 5.2 are formulated for the family S of general smooth measures. Most of them have been presented in the book [73, §6.1, §6.2] but only for a subfamily of S consisting of positive Radon measures charging no $\mathcal{E}$-polar sets. Among other things, the independence of the extended Dirichlet space of the trace Dirichlet form on the choice of measures $\mu \in S$ quasi supported by a common quasi closed set F is presented in Theorem 5.2.15 for the first time in this generality.

The last part of Section 5.2 contains two important applications of the invariance properties under the time change by a fully supported PCAF. Especially construction of the resurrected Dirichlet form and the resurrected Hunt process is well carried out for a general regular Dirichlet form by removing its killing measure. This enables us to identify in the last example of Section 5.3 the Dirichlet form of a general one-dimensional minimal diffusion with possible killings inside.

The complete characterization of the trace Dirichlet form in Sections 5.5 and 5.6 in terms of the Feller measures is based on Chen-Fukushima-Ying [26]. However, the proof of a main result (Theorem 2.6) in [26] (which corresponds to Theorem 5.5.8 of this monograph) contains a serious gap (the dual predictable projection was used incorrectly in the displays preceding [26, (2.29)]). This gap has been fixed through Lemma 5.5.4, Lemma 5.5.5, and Theorem 5.5.6 of this book. These three results as well as Exercise 5.5.7 are taken from [25].

Section 5.7 is based on Chen-Fukushima-Ying [27]. This paper also establishes the identification of the Lévy system of the time-changed process as is presented in Theorem 5.6.3 using a counterpart of Theorem 5.7.6 for a more general standard process X possessing a weak dual process under the condition that every semi-polar set is m-polar. Fitzsimmons-Getoor [57] has extended the contents of [27] to much more general right processes. However, the identification of the strongly local part of the trace Dirichlet form as is formulated in Theorem 5.6.2 is only achieved by the present approach.

The first two examples of Section 5.8 are taken from Fukushima-He-Ying [71] and [26], respectively. Similar explicit expressions of the Feller kernel and related quantities along with their estimates for symmetric reflecting Brownian motions had been presented in P. Hsu [90]. See J. Kigami [106] for an explicit expression of the Douglas integral on the Cantor set induced by random walks on trees.

The trace Dirichlet form on a quasi closed set F also arises naturally as a limit of a sequence of Dirichlet forms when their underlying measures are changing and convergent to a degenerate measure supported by F. See A. Kasue [101, Example 2.1] and Ogura-Tomisaki-Tsuchiya [125] for explicit expressions of trace Dirichlet forms obtained in this way, and Kuwae-Shioya [114] and Kim [109] for a general formulation of Mosco convergence of Dirichlet forms with varying state spaces.

Section 5.4 is an adoption from R. K. Getoor [81] into the present symmetric setting.

NOTES ON CHAPTER 6

Two variant notions of reflected Dirichlet forms were introduced by M. L. Silverstein in [138] and [139]. A fundamental tool used in both [138] and [139] is the "extended Markov process," a continuous version of Hunt's approximate Markov chain that has random birth time and death time. These two variant notions are proved to be equivalent in Z.-Q. Chen [16], using a direct martingale approach. Sections 6.1–6.3 are based on Chen [16] but the present formulation using the energy functional introduced in Section 5.4 is new. Section 6.4 is based on an unpublished paper by Z.-Q. Chen. Among the examples of reflected Dirichlet spaces presented in Section 6.5, (**5**°) and (**6**°) are taken from Chen-Fukushima [24] and Bogdan-Burdzy-Chen [14], respectively. All other examples appear here for the first time. Section 6.6 is essentially based on the book by M. L. Silverstein [138], but the presentation here is new. Section 6.7 is taken from Chen [18]. Theorem 6.7.2 was first established in Lyons and Zheng [118] for symmetric diffusions on $\mathbb{R}^n$. The most general form of such a decomposition for symmetric Markov processes can be found in [19].

Röckner-Zhang [129, 130], Takeda [147], Kawabata-Takeda [102], and Eberle [46] have studied Markovian self-adjoint extensions of some local symmetric operators in relation to the Silverstein extensions.

In [113], K. Kuwae gave a proof of the closedness of the active reflected Dirichlet form $(\mathcal{E}^{\rm ref}, \mathcal{F}^{\rm ref}_a)$ of a quasi-regular Dirichlet form $(\mathcal{E}, \mathcal{F})$ without decomposing $\mathcal{F}^{\rm ref}$ into the sum of $\mathcal{F}_e$ and **HN**. Along the way, he proved the results in Proposition 6.6.4 using a different method.

NOTES ON CHAPTER 7

Section 7.1 is basically taken from Chen-Fukushima [23]. Theorem 7.2.1 is taken from Chen-Fukushima-Ying [26, §3]. The identification of the trace Dirichlet space with the space of finite Douglas integrals for the reflecting extension (Corollary 7.2.8) and for the reflected Dirichlet space (Theorem 7.2.9) appears here for the first time. Section 7.3 is based on Chen-Fukushima [23]. The current formulation of the lateral conditions goes back to W. Feller [50] and the second author [60].

When F is countable, the description of the resolvent in Theorem 7.4.4 has its counterpart in Chen-Fukushima [22] formulated for a more general non-symmetric Markov process possessing a weak dual. Constructions of one-point and several-point extensions in Sections 7.5 and 7.7 have been dealt with by Chen-Fukushima-Ying [28] and Chen-Fukushima [22], respectively, under the weak duality setting. The uniqueness of the one-point extension stated in Theorem 7.5.4 is based on Fukushima-Tanaka [75, §6]. Its extension to the countably many-point case in Theorem 7.7.3 is new. Most examples in Section 7.6 are taken from Chen-Fukushima [21], while the last three examples in Section 7.8 are based on Chen-Fukushima [23, 24] and the second author [69].

There is a host of literature on boundary theory for Markov processes. For countable state Markov processes, see W. Feller [50], E. B. Dynkin [45],

K. L. Chung [32], Hou-Guo [89], and X.-Q. Yang [152], and the references therein. The investigation of the boundary theory for the one-dimensional diffusions was accomplished by W. Feller [49], Itô-McKean [96, 97], K. Itô [95], and E. B. Dynkin [44] among others.

Approaches to one-point extensions via excursion theory have been carried out by K. Itô [94], T. S. Salisbury [131], R. M. Blumenthal [12], Fukushima-Tanaka [75], Fitzsimmons-Getoor [56], and Chen-Fukushima-Ying [28], to name a few.

As for higher dimensional diffusions with Wentzell's boundary condition ([151]), see K. Taira [145] and the references therein for analytic approaches and S. Watanabe [150] and the references therein for probabilistic approaches.

Bibliography

[1] D. R. Adams and L. I. Hedberg, *Function Spaces and Potential Theory*. Springer, 1996.

[2] S. Albeverio and Z. M. Ma, Necessary and sufficient conditions for the existence of m-perfect processes associated with Dirichlet forms. *Séminaire de Probabilités* **25**, 374–406; Lecture Notes in Math., vol. 1485. Springer, 1991.

[3] S. Albeverio and Z. M. Ma, Additive functionals, nowhere Radon and Kato class smooth measures associated with Dirichlet forms. *Osaka J. Math.* **29** (1992), 247–265.

[4] S. Albeverio, Z. M. Ma, and M. Röckner, Quasi-regular Dirichlet forms and Markov processes. *J. Func. Anal.* **111** (1993), 118–154.

[5] R. F. Bass and P. Hsu, Some potential theory for reflecting Brownian motion in Hölder and Lipschitz domains. *Ann. Probab.* **19** (1991), 486–508.

[6] M. T. Barlow, R. F. Bass, T. Kumagai, and A. Teplyaev, Uniqueness of Brownian motion on Sierpinski carpets. *J. Europ. Math. Soc.* **12** (2010), 655–701.

[7] M. T. Barlow, J. Pitman, and M. Yor, On Walsh's Brownian motions. *Séminaire de Probabilités* **23**, 275–293; *Lecture Notes in Math.* **1372**, Springer, 1989.

[8] A. Beurling and J. Deny, Dirichlet spaces. *Proc. Nat. Acad. Sci. U.S.A.* **45** (1959), 208–215.

[9] A. Benveniste and J. Jacod, Systèmes de Lévy des processus de Markov. *Invent. Math.* **21** (1973), 183–198.

[10] L. Beznea and N. Boboc, *Potential Theory and Right Processes*. Kluwer Academic Publishers, 2004.

[11] J. Bliedtner and W. Hansen, *Potential Theory: An Analytic and Probabilistic Approach to Balayage*. Springer-Verlag, 1986.

[12] R. M. Blumenthal, *Excursions of Markov Processes*. Birkhäuser, 1992.

[13] R. M. Blumenthal and R. K. Getoor, *Markov Processes and Potential Theory*. Dover, 2007; republication of 1968 edition (Academic Press).

[14] K. Bogdan, K. Burdzy, and Z.-Q. Chen, Censored stable processes. *Probab. Theory Relat. Fields,* **127** (2003), 89–152.

[15] N. Bouleau and F. Hirsch, *Dirichlet Forms and Analysis on Wiener Space.* De Gruyter, 1991.

[16] Z.-Q. Chen, On reflected Dirichlet spaces. *Probab. Theory Relat. Fields* **94** (1992), 135–162.

[17] Z.-Q. Chen, On reflecting diffusion processes and Skorokhod decompositions. *Probab. Theory Relat. Fields* **94** (1993), 281–316.

[18] Z.-Q. Chen, On notions of harmonicity. *Proc. Amer. Math. Soc.* **137** (2009), 3497–3510.

[19] Z.-Q. Chen, P. J. Fitzsimmons, M. Takeda, J. Ying, and T.-S. Zhang, Absolute continuity of symmetric Markov processes. *Ann. Probab.* **32** (2004), 2067–2098.

[20] Z.-Q. Chen, P. J. Fitzsimmons, and R. J. Williams, Reflecting Brownian motions: Quasimartingales and strong Caccioppoli sets. *Potential Analysis* **2** (1993), 219–243.

[21] Z.-Q. Chen and M. Fukushima, One-point extensions of Markov processes by darning. *Probab. Theory Relat. Fields* **141** (2008), 61–112.

[22] Z.-Q. Chen and M. Fukushima, On Feller's boundary problem for Markov processes in weak duality. *J. Funct. Anal.* **252** (2007), 710–733.

[23] Z.-Q. Chen and M. Fukushima, Flux and lateral conditions for symmetric Markov processes. *Potential Analysis* **29** (2008), 241–269.

[24] Z.-Q. Chen and M. Fukushima, On unique extension of time-changed reflecting Brownian motions. *Ann. Inst. Henri Poincaré Probab. Statist.* **45** (2009), 864–875.

[25] Z.-Q. Chen and M. Fukushima, A localization formula in Dirichlet form theory. To appear in *Proc. Amer. Math. Soc.*

[26] Z.-Q. Chen, M. Fukushima, and J. Ying, Traces of symmetric Markov processes and their characterizations. *Ann. Probab.* **34** (2006), 1052–1102.

[27] Z.-Q. Chen, M. Fukushima, and J. Ying, Entrance law, exit system and Lévy system of time-changed processes. *Illinois J. Math.* **50** (2006), 269–312. (Special volume in memory of Joseph Doob.)

[28] Z.-Q. Chen, M. Fukushima, and J. Ying, Extending Markov processes in weak duality by Poisson point processes of excursions. Proceedings of the Abel Symposium 2005, *Stochastic Analysis and Applications—A Symposium in Honor of Kiyosi Itô*, Eds. F. E. Benth, G. Di Nunno, T. Lindstrom, B. Oksendal and T. Zhang. Springer, 2007, pp. 153–196.

[29] Z.-Q. Chen and T. Kumagai, Heat kernel estimates for stable-like processes on d-sets. *Stochastic Process Appl.* **108** (2003), 27–62.

[30] Z.-Q. Chen and T. Kumagai, A priori Hölder estimate, parabolic Harnack principle and heat kernel estimates for diffusions with jumps. *Revista Matematica Iberoamericana* **26** (2010), 551–589.

[31] Z.-Q. Chen, Z.-M. Ma, and M. Röckner, Quasi-homeomorphisms of Dirichlet forms. *Nagoya Math. J.* **136** (1994), 1–15.

[32] K. L. Chung, with the cooperation of Paul-André Meyer, *Lectures on Boundary Theory for Markov Chains*. Ann. Math. Lect. Series, vol. 65, Princeton University Press, 1970.

[33] K. L. Chung, Greenian bounds for Markov processes. *Potential Analysis* **1** (1992), 83–92.

[34] E. Cinlar, J. Jacod, P. Protter, and M. J. Sharpe, Semimartingales and Markov processes. *Z. Wahrsch. verw. Gebiete* **54** (1980), 161–219.

[35] B. Dahlberg, Estimates of harmonic measure. *Arch. Rat. Mech. Anal.* **65** (1977), 275–288.

[36] E. B. Davies, *Heat Kernels and Spectral Theory*. Cambridge University Press, 1990.

[37] C. Dellacherie and P. A. Meyer, *Probabilités et Potentiel*, Chap. I–IV, 1975; Chap. V–VIII, 1980; Chap. IX–XI, 1983; Chap. XII–XVI, 1987, Hermann.

[38] C. Dellacherie, B. Maisonneuve, and P. A. Meyer, *Probabilités et Potentiel*, Chap. XVII–XXIV. Hermann, 1992.

[39] J. Deny, Méthods Hilbertiennes en théorie du potentiel. *Potential Theory*, Centro Internazionale Matematico Estivo, Edizioni Cremonese, pp. 121–201, 1970.

[40] J. Deny and J. L. Lions, Les espaces du type de Beppo Levi. *Ann. Inst. Fourier* **5** (1953–54), 305–370.

[41] J. L. Doob, Boundary properties of functions with finite Dirichlet integrals. *Ann. Inst. Fourier* **12** (1962), 573–621.

[42] J. Douglas, Solution of the problem of Plateau. *Trans. Amer. Math. Soc.* **33** (1931), 263–321.

[43] B. K. Driver and M. Röckner, Construction of diffusions on path and loop spaces of compact Riemannian manifolds. *C. R. Acad. Sci. Paris Sér. I Math.* **315** (1992), 603–608.

[44] E. B. Dynkin, *Markov Processes, I, II*. Springer, 1965.

[45] E. B. Dynkin, General boundary conditions for denumerable Markov proceses. *Theory Probab. Appl.* **12** (1967), 187–221.

[46] A. Eberle, *Uniqueness and non-uniqueness of semigroups generated by singular diffusion operators*, Lecture Notes in Math., vol. 1718, Springer, 1999.

[47] E. Fabes, M. Fukushima, L. Gross, C. Kenig, M. Röckner, and D. W. Stroock, *Dirichlet forms*. Lectures at C.I.M.E. 1992, eds. G. Dell'Antonio, U. Mosco, Lecture Notes in Math. 1563, Springer, 1993.

[48] X. Fang, M. Fukushima, and J. Ying, On regular Dirichlet subspaces of $H^1(I)$ and associated linear diffusions. *Osaka J. Math.* **42** (2005), 1–15.

[49] W. Feller, The parabolic differential equations and the associated semi-groups of transformations. *Ann. Math.* **55** (1952), 468–519.

[50] W. Feller, On boundaries and lateral conditions for the Kolmogorov differential equations. *Ann. Math.* **65** (1957), 527–570.

[51] P. J. Fitzsimmons, Time changes of symmetric Markov processes and a Feynman-Kac formula. *J. Theor. Probab.* **2** (1989), 485–501.

[52] P. J. Fitzsimmons, Markov processes and non-symmetric Dirichlet forms without regularity. *J. Func. Anal.* **85** (1989), 287–306.

[53] P. J. Fitzsimmons, On the quasi-regularity of semi-Dirichlet forms. *Potential Analysis* **15** (2001), 151–185.

[54] P. J. Fitzsimmons and R. K. Getoor, Revuz measures and time changes. *Math. Z.* **199** (1988), 233–256.

[55] P. J. Fitzsimmons and R. K. Getoor, Smooth measures and continuous additive functionals of right Markov processes. In *Itô's Stochastic Calculus and Probobability Theory*, eds. N. Ikeda, S. Watanabe, and H. Kunita. Springer, 1996, pp. 31–49.

[56] P. J. Fitzsimmons and R. K. Getoor, Excursion theory revisited. *Illinois J. Math.* **50** (2006), 413–437. (Special volume in memory of Joseph Doob.)

[57] P. J. Fitzsimmons and R. K. Getoor, Lévy systems and time changes, in *Séminaire de Probabilités* **42**; Lecture Notes in Math., vol. 1979, Springer, 2009.

[58] G. B. Folland, *Real Analysis*, 2nd ed. John Wiley & Sons, 1999.

[59] M. Fukushima, A construction of reflecting barrier Brownian motions for bounded domains. *Osaka J. Math.* **4** (1967), 183–215.

[60] M. Fukushima, On boundary conditions for multi-dimensional Brownian motions with symmetric resolvent densities. *J. Math. Soc. Japan* **21** (1969), 58–93.

[61] M. Fukushima, Regular representations of Dirichlet spaces. *Trans. Amer. Math. Soc.* **155** (1971), 455–473.

[62] M. Fukushima, Dirichlet spaces and strong Markov processes. *Trans. Amer. Math. Soc.* **162** (1971), 185–224.

[63] M. Fukushima, *Dirichlet Forms and Markov Processes* (in Japanese). Kinokuniya, 1975.

[64] M. Fukushima, *Dirichlet Forms and Markov Processes*. Kodansha and North-Holland, 1980.

[65] M. Fukushima, Capacitary maximal inequalities and an ergodic theorem, in *Probability Theory and Mathematical Statistics*, eds. K. Itô and I. V. Prokhorov, Lecture Notes in Math. **1021**, Springer-Verlag, 1983.

[66] M. Fukushima, On a strict decomposition of additive functionals for symmetric diffusion processes. *Proc. Japan Acad.* **70**. Ser. A (1994), 277–281.

[67] M. Fukushima, On semimartingale characterizations of functionals of symmetric Markov processes. *Elect. J. Probab.* **4** (1999), Paper 18, 1–32. http://www.math.washington.edu/~ejpecp

[68] M. Fukushima, BV functions and distorted Ornstein Uhlenbeck processes over the abstract Wiener space. *J. Func. Anal.* **174** (2000), 227–249.

[69] M. Fukushima, From one dimensional diffusions to symmetric Markov processes. *Stochastic Process Appl.* **120** (2010), 590–604. (Special issue: A tribute to Kiyosi Itô.)

[70] M. Fukushima, *Selecta*, eds. N. Jacob, Y. Oshima, and M. Takeda. De Gruyter, 2010.

[71] M. Fukushima, P. He, and J. Ying, Time changes of symmetric diffusions and Feller measures. *Ann. Probab.* **32** (2004), 3138–3166.

[72] M. Fukushima and M. Hino, On the space of BV functions and a related stochastic calculus in infinite dimensions. *J. Func. Anal.* **183** (2001), 245–268.

[73] M. Fukushima, Y. Oshima, and M. Takeda, *Dirichlet Forms and Symmetric Markov Processes*. De Gruyter, 1994; 2nd Edition, 2011.

[74] M. Fukushima and M. Takeda, *Markov Processes* (in Japanese). Baifukan, Tokyo, 2008; Chinese translation by P. He, ed. by J. Ying, Science Press, 2011.

[75] M. Fukushima and H. Tanaka, Poisson point processes attached to symmetric diffuions. *Ann. Inst. Henri Poincaré Probab. Statist.* **41** (2005), 419–459.

[76] M. Fukushima and M. Tomisaki, Construction and decomposition of reflecting diffusions on Lipschitz domains with Hölder cusps. *Probab. Theory Relat. Fields* **106** (1996), 521–557.

[77] M. Fukushima and T. Uemura, Capacitary bounds of measures and ultracontractivity of time-changed processes. *J. Math. Pures Appl.* **82** (2003), 553–572.

[78] R. Gangolli, Isotropic infinitely divisible measures on symmetric spaces. *Acta Math.* **111** (1964), 213–246.

[79] R. Gangolli, On the construction of certain diffusions on a differentiable manifold. *Z. Wahrsch. Verw. Gebiete* **2** (1964), 406–419.

[80] R. K. Getoor, *Markov Processes: Ray Proceses and Right Processes*. Lecture Notes in Math. **440**, Springer, 1975.

[81] R. K. Getoor, *Excessive Measures*. Birkhäuser, 1990.

[82] R. K. Getoor and M. J. Sharpe, Naturality, standardness, and weak duality for Markov processes. *Z. Wahrsch.Verw. Gebiete* **67** (1984), 1–62.

[83] D. Gilbarg and N. S. Trudinger, *Elliptic Partial Differential Equations of Second Order*, 2nd ed. Springer, 1983.

[84] J. M. Harrison and L. A. Shepp, On skew Brownian motion. *Ann. Prob.* **9** (1981), 309–313.

[85] S. W. He, J. G. Wang, and J. A. Yan, *Semimartingale Theory and Stochastic Calculus*. Science Press, 1992.

[86] M. Hino, On singularity of energy measures on self-similar sets. *Probab. Theory Relat. Fields* **132** (2005), 265–290.

[87] M. Hino, Set of finite perimeter and the Hausdorff-Gauss measure on the Wiener space. *J. Func. Anal.* **253** (2010), 1656–1681.

[88] M. Hino and J. A. Ramírez, Small-time Gaussian behavior of symmetric diffusion semi-groups. *Annals of Prob.* **31** (2003), 1254–1295.

[89] Z. T. Hou and Q. F. Guo, *Homogeneous Denumerable Markov Processes*. Translated from the Chinese. Springer-Verlag and Science Press, 1988.

[90] P. Hsu, On excursions of reflecting Brownian motions. *Trans. Amer. Math. Soc.* **296** (1986), 239–264.

[91] N. Ikeda and S. Watanabe, *Stochastic Differential Equations and Diffusion Processes*. Kodansha and North-Holland, 1980 (2nd ed. 1989).

[92] K. Itô, *Lectures on Stochastic Processes*. Tata Institute of Fundamental Research, Bombay, 1960.

[93] K. Itô, Poisson point processes and their application to Markov processes. Lecture notes of Mathematics Department, Kyoto University, September 1969 (unpublished).

[94] K. Itô, Poisson point processes attached to Markov processes. *Proc. Sixth Berkeley Symp. Math. Stat. Probab.*, vol. 3, 1970, p. 225–239.

[95] K. Itô, *Essentials of Stochastic Processes*. Translation of Mathematical Monographs, Amer. Math. Soc., 2006; in Japanese 1957 (Iwanami Shoten).

[96] K. Itô and H. P. McKean Jr., Brownian motions on a half line. *Illinois J. Math.* **7** (1963), 181–231.

[97] K. Itô and H. P. McKean Jr., *Diffusion Processes and Their Sample Paths*. Springer, 1965; Springer's Classics in Mathematics Series, 1996.

[98] N. Jacob, *Pseudo Differential Operators and Markov Processes*. Imperial College Press, vol. 1, 2001; vol. 2, 2002; vol. 3, 2005.

[99] P. A. Jacobs, Excursions of a Markov process induced by continuous additive functionals. *Z. Wahrsch. verw. Gebiete* **44** (1978), 325–336.

[100] J. Jost, W. Kendall, U. Mosco, M. Röckner, and K-T. Sturm, *New Directions in Dirichlet Forms*. AMS International Press, 1998.

[101] A. Kasue, Convergence of Riemannian manifolds and Laplace operators. I. *Ann. Inst. Fourier, Grenoble* **52** (2002), 1219–1257.

[102] T. Kawabata and M. Takeda, On uniqueness problem for local Dirichlet forms. *Osaka J. Math.* **33** (1996), 881–893.

[103] J. L. Kelley, *General Topology*. Springer 1975; reprint of the 1955 edition (Van Nostrand).

[104] O. D. Kellogg, *Foundations of Potential Theory*. Dover 1954; republication of 1929 edition (J. Springer).

[105] J. Kigami, *Analysis on Fractals*. Cambridge University Press, 2001.

[106] J. Kigami, Dirichlet forms and associated heat kernels on the Cantor set induced by random walks on trees. *Advances in Mathematics* **225** (2010), 2674–2730.

[107] D. Kim, On spectral gaps and exit time distributions for a non-smooth domain. *Forum Math.* **18** (2006), 571–583.

[108] P. Kim, Fatou's theorem for censored stable processes. *Stochastic Process. Appl.* **108** (2003), 63–92.

[109] P. Kim, Weak convergence of censored and reflected stable processes. *Stochastic Process. Appl.* **116** (2006), 1792–1814.

[110] H. Kunita, Sub-Markov semi-groups in Banach lattice. In *Proceedings of the International Conference on Functional Analysis and Related Topics*, Tokyo, 1969, pp. 332–343.

[111] H. Kunita, General boundary conditions for multi-dimensional diffusion processes. *J. Math. Kyoto Univ.* **10** (1970), 273–335.

[112] H. Kunita and S. Watanabe, On square integrable martingales. *Nagoya Math. J.* **30** (1967), 209–245.

[113] K. Kuwae, Reflected Dirichlet forms and the uniqueness of Silverstein's extension. *Potential Analysis* **16** (2002), 221–247.

[114] K. Kuwae and T. Shioya, Convergence of spectral structures: A functional analytic theory and its applications to spectral geometry. *Communications in Analysis and Geometry* **11** (2003), 599–673.

[115] G. F. Lawler, The Laplacian-b random walk and the Schramm-Loewner evolution. *Illinois J. Math.* **50** (2006), 701–746. (Special volume in memory of Joseph Doob.)

[116] Y. LeJan, Balayage et formes de Dirichlet. *Z. Wahrsch. Verw. Gebiete* **37** (1977), 297–319.

[117] Y. LeJan, Mesures associées à une forme de Dirichlet. Applications. *Bull. Soc. Math. France* **106** (1978), 61–112.

[118] T. J. Lyons and W. A. Zheng, A crossing estimate for the canonical process on a Dirichlet space and a tightness result. Colloque Paul Lévy sur les Processus Stochastiques (Palaiseau, 1987). *Astérisque* **157–158** (1988), 249–271.

[119] Z. M. Ma and M. Röckner, *Introduction to the Theory of (Non-Symmetric) Dirichlet Forms*. Springer, 1992.

[120] B. Maisonneuve, Exit systems. *Ann. Probab.* **3** (1975), 399–411.

[121] V. G. Maz'ja, *Sobolev Spaces*. Springer, 1985.

[122] H. P. McKean and H. Tanaka, Additive functionals of the Brownian path. *Memoirs Coll. Sci. Kyoto, A. Math.* **33** (1961), 479–506.

[123] M. Motoo, Application of additive functionals to the boundary problem of Markov processes, Lévy's system of U-processes. In *Proc. Fifth Berkeley Symp. Math. Stat. II*, 1967, pp. 75–110.

[124] M. Motoo and S. Watanabe, On a class of additive functionals of Markov processes. *J. Math. Kyoto Univ.* **4** (1965), 429–469.

[125] Y. Ogura, M. Tomisaki, and M. Tsuchiya, Convergence of local type Dirichlet forms to a non-local one. *Ann. I. H. Poincaré* **38** (2002), 505–556.

[126] H. Osada, Dirichlet form approach to infinite-dimensional Wiener processes with singular interactions. *Commun. Math. Phys.* **176** (1996), 117–131.

[127] D. Ray, Resolvents, transition functions, and strongly Markovian processes. *Ann. Math.* **70** (1959), 43–72.

[128] D. Revuz, Mesures associées aux fonctionneles additives de Markov I. *Trans. Amer. Math. Soc.* **148** (1970), 501–531.

[129] M. Röckner and T. S. Zhang, Uniqueness of generalized Schrödinger operators and applications. *J. Funct. Anal.* **105** (1992), 187–231.

[130] M. Röckner and T. S. Zhang, Uniqueness of generalized Schrödinger operators. II. *J. Funct. Anal.* **119** (1994), 455–467.

[131] T. S. Salisbury, Construction of right processes from excursions. *Probab. Theory Related Fields* **73** (1986), 351–367.

[132] K. Sato, *Lévy Processes and Infinitely Divisible Distributions*. Cambridge University Press, 1999.

[133] B. Schmuland, On the local property for positivity preserving coercive forms. In *Dirichlet Forms and Stochastic Processes*, Proceedings of the International Conference held in Beijing, 1993, eds. Z.-M. Ma, M. Röckner, and J.-A. Yan. De Gruyter, 1995, pp. 345–354.

[134] B. Schmuland, Positivity preserving forms have the Fatou property. *Potential Analysis* **10** (1999), 373–378.

[135] L. Schwartz, *Théorie des Distributions*. Hermann, vol. 1, 1950; vol. 2, 1951.

[136] L. Schwartz, *Radon Measures on Arbitrary Topological Spaces and Cylindrical Measures*. Published for the Tata Institute of Fundamental Research Studies in Mathematics by Oxford University Press, 1973.

[137] M. J. Sharpe, *General Theory of Markov Processes*. Academic Press, 1988.

[138] M. L. Silverstein, *Symmetric Markov Processes*. Lecture Notes in Math., vol. 426, Springer, 1974.

[139] M. L. Silverstein, The reflected Dirichlet space. *Illinois J. Math.* **18** (1974), 310–355.

[140] M. L. Silverstein, *Boundary Theory for Symmetric Markov Processes*. Lecture Notes in Math., vol. 516, Springer, 1976.

[141] W. Stannat, The theory of generalized Dirichlet forms and its applications in analysis and stochastics. *Memoirs Amer. Math. Soc.* **678** (1999).

[142] E. M. Stein, *Singular Integrals and Differentiability Properties of Functions*. Princeton University Press, 1970.

[143] D. W. Stroock and B. Zegarlin'ski, The equivalence of the logarithmic Sobolev inequality and the Dobrushin-Shlosman mixing condition. *Comm. Math. Phys.* **144** (1992), 303–323.

[144] A.-S. Sznitman, *Brownian Motion, Obstacles and Random Media*. Springer-Verlag, 1998.

[145] K. Taira, *Semigroups, Boundary Value Problems and Markov Processes*. Springer, 2004.

[146] M. Takeda, On a martingale method for symmetric diffusion processes and its application. *Osaka J. Math.* **26** (1989), 605–623.

[147] M. Takeda, Two classes of extensions for generalized Schrödinger operators. *Potential Analysis* **5** (1996), 1–13.

[148] H. Tanemura, Uniqueness of Dirichlet forms associated with systems of infinitely many Brownian balls in $\mathbb{R}^d$. *Probab. Theory Relat. Fields* **109** (1997), 275–299.

[149] S. Watanabe, On discontinuous additive functionals and Lévy measures of Markov processes. *Japanese J. Math.* **34** (1964), 53–70.

[150] S. Watanabe, Itô's theory of excursion point processes and its developments. *Stochastic Process Appl.* **120** (2010), 653–677. (Special issue: A tribute to Kiyosi Itô.)

[151] A. D. Wentzell, On boundary conditions for multidimensional diffusion processes. *Theory Probab. Appl.* **4** (1959), 164–177.

[152] X.-Q. Yang, *The Construction Theory of Denumerable Markov Processes*. Hunan Science and Technology Publishing House and John Wiley & Sons, 1990.

[153] J. Ying and M. Zhao, The uniqueness of symmetrizing measure of Markov processes. *Proc. Amer. Math. Soc.* **138** (2010), 2181–2185.

[154] K. Yosida, *Functional Analysis*. Springer, 1968.

[155] L. Zambotti, Integration by parts formulae on convex sets of paths and application to SPDE's with reflection. *Probab. Theory Relat. Fields* **123** (2002), 579–600.

Catalogue of Some Useful Theorems

For readers' convenience, we present an index of useful theorems, some of which may not be easily located by just using the table of contents of the book.

- ◇ Theorem 1.1.5: Extended Dirichlet space on a σ-finite measure space without topological assumption.
- ◇ (1.1.22) and (1.5.1): Useful approximating expressions of $\mathcal{E}(f,f)$ for $f \in \mathcal{F}_e$.
- ◇ Corollary 1.1.9: Fatou's lemma for extended Dirichlet form $(\mathcal{F}_e, \mathcal{E})$.
- ◇ Lemma 1.2.3: If $u \in L^2_+(E; m)$ is α-excessive and $v \in \mathcal{F}$ with $u \leq v$, then $u \in \mathcal{F}$ and $\mathcal{E}_\alpha(u, u) \leq \mathcal{E}_\alpha(v, v)$.
- ◇ Theorem 1.3.14: Relation between $\mathcal{E}$-polar set and Cap_1-polar set; f is quasi continuous if and only if $f \in C(\{F_k\})$ for some Cap_1-nest $\{F_k\}$.
- ◇ Lemma 1.3.15: Every $f \in \mathcal{F}$ admits a quasi continuous version in the restricted sense with respect to Cap_1-nest.
- ◇ Theorem 1.4.3: A Dirichlet form is quasi-regular if and only if it is quasi-homeomorphic to a regular Dirichlet form.
- ◇ Theorem 1.5.2: Association of a nice Markov process to a quasi-regular Dirichlet form.
- ◇ Proposition 2.1.3: An irreducible symmetric Markovian semigroup is either transient or recurrent.
- ◇ Theorem 2.1.5: Dirichlet form characterization of transience.
- ◇ Theorem 2.1.8: Equivalent characterizations of recurrence.
- ◇ Theorem 2.1.9: Equivalent characterizations of transience.
- ◇ Proposition 2.1.10: $\mathbf{P}_x(\zeta < \infty) > 0$ on E implies that the symmetric process X is transient in the Dirichlet form sense.
- ◇ Theorem 2.1.11 and Theorem 5.2.16: A criterion for a recurrent Dirichlet form to be irreducible.
- ◇ Theorem 2.3.2: For a transient Dirichlet form $(\mathcal{E}, \mathcal{F})$, a set N is $\text{Cap}^{(0)}$-polar if and only if it is $\mathcal{E}$-polar; every $f \in \mathcal{F}_e$ admits a $\text{Cap}^{(0)}$-nest $\{F_k\}$ so that $f \in C_\infty(\{F_k\})$.

- ◇ Theorem 2.3.4: Every $f \in \mathcal{F}_e$ admits a quasi continuous version. Here $(\mathcal{E}, \mathcal{F})$ may not be transient.
- ◇ Corollary 2.3.11: For a Borel set B, $\mathrm{Cap}_1(B) = 0$ if and only if $\mu(B) = 0$ for every $\mu \in S_{00}$.
- ◇ Theorem 2.3.15: Structure of smooth measure.
- ◇ Theorem 2.4.2: Equivalent characterization of local property of a Dirichlet form that is invariant under quasi-homeomorphism.
- ◇ Theorem 2.4.3: Equivalent characterization of strongly local property of a Dirichlet form that is invariant under quasi-homeomorphism.
- ◇ Theorem 3.1.3: A set N is $\mathcal{E}$-polar if and only if it is m-polar.
- ◇ Theorem 3.1.4: Cap_1-nest vs. strong nest; probabilistic characterization of $\mathcal{E}$-nest under $\mathbf{P}_x$.
- ◇ Theorem 3.1.5: Any m-polar set is contained in a Borel properly exceptional set.
- ◇ Theorem 3.1.7: Finely continuous property for $f \in \mathcal{F}$.
- ◇ Theorem 3.1.10: Any semipolar set is $\mathcal{E}$-polar.
- ◇ Proposition 3.1.11, Corollary 3.1.14(ii): Equivalence of the absolute continuity of transition function and resolvent kernel.
- ◇ Theorem 3.1.12: Uniqueness of m-symmetric Hunt process associated with a regular Dirichlet form.
- ◇ Theorem 3.1.13: Uniqueness of m-symmetric right process properly associated with a quasi-regular Dirichlet form.
- ◇ Corollary 3.1.14(i): Any m-tight m-symmetric right process on a Radon space or any m-symmetric right process on a Lusin space is a special Borel standard process outside an m-inessential set.
- ◇ Lemma 3.1.17: Probabilistic characterization of $\mathcal{E}$-nest under $\mathbf{P}_m$.
- ◇ Theorem 3.2.2: Probabilistic representation $\mathbf{H}_B^\alpha f$ of $\mathcal{E}_\alpha$-orthogonal projection $P_{\mathcal{H}_B^\alpha} f$ of $f \in \mathcal{F}$ for nearly Borel measurable set B.
- ◇ Theorem 3.3.3: Relation between quasi notion and fine notion such as quasi open vs. finely open and quasi continuity vs. fine continuity.
- ◇ Theorem 3.3.5: Equivalent characterization of quasi support of a measure.
- ◇ Corollary 3.3.6: If $f > 0$ m-a.e., then the measure $f \cdot m$ has full quasi support E.
- ◇ Theorem 3.3.8: For a nearly Borel quasi open set D, $(\mathcal{E}^D, \mathcal{F}_D)$ is a quasi-regular Dirichlet form; relation between $\mathcal{E}^D$-polarity and $\mathcal{E}$-polarity for a set $N \subset D$.
- ◇ Theorem 3.3.9: For a regular Dirichlet form $(\mathcal{E}, \mathcal{F})$ on E and an open set $D \subset E$, $(\mathcal{E}, \mathcal{F}_D)$ is a regular Dirichlet form on $L^2(D; m)$.

- ◇ Theorem 3.4.8: For every $u \in \mathcal{F}_e$ (($\mathcal{E}, \mathcal{F}$) not necessarily transient), $\mathbf{H}_B u$ is a quasi continuous element in $\mathcal{F}_e$ that is $\mathcal{E}$-orthogonal to $\mathcal{F}_{e,E\setminus B}$.
- ◇ Theorem 3.4.9: For a nearly Borel fine open set D, the extended Dirichlet space $\mathcal{F}_{D,e}$ of $(\mathcal{E}, \mathcal{F}_D)$ can be identified with $\mathcal{F}_{e,D} = \{u \in \mathcal{F}_e : u = 0$ $\mathcal{E}$-q.e. on $D^c\}$. So we can use $\mathcal{F}_e^D$ to denote either of them.
- ◇ Lemma 3.5.1: For a transient process, a $\text{Cap}^{(0)}$-nest is a strong nest.
- ◇ Theorem 3.5.2: Sample paths wander out to infinity on $\{\zeta = \infty\}$ for transient process.
- ◇ Corollary 3.5.3: For transient process X and $f \in \mathcal{F}_e$, $\lim_{t \to \zeta-} f(X_t) = 0$ along the paths that admit no killings inside E.
- ◇ Theorem 3.5.4: If $\{f_n\} \subset \mathcal{F}_e$ is both m-a.e. convergent and $\mathcal{E}$-convergent to some $f \in \mathcal{F}_e$, then there is a subsequence of $\{\widetilde{f}_n\}$ that is uniform convergent to $\widetilde{f}$ on a strong nest. Here $(\mathcal{E}, \mathcal{F})$ may not be transient.
- ◇ Lemma 3.5.5: Consequences of recurrence for bounded excessive functions.
- ◇ Theorem 3.5.6: Probabilistic interpretation of a non-m-polar set and irreducible recurrence.
- ◇ Theorem 4.1.1: One-to-one correspondence between PCAFs and smooth measures.
- ◇ Theorem 4.2.1: Killing measure in terms of a Lévy system.
- ◇ Theorem 4.2.5: Completeness of $(\overset{\circ}{\mathcal{M}}, \mathbf{e})$ and the locally uniform convergence of $M_n \in \overset{\circ}{\mathcal{M}}$.
- ◇ Theorem 4.2.6: Fukushima's decomposition for $u \in \mathcal{F}_e$.
- ◇ Proposition 4.3.1: Strongly local property of $\mu^c_{\langle u \rangle}$ for $u \in \mathcal{F}_e$ in terms of quasi open sets.
- ◇ Theorem 4.3.3: Beurling-Deny decomposition for extended Dirichlet space $(\mathcal{F}_e, \mathcal{E})$ of a regular Dirichlet form.
- ◇ Theorem 4.3.4: Probabilistic characterizations of the local property and the strongly local one for a regular Dirichlet form.
- ◇ Theorem 4.3.7: Derivative formula for energy measure $\mu^c_{\langle u \rangle}$.
- ◇ Theorem 4.3.8: For $u \in b\mathcal{F}$, $\mu^c_{\langle u \rangle} \circ u^{-1}$ is absolutely continuous with respect to the Lebesgue measure on $\mathbb{R}$.
- ◇ Theorem 4.3.10: Normal contraction property in u for $\mu^c_{\langle u \rangle}$.
- ◇ Theorem 5.1.3: Feynman-Kac transform and perturbed Dirichlet form $(\mathcal{E}^\mu, \mathcal{F}^\mu)$, where $\mathcal{E}^\mu = \mathcal{E} + (\cdot, \cdot)_\mu$.
- ◇ Theorem 5.1.4: Equivalence of $\mathcal{E}^\mu$-nest and $\mathcal{E}$-nest.
- ◇ Theorem 5.1.5: Quasi-regularity of $(\mathcal{E}^\mu, \mathcal{F}^\mu)$ for smooth measure $\mu \in S$.

- ◇ Theorem 5.1.6: For regular Dirichlet form $(\mathcal{E}, \mathcal{F})$ and positive Radon measure μ charging no $\mathcal{E}$-polar set, $(\mathcal{E}^\mu, \mathcal{F}^\mu)$ is a regular Dirichlet form on $L^2(E; m)$.
- ◇ Proposition 5.1.9: $((\mathcal{F}^\mu)_e, \mathcal{E}^\mu) = ((\mathcal{F}_e)^\mu, \mathcal{E}^\mu)$.
- ◇ Theorems 5.2.2 and 5.2.15: Characterization of the Dirichlet form $(\check{\mathcal{E}}, \check{\mathcal{F}})$ of a time-changed process and its extended Dirichlet space $\check{\mathcal{F}}_e$.
- ◇ Theorem 5.2.5: Transience/recurrence is invariant under time-change.
- ◇ Theorem 5.2.6: Restrictions of $\mathcal{E}$-nest and $\mathcal{E}$-quasi-continuity to the quasi support F of μ yield $\check{\mathcal{E}}$-nest and $\check{\mathcal{E}}$-quasi-continuity.
- ◇ Theorem 5.2.7: Quasi-regularity of $(\check{\mathcal{E}}, \check{\mathcal{F}})$ and the proper association of time-changed process.
- ◇ Theorem 5.2.8: Let F be the quasi support of a Radon measure μ. $N \subset F$ is $\check{\mathcal{E}}$-polar if and only if it is $\mathcal{E}$-polar.
- ◇ Theorem 5.2.11: Equivalence of quasi notions for time-changed processes with full support.
- ◇ Theorem 5.2.13: For Radon measure μ, its time-changed process, properly restricted, is a Hunt process and its associated Dirichlet form is regular on the topological support F^* of μ. Moreover, $F^* \setminus F_\mu$ is $\check{\mathcal{E}}$-polar.
- ◇ Proposition 5.2.14: Reduction of a right process properly associated with a regular Dirichlet form into a Hunt process.
- ◇ Theorem 5.2.16: If a regular Dirichlet form $(\mathcal{E}, \mathcal{F})$ is irreducible, then $u \in \mathcal{F}_e$, $\mathcal{E}(u, u) = 0$ implies $u =$ constant.
- ◇ Theorem 5.2.17: Resurrected Dirichlet form and resurrected Hunt process obtained by the removal of killing measure.
- ◇ Theorem 5.3.1: Conformal invariance of two-dimensional absorbing Brownian motion.
- ◇ Example (**2°**) of Section 5.3: Conformal invariance of two-dimensional reflecting Brownian motion.
- ◇ Theorem 5.3.4: Identification of the Dirichlet form and the extended Dirichlet space of a general one-dimensional minimal diffusion (possibly with killings inside).
- ◇ Theorem 5.5.6 and Exercise 5.5.7: For quasi open subset $D \subset E$ and $v \in \mathcal{F}_e$, with $m_0 := m|_D$,

$$\lim_{t\to 0} \frac{1}{t}\mathbf{E}_{m_0}\left[(v(X_t) - v(X_0))^2; t < \tau_D\right]$$
$$= \mu^c_{\langle v\rangle}(D) + \int_{D\times D} (v(x) - v(y))^2 J(dx, dy).$$

- ◇ Theorems 5.6.2 and 5.6.3: Beurling-Deny decomposition of the extended Dirichlet space of a time-changed process.
- ◇ Lemma 5.7.2: $\sigma_F = \inf\{t > 0 : X_t \in F \text{ or } X_{t-} \in F\}$ a.s.
- ◇ Theorem 5.7.5: Characterization of the entrance law induced by an exit system.
- ◇ Theorem 5.7.6: Feller measures as joint distributions of starting and end points of excursions.
- ◇ Theorem 6.1.2: Every $M \in \overset{\circ}{\mathcal{M}}$ is uniformly $\mathbf{P}_x$-square integrable for q.e. x.
- ◇ Theorem 6.1.8: Beurling-Deny type formula for $\mathbf{e}(M^h)$ for $h \in \mathbf{HN}$.
- ◇ Theorem 6.2.13: A characterization of $\mathcal{F}^{\text{ref}}$ in terms of finite $\mathcal{E}$-energy.
- ◇ Theorem 6.2.14: $(\mathcal{E}^{\text{ref}}, \mathcal{F}^{\text{ref}}_a)$ is a Dirichlet form.
- ◇ Theorem 6.3.2: For a recurrent Dirichlet form $(\mathcal{E}, \mathcal{F})$, $(\mathcal{E}^{\text{ref}}, \mathcal{F}^{\text{ref}}) = (\mathcal{E}, \mathcal{F}_e)$.
- ◇ Theorem 6.4.2: A simpler characterization of $\mathcal{F}^{\text{ref}}$ in terms of quasi open sets.
- ◇ Proposition 6.4.6: Reflected Dirichlet space $(\mathcal{E}^{\text{ref}}, \mathcal{F}^{\text{ref}})$ is invariant under a full support time change.
- ◇ Theorem 6.4.12: A characterization of $\mathcal{F}^{\text{ref}}$ for transient quasi-regular Dirichlet form $(\mathcal{E}, \mathcal{F})$.
- ◇ Theorem 6.6.5: Probabilistic meaning of a Silverstein extension.
- ◇ Corollary 6.6.6: If $(\mathcal{E}, \mathcal{F})$ is conservative, then $(\mathcal{E}^{\text{ref}}, \mathcal{F}^{\text{ref}}_a) = (\mathcal{E}, \mathcal{F})$.
- ◇ Theorem 6.6.9: $(\mathcal{E}^{\text{ref}}, \mathcal{F}^{\text{ref}}_a)$ is the maximal Silverstein extension of $(\mathcal{E}, \mathcal{F})$.
- ◇ Theorem 6.6.10 and Remark 6.6.11(ii): Relation between $(\mathcal{E}^{\text{ref}}, \mathcal{F}^{\text{ref}})$ and the extended Dirichlet space of $(\mathcal{E}^{\text{ref}}, \mathcal{F}^{\text{ref}}_a)$.
- ◇ Remark 6.6.11(i): Reflected Dirichlet space $(\mathcal{E}^{\text{ref}}, \mathcal{F}^{\text{ref}})$ of $(\mathcal{E}, \mathcal{F})$ is also the reflected Dirichlet space of $(\mathcal{E}^{\text{ref}}, \mathcal{F}^{\text{ref}}_a)$.
- ◇ Theorem 6.7.2: Lyons-Zheng's forward-backward martingale decomposition for $u(X_t) - u(X_0)$.
- ◇ Theorem 6.7.4: For a smooth measure ν and $U\nu := \mathbf{E}_x[A^\nu_\zeta]$, $U\nu \in \mathcal{F}_e$ if and only if $\int_E U\nu(x)\nu(dx) < \infty$. In this case, $\mathcal{E}(U\nu, v) = \int_E v(x)\nu(dx)$ for $v \in \mathcal{F}_e$. Here the Dirichlet form $(\mathcal{E}, \mathcal{F})$ is not assumed a priori to be transient.
- ◇ Theorem 6.7.13: Equivalence between probabilistic and analytic notions of harmonicity.
- ◇ Theorem 7.1.6: Relation between the extended Dirichlet space of X on E and the reflected Dirichlet space of the part process X^0 of X on a quasi open set $E_0 = E \setminus F$.

- ◇ Theorem 7.1.8: Relation between the Dirichlet space of X on E and the active reflected Dirichlet space of X^0 on E_0.
- ◇ Corollary 7.2.8: When X is a reflecting extension of X^0, the trace on F of the Dirichlet space of X is identified with the space of functions on F of finite Douglas integrals.
- ◇ Theorem 7.3.5: Characterization of the L^2-infinitesimal generator of X in terms of a lateral condition on F involving the flux.
- ◇ Theorem 7.4.5: When F is countable and X admits no jumps from F to F or killings on F, the trace Dirichlet space of X is the closure of finitely supported functions with respect to the Douglas integral and X is uniquely determined by X^0. The L^2-generator of X is characterized by the zero flux condition.
- ◇ Theorem 7.5.4: Uniqueness of the one-point extension and its equivalent characterizations without assuming the regularity of the associated Dirichlet form.
- ◇ Theorem 7.5.6: Construction of a one-point extension by a Poisson point process of excursions.
- ◇ Theorem 7.5.9: Construction of a one-point extension by darning a hole.
- ◇ Theorem 7.5.10: Expression of the entrance law for a darning of a hole in terms of the entrance law induced by an exit system.
- ◇ Theorem 7.5.12: Construction of a skew one-point extension.
- ◇ Theorem 7.6.3: Conformal invariance of the one-point extension of two-dimensional Brownian motion by darning a compact hole.
- ◇ Theorem 7.7.3: Uniqueness of the countably many-point extension and its equivalent characterizations without assuming the regularity of the associated Dirichlet form.
- ◇ Theorem 7.7.4: Construction of a countably many-point extension by repeating one-point extensions.
- ◇ Theorem 7.7.5: Construction of a many-point extension by darning countably many holes.
- ◇ Theorem 7.8.1: Conformal invariance of the many-point extension of two-dimensional Brownian motion by darning finitely many compact holes.
- ◇ Proposition 7.8.5: Two-point reflecting extension of a time-changed RBM on a closed domain with two branches of infinite cones.
- ◇ Proposition 7.8.7: A skew extension of a one-dimensional absorbing Brownian motion with countable boundary.
- ◇ Theorem A.4.1: Banach-Saks Theorem.

Index